STUDY GUIDE

DIXIE J. GOSS
Hunter College

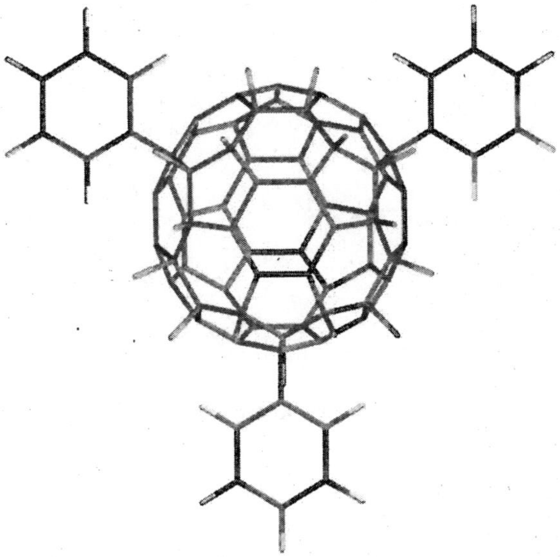

PETRUCCI HARWOOD HERRING MADURA

GENERAL CHEMISTRY

PRINCIPLES & MODERN APPLICATIONS

NINTH EDITION

courtney Hopking

PEARSON
Prentice Hall

Upper Saddle River, NJ 07458

Project Manager: Kristen Kaiser
Senior Acquisitions Editor: Kent Porter-Hamann
Editor-in-Chief, Science: Dan Kaveney
Executive Managing Editor: Kathleen Schiaparelli
Assistant Managing Editor: Karen Bosch
Production Editor: Jennifer Zisa
Supplement Cover Manager: Paul Gourhan
Supplement Cover Designer: Christopher Kossa
Manufacturing Buyer: Ilene Kahn
Manufacturing Manager: Alexis Heydt-Long

Pearson Prentice Hall
Pearson Education, Inc.
Upper Saddle River, NJ 07458

Printed in Canada

1 2 3 DPC 11 10 09

ISBN 0-13-149387-6

Pearson Education Ltd., *London*
Pearson Education Australia Pty. Ltd., *Sydney*
Pearson Education Singapore, Pte. Ltd.
Pearson Education North Asia Ltd., *Hong Kong*
Pearson Education Canada, Inc., *Toronto*
Pearson Educación de Mexico, S.A. de C.V.
Pearson Education—Japan, *Tokyo*
Pearson Education Malaysia, Pte. Ltd.

TABLE OF CONTENTS

How to Use This Book

The purpose of this study guide is to help you learn chemistry. Students of chemistry have a wide variety of back-grounds and different reasons for taking chemistry. A study guide is designed mainly for those students who have some difficulty learning a subject. But there is something of value here for all—because all of us have trouble at some point in learning new material. Because of this, you may not find all of the material useful to you. However, each suggestion is worth an honest effort on your part. It is the result of more than twenty-five years of experience helping students with what many of them consider a very difficult subject—chemistry.

Chemistry is both an art and a science. In fact, often students complain both that "chemistry has too many rules," and "there are too many exceptions to the rules." Its study often requires a different approach than does the study of many other subjects. It is not sufficient to read the text once, twice, or even several times. If you follow that approach, you will spend a great deal of time on chemistry but probably will not learn it. And it is essential that you learn chemistry well, particularly the basics. Chemistry is a vertical discipline; it builds on itself. And just as you would have great difficulty with arithmetic if you had never learned the "seven times table," so too chemistry will more than live up to its brutal reputation for you unless you learn its fundamentals well.

Learning chemistry requires that you do three things: write, practice, and test yourself. Writing helps you firm up your ideas; it makes you aware of what you really know and what you don't know. Francis Bacon stated it well: "Reading maketh a full man, conference a ready man, and writing an exact man."

Practice in the study of chemistry means that you practice solving problems. You certainly would not expect to be able to drive a car without first practicing how to do so; most of us get learner's permits so we can practice. Neither can you expect to be able to solve problems without practice, lots of practice. That also means memorization, about which more in a bit.

Finally, you need to test yourself. This has two purposes. First, it helps you find out how well you have learned what you have studied. Second, you should feel more at ease during class examinations, since you have examined yourself already.

Organization by objectives divides the material of each chapter into more easily comprehended parts Each chapter begins with short essays on each objective, grouped under "Chapter Objectives." These essays explain the material, describe how to work problems, and give detailed examples of worked-out problems. The material in the essays is taken from the author's classroom experience and includes many analogies, examples, and illustrations. The analogies are often fanciful, sometimes ridiculous. This is necessary to have them illustrate the intended point and it does make them more memorable. Sometimes topics are explained from a different viewpoint from that of the text. There is no contradiction here but mutual reinforcement. If you have trouble with a topic, a different viewpoint may be enough to make it clear. Even if you have no difficulty, the alternate view may highlight aspects that had not occurred to you and thus increase your appreciation and understanding of the topic.

For objectives involving memorization, and especially for learning new terms, 3×5 cards can be very useful. Write the term on one side of the card and your version of the definition on the other side. You can carry the cards with you and study them during spare moments: waiting in line, before classes start, and so on. In addition, you can shuffle the cards to make sure you know the terms in any order, pick out the hard ones and concentrate only on those, and turn the cards over to look at the definition and then recall the term.

There is another way to memorize that you may find easier for your style of learning. Make a list of the items you need to learn and record them on a cassette tape, with a pause after each one. T

you might say "sodium" (pause) "N...A. plus one" [Na^+], and in another place: "N...A...plus one" (pause) "sodium ion" to learn it the other way. You may now play this recorded cassette on your portable personal player as you walk between classes or on the tape player in your car if you drive a lot, such as to classes. Try to fill in the blanks as you listen. You have created a "language tape" to help you learn introductory chemistry, a course in which some say you will learn as much new vocabulary as in the first year of the study of a foreign language.

Most objectives require learning a new concept or how to solve a new type of problem. For those involving a new concept, first read the essay carefully and then summarize the concept in your words. Since a 3×5 card is too small for most concepts, this summary should be written in your lecture notebook (see "How to Take Good Notes"). Notice that both definitions and new concepts should be stated in your words. Copying from a book will help you very little. Putting the definition or concepts in your words ensures that you understand it. This is what you want. After all, *you* are the one who has to learn it and be able to use it.

Each problem-solving objective includes a detailed example of the type of problem being considered. The "tricks of the trade" are included in these examples, and often one type of problem is solved by two or more methods so that you will understand better how the answer is obtained. Sometimes there is only one example, solved by a method different from that of the text. Keep in mind that there are many different ways to solve problem and choose the one that is easiest to you. When a method of solving a problem is broken down into several steps, each step is explained.

You gain practice in problem solving by doing, so the "Chapter Objectives" section is followed by "Drill Problems." Drill problems are grouped by objective, and their presence is indicated in the essay section by an asterisk in the margin by the objective number. Occasionally there are more than 26 drill problems for a given objective, and thus they could not all be referred to with the letters of the alphabet. In these cases, those Greek alphabet capital letters that are different from Latin capital letters are used. (The Scandinavian letter Ø also is occasionally used.) Greek letters often are used in scientific equations, and you thus may become familiar with some of them. The complete Greek alphabet is given in the list that follows.

The Greek Alphabet (capital letter first in each case, followed by lower case letter, and name)

A α alpha	B β beta	$\Gamma\,\gamma$ gamma	$\Delta\,\delta$ delta	E ε epsilon	Z ζ zeta
H η eta	$\Theta\,\theta$ theta	I ι iota	K κ kappa	$\Lambda\,\lambda$ lambda	M μ mu
N ν nu	$\Xi\,\xi$ xi	O o omicron	$\Pi\,\pi$ pi	P ρ rho	$\Sigma\,\sigma$ sigma
T τ tau	$\Upsilon\,\upsilon$ upsilon	$\Phi\,\phi$ phi	X χ chi	$\Psi\,\psi$ psi	$\Omega\,\omega$ omega

The drill problems are straightforward questions based on each objective. By working them, you are practicing your mastery of that objective. If you drill yourself and become confident in these basic techniques, solving problems in which several are combined will be much easier. Answers to all drill problems are given in the Appendix. The drill problems all use actual data, not "made-up" numbers. This should give you a better feeling for the "real world" of chemistry.

When you can solve the drill problems, go on to the Exercises in the text. The drill problems give practice in the techniques of solving problems. The Review Questions and the Exercises in the text require that you be able to use these techniques in solving more involved problems.

For testing yourself, you will find in each chapter four "Quizzes" and one "Sample Test." The quizzes are designed to be a quick way to check your knowledge. There are four so that you can check you progress at several points: before you intensively study the chapter, during your study, and just before a class examination. The tests are included to give you some idea of the kinds of questions y might encounter on a class examination. The test questions differ in style within each chapter a between chapters. Different instructors use different style questions and knowing many styles can be of great help on examinations. Answers to the quizzes are in Appendix I, as are detailed solutions to the tests. Here's hoping your study of chemistry is both successful and enjoyable!

How to Take Good Notes

There are several steps involved in producing a good notebook—one that will help you in your study of a subject. Although what follows is written for a chemistry lecture notebook, it can be adapted with small changes for the lecture notebook of any subject.

The first step is prior preparation. Before you go to lecture, carefully read the chapter summary. Then skim the material in the text that will be covered. Do not read in detail. You are trying to get an overview of what is to be taught so that you understand how each topic relates to the others. Write a brief outline in your lecture notebook. You should also write out questions about points that seem unclear. Finally, review the notes you took during the previous lecture. In this way, you will understand how each point is related to the whole subject.

During lecture, take notes on only one side of each page. You will need the other side later. Many professors write all important material on the board and, of course, you will copy this in your notebook. Think about your brief outline and your questions as you take notes, and answer those questions that you can. Leave a blank line after each major topic in your lecture notes so that you can more easily see the main points when you review your notes.

Immediately after lecture, review the notes you have taken. Correct the items that are wrong, add material you remember, and emphasize important points. This should take five minutes at most; it is absolutely essential that you do it immediately, while the lecture is fresh in your mind. The lon you delay, the less value this process will have.

The evening after lecture, read the portions of the text and the study guide that were covered in lecture. Use the blank sides of the notebook pages to write clarifications and any questions that you might have. Often, just the process of writing down a question will clarify the concept enough that you will understand it.

Now try to work the problems. Start with the drill problems and work some from each objective. If you can't solve a certain type of problem, reread the essay on that objective and the appropriate part of the text. If you still can't solve it, write down a specific question in your notebook. Try to identify which part of the problem you cannot solve and solve the rest of the problem or at least detail how you will solve it.

Ask other students in the class for answers to your questions. Sometimes they will understand concepts that you do not; often you will be able to help them. Many students form study groups that meet regularly. But do not just work problems to get the right answer. Make sure that you understand the technique well enough so that you can easily solve other problems of the same type by yourself. Write summaries of techniques in your notebook so you can refresh your skills at a later time.

Finally, ask your instructor for help with your unanswered questions. By working on your own and writing down your questions, you will know exactly what to ask and you will understand the answer much better. Do not wait until the day before the exam to question your instructor. Ask daily as new questions arise. Write the answers to your questions in your notebook. Also write a solved example of each type of problem that gave you trouble.

Now you have a notebook that really is helpful. It not only contains your lecture notes but also has questions and answers on every point that gave you trouble. Such a notebook is invaluable wh you study for class exams and also when, in a later course, you want to refresh your knowledge.

1 MATTER—ITS PROPERTIES AND MEASUREMENT

CHAPTER OBJECTIVES

1. Define and use the terms in bold in the summary.

This is an objective of every chapter. You should write each term and its definition on a 3×5-inch card. (Another option is to use a cassette recorder; see Objective 1-2.) A very serious mistake is to copy definitions from the book. The definitions should be in *your* words so they make sense to *you*. After all, *you* will be the one using these terms in the future. Consider the term *heterogeneous mixture*. The definition given is that such a mixture separates "into physically distinct regions of differing properties." How might this be restated? Perhaps as "a mixture that, by itself, separates into parts that I can easily tell apart." Now think of several examples. Fruit salad is one. You can easily distinguish the pears from the juice, the pineapple, and the grapes. Another example is Italian salad dressing that has been shaken. On standing, the oil separates from the vinegar and the herbs are clearly seen as separate pieces. Finally, use the term in a sentence: "Hash is a heterogeneous mixture because I can see two parts—meat and potatoes."

2. Use the terms *element, compound, homogeneous mixture,* and *heterogeneous mixture,* to describe common materials.

It is most important you know that pure substances are elements (incapable of being broken down chemically) and compounds (having fixed proportions of elements), while mixtures have varying proportions of substances. You could regard this as an extension of Objective 1. Not only should you know definitions of terms, but you should be able to apply these definitions. Pick everyday items, classify them, and write your choices on paper. You should choose simple items: the graphite in your pencil, wood, a penny, a stone, aluminum foil, air, vinegar, water, milk, glass, and so on.

3. Write the names and chemical symbols of the more common elements.

This is a memorization task: to know the symbol that goes with the name and vice versa. This can be done by studying 3×5-inch cards with the names of elements on one side and the symbols on the other. Another way to memorize material is to use a cassette recorder to make an audiocassette of material you wish to memorize, rather like a foreign language cassette. For instance, you might say "H is the symbol of the element…," and then pause for about 2 seconds before giving the answer: "hydrogen." Then, when you play the tape, you can attempt to fill in the pauses with the correct answers. Producing such a tape has the benefit of compelling you to organize the material you need to memorize. Listening to it and responding requires that you become actively involved in memorizing, which is more effective than simply reading over the material several times.

Common elements means different things to different people. A reasonable goal would be the main group (or representative elements), along with those in the first transition series (Sc through Zn) plus Ag, Au, Cd, and Hg. These are elements with atomic numbers $1-38$, $47-56$, and $79-88$. The atomic number is the whole number in each box in the periodic table, often given the symbol Z. The best way to learn names is by families (vertical columns) in the periodic table. For example, learn the elements Li, Na, K, Rb, Cs, Fr together. Elements in the same family behave similarly chemically. By learning them together, you are also learning chemistry, not just memorizing symbols.

4. Distinguish between physical and chemical properties and simple physical and chemical changes.

Choose a few simple items and write facts you know about each one. Then decide which property is chemical (depends on the reaction of one item with another to produce a new substance) and which is physical (causes no change in composition). For example, wood is brown or tan (physical), breaks more easily with than across the grain (physical), floats in water (physical), burns in air (chemical), blackens in sulfuric acid (chemical), and is somewhat flexible (physical). Note carefully that chemical changes produce new substances; physical ones do not.

5. Describe the principal features of the scientific method and its limitations.

Start by writing a summary of the scientific method, using the terms *law*, *hypothesis*, *experiment*, and *theory*. Now construct an example of the scientific method. A fanciful one follows:

Law: Nordic children have fair skin, blond hair, and blue eyes.

Hypotheses: Trolls eat dark-haired, -eyed, or -skinned children (mythological).

The lack of sun prevents the skin, eyes, and hair of young children from "tanning" (environmental). Children inherit these traits from their parents (genetic).

Experiments: The environmental hypothesis can be disproved by moving Nordic children to sunnier climates.

Their hair, eyes, and skin do not darken.

The principle of "Occam's razor" (the explanation that makes the fewest assumptions is the best) is used to discard the mythological hypothesis. After all, trolls are imaginary—they never have been observed.

Just because the environmental and mythological hypotheses are invalid does not prove the genetic one. Experiments are needed to test it, such as observing the children produced by one Nordic parent and one Mediterranean parent. But no amount of testing "proves" a hypothesis. A hypothesis always *may* fail a future experimental test. Then a new explanation for the law must be devised. Alternatively, a more complete theoretical explanation may be developed even though there has been no contradictory experiment.

What are the limitations of the scientific method? One criticism is that it is too rigid. It does not permit conclusions to be reached from insufficient data—a process that often is a mark of genius. Can you think of others? You develop a better appreciation for any method or theory if you know what it can and what it cannot do.

6. Know common units in the English system, and the relationships between them.

Common English units, their abbreviations, and the relationships between them follow. They should be memorized (a 3×5 -inch card or cassette recorder project) if you do not know them already. Your instructor may not require that you know these at all, but most scientists (and practically all engineers) in the United States do.

Volume measure: gallon (gal) = 4 quarts (qt.)   pint (pt.) = 16 fluid ounces (fl. oz.) = 2 cups (c)

quart (qt.) = 2 pt. = 32 fl. oz.   tablespoon (Tbs.) = 3 teaspoons (tsp.) = 1/2 fl. oz.

Linear measure: mile (mi.) = 1760 yd. = 5280 ft.   foot = 12 in.

yard (yd.) = 3 feet (ft.) = 36 inches (in.)

Mass measure: ton (t) = short ton = 2000 lb.

pound (lb. or #) = 16 ounces (oz.)

7. For the metric system, state the basic units of mass, length, and volume, as well as the common prefixes.

At present there are three units to be learned: gram (g) for mass, meter (m) for length, and liter (L) for volume. The following prefixes should be memorized (a 3×5 -inch card or cassette recorder project).

tera	(T)	10^{+12}	or	1,000,000,000,000	one trillion of
giga	(G)	10^{+9}	or	1,000,000,000	one billion of
mega	(M)	10^{+6}	or	1,000,000	one million of
kilo	(k)	10^{+3}	or	1000	one thousand of
deci	(d)	10^{-1}	or	1/10	one tenth of
centi	(c)	10^{-2}	or	1/100	one hundredth of
milli	(m)	10^{-3}	or	1/1000	one thousandth of
micro	(μ)	10^{-6}	or	1/1,000,000	one millionth of
nano	(n)	10^{-9}	or	1/1,000,000,000	one billionth of
pico	(p)	10^{-12}	or	1/1,000,000,000,000	one trillionth of

You should also know how to combine the prefixes and the units and be able to write the abbreviation of the resulting unit. For example, mL is the abbreviation for milliliter, 1/1000 liter; and Mg is the abbreviation for megagram, 10^6 gram (in addition to being the chemical symbol for the element magnesium).

8. State the relationships between English and metric units.

This is a 3 x 5 –inch card or cassette recorder project. There are many relationships, but they can be reduced to only three if you know the interrelationships within each system (Objectives 1-5 and 1-7). These three can be one from each of the following columns.

EXAMPLE 1.1. A professional tennis player's serve was clocked by radar at 128 mi./hr. What was the speed in meters per second?

Convert first from miles to meters using conversion factors available and then continue to convert from hours to seconds. The same result is achieved regardless of which conversion is done first.

$$\text{speed} = \frac{128 \text{ mi.}}{\text{hr.}} \times \frac{5280 \text{ ft.}}{\text{mi.}} \times \frac{30.48 \text{ cm}}{\text{ft.}} \times \frac{1 \text{ m}}{100 \text{ cm}} \times \frac{1 \text{ hr.}}{60 \text{ min.}} \times \frac{1 \text{ min.}}{60 \text{ sec.}} = 57.2 \text{ m/sec.}$$

Notice that the conversion can be shortened, depending on the conversion factors that one has memorized. (For example, if you know the number of seconds/hr you can eliminate one step.) The conversion factors are exact and do not limit the number of significant figures in the answer.

Mass	*Length*	*Volume*
$453.6 \text{ g} = 1 \text{ lb.}$	$2.540 \text{ cm} = 1 \text{ in.}$	$0.9464 \text{ L} = 1 \text{ qt.}$
$1 \text{ kg} = 2.205 \text{ lb.}$	$1 \text{ m} = 39.37 \text{ in.}$	$1 \text{ L} = 1.057 \text{ qt.}$
$28.35 \text{ g} = 1 \text{ oz.}$	$30.48 \text{ cm} = 1 \text{ ft.}$	$29.57 \text{ mL} = 1 \text{ fl.oz.}$

9. Describe the distinction between *precision* and *accuracy*.

Scientists express the care with which measurements have been made in two ways: (1) *Precision* describes whether repeated determinations of the same quantity (multiple measurements of the same thing) agree with each other; it is a measure of reproducibility. (2) *Accuracy* describes how closely a measurement agrees with some standard or accepted value of the same quantity; it can be thought of as a measure of closeness to an ideal value. *Imprecision*—lack of precision—is expressed as the deviation of the measurement from the average of many measurements. *Inaccuracy*—lack of accuracy—is expressed as the error from the accepted value.

10. Determine the number of significant figures in a numerical calculation.

11. Express the result of a calculation with the appropriate number of significant figures.

Here is another 3×5 -inch card or cassette recorder project. Write down, in your own words, the following rules.

(1) Significant digits include all non-zero digits, zeros located between significant digits (*interior zeros*), and zeros located after non-zero digits (*trailing zeros*) and to the right of the decimal point.

(2) Zeros that precede non-zero digits (*leading zeros*) are not significant. Zeros that end a number with no decimal point may or may not be significant.

(3) The result of a multiplication or a division should have as many significant figures as the factor with the *fewest* number of significant figures.

(4) The result of an addition or a subtraction is rounded to the same number of decimal places as the term with the *fewest* number of decimal places.

(5) When rounding the digits 0, 1, 2, 3, or 4, round down (that is, truncate the digit). When rounding the digits 5, 6, 7, 8, or 9, round up.

EXAMPLE 1-2 Express the result of the following calculation with the appropriate number of significant figures:

$$\frac{[(725.6-19.1)/760.00]\ 75\times10^{-3}}{0.082057\times293.2}=?$$

The result is 2.9×10^{-3}. The factor 7.5×10^{-3} determines the number of significant figures.

12. Express numbers in scientific notation.

This is closely related to being able to work with significant figures. A reasonable pair of rules is:

(a) A number in scientific notation is written as a many-digit number, with only one significant figure to the left of the decimal point. This many-digit number is multiplied by 10 to some power.

(b) The power of 10 is determined by starting with the original number, the number being converted to scientific notation. Count the number of places that the decimal point must be moved in order to have only one significant figure to the left of the decimal point. For each place that the decimal point is moved to the left, the power of 10 is increased by one. For each place that the decimal point is moved to the right, the power of 10 is decreased by one.

EXAMPLE 1–3 Express each of the following numbers in scientific notation:

A. 1742 B. 0.0008649 C. 114.692×10^{6}

A. $1742=1.742\times10^{3}$ Since the decimal point was moved three places to the left to transform 1742 to 1.742, the power of 10 is 3.

B. $0.0008649=8.649\times10^{-4}$ Since the decimal point was moved four places to the right to transform 0.0008649 to 8.649, the power of 10 is -4.

C. $114.692\times10^{6}=1.14692\times10^{8}$ Since the decimal point was moved two places to the left to transform 114.692 to 1.14692, the power of 10 is $6+2=8$.

13. Write a conversion factor from a relationship between two quantities, and use conversion factors to solve problems.

Probably the most powerful problem-solving method you can learn is the *conversion factor method*, also known by other names, such as the *factor units method*. The method is valuable not only in general chemistry but in any numerical problem-solving course. There are many ways to use the method. One of these follows:

(1) Identify what is wanted and what units it should have. Write this down.
(2) Identify what is given, or what you start with. Write this down with its given units.
(3) Multiply what is given by one or more conversion factors to change the units into those of the answer.

EXAMPLE 1–4 How many teaspoons are there in 5.00 gallons?

Apply the steps in the order given above:

(1) ? teaspoons (tsp.)

(2) 5.00 gallons (gal.)

$$(3)\ \text{volume} = 5.00\ \text{gal.} \times \frac{4\ \text{qt.}}{1\ \text{gal.}} \times \frac{32\ \text{fl.oz.}}{1\ \text{qt.}} \times \frac{3\ \text{tsp.}}{0.5\ \text{fl.oz.}} = 3.84 \times 10^3\ \text{tsp.}$$

There is no requirement that you remember or discover the factors from right to left. You might have remembered 3 tsp. = 0.5 fl.oz. and have started with that, building the string of factors leftward from the right.

EXAMPLE 1–5 How many gallons are in 1.00 cubic foot (ft.^3)?

There is no simple relationship between gallons and cubic feet in the English system, but there is a relationship between volume and cubic length in the metric system $(\text{L} = 1000\ \text{cm}^3)$. The steps in the solution follow:

(1) ? gal .

(2) 1.00 ft.^3

(3)

$$\text{volume} = 1.00\,\text{ft.}^3 \times \left(\frac{12\ \text{in.}}{1\ \text{f.t}}\right)^3 \times \left(\frac{2.54\ \text{cm}}{1\ \text{in.}}\right)^3 \times \frac{1\ \text{L}}{1000\ \text{cm}^3} \times \frac{1\ \text{qt.}}{0.946\ \text{L}} \times \frac{1\ \text{gal.}}{4\ \text{qt.}} = 7.48\ \text{gal.}$$

In this example, we started in the center $(\text{L} = 1000\ \text{cm}^3)$ of the string of conversion factors and worked outward. Note also that two relationships (12 in. = ft. and 2.54 cm = in.) are cubed to get the needed cubic units of length. Finally, recognize that it is quite helpful to memorize the relationships in the English system, those in the metric system, and those between the two systems. In this way you have a sufficient pool of relationships to use in constructing conversion factors.

EXAMPLE 1–6 A cylinder has a diameter of 0.48 in. and a length of 2.15 in. What is its volume in cm^3?

(1) ? cm^3

(2) $d = 0.48\ \text{in.} = 2r, i = 2.15\ \text{in.}\ V = \pi r^2 h$

$$(3)\qquad V = 3.142\left(\frac{0.48\ \text{in.}}{2}\right)^2 2.15\ \text{in.}\left(\frac{2.54\ \text{cm}}{1\ \text{in.}}\right)^3 = 6.4\ \text{cm}^3$$

We should note that there is another method for solving problems: *ratios*, or *proportions*. To convert 5.00 in. to cm using this technique, one sets up the ratio:

1.00 in. is to 2.54 cm as 5.00 in. is to ? cm

or

1.00 in.: 2.54 cm :: 5.00 in. : ? cm

or

$$\frac{1.00\ \text{in.}}{2.54\ \text{cm}} = \frac{5.00\ \text{in.}}{?\ \text{cm}}$$

A little algebra gives the following equation,

$$?\text{cm} = 5.00\ \text{in.} \times \frac{2.54\ \text{cm}}{1.00\ \text{in.}} = 12.7\ \text{cm}$$

which is the conversion factor method. Therefore, the two methods *are* equivalent. But the ratio technique is cumbersome (a new ratio is needed for each change of unit): it does not allow you to work backward or from the middle in solving the problem, and it makes squaring or cubing units very difficult.

14. Express and use density in the form of conversion factors.

Density is both a physical property of a substance and the means of interconverting mass and volume of that substance. The defining equation $(d = m/V)$ has three variables: density, mass, and volume. There are only three general types of problems involving density, illustrated in Examples 1-7, 1-8, and 1-9.

EXAMPLE 1–7 Given mass and volume, determine density.

An empty container weighs 206 g. Filled with 242 mL of liquid, it weighs 938 g. What is the density of the liquid?

The mass of liquid is the difference between full and empty masses. Apply the definition to determine density.

$$\text{density} = \frac{\text{liquid mass}}{\text{volume}} = \frac{(938\text{ g} - 206\text{ g})}{242\text{ mL}} = 3.02\text{ g/mL}$$

EXAMPLE 1–8 Given volume and density, determine mass.

A 27.4 mL gold (19.3 g/mL) object has what mass?

Here we use the conversion factor method:

(1) ? g Au

(2) 27.4 mL Au

$$(3)\ \text{mass} = 27.4\text{ mL} \times \frac{19.3\text{ g Au}}{1\text{ mL}} = 529\text{ g Au}$$

EXAMPLE 1–9 Given mass and density, determine volume.

A 75.2 g piece of zirconium (6.42 g/mL) has what volume?

Again we use the conversion factor method:

(1) ? mL Zr

(2) 75.2 g Zr

$$(3)\ \text{volume} = 75.2\text{ g Zr} \times \frac{1\text{ mL Zr}}{6.49\text{ g}} = 11.6\text{ mL Zr}$$

15. Express and use percent composition in terms of conversion factors.

Percent means parts per hundred. Thus, 40.0% C in acetic acid means 40.0 g C in 100.0 g of acetic acid $(40.0\%\text{C} = 40.0\text{g C}/100.0\text{g acetic acid})$. Again, three types of problems can be expected. Can you identify all three types? One of them is illustrated in the following example.

EXAMPLE 1-10 What mass of acetic acid contains 247 g C?

We proceed via the conversion factor method:

(1) ? g acetic acid

(2) 247 g C

(3) $\text{mass} = 247\text{ g C} \times \dfrac{100.0\text{ g acetic acid}}{40.0\text{ g C}} = 618\text{ acetic acid}$

16. Be able to convert between Fahrenheit and Celsius temperatures.

You need to memorize and be able to use two relationships.

$°C = (°F - 32) \times \frac{5}{9}$ AND $°F = (°C \times \frac{9}{5}) + 32$

EXAMPLE 1-11 Perform the following temperature conversions:

A. $112\ °F = ?\ °C$ B. $25\ °C = ?\ °F$ C. $-30\ °C = ?\ °F$

A. $°C = (112\ °F - 32)(\frac{5}{9}) = 80.(\frac{5}{9}) = 44\ °C$

B. $°F = (25\ °C \times \frac{9}{5}) + 32 = 45 + 32 = 77\ °F$

C. $°F = (-30\ °C \times \frac{9}{5}) + 32 = -54 + 32 = -22\ °F.$

The following list may help you develop a useful "feel" for various Celsius temperatures. Notice that a 5°C temperature rise corresponds to a 9 °F increase.

0 °C = 32 °F	water freezes, winter day	5 °C = 41 °F	early spring morning
10 °C = 50 °F	early spring afternoon	15 °C = 59 °F	mid-spring afternoon
20 °C = 68 °F	heated home in winter	25 °C = 77 °F	cooled home in summer
30 °C = 86 °F	summer afternoon	35 °C = 95 °F	hot summer day
37 °C = 98.6 °F	normal body temperature		

17. Solve algebraic equations that arise in the course of working chemistry problems.

Solving an algebraic equation generally means obtaining a new equation, with the symbol for one variable isolated on one side (generally the left-hand side) and the remainder of the equation on the other side. We then say that we have "solved the equation" for the variable that stands alone. (Of course, that variable does not appear on the other side of the equation.) A very powerful method of solving equations is to perform the same operation on both sides of the equation. This is done several times with different operations, until the symbol for one variable *is* isolated. Keep in mind the following list of things that you can do to both sides of any equation:

(a) Add the same thing to both sides *or* subtract the same thing from both sides.

(b) Multiply both sides by the same thing *or* divide both sides by the same thing.

(c) Raise both sides to the same power. This includes the power -1, so both sides can be inverted or flipped over. It also includes fractional powers, so that you can take the square root, for instance, of both sides.

Often solving an equation will require a lot of trial and error. *Do not become discouraged!* A reason why some people can solve algebraic equations more quickly than others is that they have had more practice. Study the following example and practice on the drill problems:

EXAMPLE 1-12 Solve the following equation for *P*:

$$\left(P+\frac{an^2}{V^2}\right)(V-nb)=nRT.$$

First, divide each side by :

$$P+\frac{an^2}{V^2}=\frac{nRT}{(V-nb)}$$

Then, subtract an^2/V^2 from each side.

$$P=\frac{nRT}{(V-nb)}-\frac{an^2}{V^2}$$

SELF-ASSESSMENT EXERCISES

Comments and additional explanations are provided for selected Self-Assessment Exercises found at the end of the chapter in the text.

94. Correct: (a) cola drink. A cola drink has a number of components (sweeteners, coloring, flavoring, etc.), but they are uniformly distributed throughout the drink. Incorrect: (b) distilled water is a pure substance; the same is true for oxygen (c) in a cylinder. The material in a kitchen blender (d) will vary in composition. The composition will depend on where the sample is taken, with the heavier material in the bottom.

95. Correct: (c). Mass is the amount of substance and does not vary with the force of gravity. Weight will depend on the force of gravity and will be less on the moon.

96. Correct: (c) and (e). Because there are 1000 mg/g, if one converted each of these numbers to mg, (c) and (e) are expressed to the nearest mg. For example, (a) is 32700 mg (the last two zeros are uncertain). (d) is 32707 mg (the last 7 is uncertain).

97. Correct (d) 105^0C. Without doing any calculations, we know that Kelvin = 273 + C. Therefore, (a) and (b) are less than 105 ^{0}C. We compare (c) and (d) to the respective boiling points of water: 100 ^{0}C and 212 ^{0}F. Answer (c) is 5 ^{0}F above the boiling point and answer (d) is 5 ^{0}C above the boiling point. We know that a degree C is larger than a degree F. Therefore, (d) is the higher temperature.

98. Correct (b). The density of water is 1 g/mL, so answer (a) is 752 g. We know that 1050 ml x 0.789 g/ml will be greater than 752 g. Answer (c) is given in grams. For answer (d), converting to cm^3, 0.25 ft^3 x $(12\ in/ft)^3$ x $(2.54\ cm/in)^3$ and using density gives a mass of 778 g, which is less than (b).

99. Correct (b). The answer can have no more significant figures than the least precise number, in this case 9 mm.

100. (e) 2000 mg, (a) 1400 km, (c) 0.00304 g, (e) 1516 g, and (d) 125.34 cm. These numbers have 1, 2, 3, 4, and 5 significant figures, respectively. The precision of (e) and (a) is uncertain. To indicate precise numbers, scientific notation is used. For example, 2.000×10^3 mg has four significant figures as compared to one significant figure for 2000.

101. (c) will have the largest mass of iron. The estimates are as follows: (a) 1.0 kg of iron; (b) slightly less than 1 kg because the volume of the cube is 125 cm^3; (c) volume is 9000 cm^3, and the mass is, therefore, about 70% of 7.7 x 9000, or approximately 72 kg. For answer (d), the mass or iron is approximately 1000 ml x 1.3 g/ml x 0.3 x 0.34, or approximately 1.3 kg .

102. The density is 2.819 g/cm^3. The foil will be volume x density = 25.305 g, and because the area is 248 cm^2, the thickness will be 0.0362 cm or 0.362 mm.

Drill Problems

1. (1) Determine the number of significant figures in the following numbers.

A. 1837 B. 3.14145×10^4 C. 6005 D. 0.08206 E. 0.000014
F. 302400 G. 632 H. 8.732 I. 14.163000 J. 14.000
K. 19.7324 L. 302400.0 M. 149356 N. 205.8 O. 0.0019872
P. 8.7300 Q. 1900 R. 20000 S. 426.1 T. 1200.43
U. 1900.00 V. 0.00743 W. 6000 X. 60.0

(2) Round off each of the following numbers to four significant figures.

Y. 6.16782 Z. 213.25 Γ. 1200.43 Δ. 3135.69
Θ. 6.19648 Λ. 14.163000 Ξ. 0.0022457 Π. 152.00
Σ. 0.0019872 Υ. 302400 Φ. 14.16300 Ψ. 3.14145×10^4

2. Determine the value of each of the following expressions, with the correct number of significant figures:

(1) Multiplication and division, following the rules for significant figures.

A. 1.86/3.14 B. $(6.6262 \times 10^{-27})(2567)$ C. (37.2)(1.5)
D. (200)(87.45) E. (998)(32.157)/36 F. 4.51545/0.15
G. $(5.00 \times 10^2)/36.72$ H. 374.1/1800 I. 31.11/2.04
J. $(1.40 \times 10^8)(37.842)/147.3562$

(2) Addition and subtraction, following the rules for significant figures.

K. $104 + 37.2 - 18.57$ L. $87.6 - 0.005$
M. $6.23 + 915 - 1012.7$ N. $87.9 + 11.3 + 9.6$
O. $36.516 + 0.00258 - 32.157$ P. $6.47 \times 10^2 + 4.2 \times 10^1 + 6.8$
Q. $4.30 + 29.1 + 100.3452$ R. $204.5 - 96.5 - 32.1$

(3) Combined operations, following the rules for significant figures.

S. $(94.3)(12) - 7.62 + 300.0$ T. $(5.19 \times 10^{-2} + 1.83)(2.19 \times 10^2)$
U. $0.318 + 1.6 \times 10^{-2}(6.40/12.1) - 2.19$ V. $(3.18)(2.4)/(1.92) - 0.17$

QUIZZES

Each of the following short quizzes is designed to test your mastery of the objectives of this chapter. Limit yourself to 20 minutes for each one. The questions are multiple choice; choose the best answer for each question.

QUIZ A

1. A 6.0-in. ruler is how long in centimeters? (a) 6.0/2.54; (b) 6.0/39.37; (c) (6.0)(36)(39.37); (d) (6.0)(2.54); (e) none of these.
2. What is the mass in pounds of a small child who weighs 23.0 kg? (a) (23.0)(1000/454); (b) (23.0/2.20); (c) (23.0/1000)454; (d) (23.0)(1000)(454); (e) none of these.
3. 742 mL is how many fluid ounces? (a) 742/946; (b) 742/[(946)(32)]; (c) (742/32)(946); (d) (742)(946)(32); (e) none of these.
4. An object weighs 946 g and occupies 241 mL. What is its density? (a) 241/946; (b) 241 + 946; (c) (241)(946); (d) 1/[(241)(946)]; (e) none of these.
5. Italian salad dressing is an example of (a) an element; (b) a compound; (c) a homogeneous mixture; (d) a heterogeneous mixture; (e) none of these.
6. The density of margarine is 0.96 g/mL. What is the mass of 250 mL? (a) 1/[(0.96)(250)]; (b) 0.96/250; (c) 250/0.96; (d) (250)(0.96); (e) none of these.
7. 88 ft/s is how fast in meters/min? (a) (88/12)(2.54/100)(60); (b) (88)(12)(2.54)(60/100); (c) (88)(12)(100/2.54)(60); (d) (88/12)(100/2.54)/60; (e) none of these.
8. Matter that cannot be further separated by physical means but can be separated by chemical means is called a(n) (a) element; (b) compound; (c) homogeneous mixture; (d) heterogeneous mixture; (e) none of these.
9. A concise summary of many observations is called a(n) (a) law; (b) theory; (c) experiment; (d) hypothesis; (e) none of these.
10. 32 °C is equivalent to (a) 0 °F; (b) 58 °F; (c) 18 °F; (d) 50 °F; (e) none of these.
11. Phosphorus, sodium, and helium, respectively, have the symbols (a) Ph, Na, H; (b) P, So, He; (c) P, Na, He; (d) Ps, Na, He; (e) none of these.
12. Which answer has the correct number of significant figures? (a) $(14.7 + 24.312) \times 87.27 = 3402$; (b) $(58 + 18 + 51) / 3.000 = 36$; (c) $[97.2/(114 - 37)] = 1.26$; (d) $(1.172 - 0.4963)(4.193) = 2.83$; (e) none of these.

QUIZ B

1. The density of gold is 19.3 g/mL. What is the volume in mL of 231 g of Au? (a) 231/19.3; (b) 19.3/231; (c) (19.3)(231); (d) 1/[(19.3)(231)]; (e) none of these.
2. What is the volume in milliliters of one cup? (a) (1/2)(946/2); (b) (1)(2)(946)(2); (c) (1)(2)(946/2); (d) (1/2)(946); (e) none of these.

3. What is the height of a 6-ft. tall person in centimeters? (a) (6)(12)(2.54); (b) (6/12)(2.54); (c) (6)(2.54/12); (d) (6)(12/2.54); (e) none of these.

4. One acre equals how many square meters $(640 \text{ acres} = 1 \text{ mi.}^2)$? (a) (1.0)(640)/[(1760)(36)(0.0254)]; (b) (1.0/640)/[(1760)(36)(0.0254)]; (c) $(1.0/640)/[1760)(36)(0.0254)]^2$; (d) $(1.0)[(1760)(36)(0.0254)]^2$; (e) none of these.

5. Matter that cannot be further broken down by chemical means is called a(n) (a) element; (b) compound; (c) homogeneous mixture; (d) heterogeneous mixture; (e) none of these.

6. "A less dense solid will float in a more dense liquid" is an example of a(n) (a) hypothesis; (b) theory; (c) experiment; (d) law; (e) none of these.

7. An object weighs 314 g and occupies 432 mL. What is its density? (a) (314)(432); (b) 432/314; (c) 314/432; (d) 1/[(314)(432)]; (e) none of these.

8. 200 lb equals how many kilograms? (a) (200/454)(1000); (b) (200)(2.20); (c) (200)(454/1000); (d) (200)(1000/454); (e) none of these.

9. Sound is an example of a(n) (a) element; (b) compound; (c) homogeneous mixture; (d) heterogeneous mixture; (e) none of these.

10. Carbon, fluorine, and sodium, respectively, have the symbols (a) C, F, Na; (b) C, Fl, Na; (c) C, F, Sm; (d) Ca, F, Na; (e) none of these.

11. Which answer has the correct number of significant figures? (a) $(37 + 54 + 42)/3.000 = 44.3$; (b) $(726 + 835)/24.0 = 65.04$; (c) $(202 + 74)/73.1 = 3.8$; (d) $(1.43 + 7.26 + 9.1)/14.23 = 1.250$; (e) none of these.

12. 432 °C is equivalent to (a) 720 °F; (b) 272 °F; (c) 835 °F; (d) 810 °F; (e) none of these.

Quiz C

1. A procedure designed to test the truth or the validity of an explanation about many observations is called a(n) (a) law; (b) theory; (c) experiment; (d) hypothesis; (e) none of these.

2. An object has a mass of 764 g and occupies 265 mL. What is its density? (a) 265/764; (b) 764/265; (c) (265)(764); (d) 1/[(764)(265)]; (e) none of these.

3. What is the volume in quarts of a 250-mL beaker? (a) (250)(1000)(1.057); (b) (250/1000)(0.946); (c) (250/1000)(1.057); (d) (250)(1000/1.057); (e) none of these.

4. What is the mass in kilograms of an object that weighs 29.0 lb.? (a) (29.0/454)(1000); (b) (29.0)(2.20); (c) (29.0)(1000/454); (d) (29.0)(454/1000); (e) none of these.

5. If a room is 10.0 m wide, its width in feet is (a) (10.0)(100/2.54); (b) (10.0/1000)(2.54); (c) (10.0)(1200/2.54); (d) (10.0)(100)(2.54)(12); (e) none of these.

6. The density of copper is 8.92 g/cm^2. What is the volume of 651 g? (a) (651)(8.92); (b) 8.92/651; (c) 651/8.92; (d) $651 + 892$; (e) none of these.

7. 14.7 lb/in.^{2} equals how many kg/cm^2? (a) $14.7(0.454)(2.54)^2$; (b) $(14.7)(454)(2.54)^2$; (c) (14.7/2.20)(2.54); (d) $(14.7)(2.20)/(2.54)^2$; (e) none of these.

8. Red food coloring is an example of a(n) (a) element; (b) compound; (c) homogeneous mixture; (d) heterogeneous mixture; (e) none of these.

9. Matter that can be separated by physical means but *not* by chemical means is called a(n) (a) element; (b) compound; (c) homogeneous mixture; (d) heterogeneous mixture; (e) none of these.

10. Bromine, magnesium, and beryllium, respectively, have the symbols (a) B, Ma, Be; (b) Br, Mg, B; (c) Br, Mn, Be; (d) Br, Mn, B; (e) none of these.

11. Which answer has the correct number of significant figures? (a) $(107.4-17.2)/9.164=9.843$; (b) $9.843\times(44.7-65.2)=1082$; (c) $(1.172+0.4963)(3.145)=5.25$; (d) $(651+8.92)/14.174=46.558$; (e) none of these.

12. 450 °F is equivalent to (a) 250 °C; (b) 232 °C; (c) 218 °C; (d) 778 °C; (e) none of these.

QUIZ D

1. An object weighs 876 g and occupies 923 mL. What is its density? (a) 923/876; (b) 876/923; (c) (923)(876); (d) $923+876$; (e) none of these.

2. An automobile weighs 2750 lb. What is its mass in Mg? (a) (2750/2.2)(1000); (b) (2750)(2.20)(1000); (c) (2750)(2.2)/1000; (d) (2750/2.20)/1000; (e) none of these.

3. 100 yd. is how long in m? (36 in. = 1 yd)? (a) (100)(36)(2.54); (b) (100)(36)(2.54/100) (c) (100/36)(100/2.54); (d) (100/36)(2.54/100); (e) none of these.

4. What is the volume in mL of 27.5 fl. oz. (32 fl. oz. = 1 qt.)? (a) (37.5)(32946); (b) (37.5/32)(946); (c) (37.5)(32)(946); (d) 37.5/[(32)(946)]; (e) none of these.

5. 13.6 g/cm^3 equals how many lb/qt.? (a) (13.6/454)(1000/0.946); (b) (13.6)(454/1000)(1.057); (c) (13.6)(454)(1.057/1000); (d) (13.6/454)(1000/1.057); (e) none of these.

6. Gold is a(n) (a) element; (b) compound; (c) homogeneous mixture; (d) heterogeneous mixture; (e) none of these.

7. Matter that can be further separated by physical means but appears uniform to the eye is called a(n) (a) element; (b) compound; (c) homogeneous mixture; (d) heterogeneous mixture; (e) none of these.

8. An explanation of many observations, which has been proven true by many tests, is called a(n) (a) law; (b) theory; (c) experiment; (d) hypothesis; (e) none of these.

9. The density of silver is 10.5 g/mL. What is the volume of 475 g? (a) 475/10.5; (b) 10.5/475; (c) 1/[(10.5)(475)]; (d) (475)(10.5); (e) none of these.

10. Hydrogen, potassium, and chlorine, respectively, have the symbols (a) H, Po, Cl; (b) Hy, K, C; (c) H, K, Cl; (d) H, P, Ch; (e) none of these.

11. Which answer has the correct number of significant figures? (a) $(10.46 + 1.7394)/12.2 = 1.000$; (b) $(14.234 + 0.005169)/0.1243 = 115$; (c) $(0.723 + 9.746)/15.493 = 0.6757$; (d) $(300.4 + 0.00216)/9.18745 = 32.697$; (e) none of these.

12. 98.6 °F is equivalent to (a) 54.7 °C; (b) 37.0 °C; (c) 120 °C; (d) 86.8 °C; (e) none of these.

SAMPLE TEST

This test is designed to take 30 minutes. Treat it as if it were an in-class examination. Only allow yourself a periodic table for reference and a calculator. Express your answers with the correct number of significant figures and with proper units.

1. A 10-karat gold ring contains 41.7% gold and weighs 10.72 g. If gold sells for $652.50/oz, what is the value of the gold in the ring?

2. In the production of ammonia, the following processes occur:

   (a) *Air* is liquefied at low temperature and high *pressure.*

   (b) The *temperature* of the *liquid air* is raised gradually until the oxygen boils off. Essentially pure *liquid nitrogen* remains.

   (c) *Natural gas* is reacted with *steam* to produce *carbon dioxide* and *hydrogen.* The proportions of the products vary with the composition of the natural gas.

   (d) Hydrogen and nitrogen (both gases) are combined at high temperatures and pressures. They react to form *ammonia gas*. Some unreacted hydrogen and nitrogen remain.

   (e) The gases are cooled until the ammonia liquefies, then the gaseous hydrogen and nitrogen are re-circulated to react again.

   Each of the steps described above involves a physical change or a chemical reaction. Indicate which occurs in each step and briefly explain why you chose as you did.

   Each of the *italicized* words or phrases in the descriptions above refers to an element, a compound, a mixture, or none of these. Identify each one and explain your choice.

3. Perform the following conversions:

   A. 12.5 lb./in.^{2} = ____ oz./cm^2 = ____ kg/m^2

   B. 77.4 °F = ____ °C

   C. 5.2 m^3 = ____ L = ____ qt.

2 Atoms and the Atomic Theory

CHAPTER OBJECTIVES

*** 1. State and apply the laws of conservation of mass and definite composition.**

The first goal is to memorize your versions of these two laws: (a) mass is neither created nor destroyed; and (b) a compound has the same composition, no matter what its source. Application involves recognizing when the laws are violated and when they are obeyed. Examples of the application of these laws follow.

EXAMPLE 2-1 (The law of conservation of mass) A quantity of 4.00 g of methane (the main component of natural gas) burned in 16.00 g of oxygen produces 11.00 g of carbon dioxide and what mass of water?

If water and carbon dioxide are the only products, 9.00 g of water must be produced. This is because there are 20.00 g of reactants (4.00 g methane and 16.00 g oxygen) and there must be the same mass of products.

EXAMPLE 2-2 (The law of definite composition) A sample of methane taken from a mine (mine gas) was found to contain only C and H with 3.00 g C for every 1.00 g of H. How much hydrogen should be present in 100.0 g of methane taken from a natural gas well?

Note that 4.00 g of the mine gas contains 3.00 g C and 1.00 g H. These proportions should be the same for the natural gas methane. Thus, we can use the conversion factor method.

(1) ? g H

(2) 100.0 g methane

(3) $\text{mass} = 100.0 \text{ g methane} \times \dfrac{1.00 \text{ g H}}{4.00 \text{ g methane}} = 25.0 \text{ g H}$

2. State the basic assumptions of Dalton's atomic theory.

You should write down and memorize your versions of Dalton's assumptions. Shortened versions are: (a) Elements are composed of indestructible atoms. (b) Atoms of a given element are alike; atoms differ between elements. (c) Atoms unite in small whole number ratios to form compounds. Although the assumptions are not exactly true, they are close enough that they can help us remember how to use the atomic theory.

3. List some of the characteristic properties of cathode rays and of canal (anode) rays.

A brief list of cathode ray properties includes: (a) straight–line travel; (b) travel from cathode when current flows; (c) deflected as if negatively charged; (d) properties are independent of current source, tube material, cáthode material, and the gas that filled the tube; (e) invisible, but give off light when they strike the glass; (f) have mass.

Canal rays (positive rays or anode rays) form when *electrons* (cathode rays) knock electrons out of neutral atoms. Canal rays are, thus, positive ions or cations (cat′ - eye - uns). Their properties are: (a) straight-line travel; (b) travel toward cathode when current flows; (c) deflected as if positively charged; (d) properties depend on the gas that filled the tube, but *not* on other factors; (e) have mass but are invisible until they strike certain other objects. Canal rays are not fundamental particles since their properties are not the same under all circumstances but depend on the gas that originally filled the tube. The properties of these two types of rays lead to the idea that the atom is composed of positive and negative parts, and the negative parts (the electrons) are easily removed. You should memorize your versions of the properties of these two types of rays.

4. Describe the production of X-rays, the phenomenon of radioactivity, and the characteristics of α, β, and γ radiation.

X-rays are given off when cathode rays (electrons) are stopped by an object. X-rays penetrate matter easily and have no mass or charge. The degree to which they penetrate matter depends on both their energy and the type of matter, principally the matter's density.

Becquerel discovered radioactivity by chance. He observed that certain materials produce penetrating radiation that is unaffected by attempts to enhance or diminish it. Radioactivity is a characteristic property of certain elements. α (alpha) particles are $^4He^{2+}$ cations. They are massive and have poor penetrating power, yet they have high ionizing power. β (beta) particles are electrons. They have moderate penetrating and ionizing powers. When α or β particles are emitted by an atom, the atom changes to one of a different element (as is described in Objective 26-1). γ (gamma) rays are similar to X-rays, but they are of very high energy. They have a high penetrating power, but poor ionizing power. Except for carrying away energy, they leave unchanged the atom from which they are emitted.

5. Describe Thomson's *m/e* experiment, Millikan's oil drop experiment (to measure the charge on the electron), and Rutherford's gold–foil experiment (to establish the existence of the atomic nucleus).

In Thomson's experiment, a beam of electrons of known energy is deflected by a magnetic field, curving the beam. The radius of the curve depends on the energy of the electrons, the mass of the electrons, and the strength of the magnetic field. The energy of the electrons, in turn, depends on their charge. Thomson knew the magnetic field strength but neither the mass nor the charge of the electron. To make his experiment clearer, consider another experiment very similar in principle. Attach a weight to a spring and, holding the other end of the spring, whirl the weight about you over your head. The distance between your head and the mass (that is, the radius of the curve) depends on three things, as shown in Table 2-1.

TABLE 2-1 Relation Between Thomson's Experiment and Its Analog

Depends on	To make radius of curve greater	Analog in Thomson's experiment
mass of weight	increase weight	electron mass
speed of weight	increase speed	electron energy
stiffness of spring	use a slacker spring	magnetic field strength

Millikan's experiment is designed to measure the fundamental unit of charge. Oil drops are charged with X-rays and allowed to fall between two charged plates. The charge on each drop can be found by varying the voltage between the plates until the electrical force felt by the charged drop just balances the force of gravity. Then the drop stands still. The charge on each drop always is found to be an exact multiple of 1.602×10^{-19} coulomb.

Geiger and Marsden beamed α particles at thin gold foil. Most of the particles passed straight through, but some were deflected—a few by great angles. The experiment is similar to shooting BBs through a hole into a hot air balloon, in the center of which is a metal object. If enough BBs are shot, the size and shape of the metal object can be deduced from the pattern produced by the BBs striking the inner surface of the balloon. Rutherford interpreted this data as evidence for a nuclear atom. *Your* descriptions of these three experiments should be learned firmly.

6. State the features of Rutherford's nuclear atom and how it differs from Thomson's model of the atom.

The Rutherford model of the atom consists of a very small nucleus (about 10^{-13} cm in diameter), which contains all of the positive charge and more than 99.9% of the mass, surrounded by a tenuous cloud of electrons. The electrons possess all of the negative charge, less than 0.1% of the mass, and fill most of the space in the atom (which is about 10^{-8} cm in diameter).

Thomson's model lacks the *nuclear kernel* (the small, dense nucleus). The negative electrons are embedded in a uniform sphere of positive charge, rather like plums in a pudding. The model is sometimes called the "plum pudding" model.

7. Perform calculations involving the masses and charges of the proton, neutron, and electron.

These calculations involve determining mass-to-charge ratios. In order to do them successfully, you need to know well the charge and mass of the proton (charge = +1; mass = 1.0073 u), neutron (charge = 0; mass = 1.0087 amu), and electron (charge = −1; mass = 0.0005486 u). The unit of charge is that of the electron: 1.602×10^{-19} The atomic mass unit (u), sometimes called the dalton (d), equals 1.6605×10^{-24} g. Note that the mass of the electron is about 1/1836 that of the proton.

*** 8. List the numbers of protons, neutrons, and electrons present in atoms and ions, using the symbolism $^{A}_{Z}E$.**

The complete symbol for an atom or ion consists of the elemental symbol surrounded by subscripts and superscripts. Two of the four "corners" thus created are used for an atomic species and the third for an ionic species. (The fourth is used when writing formulas of compounds.)

(1) The leading superscript (upper left) is the mass number. This is also the number of nucleons (a *nucleon* is a proton or a neutron). Older books may show the mass number as a trailing superscript (upper right), as this was correct until about 1960. Thus ^{235}U was U^{235}. Often we see U-235 instead of ^{235}U, especially in newspapers and magazines.

(2) The leading subscript (lower left) is the atomic number, or proton number. It truly is unnecessary because the elemental symbol determines the atomic number, but often it is included for emphasis, as in $^{235}_{92}U$.

(3) The trailing superscript (upper right) is the charge or the number of protons (atomic number) minus the number of electrons. The sign $(+ \text{ or } -)$ always must be included: $^{235}_{92}U^{3+}$. The number is zero for a neutral atom, but the zero is written only for emphasis.

EXAMPLE 2-3 What is the atomic number, mass number, and charge of the nuclide $^{19}F^-$?

$^{19}F^-$ has an atomic number of 9 (F has 9 protons); a mass number of 19 $(19\text{ nucleons} = \text{protons} + x\text{ neutrons}$; therefore, $x = 10$ neutrons); and a charge of $-1\,(9\text{ protons} - y\text{ electrons} = -1$; therefore, $y = 10\text{ electrons})$.

EXAMPLE 2–4. Consider the following:

Neutrons	Protons	Electrons
7	7	7
8	6	6
7	7	8
6	6	5

Which of these are isotopes of nitrogen?

The number of protons does not change for the isotopes of the same element. Isotopes of nitrogen will have seven protons. Line 1 is nitrogen-14 and is neutral. Line 3 is nitrogen-14 with a -1 charge (an ion). Lines 2 and 4 are carbon-14 (neutral) and carbon-12 $(+1\,\text{ion})$, respectively.

*** 9. Describe how atomic mass ratios are determined by mass spectrometry and use these ratios to determine relative atomic masses.**

The mass spectrometer is an elegant version of Thomson's experiment for determining the mass-to-charge ratio of electrons, as described in Objective 2–5. This apparatus was modified to determine the properties of canal rays by the addition of a velocity selector (electric and magnetic fields just after the collimator) to ensure that all ions have the same velocity. The instrument determines ratios of atomic masses from which we compute relative atomic masses.

EXAMPLE 2-5 Naturally occurring carbon is analyzed with a mass spectrometer and the ratio of the mass of carbon-13 to that of carbon-12 is determined to be 1.083613. What is the relative mass of carbon-13?

The relative mass of carbon-13 is determined from the mass ratio and the known mass of carbon-12. Mass of $^{13}C = 1.083613 \times 12.00000 = 13.00335$.

*** 10. Calculate the atomic mass of an element from the known masses and relative abundances of its naturally occurring isotopes.**

The atomic weight of an element is the decimal number appearing in the element's block in the periodic table. For example, the atomic weight of carbon is 12.011. This atomic weight represents an average atomic mass. It can be determined from the combining weights of elements in chemical reactions. These combining weights represent the average mass of many atoms of the same element. The individual atoms are of different masses; all of the stable isotopes are present.

An example should make the concept and the calculation of atomic weights clearer. Suppose the average weight of an athlete in an 80-member track team is needed. This unusual team consists of 52 runners weighing 67 kg each and 28 shot-putters weighing 110 kg each. The coach obtains the total team weight as follows.

$$\begin{aligned} &52 \text{ runners} \times 67 \text{ kg/runner} = && 3.5 \text{ Mg} \\ &28 \text{ shot - putters} \times 110 \text{ kg/shot - putter} = && \underline{3.1 \text{ Mg}} \\ &\text{Total} = && 6.6 \text{ Mg} \end{aligned}$$

(Remember: $\text{Mg} = 10^6$; $\text{g} = 1000$ kg.) The average weight of an athlete is

$$\frac{3.5 \text{ Mg} + 3.1 \text{ Mg}}{80 \text{ athetes}} = \frac{6.6 \text{ Mg}}{80 \text{ athletes}} = 82 \text{ kg/athlete}$$

The following expressions also may be used to calculate the average weight of an athlete:

$$\frac{[52 \text{ runners} \times 67 \text{ kg/runner}] + [28 \text{ shot} - \text{putters} \times 110 \text{ kg/shot} - \text{putter}]}{80 \text{ athletes}} = 82 \text{ kg}$$

or

$$\left[\frac{52 \text{ runners}}{80 \text{ athletes}} \times \frac{67 \text{ kg}}{\text{runner}}\right] + \left[\frac{28 \text{ shot} - \text{putters}}{80 \text{ athletes}} \times \frac{110 \text{ kg}}{\text{shot} - \text{putter}}\right] = 82 \text{ kg}$$

EXAMPLE 2-6 The two isotopes of copper have masses of 62.939598 u for ^{63}Cu and 64.927793 u for ^{65}Cu. What is the percent abundance of ^{63}Cu? (The average atomic mass of copper is 63.546 u.)

We let x be the fractional abundance of ^{63}Cu. The product of each isotopic mass times its fractional abundance is added to similar products for the other isotopes to obtain the average atomic mass.

$$63.546 \text{ u} = [x \times 62.939598 \text{ u}] + [(1-x) \times 64.927793 \text{ u}] = 62.939598\ x \text{ u} + 64.927793 \text{ u} - 64.927793\ x \text{ u}$$

$$63.546 \text{ u} - 64.927793 \text{ u} = 62.939598\ x \text{ u} - 64.927793\ x \text{ u} = -1.988195\ x \text{ u}$$

$$x = \frac{63.546 \text{ u} - 64.927793 \text{ u}}{-1.988195 \text{ u}} = 0.6950 \qquad 69.50\% \text{ copper - 63}$$

11. Use the periodic table in fundamental ways, including locating elements with certain properties and predicting the charges of ions of representative elements.

As we noted in Objective 1-2, elements in the same column of the periodic table have similar properties. Each column is referred to as a periodic family or group. The horizontal rows of the periodic table are called *periods*. Elements on the right-hand side of the periodic table are nonmetals; they form *anions*, or negatively charged ions. Elements on the left-hand side of the periodic table are metals; they form *cations*, or positively charged ions. Elements within the same group will form ions with the same charge. The common ions of Groups 1A, 2A, 6A, and 7A are given in Table 2–2.

TABLE 2-2 Common Ions of Some Representative Elements, Arranged by Groups

Ion	Name	Ion	Name	Ion	Name	Ion	Name
Li^+	lithium ion	Be^{2+}	beryllium ion	O^{2-}	oxide ion	F^-	fluoride ion
Na^+	sodium ion	Mg^{2+}	magnesium ion	S^{2-}	sulfide ion	Cl^-	chloride ion
K^+	potassium ion	Ca^{2+}	calcium ion	Se^{2-}	Sellenide ion	Br^-	bromide ion
Rb^+	rubidium ion	Sr^{2+}	strontium ion	Te^{2-}	Telluride ion	I^-	iodide ion
Cs^+	cesium ion	Ba^{2+}	barium ion				

***12. Obtain and use relationships between the mole, the Avogadro constant (Avogadro's number), and the molar mass of an element.**

The *atomic weight* (or atomic mass) is the nonintegral number appearing with each element in the periodic table. Up until now, we have thought of the atomic weight as the average weight of an atom of a particular element. However, it is very inconvenient (some might say impossible) to work with individual atoms. Therefore, chemists have chosen a larger quantity—Avogadro's number—of atoms to work with. If the atomic weight of an element is expressed in grams, that mass—the molar mass—will contain Avogadro's number of atoms of that element.

Avogadro's number is truly huge—6.02214×10^{23}. The number is so large that if one million workers attempted to move 6.02214×10^{23} grains of sand with shovels, the task would require 450 years, assuming 10 shovelfuls per worker each minute and 250 million grains of sand (about 15 pounds) per shovelful. Thus, instead of counting atoms, we weigh out an amount of material and use the mass of Avogadro's number of atoms (the molar mass) to determine how many atoms are present.

We often weigh small objects instead of counting them. We buy grass seed, nails, and pieces of candy by the pound or kilogram. The merchant knows the number of items per pound or per kilogram. For atoms, in contrast, we choose a fixed number (a mole of them) and use a mass containing that number (the atomic mass). This is rather like choosing 364 g of upholstery tacks, 795 g of 4 d finishing nails, or 4288 g of 10 d common nails, all containing 1000 nails.

EXAMPLE 2-7 87.4 g Mn is equivalent to how many moles of Mn?

(1) ? moles Mn

(2) 87.4 g Mn

(3) $\text{amount Mn} = 87.4\ \text{g Mn} \times \dfrac{\text{mol Mn}}{54.9\ \text{g Mn}} = 1.59\ \text{mol Mn}$

Usually we are not concerned with how many atoms are present. We simply need to know that there are enough atoms of each element for a particular chemical reaction. Hence, we speak of moles of atoms. A mole of an element contains Avogadro's number of atoms. To make the concept clearer, imagine that a packer of machinery must put 500 bolts into each box. For each bolt there must be one nut and two washers, for a total of 500 nuts and 1000 washers. Counting these out would take quite a long time. Knowing the masses of 100 pieces of each type $(100\ \text{bolts} = 99.1\ \text{g}, 100\ \text{nuts} = 42.7\ \text{g}, 100\ \text{washers} = 41.2\ \text{g})$ makes the job faster and just as accurate as counting. In fact, the instructions to the packer might be phrased: "Put in five dops of bolts (one dop weighs 99.1 g)." [*Dop* is a made–up unit that contains 100 pieces.] Now the packer is not concerned with the number of pieces in a dop. In similar fashion, chemists are not concerned very often with the number of atoms in a mole. Moles are convenient amounts of material to use—much more so than individual atoms.

The abbreviation for mole is mol. Other symbols such as m, $\overline{m}$, and M are incorrect and should not be used as abbreviations for mole, for they will cause confusion.

One mole of an element (a) contains Avogadro's number of atoms, (b) contains 6.02214×10^{23} atoms, and (c) has a mass in grams equal to the molar mass of the element. Each of these relations can be used to construct a conversion factor. Sample factors are

$$\frac{6.02214 \times 10^{23}\ \text{atoms}}{\text{mol}} \qquad \frac{6.02214 \times 10^{23}\ \text{Mg atoms}}{24.305\ \text{g Mg}} \qquad \frac{24.305\ \text{g Mg}}{\text{mol Mg}}$$

EXAMPLE 2-8 How many moles are present in 5.00×10^{9} atoms? (Notice that we do not have to specify which element since there is the same number of atoms—Avogadro's number—in a mole of any element.)

(1) ? moles

(2) 5.00×10^{9} atoms

(3) $5.00 \times 10^{9}\ \text{atoms} \times \dfrac{\text{mole of atoms}}{6.02214 \times 10^{23}\ \text{atoms}} = 8.30 \times 10^{-15}\ \text{mol}$

SELF-ASSESSMENT EXERCISES

91. (b) 3.0 g. All of the zinc is consumed, so 10.0 g of the zinc sulfide must come from zinc. The remaining 5 g comes from sulfur, leaving 3.0 g of the original 8 g of sulfur unreacted.

92. (c) 0.374:1. Answers (a) and (b) have the same mass ratio. For (a), divide the ratio by 85.5 and one obtains 0.187:1; the same applies to (b). Answer (c) is a second compound because the mass ratios are different.

95. (d) $^{40}Ar^{+2}$. Neutral sulfur has 16 electrons (the same number as the number of protons, or the atomic number). The number in the upper left indicates the mass—the number of protons and neutrons. Changes in mass occur by addition or loss of neutrons. The number of protons determines the identity of the element. Neutral Ar has 18 electrons, but since this is Ar^{+2}, this atom has 16 electrons.

96. (c) 47.88 u. The amu is defined as 1/12 the mass of a carbon-12 atom. An individual carbon atom has the mass of 12.00000 u. Similarly for other elements, an individual atom must have a nearly whole number atomic mass.

97. (c) The isotopic mass is defined relative to carbon. Therefore, the ratio of 83.9115/12.000 would change to 84.00/x. Setting the two ratios equal and solving, x = 12.0127 u.

98. (a) 558.5 g of Fe contains 10 moles of Fe. 558.5 g x 1 mol/55.85 g = 10 mol. This eliminates answer (d). Cr has a mass of 52.0 u, so that 0.520 g x 1 mol/52.0 g = 0.010 mol, or 1/1000 the number of moles (and atoms) as the Fe sample. 600.06 g of carbon contains 50 moles. The Fe sample contains one fifth as many moles.

99. 2.312 g of Fe contains 2.312/55.847 = 0.0414 moles and 1.0 g of oxygen contains 0.0625 moles. To write the formula, divide by the smaller number (1 mol Fe: 1.5 mol O) and multiply to get whole numbers. The formula is Fe_2O_3. The second sample contains 0.0465 mol of Fe. The ratio is 1 mol Fe:1.34 mol O or 3 mol Fe:4 mol O, and the formula is Fe_3O_4.

100. The atomic mass of Sr is 87.62, which is the sum of atomic mass x abudance for all of the isotopes. The abundance of the second and third isotopes must total 16.86%, so that all of the isotopes add up to 100%. If one of the unknown isotopic abundancies is x, the other is 16.86 – x. The equation to solve for x is:

$$(83.9134)(0.0056) + (85.9134)(0.1686 - x) + (86.9089)(x) + (87.9056)(0.8258) = 87.62$$

Solving for x, x = 0.1105, or there is 11.05% abundance of the 86.9089 isomer and 5.81% abundance of 85.9134 isomer.

101. 0.250 L must be converted to tons. This can be done using a number of conversion factors, one example is given below:

$$0.0250 \text{ L x } \frac{1.03\text{ g}}{\text{ml}} \text{ x } \frac{1000\text{ ml}}{\text{L}} \text{ x } \frac{1\text{ kg}}{1000\text{ g}} \text{ x } \frac{2.205\text{ lb.}}{\text{kg}} \text{ x } \frac{1\text{ ton}}{2000\text{ lb.}} = 0.283\text{ ton}$$

The mass of Au will be 0.15 mg/ton x 0.0283 ton = 4.25×10^{-3} mg.

The number of gold atoms will be

$$4.25 \times 10^{-3}\text{ mg x } \frac{1\text{ g}}{1000\text{ mg}} \text{ x } \frac{1\text{ mol}}{197.0\text{ g}} \text{ x } \frac{6.02 \times 10^{23}\text{ atoms}}{\text{mol}} = 1.34 \times 10^{16}\text{ atoms}$$

DRILL PROBLEMS

1. (1) Law of conservation of mass.

A. If 3.41 g of hydrogen sulfide is combined with 4.80 g of oxygen, what should the products weigh? If the products are 21.9% water, what mass of water is produced?

B. 7.95 g of copper(II) oxide is combined with 0.20 g of hydrogen. The products have what mass? The products are 77.9% Cu. What mass of Cu is formed? What is the percentage of Cu in copper(II) oxide?

C. Sugar, a carbohydrate, can be decomposed by heating into only carbon and water vapor. (You do this when you bake cookies.) If 18.0 g of sugar is heated, the carbon that remains has a mass of 7.2 g. What mass of water vapor is driven off? What is the percentage of water in sugar?

D. When reacted with carbon at high temperatures, iron(III) oxide produces solely molten iron and carbon dioxide. 3.192 g of iron(III) oxide and 0.36 g of carbon produce 1.32 g of carbon dioxide. What mass of molten iron is produced? What is the percent of iron in iron(III) oxide? What mass of carbon dioxide is produced when 1.00 g of molten iron is made? How many tons of carbon dioxide are produced when 1.00 ton of molten iron is made?

E. When 9.56 g of copper(II) sulfide is heated in the presence of 4.80 g of oxygen, the products contain 55.4% copper(II) oxide. What is the total mass of products? What mass of copper(II) oxide is formed?

F. When 18.0 g of wood is burned in oxygen, the products weigh 37.2 g and contain 29.0% water. What mass of water is produced? What mass of oxygen is needed? What mass of oxygen is needed to burn 4.00 lb. of wood?

(2) Law of constant composition.

G. Sulfur dioxide, produced by heating copper(II) sulfide and oxygen, contains 50.1% S. If 57.2 g of sulfur dioxide is made by burning sulfur in oxygen, what mass of S burns? What mass of oxygen is used?

H. Carbon dioxide, produced by reacting iron(III) oxide with C, contains 27.3% C. What mass of carbon must be burned in oxygen to produce 100.0 g of carbon dioxide? What mass of oxygen is needed?

I. Water, produced by reacting hydrogen with oxygen, contains 11.1% H. What mass of hydrogen must be reacted with hot copper(II) oxide to produce 75.0 g of water? What mass of oxygen is present in 75.0 g of water?

J. When calcium carbonate (limestone) is heated, the calcium oxide (lime) produced contains 28.5% oxygen. What mass of oxygen is needed to produce 74.2 g of lime when Ca is reacted directly with oxygen (lime is the only product)? What mass of Ca is needed?

K. When pure table salt is obtained from sea water, it contains 39.3% Na. 14.0 g of chlorine reacted with excess sodium produces table salt as the only product. What mass of table salt is produced? What mass of sodium will react?

L. Ammonia, produced by reacting magnesium nitride with water, contains 17.6% H. The direct reaction of hydrogen and nitrogen produces only ammonia. What mass of ammonia can be made from 32.1 g of hydrogen? What mass of nitrogen is needed?

2. Fill in the blanks in each line that follows. There may not be enough information supplied. If so, write "IMPOSSIBLE" across the line. The first line is completed as an example.

	Symbol	*Ionic charge*	*Mass number*	*Atomic number*	*No. of electrons*	*No. of neutrons*
Ex.	$^{122}Sn^{2+}$	+2	122	50	48	72
A.	$^{81}Br^{-}$	−1	81	35	36	46
B.	____	+3	59	____	____	32
C.	$^{43}Ca^{2+}$	____	____	____	____	____
D.	____	−3	____	7	____	8
E.	____	0	20	10	____	____
F.	____	____	127	53	54	____
G.	____	+1	23	____	____	12
H.	$^{192}Os^{4+}$	____	____	____	____	____
I.	____	+3	26	____	____	30
J.	____	+2	52	24	____	____
K.	____	____	60	27	25	____

L.	____	−1	17	____	____	8
M.	$^{80}Se^{2-}$	____	____	____	____	____
N.	____	−4	14	____	____	8
O.	____	+4	118	50	____	____

3. Fill in the blank in each part that follows. The first problem is completed as an example.

	Mass of 1st isotope	*Ratio of isotope masses: 1st to 2nd*	*Mass of 2nd isotope*		*Mass of 1st isotope*	*Ratio of isotope masses: 1st to 2nd*	*Mass of 2nd isotope*
Ex.	13.00335	1.083613	12.00000	A.	190.9609	0.7745905	______
B.	77.9204	______	69.9243	C.	70.9249	______	14.0067
D.	68.9257	______	18.9984	E.	45.9537	0.283281	______
F.	______	4.80336	52.9407	G.	______	4.98818	49.9461
H.	______	1.50951	105.907	I.	63.9280	______	6.01888
J.	53.9389	1.29638	______	K.	77.9204	1.43613	______
L.	57.9353	______	150.923				

4. Fill in the blanks in each line that follows. Make use of the periodic table to fill in the first two columns. Assume that each of these elements has only two isotopes.

	Element		*Isotope A*		*Isotope B*	
	Symbol	*Atomic weight*	*Atomic weight*	*% Abundance*	*Atomic weight*	*% Abundance*
A.	Li	6.941u	6.015	7.42	7.016	92.58%
B.	B	10.811	10.013	19.88%	11.009	80.12%
C.	C	12.011	12.000	98.89	13.00u	1.11
D.	Ne	20.19	19.992	90.2	21.991	9.8
E.	Cu	63.54	62.930	69.47	64.928	30.53%
F.	Cl	35.45	34.969	75.53	36.93u	24.47
G.	K	39.102	38.964	93.1	40.964	6.9
H.	Ga	69.7	68.926	60.4	70.935	39.6
I.	Br	79.904	78.918	50.651%	80.916	49.349%
J.	Rb	85.47	84.912	72.15	86.92	27.85

5. Fill in the blanks in each line that follows. The first line is an example. Use the periodic table as needed.

	Element	*Atomic weight*	*Mass of element, g*	*Amount, mol*	*No. of atoms*
Ex.	Cs	132.9	47.2	0.355	2.14×10^{23}
A.	H	1.01	0.416	0.412	2.48×10^{23}
B.	S	32.1	87.4	2.72	1.64×10^{24}
C.	O	16.0	25.4	1.59	9.57×10^{23}
D.	C	12.0	6.02	0.501	3.02×10^{23}
E.	Cl	35.5	1.52	0.0427	2.57×10^{22}
F.	N	14.0	2.14	0.153	9.21×10^{22}
G.	Mg	24.3	99.1	4.08	2.46×10^{24}
H.	P	31.0	~~184~~ 155	5.00	3.01×10^{24}
I.	Br	____	____	6.12×10^{-5}	____
J.	K	____	____	____	8.69×10^{19}
K.	Be	____	____	5.02×10^{10}	____
L.	____	19.0	0.000302	____	____

QUIZZES (20 minutes each) Select the best answer to each question.

QUIZ A

1. There are two stable isotopes of chlorine: $Cl\text{-}35 = 34.9689$ amu (75.53%) *and* $Cl\text{-}37 = 36.9659$ amu. What is the atomic weight of chlorine? (a) $(0.7553)(34.9689) + (0.2447)(36.9659)$; (b) $(34.9689 + 36.9659)/2$; (c) $(0.7553)(36.9659) + (0.2447)(34.9689)$; (d) $(75.53)(34.9689) + (24.47)(36.9659)$; (e) none of these.

2. ${}^{35}Cl^-$, ${}^{40}Ar$, and ${}^{39}K^+$ all have the same (a) mass number; (b) atomic number; (c) number of electrons; (d) number of neutrons; (e) none of these.

3. When 16.0 g of methane is burned in 64.0 g of oxygen, 44.0 g of carbon dioxide and 36.0 g of water are produced. This is an example of the law of (a) conservation of mass; (b) atomic masses; (c) definite composition; (d) relative proportions; (e) none of these.

4. Isotopes always have the same (a) mass number; (b) atomic number; (c) number of electrons; (d) number of neutrons; (e) none of these.

5. A subatomic particle that has a very small mass and a negative charge is called (a) a proton; (b) a neutron; (c) an electron; (d) an isotope; (e) none of these.

6. According to experimental evidence, cathode rays do *not* possess which property? (a) negative charge; (b) fundamental particle; (c) radioactivity; (d) mass; (e) they possess all these properties.

7. The combination that led to the determination of the charge on the electron was (a) Dalton–oil drop; (b) Rutherford-gold foil; (c) Millikan–oil drop; (d) Thomson-magnetic field; (e) Becquerel-gold foil.

8. Dalton's assumptions included (a) isotopes; (b) the nuclear atom; (c) indestructible atoms; (d) electrons; (e) none of these.

9. 1.60×10^{22} Cu atoms (a) is 0.0531 mol Cu; (b) is 1.69 g Cu; (c) contains 2.32×10^{23} protons; (d) contains 4.64×10^{23} neutrons; (e) none of these.

QUIZ B

1. There are two stable isotopes of silver: $Ag\text{-}107 = 106.9041$ amu (51.82%) *and* $Ag\text{-}108 = 107.9047$ amu. What is the atomic weight of silver? (a) $(0.5182)(108.9047) + (0.4818)(106.9041)$;(b) $(108.9047/0.5182) + (106.9041/0.4818)$; (c) $(0.5182)(106.9041) + (0.4818)(108.9047)$; (d) $(106.9041 + 108.9047)/2$; (e) none of these.

2. When two carbon oxide samples are analyzed, one contains 36.0 g of C and 32.0 g of O, while the other contains 12.0 g of C *in* 28.0 g of the oxide. This is an example of the law of (a) conservation of mass; (b) compound proportions; (c) constant composition; (d) fixed percentages; (e) none of these.

3. $^{19}F^-, ^{20}Ne$, and $^{24}Mg^{2+}$ all have the same (a) mass number; (b) atomic number; (c) number of electrons; (d) neutron number; (e) none of these.

4. A subatomic particle that has the same mass as the hydrogen nucleus and a positive charge is called (a) a proton; (b) a neutron; (c) an electron; (d) an isotope; (e) none of these.

5. Two atoms with the same number of neutrons are called (a) protons; (b) daltons; (c) electrons; (d) isotopes; (e) none of these.

6. A fundamental particle of the atom is the (a) α particle; (b) β particle; (c) X-ray; (d) canal ray; (e) none of these.

7. The combination that lead to the theory of the nuclear atom was (a) Rutherford–gold foil; (b) Dalton–plum pudding; (c) Thomson–curved electron beam; (d) Millikan–oil drop; (e) Becquerel–radioactivity.

8. Canal rays are (a) positive ions; (b) negative ions; (c) electrons; (d) neutrons; (e) none of these.

9. There is 4.024 mol of protons (a) in 0.1006 mol of Zr; (b) in 0.9177 g of Zr; (c) associated with 2423×10^{24} electrons in Zr atoms; (d) associated with 9.693×10^{25} neutrons in Zr atoms; (e) none of these.

QUIZ C

1. There are two stable isotopes of gallium: $Ga\text{-}69 = 68.9257$ amu (60.4%) *and* $Ga\text{-}71 = 70.9249$ amu. What is the atomic weight of gallium?

   (a) $(0.604)(68.9257) + (0.396)(70.9249)$; (b) $(68.9257 + 70.9294)/2$;

(c) $(0.396)(68.9257)+(0.396)(70.9249)$; (d) $(60.4)(68.9257)+(39.6)(70.9249)$;

(e) none of these.

2. Both HCl and HBr are strong acids with similar properties. It also is true that most chlorine compounds are similar chemically to the corresponding bromine compounds. This is an example of the law of (a) conservation of mass; (b) isotopic abundance; (c) constant composition; (d) average reactivity; (e) none of these.

3. The mass of an element, expressed on a scale where carbon-12 has a mass of 12.00000, is called its (a) mass number; (b) atomic weight; (c) natural abundance; (d) isotope mass; (e) none of these.

4. An entity that contains only protons and neutrons is called (a) an atom; (b) a molecule; (c) an ion; (d) a nucleus; (e) none of these.

5. $^{40}Ca^{2+}$, $^{39}K^{+}$, and $^{41}Sc^{+}$ all have the same (a) number of electrons; (b) mass number; (c) atomic number; (d) number of neutrons; (e) none of these.

6. All of the following scientists contributed to determining the *structure* of the atom *except* (a) Thomson; (b) Rutherford; (c) Millikan; (d) Dalton; (e) Becquerel.

7. Which of the following is *not* a property of canal rays? (a) straight-line travel; (b) fundamental particle; (c) mass; (d) have a positive charge; (e) depend on the gas in the tube.

8. Which of the following is *not* a fundamental particle of the atom? (a) proton; (b) neutron; (c) beta particle; (d) alpha particle; (e) none, all are fundamental particles.

9. 91.84 g of Ti is (a) 4.175 mol of Ti; (b) contains Ti atoms; (c) contains 1.155×10^{24} protons; (d) contains 2.541×10^{25} electrons; (e) none of these.

Quiz D

1. There are two stable isotopes of europium. The element has an atomic weight of 151.96 u, and one isotope has a mass of 150.9196 u and a percent abundance of 47.820%. What is the isotopic mass of the other isotope in u? (a) 152.92; (b) 153.00; (c) 153.09; (d) 149.97; (c) none of these within 0.02 u.

2. When decomposed chemically, a 73.0 g sample of HCl produces $71.0\,g\,Cl_2$ and $2.0\,g\,H_2$, while a 34.0 g sample of H_2S produces 32.0 g S and $2.0\,g\,H_2$. This is an example of the law of (a) conservation of mass; (b) multiple variation; (c) constant composition; (d) hydrogen conservation; (e) none of these.

3. A subatomic particle that has about the same mass as the hydrogen atom and a negative charge is called (a) a proton; (b) a neutron; (c) an electron; (d) an isotope; (e) none of these.

4. ^{104}Pd, ^{105}Pd, and ^{108}Pd all have the same (a) mass number; (b) atomic number; (c) number of electrons; (d) number of neutrons; (e) none of these.

5. "Compounds are composed of atoms combined in small whole number ratios. All atoms of the same element have the same weight" is (a) a law; (b) a theory; (c) an experiment; (d) an observation; (e) a hypothesis.

6. Choose the incorrect pair. (a) alpha–radioactivity; (b) Rutherford–nuclear atom; (c) Millikan–electric charge; (d) Dalton–isotope; (e) cathode rays–electrons.

7. Which of the following is a form of natural radioactivity under a different name? (a) proton; (b) neutron; (c) isotope; (d) hydrogen nucleus; (e) none of these.

8. The mass of an atom is largely determined by the number of (a) protons; (b) neutrons; (c) electrons; (d) isotopes; (e) nucleons.

9. 1.774 mol of Ca (a) contains 2.100×10^{25} Ca atoms; (b) weighs 35.48 g; (c) contains 70.96 mol of protons; (d) contains 1.068×10^{24} protons; (e) none of these.

SAMPLE TEST (30 minutes)

1. An attempt was made to determine the atomic mass of element X. X forms a compound with oxygen that contains 46.7% X and has the formula XO. Oxygen's atomic mass is 16.00 u. What is the atomic mass of X?

2. Match the lettered terms one–for–one with the other terms by writing the correct letter in each blank.

A. same atomic number	B. alpha particle	F Millikan	G Rutherford
C. proton number	D. atomic theory	I Thomson	A isotope
E. X-ray	F. electron charge	C atomic number	J mass number
G. nuclear atom	H. proton	D Dalton	H fundamental particle
I. isotope	J. nucleon number	E radioactivity	B cathode tube emission

3. A certain element contains one atom of mass 10.013 u for every four atoms of mass 11.009 u. Compute the atomic weight of this element.

3 CHEMICAL COMPOUNDS

CHAPTER OBJECTIVES

1. Distinguish between a mole of atoms and a mole of molecules.

A molecule is simply a cluster of atoms bound together. To continue our nut–bolt–washer analogy of Objective 2-11, a bolt with two washers on it and the nut screwed on is similar to a molecule (Fig. 3-1). If these parts were preassembled, the packers would weigh out "dops" of assemblies (224.2 g/dop). Similarly, chemists find it convenient to work with moles of molecules, since molecules are "preassembled" groups of atoms.

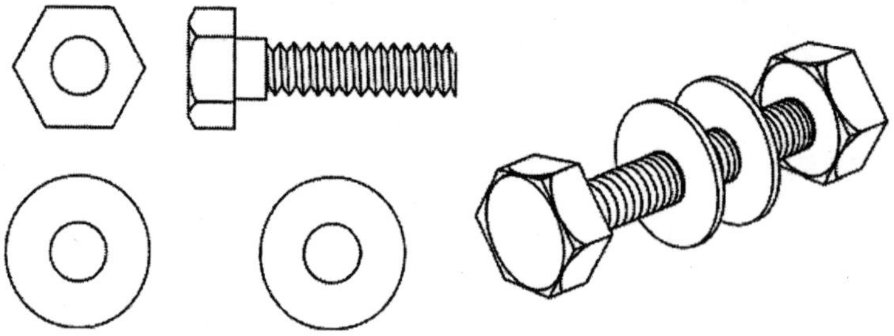

Figure 3-1 Two Formula Units—the "dop"

Some elements do not exist as individual atoms or as a large mass of somewhat attracted atoms, but rather as distinct molecules. You should firmly memorize the following forms: $H_2, F_2, Cl_2, Br_2, I_2, O_2, N_2, S_8$, and P_4. Knowing where they are located on the periodic table may help you remember them.

*2. Distinguish between formula unit and molecule, empirical formula and molecular formula, and formula mass and molecular mass.

As we have noted, a molecule is a group of atoms bound together. The atoms in a formula unit are not necessarily bound together. In our nut–bolt–washer analogy, a formula unit consists of a nut, a bolt, and two washers, whether or not they are bound together. The grouping of 1 nut +1 bolt +2 washers is the smallest collection of pieces that contains the pieces in the same ratio as when packed for shipment. Likewise, the formula unit is a small collection of atoms that contains the atoms in the same ratio they have in the compound.

A molecular formula counts the actual number of atoms of each type in the compound. Thus, the molecular formula of hydrogen peroxide is H_2O_2, and each molecule contains four atoms: 2 H and 2 O. An empirical formula expresses the *smallest* combining ratio of the atoms in the compound. For hydrogen peroxide, the empirical formula is HO. To change the molecular formula into an empirical one, divide the subscripts by the largest whole number that will go evenly into all of them. In the case of H_2O_2, this number is 2. Some examples follow.

Compound name	*Molecular formula*	*Divide by*	*Empirical formula*
Hydrazine	N_2H_4	2	NH_2
Propene	C_3H_6	3	CH_2
Diborane	B_2H_6	2	BH_3
Acetic acid	$C_2H_4O_2$	2	CH_2O

Just as the atomic mass in grams is the mass of a mole of atoms, so the formula mass in grams is the mass of a mole of formula units, and the molecular mass in grams is the mass of a mole of molecules. It is proper to refer to any of these three masses as the molar mass ($\mathcal{M}$). To determine the molar mass (sometimes called the *mole weight*), we start with the chemical formula of the compound. We interpret each subscript as the number of moles of each element in each mole of the compound. Thus, a mole of N_2O_5 contains 2 moles of nitrogen and 5 moles of oxygen. Remember the mass of a mole of an element equals the atomic weight of that element in grams.

EXAMPLE 3–1 Determine the molar mass of $HC_2H_3O_2$.

$$\begin{aligned} 4\text{ mol H} \times 1.01\text{ g H/mol H} &= \quad 4.04\text{ g H} \\ 2\text{ mol C} \times 12.0\text{ g C/mol C} &= \quad 24.0\text{ g C} \\ 2\text{ mol O} \times 16.0\text{ g O/mol O} &= \quad \underline{32.0\text{ g O}} \\ & \quad 60.0\text{ g/mol } HC_2H_3O_2 \end{aligned}$$

EXAMPLE 3-2 Determine the molar mass of $Al_2(SO_4)_3$ [$Al_2(SO_4)_3$].

A mole of the compound contains 2 moles of aluminum ions and 3 moles of sulfate ions. Since *each* moles of sulfate ions contains 1 mole of S and 4 moles of O, 3 moles of sulfate ions contain 3 moles of S and 12 moles of O (3 moles of sulfate ions $\times$ 4 moles O/ mol sulfate ions). Notice in the calculation below that *atomic*, rather than *ionic* masses are used; there are as many electrons "missing" from the cations as are present in excess among the anions.

$$\begin{aligned} 2\text{ mol Al} \times 27.0\text{ g Al/mol Al} &= \quad 54.0\text{ g Al} \\ 3\text{ mol S} \times 32.1\text{ g S/mol S} &= \quad 96.3\text{ g S} \\ 12\text{ mol O} \times 16.00\text{ g O/mol O} &= \quad \underline{192.0\text{ g O}} \\ & \quad 342.3\text{ g/mol } Al_2(SO_4)_3 \end{aligned}$$

***3. Calculate the number of atoms, ions, formula units, or molecules in a substance from a given mass, or vice versa.**

Once we can determine molar masses (Objective 3–2), we can find the amount in moles of a given mass of substance. We must know something about the substance before we can state whether or not these are moles of molecules. Only covalent compounds exist as molecules. *Binary* (two-element) compounds of two nonmetals are covalent. (See Objective 3–13 for definitions of *nonmetals* and *metals*.) Thus HCl, H_2O, NH_3, and PCl_5 are covalent compounds and composed of molecules.

In contrast, all binary compounds composed of a metal and a nonmetal are ionic. $NaCl, FeCl_3,$ and K_2S are ionic. They do not contain molecules, but consist of ions. A group of ions identical in number and type to that given by the formula is a formula unit (sometime abbreviated f.u.), not a molecule. A Na^+ cation and a Cl^- anion compose the formula unit of NaCl, while the formula unit of $FeCl_3$ consists of 1 Fe^{3+} cation and 3 Cl^- anions.

EXAMPLE 3–3 11.34 g of In_2S_3 contains **(a)** how many moles of the compound? **(b)** how many formula units of the compound? **(c)** how many molecules? **(d)** how many In^{3+} ions?

The molar mass of In_2S_3 is determined first.

$$2 \text{ mol In} \times 114.8 \text{ g In/mol In} = 229.6 \text{ g In}$$
$$3 \text{ mol S} \times 32.1 \text{ g S/mol S} = \underline{96.3 \text{ g S}}$$
$$325.9 \text{ g/mol } In_2S_3$$

The various parts of the problem can be solved with the conversion factor method.

(a) (1) ? mol In_2S_3

(2) 11.34 g In_2S_3

$$(3) \text{ amount } In_2S_3 = 11.35 \text{ g } In_2S_3 \times \frac{\text{mol } In_2S_3}{325.9 \text{ g } In_2S_3} = 0.03480 \text{ mol } In_2S_3$$

(b) (1) ? formula units In_2S_3

(2) 0.03480 mol In_2S_3

(3)

$$\text{no. f.u. } In_2S_3 = 0.03480 \text{ mol } In_2S_3 \times \frac{6.022 \times 10^{23} \text{ f.u. } In_2S_3}{\text{mol } In_2S_3} = 2.096 \times 10^{22} \text{ } In_2S_3 \text{ f. u.}$$

(c) There are no molecules, since In_2S_3 is an ionic compound, not a covalent compound.

(d) (1) ? In^{3+} ions

(2) 11.34 g In_2S_3

(3)

$$\text{no. In}^{3+}\text{ions} = 11.34 \text{ g In}_2\text{S}_3 \times \frac{\text{mol In}_2\text{S}_3}{325.9 \text{ g In}_2\text{S}_3} \times \frac{6.022 \times 10^{23} \text{ f.u. In}_2\text{S}_3}{\text{mol In}_2\text{S}_3} \times \frac{2 \text{ In}^{3+} \text{ ions}}{\text{f.u. In}_2\text{S}_3}$$

$$= 4.191 \times 10^{22} \text{ In}^{3+} \text{ ions}$$

***4. Use chemical formulas as a source of conversion factors for stoichiometric calculations.**

A chemical formula is not only "a recipe for something you wouldn't want to eat," as a sixth grader said, but also contains many relationships. $CuSO_4 \cdot 5H_2O$ contains 1 copper(II) ion, 1 sulfate ion, and 5 water molecules *or* 1 Cu atom, 1 S atom, 9 O atoms, and 10 H atoms. A *mole* of $CuSO_4 \cdot 5H_2O$ contains 1 mol of copper(II) ions, 1 mol of sulfate ions, and 5 mol of water molecules *or* 1 mol Cu atoms, 1 mol S atoms, 9 mol S atoms, and 10 mol H atoms. We already have used these relationships in obtaining the molar mass of a compound. The molar mass of $CuSO_4 \cdot 5H_2O$ is 249.6 g/mol. We can use the above relationships in other ways, as Example 3–4 illustrates.

EXAMPLE 3-4 How many moles of oxygen atoms are in $29.95 \text{ g } CuSO_4 \cdot 5H_2O$?

$$\text{amount of O} = 29.95 \text{ g cmpd} \times \frac{\text{mol cmpd}}{249.6 \text{ g cmpd}} \times \frac{9 \text{ mol O atoms}}{\text{mol cmpd}} = 1.08 \text{ mol O atoms}$$

***5. State and apply the law of multiple proportions.**

The law of multiple proportions describes the relative masses of elements in different compounds. It states that the ratio of the masses of one element combined with a constant mass of another element is a small whole number. This law is a consequence of the fact that compounds are made up of atoms combining and all atoms having the same (or at least the same average) atomic mass. Application of the law is demonstrated in Example 3-5.

EXAMPLE 3–5 Four compounds contain hydrogen, chlorine, and oxygen in the following proportions. Demonstrate that these compounds obey the law of multiple proportions.

(a) Hypocholorous acid:	$\%H = 1.91\%$	$\%Cl = 67.7\%$	$\%O = 30.5\%$
(b) Chlorous acid:	$= 1.46\%$	$= 51.9\%$	$= 46.8\%$
(c) Chloric acid:	$= 1.18\%$	$= 42.0\%$	$= 56.8\%$
(d) Perchloric acid:	$= 0.99\%$	$= 35.3\%$	$= 63.6\%$

(We discuss how to obtain these percentages in Objective 3-6.) Consider a fixed mass, such as 10.0 g of chlorine (any fixed mass will do). The combining masses of oxygen with 10.0 g Cl in order are

(a) $10.0 \text{ g Cl} \times (30.5 \text{ g O}/67.7 \text{ g Cl}) = 4.51 \text{ g O}$	$4.51/4.51 = 1.00$
(b) $10.0 \text{ g Cl} \times (46.8 \text{ g O}/51.9 \text{ g Cl}) = 9.02 \text{ g O}$	$9.02/4.51 = 2.00$
(c) $10.0 \text{ g Cl} \times (56.8 \text{ g O}/42.0 \text{ g Cl}) = 13.5 \text{ g O}$	$13.5/4.51 = 3.00$
(d) $10.0 \text{ g Cl} \times (63.6 \text{ g O}/35.3 \text{ g Cl}) = 18.0 \text{ g O}$	$18.0/4.51 = 4.00$

If each combining mass is divided by 4.51 g, they are seen to be related as 1:2:3:4. In other words, they are related as small whole numbers.

Notice that percents are converted to masses in grams here, as was done in Objective 1-13. These are the masses present in 100.0 g of each compound. Although there is nothing special about 100.0 g, choosing that mass makes the calculation simply a matter of replacing % with g.

***6. Use a compound's formula to determine its percent composition.**

The percent of one element in a compound is found from the mass of one element present divided by the mass of the compound. Both masses—numerator and denominator—are determined in the course of computing the molar mass of the compound. The percent composition states the relative mass of each element in the compound. Here we determine it from the compound's formula, which states the relative amount in moles of each element in the compound.

$$\frac{\text{mass of element}}{\text{mass of compound}} \times 100\% = \text{percent of element}$$

EXAMPLE 3-6 What is the percent of H in $(NH_4)_3PO_4$?

First, compute the molar mass.

$$\begin{array}{ll} 3\text{ mol N} \times 14.0\text{ g N}/2\text{ mol N} & = 42.0\text{ g N} \\ 12\text{ mol H} \times 1.01\text{ g H}/1\text{ mol} & = 12.1\text{ g H} \\ 1\text{ mol P} \times 31.0\text{ g P}/1\text{ mol P} & = 31.0\text{ g P} \\ 4\text{ mol O} \times 16.0\text{ g O}/1\text{ mol} & = \underline{64.0\text{ g O}} \\ & 149.1\text{ g/mol }(NH_4)_3PO_4 \end{array}$$

Then, use the masses computed in the course of determining the molar mass to calculate the percent by mass of the desired element, in this case hydrogen.

$$\%H = \frac{12.1\text{ g H in a mole of }(NH_4)_3PO_4}{149.1\text{ g }(NH_4)_3PO_4/\text{mole }(NH_4)_3PO_4} \times 100 = 8.12\%\text{ H}$$

***7. Use the percent composition of a compound to determine its empirical formula.**

The percent composition of a compound gives the relative *masses* of the elements. The empirical formula gives the combining amounts in moles of the elements in the compound. Obtaining the empirical formula involves first finding the amount (in moles) of each element present from its relative mass. These amounts are then reduced to the simplest integers.

EXAMPLE 3-7 A compound contains 27.3% C and 72.7% O. What is its empirical formula?

First, find the amount in moles of each element, starting from its mass in 100 g.

$$\text{amount C} = 27.3\ \text{g C} \times \frac{\text{mol C}}{12.0\ \text{g C}} = 2.28\ \text{mol C}$$

$$\text{amount O} = 72.7\ \text{g O} \times \frac{\text{mol O}}{16.0\ \text{g O}} = 4.54\ \text{mol O}$$

This gives $C_{2.28}H_{4.54}$ as the formula, but we wish to have whole numbers as subscripts. One way to achieve whole numbers is to divide all mole numbers by the smallest. Thus 2.20 mol C and 4.54 mol O becomes (2.28/2.28 =) 1.00 mol C and (4.54/2.28 =) 1.99 mol O. As a result, we obtain the formula $CO_{1.99}$ or CO_2.

EXAMPLE 3-8 A compound contains 47.0% C, 5.9% H, and 47.0% O. It has a molar mass of 204 g/mol. What is its **(a)** empirical formula and **(b)** molecular formula?

(a) First, find the amount of each element in moles contained in 100.0 g of the compound.

$$\text{amount C} = 47.0\ \text{g C} \times \frac{\text{mol}}{12.0\ \text{g}} = 3.92\ \text{mol C} \quad \div 2.94 \longrightarrow \quad 1.34\ \text{mol C}$$

$$\text{amount H} = 5.9\ \text{g H} \times \frac{\text{mol H}}{1.0\ \text{g}} = 5.9\ \text{mol H} \quad \div 2.94 \longrightarrow \quad 2.0\ \text{mol H}$$

$$\text{amount O} = 47.0\ \text{g O} \times \frac{\text{mol O}}{16.0\ \text{g}} = 2.94\ \text{mol O} \quad \div 2.94 \longrightarrow \quad 1.00\ \text{mol O}$$

We round any of the final mole numbers that is within 0.05 of a whole number. Here, we recognize that 0.34 is about 1/3. Thus, to obtain whole numbers, we multiply each final amount in moles by 3.

$$1.34\ \text{mol C} \times 3 = 4.02\ \text{mol C (rounded to 4.00 mol C)}$$

$$2.0\ \text{mol H} \times 3 = 6.0\ \text{mol H}$$

$$1.00\ \text{mol O} \times 3 = 3.00\ \text{mol O}$$

The empirical formula is $C_4H_6O_3$

(b) The molar mass of $C_4H_6O_3$ is

$(4 \times 12.01\ \text{g C}) + (6 \times 1.008\ \text{g H}) + (3 \times 16.00\ \text{g O}) = 48.04 + 6.048 + 48.00 = 102.09\ \text{g/mol}$.

This is almost exactly one-half the molar mass of the compound. Thus, the compound's molecular formula is $C_8H_{12}O_6$.

***8. Use the masses of the products of the complete combustion of a compound to determine its percent composition or empirical formula.**

The combustion analysis of a compound containing only C, H, and O requires the chemist to weigh the compound, and then weigh the two products of combustion: carbon dioxide and water. *All* of the carbon in the compound is in the carbon dioxide and all of the compound's hydrogen is in the water. The masses of these two elements in the products are computed first. The mass of oxygen then is computed by subtraction, if necessary. The percent of oxygen is determined by difference because the oxygen in the products is from two sources: the compound that was burned and the oxygen in which it was burned. This, of course, is not the case with either the carbon or the hydrogen.

EXAMPLE 3-9 4.27 g of a C–H–O compound produces 10.78 g CO_2 and 1.89 g H_2O when burned in air. What is the **(a)** percent composition and **(b)** the empirical formula of the compound?

The masses of C and H are found as follows.

$$\text{mass C} = 10.78\text{ g CO}_2 \times \frac{\text{mol CO}_2}{44.01\text{ g CO}_2} \times \frac{\text{mol C}}{\text{mol CO}_2}$$

$$= 0.2449\text{ mol C} \times \frac{12.01\text{ g C}}{\text{mol C}} = 2.941\text{ g C}$$

$$\text{mass H} = 1.89\text{ g H}_2\text{O} \times \frac{\text{mol H}_2\text{O}}{18.0\text{ g H}_2\text{O}} \times \frac{2\text{ mol H}}{\text{mol H}_2\text{O}}$$

$$= 0.210\text{ mol H} \times \frac{1.01\text{ g H}}{\text{mol H}} = \frac{0.212\text{ g H}}{3.153\text{ g C, H}}$$

The mass of oxygen then is found by difference, mass $\text{O} = 4.27\text{ g}$ compound $-3.153\text{ g C} \,\&\, \text{H} = 1.12\text{ g O}$

(a) The mass percent of each element then is determined.

$$\%\text{O} = \frac{1.12\text{ g O}}{4.27\text{ g cmpd}} \times 100 = 26.2\%\text{ O}$$

$$\%\text{H} = \frac{0.212\text{ g H}}{4.27\text{ g cmpd}} \times 100 = 4.96\%\text{ H}$$

$$\%\text{C} = \frac{2.94\text{ g O}}{4.27\text{ g cmpd}} \times 100 = 68.9\%\text{ C}$$

(b) The empirical formula can be determined directly from the amount of each element in moles (in a certain sample of the compound), without determining the mass percent composition.

$$\text{amount C} = 0.2449\text{ mol C} \quad \div 0.0700 \longrightarrow 3.50\text{ mol C} \times 2 \longrightarrow 7.00\text{ mol C}$$

$$\text{amount H} = 0.210\text{ mol H} \quad \div 0.0700 \longrightarrow 3.00\text{ mol H} \times 2 \longrightarrow 6.00\text{ mol H}$$

$$\text{amount O} = 1.12\text{ g O} \times \frac{1\text{ mol O}}{16.0\text{ g O}} = 0.0700\text{ mol O} \div 0.0700 \longrightarrow 1.00\text{ mol O} \times 2 \longrightarrow 2.00\text{ mol O}$$

The empirical formula of the compound is $C_7H_6O_2$.

***8a. Use the results of precipitation analysis to determine the percent of a compound in a mixture or of an element in a compound.**

A very powerful technique is that of isolating one element or compound within a solid that settles to the bottom of a liquid solution. Such a solid is called a *precipitate.*

Precipitates are separated from the solution, dried to a constant composition, and weighed. The following example gives a sample of the type of calculations required.

EXAMPLE 3-10 A 14.21-g sample of a silver-containing alloy is dissolved in nitric acid. Sodium chloride solution is added in slight excess and a silver chloride precipitate is obtained. The solid is washed, dried, and weighed. Its mass is 15.32 g. Determine the percent of silver in the alloy.

$$\text{mass Ag} = 15.32\text{ g AgCl} \times \frac{\text{mol AgCl}}{143.32\text{ g}} \times \frac{\text{mol Ag}}{\text{mol AgCl}} \times \frac{107.87\text{ g Ag}}{\text{mol Ag}} = 11.53\text{ g Ag}$$

$$\%\text{Ag in alloy} = \frac{\text{mass Ag}}{\text{mass of alloy}} \times 100 = \frac{11.53\text{ g Ag}}{14.21\text{ g alloy}} \times 100 = 81.14\%\text{ Ag}$$

*8b. Apply precipitation analysis to determine atomic weights.

Dalton's most serious handicap in determining atomic weight was that he did not know the formulas of the compounds he analyzed. But if the chemical formula is known, the determination of atomic weight is straightforward.

EXAMPLE 3-11 The sulfur present in a 11.78-g sample of M_2S is precipitated as 17.27 g of $BaSO_4$. What is the atomic weight of element M?

We first compute the mass of S in the $BaSO_4$ (and, thus, also the mass of S in M_2S.

$$\text{mass S} = 17.27\text{ g BaSO}_4 \times \frac{\text{mol BaSO}_4}{233.40\text{ g BaSO}_4} \times \frac{\text{mol S}}{\text{mol BaSO}_4} \times \frac{32.06\text{ g S}}{\text{mol S}} = 2.372\text{ g}$$

In the 11.78 g of M_2S, there are $1.78\text{ g M}_2\text{S} - 2.372\text{ g S} = 9.41\text{ g M}$. We now determine the amount of S.

$$\text{amount S} = 2.372\text{ g S} \times \frac{\text{mol S}}{32.06\text{ g S}} = 0.07399\text{ mol S}$$

and, since 1 mol S unites with 2 mol M in M_2S, the 11.78 g of M_2S contains

$$\text{amount M} = 0.07399\text{ mol S} \times \frac{2\text{mol M}}{\text{mol S}} = 0.1480\text{ mol M}$$

Now, we compute the molar mass of element M.

$$\text{molar mass} = \frac{\text{mass M}}{\text{mol M}} = \frac{9.41\text{ g M}}{0.1480\text{ mol M}} = 63.5\text{ g/mol M}$$

Note how crucial molar relationships are. In fact, an oft-repeated advice is: "When in doubt, convert to moles." Chemical calculations are based on amounts in moles.

*9. Know and apply the conventions used in determining oxidation states.

Because *oxidation state* is a formal rather than an experimental concept, it is possible to devise a rigid set of rules that work in all but the most unusual circumstances. One set of rules, slightly expanded from those of the text, is given in Table 3-1. Note very carefully that oxidation state often equals periodic table family number, or family number minus eight. This is brought out in Table 3-1a, where the relationship between location in the periodic table and the rule-assigned oxidation state is given. What is really important is the *order* in which the rules are applied.

Also make a clear distinction between an atom's oxidation state in a certain compound, and the charge on an element's most common ion (given in Table 3-2). They may be the same, but they do not have to be. In Example 3-12, the oxidation states are determined in the order in which they are listed; the letter of the applicable rule is given in parentheses before the oxidation state. Notice how often the oxidation state does not equal the charge of the most common ion. Confusing oxidation state with the charge of the common ion or forgetting to write an ion with its charge are common mistakes of beginning chemists.

TABLE 3-1 Determination of Oxidation States

	Method of applying the rules
1.	Apply the rules from the top to the bottom of the list.
2.	Search the list to find a rule that fits. Apply it.
3.	Start again at the top of the list to find the next rule that fits.
	Oxidation state rules
a.	The O.S. (oxidation state) of all uncombined elements = 0.
b.	The sum of the O.S. in a compound = 0.
c.	The sum of the O.S. in an ion = ionic charge.
d.	Alkali metals (family 1A) have O.S. = +1.
e.	Alkaline earths (2A) have O.S. = +2.
f.	F has O.S. = −1 and H has O.S. = +1.
g.	O has O.S. = −2.
h.	Cl, Br, I (in order) have O.S. = −1.
i.	S, Se, Te (in order) have O.S. = −2.
j.	N, P; As (in order) have O.S. = −3.
k.	Others have O.S. = group number, or O.S. = group number − 8, with the more electronegative element having the negative oxidation state.

TABLE 3-2 Charges of Common Ions

H^+					
Li^+	Be^{2+}	…	N^{3-}	O^{2-}	F^-
Na^+	Mg^{2+}	…	P^{3-}	S^{2-}	Cl^-
K^+	Ca^{2+}	…	As^{3-}	Se^{2-}	Br^-
Rb^+	Sr^2	…	…	Te^{2-}	I^-
Cs^+	Ba^{2+}				

EXAMPLE 3-12 Determine the oxidation state of each element in $F_2, O_2^{2-}, Na_2S, CaC_2, HCl, ClF_3, CS_2$, FeN, SiC, and SO_3^{2-} HBrO

F_2	(a) F = 0		O_2^{2-}	(c) O = −1	
Na_2S	(d) Na = +1	(b) Na = +1	CaC_2	(e) Ca = +2	(b) C = −1
HCl	(f) H = +1	(b) Cl = −1	ClF_3	(h) F = −1	(b) Cl = +3
CS_2	(i) S = −2	(b) C = +4	FeN	(j) N = −3	(b) Fe = +3
SiC	(k) C = −4	(b) Si = +4	SO_3^{2-}	(g) O = −2	(b) S = +4
HBrO	(f) H = +1	(g) O = −2	(b) Br = +1		

***10. Know the names, formulas, and charges of the ions in Tables 3-3 and 3-4, and be able to write formulas and names of the compounds formed from these ions.**

The first goal is straightforward and is a 3×5-inch card or cassette recorder project. If using cards, write the name of the ion on one side and its symbol and charge on the other side. Then practice until you can recall one while looking at the other. It may seem early to learn chemical nomenclature and how to write formulas, but these form the basis of the language of chemistry. Chemists speak to each other with the names of compounds, but chemical formulas are most useful for calculations. Thus to understand chemists, you must have a firm grasp of nomenclature, and to work problems you need to be good at writing formulas. Far too many students find chemistry difficult simply because they have not mastered nomenclature and formula-writing. Spend a couple of hours here and save yourself weeks of frustration.

TABLE 3-3 Names, Formulas, and Charges of Some Common Ions

H^+	hydrogen	Mn^{2+}	manganese(II)	Cr^{2+}	chromium(II) *or* chromous
Li^+	lithium	Ni^{2+}	nickel(II)	Cr^{3+}	chromium(III) *or* chromic
Na^+	sodium	As^{3+}	arsenic(III)	Fe^{2+}	iron(II) *or* ferrous
K^+	potassium	Zn^{2+}	zinc	Fe^{3+}	iron(III) *or* ferric
Rb^+	rubidium	Cd^{2+}	cadmium	Co^{2+}	cobalt(II) *or* cobaltous
Cs^+	cesium	Ag^+	silver	Co^{3+}	cobalt(III) *or* cobaltic
Mg^{2+}	magnesium	Au^+	gold(I) *or* aurous	Cu^+	copper(I) *or* cuprous
Ca^{2+}	calcium	Au^{3+}	gold(III) *or* auric	Cu^{2+}	copper(II) *or* cupric
Sr^{2+}	strontium	Sn^{2+}	tin(II) *or* stannous	Hg_2^{2+}	mercury(I) *or* mercurous
Ba^{2+}	barium	Sn^{4+}	tin(IV) *or* stannic	Hg^{2+}	mercury(II) *or* mercuric
Al^{3+}	aluminum	Pb^{2+}	lead(II) *or* plumbous	Tl^+	thallium(I) *or* thallous
		Pb^{4+}	lead(IV) *or* plumbic	Tl^{3+}	thallium(III) *or* thallic
H^-	hydride	F^-	fluoride	Cl^-	chloride
Br^-	bromide	I^-	iodide	N^{3-}	nitride
O^{2-}	oxide	S^{2-}	sulfide	Se^{2-}	selenide

TABLE 3-4 Names, Formulas, and Charges of Some Common Polyatomic Ions

NH_4^+	ammonium	SO_4^{2-}	sulfate	FO^-	hypofluorite
$C_2H_3O_2^-$	acetate	HSO_4^-	hydrogen sulfate	CLO^-	hypochlorite
CO_3^{2-}	carbonate	SO_3^{2-}	sulfite	ClO_2^-	chlorite
HCO_3^-	hydrogen carbonate	HSO_3^-	hydrogen sulfite	ClO_3^-	chlorate
$C_2O_4^{2-}$	oxalate	$S_2O_3^{2-}$	thiosulfate	ClO_4^-	perchlorate
CN^-	cyanide	HS^-	hydrogen sulfide	BrO^-	hypobromite
OCN^-	cyanate	OH^-	hydroxide	BrO_3^-	bromate
SCN^-	thiocyanate	O_2^{2-}	peroxide	BrO_4^-	perbromate
NO_2^-	nitrite	CrO_4^{2-}	chromate	IO^-	hypoiodite
NO_3^-	nitrate	$Cr_2O_7^{2-}$	dichromate	IO_3^-	iodate
PO_4^{3-}	phosphate	MnO_4^-	permanganate	IO_4^-	periodate
HPO_4^{2-}	hydrogen phosphate	MnO_4^{2-}	manganate		
$H_2PO_4^-$	dihydrogen phosphate				

The lists of ions presented in Tables 3-3 and 3-4 may seem rather extensive, but they contain practically all the ions you are likely to encounter in general chemistry. To help you memorize charges, note how often the charge on a cation equals its group number in the periodic table, and how often an anion's charge equals its group number minus 8. (See Table 3-2). Some of the polyatomic ions in Table 3-4 are named systematically. (See Objective 3-12.) Older names of cations, using the *-ic/-ous* convention are given second in Table 3-3; you may see them on bottles purchased from chemical companies.

Naming an ionic compound is surprisingly simple. Write down the name of the cation (positive ion), followed by a space and the name of the anion (negative ion). Examples are iron(III) chloride, sodium nitrate, and mercury(I) phosphate.

Writing formulas from names is not quite so simple. The formula contains more than just the symbols for the cation and the anion. The cation and anion symbols are multiplied so that the total charge from the cations just balances the total charge of the anions. The total cation charge plus the total anion charge thus equals zero.

Fortunately, there are not very many ways to combine cations and anions. A look at Tables 3-3 and 3-4 reveals that cations have charges of $+1$ (like Li^+), $+2$ (Ca^{2+}), and $+3$ (Al^{3+}). Cations also exist with higher charges, but $+4$ (Ce^{4+}) is the highest positive charge you are likely to encounter. Anions seem to come in only three varieties: -1 (like F^- or MnO_4^-), -2 (S^- or CrO_4^{2-}), and -3 (PO_4^{3-}). Thus, there are only 12 (4×3) different common types of ionic compounds.

The easiest type of formula to write occurs in those compounds in which the anion and cation charges are equal but of opposite signs. Sodium chloride is composed of Na^+ and Cl^- ions and the total anion and cation charges sum to zero if we take one cation and one anion:

NaCl. In like manner, magnesium sulfate is composed of Mg^{2+} and $SO_4^{\ 2-}$ ions, and has the formula $MgSO_4$. Finally, aluminum phosphate is $AlPO_4$. These compounds lie along the diagonal in Table 3-5 [from the top line of the X^- column to the third line of the X^3-column].

The next easiest type of compound has one ion with a charge of $+1$ or -1. The charge on the other ion indicates how many of the $+1$ or -1 ions are needed. For example, sodium phosphate is composed of Na^+ cations (charge $= +1$) and $PO_4^{\ 3-}$ anions (charge $= -3$). To balance the -3 charge of the phosphate ion requires a total cation charge of $+3$, which is supplied by $3\,Na^+$ cations. Compounds of this general type lie across the top line and down the X^-s column of Table 3-5.

The remaining compounds in the table are similar to aluminum sulfate, which is composed of Al^{3+} and $SO_4^{\ 2-}$ ions. One method of finding the number of each type of ion in the formula follows.

(1) Use as many anions as the value of the cation charge. In this case, the charge of Al^{3+} is $+3$, so three anions are needed.

(2) Use as many cations as the negative of the anion charge. Sulfate's charge is -2, so use two cations.

(3) If the two numbers you have produced (number of cations and number of anions) both can be divided evenly by the same number, perform this division before your write down the formula.

EXAMPLE 3-13 What is the formula of cerium(IV) oxalate?

The formula for cerium(IV) ion is Ce^{4+}, and that for oxalate ion is $C_2O_4^{\ 2-}$.

(1) Ce^{4+} has a charge = +4. Thus, use 4 $C_2O_4^{\ 2-}$ anions.

(2) $C_2O_4^{\ 2-}$ has charge = −2. Thus, use 2 Ce^{4+} cations.

(3) But 4 and 2 both can be divided evenly by 2. This yields 2 $C_2O_4^{\ 2-}$ anions and 1 Ce^{4+} cation.

The formula of cerium(IV) oxalate is $Ce(C_2O_4)_2$.

TABLE 3-5 Formulas for Twelve Common Types of Ionic Compounds [Examples are given in brackets.]

Cations		*Anions*					
		X^-	[F^-]	X^{2-}	[$SO_4^{\ 2-}$]	X^{3-}	[$PO_4^{\ 3-}$]
M^+	[$NH_4^{\ +}$]	MX	[NH_4F]	M_2X	[$(NH_4)_2SO_4$]	M_3X	$(NH_4)_3PO_4$]
M^{2+}	[Ca^{2+}]	MX_2	[CaF_2]	MX	[$CaSO_4$]	M_3X_2	[$Ca_3(PO_4)_2$]
M^{3+}	[Al^{3+}]	MX_3	[AlF_3]	M_2X_3	[$Al_2(SO_4)_3$]	MX	[$AlPO_4$]
M^{4+}	[Ce^{4+}]	MX_4	[CaF_4]	MX_2	[$Ce(SO_4)_2$]	M_3X_4	[$Ce_3(PO_4)_4$]

Remember: *Simply because a compound's name and formula can be written does not mean that the compound exists.* (The ability to name something does not guarantee that it exists. After all, the headless horseman doesn't exist, does he?)

***11. Be able to write formulas and names of simple binary covalent compounds and of binary acids.**

As the text states, covalent compounds are formed between nonmetallic elements. The easiest way to identify the nonmetals involves using the periodic table. The nonmetals are the elements in the area of the table above and to the right of a zigzag line running from between B and Al to between Po and At. All other elements are metals.

Although there are a few exceptions (such as water, H_2O; ammonia, NH_3; and methane, CH_4), the names of the binary covalent compounds are obtained from the names of the two elements. The elements are named in the same order as they appear in the formula. The first element name is unchanged; the ending of the second becomes *-ide*. The element names have prefixes depending on the subscript of that element in the formula, except that the prefix *mono-* (meaning "one of") is rarely used for the first element in the formula. Other prefixes are:

di = 2, tri = 3, tetra = 4, penta = 5, hexa = 6, hepta = 7, and octa = 8.

Yet another method of naming binary covalent compounds is the Stock system. This system uses the oxidation state of the first element, written in parentheses following its name. Thus, it is similar to binary ionic compound nomenclature. For example, N_2O_3 is nitrogen(III) oxide. The Stock system can be awkward in naming some compounds. For example, nitrogen(IV) oxide can be either NO_2 or N_2O_4.

Binary acids consist of hydrogen and a nonmetal. HCl and H_2S both are binary acids. The name of a binary acid has the prefix *hydro-* and the suffix *-ic* surrounding the root name of the element. HCl is hydrochloric acid and H_2S is hydrosulfuric acid. These binary acid names are used when the compounds are dissolved in water; that is, in aqueous solution: HCl(aq) and H_2S(aq). When they are present as pure compound (usually gases), their names follow the conventions for ionic compounds: hydrogen chloride for HCl(g) and hydrogen sulfide for H_2S(g).

***12. Use oxidation states to name oxoacids and oxoanions.**

Oxidation states are helpful in balancing some chemical equations (see Objective 4–14) and in naming many compounds. In addition to the oxidation state nomenclature that we have already studied, oxidation states are helpful in naming oxoacids and oxoanions. The general formula of an oxoanion is XO_n^{m-}, where X is the central atom, *n* the number of oxygens, and *m*– the charge on the anion. For a given representative element X, the oxidation state of X varies between oxoanions in steps of 2, as shown in Table 3–6 for chlorine oxoanions.

TABLE 3-6 Salts and acids of chlorine oxoanions

Oxidation state	*Salt*		*Example*	*Acid*		*Example*
+1	hypo-	-ite	NaClO sodium hypochlorite	hypo-	-ous	HClO hypochlorous acid
+3		-ite	$NaClO_2$ sodium chlorite		-ous	$HClO_2$ chlorous acid
+5		-ate	$NaClO_3$ sodium chlorate		-ic	$HClO_3$ chloric acid
+7	per-	-ate	$NaClO_4$ sodium perchlorate	per-	-ic	$HClO_4$ perchloric acid

All oxoanions of the same family with the same oxidation state have similar names. Thus, $KBrO_3$ is potassium bromate and HFO is hypofluorous acid. Another generality for the representative elements is that the *-ate* anion and *-ic* acid endings are used when the oxidation state of the central atom equals the periodic table family number.

The only exceptions to this occur in the halogens, where the *-ate* and *-ic* endings correspond to a $+5$ oxidation state, and in the noble gases where they correspond to $+6$. Thus, H_2SO_4 is sulfuric acid because sulfur is in Family 6A and has an oxidation state of $+6$ in this compound. Likewise, $Na_2B_4O_7$ has boron (a member of Family 3A) in an oxidation state of $+3$ and is sodium borate. Notice that the formulas of the compounds do not determine the names, only the oxidation state of the central atom. H_3PO_4 (phosphor*ic* acid) and HNO_3 (nitr*ic* acid) both have central atoms from Family 5A, both with oxidation states of $+5$.

*13. Be able to use the formulas for hydrates and other complex compounds in the same ways as those of simpler compounds

In hydrates, water is associated with another compound. An example is $CaSO_4 \cdot 2H_2O$ (calcium sulfate dihydrate, commonly called gypsum). Some or all of the associated water, known as water of hydration, can be removed by heating. For example, at $128\,°C, 1\frac{1}{2}$ mol H_2O is removed from $CaSO_4 \cdot 2H_2O$, producing $CaSO_4 \cdot \frac{1}{2}H_2O$ (also written as $2CaSO_4 \cdot H_2O$)– calcium sulfate hemihydrate, or plaster of Paris. And at $163\,°C, CaSO_4 \cdot \frac{1}{2}H_2O$ becomes $CaSO_4$, the anhydrous (literally "without water") compound. Interpreting the formula of a hydrate follows the procedure given in Objective 3-4, where the example is $CuSO_4 \cdot 5H_2O$, copper(II) sulfate pentahydrate.

Coordination compounds are similar to simpler compounds, except that either the anion or the cation (or both) is a complex ion. An example is $Co[Fe(CN)_6]_2$ (the formula of the complex ion is enclosed in brackets):

Each mole of $Co_3[Fe(CN)_6]_2$ contains

$$3 \text{ mol } Co^{2+} \text{ cations and } 2 \text{ mol } [Fe(CN)_6]^{3-} \text{ anions}$$

or

3 mol Co, 2 mol Fe, 12 mol C, and 12 mol N.

The nomenclature of complex ions is discussed in Objective 25-3.

14. Be able to recognize and name simple organic compounds.

Hydrocarbons contain only carbon and hydrogen. Saturated hydrocarbons, the alkanes, contain the maximum number of hydrogen for each carbon. The general formula for alkanes is C_nH_{2n+2}. Hydrocarbons containing a double bond are alkenes and have the general formula C_nH_{2n}.

EXAMPLE 3-14. Write the formula for, or name, the following compounds:

(a) C_2H_6 **(b)** butane **(c)** hexane **(d)** CH_4

Solutions: **(a)** two carbons, therefore ethane; **(b)** C_4H_{10} ; **(c)** C_6H_{14} ; **(d)** methane

15. Recognize organic functional groups and name the compounds containing these groups.

Alcohols contain one or more hydroxyl (OH) functional groups. Organic acids contain a carboxyl group (COOH). Halogens can also serve as organic functional groups. Chloromethane ($ClCH_3$) has one hydrogen replaced by a chlorine atom.

EXAMPLE 3-15. What type of compound is each of the following?

(a) $CH_3CH_2CH_3$ **(b)** $ClCH_2CH_3$ **(c)** $CH_3CH_2CH_2COOH$ **(d)** CH_3CH_2OH **(e)** $CH_3CHCHCH_3$

Solutions: **(a)** hydrocarbon, propane; **(b)** halohydrocarbon, chloroethane; **(c)** carboxyllic acid, butyric acid; **(d)** alcohol, ethanol; **(e)** alkene, 2-butene.

SELF-ASSESSMENT EXERCISES

106. (a) Incorrect. The mass of one mole of Br_2 will be twice the molar mass of Br, 159.808 g. (b) Incorrect. A mole contains 2 x 6.022 x 10^{23} atoms. (c) Incorrect. 9.01 g H_2O contains 9.01/18.0 g/mol = 0.50 mole and there are three atoms/molecule. (d) Correct. Br has approximately twice the atomic mass of Cl, so the molar mass of Br_2 will be twice Cl_2.

107. The empirical formula is the simplest whole number ratio of elements. Answer (b) must be a molecular formula because the empirical formula would be NH_2 (divide the molecular formula by two). None of the other formulas can be divided by a whole number to give a smaller ratio of whole numbers.

108. (d) Correct. There are seven H atoms (1.0 amu) for every N atom (14.0 amu). Therefore, the mass of N will be approximately twice that of H. Answer (a) is incorrect because

there will be 17 x 6.02 x 10^{23} atoms/mole, or 6.02 x 10^{23} molecules/mole. (b) C is heavier than H, so even though there are the same number of atoms, the mass % will be different. (c) There will be twice the number of O atoms as N, but the mass is more than twice that of N.

109. (a) Correct. 50.0 g N_2O contains slightly more than 1 mole of N_2O and twice that number of N atoms. Answer (b) contains approximately 1 mole of N. Answer (c) contains less than two moles. Multiply the vol x density to get grams and divide by the g/mol for the compound. Each mole of compound contains 1 mol of N. Answer (d) contains two moles of N atoms.

110. (c) Correct. Assume you have 100g of compound. 65 g will be F and 35g will be X. Convert to moles: 65 g x 1 mol F/19.0 g x 1 mol compound/3 mol F = 1.14 mol compound. Each mole of compound contains 1 mole of X. Therefore,

$$1.14 \text{ mol X} = 35 \text{ g and X} = 35 \text{ g}/1.14 \text{ mol} = 31 \text{ u.}$$

111. (c) Correct. The oxidation state of H is +1 and O is -2. Therefore, oxidation for the compound is the sum of the oxidation states of the elements and totals -1. If x is the oxidation state of I, +4 – 12 + x = -1 and x = +7.

112. (b). Calcium ion has a +2 charge and chlorite is ClO_2^- . The compound must be neutral.

113. (a) Incorrect. There are 4 O for every S, so the mass of oxygen is approximately twice that of sulfur. Answer (b) is also incorrect. There will be the same number of atoms, but the mass of N and O are different. Answer (c) is incorrect because there will be 12 H for every 4 N or 3 H: 1 N. Compare the masses. Answer (d) is the correct answer. There will be 1 Cu: 4 O. The mass of Cu is 63.546 u and the mass of oxygen is 16.00 u. 63.546: (4) x 16.00 ~ 1.0.

114. An isomer has the same molecular formula, but a different structure. Answer (a) may or may not be an isomer because we don't know the structure. Answer (b) is the same structure, just rotate the molecule 180^0. Answer (c) is incorrect because the molecule has a different formula. Answer (d) is correct.

115. The mass of Na_2SO_3 is 126.0 g/mole. If the compound is 50% H_2O by mass, then it contain 126.0 g/mole of water. Dividing by the molar mass of H_2O gives 7 mol of H_2O and formula is $Na_2SO_3 \cdot 7H_2O$.

116. (a) The mass % copper can be found by dividing the mass of copper by the mass of the molecule and then multiplying by 100%. The mass of Cu is 2 x 63.548 and the molecular mass is 221.13 u. The mass % is 127.1/221.1 x 100% = 57.48%. (b) Malachite yields 2 mol of copper(II)oxide, CuO, for each mol of malachite. 1 kg of

malachite contains 1000 x 1 mol/221.1 g = 4.52 mol. Therefore, 9.04 mol of CuO are formed (twice the moles of malachite) and 9.04 mol x 79.5 g/mol = 719 g.

117. To find the molecular formula, first find the empirical formula. Assume there are 100 g of sample. 63.56 g are carbon, or 5.30 mol. For the other elements, using the same calculation, H = 6.0 mol, N = 0.662 mol, and O = 1.32 mol. Divide by the smallest number (0.662) to yield C = 8.0, H = 9.0, N = 1.0, and O = 2.0. The empirical formula is $C_8H_9NO_2$. The molecular formula is found by dividing the molecular mass (151.2 u) by the empirical mass (151.2 u), which gives a value of 1. There is one empirical formula unit for each molecular formula unit and the empirical formula and molecular formula are the same.

118. In a combustion reaction, excess oxygen is used. The carbon from CO_2 comes from ibuprofen. Similarly, the water comes from the H in ibuprofen. Oxygen can only be obtained by subtracting the mass of C and H from the sample. Find the moles of CO_2, and that gives the moles of carbon:

$6.029 \text{ g } CO_2 \text{ x } \dfrac{1 \text{ mol}}{44.0 \text{ g}} = 0.1370 \text{ mol } CO_2$ and the same number of moles of carbon.

Next, find the grams of carbon by multiplying by the atomic mass (0.1370 x 12.01 = 1.646 g C). To find the mass %, divide the mass of carbon by the mass of the sample to give 75.71% C.

Use a similar procedure for H, which is the source of H in the water sample. Here, remember that water has two H molecules:

$1.709 \text{ g } H_2O \text{ x } \dfrac{1 \text{ mol}}{18.0 \text{ g}} x \dfrac{2 \text{ mol H}}{1 \text{ mol } H_2O} x \dfrac{1.01 \text{g}}{\text{mol}} = 0.1916 \text{ g H}$ and the mass % is 0.1916 g/2.174 g x 100% = 8.812 % H.

The remaining sample is oxygen, which is 15.48%. To find the empirical formula divide the % mass (assume a 100 g sample so that the % mass is the number of grams) by the atomic mass of each element to get the number of moles. We calculate 6.31 mol C, 8.72 mol H, and 0.968 mol O. Divide by the smallest number of moles to 6.52 C, 9.01 H, and 1.0 O. Multiply by 2 to give whole numbers: $C_{13}H_{18}O_2$.

DRILL PROBLEMS

1. (1) Determine the molar mass of each of these compounds.

A. Hg_2Cl_2 B. NaCl C. $SnCl_4$ D. $Ca(HS)_2$ E. $NaHCO_3$
F. $Al(NO_3)_3$ G. $BaCl_2$ H. H_2O_2 I. HCN J. $C_2Cl_6S_2$
K. $(NH_4)_2CrO_4$ L. B_2H_6

(2) Determine the molar mass of each compound. Fill in the blanks in this table.

Chemical *Mass of* *Moles of* *Chemical* *Mass of* *Moles of*

	Chem Formula	mass of compound	moles of Compound		Chem formula	mass of Compound	moles of Compound
M.	$(NH_4)_2CrO_4$	174 g	1.14	N.	Li_2SO_3	232 g	2.47
O.	NaCl	8.21 oz	______	P.	AgClO	______ g	0.0694
Q.	$FeCl_3$	0.124 lb	______	R.	H_2O_2	5.00 lb	______
S.	NaOH	______ g	1.47	T.	$Mg(CN)_2$	______ g	0.00836
U.	$Hg(NO_3)_2$	4.12 g	______	V.	CuCl	______ g	7.97
W.	$NaNO_3$	1.43 lb	______	X.	$HBrO_4$	______ g	2.18

2. & 3. Use the concept of a mole of substance to answer the following questions.

A. How many C atoms are in 14.7 g $NaHCO_3$?

B. What is the mass of 1.42×10^{23} NH_3 molecules?

C. What is the mass of 64.2 mol $CsIO_3$?

D. How many O atoms are in 74.2 g $Al_2(SO_4)_3$?

E. How many molecules are in 163.7 g HCl?

F. How many atoms are in 82.3 g H_2O_2?

G. What is the mass of 7.42×10^{27} H_2SO_4 molecules?

H. What is the mass of 27.32 mol $(NH_4)_2Cr_2O_7$?

I. How many atoms are in 64.2 g NH_4MnO_4?

J. What mass of $BaCl_2$ contains 23.1×10^{19} ions?

K. How many O atoms are in 2.68 lb of $NaClO_3$?

L. In $C_4H_{10}O$, how many moles of C are present with 18.4 mol H?

M. In NH_4CN, how many moles of H are present with 0.125 mol C?

N. In $K_4Fe(CN)_6$, how many moles of C are present with 74.2 mol K?

O. In $Al_2(SO_4)_3$, how many moles of Al are present with 16.2 mol O?

4. Each problem that follows gives the percents of elements in several compounds formed among the same elements. Demonstrate that the law of multiple proportions is valid in each case.

A.	carbon monoxide	42.9% C	57.1% O	carbon suboxide	52.9% C	47.1% O
	carbon dioxide	27.3% C	72.7% O			
B.	potassium oxide	83.0% K	17.0% O	potassium peroxide	71.0% K	29.0% O
	potassium trioxide	62.0% K	38.0% O			
C.	acetylene	7.7% H	92.3% C	methane	25.0% H	75.0% C
	ethane	20.0% H	80.0% C	ethene	14.3% H	85.7% C
D.	chlorine	18.4%	81.6% Cl	chlorine	31.1%	68.9%

	monoxide	O			dioxide	O	Cl
	chlorine heptoxide	61.2% O	38.8% Cl				
E.	sulfur monochloride	52.5% Cl	47.4% S		sulfur dichloride	68.9% Cl	31.1% S
	sulfur tetrachloride	81.6% Cl	18.4% S				
F.	ammonia	82.4% N	17.6% H		hydrazine	87.5% N	12.5% H
	hydrazoic acid	97.7% N	2.3% H		diimine	93.3% N	6.7% H
G.	orthophosphoric acid	3.1% H	31.6% P	65.3% O			
	phosphorous acid	3.7% H	37.8% P	58.5% O			
	hypophosphorous acid	4.5% H	47.0% P	48.5% O			
H.	sulfurous acid	2.4% H	39.1% S	58.5% O			
	sulfuric acid	2.0% H	32.7% S	65.2% O			
	peroxomono-sulfuric acid	1.8% H	28.1% S	71.0% O			

5. & 6. Fill in the blanks in the table that follows. The first line is an example.

	Empirical formula	*"Empirical weight"*	*Elements and percent by weight*					
Ex,	NH_4Cl	53.5	N	26.2	H	7.5	Cl	66.4
A	H_2SO_3	82.086	H	2.44	S	39.02	O	58.54
B.	NaOH	40.00	Na	57.490	O	40.0%	H	2.52
C.	$MgCO_3$	84.32	Mg	28.9%	C	14.2	O	56.9
D.	$Al(NO_3)_3$	213.01	Al	12.7	N	19.7%	O	67.6
E.	$LiClO_4$	106.93	Li	6.5	Cl	33.3	O	60.2
F.	ZnC_2O_4	153.4	Zn	42.6	C	15.6	O	41.7
G.	_____	_____	K	39.1	Mn	27.9	O	33
H.	$Mg(CN)_2$	_____	Mg	_____	C	_____	N	_____
I.	_____	_____	Cu	51.4	C	_____	O	38.9
J.	H_3PO_4	_____	H	_____	P	_____	O	_____
K.	_____	_____	Hg	_____	Br	35.0	O	21.0

7. Each line that follows gives the mass of a carbon–hydrogen–oxygen compound that has been burned completely in oxygen to give the specified masses of CO_2 and H_2O. Based on this data, determine the empirical formula of each compound.

	Mass of compound	*CO_2*	*H_2O*		*Mass of compound*	*CO_2*	*H_2O*
A.	1.320 g	1.489 g	0.914 g	B	74.6 g	226.3 g	115.8 g
C.	8.14 g	11.19 g	9.16 g	D.	0.0152 g	0.0514 g	0.0105 g
E.	0.943 g	2.412 g	0.987 g	F.	11.7 g	22.4 g	13.7 g

G.	19.4 g	28.5 g	11.6 g	H.	7.32 g	11.84 g	5.81 g
I.	0.0255 g	0.0244 g	0.0100 g	J.	0.556 g	0.388 g	0.159 g
K.	146 g	222 g	45 g	L.	92.4 g	317.6 g	52.0 g

8a. The given mass of sample is treated to form a precipitate of specified mass and identity. Determine the percent of the element being determined in the sample. The first line is an example.

	Element being determined	*Sample mass, g*	*Precipitate* Formula	*Precipitate* Mass, g	*% of element in sample*	
Ex	Ag	14.21	AgCl	15.32	81.14	(15.32 g AgCl contains 11.53 g Ag.)
A.	Ba	13.20	$BaSO_4$	15.14	_____	
B.	Mg	96.41	MgO	17.46	_____	
C.	Pb	9.915	$PbCrO_4$	11.321	_____	
D.	Ca	82.44	$Ca_3(PO_4)_2$	1.42	_____	
E.	Al	19.00	Al_2O_3	10.56	_____	
F.	Cr	14.32	Cr_2O_3	5.17	_____	
G.	Li	0.847	Li_3PO_4	0 201	_____	
H.	F	37.32	CaF_2	40.03	_____	
I.	S	1.755	$BaSO_4$	1.972	_____	
J.	C	1.650	$CaCO_3$	13.124	_____	

8b. The mass and formula of an unknown compound follow, in the first two columns. The compound is dissolved in water, The anion of the compound reacts with another cation to form a precipitate of known identity. The formula and the mass of this precipitate are given in the next two columns. Use this data to determine the atomic mass of the metallic element (that of the cation) present in the original compound. (*Hint:* Recall Example 3-11.)

Unknown compound Formula	Unknown compound Mass, g	Precipitate Formula	Precipitate Mass, g	Unknown compound Formula	Unknown compound Mass, g	Precipitate Formula	Precipitate Mass, g
A. MS	44.72	$BaSO_4$	114.99	B. MCl_3	13.27	AgCl	35.17
C. MCl_2	6.34	AgCl	21.00	D. MF	13.62	CaF_2	12.66
E. $M(SO_4)_2$	20.10	$PbSO_4$	36.70	F. $M_2(SO_4)_3$	59.87	$BaSO_4$	122.53
G. M_2CO_3	24.55	$CaCO_3$	17.76	H. $M_2(CrO_4)_3$	54.39	$PbCrO_4$	82.87

9. Give the oxidation state of each element in each compound in the drill problems for Objective 3-2.

10. Give the name or formula, as appropriate, of each of the following compounds.

A. NaCl	B. $Al_2(SO_4)_3$	C. $SrCl_2$	D. $CrHPO_4$	E. $Na_2S_2O_3$
F. CuCl	G. $FePO_4$	H. Li_2SO_3	I. $(NH_4)_2CrO_4$	J. $KHSO_4$
K. $Ca(C_2H_3O_2)_2$	L. $Cu(HS)_2$	M. $MgCO_3$	N. $Cr(CN)_3$	O. $BaSO_4$
P. potassium permanganate		Q. sodium acetate		R. iron(III) chloride

S. strontium nitrate
T. silver chlorate
U. copper(I) oxide
V. mercury(I) nitrate
W. iron(III) nitrate
X. strontium bicarbonate
Y. zinc phosphate
Z. potassium chromate
Γ. magnesium cyanide
Δ. aluminum oxide
Θ. sodium hydroxide
Λ. calcium hypochlorite

11. Give the name or formula, as appropriate, of each compound.

A. H_2O .
B. H_2S
C. CO_2
D. N_2O_3
E. P_2O_5
F. ICl_5
G. NCl_3
H. SCl_4
I. ClO_2
J. SO_3
K. nitrogen monoxide
L. diphosphorus pentoxide
M. ammonia
N. hydrogen fluoride
O. silicon tetrafluoride
P. xenon(VI) fluoride
Q. phosphorus tribromide
R. silicon(IV) sulfide
S. methane
T. boron trichloride
U. nitrogen(III) oxide
V. antimony(V) sulfide

12. For each compound, supply the correct name or formula, and give the oxidation state of each element.

A. $NaClOl_4$
B. selenic acid
C. $(NH_4)_2S$
D. $CuIO_3$
E. H_2CO_3
F. phosphoric acid
G. sodium borate
H. $NaNO_3$
I. LiClO
J. hyposulfurous acid
K. $AlPO_4$
L. boric acid
M. $HClO_4$
N. nitrous acid
O. magnesium chlorite
P. $ScAlO_3$
Q. Mg_2GeO_4
R. NH_4NO_3
S. HFO
T. hyponitrous acid
U. H_4TeO_4

13. Correctly interpret the formula of each hydrate or coordination compound to answer the following questions.

A. 186 g of $CoCl_2 \cdot 6H_2O$ contains how many moles of H?

B. There is 21.7 g of O in how many moles of $Tl(NO_3)_3 \cdot 3H_2O$?

C. How many moles of S are in 14.2 g of $Na_3[Au(S_2O_3)_2]$?

D. There is 1.48 g of H in how many moles of $[Ag(NH_3)_2]_2SO_4$?

E. What amount in moles of O is in 57.2 g of $Rb_2S \cdot 4H_2O$?

F. There is 7.62 moles of O in how many grams of $3K_2S_2O_3 \cdot 5H_2O$?

G. How many moles of C are in 91.6 g of $K_2[Pt(SCN)_6]$?

H. There is 34.0 g of N in how many grams of $[Cr(NH_3)_6]_2(SO_4)_3$?

I. There are how many moles of O in 67.9 g of $Co(IO_3)_2 \cdot 6H_2O$?

J. What mass in grams of O is in 1.88 moles of $CrSO_4 \cdot 7H_2O$?

K. What amount in moles of N is in 44.3 g of $NH_4[Au(CN)_2]$?

L. There is 3.11 mol of S in how many moles of $Al_2[Pt(SCN)_6]_3$?

14. What type of molecules correspond to the formulas:

A. CH_3COOH

B. $CH_3CHClCH_3$

C. $CH_3CBrCHCH_2CH_3$

D. $CH_3CH_2CH_2OH$

E. $CH_3CHCHCOOH$

15. Write the formulas for the following compounds:

A. propanoic acid

B. methanol

C. hexane

D. fluoroethane

E. heptene

F. benzene

QUIZZES (20 minutes each.) Choose the best answer for each question or fill in the blank.

QUIZ A

1. A compound contains 37% N and 63% O. Its empirical formula is (a) NO_2; (b) NO; (c) N_3O_2; (d) N_2O_3; (e) none of these.

2. The molar mass of $CuSO_4$ (Cu = 63.5, S = 32.1, O = 16.0 g/mol) is (a) $63.5 + 32.1 + 16.0$; (b) $(2 \times 63.5) + 32.1 + 16.0)$; (c) $63.5 + 32.1 + (16.0 \times 4)$; (d) $(2 \times 63.5) + 32.1 + (4 \times 16.0)$; (e) none of these.

3. What is the mass fraction of oxygen in $C_6H_{12}O_6$ (C = 12.0, H = 1.01, O = 16.0 g/mol)? (a) $16.0/(12.0 + 1.01 + 16.0)$; (b) $16.0[12.0 + (2 \times 1.01) + (6 \times 16.0)]$; (c) $16.0/[(6 \times 12.0) + (12 \times 1.01) + (6 \times 16.0)]$; (d) $(6 \times 16.0)/[(12.0 + (2 \times 1.01) + 16.0]$; (e) none of these.

4. Which compound has a molar mass of 34.1 g/mol (H = 1.0, Si = 28.1, P = 31.0, S = 32.1, Cl = 35.5 g/mol)? (a) HCl; (b) SiH_4; (c) H_2S; (d) PH_3; (e) none of these.

5. How many molecules are contained in $1.00\,g\,H_2O\,(H = 1.0, O = 16.0\,g/mol; N = 6.022 \times 10^{23})$? (a) (1.0/16.0)*N*; (b) (1.0/17.0)*N*; (c) (18.0/1.0)*N*; (d) (17.0/1.0)*N*; (e) none of these.

6. How many moles of atoms are contained in $15.0\,g\,H_2SO_4\,(H = 1.0, S = 32.1, O = 16.0\,g/mol)$? (a) 15.0/98.1; (b) $7 \times 15.0/98.1$; (c) 15.0/49.1; (d) $7 \times 15.0/49.1$; (e) none of these.

7. The oxidation state of N in HNO_3 is (a) $+1$; (b) $+3$; (c) $+5$; (d) -1; (e) none of these.

8. The formula of iron(III) phosphate is (a) $Fe_2(PO_4)_3$; (b) $IrPO_4$; (c) $FePO_3$; (d) $FePO_4$; (e) none of these.

9. The name of $Sr(HCO_3)_2$ is (a) strontium oxalate; (b) strontium carbonate; (c) sodium bicarbonate; (d) strontium bicarbonate; (e) none of these.

10. The name of P_2O_5 is (a) potassium peroxide; (b) phosphorus(V) oxide; (c) phosphorus peroxide; (d) dipotassium pentoxide; (e) none of these.

11. Potassium perbromate has a formula of (a) KBr; (b) K_2BrO_4; (c) $KBrO_4$; (d) $KBrO_2$; (e) none of these.

QUIZ B

1. A compound contains 46.7% N and 53.3 % O. What is the empirical formula of this compound? (a) NO; (b) NO_2; (c) N_2O_3; (d) N_2O; (e) none of these.

2. To *two significant digits*, 97 g/mol is the molar mass of $(H = 1.0, Br = 79.9, S = 32.1, O = 16.0, P = 31.0\,g/mol)$ (a) HBrO; (b) H_3PO_4; (c) H_2SO_4; (d) $H_2S_2O_2$; (e) none of these.

3. The mass fraction of carbon in $CS_2\,(C = 12.0, S = 32.1\,g/mol)$ is (a) $12.0/[12.0 + (32.1 \times 2)]$; (b) 12.0/32.1; (c) $32.1/[(12.0 \times 2) + 12.0]$; (d) $32.1/(32.1 + 12.0)$; (e) none of these.

4. The molar mass of $Mg_3(PO_4)_2\,(Mg = 24.3, P = 31.0, O = 16.0\,g/mol)$ is (a) $24.3 + 31.0 + 16.0$; (b) $24.3 + 31.0 + (6 \times 16.0)$; (c) $(24.3 \times 3) + 31.0 + (16.0 \times 4)$; (d) $72.9 + [(31.0 + 16.0 \times 4) \times 2]$; (e) none of these.

5. How many atoms are contained in $95.0\,g\,Fe_2O_3\,(Fe = 55.8, O = 16.0\,g/mol; N = 6.022 \times 10^{23})$? (a) $(2 \times 95.0)/71.8$; (b) $[(5 \times 95.0)/71.8]N$; (c) (95.0/159.6)*N*; (d) $[(5 \times 95.0)/159.6]N$; (e) none of these.

6. How many moles of formula units are contained in $45.1\,g\,Al_2O_3\,(Al = 27.0, O = 16.0\,g/mol)$? (a) 45.1/43.0; (b) 45.1/102.0; (c) $(5 \times 45.1)/43.0$; (d) $(5 \times 45.1)/102.0$; (e) none of these.

7. Mn has an oxidation state of $+4$ in (a) $MnCl_2$; (b) MnO_2; (c) $MnSO_4$; (d) $KMnO_4$; (e) none of these.

8. Mercury(I) cyanide is (a) $Hg(CN)_2$; (b) HgCN; (c) MgCN; (d) $Hg_2(CN)_2$; (e) none of these.

9. The name of $CrSO_3$ is (a) cerium sulfite; (b) chromium(II) sulfite; (c) chromium(III) sulfate; (d) chromium(II) sulfate; (e) none of these.

10. The formula of nitrogen(III) oxide is ______________________.

11. The formula of sodium chlorate is (a) $NaClO_4$; (b) $NaClO_3$; (c) $NaClO_2$; (d) NaClO; (e) NaCl.

QUIZ C

1. A compound contains 1.8% H, 42.0% O, and 56.0% S. What is the empirical formula of the compound $(H = 1.0, O = 16.0, S = 32.1\ g/mol)$? (a) H_2SO_4; (b) H_2SO_3; (c) HSO; (d) $H_2S_2O_3$; (e) none of these.

2. Which compound has a molar mass of $51.7\ g/mol\ (Li = 6.9; F = 19.0; Na = 23.0; P = 31.0; S = 32.1; Cl = 35.5; K = 39.1\ g/mol)$? (a) NaCl; (b) KF; (c) MgS; (d) Li_3P; (e) none of these.

3. What is the mass fraction of Mg in $Mg(ClO_3)_2\ (O = 16.0, Mg = 24.3, Cl = 35.5\ g/mol)$? (a) $24.3/(24.3 + 35.5 + 16.0)$; (b) $24.3/[24.3 + 35.5 + (3 \times 16.0)]$; (c) $24.3/[24.3 + (2 \times 35.5) + 2 \times 16.0)]$; (d) $24.3/[24.3 + (2 \times 35.5) + (6 \times 16.0)]$; (e) none of these.

4. The molar mass of $Al(ClO_4)_3\ (O = 16.0, Al = 27.0, Cl = 35.5\ g/mol)$ is (a) $27.0 + [3 \times (35.5 + 16.0)]$; (b) $27.0 + (3 \times 35.5) + (12 \times 16.0)$; (c) $27.0 + 35.5 + 16.0$; (d) $27.0 + [(35.5 + 16.0) \times 3]$; (e) none of these.

5. How many atoms of all kinds are there in 42.1 g of $CaC_2\ (C = 12.0, Ca = 40.1\ g/mol; N = 6.022 \times 10^{23})$? (a) (42.1/52.1)(2) *N*; (b) (42.1/52.1)(3) *N*; (c) (42.1/64.1)(3) *N*; (d) (42.1/64.1)(2) *N*; (e) none of these.

6. How many oxygen atoms are contained in 63.5 g of $Cl_2O_7\ (O = 16.0, Cl = 35.5\ g/mol; N = 6.022 \times 10^{23})$? (a) (63.5/183) *N*; (b) $[(7 \times 63.5)/167]N$; (c) $[(7 \times 63.5)/183]N$; (d) $[(7 \times 63.5)/(167]N$; (e) none of these.

7. Iodine has an oxidation state of $+7$ in (a) HIO; (b) $NaIO_2$; (c) $Mg(IO_3)_2$; (d) $Al(IO_4)_3$; (e) KI.

8. The name of $FePO_4$ is (a) fermium phosphate; (b) iron(II) phosphite; (c) iron(III) phosphate; (d) iron(II) phosphate; (e) none of these.

9. Magnesium chlorate is (a) $Mg(ClO_3)_2$; (b) $Mn(ClO_3)_2$; (c) $MnClO_2$; (d) $MnClO_4$; (e) none of these.

10. The name of Cl_2O_7 is ______________________.

11. The name of which compound ends with *ite*? (a) $NaHCO_3$; (b) HNO_2; (c) $NaNO_3$; (d) $AlPO_3$; (e) HFO.

QUIZ D

1. What is the mass fraction of carbon in CH_3CN $(H = 1.0, C = 12.0, N = 14.0)$? (a) $(2\times12.0)/[(2\times12.0)+(3\times1.0)14.0]$; (b) $12.0/(12.0+1.0+14.0)$; (c) $12.0/[12.0+(3\times1.0)+14.0]$; (d) $12.0/[12.0+(3\times1.0)+12.0+14.0]$; (e) none of these.

2. The molar mass of $Na_2S_2O_3$ $(O = 6.0, S = 23.0, S = 32.1\,g/mol)$ is (a) $[2\times(23.0+32.1+16.0)+16.0]$; (b) $23.0+32.1+16.0$; (c) $2\times(23.0+32.1+16.0)$; (d) $(2\times23.0)+32.1(3\times16.0)$; (e) none of these.

3. A compound contains 46.7% O, 51.8% Cl, and 1.5% H $(H = 1.0, O = 16.0; Cl = 35.4\,g/mol)$. This compound's empirical formula is (a) HClO; (b) $HClO_2$; (c) $HClO_3$; (d) $HClO_4$; (e) none of these.

4. Which compound has a molar mass of 102.5 g/mol (O = 16.0, Na = 23.0, Cl = 35.5 g/mol)? (a) NaCl; (b) NaClO; (c) $NaClO_2$; (d) $NaClO_3$; (e) none of these.

5. How many hydrogen atoms are contained in 14.8 g of C_3H_8 $(H = 1.0, C = 12.0; N = 6.022\times10^{23})$? (a) (14.8/13.0)(3) *N*; (b) (14.8/13.0)(8) *N*; (c) (14.8/44.0) *N*; (d) (14.8/44.0) *N*; (e) none of these.

6. 8.75 g of $AlCl_3$ contains how many moles of ions $(Al = 27.0, Cl = 35.5\,g/mol)$? (a) $8.75/(27.0+35.5)$; (b) $8.75/[3\times27.0)+35.5]$; (c) $8.75\times(27.0+35.5)$; (d) $8.75/[27.0+(3\times35.5)]$; (e) none of these.

7. The oxidation state of Mn in $MgMnO_4$ is (a) $+2$; (b) $+7$; (c) $+6$; (d) $+4$; (e) $+3$.

8. The name of Hg_2SO_3 is (a) mercury(I) sulfite; (b) mercury(II) sulfite; (c) mercury(I) sulfate; (d) mercury(II) sulfate; (e) none of these.

9. The formula of copper(II) nitrite is (a) $CuNO_2$; (b) $Cu(NO_3)_2$; (c) $Co(NO_2)_2$; (d) $Cu(NO_2)_2$; (e) Cu_2NO_2.

10. The formula of phosphorus(III) oxide is __________.

11. The name of which compound ends with *ate*? (a) HIO_4; (b) Na_4SiO_4; (c) $KClO_2$; (d) HFO; (e) NO_2.

SAMPLE TEST (30 minutes)

1. A synthetic food company is attempting to make synthetic starch. Before the product is tested on laboratory animals, it is chemically analyzed. In determining the compound's empirical formula, a chemist burns 9.72 g of the compound in oxygen. The products are $16.88\,g\,CO_2$ and $9.21\,g\,H_2O$. Determine the empirical formula of the compound, assuming that it contains only the elements H, C, and O.

2. Iron is important in the body primarily because it is present in red blood cells and acts to carry oxygen to the various organs. Without oxygen, these organs will die. There are about 2.6×10^{13} red blood cells in the blood of an adult human, and the blood contains a total of 2.9 g of iron. How many iron atoms are there in each red blood cell?

3. Give the name or formula, as appropriate, of each compound, as well as the oxidation state of each atom.

A. copper(II) sulfate	B. $Na_2S_2O_3$	C. magnesium chlorite
D. Na_3BO_4	E. ammonium oxide	F. NI_3
G. $CaXeO_4$	H. barium peroxide	I. antimony(III) sulfide

4 CHEMICAL REACTIONS

CHAPTER OBJECTIVES

*1. Write word equations and symbolic equations for chemical reactions.

Chemical equations are summaries of experimental results. Consider the description: "A solution of lead(II) nitrate is added to a solution of potassium chromate. A bright yellow precipitate forms that later is identified as lead(II) chromate. Evaporation of the remaining solution produces crystals of potassium nitrate." The word equation for this reaction is

$$\text{lead (II) nitrate solution} + \text{potassium chromate solution} =$$
$$\text{solid lead(II) chromate} + \text{potassium nitrate solution}$$

Even more compact is the symbolic equation.

$$Pb(NO_3)_2(aq) + K_2CrO_4(aq) \longrightarrow PbCrO_4(s) + 2KNO_3(aq)$$

In the symbolic equation, the substances on the left are known as *reactants*; those on the right are the *products*. The form of each substance is indicated in parentheses: (g) for a gas, (l) for a liquid, (s) for a solid, (c) for crystalline solid, and (aq) for a substance dissolved in water—an aqueous solution. Be careful not to confuse (aq) with (l). NaCl(aq) means sodium chloride (table salt) dissolved in water. This is salt water, a mixture that can exist at room temperature. NaCl(l), on the other hand, means molten sodium chloride, which requires a temperature of at least 180°C to achieve.

*1a. Practice chemical nomenclature.

You may need to review nomenclature (Objectives 3–12 through 3–14) in order to transform a word equation into a symbolic equation. In fact, now that you know you will need to be able to turn names into formulas (and vice versa) in order to write and understand chemical equations, you have even more incentive to learn nomenclature. Some nomenclature problems are included for review in the Drill Problems. The solubility rules of Objective 5–3 will enable you to predict when an insoluble compound—designated (s)—is formed in an aqueous solution.

*2. Balance chemical equations by inspection.

An unbalanced chemical expression gives the identities of the reactants and the products, but not how much of each is consumed or produced. For that we need the balanced chemical equation. Balancing by inspection means that one multiplies each species by a whole number placed before the chemical formula of that species. The whole numbers are chosen so that there are the same number of atoms of each element on both sides of the balanced chemical equation. These inserted whole numbers are called *stoichiometric coefficients. Do not change the chemical formulas of the reactants or products!* It is often easiest to balance first any element that appears in only one compound in reactants and products, and to balance last elements that are uncombined

EXAMPLE 4-1 Balance the expression $H_2(g) + O_2(g) \longrightarrow H_2O(l)$.

The formula of water shows that hydrogen and oxygen unite in a ratio of 2:1, but the left–hand side of the expression shows 1 mole of H_2 uniting with 1 mole of O_2. The first step in balancing the expression is to multiply $H_2(g)$ by 2—that is, assign $H_2(g)$ the stoichiometric coefficient of 2. This produces $2H_2(g) + O_2(g) \longrightarrow H_2O(l)$ Counting atoms, we see 4 H and 2 O on the left-hand side and 2 H and 1 O on the right-hand side. So we multiply $H_2O(l)$ by 2. Thus, 2 becomes the stoichiometric coefficient of $H_2O(l)$. $2H_2(g) + O_2(g) \longrightarrow 2H_2O(l)$

Alternatively, we could have noticed that in the original unbalanced expression there are 2 O atoms on the left–hand side and only 1 O atom on the right–hand side. To balance this, 2 becomes the stoichiometric coefficient of water, producing $H_2(g) + O_2(g) \longrightarrow 2H_2O(l)$ Then, we would observe that there are 2 H atoms on the left–hand side and 4 H atoms on the right-hand side. This would have resulted in 2 as the stoichiometric coefficient for $H_2(g)$, as before. (We also could have incorrectly "balanced" this expression by changing H_2O to H_2O_2. This is incorrect because we changed the formula of the product.)

EXAMPLE 4-2 Balance the expression
$Pb(NO_3)_2(aq) + K_2CrO_4(aq) \longrightarrow PbCrO_4(S) + KNO_3(aq)$.

This expression is easier to balance if we recognize that only ionic compounds are involved and count the number of *ions* on each side. There are the same number of Pb^{2+} and $CrO_4^{\ 2-}$ ions on each side, but there are 2 K^+ and $2\,NO_3^{\ -}$ ions on the left–hand side and only one of each on the right–hand side. To make the number of K^+ and $NO_3^{\ -}$ ions the same on each side, we merely need to assign 2 as the stoichiometric coefficient of $KNO_3(aq)$.

$$Pb(NO_3)_2(aq) + K_2CrO_4(aq) \longrightarrow PbCrO_4(s) + 2KNO_3(aq)$$

EXAMPLE 4-3 Balance the expression $NH_3(g) + O_2(g) \longrightarrow NO(g) + H_2O(l)$

We first notice that N and H are combined in a ratio of 1:3 in NH_3 on the left-hand side and thus they must be produced in that ratio on the right–hand side. A complication arises because of H being present in H_2O where there are 2 H atoms in the formula. This requires an even number of H on the left–hand side of the expression. Thus the coefficient of NH_3 is an even number; we begin with 2. $2NH_3(g) + O_2(g) \longrightarrow NO(g) + H_2O(l)$ We now balance N and H atoms with coefficients for NO(g) and $H_2O(l)$. Remember, 3 H for each N.

$$2\,NH_3(g) + O_2(g) \longrightarrow 2\,NO(g) + 3\,H_2O(l)$$

There are now two O atoms on the left–hand side, but 5 O atoms on the right-hand side. We could use $\frac{5}{2}$ as the coefficient of $O_2(g)$,

$2\,NH_3(g) + \frac{5}{2}O_2(g) \longrightarrow 2\,NO(g) + 3\,H_2O(l)$, but it is preferable to use whole numbers (integers) as stoichiometric coefficients. Thus we multiply all coefficients by 2 for $4\,NH_3(g) + 5\,O_2(g) \longrightarrow 4\,NO(g) + 6\,H_2O(l)$.

There are only two rules to follow when balancing by inspection.

(1) *Never* change the formulas of the compounds.

(2) Be certain there are the same number of atoms of each element on both sides of the equation.

*** 2a. Balance "ionic equations" by inspection.**

The technique of balancing is the same as in Objective 4–2, but there is one additional rule: Make sure that the total charge on the left–hand side equals the charge on the right-hand side.

EXAMPLE 4-4 Balance the expression $Al + Cu^{2+} \longrightarrow Al^{3+} + Cu$

There is one atom of each element on each side. But the charge on the left is $+2$ while that on the right is $+3$. In the balanced equation, atomic balance is maintained by using the same coefficient for both forms of aluminum (Al and Al^{3+}), and likewise for both forms of copper (Cu^{2+} and Cu): $2\,Al + 3\,Cu^{2+} \longrightarrow 2Al^{3+} + 3\,Cu$. Note that the coefficient for copper (3) and the charge of aluminum (+3) are the same.

***3. Predict the combustion products—particularly of carbon-hydrogen and carbon-hydrogen-oxygen compounds—and write a balanced equation.**

Combustion means that the compound burns in oxygen, $O_2(g)$. When $C-H-O$ or $C-H$ compounds burn completely in oxygen, the *only* products formed are $CO_2(g)$ and $H_2O(l)$. It helps to know that $SO_2(g)$ is produced when a sulfur-containing compound is burned. The resulting equation is balanced by inspection.

EXAMPLE 4-5 Write the balanced equation for the combustion of $C_3H_7OH(l)$.

First, write the formula expression:

$$C_3H_7OH(l) + O_2(g) \longrightarrow CO_2(g) + H_2O(l)$$

Then balance by inspection (Objective 4–2):

$$2\;C_3H_7OH(l) + 9\;O_2(g) \longrightarrow 6\;CO_2(g) + 8\;H_2O(l)$$

*** 3a. Predict the products of some simple types of chemical reactions: combination, decomposition, displacement, and metathesis reactions.**

A major accomplishment of chemistry is the prediction of reaction products. The number and type of products depend on the type of the reaction. In an old, but still useful, system of classification, there are four simple types of reactions: combination, decomposition, displacement, and metathesis (or double displacement). Another scheme is presented in Chapter 5.

Combination: In a combination reaction, two substances unite to produce a single substance. The two substances may be two elements:

$$C(s) + O_2(g) \longrightarrow CO_2(g) \qquad Cu(s) + S(s) \longrightarrow CuS(s)$$

or two compounds:

$$SO_3(g) + H_2O(l) \longrightarrow H_2SO_4(l) \qquad CaO(s) + H_2O(l) \longrightarrow Ca(OH)_2(s)$$

$$Li_2O(s) + CO_2(g) \longrightarrow Li_2CO_3(s)$$

or an element and a compound:

$$2\,FeCl_2(s) + Cl_2(g) \longrightarrow 2\,FeCl_3(s) \qquad 2\,Cu_2O(s) + O_2(g) \longrightarrow 4CuO(s)$$

Notice that the products are composed of ions that you already have learned, $(Cu^{2+}, S^{2-}, Ca^{2+}, OH^-, Li^+, CO_3^{2-}, Fe^{3+}, Cl^-, O^{2-})$ or are compounds that should be familiar (CO_2, NH_3, H_2SO_4). This will be true for practically all reactions you encounter in this course.

Decomposition: In decomposition reactions, a single compound produces two or more substances. These substances may be compounds:

$$CaCO_3 \xrightarrow{\text{heat}} CaO(s) + CO_2(g) \qquad NH_4HCO_3(s) \xrightarrow{\text{heat}} NH_3(g) + H_2O(l) + CO_2(g)$$

or they may be elements:

$$2\,HgO(s) \xrightarrow{\text{heat}} 2Hg(l) + O_2(g) \qquad MgCl_2(l) \xrightarrow{\text{heat \& electricity}} Mg(l) + Cl_2(g)$$

or they may be elements and compounds:

$$2KClO_3(s) \xrightarrow{\text{heat}} 2\,KCl(s) + 3\,O_2(g) \qquad 2\,NaNO_3(s) \xrightarrow{\text{heat}} 2\,NaNO_2(s) + O_2(g)$$

Again, notice that the compounds produced either are familiar (NH_3, CO_2, H_2O) or are composed of ions that you have memorized $(Ca^{2+}, O^{2-}, K^+, Cl^-, Na^+, NO_2^-)$. Notice also that heat or electricity is normally required to decompose compounds. Few compounds decompose spontaneously. Finally, notice that this is the only one of the four types of reactions that has but one reactant.

Displacement: In a displacement reaction, normally one element reacts with a compound to take the place of one of the elements in that compound. Sometimes, however, one compound displaces another from a third compound. Some examples follow.

$$Zn(s) + CuSO_4(aq) \longrightarrow ZnSO_4(aq) + Cu(s) \qquad 2\,Na(s) + 2\,H_2O(l) \longrightarrow 2\,NaOH(aq) + H_2(g)$$

$$Mg(s) + 2\,HCl(aq) \longrightarrow MgCl_2(aq) + H_2(g) \qquad 3\,H_2(g) + Fe_2O_3(s) \xrightarrow{\text{heat}} 3\,H_2O(g) + 2\,Fe(s)$$

$$Cl_2(g) + 2\,KI(aq) \longrightarrow 2\,KCl(aq) + I_2(s) \qquad CO_2(g) + Ca(OH)_2(s) \longrightarrow CaCO_3(s) + H_2O(g)$$

In each of the previous equations, the "displacing element" (or compound) is written first and the element (or compound) that is displaced is written last. Again, notice that all of the products are substances that you can name.

Metathesis: Metathesis (or double displacements) is when two ionic compounds "change partners" or swap ions.

$BaCl_2(aq) + K_2SO_4(aq) \longrightarrow BaSO_4(s) + 2\,KCl(aq)$

$NaCN(aq) + HCl(aq) \longrightarrow HCN(g) + NaCl(aq)$

$2\,KOH(aq) + H_2SO_4(aq) \longrightarrow 2\,H_2O(l) + K_2SO_4(aq)$

$MgCO_3(s) + 2\,HCl(aq) \longrightarrow H_2O(l) + CO_2(g) + MgCl_2(aq)$

$Na_2SO_3(aq) + 2\,HI(aq) \longrightarrow H_2O(l) + SO_2(g) + 2\,NaI(aq)$

In the first three equations above, it is clear that the cations have exchanged anions. For example, Ba^{2+} is associated with Cl^- in the reactants of the first equation and with $SO_4^{\ 2-}$ in the products. This exchange is not so evident in the last two equations until we recognize that the first two products in each case are the result of decomposition reactions. ["$H_2CO_3(aq)$" is enclosed in quotes because this compound has never been isolated.]

$``H_2CO_3(aq)" \longrightarrow H_2O(l) + CO_2(g) \qquad H_2SO_3(aq) = SO_2 \cdot H_2O \longrightarrow H_2O(l) + SO_2(g)$

Again, notice that all the products are substances you are able to name. This provides another good reason for learning chemical nomenclature: it makes it much easier to predict the products of chemical reactions. Also, because you know how elements combine, you can recognize unusually, and most likely incorrect, chemical formulas, such as KSO_4.

*** 4. From balanced chemical equations derive conversion factors for use in stoichiometric calculations.**

A chemical equation can be interpreted either in terms of atoms, ions, molecules, and formula units, or in terms of moles of these entities. Thus, there are two ways of interpreting the following equation.

$$2\,Na(s) + 2\,H_2O(l) \longrightarrow 2\,NaOH(aq) + H_2(g)$$

The first interpretation is that 2 sodium atoms react with two water molecules to produce 2 formula units of sodium hydroxide (or 2 sodium ions and 2 hydroxide ions) and 1 hydrogen molecule. The second is that 2 moles of Na(s) react with 2 moles of $H_2O(l)$ to produce 2 moles of NaOH(aq) and 1 mole of $H_2(g)$. Therefore, if we are given the amount in moles of one reactant or product and the balanced chemical equation, we can determine the amounts of the other species involved.

EXAMPLE 4-6 Determine the amounts of Sr(s) and $H_2O(l)$ that reacted and the amount of $Sr(OH)_2(aq)$ produced when 3.4 mol $H_2(g)$ is produced by the reaction $Sr(s) + 2\,H_2O(l) \longrightarrow Sr(OH)_2(aq) + H_2(g)$

The balanced equation states that 1 mol $Sr(OH)_2(aq)$ is produced for every 1 mol $H_2(g)$ produced.

$$\text{amount Sr(OH)}_2\text{(aq)} = 3.4 \text{ mol H}_2\text{(g)} \times \frac{1 \text{ mol Sr(OH)}_2\text{(aq)}}{1 \text{ mol H}_2\text{(g)}} = 3.4 \text{ mol Sr(OH)}_2\text{(aq)}$$

Note carefully that 1 mol $H_2(g)$ is not *equal* to 1 mol $Sr(OH)_2(aq)$. They are equivalent to each other in that they are produced together. In like fashion, we determine the amounts of reactants consumed.

$$\text{amount Sr(s)} = 3.4 \text{ mol H}_2\text{(g)} \times \frac{1 \text{ mol Sr(s)}}{1 \text{ mol H}_2\text{(g)}} = 3.4 \text{ mol Sr(s)}$$

$$\text{amount H}_2\text{O(l)} = 3.4 \text{ mol H}_2\text{(g)} \times \frac{2 \text{ mol H}_2\text{O(l)}}{1 \text{ mol H}_2\text{(g)}} = 6.8 \text{ mol H}_2\text{O(l)}$$

The stoichiometric coefficient for each species is used in the conversion factor. Quantities are converted two and from amounts in moles. It is essential that you thoroughly master this technique. All of the material in this chapter either leads up to or depends on this objective. In addition, you will use this technique repeatedly in subsequent chapters.

*** 4a. Solve problems based on balanced chemical equations with quantities given or sought in a variety of units.**

This objective combines the previous objective with the techniques of using conversion factors that you mastered when studying Chapters 1 and 3. You may wish to review the conversion factor method (Objective 1–11). One or more of the conversion factors is between amounts in moles, obtained from a balanced chemical equation.

EXAMPLE 4-7 Under normal conditions, hydrogen gas has a density of 0.0824 g/L. What volume of hydrogen gas in liters is needed to produce 2.50 lb. of iron from $Fe_2O_3(s)$? Water is the other reaction product.

$$\text{Fe}_2\text{O}_3\text{(s)} + \text{H}_2\text{(g)} \longrightarrow \text{Fe(s)} + \text{H}_2\text{O(l)}$$

Review Objective 4–1 if you have trouble writing the unbalanced expression. The expression then is balanced by inspection (Objective 4–2),

$$\text{Fe}_2\text{O}_3\text{(s)} + 3\,\text{H}_2\text{(g)} \longrightarrow \text{Fe(s)} + 3\,\text{H}_2\text{O(l)}.$$

Now can apply the conversion factor method. (1) ? L $H_2(g)$ (2) 2.50 lb Fe(s) (3)

$$\text{volume H}_2 = 2.50 \text{ lb Fe} \times \frac{454 \text{ g Fe}}{1 \text{ lb.}} \times \frac{1 \text{ mol Fe}}{55.8 \text{ g}} \times \frac{3 \text{ mol H}_2}{2 \text{ mol Fe}} \times \frac{2.02 \text{g H}_2}{1 \text{ mol H}_2} \times \frac{1 \text{ L H}_2\text{(g)}}{0.0824 \text{g H}_2}$$

$$= 748 \text{ L H}_2$$

Note that *only one* conversion factor (3 mol H_2 / 2 mol Fe) is obtained from the balanced chemical reaction. It is very unusual that more than one conversion factor will be taken from a given chemical equation. Also, notice that moles (not grams or liters) of different substances are related. Therefore, when you wish to convert from one substance to another, you must convert to (and then from) moles.

*** 5. Define the terms associated with solutions, including *molarity*; compute molarities and solution volumes.**

The first goal is to define in your words and commit to memory the terms *solute, solution, solvent, concentration, aqueous solution*, and *molarity*. You should also know the common symbol for molarity (brackets around the chemical formula of the species in solution). Thus, $[NH_3]$ means "the molar concentration of ammonia." The formula in brackets is that of the species actually present in solution. Therefore, for an aqueous solution of magnesium chloride, we can state $[Mg^{2+}] = 0.320\text{ M}$ and $[Cl^-] = 0.640\text{ M}$, but not $[MgCl_2] = 0.320\text{ M}$. This is because the ionic solute exists as ions in solution and not grouped as formula units. Instead, we say that the $MgCl_2$ molarity is 0.320 M. (Be aware that some chemists do not make this distinction. They will write $[MgCl_2] = 0.320\text{ M}$.)

We need to be careful in another regard. Chemists often refer to a solution by the name of the solute. Thus, an aqueous sodium hydroxide solution often is spoken of as simply "sodium hydroxide." This is familiar to us in other areas: "coffee" refers to both the solution we drink and the beans from which it is brewed. The definition of *molarity* (moles of solute/liters of solution) suggests three types of problems:

EXAMPLE 4-8 (Determine molarity given moles of solute and volume of solution.) A 205-mL volume of solution contains 4.1 g NaCl. What is the concentration of the solution in moles per liter?

To determine molarity, we divide the *amount* of solute (in moles) by the *volume* of solution (in liters).

$$\text{NaCl molarity} = \frac{4.1\text{ g NaCl} \times \dfrac{1\text{ mol NaCl}}{58.5\text{ g NaCl}}}{205\text{ mL solution} \times \dfrac{1\text{ L}}{1000\text{ mL}}} = \frac{0.070\text{ mol NaCl}}{0.205\text{ L soln}} = 0.34\text{ M}$$

EXAMPLE 4-9 (Determine moles of solute given solution volume and concentration.) A 114-mL sample of $0.357\text{ M Al(NO}_3)_3$ contains what amount of solute in moles?

The solution molarity is a conversion factor:

(1) $?\text{ mol Al(NO}_3)_3$

(2) 114 mL solution

(3)

$$\text{Al(NO}_3)_3\text{ amount} = 114\text{ mL soln} \times \frac{1\text{L}}{1000\text{ mL}} \times \frac{0.357\text{ mol Al(NO}_3)_3}{1\text{ L soln}}$$
$$= 0.0407\text{ mol Al(NO}_3)_3$$

EXAMPLE 4-10 (Determine volume of solution given solution concentration and moles of solute.) What volume in mL of 0.628 M AgNO_3 solution contains 55.1 g $AgNO_3$?

(1) ? mL solution

(2) 55.1 g AgNO_3

(3) solution

$$\text{volume} = 55.1\,\text{g AgNO}_3 \times \frac{1\,\text{mol AgNO}}{169.9\,\text{g}} \times \frac{1\,\text{L soln}}{0.628\,\text{mol AgNO}_3} \times \frac{1000\,\text{mL}}{1\,\text{L}} = 516\,\text{mL}$$

*** 6. Solve dilution problems and those involving the mixing of two solutions.**

There is yet another type of problem, which concerns the dilution of solutions. We first determine the amount of solute in the original solution with the technique of Example 4-9, and then compute the molarity with the new solution volume and the method of Example 4-8.

EXAMPLE 4-11 A volume of 54.0 mL of water is mixed with 114 mL of 0.357 M $NaNO_3$ solution. What is the final concentration of sodium nitrate, the $NaNO_3$ molarity?

$$\text{amount of solute} = 114\,\text{mL soln} \times \frac{1\,\text{L}}{1000\,\text{mL}} \times \frac{0.357\,\text{mol NaNO}_3}{1\,\text{L soln}} = 0.0407\,\text{mol NaNO}_3$$

$$\text{Total solution volume} = 54.0\,\text{mL} + 114\,\text{mL} = 168\,\text{mL}.$$

$$\text{NaNO}_3\text{ molarity} = \frac{\text{mol NaNO}_3}{\text{L solution}} = \frac{0.0407\,\text{mol NaNO}_3}{168\,\text{mL} \times \frac{1\,\text{L}}{1000\,\text{mL}}} = 0.242\,\text{M}$$

Often, an alternate definition of molarity (= millimoles of solute/milliliters of solution) is more convenient because most solution volumes are expressed in milliliters. The solution for Example 4-11 then becomes

$$\text{amount of solute} = 114\,\text{mL soln} \times \frac{0.357\,\text{mmol NaNO}_3}{1\,\text{mL soln}} = 40.7\,\text{mmol NaNO}_3$$

$$\text{NaNO}_3\text{ molarity} = \frac{40.7\,\text{mmol NaNO}_3}{168\,\text{mL}} = 0.242\,\text{M}$$

When two solutions are mixed, the concentration of a solute (or a solute ion) that is present in only one of the solutions is determined with the method of Example 4-11. If a solute (or a solute ion) is present in both solutions, the amount of solute in each solution is determined before its concentration in the final solution can be calculated, as shown in Example 4-12.

EXAMPLE 4-12 A volume of 54.0 mL of 0.628 M $NaNO_3$ is mixed with 114 mL of 0.357 M $NaNO_3$ solution. What is the final concentration of sodium nitrate?

As in Example 4-11, the total solution volume is 168 mL. First the amount of sodium nitrate in each of the two solutions is determined. These amounts are added when determining the concentration of $NaNO_3$.

$$\text{amount NaNO}_3 = 54.0\,\text{mL} \times \frac{0.628\,\text{mmol NaNO}_3}{\text{mL soln}} + 114\,\text{mL} \times \frac{0.357\,\text{mmol NaNO}_3}{\text{mL soln}}$$
$$= 33.9\,\text{mmol NaNO}_3 + 40.7\,\text{mmol NaNO}_3$$

$$\text{NaNO}_3 \text{ molarity} = \frac{(33.9 + 40.7) \text{ mmol NaNO}_3}{168 \text{ mL soln}} = 0.444 \text{ M}$$

*** 7. Solve stoichiometry problems when either the reactants or the products are species in solution and concentration and volume data are given.**

This objective is a combination of Objectives 4-5 and 4-7. Therefore, it only requires the techniques you have mastered already. Apply these techniques to somewhat more involved problems. Be sure that you work on some of the drill problems of these two previous objectives before you attempt the drill problems of this objective. Remember that, when changing from one substance to another, you must convert to (and from) moles.

EXAMPLE 4-13 What volume of 0.154 M $Ca(NO_3)_2$, in mL, must be added to an excess of 0.500 M $Na_3PO_4(aq)$ to produce 65.1 mg of $Ca_3(PO_4)_2$ (310.2 mg/mmol)?

The balanced chemical equation is:

$$3\,Ca(NO_3)_2(aq) + 2\,Na_3PO_4(aq) \longrightarrow Ca_3(PO_4)_2(s) + 6\,NaNO_3(aq).$$

The conversions we need are mass

$$Ca_3(PO_4)_2 \longrightarrow \text{amount } Ca_3(PO_4)_2 \longrightarrow \text{amount } Ca(NO_3)_2 \longrightarrow \text{ volume } Ca(NO_3)_2.$$

In the calculation, each arrow in the previous sentence is replaced by a conversion factor. Because information is supplied in mg and requested in mL, we use concentrations in mmol/mL

solution volume =

$$65.1 \text{ mg of } Ca_3(PO_4)_2 \times \frac{1 \text{ mmol } Ca_3(PO_4)_2}{310.2 \text{ mg } Ca_3(PO_4)_2} \times \frac{3 \text{ mmol } Ca(NO_3)_2}{1 \text{ mmol } Ca_3(PO_4)_2} \times \frac{1 \text{ mL soln}}{0.154 \text{ mol } Ca(NO_3)_2}$$

$$= 4.09 \text{ mL } Ca(NO_3)_2(aq)$$

*** 8. Determine the reactant(s) in excess, the limiting reactant, and the amounts of products obtained in a chemical reaction.**

Limiting reactant (or limiting reagent) problems, as they commonly are called, are based on the fact that reactants combine to yield definite amounts of products. If there is an excess of one reactant, that excess will undergo no chemical change. The reactant that determines the final amount of product is the limiting reactant. Chemical reactions are quite different in this respect from cookie recipes. Using 12 oz .of chocolate chips in a recipe calling for 16 oz. only causes the cookies to have a scant number of chocolate chips.

Chemical reactions are more similar to automobile production. If a manufacturer has 7200 spark plugs, only 1200 six-cylinder cars can be produced. Likewise, 4000 wheels limits production to 1000 cars. It makes no difference that there are 9000 engines and 4000 car bodies on hand. Limiting reactant problems often are most easily solved by determining the amount of each product that can be produced from each reactant. The limiting reagent is the one that produces the least amount of product.

$$7200 \text{ spark plugs} \times \frac{1 \text{ car}}{6 \text{ spark plugs}} = 1200 \text{ cars} \qquad 4000 \text{ wheels} \times \frac{1 \text{ car}}{4 \text{ wheels}} = 1000 \text{ cars}$$

$$3000 \text{ bodies} \times \frac{1 \text{ car}}{1 \text{ body}} = 3000 \text{ cars} \qquad 9000 \text{ engines} \times \frac{1 \text{ car}}{1 \text{ engine}} = 9000 \text{ cars}$$

Hence, only 1000 cars can be produced. Notice the limiting reactant is *not* the smallest number of parts (3000 bodies), but the number of parts that *produces the smallest number* of cars. A key to spotting a limiting reactant problem is that the quantities of two or more reactants must be stated. Although these quantities may be stated in different units (such as mass of a solid along with volume of a solution), they both will be given.

EXAMPLE 4-14 A sample of 1.50 g of Mg reacts with 342 mL of 0.350 M HCl solution to produce what mass in grams of $MgCl_2$? What mass of the excess reactant is left over?

The reaction is a displacement (See Objective 4-4):

$$Mg(s) + 2\,HCl(aq) \longrightarrow MgCl_2(aq) + H_2(g)$$

$$\text{amount } MgCl_2 = 1.50 \text{ g Mg} \times \frac{1 \text{ mol Mg}}{24.3 \text{ g Mg}} \times \frac{1 \text{ mol } MgCl_2}{1 \text{ mol Mg}} = 0.0617 \text{ mol } MgCl_2$$

$$\text{amount } MgCl_2 = 342 \text{ mL soln} \times \frac{1 \text{ L}}{1000 \text{ mL}} \times \frac{0.350 \text{ mol HCl}}{1 \text{ L soln}} \times \frac{1 \text{ mol } MgCl_2}{2 \text{ mol HCl}}$$

$$= 0.0599 \text{ mol } MgCl_2$$

HCl is the limiting reactant, and only 0.0599 mol $MgCl_2$ is produced. We convert this amount to a mass.

$$\text{mass } MgCl_2 = 0.0599 \text{ mol } MgCl_2 \times \frac{95.3 \text{ g } MgCl_2}{1 \text{ mol } MgCl_2} = 5.71 \text{ g } MgCl_2$$

To determine the mass of Mg left over (the excess reactant), first compute the mass of Mg that reacted and then subtract that mass from the 1.50 g provided.

$$342 \text{ mL soln} \times \frac{1 \text{ L}}{1000 \text{ mL}} \times \frac{0.350 \text{ mol HCl}}{1 \text{ L soln}} \times \frac{1 \text{ mol Mg}}{2 \text{ mol HCl}} \times \frac{24.3 \text{ g Mg}}{1 \text{ mol Mg}} = 1.45 \text{ g Mg}$$

Mass in excess $= 1.50 \text{ g} - 1.45 \text{ g} = 0.05 \text{ g Mg}$.

*** 9. Define the terms *actual yield*, *theoretical yield*, and *percent yield*. Compute these quantities for a given reaction.**

The *theoretical yield*, or calculated yield, of a reaction is the computed quantity of product based on the quantities of reagents used. This quantity may have to be determined by solving a limiting reagent problem. The actual yield (often called the experimental yield) is the quantity of product actually obtained. The percent yield is calculated from their ratio:

$$\text{percent yield} = \frac{\text{actual yield}}{\text{theoretical yield}} \times 100\%$$

EXAMPLE 4-15 The reaction $N_2 + 3\,H_2 \longrightarrow 2\,N_3$ has a percent yield of 62.1% under certain conditions. **(a)** What mass of NH_3 is produced from $74.8\text{ g }N_2$? **(b)** What mass of H_2 is required to produce $36.5\text{ g }NH_3$?

We use the conversion factor method in each case, labeling the actual yield as "g NH_3 produced" or "g NH_3 formed" and the theoretical yield as "g g NH_3 calculated."

(a)

$$\text{mass NH}_3 = 74.8\text{ g N}_2 \times \frac{1\text{ mol N}_2}{28.0\text{ g N}_2} \times \frac{2\text{ mol NH}_3}{1\text{ mol N}_2} \times \frac{17.0\text{ g NH}_3}{1\text{ mol NH}_2} \times \frac{62.1\text{ g NH}_3\text{ produced}}{100.0\text{ g NH}_3\text{ calculated}}$$

$$= 56.4\text{ g NH}_3\text{ produced}$$

(b)

$$\text{mass H}_2 = 36.5\text{ g NH}_3\text{ formed} \times \frac{100.0\text{ g NH}_3\text{ calc'd}}{62.1\text{ g NH}_3\text{ formed}} \times \frac{1\text{ mol NH}_3}{17.0\text{ g NH}_3} \times \frac{3\text{ mol H}_2}{2\text{ mol NH}_3} \times \frac{2.02\text{ g H}_2}{1\text{ mol H}_2}$$

$$= 10.5\text{ g H}_2\text{ required}$$

Notice that the correct use of percent yield as a conversion factor requires careful labeling of both numerator and denominator. Often, problems involving consecutive reactions (Objective 4-11) will be complicated by including the percent yield of each reaction.

*** 10. Compute the amount of product produced or reactant consumed by two or more simultaneous reactions.**

Simultaneous reactions either consume an identical reactant or yield an identical product. We are asked to determine the amount of the common species involved. The techniques of solving stoichiometric problems (Objective 4-5) are used for each reactant, in turn, and the results are added.

EXAMPLE 4-16 What amount of $AgNO_3$ is consumed when 406 mL of a solution that is both 0.100 M in KCl and 0.250 M in Na_2CrO_4 reacts completely?

Both reactions are metatheses. $KCl(aq) + AgNO_3(aq) \longrightarrow AgCl(s) + KNO_3(aq)$

The amount of $AgNO_3$ required to react with the KCl in solution is

$$\text{amount AgNO}_3 = 406\text{ mL} \times \frac{1\text{ L}}{1000\text{ mL}} \times \frac{0.100\text{ mol KCl}}{1\text{ L soln}} \times \frac{1\text{ mol AgNO}_3}{1\text{ mol KCl}}$$

$$= 0.046\text{ mol AgNO}_3$$

The amount of $AgNO_3$ required to react with the Na_2CrO_4 in solution is

$$\text{amount AgNO}_3 = 406\text{ mL} \times \frac{1\text{ L}}{1000\text{ mL}} \times \frac{0.250\text{ mol Na}_2\text{CrO}_4}{1\text{ L soln}} \times \frac{2\text{ mol AgNO}_3}{1\text{ mol Na}_2\text{CrO}_4}$$

$$= 0.203\text{ mol AgNO}_3$$

Total $AgNO_3 = 0.0406\text{ mol} + 0.203\text{ mol} = 0.244\text{ mol}$.

*** 11. Compute the amount of product produced by two or more consecutive reactions.**

Consecutive reactions yield products from reactants in a series of steps rather than in a single reaction. Thus the solution of a stoichiometric problem will contain not just one conversion factor obtained from a chemical reaction, but several—one from each step. An example of a series of reactions is the Solvay process for making sodium carbonate. It consists of six steps.

[1 & 2] $CaCO_3(s) \xrightarrow{\text{heat}} CaO(s) + CO_2(g)$ $\qquad$ $CaO(s) + H_2O(l) \longrightarrow Ca(OH)_2(aq)$

[3] $Ca(OH)_2(aq) + 2\,NH_4Cl(aq) \longrightarrow 2\,NH_3(g) + 2\,H_2O(l) + CaCl_2(aq)$

[4] $NH_3(g) + CO_2(g) + H_2O(l) \longrightarrow NH_4HCO_3(aq)$

[5] $NaCl(aq) + NH_4HCO_3(aq) \longrightarrow NaHCO_3(s) + NH_4Cl(aq)$

[6] $2\,NaHCO_3(s) \xrightarrow{\text{heat}} Na_2CO_3(s) + H_2O(l) + CO_2(g)$

EXAMPLE 4-17 What mass of sodium carbonate in grams can be produced from 64.0 g of $CaCO_3$ using the Solvay process?

$$\text{mass } Na_2CO_3 = 64.0\text{ g } CaCO_3 \times \frac{1\text{ mol } CaCO_3}{100.0\text{ g}} \times \frac{1\text{ mol } CO_2}{1\text{ mol } CaCO_3} \times \frac{1\text{ mol } NH_4HCO_3}{1\text{ mol } CO_2}$$

$$\times \frac{1\text{ mol } NaHCO_3}{1\text{ mol } NH_4HCO_3} \times \frac{1\text{ mol } Na_2CO_3}{2\text{ mol } NaHCO_3} \times \frac{106.0\text{ g } Na_2CO_3}{1\text{ mol } Na_2CO_3} = 33.9\text{ g } Na_2CO_3$$

The hardest part of solving this type of problem is in tracing the product (or a part, thereof) through the series of reactions. In this case the element carbon was used. Sometimes this tracing is more easily done if one works backward from the products. Notice that each chemical equation furnishes only one conversion factor. In Example 4-17, only conversion factors from equations [1], [4], [5], and [6] are needed.

*** 12. Determine the overall or net reaction for a process consisting of several steps.**

The easiest way to find the overall reaction is by combining the chemical reactions as if they were algebraic equations, eliminating those species that appear on both sides. The equations are manipulated so that as many species cancel as possible. For the Solvay process, equations [1] through [6] of Objective 4-12, we proceed as follows.

Combination: $\qquad [6] + 2 \times [5] = [7]$

$$\cancel{2\,NaHCO_3} + 2\,NaCl + 2\,NH_4HCO_3 \longrightarrow Na_2CO_3 + H_2O_2 + CO_2 + \cancel{2\,NaHCO_3} + 2\,NH_4Cl$$

Result: $2\,NaCl + 2\,NH_4HCO_3 \longrightarrow Na_2CO_3 + H_2O + CO_2 + 2\,NH_4Cl$ $\qquad$ [7]

Combination: $\qquad [7] + 2 \times [4] = [8]$

$$2\,NaCl + \cancel{2\,NH_4HCO_3} + 2\,NH_3 + \cancel{2}\,CO_2 + \cancel{2}\,CO_2 + \cancel{2}\,H_2O \longrightarrow Na_2CO_3 + \cancel{H_2O} + \cancel{CO_2} + 2\,NH_4Cl + \cancel{2NH_4HCO_3}$$

Result: $2\,NaCl + 2\,NH_3 + CO_2 + H_2O \longrightarrow Na_2CO_3 + 2\,NH_4Cl$ [8]

Combination: $[8] + [3] = [9]$

$$2\,NaCl + \cancel{2\,NH_3} + CO_2 + \cancel{H_2}O + Ca(OH)_2 + \cancel{2NH_4Cl} \longrightarrow Na_2CO_3 + \cancel{2NH_4Cl} + \cancel{2NH_3} + \cancel{2}H_2O + CaCl_2$$

Result: $2\,NaCl + CO_2 + Ca(OH)_2 \longrightarrow Na_2CO_3 + H_2O + CaCl_2$ [9]

Combination: $[1] + [2] = [10]$

Result: $CaCO_3 + \cancel{CaO} + H_2O \longrightarrow \cancel{CaO} + CO_2 + Ca(OH)_2$ [10]

Combination: $[9] + [10] = \text{overall reaction} = [11]$

$$2\,NaCl + \cancel{CO_2} + \cancel{Ca(OH)_2} + CaCO_3 + \cancel{H_2O} \longrightarrow Na_2CO_3 + \cancel{H_2O} + CaCl_2 + \cancel{CO_2} + \cancel{Ca(OH)_2}$$

Result: $2\,NaCl + CaCO_3 \longrightarrow Na_2CO_3 + CaCl_2$ overall reaction [11]

SELF- ASSESSMENT EXERCISES

114. (d) Correct. This answer gives on the left side of the equation 3 Cu, 8H, 8 N, 24 O, on the right side of the equation 3 Cu, 6 N, 18 O (from $Cu(NO_3)_2$) and 8 H, 4 O, 2 N, and 2 O for a total left-hand side. 3 Cu, 8 H, 8 N, 24 O and the same on the right-hand side. The equation is balanced.

115. (d) Correct. For each 3 moles of H_2O, 2 moles of NH_3 are formed. Therefore,

$$1 \text{ mol } H_2O \text{ x } \frac{2 \text{ mol } NH_3}{3 \text{ mol } H_2O} = 2/3 \text{ mol } NH_3.$$

116. (a) Write the balanced equation:

$$2\,Ag_2CO_3 \rightarrow 4\,Ag + 2\,CO_2 + O_2$$

The correct ratio is 1 mol O_2 : 2 mol CO_2. There are 4 mol of Ag formed for each 2 mol of silver carbonate.

117. (c) A 1.00 M solution of $NaNO_3$ contains 1 gram mol wt of $NaNO_3$ in 1.00 L of solution. The molar mass of $NaNO_3$ is 85.0 u. Answer (b) is incorrect because 1 kg of aqueous

solution will be more than 1.0 L. Answer (c) 425 g/5.0 L is 85.0 g/L and Answer (d) is 8.5 g/L when converted.

118. The best answer is to evaporate 20 mL. Answer (b) will give the correct number of moles, but will also increase the volume slightly.

119. (d) The theoretical yield is the amount of product formed if all of the available reactants formed products. CCl_4 is limiting. Therefore, 2 mol of CCl_2F_2 could, theoretically, be formed. The % yield is the actual yield divided by the theoretical yield x 100%. The yield is 1.70/2.00 x 100% = 85%.

120. (a) Complete combustion:

$$C_8H_{18} + 12\ \tfrac{1}{2}\ O_2 \rightarrow 8\ CO_2 + 9\ H_2O$$

(b) Incomplete combustion produces 25% CO. Therefore, 25% of the carbon ends up as CO, or 2 of the 8 mol CO_2 will be CO.

$$C_8H_{18} + 11\ \tfrac{1}{2}\ O_2 \rightarrow 2CO + 6\ CO_2 + 9\ H_2O$$

121. The number of moles of C in the sample is 0.477 g/ 44.0 g/mol = 0.0108 mol. 0.0108 mol of C = 0.1301 g or 13.0% since a 1 g sample was used. Calcium carbonate contains 12% carbon (atomic mass of C/molar mass of calcium carbonate) and magnesium carbonate contains 14.2% C. Therefore, dolomite will give 13 % C.

DRILL PROBLEMS

1. Write a balanced equation for each of the following word equations.

A. aluminum + magnesium oxide ⟶ magnesium + aluminum oxide

B.
aluminum chloride + sodium hydroxide ⟶ aluminum hydroxide + sodium chloride

C. silver nitrate + sodium phosphate ⟶ silver phosphate + sodium nitrate

D. chlorine + potassium iodide ⟶ potassium chloride + iodide

E. iron(III) sulfate + calcium hydroxide ⟶ iron(III) hydroxide + calcium sulfate

F. barium nitrate + ammonium carbonate ⟶ ammonium nitrate + barium carbonate

G. potassium hydroxide + sulfuric acid ⟶ potassium sulfate + water

H. iron(II) hydroxide + hydrochloric acid ⟶ iron(II) chloride + water

I. copper(II) sulfate + hydrogen sulfide ⟶ silver chromate + potassium nitrate

J. silver nitrate + potassium chromate $\longrightarrow$ silver chromate + potassium nitrate

1a. (1) Give the correct name of each of the following compounds.

A.	$K_2Cr_2O_7$	B.	$AgBrO_3$	C.	$Mg(HSO_3)_2$	D.	KIO
E.	$NaClO_4$	F.	KCNS	G.	$Pb(CO_3)_2$	H.	CuC_2O_4
I.	K_2O	J.	Tl_2S	K.	$Sn(MnO_4)_2$	L.	$Mn(IO_3)_2$
M.	$Ni(CN)_2$	N.	$Mn(CNO)_2$	O.	$CdSO_4$	P.	$MnSO_3$
Q.	$(NH_4)_2Cr_2O_7$	R.	$CoMnO_4$	S.	AgCNS	T.	Li_2O_2
U.	$Hg_2(C_2H_3O_2)_2$	V.	Tl_2S_3	W.	$Ag_2Cr_2O_7$	X.	$Au_2(SO_4)_3$
Y.	$Mg(IO_3)_2$	Z.	$Cd(HSO_3)_2$				

(2) Give the chemical formula of each of the following compounds.

A.	strontium peroxide	B.	manganese(II) sulfite	C.	lead(II) oxalate
D.	rubidium bromate	E.	magnesium cyanate	F.	sodium bisulfide
G.	calcium hypoiodite	H.	iron(II) peroxide	I.	silver sulfide
J.	barium iodate	K.	cadmium sulfate	L.	potassium manganate
M.	arsenic(III) chloride	N.	chromium(II) bromite	O.	tin(IV) carbonate
P.	arsenic(III) sulfide	Q.	manganese(II) bicarbonate	R.	lead(IV) carbonate
S.	gold(I) dichromate	T.	potassium oxalate	U.	manganese(II) thiosulfate
V.	hydrogen peroxide	W.	gold(III) iodate	X.	sodium cyanide
Y.	copper(I) cyanate	Z.	tin(II) fluoride		

2. Balance the following equations by inspection.

A. $P_4 + O_2 \longrightarrow P_2O_5$

B. $Na + O_2 \longrightarrow Na_2O_2$

C. $Al + HCl \longrightarrow AlCl_3 + H_2$

D. $Ca + H_2O \longrightarrow Ca(OH)_2 + H_2$

E. $FeCl_3 + Ca(OH)_2 \rightarrow Fe(OH)_3 + CaCl_2$

F. $Al + N_2 \longrightarrow AlN$

G. $HCl + Fe_2O_3 \longrightarrow FeCl_3 + H_2O$

H. $Cl_2 + H_2O \longrightarrow HCl + HClO_3$

I. $Al(OH)_3 + HCl \longrightarrow AlCl_3 + H_2O$

J. $CaSO_3 + H_2SO_4 \rightarrow CaSO_4 + H_2O + SO_2$

K. $NaCl + H_2SO_4 \longrightarrow Na_2SO_4 + HCl$

L. $Pb(NO_3)_2 \longrightarrow PbO + NO + O_2$

M. $HNO_2 \longrightarrow HNO_3 + NO + H_2O$

N. $Ca(OH)_2 + H_3PO_4 \rightarrow Ca_3(PO_4)_2 + H_2O$

O. $SiF_4 + H_2O \longrightarrow HF + SiO_2$

2a. Determine the proper stoichiometric coefficients to balance the following ionic equations.

A. $Cu + Fe^{3+} \longrightarrow Cu^{2+} + Fe$

B. $Zn + H^+ \longrightarrow Zn^{2+} + H_2$

C. $H^+ + S^{2-} \longrightarrow H_2S(g)$

D. $Pb^{2+} + Hg^{2+} \longrightarrow Pb^{4+} + Hg_2{}^{2+}$

E. $Fe^{3+} + Sn^{2+} \longrightarrow Fe(CN)_6{}^{4+}$

F. $H^+ + N^{3-} \longrightarrow NH_3$

G. $CN^- + Fe^{2+} \longrightarrow Fe(CN)_6{}^{4-}$

H. $Bi^{3+} + S^{2-} \longrightarrow Bi_2S_3$

3. Write the balanced equation for the combustion of each of the following compounds.

A. methanol, CH_3OH B. benzene, C_6H_6 C. pentane, C_5H_{12}

D. sucrose, $C_{12}H_{22}O_{11}$ E. carbon monoxide, CO F. acetic acid, $HC_2H_3O_2$

G. naphthalene, $C_{10}H_8$ H. ninhydrin, $C_9H_6O_4$ I. benzoic acid, C_6H_5COOH

J. testosterone, $C_{19}H_{28}O_2$

3a. Classify each of the following reactions as *combination, decomposition, displacement,* or *metathesis*. Predict the products. Balance the resulting equation.

A. $Al + Br_2 \longrightarrow$ B. $Al + Fe_2O_3 \longrightarrow$ C. $AlCl_3 + KOH \longrightarrow$

D. $BaCl_2 + Na_2CO_3 \longrightarrow$ E. $BaCl_2 + AgNO_3 \longrightarrow$ F. $BaO + SO_3 \longrightarrow$

G. $Ba(OH)_2 + \text{``}H_2CO_3\text{''} \rightarrow$ H. $Br_2 + H_2 \longrightarrow$ I. $Cd(NO_3)_2 + (NH_4)_2S \rightarrow$

J. $Ca + H_2O \longrightarrow$ K. $HCl + Al \longrightarrow$ L. $H_2O + Li_2O \longrightarrow$

M. $Ba(OH)_2 + H_3PO_4 \longrightarrow$ N. $BaO + H_2O \longrightarrow$ O. $CaO + HCl \longrightarrow$

P. $CaO + CO_2 \longrightarrow$ Q. $Ca(OH)_2 + HNO_3 \longrightarrow$ R. $CaO + HNO_3 \longrightarrow$

S. $NH_3(g) + HCl(g) \longrightarrow$ T. $SO_2 + O_2 \longrightarrow$ U. $AgNO_3 + Cu \longrightarrow$

V. $Cl_2 + FeCl_2 \longrightarrow$ W. $Cl_2 + KI \longrightarrow$ X. $N_2 + H_2 \longrightarrow$

Y. $Al(OH)_3 + H_2SO_4 \longrightarrow$ Z. $CuBr + Br_2 \longrightarrow$

4. & 4a. Beneath a reactant or a product of each balanced equation is given the quantity of material produced or consumed. Densities are given at right. Fill in the blanks under each equation. Each lettered part represents a different situation.

$$Fe(OH)_3(s) + 3\,HCl(aq) \longrightarrow FeCl_3(s) + 3\,H_2O$$

Density

A. 34.5 g	____ g	____ g	____ cm^3	H_2O 1.00 g/cm^3
B. ____ g		21.2 cm^3	____ g	$FeCl_3$ 2.90 g/cm^3
C. ____ g	18.2 g	____ g		

$$C_5H_{12}(g) + 8\,O_2(g) \longrightarrow 5\,CO_2(g) + 6\,H_2O(l)$$

D. 14.3 g	____ g	____ g	____ cm^3	H_2O 1.00 g/cm^3
E. ____ g		1.26 L	____ g	CO_2 1.96 g/L
F. ____ g	1.46 g	____ L	____ cm^3	

$$N_2(g) + 3\,H_2(g) \longrightarrow 2\,NH_3(g)$$

G. 1.46 g	____ g	____ g	NH_3 0.625 g/L
H. ____ g	____ g	21.4 L	
I. ____ g	19.4 g	____ L	

$$6\,HCl(g) + 2\,HNO_3(aq) \longrightarrow 4\,H_2O(g) + 2\,NO(g) + 3\,Cl_2(g)$$

J. 18.4 g	____ g	____ g	____ L	NO 1.34 g/L
K. 14.2 L	____ g	____ g	____ g	HCl 1.63 g/L
L. ____ L	____ g	____ g	19.4 g

$$3\,Cu(s) + 8\,HNO_3(aq) \longrightarrow 3\,Cu(NO_3)_2(aq) + 2\,NO(g) + 4\,H_2O(l)$$

M. 3.14 g	____g		____g		NO 1.34 g/L
N. ____g		____g	2.16 L		Cu 8.96 g/cm³
O. 19.4 cm³	____g	____g	____L		
P. ____ cm³		21.2 g	____L	____ _g	

5. Each line that follows represents a different solution. Fill in the blanks, based on the numbers given.

	Solute			*Solution*	
	Identity	*Mass, g*	*Moles*	*Volume, mL*	*Concentration, M*
A.	$HC_2H_3O_2$	____	____	250.0	3.414
B.	HCl	____	1.270	400.0	____
C.	$Pb(NO_3)_2$	____	0.221	____	1.104
D.	$BaCl_2$	____	____	275.0	0.888
E.	NaCl	14.3	____	____	1.462
F.	KNO_3	49.7	____	174.0	____
G.	$KHCO_3$	____	____	200.0	1.064
H.	KBr	13.75	____	350.0	____
I.	HNO_3	____	12.4	____	6.00
J.	CsCl	____	0.200	____	1.207
K.	H_2SO_4	____	____	400.0	13.026
L.	H_3PO_4	21.9	____	300.0	____

6. In each line that follows, two solutions (or a solution and water, a solvent) are mixed to produce a new solution. Assume the new solution's volume is the sum of the volumes that were mixed. Fill in the blanks.

Solution A			*Solution B*			*Final Solution*		
Solute	*M*	*Volume*	*Solute*	*M*	*Volume*	*Solute*	*M*	*Volume*
A. KNO_3	0.250	120 mL	water		450 mL		KNO_3	____
B. HCl	0.250	____mL	water		____mL	HCl	0.100	600 mL
C. $MgSO_4$	1.74	87.5 mL	water		____mL	$MgSO_4$	1.00	____ mL
D. NaCl	1.40	120 mL	water		500 mL	NaCl	____	____ mL
E. $MgSO_4$	2.40	____mL	water		____mL	Mg^{2+}	0.250	300 mL
F. Na_2SO_4	3.00	140 mL	water		____mL	Na_2SO_4	0.50	____ mL
G.	0.350	120 mL	water		600 mL	K_2SO_4	____	____ mL

K_2SO_4								
H. AlI_3	0.750	____mL	water		____mL	AlI_3	0.250	400 mL
I. KNO_3	0.642	100 mL	NaCl	0.342	200 mL	NaCl	____	____ mL
J. Na_2SO_4	0.578	50. mL	KCl	0.618	____mL	Na_2SO_4	0.214	____ mL
K. $NaNO_3$	0.745	100 mL	$NaNO_3$	0.312	200 mL	$NaNO_3$	____	____ mL
L. $MgSO_4$	0.134	____mL	$MgSO_4$	0.817	____mL	$MgSO_4$	0.400	500 mL
M. HCl	____	____mL	HCl	1.00	300 mL	HCl	1.45	600 mL
N. $NaNO_3$	1.50	200 mL	$NaNO_3$	4.00	300 mL	$NaNO_3$	____	____ mL

7. Beneath a reactant or a product in each equation is given either the mass of the species or the volume and concentration of the species in solution. Fill in the blanks with the data provided. To simplify calculations, assume that any water produced does not dilute the final solution. Each lettered part represents a different situation.

$3\,AgNO_3(aq) +$	$Na_3PO_4(aq) \longrightarrow$	$Ag_3PO_4(s)\ +$	$3\,NaNO_3(aq)$
A. 25.0 mL, 1.20 M		____g	100 mL, ____M
B.	____mL, 1.00 M	13.2 g	200 mL, ____M
C. ____mL, 2.00 M		____g	150 mL, 1.75 M
D.	84.0 mL, ____M	6.25 g	____mL, 1.40 M

$Mg(CN)_2(s)\ +$	$H_2SO_4(aq) \longrightarrow$	$2\,HCN(g)\ +$	$MgSO_4(aq)$
E. 13.4 g		____g	____mL, 1.00 M
F. ____g	____mL, 6.00 M	9.47 g	
G. ____g	172 mL, 3.50 M	____g	172 mL, ____M
H. ____g		____g	200 mL, 1.75 M

$Al_2O_3(s)\ +$	$3\,H_2SO_4(aq) \longrightarrow$	$Al_2(SO_4)_3(aq)\ +$	$3\,H_2O(l)$
I. 2.31 g	____mL, 6.00 M		
J. ____g		174 mL, 3.12 M	____g
K.	196 mL, 1.00 M	196 mL, ____M	____g
L. ____g	____mL, 1.00 M		7.42 g

$2\,KOH(aq)\ +$	$CO_2(g) \longrightarrow$	$K_2CO_3(aq)\ +$	$H_2O(l)$
M. 274 mL, ____M	____g	274 mL, ____M	17.4 g
N. 194 mL, 1.00 M	____g	____mL, 0.50 M	____g
O. 142 mL, ____M	1.34 g		____g
P.	____g	86.5 mL, 1.00 M	____g

$2\,KOH(aq)\ +$	$H_2SO_3(aq) \longrightarrow$	$K_2SO_3(aq)\ +$	$2\,H_2O(l)$
Q. 182 mL, 1.00 M	50.0 mL, ____M	____mL, 0.750 M	____g
R. ____mL, 1.32 M		143 mL, 1.00 M	____g
S.	____mL, 4.50 M	____mL, 3.21 M	14.3 g

8. After each formula is given the mass of each reactant. Determine the mass in grams of the first-listed product. Also, determine the mass in grams of each reactant that is left over.

A. $3\,AgNO_3(102\,g, aq) + Na_3PO_4(25.4\,g, aq) \longrightarrow Ag_3PO_4(s) + 3\,NaNO_3(aq)$

B. $Ba(OH)_2(105\,g, aq) + H_2SO_4(83.1\,g, aq) \longrightarrow 2\,H_2O(l) + BaSO_4(s)$

C. $4\,C(12.4\,g, s) + BaSO_4(13.2\,g, s) \longrightarrow 4\,CO(g) + BaS(s)$

D. $Zn(7.15\,g, s) + 2\,HCl(14.2\,g, aq) \longrightarrow H_2(g) + ZnCl_2(aq)$

E. $Mg(21.3\,g, s) + S(19.4\,g, s) \longrightarrow MgS(s)$

F. $Cu_2O(49.7\,g, s) + 2\,HCl(32.1\,g, g) \longrightarrow 2\,CuCl(s) + H_2O(g)$

G. $4\,NaOH(146\,g, aq) + SiO_2(84.3\,g, s) \longrightarrow 2\,H_2O(l) + Na_4SiO_4(s)$

H. $2\,Al(OH)_3(242\,g, s) + 3\,H_2SO_4(127\,g, aq) \longrightarrow 6\,H_2O(l) + Al_2(SO_4)_3(aq)$

I. $SO_3(47.2\,g, g) + 2\,HNO_3(195\,g, aq) \longrightarrow H_2SO_4(aq) + N_2O_5(aq)$

J. $6\,HCl(143\,g, aq) + Fe_2O_3(204\,g, s) \longrightarrow 2\,FeCl_3(aq) + 3\,H_2O(l)$

K. $2\,HNO_3(123\,g, aq) + 3\,H_2S(164\,g, g) \longrightarrow 4\,H_2O(l) + 2\,NO(g) + 3\,S(s)$

L. $Na_2TeO_4(174\,g, s) + 4\,NaI(157.1\,g, s) + 6\,HCl(63.2\,g, aq) \longrightarrow 6\,NaCl(s) + Te(s) + 2\,I_2(s) + 3\,H_2O(l)$

M. $4\,Zn(105\,g, s) + H_3AsO_4(194\,g, aq) + 8\,HCl(128\,g, aq) \longrightarrow 4\,ZnCl_2(aq) + AsH_3(g) + 4\,H_2O(l)$

N. $2\,KNO_3(48.7\,g) + 4\,H_2SO_4(102\,g) + 3\,Hg(193\,g) \longrightarrow K_2SO_4 + 3\,HgSO_4 + 4\,H_2O + 2\,NO$

O. $3\,K_2Cr_2O_7\,(247\,g) + 2\,C_2H_3OCl\,(92.1\,g) + 12\,H_2SO_4\,(374\,g) \longrightarrow Cl_2 + 6\,CrSO_4 + 4\,CO_2 + 15\,H_2O$

9. Determine the mass of reactant, the mass of product, or the percent yield as requested, and fill in the blank in each line below each reaction.

$P_4(s) + 6\,Cl_2(g) \longrightarrow 4\,PCl_3(g)$

A. 24.6 g	100.0 g	____%
B. 9.3 g	____g	74.2%
C. ____g	81.2 g	63.0%

$CaCO_3(s) \xrightarrow{heat} CaO(s) + CO_2(g)$

D. 87.6 g	30.7 g	____%
E. 143 g	____g	74.1%

F. ____ g		45.3 g	82.4%

$$4\,NH_3(g) + 5\,O_2(g) \longrightarrow 4\,NO(g) + 6\,H_2O(l)$$

G.		48.2 g	30.7 g	____ %
H.		____ g	137 g	73.5 %
I.	14.6 g		____ g	89.7%

$$Fe_2O_3(s) + 3\,H_2(g) \xrightarrow{heat} 2\,Fe(s) + 3\,H_2O(l)$$

J.	64.7 g			14.0 g	____ %
K.		18.9 g	____ g		94.6%
L.		____ g		14.3 g	79.6%

$$2\,HNO_3(aq) + 3\,H_2S(g) \longrightarrow 4\,H_2O(g) + 2\,NO(g) + 3\,S(s)$$

M.	124 g				85.1 g	____ %
N.		8.43 g	____ g			93.1%
O.		____ g		1.74 g		90.2%

10. Answer each question based on the simultaneous reactions that follow it.

A. 14.5 g of a $AgNO_3 - Pb(NO_3)_2$ mixture that is 74.0% $AgNO_3$ consumes how many grams of NaCl? ...how many mL of 0.500 M NaCl(aq)?

$$AgNO_3(aq) + NaCl(aq) \longrightarrow AgCl(s) + NaNO_3(aq)$$

$$Pb(NO_3)_2(aq) + 2\,NaCl(aq) \longrightarrow PbCl_2(s) + 2\,NaNO_3(aq)$$

B. 17.4 g of a $CaCO_3 - NaHCO_3$ mixture that is 45.0% $CaCO_3$ produces what mass of $CO_2(g)$? ...consumes how many mL of 0.100 M HCl?

$$CaCO_3(s) + 2\,HCl(aq) \longrightarrow CaCl_2(aq) + H_2O(l) + CO_2(g)$$

$$NaHCO_3(s) + HCl(aq) \longrightarrow NaCl(aq) + H_2O(l) + CO_2(g)$$

C. 74.2 g of a $KOH - LiOH$ mixture that is 17.2% KOH consumes what mass of $CO_2(g)$? ...produces what total mass of solid?

$$2\,KOH(s) + CO_2(g) \longrightarrow K_2CO_3(s) + H_2O(l)$$

$$2\,LiOH(s) + CO_2(g) \longrightarrow Li_2CO_3(s) + H_2O(l)$$

D. 347 mL of a solution with $[H_2SO_4] = 0.640$ M and $[HCl] = 0.935$ M consumes what volume of 0.623 M KOH? ...produces what mass of $H_2O(l)$?

$$HCl(aq) + KOH(aq) \longrightarrow KCl(aq) + H_2O(l)$$

$$H_2SO_4(aq) + 2\,KOH(aq) \longrightarrow K_2SO_4(aq) + 2\,H_2O(l)$$

E. 9.52 g of a $Mg - Fe$ mixture containing 17.4% Mg produces what mass of $H_2(g)$? ...consumes what volume of 0.200 M HCl?

$$Mg(s) + 2\,HCl(aq) \longrightarrow MgCl_2(aq) + H_2(g)$$

$$2\ Fe(s) + 6\ HCl(aq) \longrightarrow 2\ FeCl_3(aq) + 3\ H_2(g)$$

F. 84.6 g of a $Zno - Al_2O_3$ mixture containing 43.6% ZnO produces what mass of $H_2O(l)$? ...consumes what volume of 3.47 M H_2SO_4?

$$Zno(s) + H_2SO_4(aq) \longrightarrow ZnSO_4(aq) + H_2O(l)$$

$$Al_2O_3(s) + 3\ H_2SO_4(aq) \longrightarrow Al_2(SO_4)_3(aq) + 3\ H_2O(l)$$

G. 17.3 g of a $Na_3PO_4 - Na_2CrO_2$ mixture that contains 26.1 % Na_2CrO_4 produces what mass of $NaNO_3(aq)$, once the solvent water is evaporated? ...produces what mass of silver-containing solid?

$$3\ AgNO_3(aq) + Na_3PO_4(aq) \longrightarrow Ag_3PO_4(s) + 3\ NaNO_3(aq)$$

$$2\ AgNO_3(aq) + Na_2CrO_4(aq) \longrightarrow Ag_2CrO_4(s) + 2\ NaNO_3(aq)$$

11. Answer each question based on the series of reactions that follows it.

A. 19.4 g of CuS(s) produces what mass of H_2SO_4?

B. 27.5 g of H_2SO_4 requires what mass of $O_2(g)$ for its production?

$$2\ CuS(s) + 3\ O_2(g) \longrightarrow 2\ CuO(s) + 2\ SO_2(s) \qquad 2\ SO_2(g) + O_2(g) \longrightarrow 2\ SO_3(g)$$

$$SO_3(g) + H_2O(l) \longrightarrow H_2SO_4(aq)$$

C. 83.1 g of Cu(s) requires what mass of NaCl(s)?

D. 34.2 g of $H_2O(l)$ requires what volume of 12.5 M H_2SO_4?

E. 28.2 g of NaCl produces what mass of copper?

$$2\ NaCl(s) + H_2SO_4(aq) \longrightarrow Na_2SO_4(aq) + 2\ HCl(g)$$
$$6\ HCl(g) + 2\ Fe(s) \longrightarrow 2\ FeCl_3(aq) + 3\ H_2(g)$$

$$CuO(s) + H_2(g) \longrightarrow Cu(s) + H_2O(l)$$

F. 85.1 g N_2 produces what mass of HNO_3? Do not recycle by-product NO(g).

G. 14.3 g HNO_3 requires what mass of $H_2(l)$?

$$N_2(g) + 3H_2(g) \longrightarrow 2NH_3(g)$$
$$4NH_3(g) + 5\ O_2(g) \longrightarrow 4\ NO(g) + 6\ H_2O(g)$$

$$2\ NO(g) + O_2(g) \longrightarrow 2\ NO_2(g)$$
$$3\ NO_2(g) + H_2O(l) \longrightarrow 2\ HNO_3(aq) + NO(g)$$

H. 19.5 g $CaCO_3$ produces what mass of $CaCl_2$?

I. 73.9 g $CaCO_3(s)$ produces what mass of $NH_3(g)$?

J. 43.1 g $H_2O(l)$ requires what mass of $CaCO_3(s)$?

$CaCO_3(s) \longrightarrow CaO(s) + CO_2(s)$ $\qquad$ $CaO(s) + H_2O(l) \longrightarrow Ca(OH)_2(s)$

$Ca(OH)_2(aq) + 2\ NH_4Cl(aq) \longrightarrow CaCl_2(aq) + 2\ NH_3(g) + 2\ H_2O(l)$

K. 21.7 g $P_4(s)$ forms what mass of AgCl(s)?

L. 34.2 g $Cl_2(g)$ produces what volume of 3.40 M HNO_3?

M. 27.4 g AgCl(s) requires what mass of $P_4(s)$? ...of $Cl_2(g)$?

$P_4(s) + 3\ Cl_2(g) \longrightarrow 4\ PCl_3(g)$

$PCl_3(g) + 3\ H_2O(l) \longrightarrow H_3PO_4(aq) + 3\ HCl(aq)$

$HCl(aq) + AgNO_3(aq) \longrightarrow AgCl(s) + HNO_3(aq)$

12. Write the overall reaction for each of the five series of reactions in the drill problems for Objective 4-11.

QUIZZES (20 minutes each). Choose the best answer or fill in the blanks for each question. Nomenclature questions are included for review.

QUIZ A

1. The name of $NaNO_2$ is ____________________.

2. The formula of sulfuric acid is ____________.

3. $AlPO_4 + Ca(OH)_2 \longrightarrow Al(OH)_3 + Ca_3(PO_4)_2$. When this equation is balanced with the smallest whole number coefficients, the coefficient of $Al(OH)_3$ is (a) 4; (b) 3; (c) 5; (d) 2; (e) none of these.

4. $Fe_2O_3 + C \longrightarrow Fe + CO$. When this equation is balanced with the smallest whole number coefficients, the sum of all the coefficients is (a) 4; (b) 8; (c) 5; (d) 7; (e) none of these.

5. $PCl_5 + 4\,H_2O \longrightarrow H_3PO_4 + 5HCl$. A 12.0-g sample of PCl_5 (208.5 g/mol) produces how many grams of HCl (36.5 g/mol)? (a) $12.0(36.5 \times 5)/208.5$; (b) 12.0(26.5/208.5); (c) $12.0 \times 36.5/208.5$; (d) 12.0/208.5; (e) none of these.

6. $Fe_2O_3 + 3\,H_2 \longrightarrow 2\,Fe + 3\,H_2O$. A mixture of 15.0 g Fe_2O_3 (159.0 g/mol) and 3.1 g H_2 (2.0 g/mol) produces how many grams of H_2O (18.0 g/mol)? (a) 15.0(18.0/159.6); (b) $15.0(18.0/159.6) \times 3$; (c) $3.1(18.0/2.0) \times 3$; (d) 3.1(18.0/2.0); (e) none of these.

7. Complete and balance $Na_2S(s) + HCl(aq) \longrightarrow$ ____________________.

8. How many mL of 14.5 M HNO_3 must be diluted to produce 5.00 L of 1.00 M solution? (a) 5000(14.5/1.00); (b) 5000(1.00/14.5); (c) 5.00(1.00/14.5); (d) 5.00(14.5/1.00); (e) none of these.

9. A mass of 56.7 g of sodium bicarbonate (84.0 g/mol) is enough solute for how many mL of 2.77 M solution? (a) (56.7/84.0)/2.77; (b) (84.0)(56.7)(2.77); (c) (84.0/56.7)(1000/2.77); (d) (56.7/84.0)(1000/2.77); (e) none of these.

10. $N_2(g) + 3H_2(g) \longrightarrow 2NH_3(g)$ has an 85% yield. What mass in grams of $NH_3(g)$ is produced from 12.0 g $N_2(g)$? (a) (12.0/28.0)(2)(17.0)(0.85); (b) (12.0/14.0)(2)(17.0/0.85); (c) (12.0/28.0)(2)(17.0/0.85); (d) (12.0/14.0)(2)(0.85)(17.0); (e) none of these.

Quiz B

1. The name of $Mg(ClO_3)_2$ is ________________.

2. The formula of sodium hypofluorite is __________.

3. $NO_2 + H_2 \longrightarrow NH_3 + H_2O$. When this equation is balanced with the smallest whole number coefficients, the coefficient of nitrogen dioxide is (a) 1; (b) 2; (c) 3; (d) 4; (e) none of these.

4. $H_2S + O_2 \longrightarrow H_2O + SO_3$. When this equation is balanced with the smallest whole number coefficients, the sum of the coefficients is (a) 4; (b) 5; (c) 6; (d) 8; (e) none of these.

5. How many moles of SO_2 are produced when 72.0 g H_2O is produced by the process $2H_2S + 3O_2 \longrightarrow 2H_2O + 2SO_2$ ($H = 1.0, O = 16.0, S = 32.1$ g/mol)? (a) (72.0/17.0)(3/2); (b) (72.0/18.0); (c) (72.0/18.0)(2/3); (d) (72.0)(18.0); (e) none of these.

6. $2Al + Fe_2O_3 \longrightarrow 2Fe + Fe_2O_3$. A mixture of 2.5 g Al (27.0 g/mol) and 7.2 g Fe_2O_3 (158.9 g/mol) produces how many grams of Fe (55.8 g/mol)? (a) 2.5(55.8/27.0); (b) 7.2(55.8/158.9); (c) $2.5(55.8 \times 2)/(27.0 \times 2)$; (d) $(7.2)(55.8 \times 2)/158.9$; (e) none of these.

7. Complete and balance $Al_2O_3(aq) + H_2SO_4(aq) \longrightarrow$ ____________________.

8. 67.5 mL of a 3.15 M CsI solution is diluted to 2.50 L with water. The molarity of the resulting solution is (a) (3.15)(67.5)(2500); (b) (3.15/67.5)/2500; (c) 3.15(67.5/2500); (d) 3.15(2500/67.5); (e) none of these.

9. A 151.6-mL volume of a 0.45 M solution of iron(II) chloride contains how many grams of solute? (a) (151.6)(0.45)(126.8); (b) (151.6/1000)(0.45)(146.1); (c) (151.6/0.45)(126.8); (d) (151.6/0.45)(146.1); (e) none of these.

10. $CO(g) + 2H_2 \longrightarrow CH_3OH(l)$ has a 72.1 % yield. To produce 14.0 g $CH_3OH(l)$ requires what mass of hydrogen? (a) (14.0/32.0)(2)(2.0/72.1); (b) (14.0/32.0)(2)(1.0/72.1); (c) (14.0/32.0)(2)(2.0/0.721); (d) (14.0/32.0)(2)(2.0)(0.721); (e) none of these.

Quiz C

1. The name of BrO_4 is ________________.

2. The formula of nitric acid is __________.

3. $H_3PO_4 + NaOH \longrightarrow H_2O + Na_3PO_4$. When this equation is balanced with the smallest whole number coefficients, the coefficient for water is (a) 1; (b) 2; (c) 3; (d) 4; (e) none of these.

4. $AgNO_3 + CuCl_2 \longrightarrow Cu(NO_3)_2 + AgCl$. When this equation is balanced with the smallest whole number coefficients, the sum of all the coefficients is (a) 8; (b) 5; (c) 6; (d) 4; (e) none of these.

5. $2\,C_3H_7OH + 9\,O_2 \longrightarrow 6\,CO_2 + 8\,H_2O$. To produce 47.2 g CO_2 (44.0 g/mol) requires what mass in grams of C_3H_7OH (60.0 g/mol)? (a) $47.2[(6\times 44.0)/(2\times 60.0)]$; (b) $47.2[2\times 60.0)/(6\times 44.0)]$; (c) 47.2(60.0/44.0); (d) 47.2(44.0/60.0); (e) none of these.

6. $Mg_3N_2 + 6\,H_2O \longrightarrow 3\,Mg(OH)_2 + 2\,NH_3$. A mixture of 2.5 mol Mg_3N_2 (100.9 g/mol) and 6.0 mol H_2O (18.0 g/mol) produce how many grams of NH_3 (17.0 g/mol)? (a) $(6.0/18.0)(2\times 17.0)$; (b) 2×17.0; (c) $2.5(2\times 17.0)$; (d) $(2.5\times 100.9)(2\times 17.0)$; (e) none of these.

7. Complete and balance $Ba(OH)_2(aq) + H_2SO_4(aq) \longrightarrow$ ____________________.

8. If 85.6 mL of a 6.75 M solution is diluted to 6.20 L with water, what is the molarity of the final solution? (a) 6.75(85.6/6.20); (b) 6.75(6.20/85.6); (c) 6.75(6200/85.6); (d) 6.75(85.6/6200); (e) none of these.

9. A 433-mL volume of a 0.675 M solution of sodium sulfite contains what mass in grams of solute? (a) (433/1000)(0.675)(126.1); (b) (433/1000)(0.675)(142.1); (c) (433)(0.675)(103.1); (d) (433)(0.675)(126.1); (e) none of these.

10. $2\,SO_2(g) + O_2(g) \longrightarrow 2\,SO_3(g)$ has a 93.7% yield. What mass in grams of $SO_3(g)$ is produced from 6.75 g of $SO_2(g)$? (a) (6.75/64.1)(80.1/93.7); (b) (6.75/64.1)(80.1/0.937); (c) 6.75/0.937; (d) (6.75/64.1)(80.1)(93.7); (e) none of these.

QUIZ D

1. The name of H_2SO_2 is ____________________________.

2. The formula of sodium bicarbonate is ___________.

3. $PCl_3H_2O \longrightarrow HCl + H_3PO_3$. When this equation is balanced with the smallest whole number coefficients, the coefficient of hydrochloric acid is (a) 6; (b) 2; (c) 3; (d) 4; (e) none of these.

4. $SCl_4 + H_2O \longrightarrow H_2SO_3 + HCl$. When this equation is balanced with the smallest whole number coefficients, the sum of the coefficients is (a) 4; (b) 6; (c) 8; (d) 10; (e) none of these.

5. What amount in moles of NH_3 is required to produce 256 g N_2H_4 via the process: $4\,NH_3 + Cl_2 \longrightarrow N_2H_4 + 2\ NH_4Cl$ (Cl = 35.5, N = 14.0, H = 1.0 g/mol)? (a) (256/16)(4); (b) (256/32)(2); (c) (256/32)(4); (d) (256/32)/4; (e) none of these.

6. $2\,Fe_2O_3 + 3\,C \longrightarrow 4\,Fe + 3\,CO_2$. A mixture of 18.0 g $Fe_2 + O \times$ (159.7 g/mol) and 2.5 g C (12.0 g/mol) produces what mass in grams of Fe (55.8 g/mol)? (a) 18.0(55.8/159.7); (b) 18.0(4 × 55.8)/(2 × 159.7); (c) 2.5(55.8/12.0); (d) 2.5(55.8 × 4)/(3 × 12.0); (e) none of these.

7. Complete and balance $CaCO_3(s) + HCL(aq) \longrightarrow$ ____________________.

8. What volume in mL of 17.0 M HCl solution is needed to produce 2.0 L of a 3.00 M HCl solution? (a) (2000.17.0)/3.00; (b) 2.000(17.0)(3.00); (c) 2000(17.0/3.00); (d) 2000(3.00/17.0); (e) none of these.

9. A 81.2-mL volume of a 14.0 M solution of nitric acid contains how many grams of solute? (a) (81.2/1000)(14.0)(81.0); (b) (81.2/1000)(14.0)(63.0); (c) (81.2)(14.0)(47.0); (d) (81.2)(14.0)(65.0); (e) none of these.

10. $2\,P(s) + 5\,Cl_2(g) \longrightarrow 2\,PCl_5(g)$ has a 89.0% yield. A 19.2-g sample of $PCl_5(g)$ requires how many grams of $Cl_2(g)$? (a) (19.2/208.5)(5/2)(71.0/0.890); (b) (19.2/66.5)(5/2)(71.0/0.890); (c) (19.2/208.5)(71.0)(0.890); (d) (19.2/66.5)(5/2)(71.0)(0.890); (e) none of these.

SAMPLE TEST (30 minutes)

1. Give the name or formula, as appropriate, of each of the following compounds.

A. $Pb(CNS)_2$	B. lead(IV) thiosulfate	C. arsenic(III) sulfide
D. $Mn(BrO)_2$	E. HNO_2	F. periodic acid
G. stannous sulfide	H. ammonium bicarbonate	

2. What volume in mL of 0.160 M KNO_3 must be mixed with 200.0 mL of 0.240 M K_2SO_4 to produce a solution having $[K^+] = 0.400$ M?

3. Complete and balance the following equations:

A. $BaCl_2 + AgNO_3 \longrightarrow$

B. $Ca(OH)_2 \xrightarrow{\text{heat}}$

C. $NH_3 + HCl \longrightarrow$

D. $Zn + CuSO_4 \longrightarrow$

E. $C_2H_6 + O_2 \longrightarrow$

4. Consider the reaction $CuO(s) + 2\,HCl(aq) \longrightarrow CuCl_2(aq) + H_2O(l)$. If 55.2 g CuO is added to 158 mL of 2.70 M HCl, what mass in grams of H_2O is formed?

5. 26.4 g of a NaOH – CaO mixture that contains 40.0% CaO reacts with HCl according to the equations:

$$NaOH(s) + HCl(aq) \longrightarrow NaCl(aq) + H_2O(l)$$

$$CaO(s) + 2\,HCl(aq) \longrightarrow CaCl_2(aq) + H_2O(l)$$

All of the mixture reacts and there is no HCl left over. The resulting solution is evaporated to dryness. What is the mass in grams of solid obtained?

5 Introduction to Reactions in Aqueous Solutions

Chapter Objectives

*** 1. Identify compounds according to whether they are non-, weak, or strong electrolytes; strong or weak acids or bases; or salts.**

As a general rule, ionic compounds are considered to be completely dissociated into ions when they dissolve in water. That is, ionic compounds are strong electrolytes. These soluble ionic compounds also are known as *salts.* Recall that an ionic compound consists of a cation and an anion. Hence, a thorough mastery of the common ions listed in Tables 3-2 and 3-3 will make it easy for you to recognize ionic compounds. If the ions of the compound are not common ones, you still can recognize an ionic compound as consisting of a metal and a nonmetal. (Recall Objective 3–13, to distinguish between a metal and a nonmetal.)

Also, it is easy to remember that practically all covalent compounds are nonelectrolytes. And we recall that covalent compounds contain only nonmetal atoms. (Ammonium ion is the only common cation that contains no metal atoms.)

There are two large classes of covalent compounds that are electrolytes: acids and bases. Strong acids and strong bases are completely dissociated into ions in aqueous solution; they are strong electrolytes. It is essential that you memorize the common strong acids and bases, listed in Table 5-1. The periodic table can help. The strong bases are the hydroxides of the Group 1A and 2A elements with the exceptions of $Be(OH)_2$ and $Mg(OH)_2$. Three of the strong acids are hydrogen halides: HCl, HBr, and HI. In contrast, HF is a weak acid.

Table 5-1 Common Strong Bases and Acids in the Periodic Table

Bases		*Acids*	*Period*
LiOH		HNO_3	2nd
NaOH		H_2SO_4 $HClO_3$ HCl	3rd
KOH	$Ca(OH)_2$	HBr	4th
RbOH	$Sr(OH)_2$	HI	5th
CsOH	$Ba(OH)_2$		6th

The formulas of all acids, weak as well as strong, begin with H, and their names end with the word *acid.* (Later, when you study organic chemistry, you will note that the formulas of organic acids end with —COOH. Thus, we shall write the formula of acetic acid as $HC_2H_3O_2$, but organic chemists write it as CH_3COOH.) Of course, if an acid is not strong (listed in Table 5-1), it must be weak and, thus, only partially dissociated in aqueous solution. (*Un*common strong acids will be explicitly indicated.) Strong bases are hydroxides of active metal cations; their formulas end with OH. Weak bases are harder to identify. Most of them are based on ammonia (NH_3).They are produced by replacing one of the H's with a group of atoms. Thus, CH_3NH_2, where $—CH_3$ replaces —H, is a weak base (methylamine).

*** 1a. Be able to write the names and formulas of acids.**

Notice that acids are named differently than salts. HNO_3 is not *hydrogen nitrate*, but *nitric acid.* In order to form the name of the acid, we begin with the name of the anion. Remove the anion name's suffix (the last three letters in each case) and replace it with the suffix of the acid as illustrated in Table 5-2. In the case of binary acids (those composed of hydrogen and only one other element) the prefix *hydro-* also is added.

TABLE 5-2 Nomenclature of Acids (*examples in right-hand two columns*)

Anion suffix	*Acid suffix*	*Anion name, formula*	*Acid name, formula*
-ite	-ous	nitrite, NO_2^-	nitrous acid, HNO_2
-ate	-ic	oxalate, $C_2O_4^{2-}$	oxalic acid, $H_2C_2O_4$
-ide	(hydro-) -ic	iodide, I^-	hydroiodic acid, HI

*** 2. Relate the molarities of ions in solutions to the concentrations of strong electrolytes from which they are derived.**

Determining the concentrations of ions in aqueous solutions is similar to determining the concentration of the solute (recall Objectives 4–6 and 4–7). There is one added feature: each mole of solute may produce more than one mole of cations or anions. The calculation thus involves an additional conversion factor, derived from the chemical formula of the solute.

EXAMPLE 5-1 How many millimoles of nitrate ion are present in 114 mL of 0.357 M $Al(NO_3)_3$ solution?

This is similar to Example 4–9, with the addition of the conversion factor mentioned above.

$$\text{amount } NO_3^- = 114 \text{ mL soln} \times \frac{0.357 \text{ mmol } Al(NO_3)_3}{1 \text{ mL soln}} \times \frac{3 \text{ mmol } NO_3^-}{1 \text{ mmol } Al(NO_3)_3} = 122 \text{ mmol } NO_3^-$$

Of course, problems similar to Examples 4–8, 4–10, 4–11, and 4–12 also can involve ionic solutes. Perhaps the most involved of these is the situation where two solutions that have one ion in common are mixed.

EXAMPLE 5-2 54.0 mL of 0.314 M $Mg(NO_3)_2$ is mixed with 114 mL of 0.357 M $Al(NO_3)_3$ solution. What is the final concentration of nitrate ion, $[NO_3^-]$?

$$\text{Total solution volume} = 54.0\ \text{mL} + 114\ \text{mL} = 168\ \text{mL}.$$

Now the amount of nitrate ion in each of the two solutions is determined. We interpret molarity as millimoles of solute per milliliter of solution.

$$\text{amount } NO_3^- = 54.0\ \text{mL} \times \frac{0.628\ \text{mmol } Mg(NO_3)_2}{1\ \text{mL soln}} \times \frac{2\ \text{mmol } NO_3^-}{1\ \text{mmol } Mg(NO_3)_2} = 33.9\ \text{mmol } NO_3^-$$

$$\text{amount } NO_3^- = 114\ \text{mL} \times \frac{0.357\ \text{mmol } Al(NO_3)_3}{1\ \text{mL soln}} \times \frac{3\ \text{mmol } NO_3^-}{1\ \text{mmol } Al(NO_3)_3} = 122\ \text{mmol } NO_3^-$$

$$[NO_3^-] = \frac{(33.9 + 122)\ \text{mmol } NO_3^-}{168\ \text{mL soln}} = 0.928\ \text{M}$$

*** 3. State general rules that apply to the aqueous solubilities of ionic compounds, and write net ionic equations based on these solubility rules.**

Solubility rules enable us to determine whether a precipitation reaction will occur. In most precipitation reactions that we will study, aqueous solutions of two soluble compounds are mixed, and their ions recombine into two new compounds, one of which is insoluble. This is the description of a metathesis reaction (recall Objective 4-3a) with one insoluble product. An example is the reaction of $Ba(NO_3)_3(aq)$ and $Al_2(SO_4)_3(aq)$, equation [1].

$$3\,Ba(NO_3)_2(aq) + Al_2(SO_4)_3(aq) \longrightarrow 3\,BaSO_4(s) + 2\,Al(NO_3)_3(aq) \qquad [1]$$

The solubility rules in the text can be restated in a fashion that might be more memorable. As you learn these rules, remember three points. First, the rules are organized by the anions of the compounds. Second, the compounds of one group of anions generally are soluble, while those of another group generally are insoluble. Third, most rules have exceptions, even though they deal only with common compounds. The restatement of the rules is in Table 5-3, with some additional anions (written in *italics*) included for completeness. Learn whichever set of rules is easiest for you to remember and apply, these or the ones in the *text.*

Ionic equations were first discussed in Objective 4-2a. To produce a net ionic equation, we can start with a whole-formula equation, written in terms of compounds. Consider the reaction of aqueous solutions of $Al_2(SO_4)_3$ and $Ba(NO_3)_2$. The products of the reaction are predicted by metathesis, switching the partners of each ion. They are $BaSO_4$ and $Al_2(NO_3)_3$. The solubility rules (rules 2.c. and 2.a., respectively) indicate that $BaSO_4$ is insoluble and $Al(NO_3)_3$ is soluble. The unbalanced chemical expression then is equation [2].

$$Al_2(SO_4)_3(aq) + Ba(NO_3)_2(aq) \longrightarrow BaSO_4(s) + Al(NO_3)_3(aq) \qquad [2]$$

We now take advantage of a simple generality: *Ionic compounds dissociate completely into ions when they dissolve in water*. We write ions for each compound that is designated (aq), and we have

$$Al^{3+}(aq) + SO_4^{2-}(aq) + Ba^{2+}(aq) + SO_4^{2-}(aq) \longrightarrow BaSO_4(s) + Al^{3+}(aq) + NO_3^-(aq) \qquad [3]$$

[We do not break into ions those compounds that are designated (s), (l), or (g) when writing net ionic equations.] Notice that we write the ions without stoichiometric coefficients. To obtain the net ionic expression, eliminate the ions that appear on both sides of the expression. These are the spectator ions: $Al^{3+}(aq)$ and $NO_3^-(aq)$..

$$SO_4^{2-}(aq) + Ba^{2+}(aq) \longrightarrow BaSO_4(s) \qquad [4]$$

TABLE 5-3 Solubility Rules for Common Cations and Anions

1. The common compounds of ammonium ion (NH_4^+) and the alkali metal (periodic group 1A) cations are SOLUBLE.
2. Most of the common compounds of the following anions are SOLUBLE.

   A The common compounds of nitrate (NO_3^-), chlorate (ClO_3^-), perchlorate (ClO_4^-), acetate ($C_2H_3O_2^-$), and *permanganate* (MnO_4^-), *nitrite* (NO_2^-) ions are SOLUBLE. Exceptions are moderately soluble $AgC_2H_3O_2$ and insoluble $AgNO_2$.

   B. The common compounds of chloride (Cl^-), bromide (Br^-), and iodide (I^-) ions are SOLUBLE. Exceptions are the insoluble compounds of these anions with Pb^{2+}, Hg_2^{2+}, and Ag^+.

   C. The common compounds of sulfate (SO_4^{2-}) and *thiosulfate* ($S_2O_3^{2-}$) ions are SOLUBLE. Exceptions are the insoluble compounds of these anions with $Sr^{2+}, Ba^{2+}, Pb^{2+}$ and Hg_2^{2+}, and their moderately soluble compounds with Ca^{2+} and Ag^+.
3. Most of the common compounds of the following anions are INSOLUBLE. Exceptions, of course, are the compounds that they form with ammonium ion and the alkali metal cations.

   A. The common compounds of the carbonate (CO_3^{2-}), chromate (CrO_4^{2-}), phosphate (PO_4^{3-}), *sulfite* (SO_3^{2-}), and *oxalate* ($C_2O_4^{2-}$) ions are INSOLUBLE. An exception is soluble $Fe_2(C_2O_4)_3$.

   B. The common compounds of the hydroxide (OH^-) and *oxide* (O^{2-}) ions are INSOLUBLE. Exceptions are their moderately soluble compounds with Ca^{2+}, Sr^{2+}, and Ba^{2+}.

   C. The common compounds of the sulfide ion (S^{2-}) are INSOLUBLE. Exceptions are the soluble sulfides of the alkaline earth (group 2A) cations.

If necessary, we balance this expression to obtain the net ionic equation. The net ionic equation is more general than any specific equation. In this case, for example, it tells us that a $BaSO_4(s)$ precipitate will form when any soluble barium compound is mixed with a soluble sulfate compound in aqueous solution.

Example 5.3. Predict whether a reaction will occur in each of the following cases. If so, write a net ionic equation for the reaction.

A. $MgCl_2 + KOH \rightarrow$

B. $CH_3COOH + NaOH \rightarrow$

C. $Ca(OH)_2 + NaCl \rightarrow$

Solution: In each case we need to determine the possible combination of ions and which, if any, of the resulting compounds are soluble. Write all the ions present from the two compounds and look to determine which anions and cations can combine. If the resulting combination of anion and cation results in an insoluble compound, a precipitate will form. Also, check if an acid-base neutralization reaction can occur.

A. The possible products are $Mg(OH)_2$ and KCl. $Mg(OH)_2$ is insoluble and KCl is soluble. The net ionic equation is $Mg^{2+} + 2OH^- \rightarrow Mg(OH)_2$.

B. A neutralization reaction occurs $CH_3COO^- + H^+ + Na^+ + OH^- \rightarrow$

The net ionic equation is: $H^+ + OH^- \rightarrow H_2O$

C. All products are soluble, therefore no reaction occurs.

4. Write net ionic equations for neutralization reactions and for reactions that result in the dissolving (dissolution) of a water-soluble substance or the evolution of a gas.

In neutralization reactions, an acid and a base react to form a salt and water. An example is the reaction of sulfuric acid with sodium hydroxide.

$$2\ NaOH(aq) + H_2SO_4(aq) \longrightarrow Na_2SO_4(aq) + 2\ H_2O(l) \qquad [5]$$

The net ionic equation for this neutralization reaction, and for all other acid-base neutralization reactions, is the reaction of hydroxide ion with hydrogen ion. (We often write hydrogen ion as H^+ for simplicity, even though we recognize that it exists in aqueous solution as H_3O^+.)

$$H^+(aq) + OH^-(aq) \longrightarrow H_2O \qquad [6]$$

Recognize that sometimes acid-base reactions may occur between an acid and a metal oxide or between a base and a nonmetal oxide. The following two equations are examples.

$$H_2SO_4(aq) + CaO(s) \longrightarrow CaSO_4(aq) + H_2O \quad [7]$$

$$2\ NaOH(aq) + SO_2(aq) \longrightarrow Na_2SO_3(aq) + H_2O \quad [8]$$

The net ionic equation for the dissolving of a water-soluble substance has that substance as the reactant, and its dissolved form as the products. In the case of strong electrolytes, the products are ions, as in the case of the dissolution of $Al_2(SO_4)_3$.

$$Al_2(SO_4)_3(s) \xrightarrow{H_2O} 2\ Al^{3+}(aq) + 3\ SO_4^{\ 2-}(aq) \quad [9]$$

A limited number of gases are produced from aqueous solutions, most through the actions of acids and bases. Two of these, sulfur dioxide and carbon dioxide, were described in Objective 4-4. These are formed by the reaction of an acid on sulfites and carbonates, respectively.

$$2\ H^+(aq) + SO_3^{\ 2-}(aq) \longrightarrow H_2SO_3(aq) = SO_2 \cdot H_2O \longrightarrow H_2O(l) + SO_2(g) \quad [10]$$

$$2\ H^+(aq) + CO_3^{\ 2-}(aq) \longrightarrow \text{``}H_2CO_3(aq)\text{''} \longrightarrow H_2O(l) + CO_2(g) \quad [11]$$

Sulfur dioxide gas also is formed by the reaction of an acid on thiosulfates. Hydrogen sulfide gas is formed by the reaction of an acid on sulfides. And ammonia gas is formed by the reaction of a base [OH^-, such as from NaOH(aq)] on ammonium ion. This last reaction is the basis of a qualitative test for ammonium ion.

$$2\ H^+(aq) + S_2O_3^{\ 2+}(aq) \longrightarrow H_2S_2O_3(aq) \longrightarrow SO_2(g) + H_2O(l) + S(s) \quad [12]$$

$$2\ H^+(aq) + S^{2-}(aq) \longrightarrow H_2S(aq) \quad [13]$$

$$OH^-(aq) + NH_4^{\ +}(aq) \longrightarrow NH_3(g) + H_2O(l) \quad [14]$$

$SO_2(g)$, $CO_2(g)$ and $H_2S(g)$ also are produced by reaction with the $HSO_3^{\ -}$, $HCO_3^{\ -}$, and HS^- ions, respectively. Gases other than these four also can be produced, many as the result of oxidation-reduction equations.

*** 4a. State and apply the conditions under which reactions between ions in aqueous solution may go to completion.**

Reactions between ions in aqueous solution will go to completion if some of the ions are removed from solution. This will occur when the reaction forms one of three types of substance.

(1) Precipitate, that is, a compound that is insoluble in the solution (a precipitation reaction).
(2) Gas that escapes from the solution (a gas evolution reaction).
(3) Nonelectrolyte, such as water, that results from the union of ions (an acid-base reaction).

This method of prediction is straightforward to apply when both reactants are strong electrolytes. But sometimes strong electrolytes are present on both sides of the chemical equation. We then cannot predict whether the reaction will go to completion or not. Such is the case with the reaction between cobalt(II) sulfate and hydrogen sulfide.

$$CoSO_4(aq) + H_2S(aq) \longrightarrow CoS(s) + H_2SO_4(aq) \qquad [15]$$

We recognize cobalt(II) sulfide as a precipitate, but we also note that H_2S normally appears as a gas escaping from solution. In this and similar cases we will be content for now to state that the reaction proceeds to an intermediate state of equilibrium. We will return to consider these cases in Chapters 16 through 19.

5. Identify the metals that readily dissolve in common mineral acids and those that do not.

Active metals dissolve in common mineral acids, such as HCl(aq). These include the alkali metals (Li, Na, K, Rb, Cs), the alkaline earth metals (Mg, Ca, Sr, Ba), and the metals Al, Zn, Fe, Sn, and Pb. In every case, hydrogen gas is evolved and a compound of the metal's cation and the acid's anion is produced.

$$2\,Li(s) + 2\,HCl(aq) \longrightarrow H_2(g) + 2\,LiCl(aq) \qquad Mg(s) + H_2SO_4(aq) \longrightarrow H_2(g) + MgSO_4(aq)$$

$$2\,Al(s) + 6\,HBr(aq) \longrightarrow 3\,H_2(g) + 2\,AlBr_3(aq) \qquad Sn(s) + 2\,HCl(aq) \longrightarrow H_2(g) + SnCl_2(aq)$$

But inactive metals—Cu, Ag, Au, Hg—do not dissolve in common mineral acids. They require an oxidizing acid (a strong acid that also is an oxidizing agent) such as $HNO_3(aq)$.

*** 6. Recognize an oxidation-reduction reaction by changes in oxidation state and identify the oxidizing and reducing agents in an oxidation-reduction reaction.**

An oxidation-reduction reaction or a *redox* reaction has occurred when some of the atoms change oxidation state in going from reactants to products. (Objective 3–11 discusses how to assign oxidation states.) An atom is *oxidized* when its oxidation state increases. An atom is *reduced* when its oxidation state decreases. Oxidation and reduction *always* occur together in a reaction, *never* one without the other.

An oxidizing agent (or oxidant) accepts electrons from the substance that is being oxidized. A reducing agent (or reductant) gives electrons to the substance that is being reduced. Thus, the oxidizing agent itself is reduced and the reducing agent is oxidized. (If this seems confusing, consider the following: If you give money to someone, you are that person's enriching agent; yet you are impoverished by the transaction.) An individual atom is *not* the oxidizing or reducing agent. The agent is the compound or ion that contains the atom that changes oxidation state.

EXAMPLE 5-4 Identify the oxidizing agent and the reducing agent in the reaction of copper with nitric acid:

$$3\,Cu(s) + 8\,HNO_3(aq) \longrightarrow 3\,Cu(NO_3)_2(aq) + 2\,NO(g) + 4\,H_2O(l)$$

Elemental copper (O.S. = 0) becomes copper(II) ion (O.S. = +2); copper is oxidized. Nitrogen has (O.S. = +5) in HNO_3 and O.S. = +2 in NO; nitrogen is reduced. Nitric acid is the oxidizing agent (it oxidizes elemental copper) and copper is the reducing agent in this reaction.

*** 7. Separate an oxidation-reduction equation into half equations, complete and balance the half equations, and recombine them into a balanced net oxidation-reduction equation.**

This method of balancing redox equations has three advantages. First, it does not rely on your ability to determine oxidation states, a fact that many beginners appreciate. Second, it tends to be more reliable than other methods. And finally, it produces two balanced half equations that can be used to determine the standard cell voltage of the reaction (Objective 20-3). We illustrate the method by balancing the following redox equation.

$$HCl(aq) + MnO_2(s) \longrightarrow Cl_2(g) + H_2O(l) + MnCl_2(aq)$$

(1) Identify the substances that are oxidized or reduced, separate them into ions (if that is how they exist in the reaction mixture), and write two *couples*. (We only separate ionic compounds that are dissolved in aqueous solution.) A couple consists of the oxidized and the reduced species of an element separated by an arrow or a bar (I), with the reactant written first. In this reaction, chlorine changes oxidation state from being combined in HCl(aq), to being uncombined in $Cl_2(g)$. Manganese changes oxidation state from $MnO_2(s)$ to $MnCl_2(aq)$. Chlorine is present as $Cl^-(aq)$ in HCl(aq) and manganese as $Mn^{2+}(aq)$ in $MnCl_2(aq)$. Because $MnO_2(s)$ is a solid, it is not separated into ions.

$$Cl^-(aq) \longrightarrow Cl_2(g) \text{ or } Cl^-(aq) \mid Cl_2(g)$$

$$\text{AND } MnO_2(s) \longrightarrow Mn^{2+}(aq) \text{ or } MnO_2(s) \mid Mn^{2+}(aq)$$

(2) Next, balance all atoms except H and O: $2\,Cl^-(aq) \longrightarrow Cl_2(g)$.

(3) Now, use H_2O and $H^+(aq)$ to balance H and O atoms. First, add one H_2O for each O atom needed. Then add one $H^+(aq)$ for each H atom needed.

$$MnO_2(s) \longrightarrow Mn^{2+}(aq) + 2\ H_2O \text{ THEN } MnO_2(s) + 4\,H^+(aq) \longrightarrow Mn^{2+}(aq) + 2\,H_2O$$

(4) Balance charge by adding electrons to reactants or products as needed:

$$2\,Cl^-(aq) \longrightarrow Cl_2(g) + 2\ e^- \qquad [16]$$

$$2\,e^- + 4\,H^+(aq) + MnO_2(s) \longrightarrow Mn^{2+}(aq) + 2\,H_2O \qquad [17]$$

These are the two balanced half equations. In Equation [16], electrons are produced; this is an oxidation. Chloride ion is oxidized to elemental chlorine and electrons are produced. In Equation [17] electrons are consumed; this is a reduction. Manganese(IV) in $MnO_2(s)$ is reduced to manganese(II), $Mn^{2+}(aq)$.

(5) Now, multiply the two half equations by whole numbers so that the total number of electrons produced in the oxidation equals the total number consumed in the reduction. Add the resulting (multiplied) half equations and eliminate substances that appear on both sides. This is the net ionic equation.

$$2\,Cl^-(aq) + MnO_2(s) + 4\,H^+(aq) \longrightarrow Cl_2(g) + Mn^{2+}(aq) + 2\,H_2O$$

(6) Finally, add spectator ions as needed to produce the original substances. In this case, we add 2 $Cl^-(aq)$ to each side of the equation.

$$MnO_2(s) + 4\,HCl(aq) \longrightarrow Cl_2(g) + MnCl_2(aq) + 2\,H_2O$$

In *alkaline* (basic) solution, we add a step after Step 3. The purpose is to eliminate the $H^+(aq)$, which appears only in extremely small concentrations in alkaline solutions, and use $OH^-(aq)$ instead. Consider this reaction.

$$KMnO_4(aq) + H_2S(g) \longrightarrow MnO_2(s) + S(s) + KOH(aq) + H_2O$$

1. & 2. Write half equations, and balance all atoms except H and O.

$$MnO_4^-(aq) \longrightarrow MnO_2(s) \text{ AND } H_2S(g) \longrightarrow S(s)$$

3. Balance O and H by adding H_2O and H^+.

$$MnO_4^-(aq) + 4\,H^+(aq) \longrightarrow MnO_2(s) + 2\,H_2O \text{ AND } H_2S(g) \longrightarrow S(s) + 2\,H^+(aq)$$

3a. Now add, to *both sides* of each half equation, a number of $OH^-(aq)$ equal to the number of $H^+(aq)$ present in that half equation.

$$MnO_4^-(aq) + 4\,H^+(aq) + 4\,OH^-(aq) \longrightarrow MnO_2(s) + 2\,H_2O + 4\,OH^-(aq)$$
$$H_2S(g) + 2\,OH^-(aq) \longrightarrow S(s) + 2\,H^+(aq) + 2\,OH^-(aq)$$

And now, when $H^+(aq)$ and $OH^-(aq)$ appear on the same side of a half equation, combine these two ions to form H_2O, and then eliminate H_2O that appears on both sides of the half equation.

$$MnO_4^-(aq) + 2\,H_2O \longrightarrow MnO_2(s) + 4\,OH^-(aq)$$
$$\text{AND } H_2S(g) + 2\,OH^-(aq) \longrightarrow S(s) + 2\,H_2O$$

4. Balance charge to produce two balanced half equations.

$$3\,e^- + MnO_4^-(aq) + 2\,H_2O \longrightarrow MnO_2(s) + 4\,OH^-(aq)$$ [18]

$$H_2S(g) + 2\,OH^-(aq) \longrightarrow S(s) + 2\,H_2O + 2\,e^-$$ [19]

5. Equation [18] is multiplied by 2, and equation [19] by 3.

$$2\,MnO_4^-(aq) + 3\,H_2S(g) + 4\,H_2O + 6\,OH^-(aq) \longrightarrow$$
$$2\,MnO_2(s) + 8\,OH^-(aq) + 3\,S(s) + 6\,H_2O$$

Then, $OH^-(aq)$ and H_2O are eliminated when they appear on both sides of the equation to produce the balanced net ionic equation.

$$2\,MnO_4^-(aq) + 3\,H_2S(g) \longrightarrow 2\,MnO_2(s) + 2\,OH^-(aq) + 3\,S(s) + 2\,H_2O$$

6. Two $K^+(aq)$ are added as spectator ions to each side to obtain the balanced oxidation-reduction equation.

$$2\,KMnO_4(aq) + 3\,H_2S(g) \longrightarrow 2\,MnO_2(s) + 2\,KOH(aq) + 3\,S(s) + 2\,H_2O$$

*** 8. Extend the stoichiometric methods of Chapter 4 to reactions involving precipitation, neutralization, or oxidation-reduction.**

Just because reactions are given different names does not mean that the techniques for dealing with them have changed. In fact, practically all precipitation, gas-evolution, and acid-base reactions are metathesis reactions. And many decomposition, displacement, and combination reactions are oxidation-reduction reactions. Of course, some reactions do not neatly fit into any category. Consider the reaction of solid lithium hydroxide with water.

$$Li_2O(s) + H_2O(l) \longrightarrow 2\,LiOH(aq)$$ [20]

This reaction clearly is a combination reaction; it does not fit into the categories of precipitation, acid-base, or oxidation-reduction. Another example is the reaction of chlorine with water.

$$3\,Cl_2(aq) + 3\,H_2O(l) \longrightarrow 5\,HCl(aq) + HClO_3(aq)$$ [21]

This is clearly an oxidation-reduction reaction, and yet it cannot be classified as a combination, decomposition, displacement, or metathesis reaction. It is difficult to devise any one classification scheme that works in all cases. The wise student uses many schemes as aids to learning. The goal is being able to predict products from a list of reactants. The drill problems for this objective provide practice classifying reactions by each of the two schemes.

*** 9. Be familiar with the technique of titration—how the experiment is performed, what data is collected, why an indicator is used, and how to use the data.**

Titrations involve reactions in solution that go to completion. *Titrant*, a solution of known concentration, is added to a solution of known volume until the endpoint of the titration is reached. The titrant is added in such a way that its total volume is known precisely. Often an indicator is added so that a color change occurs when the endpoint is reached. The data is (a) the chemical reaction that occurs, called the *titration reaction*; (b) the volumes of the solution being titrated and of the titrant; and (c) the concentration of the titrant. From this data, one determines the concentration of the titrated solution or the amount of titrated material present. Thus, the drill problems of this objective are merely applications of Objective 4–8.

EXAMPLE 5-5 200.00-mL sample of H_2SO_4 solution is completely titrated with 28.16 mL of 0.1078 M NaOH. What is the H_2SO_4 molarity?

$$H_2SO_4(aq) + 2\ NaOH(aq) \longrightarrow Na_2SO_4(aq) + 2\ H_2O$$

First, we determine the amount of H_2SO_4 present from the amount of NaOH used.

$$H_2SO_4 \text{ amount} = 28.16 \text{ mL base} \times \frac{0.1078 \text{ mmol NaOH}}{1 \text{ mL base}} \times \frac{1 \text{ mmol } H_2SO_4}{2 \text{ mmol NaOH}}$$

$$= 1.518 \text{ mmol } H_2SO_4$$

Then we used the definition of molarity.

$$H_2SO_4 \text{ molarity} = \frac{\text{amount } H_2SO_4}{\text{acid volume}} = \frac{1.518 \text{ mmol } H_2SO_4}{20.00 \text{ mL acid}} = 0.07590 \text{ M}$$

*** 9a. Draw conclusions about the presence or absence of ions in an unknown from experimental observations.**

The formation of a precipitate or the evolution of a gas can provide evidence to help you determine which ions are present and which ones are absent in a sample. If the smell of the gas is known, further conclusions can be drawn. $CO_2(g)$ is odorless, $H_2S(g)$ has the odor of rotten eggs, $SO_2(g)$ has the sharp odor of burning sulfur (the smell of a burning match), and $NH_3(aq)$ has the characteristic odor of ammonia (smelling salts or ammonia cleaning solution).

EXAMPLE 5–6 A solid is known to be $BaSO_4, Ba(NO_3)_2, CaCO_3$, or Na_2SO_3. $Na_2SO_4(aq)$ is added to an aqueous solution of this unknown solid, and no precipitate forms. When dilute HCl(aq) is added to the unknown solid, a gas forms and a sharp odor is observed. What is the identity of the unknown?

Because the unknown is soluble in water, it cannot be $BaSO_4$. Likewise, it cannot be $Ba(NO_3)_2$ because no precipitate (of $BaSO_4$) forms when a solution of sulfate ion is added. The formation of a gas also eliminates $BaSO_4$ and $Ba(NO_3)_2$, since neither anion evolves a gas on the addition of acid. The sharp odor is that of $SO_2(g)$, which is produced from the treatment of sulfite ion with an acid. Thus, the unknown solid is Na_2SO_3.

Self-Assessment Exercises

96. (b) is correct. 0.005 moles/L $Ba(OH)_2$ x 0.300 L x 2 mol (OH^-)/mol $Ba(OH)_2$ = 0.0030 mol.

97. (d) 0.10 M H_2SO_4. Both H_2SO_4 and HCl are strong acids. However, H_2SO_4 will be a strong acid for the first ionization and will contribute a small amount of H^+ from the second ionization.

98. (c) K_2CO_3 should be added. $ZnCO_3$ will form and is insoluble. $ZnCl_2$, $ZnBr_2$ and $ZnSO_4$ are all soluble.

99. (a) $BaSO_3$. SO_3^{2-} will react according to the following equation:

$$SO_3^{2-} + 2H^+ \rightarrow SO_2\,(g) + H_2O\,(l)$$

100. (b) If the charges on each side are balanced, 4 x Fe^{2+} = +8 and 4 x H^+ = 4 or a total of +12 for the left side. 4 x Fe^{+3} gives +12 for the right side.

101. (a) One electron is transferred. The half reaction is $Np^{+5} + 1\,e^- \rightarrow Np^{+4}$.

Drill Problems

1. State whether each of the following compounds is a non-, weak, or strong electrolyte. If relevant, also state whether each is a weak or strong acid, base, or a salt.

A. HCl	B. H_2O	C. NH_3	D. NaCl	E. HNO_3	F. $HC_2H_3O_2$
G. $BaCl_2$	H. SO_2	I. $MnBr_2$	J. KNO_3	K. $MgSO_4$	L. NaOH
M. FeS_2O_3	N. HCN	O. CH_4	P. $RbMnO_4$	Q. Na_2CO_3	R. $Ni(NO_3)_2$
S. NH_4Cl	T. NH_2OH	U. H_2O_2	V. SrI_2	W. K_2S	X. HF
Y. $AgClO_3$	Z. Na_3PO_4	Γ. CH_3OH	Δ. $CuSO_4$	Θ. K_3PO_4	Λ. $H_2C_2O_4$
Ξ. PCl_3	Π. SCl_4	Σ. $CaCl_2$	Υ. $AuClO_3$	Φ. $CoSO_4$	Ψ. CO_2

1a. Give the name or formula as appropriate of each of the following acids.

Common Acids	A. H_2SO_4	B. nitric acid	C. sulfuric acid	D. $HC_2H_3O_2$
E. acetic acid	F. HCl	G. H_3PO_4	H. $HClO_4$	

Uncommon Acids	I. HIO_3	J. hydrofluoric acid	K. nitrous acid
L. HFO	M. $H_2C_2O_4$	N. periodic acid	O. H_2MnO_4
P. hypobromous acid	Q. hydrocyanic acid	R. HIO	S. carbonic acid
T. permanganic acid	U. perchloric acid	V. $HClO_2$	

2. (1) Determine the indicated concentration in each of the following solutions.

A. $[NO_3^-]$ in 2.16 M $Al(NO_3)_3$

B. $[SO_4^{2-}]$ in 1.22 M Na_2SO_4

C. $[Fe^{3+}]$ in 0.233 M $Fe_2(C_2O_4)_3$

D. $[Na^+]$ in 0.210 M Na_3PO_4

E. $[Br^-]$ in 2.40 M $MgBr_2$

F. $[Cl^-]$ in 0.743 M $FeCl_3$

G. $[Mg(ClO_3)_2]$ when $[ClO_3^-] = 0.538\ M$

H. $Hg_2(NO_3)_2$ molarity when $[NO_3^-] = 0.428\ M$

I. $Al_2(SO_4)_3$ molarity when $[Al^{3+}] = 0.338\ M$

J. $BaBr_2$ molarity when $[Br^-] = 0.406\ M$

K. $Fe(NO_3)_3$ molarity when $[NO_3^-] = 0.972\ M$

L. $[PO_4^{3-}]$ in 0.665 M K_3PO_4

(2) In each line in the following table, two solutions (or a solution and water, the solvent) are mixed to produce a new solution. Fill in the blanks based on the numbers supplied.

	Solution A			*Solution*			*Final Solution*		
	Solute	*M*	*Volume*	*Solute*	*M*	*Volume*	*Solute*	*M*	*Volume*
M.	$MgSO_4$	2.40	_____ mL	water		_____ mL	Mg^{2+}	0.250	300 mL
N.	Na_2SO_4	3.00	140 mL	water		_____ mL	SO_4^{2-}	0.50	_____ mL
O.	K_2SO_4	0.350	120 mL	water		600 mL	K^+	_____	_____ mL
P.	$Al(NO_3)_3$	0.750	_____ mL	water		_____ mL	NO_3^-	0.250	400 mL
Q.	KNO_3	0.642	100 mL	NaCl	0.342	200 mL	Na^+	_____	_____ mL
R.	$NaNO_3$	0.745	100	NaCl	0.312	200 mL	Na^+	_____	_____ mL
S.	$MgSO_4$	0.134	_____ mL	Na_2SO_4	0.817	_____ mL	SO_4^{2-}	0.400	500 mL
T.	KCl	_____	_____ mL	HCl	1.00	300 mL	Cl^-	1.45	600 mL
U.	NaCl	1.50	200 mL	$MgCl_2$	2.00	300 mL	Cl^-	_____	_____ mL

3. Use the solubility rules to determine whether or not each of the following compounds is soluble in water.

A. Ag_2CrO_4	B. CdS	C. $(NH_4)_2S$	D. Cr_2O_3	E. $AlCl_3$	F. $Pb(NO_3)_2$
G. K_3PO_4	H. $Sr(OH)_2$	I. $FeCl_3$	J. CaS	K. $BaCO_3$	L. $Cr_2(SO_4)_3$
M. $AgNO_2$	N. Hg_2Cl_2	O. $CdSO_4$	P. $ZnSl_2$	Q. $MgSO_4$	R. $Al(NO_3)_3$
S. Rb_2SO_3	T. $Zn(OH)_2$	U. $FeCl_3$	V. SrS	W. $CuSO_4$	X. $Hg_6(PO_4)_2$
Y. $MnBr_2$	Z. CaO	Γ. Ag_2SO_4	Δ. NiS_2O_3	Θ. $FeCrO_4$	Λ. $Sr(NO_2)_2$
Ξ. $AgClO_3$	Π. CaC_2O_4	Σ. $Ba(OH)_2$	Υ. $AlPO_4$	Φ. Bi_2S_3	Ψ. $(NH_4)_2CO_3$

4. Write each of the following as a net ionic expression, then balance the result. (A word about physical states: All substances are dissolved in aqueous solution, (aq), unless otherwise indicated, with three exceptions. (1) Water, of course, is a liquid. (2) Insoluble compounds are determined by application of the solubility rules. (3) Sulfur dioxide, carbon dioxide, hydrogen sulfide, and ammonia are gases.)

A. $HCl + Pb(NO_3)_2 \longrightarrow PbCl_2 + HNO_3$

B. $Ca(OH)_2 + K_3PO_4 \longrightarrow Ca_3(PO_4)_2 + KOH$

C. $BaCl_2 + K_2CO_3 \longrightarrow BaCO_3 + KCl$

D. $NaCl + H_3PO_4 \longrightarrow Na_3PO_4 + HCl(g)$

E. $KOH + H_2SO_4 \longrightarrow K_2SO_4 + H_2O$

F. $CaS + HCl \longrightarrow CaCl_2 + H_2S$

G. $BaI_2 + Na_2SO_4 \longrightarrow BaSO_4 + NaI$

H. $AgNO_3 + K_2CrO_4 \longrightarrow Ag_2CrO_4 + KNO_3$

I. $Na_2CO_3 + HCl \longrightarrow NaCl + H_2O + CO_2$

J. $NH_4Br + KOH \longrightarrow NH_3 + H_2O + KBr$

K. $CdSO_4 + K_2S \longrightarrow CdS + K_2SO_4$

L. $Mg(HO)_2 + H_3PO_4 \longrightarrow H_2O + Mg_3(PO_4)_2$

M. $FeCl_3 + NaOH \longrightarrow Fe(OH)_3 + NaCl$

N. $AgNO_3 + K_3PO_4 \longrightarrow Ag_3PO_4 + KNO_3$

O. $H_3PO_4 + Sr(OH)_2 \longrightarrow Sr_3(PO_4)_2 + H_2O$

P. $Al(OH)_3 + HCl \longrightarrow AlCl_3 + H_2O$

5. Complete and balance each of the following equations. Determine whether you expect each reaction to go to completion and state why.

A. $Hg_2(NO_3)_2(aq) + H_3PO_4(aq) \longrightarrow$

B. $Pb(NO_3)_2(aq) + HCl(aq) \longrightarrow$

C. $Sr(NO_3)_2(aq) + H_2SO_4(aq) \longrightarrow$

D. $RbOH(aq) + H_2SO_3(aq) \longrightarrow$

E. $AsCl_3(aq) + H_2S(aq) \longrightarrow$

F. $MnCl_2(aq) + H_2S(aq) \longrightarrow$

G. $Ba(OH)_2(aq) + H_2SO_4(aq) \longrightarrow$

H. $Zn(OH)_2(s) + HCl(aq) \longrightarrow$

I. $Al_2O_3(s) + HNO_3(aq) \longrightarrow$

J. $Cu_2O(s) + HCl(aq) \longrightarrow$

K. $Cr_2O_3(s) + H_2SO_4(aq) \longrightarrow$

L. $KOH(aq) + CO_2(g) \longrightarrow$

M. $CuSO_4(aq) + H_2S(aq) \longrightarrow$

N. $ZnCl_2(aq) + H_2S(aq) \longrightarrow$

O. $ZnO(s) + H_3PO_4(aq) \longrightarrow$

P. $Al_2O_3(s) + H_2SO_4(aq) \longrightarrow$

Q. $HCl(aq) + Na_2S(s) \longrightarrow$

R. $K_2CO_3(aq) + H_2SO_4(aq) \longrightarrow$

S. $NH_4Cl(aq) + NaOH(aq) \longrightarrow$

T. $CuSO_3(s) + HBr(aq) \longrightarrow$

6. Identify the oxidizing agent and the reducing agent in each equation that follows.

A. $Ag(s) + HNO_3(aq) \longrightarrow AgNO_3(aq) + H_2O(l) + NO(g)$

B. $HCl(aq) + PbO_2(s) \longrightarrow H_2O(l) + Cl_2(l) + PbCl_2(s)$

C. $Al(s) + HBr(aq) \longrightarrow AlBr_3(aq) + H_2(g)$

D. $BaSO_4(s) + C(s) \longrightarrow BaS(s) + CO(g)$

E. $Br_2(l) + H_2O(l) + SO_2(g) \longrightarrow HBr(aq) + H_2SO_4(aq)$

F. $Br_2(l) + KOH(aq) \longrightarrow KBr(aq) + KBrO_3(aq) + H_2O(l)$

G. $C(s) + HNO_3(aq) \longrightarrow CO_2(g) + H_2O(l) + NO_2(g)$

H. $Ca(PO_3)_2 + C \longrightarrow Ca_3(PO_4)_2 + P_4 + CO(g)$

I. $ClO_2(g) + H_2O_2(aq) + KOH(aq) \longrightarrow H_2O(l) + KClO_2(aq) + O_2(g)$

J. $Cu(s) + HNO_3(aq) \longrightarrow Cu(NO_3)_2(aq) + NO_2(g) + H_2O(l)$

K $HClO_3(aq) \longrightarrow HClO_4(aq) + ClO_2(g) + H_2O(l)$

L. $H_3AsO_3(aq) + H_2O(l) + I_2(s) \longrightarrow HI(aq) + H_3AsO_4(aq)$

M. $Cr(OH)_3(s) + Na_2O_2(aq) \longrightarrow Na_2CrO_4(aq) + NaOH(aq) + H_2O(l)$

N. $FeI_2(aq) + H_2SO_4(aq) \longrightarrow Fe_2(SO_4)_3(aq) + I_2(s) + SO_2(g) + H_2O(l)$

O. $FeS(s) + HNO_3(aq) \longrightarrow Fe(NO_3)_3(aq) + S(s) + NO_2(g) + H_2O(l)$

7. (1) Write the balanced half equation for each of the following couples, in acidic solution, unless indicated (with a bullet) as being in alkaline solution.

A. $Ag(s) | Ag^+(aq)$ B. $HNO_3(aq) | NO(g)$ C. $HCl(aq) | Cl_2(g)$

D. $PbO_2(s) | Pb^{2+}(aq)$ E. $Br_2(l) | BrO_3^-(aq)$ F. $Br_2(l) | BrO_3^-(aq)$ (B)

G. $HNO_3(aq) | NO_2(g)$ H. $H_2O_2(aq) | O_2(g)$ (B) I. $HClO_3(aq) | ClO_2(g)$

J. $HClO_3(aq) | HClO_4(aq)$ K. $Cr(OH)_3(s) | CrO_4^{2-}$ (B) L. $O_2^{2-}(aq) | HO^-(aq)$ (B)

M. $H_2SO_4(aq) | SO_2(g)$ N. $NO_2^-(aq) | NO_3^-(aq)$ (B) O. $MnO_4^{2-}(aq) | MnO_2(s)$ (B)

(2) Balance each of the following oxidation-reduction equations by the ion-electron method.

A. $CuSO_4(aq) + H_2O_2(aq) \longrightarrow Cu(s) + O_2(g) + H_2SO_4(aq)$

B. $Mg(s) + AgNO_3(aq) \longrightarrow Ag(s) + Mg(NO_3)_2(aq)$

C. $Al(s) + H_2SO_4(aq) \longrightarrow Al_2(SO_4)_3(aq) + H_2(g)$

D. $Fe_2(SO_4)_3(aq) + Pb(s) \longrightarrow PbSO_4(s) + Fe(s)$

E. $NaI(aq) + F_2(g) \longrightarrow I_2(s) + NaF(aq)$

F. $H_2(g) + KClO(aq) \longrightarrow KCl(aq) + H_2O(l)$, in alkaline solution

G. $MnO_2(s) + HCl(aq) \longrightarrow MnCl_2(aq) + H_2O(l) + Cl_2(g)$

H. $I_2(s) + H_2O(l) + Br_2(l) \longrightarrow HBr(aq) + HIO_3(aq)$

I. $Ag(s) + HNO_3(aq) \longrightarrow AgNO_3(aq) + NO(g) + H_2O(l)$

J. $S(s) + H_2O(l) + Pb(NO_3)_2(aq) \longrightarrow Pb(s) + H_2SO_3(aq) + HNO_3(aq)$

K. $O_2(g) + H_2O(l) \longrightarrow O_3(g) + H_2O_2(aq)$

L. $H_2S(g) + Br_2(l) \longrightarrow S(s) + HBr(aq)$

M. $SnSO_4(aq) + FeSO_4(aq) \longrightarrow Sn(s) + Fe_2(SO_4)_3(aq)$

N. $H_2SO_4(aq) + S(s) + H_2O(l) \longrightarrow H_2SO_3(aq)$

O. $I_2(s) + H_2O(l) \longrightarrow HI(aq) + HIO_3(aq)$

8. Balance each of the oxidation-reduction equations of Objective 5–6.

9. Classify each of the reactions of the Drill Problems of Objectives 4-2 and 4-3a in one of two ways:

(1) As combination (c), decomposition (de), displacement (di), or metathesis (m).

(2) As precipitation (p), gas-evolution (g), acid-base (a), or oxidation-reduction (o) reactions.

(Use *?* if the reaction does not fit into one of these categories.)

10. Determine the concentration of the titrated solution or the purity of the solid in each reaction that follows.

A. $HCl(25.0\,mL\ of\_\_\_M) + NaOH(18.7\,mL\ of\ 0.100\,M\ NaOH) \longrightarrow NaCl(aq) + H_2O(l)$

Meaning: 25.0 mL of solution with ___M HCl solution is titrated with 18.7 mL of 0.100 M NaOH.

B. $HNO_3(20.00\,mL\ of\_\_\_M) + KOH(26.42\,mL\ of\ 0.125\,M) \longrightarrow KNO_3(aq) + H_2O(l)$

C. $HCl(17.42\,mL\ of\ 1.43\,M) + KOH(20.00\,mL\ of\ \_\_\_M) \longrightarrow KCl(aq) + H_2O(l)$

D. $HNO_3(19.74\,mL\ of\ 1.50\,M) + NaOH(25.00\,mL\ of\ \_\_\_M) \longrightarrow NaNO_3(aq) + H_2O(l)$

E..

$H_2SO_4(20.00\,mL\ of\ \_\_\_M) + 2\,KOH(20.41\,mL\ of\ 0.100\,M) \longrightarrow K_2SO_4(aq) + 2\,H_2O(l)$

F. $2\,HNO_3(aq) + CaO(s) \longrightarrow Ca(NO_3)_2(aq) + H_2O(l)$

1.75 g of pure CaO(s) reacts with 20.13 mL of solution with _____ M HNO_3.

G.

$2\,HCl(19.40\,mL\ of\ 0.100\,M) + Mg(OH)_2(25.00\,mL\ of\ \_\_\_M) \longrightarrow MgCl_2(aq) + 2\,H_2O(l)$

H. $CaCO_3(s) + 2\,HCl(aq) \longrightarrow CaCl_2(aq) + H_2O(l) + CO_2(g)$

19.41 mL of 1.00 M HCl titrates 5.00 g of $CaCO_3$-containing solid. $CaCO_3$ = _____%.

I.

$AgNO_3(17.42\,mL\ of\ 0.100\,M) + NaCl(25.00\,mL\ of\ \_\_\_M) \longrightarrow AgCl(s) + NaNO_3(aq)$

J. $2\,AgNO_3(50.00\,mL\ of\ \_\_\_M) + K_2CrO_4(19.21\,mL\ of\ 0.500\,M) \longrightarrow 2\,KNO_3(aq) + Ag_2CrO_4(s)$

K. $Sr(NO_3)_2(25.00\ mL\ of\ 0.100\ M) + Na_2SO_4(27.20\ mL\ of\ \_\_\ M)$
$\longrightarrow 2\ NaNO_3(aq) + SrSO_4(s)$

L. $Al_2O_3(s) + 6\ HNO_3(aq) \longrightarrow 2\ Al(NO_3)_3(aq) + 6\ H_2O(l)$

24.31 mL of 0.100 M HNO_3 titrates 0.100 g Al_2O_3-containing solid. Al_2O_3 = _____%

M. $Na_2CO_3(s) + 2\ HCl(aq) \longrightarrow 2\ NaCl(aq) + H_2O(l) + CO_2(g)$

13.41 mL of 0.25 M HCl titrates 4.00 g Na_2CO_3-containing solid. Na_2CO_3 = _____%

N. $Mg(CN)_2(s) + H_2SO_4(aq) \longrightarrow MgSO_4(aq) + 2\ HCN(g)$

17.21 mL of 0.500 M H_2SO_4 titrates 1.05 g $Mg(CN)_2$-containing solid. $Mg(CN)_2$ = _____%

O. $KCNS(20.00\ mL\ of\ \_\_\_\_\ M) + AgNO_3(14.32\ mL\ of\ 0.200\ M)$
$\longrightarrow AgCNS(s) + KNO_3(aq)$

11. In each of the following cases, the unknown is one of the three compounds listed first. Identify the unknown, based on the information supplied.

A. $AgNO_3$, NH_4NO_3, $PbSO_3$. When NaOH(aq) is added, a gas is evolved that has a sharp odor.

B. $BaCl_2$, $FeSO_4$, $MgCO_3$. Precipitate forms when $Na_2S(aq)$ is added to a solution of the unknown.

C. AgI, KCl, $FeBr_2$. No precipitate forms when $Na_2CrO_4(aq)$ is added to a solution of the unknown.

D. $AgClO_3$, $Ba(NO_3)_2$, $FeSO_4$. Precipitate forms when NaCl(aq) is added to a solution of the known.

E. $BaCrO_4$, $MgSO_4$, $Sr(NO_3)_2$. Precipitate forms when a small volume of NaOH(aq) is added to a solution of the unknown.

F. $PbSO_3$, Na_2CO_3, FeS. When HCl(aq) is added, a gas with a rotten-egg odor is evolved.

G. $FeSO_4$, $MgSO_4$, K_2SO_4. Precipitate forms when $Na_2CrO_4(aq)$ is added to a solution of the unknown.

H. $CoCl_2$, Hg_2Br_2, $BaCl_2$. No precipitate forms when $Na_2SO_4(aq)$ is added to a solution of the unknown.

I. $Cu(OH)_2$, $AlPO_4$, MgS. No precipitate forms when $NaNO_3(aq)$ is added to a solution of the unknown.

J. $(NH_4)_2S$, $CaCO_3$, $NaC_2H_3O_2$. No gas is evolved when HCl(aq) is added.

K. $BaBr_2$, $MgSO_4$, Na_2CO_3. Precipitate forms when $Co(ClO_3)_2(aq)$ is added to a solution of the unknown.

L. BaS, $FeSO_3$, $MgCO_3$. An odorless gas is evolved when HCl(aq) is added.

QUIZZES (20 minutes each) Choose the best answer to each question or complete and balance the equation.

Quiz A

1. Which of the following contains, in order, a nonelectrolyte, a strong electrolyte, and a base?

   (a) H_2O, NH_3, NaOH; (b) KCl, $BaBr_2$, Na_2SO_3; (c) $MnBr_2$, CH_4, H_2O; (d) CO_2, $MgSO_4$, KOH; (e) none of these.

2. $2\ KClO_3 \xrightarrow{\text{heat}}$? When this reaction is completed, it is which of the following types? (a) precipitation; (b) gas evolution; (c) acid-base; (d) oxidation-reduction; (e) none of these.

3. Which contains two insoluble compounds and a soluble one? (a) $HgBr_2$, Ag_2CrO_4, $Co(ClO_4)_2$;

   (b) $CoSO_4$, CuCl, $AgNO_3$; (c) $PbSO_4$, $Ca(ClO_3)_2$, BaS; (d) $CuCl_2$, $Ba(C_2H_3O_2)_2$, Na_2CO_3; (e) none of these.

4. 11.2 mL of 0.100 M NaOH is needed to titrate 25.00 mL of HCl(aq). The HCl molarity is (a) 0.448 M; (b) 2.23 M; (c) 0.223 M; (d) 0.112 M; (e) none of these.

5. Which of the following represents an oxidant followed by a reductant, from the equations of Questions 7 and 8 of this quiz? (a) I_2, $HClO_3$; (b) $HClO_3$, I_2; (c) I^-, $Cr_2O_7^{2-}$; (d) H^+, I^-; (e) none of these.

6. What is $[Al^{3+}]$ in an $Al_2(SO_4)_3$ solution that has $[SO_4^{2-}] = 0.342$ M? (a) 0.228 M; (b) 0.513 M; (c) 0.342 M; (d) 0.114 M; (e) none of these.

7. When the equation $HClO_3 + H_2O + I_2 \longrightarrow HIO_3 + HCl$ is balanced with the smallest integer coefficients, the coefficient of chloric acid is (a) 10; (b) 12; (c) 6; (d) 8; (e) none of these.

8. $Cr_2O_7^{2-} + I^- + H^+ \longrightarrow Cr^{3+} + I_2 + H_2O$. When the smallest whole number coefficients are used, the correct coefficient for I^- in the balanced equation is (a) 2; (b) 3; (c) 6; (d) 14; (e) none

9. Complete and balance $AgClO_3(aq) + H_2SO_4(aq) \longrightarrow$ 10. $LiOH(aq) + H_2C_2O_4(aq) \longrightarrow$

Quiz B

1. Which of the following contains, in order, a strong electrolyte, an acid, and a base? (a) KCl, $CoSO_4$, CH_4;

   (b) HCl, $HC_2H_3O_2$, NH_3; (c) $Mg(NO_3)_2$, H_2SO_4, $Ca(OH)_2$; (d) SO_2, HF, HCl; (e) none of these.

2. $AgNO_3(aq) + H_2SO_4(aq) \longrightarrow$? When this reaction is completed, it is which of the following types?

   (a) precipitation; (b) gas evolution; (c) acid-base; (d) oxidation-reduction; (e) none of these.

3. Which contains two soluble compounds and an insoluble one? (a) $HgBr_2$, $MnSO_4$, $Na_2C_2O_4$;

   (b) $Na_2S_2O_3$, NH_4Cl, CoI_2; (c) MnS, $Cu(OH)_2$, Al_2O_3; (d) $Mn(ClO_3)_2$, $Ca(MnO_4)_2$, Hg_2SO_4; (e) none of these.

4. 19.4 mL of 2.00 M HNO_3 is needed to titrate 25.00 mL of LiOH(aq). The LiOH molarity is (a) 0.776 M; (b) 1.55 M; (c) 3.10 M; (d) 1.29 M; (e) none of these.

5. Which of the following represents an oxidant followed by a reductant, from the equations of Questions 7 and 8 of this quiz? (a) KNO_2, $KMnO_4$; (b) $KMnO_4$, KNO_2; (c) NO_3^-, H_2O; (d) NO, NO_3^-; (e) none of these.

6. What is $[NO_3^-]$ in a $Ca(NO_3)_2$ solution that has $[Ca^{2+}] = 0.840$ M? (a) 0.840 M; (b) 1.68 M; (c) 0.420 M; (d) 1.26 M; (e) none of these.

7. When the equation $NO_3^- + H_2O \longrightarrow NO + OH^-$ is balanced with the smallest whole number coefficients, the coefficient of hydroxide ion is (a) 3; (b) 2; (c) 4; (d) 1; (e) none of these.

8. When the equation $KMnO_4 + KNO_2 + H_2O \longrightarrow MnO_2 + KNO_3 + KOH$ is balanced with the smallest whole number coefficients, the coefficient of potassium nitrate is (a) 1; (b) 2; (c) 3; (d) 4; (e) 6.

9. Complete and balance $MnSO_3(aq) + HCl(aq) \longrightarrow$

10. Complete and balance $ZnSO_4(aq) + BaS(aq) \longrightarrow$

Quiz C

1. Which contains a nonelectrolyte, a weak electrolyte, and a strong electrolyte? (a) CO_2, NaCl, $MnSO_4$; (b) H_2SO_4, $HC_2H_3O_2$, CuCl; (c) SO_2; HF, $FeSO_4$; (d) $Ba(ClO_3)_2$, $K_2S_2O_3$, $NaMnO_4$; (e) none of these.

2. $MnSO_3(s) + HCl(aq) \longrightarrow$? When this reaction is completed, it is which of the following types? (a) precipitation; (b) gas evolution; (c) acid-base; (d) oxidation-reduction; (e) none of these.

3. Which of the following contains three insoluble compounds? (a) FeC_2O_4, SrS, $AgC_2H_3O_2$; (b) $Ba(OH)_2$, MnS_2O_3, $Sr(ClO_3)_2$; (c) $CrCl_3$, $SnSO_4$, $(NH_4)_2S$; (d) $ZnSO_4$, AgI, CdS; (e) none of these.

4. 21.6 mL of 0.500 M CsOH is needed to titrate 20.00 mL of $HClO_3(aq)$. The $HClO_3$ molarity is (a) 0.270 M; (b) 0.540 M; (c) 1.08 M; (d) 0.926 M; (e) none of these.

5. Which of the following represents an oxidant followed by a reductant, from the equations of Questions 7 and 8 of this quiz? (a) Cs_2MnO_4, H_2O; (b) H_2O, Cs_2MnO_4; (c) H_2S, Br_2; (d) Br_2, H_2S; (e) none of these.

6. What is $[Na^+]$ in a Na_3PO_4 solution that has $[PO_4^{3-}] = 0.444$ M? (a) 0.111 M; (b) 0.148 M; (c) 0.592 M; (d) 1.33 M; (e) none of these.

7. When the equation $Br_2 + H_2S \longrightarrow S + HBr$ is balanced with the smallest whole number coefficients, the sum of the coefficients is (a) 7; (b) 8; (c) 4; (d) 5; (e) 6.

8. When the equation $Cs_2MnO_4 + H_2O \longrightarrow CsMnO_4 + CsOH + MnO_2$ is balanced with the smallest whole number coefficients, the coefficient of MnO_2 is (a) 1; (b) 2; (c) 3; (d) 4; (e) none of these.

9. Complete and balance $FeS(s) + H_2SO_4(aq) \longrightarrow$

10. Complete and balance $Al(OH)_3(s) + H_2SO_4(aq) \longrightarrow$

Quiz D

1. Which of the following contains a weak acid, a weak base, and a salt? (a) HCl, NH_3, Na_2SO_4; (b) HNO_2, NH_3, NH_4NO_2; (c) HCl, $Ca(OH)_2$, $CaSO_4$; (d) HNO_2, $HMnO_4$, Cs_2CrO_4; (e) none of these.

2. $NaI(aq) + Cl_2(aq) \longrightarrow$? When this reaction is completed, it is which of the following types? (a) precipitation; (b) gas evolution; (c) acid-base; (d) oxidation-reduction; (e) none of these.

3. Which contains two insoluble compounds and a soluble one? (a) CuS, CaS, K_2S; (b) $PbSO_4$, MnS; $Co(NO_2)_2$; (c) $AgClO_4$, $Mn(C_2H_3O_2)_2$, CaO; (d) $BaBr_2$, $Ba(BrO_3)_2$, AgBr; (e) none of these.

4. 31.2 mL of 0.200 M HI(aq) is needed to titrate 25.00 mL of KOH(aq). The KOH molarity is (a) 1.25 M; (b) 0.801 M; (c) 0.245 M; (d) 4.08 M; (e) none of these.

5. Which of the following represents an oxidant followed by a reductant, from the equations of Questions 7 and 8 of this quiz? (a) Cl^-, MnO_4^-; (b) MnO_4^-, H^+; (c) H_2O_2, KOH; (d) ClO_2, H_2O_2; (e) none of these.

6. What is $[Na^+]$ in a Na_2SO_4 solution that has $[SO_4^{2-}] = 0.344\,M$? (a) 0.0860 M; (b) 0.172 M; (c) 0.258 M; (d) 0.688 M; (e) none of these.

7. When the equation $MnO_4^- + Cl^- + H^+ \longrightarrow Mn^{2+} + Cl_2 + H_2O$ is balanced, the sum of the smallest whole number coefficients is (a) 21; (b) 24; (c) 38; (d) 43; (e) 51.

8. When the equation $ClO_2 + H_2O_2 + KOH \longrightarrow H_2O + KClO_2 + O_2$ is balanced with the smallest integer coefficients, the coefficient of water is (a) 3; (b) 5; (c) 2; (d) 4; (e) none of these.

9. Complete and balance $BaCO_3(s) + HNO_3(aq) \longrightarrow$

10. Complete and balance $Ba(OH)_2(aq) + H_2SO_4(aq) \longrightarrow$

SAMPLE TEST (20 minutes)

1. Complete and balance the following equations, and identify each one in two ways: as combination, decomposition, displacement, or metathesis; then as precipitation, oxidation-reduction, gas-evolution, or acid-base.

A. $Sr(NO_3)_2(aq) + H_2SO_4(aq) \longrightarrow$

B. $Na_2CO_3(s) + HCl(aq) \longrightarrow$

C. $CaO(s) + HNO_3(aq) \longrightarrow$

D. $NaI(aq) + Cl_2(aq) \longrightarrow$

G. $CaO(s) + CO_2(g) \longrightarrow$

H. $Al(s) + Cl_2(g) \longrightarrow$

2. Balance each of the following oxidation-reduction equations.

A. $Ag(s) + HNO_3(aq) \longrightarrow AgNO_3(aq) + NO(g) + H_2O(l)$

B. $Cl_2(aq) + KOH(aq) \longrightarrow KCl(aq) + KClO(aq) + H_2O(l)$

C. $Fe_2(SO_4)_3(aq) + SnSO_4(aq) \longrightarrow Sn(SO_4)_2(aq) + FeSO_4(aq)$

D. $K_2C_2O_4(aq) + KMnO_4(aq) + H_2SO_4(aq) \longrightarrow$
$MnSO_4(aq) + K_2SO_4(aq) + CO_2(g) + H_2O(l)$

3. An unknown, which is a pure compound, forms a solution when water is added. A precipitate forms when $Ba(NO_3)_2(aq)$ is added to this solution, but not when $Cu(NO_3)_2(aq)$ is added. When $HC_2H_3O_2(aq)$ is added to the unknown solid, no gas is produced. When NaOH(aq) is added to the solid, a gas with a pungent odor is produced. What is the identity of the unknown?

6 GASES

CHAPTER OBJECTIVES

*** 1. Be able to convert between the common units of pressure.**

Your goal is to firmly commit the following expressions to memory and use them in problems. (Notice that there is no abbreviation for torr; t or T is incorrect.) The problems are largely of the conversion factor type.

$$1.000 \text{ atmosphere (atm)} = 760 \text{ mmHg} = 760 \text{ torr} = 14.7 \text{ lb./in.}^2$$
$$= 101{,}325 \text{ N/m}^2 = 101{,}325 \text{ pascal (Pa)}$$

$$1000 \text{ pascal} = 1.000 \text{ kilopascal (kPa)}$$

2. Explain the operation of a mercury barometer, an open-end manometer, and a closed-end manometer and be able to use the data obtained with these devices.

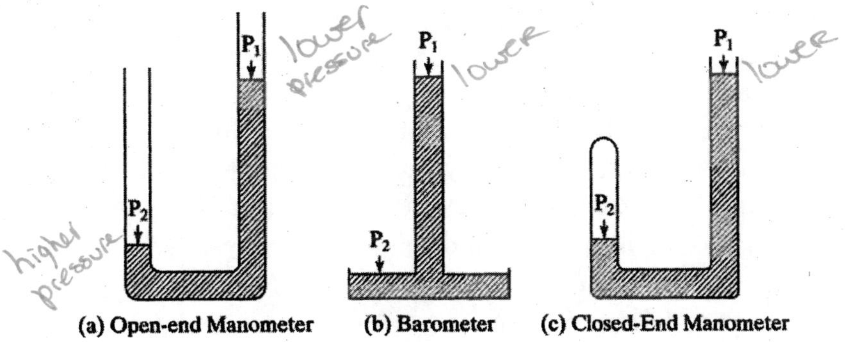

FIGURE 6-1 Liquid-Filled Pressure-Measuring Devices

All liquid-filled pressure-measuring devices operate on the same principle. A liquid is confined in a U-shaped tube, as shown in Figure 6-1a. If the pressure is different on the two ends of the liquid column—that is, if P_1 differs from P_2—the liquid levels in the two arms of the tube are different. The liquid level is higher on the low-pressure end of the tube. (Think of sucking on a soda straw to remember which side is higher: the side with lower pressure.) In all parts of Figure 6-1, P_1 is a lower pressure than P_2. The difference in liquid levels (h) is proportional to the pressure difference.

$$\Delta P = P_2 - P_1 \propto h \qquad \text{OR} \qquad \Delta P = gdh \qquad [1]$$

In Equation [1], g is the gravitational constant ($9.807\ \text{m/s}^2$) and d is the density of the liquid. Normally, we know one of the pressures (P_1 or P_2) and the density of the liquid, are given or can measure a value of h, and wish to find the other pressure. There is a simpler method of finding ΔP than Equation [1]. It depends on the fact that a denser liquid produces a smaller level difference. In fact, for two liquids, A and B:

$$h_A d_A = h_B d_B \quad \text{OR} \quad h_A = h_B d_B / d_A \qquad [2]$$

Frequently, mercury is used as liquid A ($d_A = 13.5951\ \text{g/cm}^3$) because $760\ \text{mmHg} = 1.000\ \text{atm}$.

EXAMPLE 6-1 An open-end manometer is filled with dibutyl phthalate (DBP, $d = 1.047\ \text{g/cm}^3$). If the level difference is 625 mm, what is the pressure difference in atmospheres?

$$\Delta P = h_{Hg} = 625\ \text{mm DBP} \times \frac{1.047\ \text{g/cm}^3}{13.5951\ \text{g/cm}^3} = 48.1\ \text{mmHg} \times \frac{1\ \text{atm}}{760\ \text{mmHg}} = 0.0633\ \text{atm}$$

In a closed-end manometer and in a barometer, the pressure in the low-pressure end (the closed end) is almost zero. It is the vapor pressure of the liquid being used. At 25 °C, the vapor pressure of mercury is 6×10^{-6} atm, that of glycerol is 6×10^{-7} atm, and that of dibutyl phthalate is 2×10^{-7} atm. See Objectives 13-2 through 13-5 for a discussion of vapor pressure.

*** 3. Learn Boyle's law both mathematically and graphically, and be able to use it in calculations.**

Robert Boyle observed that for a fixed amount of gas at constant temperature, the product of volume and pressure is constant. The condition of fixed amount means that the apparatus must not leak and also that no physical or chemical change that produces or consumes a gas may occur within the container. There are no units specified for pressure or volume in Boyle's law. Any units are satisfactory, but they must be consistent: all pressures must be measured in the same units (for example, inches of mercury [in. Hg]), and so must be all volumes (for example, in.^{3}). Graphically, the law is demonstrated when a plot of $1/V$ vs. P (Figure 6-2b) or of V vs. $1/P$ (Figure 6-2c) is a straight line. Other plots give curved lines. The data for Figure 6-2, given in Table 6-1, are part of Boyle's original data.

Problems based on Boyle's law, sometimes called pressure-volume problems, supply an initial pressure and volume, and a final pressure or volume. A convenient expression to use is Equation [3].

$$P_i V_i = P_f V_f \quad (i = \text{initial};\ f = \text{final}) \qquad [3]$$

TABLE 6-1 Robert Boyle's Data of Pressure and Volume

P, in. Hg		V, in.^{3}		$P \times V$	$1/P$	$1/V$
$33\frac{1}{2}$	33.50	$10\frac{1}{2}$	10.50	352.8	0.02985	0.09524
$39\frac{1}{4}$	39.25	9	9.00	353	0.02548	0.111
$54\frac{5}{16}$	54.31	$6\frac{1}{2}$	6.50	353	0.01841	0.154
$74\frac{1}{8}$	74.13	$4\frac{3}{4}$	4.75	353	0.01346	0.211
$117\frac{9}{16}$	117.56	3	3.00	353	0.0085063	0.333

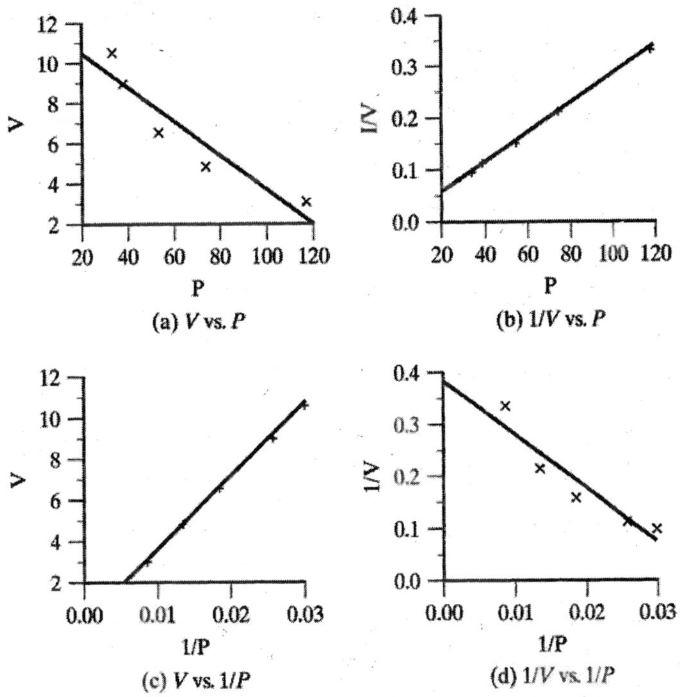

FIGURE 6-2 Graphs of Boyle's Pressure-Volume Data In each graph, the solid line is the best straight line (least squares) through the data points.

EXAMPLE 6-2 When measured at 1.25 atm, a gas occupies $247\,\text{cm}^3$. What is its volume in liters at a pressure of 76.5 mmHg?

Since the two pressures are given in different units, we convert one of them.

$$P_i = 1.25\,\text{atm} \times \frac{760\,\text{mmHg}}{1\,\text{atm}} = 950\,\text{mmHg}$$

Then equation [3] gives the final volume.

$$950\,\text{mmHg} \times 247\,\text{cm}^3 = 76.5\,\text{mmHg} \times V_f \qquad V_f = 3.07 \times 10^3\,\text{cm}^3 \times \frac{1\,\text{L}}{1000\,\text{cm}^3} = 3.07\,\text{L}$$

*** 4. Learn Charles's law both mathematically and graphically, and be able to use it in calculations.**

Charles observed that, for a fixed amount of gas at constant pressure, the ratio of volume to temperature is constant. Mathematically, this law can be expressed in three ways, as shown in Equation [4].

$$V = \text{constant} \times T \quad \text{OR} \quad \frac{V}{T} = \text{constant} \quad \text{OR} \quad \frac{V_i}{T_i} = \frac{V_f}{T_f} \qquad [4]$$

Graphically, data points fall along a straight line when volume is plotted against temperature. The only requirement for *graphical* data is that all volumes must have the same units and all temperatures must be measured on the same scale (either K, °C, or °F). When solving Charles's law *problems*, however, one *must* express temperatures on an absolute scale, such as kelvins, *not* °C or °F.

EXAMPLE 6-3 At 0.0°C, a gas has a volume of 22.414 L. What is its volume at 25.0°C?

Both temperatures must be expressed on the Kelvin scale before Charles's law can be used.

$$T_i = 0.0 + 273.15 = 273.2\text{ K} \quad T_f = 25.0 + 273.15 = 298.2\text{ K}$$

Then application of Charles's law, Equation [4], gives the final volume.

$$\frac{22.414\text{ L}}{273.2\text{ K}} = \frac{V_f}{298.2\text{ K}} \qquad V_f = 22.414\text{ L} \times \frac{298.2\text{ K}}{273.2\text{ K}} = 24.47\text{ L}$$

In addition, the pressure of a gas kept in a constant volume varies with its absolute temperature (Amonton's Law).

5. Discuss the significance of the absolute zero of temperature, and be able to convert between Celsius and Kelvin temperatures.

When Charles's law ($V = \text{constant} \times T$) is carefully considered, one concludes that a gas will occupy no volume once its Kelvin temperature has been lowered to zero. Charles's law ($V/T = \text{constant}$) is not valid on the Celsius scale. For example, consider the data of Example 6-3:

$$\frac{V_i}{t_i} = \frac{22.414\text{ L}}{0\text{ °C}} = \infty \quad \text{but} \quad \frac{V_f}{t_f} = \frac{24.47\text{ L}}{25.0\text{ °C}} = 0.979\text{ L/°C}$$

Charles's law *is* valid on the Kelvin scale.

$$\frac{V_i}{T_i} = \frac{22.414\text{ L}}{273.2\text{ K}} = 0.08204\text{ L/K} \quad \text{but} \quad \frac{V_f}{T_f} = \frac{24.47\text{ L}}{298.2\text{ K}} = 0.08206\text{ L/K}$$

It is very important to remember that *all problems involving gas laws must use Kelvin temperatures*. The relationship to be memorized is $T(\text{K}) = t(\text{°C}) + 273.15$. Notice that the degree sign (°) is not used in expressing temperatures on the Kelvin scale, which are represented by a capital *T* (and referred to as *kelvins*). Celsius temperatures, formerly called centigrade temperatures, are symbolized by a lower case *t* (and referred to as *degrees Celsius*).

6. State what is meant by STP and the STP molar volume, and be able to use the latter in calculations.

STP is the abbreviation for Standard Temperature and Pressure: $T = 273.15\text{ k}$ and $P = 1.000$ atm. The volume occupied by one mole of an ideal gas at STP is 22.414 L. This volume is known as the STP molar volume.

EXAMPLE 6-4 What volume of $H_2(g)$, measured at STP, is needed to reduce 36.1 g WO_3 according to $WO_3(s) + 3\,H_2(g) \xrightarrow{\text{heat}} W(s) + 3\,H_2O(l)$.

The problem is solved with the conversion factor method.

$$\text{volume } H_2 = 36.1\text{ g } WO_3 \times \frac{1\text{ mol } WO_3}{231.8\text{ g } WO_3} \times \frac{3\text{ mol } H_2}{1\text{ mol } WO_3} \times \frac{22.4\text{ L } H_2}{1\text{ mol } H_2} = 10.5\text{ L } H_2$$

7. State and be able to use Avogadro's hypothesis.

Avogadro's hypothesis, expressed mathematically in Equation [5], states that equal volumes of gases at the same temperature and pressure contain equal amounts (the same number of moles) of the gases, even if the gases are different. We *must* compare the gases at the *same temperature and pressure*; otherwise, the rule is invalid.

$$V = n \times \text{constant} \quad \text{OR} \quad \frac{V}{n} = \text{constant} \quad \text{OR} \quad \frac{V_1}{n_1} = \frac{V_2}{n_2} \qquad [5]$$

***8. Solve for one of *P*, *V*, *n*, or *T* when given values of the other three for an ideal gas.**

This objective involves the proper use of the ideal gas equation, $PV = nRT$. Since *R* has the units of liter-atmospheres/mole-kelvins, the pressure must be expressed in atmospheres, the volume in liters, the amount of gas in moles, and the temperature in kelvins.

EXAMPLE 6-5 What is the volume of 3.04 moles of gas confined at 741 mmHg and 14.0 °F?

We first express the pressure in atmospheres and the temperature in kelvins.

$$P = 741\text{ mmHg} \times \frac{1.00\text{ atm}}{760\text{ mmHg}} = 0.975\text{ atm}$$

$$°C = (°F - 32) \times \tfrac{5}{9} = (14.0 - 32.0)(\tfrac{5}{9}) = -10.0\text{ °C} \qquad T = -10.0 + 273.15 = 263.2\text{ K}$$

We then solve the ideal gas equation for *V*. (Notice that the units in the expression cancel appropriately.)

$$V = \frac{nRT}{P} = \frac{(3.04\text{ mol})(0.08206\text{ L atm mol}^{-1}\text{ K}^{-1})(263.2\text{ K})}{0.975\text{ atm}} = 67.4\text{ L}$$

*** 9. Obtain the value of one final variable (*P*, *V*, *n*, or *T*), given the values of the other final variables and of all the initial variables, excluding those that remain unchanged, in the ideal gas law.**

This type of problem often is called an "initial-final problem." It depends on a rearranged version of the ideal gas law, $R = PV/nT$, which produces Equation [6].

$$\frac{P_iV_i}{n_iT_i} = \frac{P_fV_f}{n_fT_f} \quad (i = \text{initial}, f = \text{final}) \qquad [6]$$

Any of the simple gas laws (Boyle's, Charles's, Guy-Lussac's, Avogadro's) can be produced from this equation. Equation [6] also can be used to solve problems for which there is no "law."

EXAMPLE 6-6 2.40 mol of an ideal gas is confined to a 26.4 L vessel at 250.0 K. If an additional 2.00 mol of gas is added to the vessel and the pressure does not vary, what is the final temperature?

Notice that this problem does not explicitly state that the volume of the container does not vary. It could be a balloon, for instance. However, the final volume is not given and, thus, one *assumes* that the volume is constant. You often will have to make such an assumption to solve a problem. Both volume and pressure are fixed and, thus, $V_i = V_f$ and $P_i = P_f$. We use Equation [6].

$$\frac{P_iV_i}{n_iT_i} = \frac{P_iV_i}{n_fT_f} \quad \text{OR} \quad n_iT_i = n_fT_f$$

We are given the following data:

$$n_i = 2.40\text{ mol} \quad T_i = 250.0\text{ K} \quad n_f = n_i + 2.0 = 2.40\text{ mol} + 2.00\text{ mol} = 4.20\text{ mol}$$

And the result is:

$$(2.40\text{ mol})(250.0\text{K}) = (4.40\text{ mol})\,T_f \quad T_f = \frac{(2.40\text{ mol})(250.0\text{ K})}{4.40\text{ mol}} = 136.4\text{ K}$$

Notice that the equation produced to solve this problem, $n_iT_i = n_fT_f$, is not one of the classical gas laws.

We conclude with a word of caution. Frequently, it is stated that a problem like that of Example 6-6 can be solved with an equation of the type $T_f = T_i \times (\text{mole ratio})$. The "mole ratio" is the quotient of the two numbers of moles (4.40 and 2.40). But is the ratio 4.40/2.40 or 2.40/4.40? In order to confine more moles in the same volume, at the same pressure, the temperature must be reduced, and, thus, the correct ratio is 2.40/4.40. This kind of intuitive or "gut" feeling about how gases behave comes with experience, which many students lack. Therefore, you might get the ratio upside down, especially in the heat of an examination. Starting with Equation [6] is safer.

*** 10. Use the alternate versions of the ideal gas law for calculating molar masses of gases and determining gas densities.**

An alternate form of the ideal gas law (Equation [7]) can be used to determine molar mass.

$$PV = \frac{m}{M}RT \qquad [7]$$

Equation [7] is obtained from the ideal gas law by substituting Expression [8] into the ideal gas equation

$$n = m/M \qquad [8]$$

where m = the mass of the gas and M = the molar mass of the gas. It is convenient, but not essential, that you know the combined Equation [7]. The two parts, the ideal gas law and Equation [8], work just as well, as Example 6-7 illustrates. Just use the technique you find easier.

EXAMPLE 6-7 What is the molar mass of a gas if 11.770 g has a volume of 5.00 L at 30°C and 742 mmHg?

With the combined Equation [7]:

$$P = 742\ \text{mmHg} \times \frac{1\ \text{atm}}{760\ \text{mmHg}} = 0.976\ \text{atm}, \quad T = 30°\text{C} + 273.15 = 303\ \text{K},$$

$$m = 11.770\ \text{g}, \quad V = 5.00\ \text{L}$$

$$PV = \frac{m}{M}RT \qquad 0.976\ \text{atm} \times 5.00\text{L} = \frac{11.770\ \text{g}}{M}(0.08206\frac{\text{L atm}}{\text{mol K}})(303\ \text{K}) \quad M = 60.0\ \text{g/mol}$$

With the ideal gas equation and Equation [8]:
$(0.976\ \text{atm})(5.00\ \text{L}) = n(0.08206\ \text{L atm mol}^{-1}\ \text{K}^{-1})(303\ \text{K})$

$$n = 0.196\ \text{mol} = \frac{\text{m}}{M} = \frac{11.770\ \text{g}}{M} \qquad M = \frac{11.770\ \text{g}}{0.196\ \text{mol}} = 60.0\ \text{g/mol}$$

*** 11. Combine gas molar masses with empirical formulas to determine molecular formulas.**

Percent composition data only enables us to determine the empirical formula of a compound. But several compounds may have the same empirical formula. The molar mass of a compound allows us to determine the molecular formula of a compound if we know its empirical formula. Thus, if the gas of Example 6-7 has an empirical formula of CH_2O, its molecular formula will be an integral multiple of its molecular formula: CH_2O, $C_2H_4O_2$, $C_3H_6O_3$, etc. The *empirical molar mass* is the mass of 1 mol C, 2 mol H, and 1 mol O or $12.0\ \text{g} + 2 \times 1.0\ \text{g} + 16.0\ \text{g} = 30.0\ \text{g/mol}$. Since the molar mass obtained in Example 6-7 (60.0 g/mol) is twice this empirical molar mass, the molecular formula must be twice the molecular formula, or $C_2H_4O_2$.

*** 12. Solve stoichiometry problems involving gases.**

A large variety of stoichiometry problems use the ideal gas law either to determine the amount of reactant or product, or to determine the volume (or occasionally the pressure) of a reactant or product. If you can solve stoichiometry problems and problems based on the ideal gas law, these combined problems should give you little trouble. However, in problems involving the law of combining volumes, the data supplied often is insufficient to allow you to compute the number of moles. Recall that equal volumes of gases at the same temperature and pressure contain equal numbers of moles (Avogadro's hypothesis). Thus, when temperature and pressure are fixed, gas volumes can be used in the same manner as gas amounts (in moles).

EXAMPLE 6-8 3.04 L of $O_2(g)$ and excess $SO_2(g)$, both at 397 K and 0.351 atm, combine to produce what volume of SO_2? $2\,SO_2(g) + O_2(g) \longrightarrow 2\,SO_3(g)$

We assume that the pressure and temperature remain fixed. Thus, we have

$$\text{volume}\,O_2 = 3.04\,\text{L}\,O_2 \times \frac{2\,\text{L}\,SO_3(g)}{1\,\text{L}\,O_2(g)} = 6.08\,\text{L}\,SO_3(g)$$

*** 13. Solve problems involving mixtures of gases with (a) the ideal gas law, (b) Dalton's law of partial pressures, or (c) Amagat's law of partial volumes.**

Since ideal gas molecules do not attract or repel each other, a mixture of several gases displays properties that are the sums of the individual gas properties. Each gas in the mixture acts as if the other gases were not present, unless a chemical reaction occurs. Thus, the total pressure is the sum of the partial pressures. (The partial pressure is the pressure that the individual gas would exert if it were confined in the same volume at the same temperature as the mixture.) In like fashion, the total gas volume is the sum of the partial volumes. (The partial volume is the volume an individual gas would occupy if confined at the same temperature and pressure as the mixture.) This additive feature of pressures (Dalton's law) and volumes (Amagat's law) is valid only for ideal gases. For other states of matter—solids and liquids—only masses and amounts (in moles) can be summed to obtain the total mass and the total amount of the mixture.

EXAMPLE 6-9 A 2.46-L volume of argon at 0.320 atm and 400 K is combined with 4.12 L of neon at 0.440 atm and 400 K in a rigid 5.00-L container at 400 K. Determine **(a)** the total final pressure, **(b)** the partial pressure of argon in the 5.00-L container, and **(c)** the partial volume of neon in the 5.00-L container.

We first compute the amount of each gas and the total moles of gas placed in the 5.00- L container.

$$n_{\text{Ar}} = \frac{PV}{RT} = \frac{(0.320\,\text{atm})(2.46\,\text{L})}{(0.08206\,\text{L atm mol}^{-1}\,\text{K}^{-1})(400\,\text{K})} = 0.0240\,\text{mol}$$

$$n_{\text{Ne}} = \frac{(0.440\,\text{atm})(4.12\,\text{L})}{(0.08206\,\text{L atm mol}^{-1}\,\text{K}^{-1})(400\,\text{K})} = 0.0552\,\text{mol}$$

$$n_{\text{total}} = n_{\text{Ar}} + n_{\text{Ne}} = 0.0240\,\text{mol} + 0.0552\,\text{mol} = 0.792\,\text{mol}$$

(a) We compute the final total pressure with the ideal gas law:

$$P_f = \frac{nRT}{V} = \frac{(0.0792)\,(0.08206\,\text{L atm mol}^{-1}\,\text{K}^{-1})(400\,\text{K})}{5.00\,\text{L}} = 0.520\ \text{atm}$$

(b) The pressure of argon in the 5.00-L container can be computed in at least two ways:

(1)

$$P_{Ar} = \frac{n_{Ar}RT}{V} = \frac{(0.0240\ \text{mol})(0.08206\ \text{L atm mol}^{-1}\ \text{K}^{-1})(400\ \text{K})}{5.00\ \text{L}} = 0.158\ \text{atm}$$

(2) $$P_{Ar} = \frac{n_{Ar}}{n_{total}} P_{total} = \frac{0.0240\ \text{mol}}{0.0792\ \text{mol}} \times 0.520\ \text{atm} = 0.158\ \text{atm}$$

(c) The partial volume of neon is computed from its mole fraction. (See Objective 13-2b.)

$$V_{Ne} = \chi_{Ne} V_{total} = \frac{n_{Ne}}{n_{total}} V_{total} = \frac{0.0552\ \text{mol}}{0.0792\ \text{mol}} \times 5.00\ \text{L} = 3.48\ \text{L}$$

14. Compute the pressure of gases collected over water.

When a gas is collected over water, it is mixed with water vapor. Thus, the total pressure is the sum of the pressure of the gas collected and the vapor pressure of water, which varies with temperature (as shown in Table 6-2).

$$P_{total} = P_{gas} + P_{water} \qquad [9]$$

EXAMPLE 6-10 A 1.05-g sample of HgO is thermally decomposed to its elements and the $O_2(g)$ is collected over water at 25°C and 753.6 mmHg barometric pressure. What volume of gas is collected in mL?

The pressure of oxygen is part of the total pressure; the rest is water vapor. From Table 6-2, $P\{H_2O\} = 23.8\ \text{mmHg}$ at 25 °C. Thus, we have

$$P_{total} = P\{O_2\} + P\{H_2O\} = 753.6\ \text{mmHg} = P\{O_2\} + 23.8\ \text{mmHg}$$

$$P\{H_2O\} = 753.6\ \text{mmHg} - 23.8\ \text{mmHg} = 729.8\ \text{mmHg}$$

The mass of HgO is related to the amount of $O_2(g)$ produced by the balanced chemical equation:

$$2\ \text{HgO(s)} \xrightarrow{\Delta} 2\ \text{Hg(l)} + O_2(g)$$

$$O_2\ \text{amount} = 1.05\ \text{g HgO} \times \frac{1\ \text{mol HgO}}{216.62\ \text{g HgO}} \times \frac{1\ \text{mol}\ O_2}{2\ \text{mol HgO}} = 0.00242\ \text{mol}\ O_2$$

Then, we used the ideal gas law:

$$\text{gas volume} = V = \frac{nRT}{P} = \frac{0.00242\ \text{mol}\ O_2 \times 0.08206 \frac{\text{L atm}}{\text{mol K}} \times (273.2 + 25)\ \text{K}}{729.8\ \text{mmHg} \times \frac{1\ \text{atm}}{760\ \text{mmHg}}} \times \frac{1000\ \text{mL}}{1\ \text{L}}$$

$$= 61.7\ \text{mL}\ O_2$$

TABLE 6-2 Vapor Pressure of Water at Various Temperatures

t, °C	P, mmHg	t, °C	P, mmHg	t, °C	P, mmHg	t, °C	P, mmHg
13.0	11.2	19.0	16.5	25.0	23.8	31.0	33.7
14.0	12.0	20.0	17.5	26.0	25.2	32.0	35.7
15.0	12.8	21.0	18.7	27.0	26.7	33.0	37.7
16.0	13.6	22.0	19.8	28.0	28.3	34.0	39.9
17.0	14.5	23.0	21.1	29.0	30.0	35.0	42.2
18.0	15.5	24.0	22.4	30.0	31.8	36.0	44.6

15. State the postulates and the basic mathematical relationships of the kinetic molecular theory of gases.

The postulates are a set of assumptions about the structure of gases and how gas molecules interact. One form is:

1. Gases are composed of molecules.
2. Gas molecules have no volume—they are point masses.
3. Gas molecules neither attract nor repel each other—there are no forces between them.
4. Gas molecules are in constant, random motion.

The kinetic energy of an average ideal gas molecule is given by two equations, [10] and [11], where $k = R/N_A = 1.38\times10^{-23}$ J molecule^{-1} K^{-1}. k is known as Boltzmann's constant.

$$\overline{e_k} = \tfrac{1}{2}m\overline{u^2} \qquad [10]$$

$$\overline{e_k} = \tfrac{3}{2}kT \qquad [11]$$

16. Deduce the simple gas laws from the kinetic molecular theory.

Consider Boyle's law as an example. A consequence of random motion of molecules is that molecules continually hit the sides of the container. The average molecular speed, and, thus, the average force exerted by each collision with the wall, depends only on the temperature (Equation [12], below). If the gas now is compressed into a smaller volume, each collision still has the same force. But there are many more collisions, simply because the walls are closer. In fact, if the volume is halved, there are twice as many collisions and the pressure will double. A particular gas molecule should hit each side twice as often. You can produce the same type of explanation for the other gas laws: Charles's, Avogadro's, and Dalton's.

*** 17. Compute molecular velocities; know and apply Graham's law.**

When the two expressions for average kinetic energy, Equations [10] and [11] of Objective 6-12, are combined and solved for $u_{\text{rms}} = \sqrt{\overline{u^2}}$, one obtains

$$u_{\text{rms}} = \sqrt{\overline{u^2}} = \sqrt{\frac{3kT}{m}} = \sqrt{\frac{3RT}{M}},$$

where m = molecular mass (kg/molecule), M = molar mass (kg/mole of molecules), and $R = 8.3145\ \text{J mol}^{-1}\ \text{K}^{-1}$, giving a velocity in m/s. (Note that the unit joule is equal to $\text{kg m}^2\ \text{s}^{-2}$.)

EXAMPLE 6-11 Determine the average velocity of a xenon molecule at 37°C (310 K)

$$u_{rms} = \sqrt{\frac{3 \times 8.3145\ \text{J mol}^{-1}\ \text{K}^{-1} \times 310\ \text{K}}{0.1313\ \text{kg/mol}}} = 243\ \text{m/s} \qquad [12]$$

Equation [12] is true for all ideal gases, specifically for two different gases at the same temperature:

$$u_{\text{rms, A}} = \sqrt{\frac{3RT}{M_\text{A}}} \qquad \text{AND} \qquad u_{\text{rms, B}} = \sqrt{\frac{3RT}{M_\text{B}}} \qquad [13]$$

When we solve the two expressions for the constant expression $\sqrt{3RT}$, we can rearrange the result:

$$\text{u}_{\text{rms, A}}\sqrt{\text{M}_\text{A}} = \text{u}_{\text{rms, B}}\sqrt{\text{M}_\text{B}} \qquad \text{REARRANGES TO} \qquad \frac{\text{u}_{\text{rms,A}}}{\text{u}_{\text{rms, B}}} = \sqrt{\frac{\text{M}_\text{B}}{\text{M}_\text{A}}} \qquad [14]$$

Equation [14], obtained from kinetic molecular theory, is similar to Graham's law, which was experimentally obtained. An example of Graham's law is the rate at which a helium-filled toy balloon deflates—far faster than a similar balloon filled with air. This occurs because the lower-molar-mass helium effuses more rapidly through the small pores in the balloon than the high-molar-mass oxygen and nitrogen of air.

EXAMPLE 6-12 How much faster than air ($M_\text{air} = 29.0\ \text{g/mol}$) does helium effuse through a pinhole?

$$\frac{\text{rate of He}}{\text{rate of air}} = \frac{u_\text{He}}{u_\text{air}} = \sqrt{\frac{M_\text{air}}{M_\text{He}}} = \sqrt{\frac{29.0\ \text{g/mol}}{4.00\ \text{g/mol}}} = 2.69$$

Helium effuses 2.69 times as fast as air.

18. Explain why real gases differ from ideal gases and how the differences lead to the van der Waals equation. Know under what conditions gases are most nearly ideal.

Two assumptions of the kinetic molecular theory clearly are incorrect: gas molecules *do* have volume and they *do* attract each other. If these assumptions were true, liquids would have no volume because the molecules composing them would be volumeless. Second, liquids would spontaneously and completely vaporize because no forces would act to hold the molecules together. The ideal gas equation can be modified to account for these two incorrect assumptions. The volume in the ideal gas equation is the *free volume*—the space within which molecules are free to move. In a real gas, the molecules take up, or exclude, some of the volume of the container.

free volume = container volume − excluded volume *or*

$$V_\text{free} = V - nb \quad (b = \text{excluded volume per mole of gas}) \qquad [15]$$

The ideal gas pressure assumes the molecules do not attract each other. But because there is an attraction, the molecules slow down as they approach the wall. They are attracted back into the body of the gas by other molecules. Hence, they strike the wall less hard and the measured pressure, P_{meas}, is less than the ideal pressure would be. The corrected pressure, P_{corr}, approximates what the ideal gas pressure would be:

$$P_{corr} = P_{meas} + \frac{an^2}{V^2} \qquad [16]$$

The constant a is multiplied by the square of the concentration of the gas, n/V The concentration is squared because molecule A attracts molecule B, but molecule B also attracts molecule A. The result of these two modifications (Equations [15] and [16]) is the van der Waals equation.

$$\left(P + \frac{an^2}{V^2}\right)(V - nb) = nRT \qquad [17]$$

The constants a and b principally depend on the gas being considered. To a small extent, they also depend on pressure and temperature. The volume correction ($-nb$) will be insignificant if it is small compared to the volume of the container. For a given quantity of the gas, the container volume is large when the pressure is low and the temperature is high. A large container volume also means that the concentration is low and, thus, the pressure correction, $(+an^2/V^2)$ will be small. Hence, a gas will behave most like an ideal gas at low pressure and high temperature.

EXAMPLE 6-13 At 273 K, 1.00 mol of an ideal gas confined to a 2.00-L container exerts a pressure of 11.2 atm. Under the same conditions, what pressure is exerted by CO_2, for which $a = 3.59\ \text{L}^2\ \text{atm mol}^{-2}$ and $b = 0.0427\ \text{L mol}^{-1}$?

Since 1 mole of gas is being considered, the value of n^2a is numerically the same as the value of a, and the value of nb is numerically the same as the value of b.

$$P = \frac{nRT}{V - nb} - \frac{n^2a}{V^2} = \frac{1.00\ \text{mol} \times \dfrac{0.08206\ \text{L atm}}{\text{mol K}} \times 273\ \text{K}}{(2.00 - 0.0427)\ \text{L}} - \frac{3.59\ \text{L}^2\ \text{atm}}{(2.00\ \text{L})^2} = 11.4_4\ \text{atm} - 0.898\ \text{atm}$$
$$= 10.5\ \text{atm}\ CO_2(g)$$

SELF-ASSESSMENT EXERCISES

114. (b) is correct. Answer (a) gives a pressure of 750 mm Hg, slightly less then 1 atmosphere. Answer (b) is 1 atm. For answer (c), the pressure, in mm Hg, is the height of the column times the ratio of the density of CCl_4 to Hg, giving a pressure of 500 mm Hg. Answer (d) would give a volume of 10 mL of H_2O. Even in a very small capillary, this would not give 1 atm of pressure.

115. (c) Decrease. 100.0 ^{0}C is 373 K. The temperature is decreasing and, therefore, the volume is also decreasing.

116. (c) SO_3. At STP, 1 mol of all gases will occupy the same volume. The density, mass/volume, will depend on the molecular mass of the gas. SO_3 has the largest molecular mass.

117. (c) The total pressure will be the sum of the partial pressures. Subtract the vapor pressure of water from the total: 751 mm Hg – 21 mm Hg = 731 mm Hg. Converting to atmospheres gives 0.96 atm.

118. (a) Using the ideal gas law (and temperature in K), the sample contains 0.10 mol of NH_3 or 6.02 x 10^{22} atoms. It also contains 1.7 g.

119. (b) To achieve the desired pressure, we need 0.20 mol of gas. Oxygen is diatomic, so we have only 0.05 mol (16.0 g/mol / 32.0 g = 0.05 mol). To determine the pressure, the identity of the gas doesn't matter. 0.06 g of He contains 0.15 mol. Added to the oxygen, this gives the desired pressure.

120. (a) The lighter gas diffuses faster.

121. (c) Gases behave more ideally at higher temperature and low pressure. Since T and P are directly related, the lower temperature is more than offset by the lower pressure.

122. To find the empirical formula, assure 100 g of compound. There are then 82.7 g C and 17.3 g H. Convert grams to mol. 82.7 g x $\frac{1 \text{ mol}}{12.01 \text{ g}}$ = 6.88 mol C and 17.33 g H x $\frac{1 \text{ mol}}{1.01 \text{ g}}$ = 17.16 mol. Taking the ratio of the number of moles 17.16/6.88 = 2.49.

There are 2.49 H for each C, but we must have whole numbers. Multiply by 2 so that the empirical formula is C_2H_5. Using the information given, assume 1.0 L of gas. This will contain 2.35 g. Using the ideal gas law ($PV = \frac{m}{M}RT$) and

rearranging, $M = \frac{mRT}{PV} = \frac{2.35 \text{ g} \times 0.082 \text{ L atm K}^{-1} \text{ mol}^{-1} \times 298 \text{ K}}{0.953 \text{ atm} \times 1.0 \text{ L}} = 59.0 \text{ g/mol}$

Divide the molar mass by the empirical mass. This gives the multiplier for the empirical formula. 59.0/29.01 ~ 2 and the molecular formula is C_4H_{10}.

Drill Problems

1. Fill in the blanks in each line that follows.

A. 30.0 atm = ___ lb./in.^{2} = ___ kPa
B. 142 mmHg = ___ lb./in.^{2} = ___ Pa
C. 242 lb./in.^{2} = ___ atm = kPa
D. 5.00×10^2 atm = ___ mmHg = ___ Pa
E. 5.00×10^2 mmHg = ___ atm = ___ Pa
F. 5.00×10^2 Pa = ___ atm = ___ mmHg
G. 5.00×10^2 lb./in.^{2} = ___ atm = ___ Pa
H. 3.52 kPa = ___ lb./in.^{2} = ___ mmHg

2. Fill in the blank in each line that follows. Temperature and amount are constant in each line.

P_i	V_i	P_f	V_f
A. 732 mmHg	6.49 L	1.02 atm	___ L
B. 30.0 mmHg	35.2 L	___ atm	35.2 mL
C. 9.75 Pa	1.46 L	923 mmHg	___ L
D. 104 kPa	22.4 L	___ atm	1.00 L
E. 247 kPa	0.972 L	143 mmHg	___ L
F. 1.43 atm	0.947 qt .	___ atm	4.06 qt.
G. 65.0 lb./in.^{2}	342 ft.^{3}	14.7 lb./in.^{2}	___ ft.^{3}
H. 175.5 kPa	12.4 m^3	___ kPa	8.46 m^3
I. 1.11 atm	85.2 L	0.506 atm	___ L
J. 4.11 atm	24.2 m^3	___ mmHg	100.0 m^3

3. Fill in the blanks in each line that follows. Both pressure and amount are constant in each line.

V_i	T_i	V_f	T_f
A. 6.49 L	273 K	___ L	94.2°C 9
B. 947 qt.	446 K	3.37 qt	___ K
C. 85.2 L	200. K	___ L	400. K
D. 24.2 m^3	37.0°C	19.4 m^3	___ °C
E. 22.4 L	273 K	___ L	298 K
F. 22.4 L	273 K	100.0 L	___ K
G. 14.7 L	50.6°C	___ L	300. K
H. 342 ft.^{3}	300. K	67.4 ft.^{3}	___ °C

4. Fill in the blank in each line that follows.

P	V	n	T
A. 1.23 atm	___ L	2.50 mol	298 K
B. 1.00 atm	22.4 L	1.46 mol	___ K
C. 2.64 atm	1.23 L	___ mol	143 K
D. ___ atm	9.61 L	8.41 mol	312 K
E. 746 mmHg	___ qt.	1.00 mol	87.6 °C
F. 1024 kPa	14.7 L	9.00 mol	___ °C

G. 5.17 atm	94.2 L	____mol	25.0 °C
H. 1726 kPa	____L	4.40 mol	37.0 °C
I.____mmHg	173 qt	1.26 mol	– 40.0 °C
J. 1.00 atm	____L	12.4 g CO_2	94.2 °C
K. 7.43 lb./in.^{2}	91.6 L	____g H_2O	293 °F
L. ____ lb./in.^{2}	123 gal	9.63 g Ne	31 °F
M. 1.46 kPa	84.1 m^3	143 g He	____°C
N. 9.47 atm	____L	22.0 g Xe	14.6 °C
O. 522 mmHg	19.2 L	____g CH_4	83.1 °C
P.____mmHg	245 L	19.6 g N_2	250.0 K

5. (1) Fill in the blank in each line that follows. All numbers have the units given in the column headings unless otherwise specified. All substances are ideal gases. The number of moles is the same for each line.

P_i, atm	V_i, L	T_i, K	P_f, atm	V_f, L	T_f, K
A. 1.42	7.26	274	3.74	2.16	____
B. 0.926	24.7	80.1 °C	1.42	____	481
C. 1.43	7.32 qt.	91.7 °C	____	3.17	260.
D. 4.60 mmHg	9.32	____	8.23	0.164	272. °C
E. 2.72	____	403	0.247	10.3	52.4 °F
F. ____	82.1 gal.	96.4 °C	1.79	47.2	294
G. 829 mmHg	500 mL	73.1 °F	0.725	1.42	____
H. 32.1	16.4	24.7 °C	7.15	____	578
I. 466 mmHg	8.83	98.6 °F	____	8.92 qt.	78.0 °F
J. 1776 mmHg	68.4	____	1086 mmHg	54.3	347
K. 5.13	____	671 °C	2.62	9.13	403
L. ____	1.34	72.4 °C	1.46	8.71	272

(2) Fill in the blank in each line that follows. Properties other than those specified are constant.

M. $T_i = 296\text{ K}$	$n_i = 14.3\text{ mol}$	$T_f = 123\text{ °C}$	$n_f =$ ____ mol
N. $P_i = 14.3\text{ atm}$	$n_i = 1.23\text{ mol}$	$P_f =$ ____ atm	$n_f = 2.16\text{ mol}$
O. $V_i = 22.4\text{ L}$	$n_i = 1.00\text{ mol}$	$V_f = 14.2\text{ L}$	$n_f =$ ____ mol
P. $T_i = 500.\text{ K}$	$n_i = 2.74\text{ mol}$	$T_f =$ ____ °C	$n_f = 1.43\text{ mol}$
Q. $P_i = 21.2\text{ atm}$	$n_i = 5.00\text{ mol}$	$P_f = 12.3\text{ atm}$	$n_f =$ ____ mol
R. $V_i = 19.4\text{ L}$	$n_i = 1.32\text{ mol}$	$V_f =$ ____ L	$n_f = 4.61\text{ mol}$

6. Fill in the blanks in each line. Numbers have the units given in the column headings unless otherwise specified. Each line represents a different situation.

	V, L	*P*, atm	*n*, mol	M, g/mol	*T*, K	*mass*, g	*density*, g/L
A.	25.0	2.00	____	____	27°C	____	1.25
B.	____	10.4	2.32	____	165	75.0	____
C.	72.1	46.8 mmHg	1.76	____	____	34.2	____
D.	34.6	____	0.832	17.4	−14°C	____	____
E.	10.4	1364 kPa	1.54	27.3	____	____	____
F.	134	72.1	____	79.2	279	____	____
G.	____	7.15	5.32	34.1	217	____	____
H.	____	3.12	1.91	____	36.0°C	76.4	____
I.	____	5.27	____	80.9	302	417	____
J.	____	____	2.73	75.2	463	____	2.14
K.	____	____	1.26	____	275	104	0.312
L.	____	____	____	44.0	350.	396	4.16
M.	____	34.6	____	____	46.3°C	4.76	54.0
N.	75.2	0.268	____	____	450.	____	0.178
O.	3.74	24.2	5.441	____	____	____	145
P.	7.42	1.74	____	8.42	____	____	0.874

7. The mass and the empirical formula of each gas used in part (1) of the drill problems of Objective 6–9 follow. They match, line-for-line. Determine the molecular formula of each gas.

A. 19.3 g CH_2	B. 101 g CHF	C. 61.5 g CF	D. 0.966 g NH_2
E. 9.16 g CH_2	F. 168 g O	G. 1.17 g CH	H. 991 g NO_2
I. 8.50 g C_2H_2N	J. 158 g CHO	K. 43.4 g CH_2O	L. 17.1 g CH_2O

8. (1) Balance each equation if necessary, then fill in the blanks. Gases are at the same *T* and *P* in each line.

$$C_5H_{12}(g) + O_2(g) \longrightarrow CO_2(g) + H_2O(g)$$

A. 7.41 L ____L ____L

B. ____L 87.9 L ____L

C. 6.75 L 73.1 L ____L ____L

$$N_2(g) + H_2(g) \longrightarrow NH_3(g)$$

D. 27.2 L ____L ____L

E. 8.63 L 25.2 L ____L

F. ____L ____L 21.4 L

$$6\,HCl(g) + 2\,HNO_3(g) \longrightarrow 4\,H_2O(g) + 2\,NO(g) + 3\,Cl_2(g)$$

G. 9.14 L ____L ____L

H.____L 2.17 L ____L ____L

I. 1.46 L 0.521 L ____ L ____ L

$$4\,NH_3(g) + 5\,O_2(g) \xrightarrow{Fe} 4\,NO(g) + 6\,H_2O(g)$$

J. 4.61 L ____ L ____ L

K. 36.0 L 42.1 L ____ L ____ L

L. ____ L 2.74 L ____ L

$$4\,NH_3(g) + 3\,O_2(g) \longrightarrow 2\,N_2(g) + 6\,H_2O(g)$$

M. ____ L ____ L 19.2 L

N. 24.9 L 18.6 L ____ L ____ L

O. ____ L ____ L ____ L 14.6 L

(2) A given quantity of a $C-H-O$ compound is burned in $O_2(g)$. Both the products and the reactants are at the pressure and the temperature specified. Fill in the blank in each line.

P. 14.2 g $C_2H_6(g) \longrightarrow$ ____ L $CO_2(g)$ 1.94 atm, 50. °C

Q. 42.1 L $C_3H_8(g)$ + 19.2 g $O_2 \longrightarrow$ ____ L H_2O 0.746 atm, 428 K

R. 7.07 g $C_2H_5OH(l)$ + ____ L $O_2(g) \longrightarrow$ ____ L $CO_2(g)$ 0.203 atm, 540. K

S. 1.43 L $C_3H_8(g)$ + ____ L $O_2(g) \longrightarrow$ ____ L $CO_2(g)$ 0.761 atm, 397 K

T. 8.16 L $H_2CO(g)$ + ____ g $O_2(g) \longrightarrow$
____ g $H_2O(l)$ + ____ L $CO_2(g)$ 0.331 atm, 100 °C

U. 1.43 g $C_6H_6(l)$ + 9.14 L $O_2(g) \longrightarrow$ ____ L $H_2O(g)$ 0.621 atm, 150. °C

V. 12.3 L $CH_4(g)$ + 123 L $O_2(g) \longrightarrow$ ____ L $CO_2(g)$ 0.714 atm, 200. °C

W. 1.13 g $C_2H_6O_2(l)$ + 4.37 L $O_2(g) \longrightarrow$ ____ L $CO_2(g)$ 0.506 atm, 350 K

X. 9.06 g $C_3H_8O(l) \longrightarrow$ ____ L $CO_2(g)$ 1.23 atm, 74°F

Y. 8.32 g $C_6H_{14}(l) \longrightarrow$ ____ L $CO_2(g)$ + ____ g $H_2O(l)$ 790 mmHg, 50. °C

9. Gas *A* is mixed with Gas *B* to produce the mixture of gases. All gases are ideal, and no chemical reactions occur. Fill in the blank in each line below. Each line represents a different situation.

	Gas *A*			Gas *B*			Mixture		
	P, atm	*V*, L	*T*, K	*P*, atm	*V*, L	*T*, K	*P*, atm	*V*, L	*T*, K
A.	1.04	27.2	298	0.96	21.2	304	1.00	____	298

B.	2.14	7.86	250.	1.32	9.42	273	______	11.6	260.
C.	1.21	8.17	308	1.41	16.2	308	1.31	25.0	______
D.	1.73	3.12	298	______	2.14	298	204	6.15	298
E.	1.96	74.2	401	1.96	4.01	______	1.96	80.0	374
F.	1.43	2.13	298	______	2.13	298	7.42	2.13	250.
G.	2.17	14.6	294	2.17	16.4	306	4.34	______	300.
H.	3.00	17.2	286	3.00	17.2	290	______	17.2	294
I.	1.62	9.46	271	2.61	8.46	271	1.23	17.9	______

10. Apply the kinetic molecular theory and Graham's law.

A. How fast is the average molecule of each of the following gases traveling at 300K?

$$\text{He, Ne}, H_2, N_2, O_2, CO_2, \text{Ar}, CH_2, NH_3, H_2O?$$

B. At what temperature is the average speed of each of the following molecules equal to 300. m/s?

$$\text{He, Ne}, H_2, N_2, O_2, CO_2, \text{Ar}, CH_4, NH_3, H_2O?$$

C. How much faster does He effuse than N_2 at 300 K?

D. How much faster does N_2 effuse than O_2 at 1.50 atm?

E. If 1.25 mol He effuses in 8.00 hr, how long is needed for 1.25 mol of Ar to effuse? (Note: Both gases are at 74°C.)

F. If 0.230 mol N_2 effuses in 12.4 hr, how long is needed for 12.4 mol of H_2O to effuse? (Note: Both gases are at 4.71 atm.)

QUIZZES (20 minutes each) Choose the best answer for each question.

QUIZ A

1. STP refers to which one of the following conditions? (a) 1.00 atm, undefined temperature; (b) 1.00 atm, 0 °C; (c) 760 mmHg, 298 K; (d) 100 mmHg, 0 °C; (e) none of these.

2. The kinetic energy of a gas is directly proportional to its (a) temperature; (b) pressure; (c) volume; (d) composition; (e) none of these.

3. A mathematical statement of Boyle's law is (a) $V/T = \text{constant}$; (b) $V/n = \text{constant}$; (c) $P/T = \text{constant}$; (d) $PV = \text{constant}$; (e) none of these.

4. At 1000 °C, the pressure of a gas is 5020 kPa. What is its pressure in kPa at 3450 °C? (a) 5020(1273/3723); (b) 5020(3723/1273); (c) 5020(3450/1000); (d) 5020(1000/3450); (e) none of these.

5. At 850 kPa and 252 K, a gas occupies 17.9 mL. What is its Kelvin temperature at 722 kPa and a volume of 625 mL? (a) 252(722/850)(625/17.9); (b) 252(722/850)(17.9/625); (c) 252(850/722)(17.9/625); (d) 252(850/722)(625/17.9); (e) none of these.

6. Into a 6.00-L container are placed 4.0 L of argon at 600 mmHg and 2.0 L of xenon at 300 mmHg. The final pressure is (a) 3000 mmHg; (b) 480 mmHg; (c) 500 mmHg; (d) 450 mmHg; (e) none of these.

7. At what Kelvin temperature does 2.30 mol of gas occupy a volume of 62 mL at 6.14 atm? (a)(6.14)(62)/[(2.30)(0.0821)]; (b) (6.14)(2.30)(62)/(0.0821); (c) (6.14)(0.062)/[(2.30)(0.0821)]; (d) (6.14)(0.062)(0.0821)/(62); (e) none of these.

8. 85.1 g of gas occupies 120 L at 0.224 atm and $450\,°C$. Its mole weight is (a) 85.1(0.0821/0.224)(450/120); (b) 85.1(0.224/0.0821)(120/450); (c) 85.1(0.224/0.0821)(120/723); (d) 85.1(0.0821/0.224)(120/723); (e) none of these.

9. In the reaction $4\,NH_3(g) + Cl_2(g) \longrightarrow N_2H_4(g) + 2\,NH_4Cl(s)$, how many liters of ammonia will combine with 14.0 L of chlorine? (a) (14.0/22.4); (b) (14.0)(22.4)(4); (c) (14.0)(4); (d) (14.0/22.4)(1/4); (e) none of these.

10. A gas has a temperature of 200 K. What is its Kelvin temperature if the average kinetic energy of the molecules doubles? (a) $\sqrt{2}(200\,K)$; (b) 200 K; (c) (2)(200 K); (d) (3/2)(200 K); (e) none of these.

QUIZ B

1. The postulates of the kinetic molecular theory of gases include all those that follow *except* (a) no forces exist between molecules; (b) molecules are point masses; (c) molecules are repelled by the walls of the container; (d) molecules are in random motion; (e) all are postulates.

2. The mutual attraction of gas molecules is an important aspect of (a) Avogadro's law; (b) Dalton's law; (c) Graham's law; (d) van der Waals's theory; (e) none of these.

3. Subtracting the vapor pressure of water from the total pressure of a gas collected over water is an example of the application of (a) Avogadro's law; (b) Dalton's law; (c) Graham's law; (d) van der Waals's theory; (e) none of these.

4. A gas's volume is 16.0 L at $37\,°C$. What is its volume in liters at $82\,°C$? (a) 16.0(82/37); (b) 16.0(37/82); (c) 16.0(310/355); (d) 16.0(355/310); (e) none of these.

5. At 1.70 atm and $27\,°C$, a gas occupies 14.0 L. What is its volume in L at 1.05 atm and $-3\,°C$? (a) 14.0(1.70/1.05)(270/300); (b) 14.0(1.05/1.70)(270/300); (c) 14.0(1.05/1.70)(300/270); (d) 14.0(1.70/1.05)(300/270); (e) none of these.

6. Into a 3.0 L-container are placed 1.0 L of hydrogen at 700 kPa and 2.0 L of nitrogen at 1000 kPa. The final pressure is (a) 900 kPa; (b) 800 kPa; (c) 850 kPa; (d) 750 kPa; (e) none of these.

7. How many moles of gas occupy 3.14 L at 272 K and 0.112 atm? (a) (0.112)(272)/[(0.0821)(3.14)]; (b) (0.0821)(272)/[(0.112)(3.14)]; (c) (0.112)(3.14)/[(0.0821)(272)]; (d) (0.0821)(3.14)/[(0.112)(272)]; (e) none of these.

8. 22.4 g of gas occupies 15.2 L at 77 °C and 3.21 atm. Its mole weight is (a) 22.4(0.0821/3.21)(350/15.2); (b) 22.4(0.0821/3.21)(77/15.2); (c) 22.4(0.0821/3.21)(15.2/350); (d) 22.4(3.21/0.0821)(350/15.2); (e) none of these.

9. In the reaction $2\,H_2S(g) + 3\,O_2(g) \longrightarrow 2\,H_2O(l) + 2\,SO_2(g)$, how many liters of $O_2(g)$ are needed to produce 16.2 L of SO_2? (a) (16.2)(2/3); (b) (16.2)(3/2); (c) 16.2; (d) (16.2)(22.4)(3/2); (e) none of these;

10. The rate of effusion of one gas (38 g/mol) is 3.14 L/s. What is the rate in L/s for a second gas (28 g/mol) (a) 3.14(38/28); (b) 3.14 (28/38); (c) $3.14\sqrt{38/28}$; (d) $3.14\sqrt{28/38}$; (e) none of these.

QUIZ C

1. If the Kelvin temperature of a gas is doubled, then (a) the average molecule moves twice as fast; (b) every molecule moves twice as fast; (c) every molecule has twice the kinetic energy; (d) the average molecule has twice the kinetic energy; (e) none of these.

2. The ideal gas constant has the value of 0.08206. The units of this number are (a) $\mathrm{L\,atm\,mol^{-1}\,K^{-1}}$; (b) $\mathrm{L\,mmHg\,mol^{-1}\,K^{-1}}$; (c) $\mathrm{mL\,mmHg\,mol^{-1}\,K^{-1}}$; (d) $\mathrm{mL\,atm\,mol^{-1}\,K^{-1}}$; (e) none of these.

3. $PV = nRT$ is a statement of (a) Avogadro's law; (b) Dalton's law; (c) Graham's law; (d) van der Waals's theory; (e) none of these.

4. At 5.12 atm, the temperature of a gas is 47 °C. What is the Celsius temperature at 7.14 atm? (a) $320(7.14/5.12)-273$; (b) $320(5.12/7.14)-273$; (c) $47(5.12/7.14)-273$; (d) $47(7.14/5.12)+273$; (e) none of these.

5. At 250 mmHg and 150 °C, a gas occupies 75.2 L. What is its volume in liters at 520 mmHg and 275 °C? (a) 75.2(250/520)(275/150); (b) 75.2(250/520)(548/423); (c) 75.2(520/250)(548/423); (d) 75.2(520/250)(275/150); (e) none of these.

6. Into a 5.0-L container are placed 3.0 L of neon at 640 kPa and 2.0 L of xenon at 250 kPa. The final pressure is (a) 445 kPa; (b) 484 kPa (c) 2420 kPa; (d) 406 kPa; (e) none of these.

7. 2.24 mol of gas exerts what pressure in atm at 25 °C and a volume of 72 L? (a) (2.24)(0.0821)(25/72); (b) (2.24)(0.0821)(72/298); (c) (2.24)(72)(298/0.0821); (d) (25)(72)(0.0821/2.24); (e) none of these.

8. 22.4 g of gas occupies 22.4 L at 273 K and 2.24 atm. Its mole weight is (a) 22.4; (b) 22.4(2.24/0.0821)(273/22.4); (c) 22.4(0.0821/2.24)(273/22.4); (d) 22.4(0.0821/2.24)(546/22.4); (e) none of these.

9. In the reaction $3\,H_2S(g) + 2\,HNO_3 \longrightarrow 3\,S + 2\,NO(g) + 4\,H_2O$, how many liters of $H_2S(g)$ are needed to produce 75.1 L of NO(g)? (a) 75.1(2/3); (b) 75.1(3/2); (c) 75.1/22.4)(3/2); (d) 75.1; (e) none of these.

10. 3.12 mol of Ne (20.2 g/mol) effuses through a pinhole in 14.2 min. How many minutes are needed for 3.12 mol of methane (CH_4, 16.0 g/mol) to effuse through the same pinhole? (a) 14.2; (b) 14.2(20.2/16.0); (c) $14.2\sqrt{20.2/16.0}$; (d) $14.2\sqrt{16.0/20.2}$; (e) none of these.

Quiz D

1. Pressure is the result of (a) molecular speed only; (b) molecular speed and mass; (c) molecular mass only; (d) repulsive forces between molecules; (e) attractive forces between molecules.

2. Gases approach ideal behavior most closely under what set of conditions? (a) high temperature, low pressure; (b) low temperature, low pressure; (c) high temperature, high pressure; (d) low temperature, high pressure; (e) no such simple statement can be made.

3. Equal volume of two gases at the same temperature and pressure have (a) equal masses; (b) Avogadro's number of molecules; (c) identical chemical compositions; (d) the same number of molecules; (e) none of these.

4. At 946 mmHg, the volume of a gas is 2.54 L. What is its volume in liters at 454 mmHg? (a) 2.54(454/946)(760); (b) 2.54(454/946); (c) 2.54(946/454)(760); (d) 2.54(946/454); (e) none of these.

5. At 8.59 atm and $25\,°C$, a gas occupies 16.2 L. What is its pressure in atm at 421 K and a volume of 14.0 L? (a) 8.59(421/25)(16.2/14.0); (b) 8.59(25/421)(14.0/16.2); (c) 8.59(298/421)(14.0/16.2); (d) 8.59(421/298)(16.2/14.0); (e) none of these.

6. Into a 5.0-L container are placed 2.0 L of argon at 400 kPa and 3.0 L of neon at 600 kPa. The final pressure is (a) 500 kPa; (b) 520 kPa; (c) 580 kPa; (d) 2600 kPa; (e) none of these.

7. 0.82 mol of gas has what pressure in mmHg at $23\,°C$ in a volume of 72 L? (a) (0.82)(0.0821)(296)(760/72); (b) (0.82)(0.0821)(296/72); (c) (0.82)(0.0821)(72/296); (d) (0.82)(0.0821)(296)/[(72)(760)]; (e) none of these.

8. 374 g of gas occupies 82.4 L at 12.4 atm and 750 K. Its mole weight is (a) 374(0.0821/12.4)(750/82.4); (b) 374(0.0821/12.4)(82.4/750); (c) 374(12.4/0.0821)(82.4/750); (d) 374(12.4/0.0821)(750/82.4); (e) none of these.

9. In the reaction $4\,NH_3(g) + 5\,O_2(g) \longrightarrow 4\,NO(g) + 6\,H_2O$, how many milliliters of O_2 are needed to form 200 mL of NO? (a) 200(4/5); (b) 200(4/6); (c) 200(5/6); (d) 200(6/4); (e) none of these.

10. 2.45 mL of methane (CH_4, 16.0 g/mol) effuses through a pinhole in 1.25 min. What is the molar mass (in g/mol) of a second gas if 2.45 mL of it (measured at the same *T*

and *P*) requires 2.61 min to effuse through the same pinhole? (a) $16.0(2.61/1.25)^2$; (b) 16.0(2.61/1.25); (c) $16.0\sqrt{1.25/2.61}$; (d) $16.0\sqrt{2.61/1.25}$; (e) none of these.

SAMPLE TEST (20 minutes)

1. Convert the units of each of the following properties as indicated.

   A. Pressure: 26.4 lb./in.^{2} = _____ atm = _____ kPa

   B. Volume: 2.0 ft.^{3} = _____ m^3 = _____ L

   C. Temperature: 92.0 °F = _____ °C = _____ K

2. A gas that has the pressure, volume, and temperature given in the previous problem has a mass of 1.305 lb. What is the mole weight, in g/mol, of this gas?

3. What is the numerical value of the ideal gas constant *R* when it has the units mL mmHg mmol^{-1}K^{-1}?

4. Two gases are to be mixed into a 5.000-L container at 291.0 K. Gas *A* originally is confined in 14.20 L at 1.067 atm and 303.1 K. Gas *B* originally is confined in 1.251 L at 26.42 atm and 327.5 K. Once the two gases are mixed in the 5.000-L container,

   A. What is the total pressure?

   B. What is the partial pressure of Gas *A*?

   C. What is the partial pressure of Gas *B*?

5. Precisely 1 mol of helium and 1 mole of neon are mixed in a container.

   A. Which gas has the larger mass?

   B. Which gas has the greater average molecular speed?

   C. Which type of molecule strikes the wall of the container more frequently?

   D. Which gas exerts the larger pressure?

7 THERMOCHEMISTRY

CHAPTER OBJECTIVES

1. Distinguish between heat and work.

Work moves an object through a distance and *heat* raises its temperature. Although work can be transformed completely into heat, heat cannot be transformed completely into work. Another difference between the two is that work is an organized form of energy and can move large objects. For instance, the work done by the expansion of a gas against a fixed pressure is given by: $\text{work} = P_{\text{ext}}(V_{\text{final}} - V_{\text{initial}})$. Heat is more random and mainly is able to make small particles—atoms and molecules—move more rapidly. Both heat and work appear only during energy changes. Also note that we cannot speak of absolute energy, but rather of energy changes.

* **2. Use specific heat to determine temperature changes and quantities of heat.**

Specific heat is the quantity of heat needed to raise the temperature of 1 gram of substance by 1 degree Celsius. Specific heat depends on the nature of the substance being heated. The specific heat of water, for example, is greater than that of iron. Specific heat has the units of joules per gram per degree Celsius ($\text{J g}^{-1}\,{}^\circ\text{C}^{-1}$) [or sometimes calories per gram per degree Celsius ($\text{cal g}^{-1}\,{}^\circ\text{C}^{-1}$)]. The quantity of heat absorbed by an object that undergoes a temperature change is given by Equation [1].

heat absorbed = (mass)(specific heat)(temperature change)

$$q = (m)(\text{sp ht})(\Delta T) \qquad [1]$$

The molar heat capacity (C_p) is the quantity of heat needed to raise the temperature of one mole of a substance by one degree Celsius. C_p has the units of $\text{J mol}^{-1}\,{}^\circ\text{C}^{-1}$ or $\text{cal }{}^\circ\text{C}^{-1}$. Thus the heat absorbed by an object also is given by equation [2].

$$q = n\,C_p\Delta T \quad (n = \text{amount of substance in moles}) \qquad [2]$$

There is a standard convention for using the symbol Δ (delta). It represents the final value minus the initial value of a property: $\Delta T = T_f - T_i, \Delta n = n_f, \Delta V = V_f - V_i$, and so forth.

EXAMPLE 7-1 What quantity of heat is needed to raise the temperature of 500.0 g Cu (for which $C_p = 24.4\ \text{J mol}^{-1}\,{}^\circ\text{C}^{-1}$) from 20.0 °C to 50.0 °C?

$$\text{heat} = n\,C_p\Delta T = 500.0\,\text{g} \times \frac{1\,\text{mol Cu}}{63.54\,\text{g}} \times 24.4\frac{\text{J}}{\text{mol}\,°\text{C}}(50.0\,°\text{C} - 20.0\,°\text{C})$$

$$= 5.78 \times 10^3\ \text{J} = 5.78\,\text{kJ}$$

One can combine Equations [1] or [2] with the law of conservation of energy to determine either specific heats of materials or final temperatures of mixtures.

EXAMPLE 7-2 $150.0\,\text{g H}_2\text{O}$ (sp ht = $4.18\,\text{J g}^{-1}\,°\text{C}^{-1}$) at $20.0\,°\text{C}$ is mixed with 100.0 g of vanadium (sp ht = $0.489\,\text{J g}^{-1}\,°\text{C}^{-1}$) at $100.0\,°\text{C}$. What is the final temperature of the mixture?

We use the law of conservation of energy first. (The negative sign is inserted before the heat gained by vanadium because the metal actually loses heat in going to a lower temperature.)

$$\text{-heat gained by V} = \text{heat gained by H}_2\text{O}$$

$$-(100.0\,\text{g})(0.489\frac{\text{J}}{\text{g}\,°\text{C}})(T_f - 100.0\,°\text{C}) = (150.0\,\text{g})(4.18\frac{\text{J}}{\text{g}\,°\text{C}})(T_f - 20.0\,°\text{C})$$

$$-48.9T_f + 48.9\times 10^3 = 627\,T_f - 1.25\times 10^4 \text{ OR } 1.75\times 10^4 = 676\,T_f \text{ OR } T_f = 26.0\,°\text{C}$$

EXAMPLE 7-3 $140.0\,\text{g H}_2\text{O}$ at $25.0\,°\text{C}$ is mixed with 100.0 g of metal at $100.0\,°\text{C}$. The final temperature of the mixture is $29.6\,°\text{C}$. What is the specific heat of the metal?

$$-\text{heat gained by the metal} = \text{heat lost by the water}$$

$$-(100.0\,\text{g})(\text{sp ht})(29.6\,°\text{C} - 100.0°\text{C}) = (140.0\,\text{g})(4.18\,\text{J g}^{-1}\,°\text{C}^{-1})(29.6\,°\text{C} - 25.0\,°\text{C})$$

$$(7040\,\text{g}\,°\text{C})(\text{sp ht}) = 2692\ \text{J} \quad \text{OR} \quad \text{sp ht} = 0.38\ \text{J g}^{-1}\,°\text{C}^{-1}$$

3. **Interconvert joules and calories. Apply the first law of thermodynamics:** $\Delta u = q + w$.

Both the English and the metric systems have two different units of energy—one for heat and the other for work.

calorie: the heat needed to raise the temperature of one gram of water from $14.5\,°\text{C}$ to $15.5\,°\text{C}$. (The Calorie, written with a capital C, is a nutritional Calorie, equal to 1000 calories.)

Btu (British thermal unit): the heat needed to raise the temperature of 1 pound of water 1 degree Fahrenheit at or near $39.1\,°\text{F}$.

Joule: the work done when a force of 1 newton acts through a distance of 1 meter.

foot-pound: the work done when a force of 1 pound acts through a distance of 1 foot.

The calorie now is defined in terms of an SI unit: one calorie = 4.184 Joule. Use of the calorie will be phased out.

EXAMPLE 7-4 Determine the number of (a) Joules in 267.2 cal; (b) calories in 794.2 J.

The conversion factor method is used in each part.

(a) $265.2\,\text{cal} \times (4.184\,\text{J/cal}) = 1117\,\text{J}$ (b) $791.2\,\text{J} \times (\text{cal}/4.184\,\text{J}) = 189.1\,\text{cal}$

The ability to phrase heat and work in the same units is needed when we use the first law of thermodynamics: $\Delta u = q + w$. In this statement of the first law, both heat and work represent energy flowing into the system. q is the heat absorbed by the system, and w is the work done on the system. The first law can be thought of as an energy bookkeeping equation. It operates in a fashion similar to determining the assets of a business, such as a gold bullion merchant. The assets may be in either cash (heat) or gold (work). The two assets may be interconverted, and either cash or gold may be lost or gained by the merchant. The comparison fails in that the "rate of exchange" between heat and work is fixed (whereas that between gold and cash varies daily). Furthermore, in energy systems, we do not know the value of u (the total assets); we can only determine its change.

EXAMPLE 7-5 A system absorbs 189.1 cal of heat and does 1117 J of work. What is the value of Δu for this system, in J?

The energy values are those used in Example 7-4. In this case, we must be careful to note that work is negative; work done by the system is energy that flows out of the system (it is negative).

$\Delta u = q + w = 791.2\,\text{J} - 1117\,\text{J} = -326\,\text{J}$

4. **Know the definition of *work* and calculate pressure-volume work for a gas.**

Work is force x distance. Pressure-volume work is work done as a result of a volume change in the system. If the system expands, the system does work on the surroundings and the energy of the system decreases.

Example 7-6. How much work is done when 1.00 mol of H_2 gas is compressed from 1.50 L to 0.80 L at a temperature of 298 K?

Solution: We are given the number of moles, temperature, and volumes for the initial and final states. The work is $-P\Delta V$, but we need to determine the final pressure and to make sure the sign of the work is consistent with the energy gain or loss.

$$P_{\text{Final}} = \frac{nRT}{V_{\text{Final}}} = \frac{0.0821\,\text{L atm mol}^{-1}\,\text{K}^{-1} \times 1.00\,\text{mol} \times 298\,\text{K}}{0.80\,\text{L}}$$
$$= 30.6\,\text{atm}$$

Work is then:

$w = -30.6\,\text{atm}\,(0.80\,\text{L} - 1.50\,\text{L})$ (note V_{Final} is smaller than V_{Initial})

$= 21\,\text{L atm} \times 101\,\text{J/1 Latm} = \mathbf{2.1 \times 10^3\,J}$

Work is a positive number in this case, so the energy of the system is increasing. Another way to look at it is that work had to be done on the system to compress the gas; therefore, the energy of the gas must increase.

5. **State the meaning of the concept *state function*, especially as demonstrated by enthalpy.**

The concept of a state function is one of the most powerful in thermodynamics. A state function, or state property, only depends on where the system starts (its initial state) and where it ends (its final state). A state function does *not* depend on how the change in the system occurred (the pathway). For example, the distance from Los Angeles to New York is a state function. Yet the actual mileage traveled depends on the route. Thus, the mileage traveled is a path function. Both heat and work are path functions; they can become state functions when the path is specified. An important path is one of constant pressure, a condition frequently encountered in the laboratory. The heat evolved at constant pressure equals a state function. It is called the *change in enthalpy of the system.*

*** 6. Calculate the heat of a reaction at constant volume, q_v, from bomb calorimetry data.**

The heat given off by a reaction occurring in a bomb calorimeter increases the temperature of both the bomb and the water that surrounds it.

$$q_v = -\text{heat gained by water} - \text{heat gained by bomb} = -(\text{mass } H_2O)(\text{sp ht})(\Delta T) - (C_B)(\Delta T) \quad [3]$$

C_B is the heat capacity of the bomb—the quantity of heat needed to raise the temperature of the bomb by 1 °C. The change in the state function, internal energy ($\Delta E = q_v$), is the heat produced at constant volume. Notice that a path function (heat) has been restricted by specifying its pathway (one of constant volume) so that it now becomes a state function.

Most often, combustion reactions are those run in a bomb calorimeter. In fact, there are extensive tabulations of heats of combustion. In order to relate these to enthalpy changes, you need to know the products of the combustion reaction. Some of the usual conventions are that $CO_2(g)$ is formed from compounds that contain carbon, $H_2O(l)$ from those containing hydrogen, $N_2(g)$ from those containing mitrogen, and $SO_2(g)$ from those containing sulfur. In addition, sometimes heats of combustion are tabulated as positive numbers, even though the enthalpy change is negative (most combustion reactions are exothermic). Consult the legend of the table you are using to see what conventions are used.

EXAMPLE 7-7 A calorimeter surrounded by 925 g of water initially at 25.00 °C has burned in it 0.640 g of a substance whose heat of combustion is 31.48 kJ/g. The temperature increases to 28.87 °C. What is the heat capacity of the calorimeter, C_B?

Part of the heat produced in the combustion is absorbed by the water, the rest by the calorimeter. Heat produced by combustion = heat absorbed by water + heat absorbed by calorimeter.

$$0.640\text{ g} \times 31.48 \times 10^3\text{ J/g} = 2.01 \times 10^4\text{ J}$$

$$= 925\text{ g} \times 4.18\text{ J g}^{-1}\text{ °C}^{-1}\ (28.87 - 25.00)\text{ °C} + C_B\Delta T$$

$$= 1.50 \times 10^4\text{ J} + C_B\Delta T$$

$$C_B = \frac{2.01 \times 10^4\text{ J} - 1.50 \times 10^4\text{ J}}{28.87\text{ °C} - 25.00\text{ °C}} = 1.3\text{ kJ/°C}$$

7. Explain how to use a Styrofoam coffee cup calorimeter and interpret the data obtained.

A nest of Styrofoam coffee cups with a cardboard lid is almost a perfectly adiabatic container—very little heat is lost or gained by the system inside the cups. All the heat produced or absorbed by any process (including a chemical reaction) occurring in the calorimeter produces a temperature change of the system. The process occurs at constant pressure and, thus, we measure $q_p = \Delta H$.

EXAMPLE **7-8** 50.0 mL each of NaOH(aq) and HCl(aq), both at a concentration of 1.00 M and a temperature of 20.5 °C, are mixed in a Styrofoam cup calorimeter. The final temperature of the system is 27.3 °C. The resulting 0.500 M NaCl(aq) has a density of 1.02 g/mL and a specific heat of $4.02\ \text{J g}^{-1}\ °\text{C}^{-1}$. Compute ΔH in kJ/mol for the reaction:

$$\text{NaOH(aq)} + \text{HCl(aq)} \longrightarrow \text{NaCl(aq)} + \text{H}_2\text{O(l)}$$

$$q_p = [(1.02\ \text{g/mL})(100.0\ \text{mL})](4.02\ \text{J g}^{-1}\ °\text{C}^{-1})(27.3\ °\text{C} - 20.5\ °\text{C}) = 2.79 \times 10^3\ \text{J}$$

This is the heat absorbed by the solution. The heat produced by the reaction is the negative of this value:

$$q_{\text{p,rxn}} = -2.79 \times 10^3\ \text{J}$$

We find the amount of NaCl:

$$\text{amount NaCl} = 0.1000\ \text{L soln} \times \frac{0.500\ \text{mol NaCl}}{1\ \text{L soln}} = 0.0500\ \text{mol NaCl}$$

Finally, we compute ΔH:

$$\Delta H = \frac{q_{\text{p rxn}}}{\text{amount NaCl}} = \frac{-2.79 \times 10^3\ \text{J}}{0.0500\ \text{mol NaCl}} = -55.7 \times 10^3\ \text{J/mol} = -55.7\ \text{kJ/mol}$$

***8. Apply Hess's law of constant heat summation.**

It is not necessary to actually *run* a chemical reaction in order to determine how much heat it will produce or consume. We merely combine (on paper) the heats of a series of consecutive reactions whose overall result is the same as the reaction we wish to study. On paper, we start with the same reactants (initial state) and end with the same products (final state), but we use (perhaps) a different pathway.

EXAMPLE **7–9** What is $\Delta H°$ for the reaction C(graphite) + 2 S(rhombic) $\longrightarrow \text{CS}_2\text{(l)}$?

The following thermochemical reactions are provided:

(a) $\text{C(graphite)} + \text{O}_2\text{(g)} \longrightarrow \text{CO}_2\text{(g)} \quad \Delta H° = -393.5\ \text{kJ/mol}$

(b) $\text{S(rhombic)} + \text{O}_2\text{(g)} \longrightarrow \text{SO}_2\text{(g)} \quad \Delta H° = -296.9\ \text{kJ/mol}$

(c) $\text{CS}_2\text{(l)} + 3\,\text{O}_2\text{(g)} \longrightarrow \text{CO}_2\text{(g)} + 2\,\text{SO}_2\text{(g)} \quad \Delta H° = -1075.2\ \text{kJ/mol}$

We combine the given reactions: $(a)+2(b)-(c)$. Whatever we do to the reactions (multiply them by a constant or change their sign [reverse them]), we also do to the associated $\Delta H°$ values:

(a)	$C(graphite) + O_2(g) \longrightarrow CO_2(g)$	$\Delta H° = -393.5$ kJ/mol
2(b)	$2\,S(rhombic) + 2\,O_2(g) \longrightarrow 2\,SO_2(g)$	$\Delta H° = -593.8$ kJ/mol
−(c)	$CO_2(g) + 2\,SO_2(g) \longrightarrow CS_2(l) + 3\,O_2(g)$	$\Delta H° = +1075.2$ kJ/mol
$C(graphite) + 2\,S(rhombic) \longrightarrow CS_2(l)$		$\Delta H° = +87.9$ kJ/mol

*** 9. State the definitions of *standard state* and *standard formation reaction*, and write the standard formation reaction for any substance.**

To use Hess's law, all reactants and products must be under the same conditions. For example, consider the $CO_2(g)$ produced by reaction (a) in Objective 7–8 and consumed by reaction. If reaction (a) produces $CO_2(g)$ at 298 K and 1.00 atm and reaction consumes $-(c)$ at 323 K and 4.00 atm, then energy is needed to heat and compress the gas. On the other hand, if both reactants and products are in a well-defined state at one temperature, no energy is needed for heating and compression (or produced by cooling and expansion). The *standard state* is a pressure of 1 atmosphere or, for a solute in solution, a concentration of 1 mole per liter of solution. The degree sign (°) on $\Delta H°$ indicates that the molar enthalpy change is measured with reactants and products in their standard states. Note that the standard state does not mention the temperature. Temperature should be written as a subscript in kelvins, ΔH°_{298} . Often, however, the subscript is omitted and 298 K is assumed.

The standard state also does not mention anything about the form of a substance. We can correctly refer to the standard state of each of the 2 allotropes of oxygen, $O_2(g)$ and $O_3(g)$, or ozone. Also, we can refer to the standard state of steam or water vapor, $H_2O(g)$; of ice, $H_2O(s)$; and of liquid water, $H_2O(l)$. It is most helpful to speak of all substances as being formed from one set of reactants. Since all matter is composed of atoms, we pick the *most stable form* (the one of lowest energy) of each element. These ideas are contained in the concept of the standard formation reaction: the reaction that produces 1 mole of a substance in its standard state from the elements in their standard states and most stable forms. The stable forms of the elements at 25 °C (298.15 K) are given in Table 7-1. The standard formation reactions for $C_2H_5OH(l)$, $P_2O_5(s)$, and $CaCO_3(s)$ follow:

$$2\,C(graphite) + 3\,H_2(g) + \tfrac{1}{2}O_2 \longrightarrow C_2H_5OH(l)$$

$$2\ P(s, white) + \tfrac{5}{2}O_2(g) \longrightarrow P_2O_5(s)$$

$$Ca(s) + C(graphite) + \tfrac{1}{2}O_2(g) \longrightarrow CaCO_3(s)$$

TABLE 7-1 Most Stable Forms of Elements at 25°C (*in periodic table arrangement*)

						$H_2(g)$	He(g)
			C(graphite)	$N_2(g)$	$O_2(g)$	$F_2(g)$	Ne(g)
			...	P_4(s, white)	S_8(s, rhombic)	$F_2(g)$	Ar(g)
...	Mn(c, α)	...	...	As(s, gray)	Se(s, gray)	$Br_2(l)$	Kr(g)
...	...	Cd(c, α)	Sn(s, white)	Sb(c, III)	...	$I_2(s)$	Xe(g)
U(c,III)	...	Hg(l)	ALL OTHERS—crystalline solid: (c) or (s)				

The standard molar enthalpy of formation is the heat evolved at constant pressure (the enthalpy change) when the standard formation reaction occurs. But what if the reaction does not occur as written? We already have answered this question in Example 7-7. The reaction

$$\text{C(graphite)} + 2\,\text{S(rhombic)} \longrightarrow CS_2(l)$$

does not occur, but it is the formation reaction for $CS_2(l)$. In Example 7-7, we determined ΔH_f°, the standard enthalpy of formation for $CS_2(l)$, by combining the ΔH° values of several reactions that do occur.

*** 10. Apply Hess's law in the special case of standard formation reactions; that is, compute ΔH_{rxn}° from ΔH_f° values.**

Standard formation reactions greatly simplify the application of Hess's law.

EXAMPLE 7–10 What is ΔH_{rxn}° in kJ/mol for the oxidation of ammonia in the presence of a platinum catalyst?

$$4\,NH_3(g) + 5\,O_2(g) \xrightarrow{Pt} 4\,NO(g) + 6\,H_2O(g)$$

$$\Delta H_f^\circ \quad -46.11 \qquad 0.00 \qquad\qquad +90.25 \quad -285.8 \text{ kJ/mol}$$

The ΔH_f° values that are given are equivalent to the following thermochemical equations:

(*d*) $\frac{1}{2}N_2(g) + \frac{3}{2}H_2(g) \longrightarrow NH_3 \qquad \Delta H^\circ = -46.11\text{ kJ/mol}$

(*e*) $O_2(g) \longrightarrow O_2(g) \qquad \Delta H^\circ = 0.00\text{ kJ/mol}$

(*f*) $\frac{1}{2}N_2(g) + \frac{1}{2}O_2(g) \longrightarrow NO(g) \qquad \Delta H^\circ = +90.25\text{ kJ/mol}$

(*g*) $\frac{1}{2}O_2(g) + H_2(g) \longrightarrow H_2O(l) \qquad \Delta H^\circ = -285.8\text{ kJ/mol}$

Notice that ΔH_f° for $O_2(g)$ is 0.00 kJ/mol since the product of the formation reaction is an element in its most stable form. [In contrast, $\Delta H_f^\circ = 142.3\text{ kJ/mol}$ for ozone,

$O_3(g)$.] Equations *(d)* through *(g)* can be combined to yield the equation for the catalytic oxidation of ammonia: $-4(d)-5(e)+4(f)+6(g)$.

$-4\ (d)$	$4\,NH_3(g) \longrightarrow 2\,N_2(g) + 6\,H_2(g)$	$\Delta H° = +184.4\ kJ/mol$
$-5\ (e)$	$5\ O_2(g) \longrightarrow 5\,O_2(g)$	$\Delta H° = \quad 0.00\ kJ/mol$
$+4\ (f)$	$2\ N_2(g) + 2\,O_2(g) \longrightarrow 4\,NO(g)$	$\Delta H° = +361.0\ kJ/mol$
$+6\ (g)$	$3\ O_2\ (g) + 6\,H_2(g) \longrightarrow 3\,H_2O(l)$	$\Delta H° = -1714.8\ kJ/mol$
	$4\,NH_3(g) + 5\,O_2(g) \longrightarrow 4\,NO(g) + 6\,H_2O(l)$	$\Delta H° = -1714.8\ kJ/mol$

Notice that the numbers we used to multiply equations (*d*) through (*g*) are the stoichiometric coefficients in the original equation! In order to determine ΔH°_{rxn} for a reaction, we first add the standard formation enthalpies for the products, each one multiplied by the stoichiometric coefficient that that substance has in the balanced chemical equation. From this sum of product ΔH°_{f} values, we subtract a similarly constructed sum of reactant ΔH°_{f} values. (Values of ΔH°_{f} are listed in Table 20-1.)

$$\Delta H°_{rxn} = \sum_{products} \nu_{prod}\, \Delta H^{\circ}_{f,prod} - \sum_{reactants} \nu_{reac}\, \Delta H_{f,reac}^{\ \circ} - \sum_{reactants} \nu_{reac}\, \Delta H_{f,reac}^{\ \circ} \qquad [4]$$

In Equation [4], ν_{prod} and ν_{reac} represent the stoichiometric coefficients in the balanced chemical equation. For the calculation of Example 7–8, Equation [4] becomes

$$\begin{aligned}\Delta H^{\circ}_{rxn} &= 4\,\Delta H^{\circ}_{f}\ [NO(g)] + 6\,\Delta H^{\circ}_{f}\ [H_2O(l)] - 4\,\Delta H^{\circ}_{f}\ [NH_3(g)] - 5\,\Delta H^{\circ}_{f}\ [O_2(g)] \\ &= 4(90.25) + 6(-285.8) - 4\,(-46.11) - 5\,(0.00)\ kJ/mol = -1169.4\ kJ/mol\end{aligned}$$

This technique works just as well for determining the enthalpies of reactions of ions in solution. Net ionic equations can be used, since spectator ions are unaltered in these reactions. To compile tables of ionic standard enthalpy changes, such as Table 7-2 that follows, it is assumed that $\Delta H^{0}_{f}\ [H^{+}9aq)] = 0\ kJ/mol$ exactly.

EXAMPLE 7–11 Write the net ionic equation for the neutralization of HCl(aq) by NaOH(aq) and H use values from Tables 20-1 and 7-2 to determine $\Delta H°$ for this neutralization reaction.

The neutralization reaction $HCl(aq) + NaOH(aq) \longrightarrow NaCl(aq) + H_2O$ has as its net ionic equation:

$$H^{+}(aq) + OH^{-}(aq) \longrightarrow H_2O(l)$$

The enthalpy change for the neutralization reaction is:

$$\begin{aligned}\Delta H^{\circ}_{rxn} &= \Delta H^{\circ}_{f}\,[H_2O(l)] - \Delta H^{\circ}_{f}\ [H^{+}(aq)] - \Delta H^{\circ}_{f}\ [OH^{-}(aq)] \\ &= -285.8\ kJ/mol - (0.0\ kJ/mol) - (-230.0\ kJ/mol) = -55.8\ kJ/mol\end{aligned}$$

This compares well with the value of $-55.7\ kJ/mol$ determined experimentally in Example 7-7.

TABLE 7-2 Some Standard Enthalpies of Formation (ΔH_f° in kJ/mol) of Ions in Aqueous Solution

H^+	0	Mg^{2+}	−466.9	OH^-	−230.0	CO_3^{2-}	−677.1	CrO_4^{2-}	−881.2
Li^+	−278.5	Ca^{2+}	−542.8	F^-	−332.6	S^{2+}	+33.1	$Cr_2O_7^{2-}$	−1490.
Na^+	−240.1	Ba^{2+}	−537.6	Cl^-	−167.2	SO_4^{2-}	−909.3	MnO_4^-	−541.4
K^+	−252.4	Zn^{2+}	−153.9	Br^-	−121.6	$S_2O_3^{2-}$	−648.5	Mn^{2+}	−220.8
NH_4^+	−132.5	Cu^{2+}	+ 64.77	I^-	−55.19	PO_4^{3-}	−1277	Fe^{2+}	−89.1
Ag^+	+105.6	Pb^{2+}	−1.7	Al^{3+}	−531	NO_3^-	−205.0	Fe^{3+}	−48.5

SELF-ASSESSMENT EXERCISES

112. (b) aluminum. The quantity of heat lost by the metal will be equal to the quantity gained by the water: $-q = m \times s.h. \times \Delta T$. The metal with the highest s.h. will cause the greatest temperature change.

113. (c) If equal volumes were mixed, the temperature would be halfway between the two starting temperatures: 50 ^{0}C. But since there is less of the warmer solution, the temperature will be less than 50 ^{0}C and correspond to the ratio of the volumes.

114. (a) If the total change in ΔU = 100 J, and all of it is given off as heat, then no work is done and no other energy is added to the system.

115. (a) Increases. The heat of solution is negative; therefore, heat is given off by the reaction and the temperature will increase.

117. (a) and (d). Answer (a) is true since the number of moles is the same for products and reactants. Answer (b) is true by definition. Answer (c) is not true; it is the most stable state of an element at standard conditions. Answer (d), in the case where no $P\Delta V$ work is done, $\Delta U = \Delta H$. For answer (e), the reaction of a strong acid and a strong base is exothermic, $\Delta H < 0$.

118. From the information given, calculate the heat gain for water and the specific heat of the iron using the relationship heat gained = heat lost. The s.h.of the iron is calculated to be 1.25 J/g ^{0}C. For the temperature change for glycerol, first convert c_p to specific heat by dividing by gram molar mass (92.0 for glycerol). The problem is now solved using the relationship heat gained = - heat lost and the final temperature is 71.1 ^{0}C.

120. The reaction desired is:

$$C(graphite) + \tfrac{1}{2} O_2 + Cl_2(g) \rightarrow COCl_2$$

The combustion reactions are:

(1) $C(graph) + O_2 \rightarrow CO_2$ $\Delta H = -393.5$ kJ

(2) $CO + ½ O_2 \rightarrow CO_2$ $\Delta H = -283$ kJ

The reaction given:

(3) $CO(g) + Cl_2(g) \rightarrow COCl_2$ $\Delta H = -108$ kJ

Using Hess's Law, if we take equation (1) – (2) + (3), we arrive at the desired equation (Reverse Equation (2), hence the minus sign). The ΔH values are treated in the same way and ΔH of formation for $COCl_2$ = (-393.5 kJ) – (-283 kJ) + (-108 kJ) = 218.5 kJ.

Drill Problems

1. (1) A substance of mass *m*, with specific heat given by (sp ht) has *q* joules of heat added and its temperature rises from t_i to t_f. Fill in the blanks in each line that follows. The numbers have the units given in the column headings, unless otherwise specified.

	q, J	m, g	sp ht, J g^{-1} °C^{-1}	t_i, °C	t_f, °C
A.	640	124	0.931	25.0	____
B.	____	75.2	0.587	20.0	30.0
C.	374	103	____	19.5	25.4
D.	1300	____	0.863	17.2	30.1
E.	194 cal	125	0.750	21.2	____
F.	____	142	1.086	66.1 °F	52.3 °F
G.	876	4.32 oz	____	22.1	30.4
H.	444	____	0.0953 cal g^{-1}°C^{-1}	19.4	26.5
	q, J	n, mol	C_p, J mol^{-1}°C^{-1}	t_i, °C	t_i, °C
I.	1962	14.0	72.0	19.3	____
J.	83.5	____	32.1	20.4	22.1
K.	572	1.65	____	21.4	35.7
L.	____	4.32	26.1	19.7	17.3
M.	1043	5.72	56.2	16.8	____

(2) Fill in the blank in each line that follows. Substance *A* of specified mass, specific heat, and initial temperature is mixed with substance *B*. The mixture has the temperature given in the right-hand column. No heat is lost from, or gained by, the system.

	Substance A			*Substance B*		
	Sp. ht.			*Sp. ht.*		*Mixture*
m, g	J g^{-1}°C^{-1}	t, °C	m, g	J g^{-1} °C^{-1}	t, °C	t, °C

N.	52.0	4.18	25.0	194	_____	35.0	31.2
O.	124	4.60	20.0	10.2	0.412	75.2	_____
P.	76.4	4.31	15.0	24.2	1.326	_____	27.3
Q.	142	3.90	24.7	_____	0.435	68.4	37.4
R.	7.51	0.682	94.1	_____	3.66	27.2	39.0
S.	9.64	1.293	85.7	274	_____	12.8	25.9
T.	7.28	3.60	106.0	15.4	4.18	_____	82.0
U.	16.3	1.188	23.1	75.6	3.90	47.5	_____
V.	12.3	1.686	64.7	105	3.06	19.4	_____

2. (1) Convert the following values that are in calories to Joules, and those that are in Joules to calories.

A. 17.64 J B. 572.3 cal C. 2455 J D. 542.0 cal E. 312.4 J F. 73.20 cal

G. 627.6 J H. 1.264 cal I. 6773 J J. 2.600 cal K. 4002 J L. 34.51 cal

(2) Determine the value of ΔE, in J or kJ, of the system for each of the following situations. For this set of problems, energy in *cal* is heat, while that in *J* is work done on or done by the system.

M. 87.5 cal absorbed, 434 J done by.
N. 96.2 cal given off, 804 J done by.
O. 134 cal given off, 506 J done on.
P. 222 cal absorbed, 734 J done on.
Q. 987 cal absorbed, 2.17 kJ done by.
R. 455 cal given off, 1.98 kJ done by.
S. 884 cal given off, 3.02 kJ done on.
T. 976 cal absorbed, 1.201 kJ done on.
U. 434 cal given off, 1.502 kJ done on.
V. 303 cal absorbed, 863 J done on.
W. 155 cal given off, 733 J done by.
X. 87.6 cal absorbed, 95.0 J done by.

3. Fill in the blank in each line below. m = the mass of substance burned in the calorimeter, and q_v is the heat of combustion in kJ/g of that substance. $m[H_2O]$ is the mass of water surrounding the calorimeter, and C_B is the calorimeter's heat capacity.

	m, g	q_v, kJ/g	$m[H_2O]$, g	C_B, kJ/°C	t_f	t_i
A.	0.750	26.45	1043.3	_____	24.1 °C	20.0 °C
B.	0.932	_____	1032.1	1.614	26.9 °C	20.5 °C
C.	0.516	17.29	974.1	1.312	_____	19.6 °C
D.	_____	4.09	836.4	1.435	77.9 °F	68.1 °C
E.	3.000	14.25	1017.0	_____	25.9 °C	17.3 °C
F.	0.816	_____	894.3	1.826	22.1 °C	15.1 °C
G.	_____	40.56	1055.2	1.414	26.4 °C	18.8 °C
H.	1.024	13.08	1000.0	1.732	_____ °F	59.1 °F

4. Fill in the blank in each line below. The solute and water, each at initial temperature t_i, are mixed together. The resulting solution has temperature t_f, and a specific heat of $4.00\,\mathrm{J\,g^{-1}\,°C^{-1}}$.

	Solute Formula	*Solute mass*, g	*Water mass*, g	t_f,°C	t_i,°C	$\Delta H^\circ_{solution}$ kJ/mol of solute
A.	$KClO_3$	9.04	91.20	22.4	29.7	____
B.	$NaNO_3$	8.61	157.98	____	18.6	– 21.38
C.	NH_4Cl	18.31	____	17.1	22.0	+15.15
D.	NH_4NO_3	15.27	473.15	22.7	25.1	____
E.	NaOH	11.73	143.06	____	20.0	– 42.89
F.	NaCl	11.98	104.32	____	20.1	+3.88
G.	RbI	74.24	569.87	25.0	28.4	____

5. For each lettered part, compute ΔH° in kJ/mol for the first reaction by appropriately combining the ΔH°s of the remaining reactions [for each of which the value of ΔH°, in kJ/mol, is written in square brackets].

A. $CO_2(g) + H_2(g) \longrightarrow H_2O(l) + CO(g)$

$H_2(g) + \frac{1}{2}O_2(g) \longrightarrow H_2O(l) [-285.9]$

$CO(g) + \frac{1}{2}O_2(g) \longrightarrow CO_2(g) [-393.5]$

B. $2C_2H_5OH(l) + O_2(g) \longrightarrow 2C_2H_4O(l) + 2\ H_2O(l)$

$C_2H_5OH(l) + 3O_2(g) \longrightarrow 2CO_2(g) + 3H_2O(l)\ [-1370.7]$

$C_2H_4O(l) + \frac{5}{2}O_2(g) \longrightarrow 2CO_2(g) + 2H_2O(g)\ [-1167.3]$

C. $2\ C(gr) + \frac{5}{2}H_2(g) + \frac{1}{2}Cl_2(g) \longrightarrow C_2H_5Cl(g)$

$H_2(g) + \frac{1}{2}O_2(g) \longrightarrow H_2O(l)\ [-285.9]$

$C_2H_4(g) + HCl(g) \longrightarrow C_2H_5Cl(g)\ [-72.0]$

$C(gr) + O_2(g) \longrightarrow CO_2(g)\ [-393.5]$

$C_2H_4(g) + 3O_2(g) \rightarrow 2CO_2(g) + 2H_2O(l)\ [-1410.8]$

$H_2(g) + Cl_2(g) \longrightarrow 2HCl(g)\ [-184.6]$

D. $Na(s) + \frac{1}{2}Cl_2(g) \longrightarrow NaCl(s)$

$2Na(s) + 2HCl(g) \longrightarrow 2NaCl(s) + H_2(g)\ [-637.4]$

$H_2(g) + Cl_2(g) \longrightarrow 2HCl(g)\ [-184.6]$

E. $2XO_2(s) + CO(g) \longrightarrow X_2O_3(s) + CO_2(s)$ (X is a metallic element)

$XO_2(s) + CO(g) \longrightarrow XO(s) + CO_2(g)$ [−83.7]
$X_3O_4(s) + CO(g) \longrightarrow 3\ XO(s) + CO_2(g)$ [+25.1]

$3\ X_2O_3(s) + CO(g) \longrightarrow 2X_3O_4(s) + CO_2(g)$ [−50.2]

F. $2\ MnO_2(s) + CO(g) \longrightarrow Mn_2O_3(s) + CO_2(s)$

$MnO_2(s) + CO(g) \longrightarrow MnO(s) + CO_2(g)$ [−150.6]
$Mn_3O_4(s) + CO(g) \rightarrow 3MNO(s) + CO_2(g)$ [−54.4]

$3\ Mn_2O_3(s) + CO(g) \longrightarrow 2Mn_3O_4(s) + CO_2(g)$ [−142.3]

G. $3\ V_2O_3(s) \longrightarrow V_2O_5(s) + 4VO(s)$

$2\ V(s) + \frac{3}{2}O_2(g) \longrightarrow V_2O_3(s)$ [−1213]

$2\ V(s) + \frac{5}{2}O_2(g) \longrightarrow V_2O_5(s)$ [−1561]

$V(s) + \frac{1}{2}O_2(g) \longrightarrow VO(s)$ [-418]

$V(s) + O_2(g) \longrightarrow VO_2(s)$ [−720]

H. $V_2O_5(s) + 3\ V(s) \longrightarrow 5\ VO(s)$

I. $Fe_2O_3(s) + 3\ CO(g) \longrightarrow 2\ Fe(s) + 3\ CO_2(g)$

$Fe_2O_3(s) + CO(g) \rightarrow 2\ FeO(s) + CO_2(g)$ [−2.9]
$Fe(s) + CO_2(g) \rightarrow FeO(s) + CO(g)$ [+11.3]

J. $2\ C(gr) + H_2(g) \longrightarrow C_2H_2(g)$

$CaC_2(s) + 2\ H_2O(l) \longrightarrow Ca(OH)_2(s) + C_2H_2(g)$ [−125.5]

$CaO(s) + H_2O(l) \longrightarrow Ca(OH)_2(s)$ [−65.3]

$2H_2O(l) \longrightarrow 2H_2(g) + O_2(g)$ [+571.5]

$CaO(s) + 3\ C(gr) \longrightarrow CaC_2(s) + CO(g)$ [+462.3]

$2\ C(gr) + O_2(g) \longrightarrow 2\ CO(g)$ [−220.9]

K. $Cu(s) + \frac{1}{2}\ O_2(g) \longrightarrow CuO(s)$

$Cu_2O(s) + \frac{1}{2}\ O_2(g) \longrightarrow 2\ CuO(s)$ [−143.9]

$CuO(s) + Cu(s) \longrightarrow Cu_2O(s)$ [−11.3]

6. Write the standard formation reaction for each substance.

A. $C_2H_5OH(l)$ B. $(NH_4)_2SbCl_5(s)$ C. $Hg_2Cl_2(s)$ D. $H_3PO_4(s)$ E. $NaIO_3(s)$

F. $SnF_2(s)$ G. $XeO_3(s)$ H. $HNO_3(l)$ I. $KBrO_3(s)$ J. $ICl_3(g)$

7. Use the enthalpies of formation (all in kJ/mol) in Table 20-1 to determine the enthalpy change of each reaction that follows.

A. $CaCO_3(s) \longrightarrow CaO(s) + CO_2(g)$

B. $NH_3(g) + HCl(g) \longrightarrow NH_4Cl(s)$

C. $NH_3(g) + H_2O(l) + CO_2(g) \longrightarrow NH_4HCO_3(s)$

D. $CaO(s) + 2\ HCl(g) \longrightarrow CaCl_2(s) + H_2O(l)$

E. $Li_2CO_3(s) \longrightarrow Li_2O(s) + CO_2(g)$

F. $Li_2O(s) + H_2O(l) \longrightarrow 2\ LiOH(s)$

G. $4\ NH_3(g) + 3\ O_2(g) \longrightarrow 2\ N_2(g) + 6\ H_2O(l)$

H. $4\ NH_3(g) + 5\ O_2(g) \longrightarrow 4\,NO(g) + 6\,H_2O(l)$

I. $2\ NO(g) + O_2(g) \longrightarrow 2\ NO_2(g)$

J. $3\ NO_2(g) + H_2O(l) \longrightarrow 2\ HNO_3(l) + NO(g$

K. $2\ HNO_3(l) + Li_2O(s) \rightarrow 2\ LiNO_3(s) + H_2O(l$

L. $Li_2O(s) + 2\ HCl(g) \longrightarrow LiCl(s) + H_2O(l)$

M. $LiOH(s) + HCl(g) \longrightarrow LiCl(s) + H_2O(l)$

N. $4\ LiNO_3(s) \longrightarrow 2\ Li_2O(s) + 4\ NO_2(g) + O$

O. $CaO(s) + H_2O(l) \longrightarrow Ca(OH)_2(s)$

QUIZZES (20 minutes each) Choose the best answer for each question. For problems with a numerical answer, an answer is correct if it is within 1% of the correctly computed value.

QUIZ A

1. The standard state of a substance is the (a) pure form at 1 atm; (b) most stable form at 25°C and 1 atm; (c) most stable form at 0°C; (d) pure gaseous form at 25°C; (e) none of these.

2. 75.0 g of water at 10°C is mixed with 125.0 g of water at 50°C. The final temperature of the mixture is (a) 30°C; (b) 40°C; (c) 35°C; (d) 25°C; (e) none of these.

3. $2\ Mg(s) + O_2(g) \longrightarrow 2\ MgO(s)$ $\Delta H° = -1203$ kJ/mol O_2. The heat given off in kJ when 1.0 g of MgO(s) is formed is (a) 29.9; (b) 1203; (c) 601.7; (d) 14.9; (e) none of these.

4. 10.0 g of H_2O (sp ht $= 4.184\ J\,g^{-1}\,°C^{-1}$) cools from 60°C to 30°C. How many joules of heat are lost to the surroundings? (a) 1393; (b) 62.8; (c) 12.6; (d) 1255; (e) none of these.

5. Which is the formation reaction for NO(g)? (a) $\frac{1}{2}\ N_2(g) + \frac{1}{3}\ O_3(g) \longrightarrow NO(g)$; (b) $N(g) + O(g) \longrightarrow NO(g)$; (c) $NO_2(g) \longrightarrow NO(g) + \frac{1}{2}O_2(g)$; (d) $N_2O_3(g) \longrightarrow NO(g) + NO_2(g)$; (e) none of these.

6. $Sn(s) + 2\ Cl_2(g) \longrightarrow SnCl_4(s)$ $\Delta H° = -545.2$ kJ/mol

$SnCl_2(s) + Cl_2(g) \longrightarrow SnCl_4(s) \quad \Delta H° = -195.4\, kJ/mol$

What is $\Delta H°$ in kJ/mol for the reaction $Sn(s) + Cl_2(g) \longrightarrow SnCl_2(g)$? (a) 349.8; (b) -349.8; (c) 740.6; (d) 195.4; (e) none of these.

7. $H_2(g) + \frac{1}{2}O_2(g) \longrightarrow H_2O(l)\ \Delta H° = -286\ kJ/mol$

$C(gr) + O_2(g) \longrightarrow CO_2(g)\ \Delta H^{\circ} = -393\ kJ/mol$

$C(gr) + O_2(g) \longrightarrow CO_2(g)\ \Delta H^{\circ\circ} = -393\ kJ/mol$

$C_2H_6(g) + \frac{7}{2}O_2(g) \longrightarrow 2\,CO_2(g) + 3\,H_2O(g)\ \ \Delta H° = -1541\ kJ/mol$

The value of $\Delta H°$ in kJ/mol for 2 C(gr) + 3 $H_2(g) \longrightarrow C_2H_6(g)$ is (a) -1541; (b) 862; (c) 469; (d) 103; (e) none of these.

8. $SnO_2(s) + 2\,H_2(g) \longrightarrow Sn(s) + 2\,H_2O(l) \quad \Delta H^{\circ}_{rxn}$ in kJ/mol is (a) 294.8; (b) 8.9;
ΔH°_{f} -580.7 0.00 0.00 -285.9 kJ/mol; (c) -294.8; (d) -571.8; (e) none of these.

QUIZ B

1. Each of the following is a stable form of an element *except* (a) $O_2(g)$; (b) $F_2(g)$; (c) C(diamond); (d) $N_2(g)$; (e) all of these are stable.

2. 15.0 g of water at 100 °C is mixed with 85.0 g of water at 20°C. The final temperature of the mixture is (a) 60°C; (b) 32 °C; (c) 50 °C; (d) 48 °C; (e) none of these.

3. The molar heat of combustion of acetylene (26 g/mol) is -1301 kJ/mol. Combustion of 0.130 g of acetylene produces how many kilojoules of heat? (a) 50.05; (b) 384.9; (c) 6.51; (d) 169.1; (e) none of these.

4. 10.00 g of HCl solution (sp ht = 4.017 J g^{-1} $°C^{-1}$) is heated from 25.0 °C to 40.0 °C. How many joules of heat does this heating require? (a) 602.6; (b) 627.6; (c) 654.0; (d) 1004; (e) none of these.

5. Which is the formation reaction for $H_2O(l)$? (a) $H^+(aq) + OH^-(aq) \longrightarrow H_2O(l)$; (b) $H_2O(g) \longrightarrow H_2O(l)$; (c) $H_2(g) + \frac{1}{2}O_2(g) \longrightarrow H_2O(l)$; (d) $H_2(g) + \frac{1}{3}O_3(g) \longrightarrow H_2O(l)$; (e) none of these.

6. $MnO_2(s) \rightarrow MnO(s) + \frac{1}{2}O_2(g)\ \Delta H° = +136.0\ kJ/mol$

$MnO_2(s) + Mn(s) \rightarrow 2\,MnO(s)\ \Delta H° = -248.9\ kJ/mol$

For the reaction $Mn(s) + O_2(g) \longrightarrow MnO_2(s)$, the value of $\Delta H°$ in kJ/mol is (a) -112.9; (b) -384.9; (c) -520.9; (d) +361.9; (e) none of these.

7. $2\,ClF + O_2 \longrightarrow Cl_2O + F_2O \quad \Delta H° = 167.4$ kJ

$2\,ClF_3 + 2\,O_2 \longrightarrow Cl_2O + 3\,F_2O \; \Delta H° = 341.4$ kJ

$\Delta H_f^\circ = -21.8$ kJ/mol for F_2O.

For the reaction $ClF + F_2 \longrightarrow ClF_3$, the value of $\Delta H°$ in kJ is (a) -217.6; (b) -435.2; (c) +223.6; (d) -130.2; (e) none of these.

8. $NH_3(g) + \frac{3}{4}O_2(g) \longrightarrow \frac{1}{2}N_2(g) + \frac{3}{2}H_2O(g)$ $\quad \Delta H^\circ_{rxn}$ in kJ/mol is (a) -195.8; (b) +316.7; ΔH_f° -46.0 0.00 0.00 -241.8 kJ/mol (c) -408.8; (d) -316.7; (e) none of these.

Quiz C

1. When an element is involved in a formation reaction, it does *not* have to be (a) pure; (b) at 1.00 M concentration; (c) at 1.00 atm pressure; (d) in its most stable form; (e) none of these.

2. 25.0 g of water at 50.0 °C is mixed with 15.0 g of water at 30.0 °C. The final temperature of the mixture is (a) 40.0 °C; (b) 42.5 °C; (c) 37.5 °C; (d) 35.0 °C; (e) none of these.

3. The molar enthalpy of combustion of $CS_2(l)$ is -897.46 kJ/mol. The products of combustion are $CO_2(g)$ and $SO_2(g)$. How many kJ of heat are given off when 6.41 g of $SO_2(g)$ is produced by the combustion of $CS_2(l)$? (a) 44.9; (b) 89.7; (c) 179; (d) 29.9; (e) none of these.

4. 1674 J of heat is absorbed by 25.0 mL of NaOH(aq) (d = 1.10 g/mL; sp ht = 4.10 J g^{-1} $°C^{-1}$). The temperature of the NaOH(aq) increases how many °C? (a) 17.2; (b) 14.2; (c) 14.8; (d) 18.0; (e) none of these.

5. Which is the formation reaction for $CO_2(g)$? (a) $C(dia) + O_2(g) \longrightarrow CO_2(g)$; (b) $C(gr) + O_2(g) \longrightarrow CO_2(g)$; (c) $CO(g) + \frac{1}{2}O_2(g) \longrightarrow CO_2(g)$; (d) $CaCO_3(s) \longrightarrow CaO(s) + CO_2(g)$; (e) none of these.

6. $CH_4 + 2\,O_2 \longrightarrow CO_2 + 2\,H_2O \quad \Delta H° = -890.4$ kJ

$CH_2O + O_2(g) \longrightarrow CO_2 + H_2O \quad \Delta H° = -563.5$ kJ

For the reaction, $CH_4 + O_2 \longrightarrow CH_2O + H_2O$, the value of $\Delta H°$ in kJ is (a) -236.6; (b) +118.3; (c) -326.9; (d) -1453.9; (e) none of these.

7. $2\,LiOH(s) \rightarrow Li_2O(s) + H_2O(l)$ $\Delta H° = +379.1$ kJ/mol

$2H_2(g) + O_2(g) \rightarrow 2\,H_2O(l)$ $\Delta H° = -285.9$ kJ/mol

$LiH(s) + H_2O(l) \longrightarrow LiOH(s) + H_2(g)$ $\Delta H° = -110.9$ kJ/mol

For the reaction 2 $LiH(s)$ + $O_2(g) \longrightarrow Li_2O(s) + H_2O(l)$, $\Delta H°$ in kJ/mol is (a) -128.6; (b) -17.7; (c) 268.2; (d) 551.4; (e) none of these.

8. $2\ H_2S(g) + 3\,O_2(g) \longrightarrow 2\,H_2O(l) + 2\,SO_2(g)$ ΔH°_{rxn} 2 in kJ/mol is (a) -562.8; (b) +562.8; ΔH°_f -20.1 0.00 -285.8 -297.1 kJ/mol (c) -3.14; (d) -477.4; (e) none of these.

QUIZ D

1. The metric system unit of energy that was originally designed to measure heat is the (a) joule; (b) Btu; (c) horsepower; (d) calorie; (e) none of these.

2. 77.0 g of water at 95.0 °C is mixed with 23.0 g of water at 5.0 °C. The final temperature of the mixture is (a) 50.0 °C; (b) 63.6 °C; (c) 23.6 °C; (d) 74.3 °C; (e) none of these.

3. $CH_3OH(l)$'s molar enthalpy of combustion is -726.51 kJ/mol. How much heat in kJ is produced by burning $CH_3OH(l)$ with 1.00 mol $O_2(g)$? (a) 726.51; (b) 363.26; (c) 1089.8; (d) 242.17; (e) none of these.

4. 40.0 g of solution has its temperature raised from 25.0 °C to 50.0 °C by the addition of 4067 J of heat. Its specific heat (in J g^{-1} °C^{-1}) is (a) 4.31; (b) 4.07; (c) 4.18; (d) 2.03; (e) none of these.

5. Which is the formation reaction for $SO_3(g)$? (a) $S(monoclinic) + \frac{3}{2}O_2(g) \longrightarrow SO_3(g)$; (b) $2\,SO_2(g) \longrightarrow SO_3(g) + SO(g)$; (c) $SO_2(g) + \frac{1}{2}O_2(g) \longrightarrow SO_3(g)$; (d) $H_2SO_4(l) \longrightarrow H_2O(l) + SO_3(g)$; (e) none of these.

6. $CO(g) + \frac{1}{2}O_2(g) \rightarrow CO_2(g)$ $\Delta H° = -282.8$ KJ/mol

$C(gr) + O_2(g) \rightarrow CO_2(g)$ $\Delta H° = -393.3$ kJ/mol

For the reaction $C(gr) + \frac{1}{2}O_2(g) \longrightarrow CO(g)$, the value of $\Delta H°$ in kJ/mol is (a) $+282.8$; (b) -676.1; (c) $+110.5$; (d) -110.5; (e) none of these.

7. $S(rhombic) + O_2(g) \rightarrow SO_2(g)$ $\Delta H° = -297.1$ kJ/mol

$H_2(g) + \frac{1}{2}O_2(g) \rightarrow H_2O(l)$ $\Delta H° = -285.8$ kJ/mol

$H_2S(g) + \frac{3}{2}O_2 \longrightarrow H_2O(l) + SO_2(g)$ $\Delta H° = -562.3$ kJ/mol

For the reaction $H_2(g) + S(\text{rhombic}) \longrightarrow H_2S(s)$, the value of $\Delta H°$ in kJ/mol is (a) 562.3; (b)265.3; (c)276.6; (d) -20.6; (e) none of these.

8. $CH_3OH(l) + \frac{3}{2}O_2(g) \longrightarrow CO_2(g) + 2\,H_2O(l)$ ΔH°_{rxn} in kJ/mol is (a) -140.8; (b) -725.6; ΔH°_f -238.5 0.00 -393.5 -285.3 kJ/mol; (c) $+154.8$; (d) -154.8; (e) none of these.

Sample Test (15 minutes)

1. Write the formation reaction for each of the following:

   A. $SnCl_4(s)$ B. $C_6H_5COOH(s)$ C. $COCl_2(g)$

2. Compute ΔH°_{rxn} for these two reactions. (The value of ΔH°_f in kJ/mol is given below each formula.)

   A. $SiO_2(s) + 4\,HF(g) \longrightarrow SiF_4(g) + 2\,H_2O(g)$

   -48.5 0.00 -155 -296.9

   B. $2\,CuS(s) + 3\,O_2(g) \longrightarrow 2\,CuO(s) + 2\,SO_2(g)$

   -859.4 -269 -1550 -285.9

3. When dissolved in water, 1.00 mole LiCl produces 37.2 kJ of heat. What is the final temperature when 5.00 g LiCl dissolves in 110.0 g of water at 20.00 °C? (The solution produced has a specific heat = 4.00 J g^{-1} °C^{-1}.)

8 ELECTRONS IN ATOMS

CHAPTER OBJECTIVES

*** 1. Apply the fundamental expression relating the frequency, wavelength, and velocity of electromagnetic radiation, with appropriate regard for units.**

Equation [1] is the relationship between the frequency of radiation ν (pronounced "nu") and its wavelength λ ("lambda"). The constant velocity of electromagnetic radiation has a value $c = 2.9979 \times 10^8$ m/s. We shall often use the three-significant-figure value, $c = 3.00 \times 10^8$ m/s.

$$c = \lambda\nu = 2.9979 \times 10^8 \text{ m/s} \qquad [1]$$

Frequency consistently is expressed in units of hertz (Hz). However, previous usage has resulted in many ways of writing hertz, including: /sec, /s, s^{-1}, cycles per second, cycles, and cps. For the units of wavelength, the situation is not as simple. First, wavelengths often are not written in scientific notation as are frequencies, but rather are expressed in the most conveniently sized unit of length. For example, red light has a wavelength of 760 nm (7.60×10^{-7} m). Second, three non-SI units are or have been used for wavelengths. They are the micron $(\mu) = 10^{-6}\text{ m} = \text{micrometer}\,(\mu\text{m})$, the millimicron $(\text{m}\mu) = 10^{-9}\text{ m} = \text{nanometer (nm)}$, and the Ångstrom $(\text{Å}) = 10^{-10}\text{m} = 10^{-8}\text{cm} = 0.1\text{ nm} = 100\text{ pm}$. To use Equation [1], however, wavelengths must be expressed in meters.

EXAMPLE 8-1 A radio station's frequency is 95 kilocycles. What is the wavelength of this radiation in km? (95 kilocycles is the same as 95×10^3 Hz.) Rearrangement of Equation [1] produces $\lambda = c/\nu$.

$$\lambda = \frac{c}{\nu} = \frac{3.00 \times 10^8 \text{ m/s}}{95 \times 10^3 \text{ s}^{-1}} = 3.2 \times 10^3 \text{ m} \times \frac{1 \text{ km}}{1000 \text{ m}} = 3.2 \text{ km}$$

EXAMPLE 8-2 Electromagnetic radiation with a wavelength of 3429 Å has what frequency?

$$\nu = \frac{c}{\lambda} = \frac{3.00 \times 10^8 \text{ m/s}}{3429 \text{ Å} \times \dfrac{1\text{m}}{10^{10} \text{ Å}}} = 8.75 \times 10^{14} \text{ Hz}$$

*** 2. List the various types of radiation and their approximate wavelengths.**

This objective involves learning the approximate wavelengths shown in Table 8-1 and Figure 8-1. The boundaries between the areas of the spectrum are not sharp divisions, but rather gradual transitions. It is helpful to associate the various types of radiation with the changes in matter they produce or that produce them, as shown in Table 8-1. These associations are mutual. For example, infrared radiation may cause changes in the vibrations of molecules when it strikes them. Or a change in how a molecule vibrates may produce infrared radiation. This is how spectroscopy is used to identify molecules. Specific parts of molecules absorb light of specific wavelengths. By detecting this absorption, we often can identify a particular compound.

3. Know how light is dispersed into a spectrum and the difference between continuous and line spectra.

When light passes through a prism, it is refracted or bent as it enters and as it leaves the prism. Different frequencies of light are refracted to different degrees. High-frequency (short-wavelength) light is refracted most; low-frequency (long-wavelength) light is refracted least. Among the colors of visible light, red light is refracted least; next orange; then yellow; green; blue; indigo; and, finally, violet, which is refracted most (see Figure 9-1). (Note that the initial letters of the colors spell (ROY G BIV.) You can remember which is bent most by imagining that high-frequency light interacts most frequently with the matter of the prism as it passes through and, thus, is refracted the most. This is only a memory aid, not what actually occurs; refraction occurs at the surfaces of the prism.

TABLE 8-1 Changes in Matter Associated with Various Types of Electromagnetic Radiation.

Type(s) of radiation	Approx. λ	Associated change in matter
Cosmic rays, γ rays	0.01 pm	Nuclear transformations (changes in nuclear structure)
X rays	0.1 nm	Inner electronic transitions in atoms
Ultraviolet rays	100 nm	Outer electron transitions in atoms and in molecules
Visible light	0.5 μm	(Outer electron transitions in atoms and in molecules)
Infrared radiation	10 μm	Vibrations of atoms in molecules
Microwave radiation	1 cm	Rotation of molecules
Radar	1 m	(Rotation of molecules)
Radio	10 m	Modifications of nuclear spins (NMR and MRI)
Television	1 km	Translations of molecules (their movement)

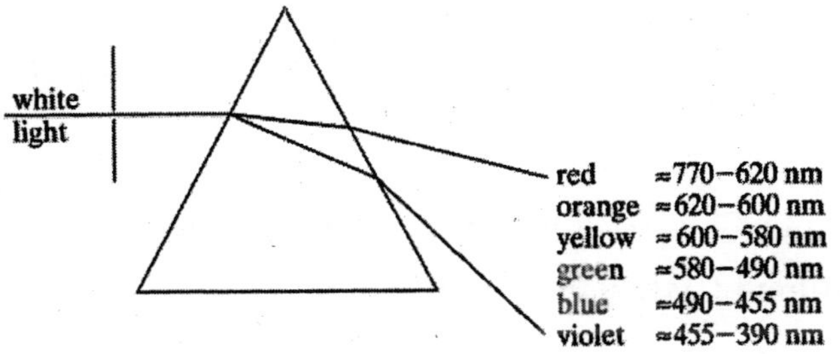

FIGURE 8-1 Dispersion of White Light by a Prism

When white light is dispersed by a prism, it produces a continuous spectrum. When an element is heated in a flame and the resulting light is dispersed by a prism, the spectrum consists of several lines. The pattern of lines is characteristic of the element being heated. This line spectrum also is called an *emission spectrum.* A neon light is a common example of a heated element giving off light, as is a sodium vapor lamp. Elements can be tentatively identified visually by the colors they emit when heated. Some of these flame colors are given in Table 8-2.

TABLE 8-2 Flame Colors of Some of the Elements

Element	Color	Element	Color
Lithium	Carmine	Calcium	Orange-red
Sodium	Orange-yellow	Strontium	Brick-red
Potassium	Violet	Barium	Yellowish green
Rubidium	Red	Copper	Azure blue
Cesium	Blue	Lead	Light blue

*** 4. Use the Rydberg equation to determine the wavelengths of lines in hydrogen's spectrum.**

The Balmer equation is a special case of the Rydberg equation, [2].

$$\frac{1}{\lambda} = R_{\mathrm{H}}\left(\frac{1}{n^2} - \frac{1}{m^2}\right) \quad \text{OR} \quad \nu = \frac{c}{\lambda} = R_{\mathrm{H}}\left(\frac{1}{n^2} - \frac{1}{m^2}\right) \quad \text{OR} \quad E = \frac{hc}{\lambda} = R_{\mathrm{H}}\left(\frac{1}{n^2} - \frac{1}{m^2}\right) \qquad [2]$$

The numbers *n* and *m* are whole numbers with *m* larger than *n*.
$R_{\mathrm{H}} = 109{,}678\ \mathrm{cm}^{-1}$; $R_{\mathrm{H}} = 3.2281 \times 10^{-15}\ \mathrm{s}^{-1}$; $R_{\mathrm{H}} = 2.179 \times 10^{-19}$ J. (It may seem strange to have the same constant, R_{H} expressed in cm^{-1}, s^{-1}, and J. But the proper value to use will be evident from the problem you are solving.) The Balmer equation results when $n = 2$. The wavelengths calculated then are those of the *Balmer series*—a set of visible lines emitted by hydrogen. When $n = 1$, the Lyman series in the ultraviolet region is produced. The remaining series all are in the infrared: Paschen ($n = 3$), Brackett ($n = 4$), Pfund ($n = 5$), and Humphreys ($n = 6$). The Rydberg equation describes these lines. It is simply a summary of data, *not* based on a theoretical model.

EXAMPLE 8-3 What is the wavelength of the Lyman series line with $m = 2$? Is this visible light?

$\frac{1}{\lambda} = 109{,}678\ \text{cm}^{-1}\left(\frac{1}{1^2} - \frac{1}{2^2}\right) = 82{,}258.5\ \text{cm}^{-1}$ $\lambda = \frac{1}{82258.5\ \text{cm}^{-1}}$ 121.6 nm is

$= 1.216\times10^{-5}\ \text{cm} = 121.6\ \text{nm}$

ultraviolet radiation. The limit of visible light is about 350 nm.

*** 5. Know and be able to use Planck's equation.**

Planck's equation, $E = h\nu$, enables us to relate the energy to the frequency of radiation.

EXAMPLE 8-4 What is the energy of a light photon with a frequency of 110 Hz? A mole of such photons?

$$E = h\nu = (6.626\times10^{-34}\ \text{J}\cdot\text{s})(110.\times10^{6}\ \text{Hz}) = 7.29\times10^{-26}\ \text{J/photon}$$

$$E_{\text{mol}} = N_A\varepsilon = (6.022\times10^{23}\ \text{photons/mol})(7.29\times10^{-26}\ \text{J/photon}) = 4.39\times10^{-2}\ \text{J/mol photons}$$

This is a small energy. Compare it to the 75.3 J required to heat 1 mole (18.0 g) of water by 1.00 °C.

Frequently, $E = h\nu$ and $c = \lambda\nu$ are combined to produce Equation [3], whose use is illustrated in Example 8-5.

$$E = hc/\lambda \qquad [3]$$

EXAMPLE 8-5 400. kJ is required to break a mole of N—H bonds. What is the wavelength of the radiation needed to break one N—H bond?

First, we determine the energy needed to break one bond from $E = N_A\varepsilon$ or $\varepsilon = E/N_A$.

$$E = \frac{400.\times10^{3}\ \text{J}}{1\ \text{mol}} \times \frac{1\ \text{mol bonds}}{6.022\times10^{23}\ \text{bonds}} = 6.64\times10^{-19}\ \text{J/bond}$$

Then, we determine the wavelength from a rearranged version of Equation [3].

$$\lambda = \frac{hc}{E} = \frac{6.626\times10^{-34}\ \text{J/s}\times2.998\times10^{8}\ \text{m/s}}{6.64\times10^{-19}\ \text{J}} = 3.00\times10^{-7}\ \text{m}\times\frac{10^{9}\ \text{nm}}{1\ \text{m}} = 300.\ \text{nm}$$

This radiation is just beyond the visible region, in the ultraviolet. This partly explains why some substances deteriorate when they are exposed to sunlight—the chemical bonds are disrupted.

6. Know and be able to apply Bohr's model of the hydrogen atom: the assumptions, the picture of the atom, the energy expression, and the energy-level diagram.

To explain the Rydberg equation, Bohr proposed a model of the hydrogen atom based on several assumptions:

1. The electron moves around the nucleus in one of several circular orbits. The electron does not spiral into the nucleus as classical physics requires.

2. In each orbit of radius r, the angular momentum of the electron, $m_e v r$ (where m_e and v are the electron mass and velocity), is restricted to values of $nh/2\pi$ (where n is a whole number and h is Planck's constant).

$$m_e vr = nh/2\pi \qquad [4]$$

3. When an electron moves from one orbit to another, the energy difference (ΔE) between the two orbits appears as light emitted (if the second orbit is of lower energy) or absorbed (if the second orbit has higher energy) by the atom.

$$\Delta E = h\nu \qquad [5]$$

After some algebraic manipulation, these equations, along with the principles of classical physics, give expressions for the energy of each orbit (E_n) and the difference in energy between two orbits (ΔE).

$$E_n = (2.179\times10^{-18}\ \text{J})/n^2 \qquad [6]$$

$$\Delta E = (2.179\times10^{-18}\ \text{J})[(1/\text{n}^2) - (1/m^2)] \qquad [7]$$

The wavelength of light absorbed or emitted is given by the Rydberg equation, [2], which now is based on a theoretical model. When the energies determined from Equation [6] are plotted, an energy-level diagram is produced (Figure 9-2). This diagram illustrates the various possible values of ΔE. Another aspect of each series in the hydrogen spectrum is the series limit. The energy levels get closer to each other as the quantum number (n) increases. Thus, the spectral lines also are closer in frequency.

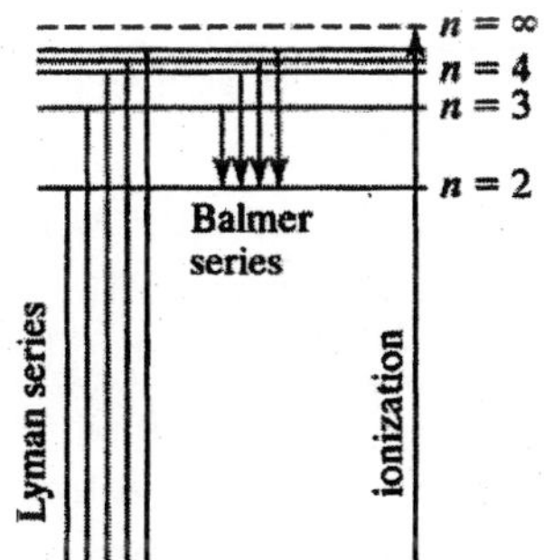

FIGURE 8-2 Energy Level Diagram for the Hydrogen Atom

7. Summarize de Broglie's and Heisenberg's ideas.

A serious problem with Bohr's model is why his second assumption, Equation [4], is true. This was explained in 1924 by Louis de Broglie (and verified experimentally by

Davisson and Germer in 1927), who stated that a particle of mass (m) and velocity (v) has a wavelength given by

$$\lambda = h/mv \qquad [8]$$

An integral number (whole number) of wavelengths must fit in a circular orbit of the atom, according to classical wave theory. Substitution of the de Broglie wavelength, and rearrangement produces Equation [10].

$$n\lambda = 2\pi r = \text{circumference} \qquad [9]$$

SUBSTITUTING $n(h/mv) = 2\pi r$ REARRANGING $nh/2\pi = mvr$ 10]

Equation [10] is Bohr's assumption (Equation [4]). But de Broglie's equation does much more than support Bohr's theory. It is a fundamental fact of nature. Particles moving at high velocities have wave properties. But even this statement is not complete. Particles *are* waves. It merely depends on how we observe them. In the same way, a woman can be both a mother to her children and a surgeon, depending on how (or when) we observe her. (Also, think of a coin twirling in the air. Is it heads or tails while it is in the air?) Another consequence of this wave-particle duality is the Heisenberg uncertainty principle.

$$\Delta x \Delta p > h/2p \ \text{(where } p = mv = \text{momentum)} \qquad [11]$$

This uncertainty is a fundamental limitation of nature, not a result of the crudeness of measuring devices. We have a different picture of the electron in the atom. Bohr's definite orbits are replaced by *orbitals*—regions of space in which there is a reasonable chance (for example, better than 90%) of finding the electron. An orbital is somewhat similar to the boundaries of a city. We believe there is a good chance (or a high probability) of finding a citizen of the city somewhere within the city limits. In the same way there is a high probability of finding the electron within the boundaries of the orbital.

8. Explain the differences between Bohr's and Schrödinger's models of the atom.

Bohr treated the electron as a classical particle fixed in certain orbits, defined by its wavelength. Schrödinger allows the electron to fully express its wave properties. Schrödinger's electron does not plod around an orbit, but totally fills the space of the orbital. It does not fill this space as would a fly in a cage, but rather as a gas fills a bottle. Yet it does not fill the space by breaking into fragments or by making itself larger, but rather in the way that sound fills the room. Unfortunately, the mathematics describing the Schrödinger atom are more complex than those needed by the Bohr model, but the results can be summarized in rules and visualized, which is what we will do next.

*** 9. Know and be able to apply the quantum number relationships of wave mechanics.**

The *principal quantum number*, designated by n, can only be a whole number larger than zero.

$$n = 1,\ 2,\ 3,\ 4,\ 5,\ldots \qquad [12]$$

Electrons with the same principal quantum number are in the same *shell* or *level.* Often the shells are designated with capital letters.

value of n:	1	2	3	4	5
shell:	K	L	M	N	O

Thus, all electrons with $n = 1$ are in the K shell. The average distance of the electron from the nucleus depends almost entirely on n. The *azimuthal quantum number,* ℓ (also called the orbital quantum number), is a nonnegative whole number less than n.

$$l = 0, 1, 2, 3, \ldots, n - 1 \qquad [13]$$

Electrons with the same value of ℓ are in the same *subshell* or *sublevel.* Subshells are designated by lower-case letters.

value of ℓ:	0	1	2	3	4	5	6	7
subshell:	*s*	*p*	*d*	*f*	*g*	*h*	*i*	*k*

(The subshell letters are the first letters of the words in: "Sober physicists don't find giraffes hiding in kitchens." This sentence is a convenient memory aid.) The energy of a many-electron atom depends on the sum $n + \ell$. Also, ℓ specifies the shape of an orbital. The *magnetic quantum number,* m_ℓ (or m, also called the orientation quantum number), ranges from $-\ell$ to $+\ell$ in whole number steps.

$$m_\ell = -\ell, -\ell + 1, -\ell + 2, \ldots, -1, 0, +1, \ldots, \ell - 2, \ell - 1, \ell \qquad [14]$$

Electrons with the same values of n, ℓ, and m_ℓ are in the same *orbital.* The orientation quantum number, m_ℓ, specifies the general direction of the orbital (the direction in which it "points").

The rules tell us possible values of quantum numbers. For example, when $n = 4$ (the N shell), the only possible values of ℓ are 0, 1, 2, 3. Therefore, the only subshells in the 4th shell are 4*s*, 4*p*, 4*d*, and 4*f*, listed in order of increasing energy. Within the $4d\,(\ell = 2)$ subshell, there are five orbitals $(m_\ell = -2, -1, 0, +1, +2)$, which is the number of orbitals in any *d* subshell. These quantum numbers designate the possible orbitals that electrons can occupy. Electrons tend to be present in the low energy orbitals, those with the smallest $n + \ell$ sum.

*** 10. Know what an orbital is and sketch the appearance of *s*, *p*, and *d* orbitals.**

An orbital is a region in space where there is a good chance or a high probability of finding an electron. All *s* orbitals are spherical in shape and increase in size as *n* increases. The *p* orbitals have a double-squashed sphere shape (see Figure 8-3). Each squashed sphere is called a *lobe*, and the region of zero electron probability between the lobes is called a *node*. A *p* orbital corresponds to $\ell = 1$ and, thus, there are three *p* orbitals since m_ℓ can equal -1, 0, or $+1$. Since $\ell = 2$ for *d* orbitals, there are five values for m_ℓ $(= -2, 0, +1, +2)$ and, thus, five *d* orbitals. These are shown in Figure

8-4. Except for d_{z^2}, each orbital has four lobes and two nodes and looks rather like a four-leaf clover or four teardrops pointing to a common center.

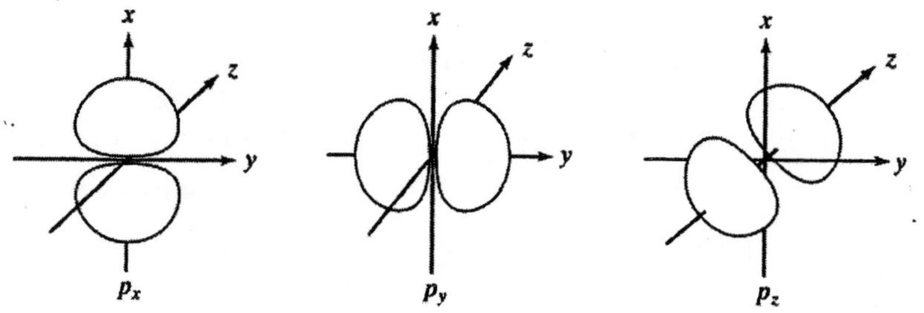

(a) The Orbitals as They Most Likely Appear

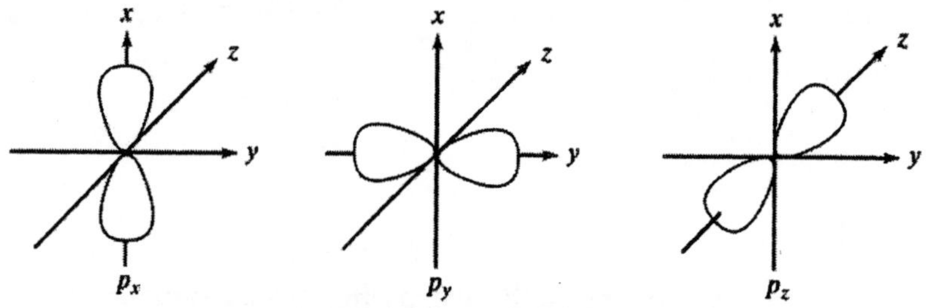

(b) The Orbitals as They Often Are Drawn

FIGURE 8-3 The Three *p* Orbitals

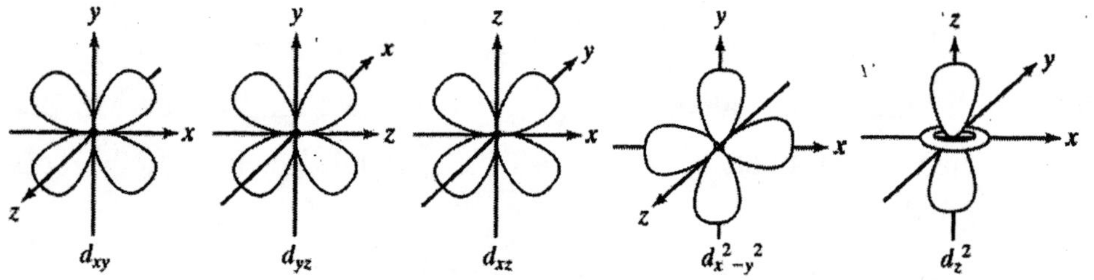

FIGURE 8-4 The Five *d* Orbitals

11. Know how orbital energies are modified when more than one electron is present in an atom.

For one-electron atoms and ions (H, He^+, Li^{2+}, and so forth), orbital energies depend on only the principal quantum number *n*. As other electrons are added, the principal levels split into sublevels of different energies. (As we shall see later in the text, sometimes, when atoms combine, the orbitals within a subshell have different energies.) Listed in order of increasing energy, the orbitals are:

$$1s, 2s, 2p, 3s, 3p, 4d, 5p, 6s, 4f, 5d, 6p, 7s, 5f, 6d, 7p \qquad [15]$$

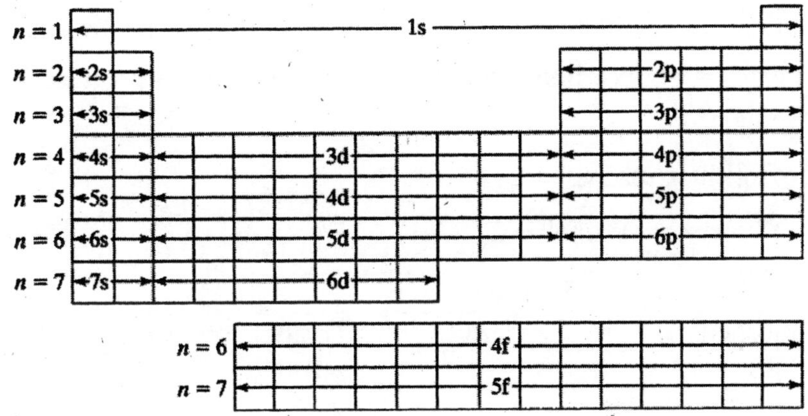

FIGURE 8-5 Order of Subshell Filling Based on the Periodic Table

There is a superb way of remembering this order, based on the periodic table. First, remember that each orbital can hold two electrons. Thus, an *s* subshell with one orbital can hold two electrons, a *p* subshell with three orbitals can hold six electrons, a *d* subshell with five orbitals can hold 10 electrons, and an *f* subshell with seven orbitals can hold 14 electrons. Now look carefully at the periodic table in the front of this book. Notice that there is a region that is two columns wide. These columns, headed 1A and 2A, represent the filling of the *s* subshell. The six columns headed 3A, 4A, 5A, 6A, 7A, and 8A, represent the filling of the *p* subshell. The "waist" of the periodic table is 10 columns wide—from 3B to 2B—and represents the filling of the five orbitals of the *d* subshell. Finally, the two long rows at the bottom are 14 elements wide and represent the filling of the seven *f* orbitals. This order of filling is shown in Figure 8-5.

We also can use this method to determine the subshell of the last electron added. For example, the last electron was added to the 4*d* subshell of Ru, because this element is in the area of the periodic table labeled 4*d*. For representative elements, the outermost electrons determine how they react chemically. These outermost electrons, or valence electrons, are the last ones added for representative elements. By simply looking at the periodic table, we can determine which electrons these are.

***12. Learn the three basic principles governing electron configuration.**

1. *The order in which orbitals are filled.* The correct order is given in expression [15] and a method of recalling it is discussed in Objective 8-11.

2. *The Pauli exclusion principle and its implications.* The Pauli exclusion principle states that no two electrons in an atom may have the same set of four quantum numbers. An orbital contains electrons that have the same value of n, ℓ, and m_ℓ. Hence m_s (or *s*, the spin quantum number) must be different for the two electrons in an orbital. Note that m_s can have only two values: either $m_s = +\frac{1}{2}$ or $m_s = -\frac{1}{2}$.

$$\text{number of electrons/orbita } l = 2 \qquad [16]$$

The number of orbitals in a subshell is governed by m_ℓ, from $m_\ell = -\ell$ to $m_\ell = +\ell$ by whole number steps. There are $2\ell+1$ values of m_ℓ for a given value of ℓ.

number of orbitals/subshell $= 2\ell+1$ [17]

number of electrons/subshell $= 2(2\ell+1)$ [18]

TABLE 8-3 Electrons in Shells and Subshells

Shell	Subshells (electron capacity)	Total electrons in each shell	
1	1*s*(2)	2	= 2
2	2*s*(2) 2*p*(6)	2 + 6	= 8
3	3*s*(2) 3*p*(6) 3*d*(10)	2 + 6 + 10	= 18
4	4*s*(2) 4*p*(6) 4*d*(10) 4*f*(14)	2 + 6 + 10 + 14	= 32
5	5*s*(2) 5*p*(6) 5*d*(10) 5*f*(14) 5*g*(18)	2 + 6 + 10 + 14 + 18	= 50

TABLE 8-4 Possible Electron Configurations of Six *d* Electrons

Configuration	*Number of electrons with* $m_s = +\frac{1}{2}$	*Number of electrons with* $m_s = -\frac{1}{2}$	Total spin
[↑↓] [↑↓] [↑↓] [ ] [ ]	3	3	0
[↑↓] [↑↓] [↑] [↑] [ ]	4	2	1
[↑↓] [↑] [↑] [↑] [↑]	5	1	2

The number of subshells in a shell is determined by the value of *n*. There are *n* subshells in each shell, from $\ell = 0$ to $\ell = n-1$. In Table 8-3, the subshells of each shell are listed and the capacity of each subshell is given in parentheses. Note that each *s* subshell can hold 2 electrons, each *p* subshell can hold 6, each *d* subshell can hold 10 electrons, and each *f* subshell can hold 14 electrons.

number of subshells/shell $= n$ [19]

3. *Hund's rule of maximum multiplicity.* The multiplicity of an atom is related to its total spin or to the sum of the m_s quantum numbers. Hund's rule states that when electrons fill orbitals of the same energy, they do so with the same values of m_s if possible. Suppose, for example, that there are six electrons to place in the five *d* orbitals. The five *d* orbitals can be represented by five boxes and the electrons by arrows—pointing up (↑) if $m_s = +\frac{1}{2}$, and pointing down (↓) if $m_s = -\frac{1}{2}$. The three possibilities for six *d* electrons are summarized in Table 8-4. Hund's rule tells us to choose the last of these three.

* **13. Apply the Aufbau principle; write electron configurations with many different methods.**

With the Aufbau process, we build the electron configuration of an atom upon the configuration of the previous atom (that of next lowest atomic number). There are many ways to write electron configurations. One is to give the four quantum numbers of the last electron added. Table 8-5 lists these values for the elements with atomic numbers up to $Z = 10$. This is a cumbersome technique and does not account for those cases (of which chromium is one) in which the quantum numbers of one of the previous electrons change. Furthermore, note in Table 8-5, that there is no good reason for assigning $m_s = +\frac{1}{2}$ before $m_s = -\frac{1}{2}$. If we would have assigned them in the opposite order, the configuration for H would be: $n = 1, \ell = 0, m_\ell = 0, m_s = -\frac{1}{2}$. You should be consistent, assigning either $m_s = +\frac{1}{2}$ or $m_s = -\frac{1}{2}$ first. There also is no preferred order of assigning m_l values. Table 8-5 shows them assigned in increasing order $(m_\ell = -1, 0, +1)$, but they could have been assigned in decreasing order $(m_\ell = +1, 0, -1)$. Again, either order is correct as long as you are consistent.

TABLE 8-5 Quantum Numbers of the Last Electron Added for the First 13 Elements

Quantum number	H	He	Li	Be	B	C	N	O	F	Ne	Na	Mg	Al
n	1	1	2	2	2	2	2	2	2	2	3	3	3
ℓ	0	0	0	0	1	1	1	1	1	1	0	0	1
m_ℓ	0	0	0	0	-1	0	$+1$	-1	0	$+1$	0	0	-1
m_s	$+\frac{1}{2}$	$-\frac{1}{2}$	$+\frac{1}{2}$	$-\frac{1}{2}$	$+\frac{1}{2}$	$+\frac{1}{2}$	$+\frac{1}{2}$	$-\frac{1}{2}$	$-\frac{1}{2}$	$-\frac{1}{2}$	$+\frac{1}{2}$	$-\frac{1}{2}$	$+\frac{1}{2}$

Electron configuration is written more compactly as an energy-level diagram, so called because the vertical direction indicates approximate orbital energy. The basic diagram is shown in Figure 8-6a. Each box represents an orbital to be filled with two electrons, drawn as arrows. The electron configurations of C, Mg, and P are shown in Figure 8-6. When energy-level diagrams are written on one line, as in Figure 8-7, they are called orbital diagrams.

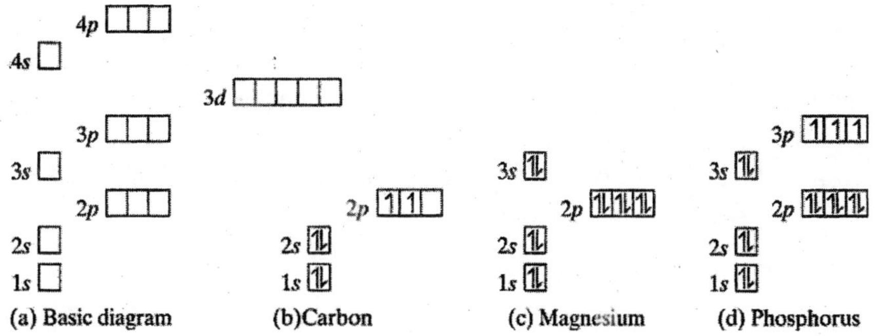

FIGURE 8-6 Energy Level Diagrams for C, Mg, and P

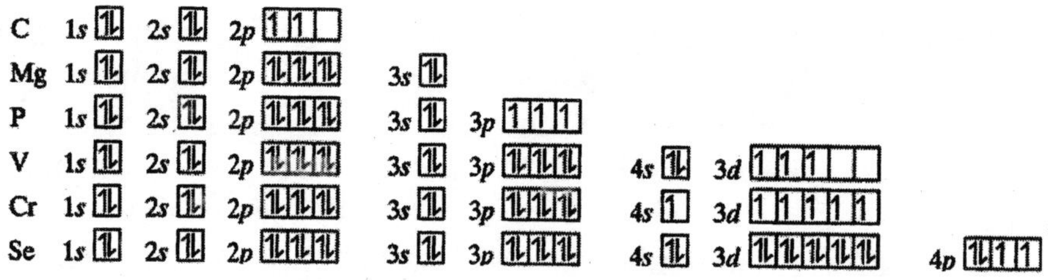

FIGURE 8-7 Orbital Diagrams for C, Mg, P, V, Cr, and Se

The electron configuration of Cr is not built upon that of V. A useful generalization is that half-filled and filled subshells are unusually stable, and that if they can be obtained by "moving" only one electron between very closely spaced energy subshells, one should do so. The closely spaced subshells are (4*s*, 3*d*), (5*s*, 4*d*), (6*s*, 4*f*, 5*d*), and (7*s*, 5*f*, 6*d*). This is an extension of Hund's rule. It works quite well for elements up to atomic number 60, but not very well after that.

TABLE 8-6 *spdf* Notation for C, Ca, P, V, Cr, Se

	Ordered by energy	Abbreviated	Ordered by *n*
C	$1s^2\ 2s^2\ 2p^2$	[He] $2s^2\ 2p^2$	$1s^2\ 2s^2\ 2p^2$
Mg	$1s^2\ 2s^2\ 2p^6\ 3s^2$	[Ne] $3s^2$	$1s^2\ 2s^2\ 2p^6\ 3s^2$
P	$1s^2\ 2s^2\ 2p^6\ 3s^2\ 3p^3$	[Ne] $3s^2\ 3p^3$	$1s^2\ 2s^2\ 2p^6 3s^2 3p^3$
V	$1s^2\ 2s^2\ 2p^6\ 3s^2 3p^6 4s^2 3d^3$	[Ar] $4s^2\ 3d^3$	$1s^2\ 2s^2\ 2p^6\ 3s^2\ 3p^6\ 3d^3\ 4s^2$
Cr	$1s^2\ 2s^2\ 2p^6\ 3s^2 3p^6 4s^1 3d^5$	[Ar] $4s^1\ 3d^5$	$1s^2\ 2s^2\ 2p^6\ 3s^2\ 3p^6 3d^5 4s^1$
Se	$1s^2\ 2s^2\ 2p^6\ 3s^2 3p^6 4s^2 3d^{10} 4p^4$	[Ar] $4s^2\ 3d^{10}\ 4p^4$	$1s^2\ \ 2s^3\ 2p^6\ 3s^6\ 3p^6\ 3d^{10}\ 4s^2\ 4p^4$

A more compact notation is *spdf* or spectroscopic notation. (The letters *s, p, d,* and *f* originally described emission spectrum lines: "sharp," "principal," "diffuse," and "fundamental.") One writes the subshells in order of increasing energy, as in expression [15], and indicates the number of electrons in each subshell with a superscript. Sometimes the orbitals are written in order of increasing *n* value. Both methods are shown in Table 8-6. One often abbreviates the *spdf* notation by writing the symbol for the last noble gas in brackets to represent its electron configuration. But the *spdf* notation does not tell us how electrons are paired in the unfilled subshells. Often, this information is expressed by writing the designation for the unfilled subshells, as shown in Table 8-7.

TABLE 8-7 Extended *spdf* Notation

C	[He] $2s^2\ 2p_x^1\ 2p_y^1$	Mg	[Ne] $3s^2$
P	[Ne] $3s^2\ 2p_x^1\ 2p_y^1 2p_z^1$	V	[Ar] $4s^2\ 3d_{xy}^1\ 3d_{xz}^1\ 3d_{yz}^1$
Cr	[Ar] $4s^1\ 3d_{xy}^1\ 3d_{xz}^1\ 3d_{yz}^1\ 3d_{z^2}^1\ 3d_{x^2-y^2}^1$	Se	[Ar] $4s^2\ 3d^{10}\ 4p_x^2\ 4p_x^2 4p_y^1 4p_z^1$

To summarize, the electron configurations of the first 36 elements are given in Table 8-8 in several forms: *spdf* notation, extended *spdf* notation, orbital diagrams, and energy-level diagrams.

***14. Use the periodic table to describe the Aufbau process. Explain the basic features of the electron configurations of the representative and transition elements, especially the number of valence electrons in each representative group.**

We used the periodic table to help remember the order of subshell-filling (Objective 8–11 and Figure 8-5). The electron configuration of an element determines its physical and chemical properties. An atom displays mainly its outer electrons—those in the shell of highest principal quantum number. Therefore, there is a major difference in chemical and physical properties between elements in different representative families, since each family differs in the number of outer-shell electrons. There is not so large a difference between transition families, because these families differ by inner-shell electrons. The number of outer-shell (that is, *s* and *p*) electrons of a representative element equals its family number. These outer-shell electrons are called *valence electrons*, and are said to be in the valence shell.

TABLE 8-8 Electron Configurations of the First 36 Elements

1. Electron configurations for H through Ne

	1. Quantum numbers of last electron added					2. *spdf* notation	3. Extended *spdf* notation	4. Orbital diagram
	Z	n	ℓ	m_ℓ	m_s			
H	1	1	0	0	$+\frac{1}{2}$	$1s$	$1s$	[↑] $1s$
He	2	1	0	0	$-\frac{1}{2}$	$1s^2$	$1s^2$	[↑↓] $1s$
Li	3	2	0	0	$+\frac{1}{2}$	$1s^2 2s$	[He]$2s$	[He] [↑] $2s$
Be	4	2	0	0	$-\frac{1}{2}$	$1s^2 2s^2$	[He] $2s^2$	[He] [↑↓] $2s$
B	5	2	1	+1	$+\frac{1}{2}$	[He] $2s^2 2p^1$	[He] $2s^2 2p_x$	[He] [↑↓] $2s$ [↑][ ][ ] $2p$
C	6	2	1	0	$+\frac{1}{2}$	[He] $2s^2 2p^2$	[He] $2s^2 2p_x 2p_y$	[He] [↑↓] $2s$ [↑][↑][ ] $2p$
N	7	2	1	−1	$+\frac{1}{2}$	[He] $2s^2 2p^3$	[He] $2s^2 2p_x 2p_y 2p_z$	[He] [↑↓] $2s$ [↑][↑][↑] $2p$
O	8	2	1	+1	$-\frac{1}{2}$	[He] $2s^2 2p^4$	[He] $2s^2 2p_x^2 2p_y 2p_z$	[He] [↑↓] $2s$ [↑↓][↑][↑] $2p$
F	9	2	1	0	$-\frac{1}{2}$	[He] $2s^2 2p^5$	[He] $2s^2 2p_x^2 2p_y^2 2p_z$	[He] [↑↓] $2s$ [↑↓][↑↓][↑] $2p$
Ne	10	2	1	−1	$-\frac{1}{2}$	[He] $2s^2 2p^6$	[He] $2s^2 2p^6$	[He] [↑↓] $2s$ [↑↓][↑↓][↑↓] $2p$

2. The electron configurations for Na through Ar. The principal quantum number has changed: $n = 2 \rightarrow n = 3$.

		2. *spdf notation*	3. *Extended spdf*	4. Orbital diagram
Na	$Z = 11$	[Ne]$3s$	[Ne]$3s$	[Ne] [↑]$_{3s}$
Mg	$Z = 12$	[Ne] $3s^2$	[Ne] $3s^2$	[Ne] [↑↓]$_{3s}$
Al	$Z = 13$	[Ne] $3s^2 3p^1$	[Ne] $3s^2 3p_x$	[Ne] [↑↓]$_{3s}$ [↑][ ][ ]$_{3p}$
Si	$Z = 14$	[Ne] $3s^2 3p^2$	[Ne] $3s^2 3p_x 3p_y$	[Ne] [↑↓]$_{3s}$ [↑][↑][ ]$_{3p}$
P	$Z = 15$	[Ne] $3s^2 3p^3$	[Ne] $3s^2 3p_x 3p_y 3p_z$	[Ne] [↑↓]$_{3s}$ [↑][↑][↑]$_{3p}$
S	$Z = 16$	[Ne] $3s^2 3p^4$	[Ne] $3s_2 3p_x^2 3p_y 3p_z$	[Ne] [↑↓]$_{3s}$ [↑↓][↑][↑]$_{3p}$
Cl	$Z = 17$	[Ne]$3s^2 3p^5$	[Ne] $3s^2 3p_x^2 3p_y^2 3p_z$	[Ne] [↑↓]$_{3s}$ [↑↓][↑↓][↑]$_{3p}$
Ar	$Z = 18$	[Ne]$3s^2 3p^6$	[Ne]$3s^2 3p^6$	[Ne] [↑↓]$_{3s}$ [↑↓][↑↓][↑↓]$_{3p}$

3. Electron configurations for K through Kr. Note Hund's rule extended for Cr and Cu.

	Z	2. *spdf* notation	4. Orbital diagram
K	19	[Ar]4s	[Ar] [↑]$_{4s}$
Ca	20	[Ar]$4s^2$	[Ar] [↑↓]$_{4s}$
Sc	21	[Ar]$4s^2 3d^1$	[Ar] [↑][ ][ ][ ][ ]$_{3d}$ [↑↓]$_{4s}$
Ti	22	[Ar]$4s^2 3d^2$	[Ar] [↑][↑][ ][ ][ ]$_{3d}$ [↑↓]$_{4s}$
V	23	[Ar]$4s^2 3d^3$	[Ar] [↑][↑][↑][ ][ ]$_{3d}$ [↑↓]$_{4s}$
Cr	24	[Ar]$4s^1 3d^5$	[Ar] [↑][↑][↑][↑][↑]$_{3d}$ [↑]$_{4s}$
Mn	25	[Ar]$4s^2 3d^5$	[Ar] [↑][↑][↑][↑][↑]$_{3d}$ [↑↓]$_{4s}$
Fe	26	[Ar]$4s^2 3d^6$	[Ar] [↑↓][↑][↑][↑][↑]$_{3d}$ [↑↓]$_{4s}$
Co	27	[Ar]$4s^2 3d^7$	[Ar] [↑↓][↑↓][↑][↑][↑]$_{3d}$ [↑↓]$_{4s}$
Ni	28	[Ar]$4s^2 3d^8$	[Ar] [↑↓][↑↓][↑↓][↑][↑]$_{3d}$ [↑↓]$_{4s}$
Cu	29	[Ar]$4s^1 3d^{10}$	[Ar] [↑↓][↑↓][↑↓][↑↓][↑↓]$_{3d}$ [↑]$_{4s}$
Zn	30	[Ar]$4s^2 3d^{10}$	[Ar] [↑↓][↑↓][↑↓][↑↓][↑↓]$_{3d}$ [↑↓]$_{4s}$
Ga	31	[Ar]$4s^2 3d^{10} 4p^1$	[Ar] [↑↓][↑↓][↑↓][↑↓][↑↓]$_{3d}$ [↑↓]$_{4s}$ [↑][ ][ ]$_{4p}$
Ge	32	[Ar]$4s^2 3d^{10} 4p^2$	[Ar] [↑↓][↑↓][↑↓][↑↓][↑↓]$_{3d}$ [↑↓]$_{4s}$ [↑][↑][ ]$_{4p}$
As	33	[Ar]$4s^2 3d^{10} 4p^3$	[Ar] [↑↓][↑↓][↑↓][↑↓][↑↓]$_{3d}$ [↑↓]$_{4s}$ [↑][↑][↑]$_{4p}$
Se	34	[Ar]$4s^2 3d^{10} 4p^4$	[Ar] [↑↓][↑↓][↑↓][↑↓][↑↓]$_{3d}$ [↑↓]$_{4s}$ [↑↓][↑][↑]$_{4p}$
Br	35	[Ar]$4s^2 3d^{10} 4p^5$	[Ar] [↑↓][↑↓][↑↓][↑↓][↑↓]$_{3d}$ [↑↓]$_{4s}$ [↑↓][↑↓][↑]$_{4p}$
Kr	36	[Ar]$4s^2 3d^{10} 4p^6$	[Ar] [↑↓][↑↓][↑↓][↑↓][↑↓]$_{3d}$ [↑↓]$_{4s}$ [↑↓][↑↓][↑↓]$_{4p}$

ENERGY LEVEL DIAGRAM

Subscripts and superscripts are atomic numbers.

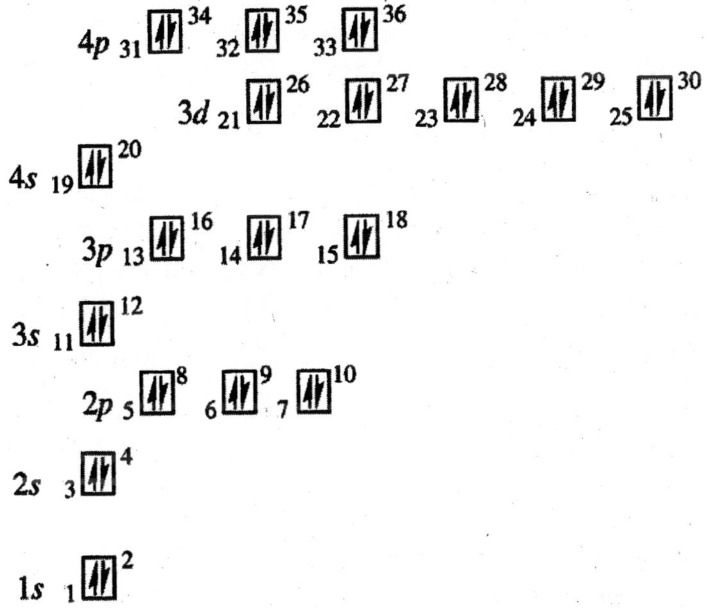

EXAMPLE 8-6 What is the electron configuration of **(a)** the valence shell of elements in Group 2A? **(b)** the element Ca?

(a) By reference to Figure 8-6, we see that each element in family 2A has two *s* electrons. The valence electron configuration is ns^2.

(b) Ca is in the fourth period; thus, its valence electron configuration is $4s^2$. Ca follows the noble gas, argon. Thus, the electron configuration of Ca is [Ar] $4s^2$.

SELF-ASSESSMENT EXERCISE

127. (a) Velocity-independent because all electromagnetic radiation has a constant velocity in a vacuum. (b) Wavelength is inversely proportional $\upsilon = c/\lambda$. The speed of light is a constant and, therefore, $\upsilon\lambda$ must be constant. A larger υ value leads to a smaller λ and vice versa. (c) Energy per mole is directly proportional to the frequency, $E = h\upsilon$. Whether this is per photon or per mole, the relationship is the same.

DRILL PROBLEMS

1. & 2. Determine the wavelength or frequency as appropriate of each of the following. Also, classify each according to its region: gamma (γ) rays, X-rays, ultraviolet (UV) radiation, visible light, infrared radiation, microwaves, radar waves, television (TV) waves, or radio waves.

A. $\nu = 2.7\times10^{14}$ Hz	B. $\lambda = 1.54$ Å	C. $\nu = 25$ MHz
D. $\lambda = 1.50$ km	E. $\nu = 1.75\times10^{22}$ Hz	F. $\lambda = 7256\ \mu m$
G. $\nu = 51.5$kHz	H. $\lambda = 124$ cm	I. $\nu = 7.42\times10^{11}$ MHz
J. $\lambda = 5.04\times10^{-5}$ cm	K. $\nu = 3.15\times10^{12}$ kHz	L. $\lambda = 12.45$ m
M. $\nu = 8.46\times10^{16}$ MHz	N. $\lambda = 1.359$ mm	O. $\nu = 5.02\times10^{11}$ kHz
P. $\lambda = 82.5$ nm	Q. $\nu = 14.6\times10^{3}$ MHz	R. $\lambda = 0.132$ pm
S. $\nu = 8.46\times10^{10}$Hz	T. $\lambda = 4.40\times10^{-6}$ pm	U. $\nu = 525$ MHz

3. For each various line of the hydrogen spectrum given, find n (the smaller integer), m (the larger integer), or λ (the wavelength of the line). Also determine the series (Lyman, Balmer, Paschen, Brackett, Pfund, or Humphreys), and identify the region of the spectrum (cosmic, visible, IR, etc.) to which the line belongs.

A. $m = 3, \lambda = 656.5$ nm	B. $n = 1, \lambda = 102.6$ nm	C. $n = 2, m = 10$
D. $n = 3, m = 4$	E. $m = 4, \lambda = 486.3$ nm	F. $n = 2, \lambda = 434.2$ nm
G. $n = 4, m = 8$	H. $n = 6,\ m = 7$	I. $m = 6,\ \lambda = 7459.9$ nm
J. $n = 1,\ \lambda = 121.57$	K. $n = 2, m = 8$	L. $n = 1,\ m = 9$

4. Determine three of the following, given one of them: (1) frequency, ν; (2) wavelength, λ; (3) energy per photon, E (in ergs for A through H, and in joules for the remainder); (4) energy per mole of photons, E_{mol} (in kJ/mol). (Note: $1\,J = 10^7$ erg.)

A. $\lambda = 1.54$ Å	B. $E_{mol} = 418$ kJ/mol	C. $\nu = 105$ MHz	D. $E = 1.07\times10^{-15}$ erg
E. $\lambda = 527$ nm	F. $E_{mol} = 28.5$ kJ/mol	G. $\nu = 6.12\times10^{12}$ Hz	H. $E = 8.13\times10^{-20}$ erg
I. $\lambda = 1.34$ km	J. $E_{mol} = 526$ J/mol	K. $\nu = 512$ kHz	L. $E = 3.14\times10^{-17}$ J
M. $\lambda = 6.15$ cm	N. $E_{mol} = 725$ J/mol	O. $\nu = 34.2\times10^{15}$ Hz	P. $E = 1.24\times10^{-13}$ J

5. (1) Supply the letter designations (such as p_x, p_y, and p_z) and the set of magnetic quantum numbers (such as $-1, 0$ and $+1$) for the orbitals in each of the following subshells:

A. 1*s*	B. 3*p*	C. 4*d*	D. 5*s*	E. 4*f*
F. 4*p*	G. 3*s*	H. 2*p*	I. 5*d*	J. 3*d*

(2) Give the number and letter designations of all subshells (such as 2*s* and 2*p*), as well as the total number of electrons in each of the following shells.

K. *K* shell	L. *M* shell	M. 2nd shell	N. $n = 4$	O. 5th level

6. Sketch each of the following sets of orbitals, each set on the same axes and to approximate scale.

A. $1s, 2s, 3s$ B. $2p_x, 2p_y$ C. $3d_{xy}, 3d_{x^2-y^2}$ D. $2p_x, 3p_x$

E. $3p_z, 3d_{z^2}$ F. $3d_{xz}, 3d_{z^2}$ G. $3s, 3p_x$ H. $3d_{xy}, 4d_{xy}$

7. The symbol for each element is followed by an electron configuration that violates one of the following principles: (1) the order of filling, or the Aufbau principle; (2) the Pauli exclusion principle; (3) Hund's rule of maximum multiplicity; (4) Hund's rule extended—stability of full and half-full subshells; (5) allowed quantum numbers (Objective 8-10). State which principle is violated by each configuration and give the correct configuration for each atom.

A. N $1s^2\ 2s^2 2p_x^2 2p_y^1$ B. Al $1s^2\ 2s^2\ 2p^6 2d^3$ C. B $1s^2\ 2s^3$

D. P $1s^2 2s^2\ 2p^6\ 3p^5$ E. Cu $[Ar]4s^2 3d^9$ F. Be $1s^2\ 1p^2$

G. Mg [Ne] [↑↑] H. C $1s^2\ 2s^1\ 2p_x^1\ 2p_x^1\ 2p_y^1 2p_z^1$ I. V $1s^2\ 2s^2\ 2p^6 3s^2 3p^6 3d^5$

J. S [Ne] $3s^2 3p_x^2 3p_y^2$ K. Mn $[Ar]4s^1 3d^6$ L. N $1s^2\ 1p^5$

M. Ag [Kr] $5s^2\ 4d^9$ N. Ni [Kr] $5s^2 4d_{xy}^2\ 4d_{xz}^2\ 4d_{yz}^2\ 4d_{z^2}^2$

O. Na $1s^2 1p^9$ P. Sc [Ne] $3s^2 3p^6 3d^3$ Q. Cl $1s^2\ 1p^6\ 2p^6 3s^1$

R. C $1s^2\ 2s^2 2p_x^2$ S. Cl [Ne] $3s^2$ [↑↑] [↑↑] [↑] T. B $1s^1\ 2s^1\ 2p_x^1\ 2p_y^1\ 2p_z^1$

In your own words, state each of the following principles or rules and give two examples illustrating each.

U. Hund's rule extended V. Hund's rule of maximum multiplicity

W. The Aufbau principle X. Pauli exclusion principle.

8. (1) Give the symbol of the element of lowest atomic number that has the characteristics listed in each of the following parts. Also, give the electron configuration of that element.

A. one electron with $m_\ell = 2$ B. two unpaired electrons

C. four pairs of electrons D. two electrons with $n = 3$ and $\ell = 2$

E. three electrons with $n = 3$ F. 11 p electrons

G. five d electrons H. three electrons with $m_\ell = 2$

I. five electrons with $m_\ell = +1$ J. five electrons with $m_\ell = 0$

K. five electrons with $m_s = \frac{1}{2}$

(2) The last few quantum numbers of the electrons or the last part of the electron configuration of different elements is given in each part below. Identify each element based on its electron configuration. (For example, boron might be $n = 2, \ell = 0,\ m_\ell = 0, m_s = +\frac{1}{2}; n = 2, \ell = 1, m_\ell = -1, m_s = -\frac{1}{2}$)

L. $3s^2\ 3p^4$ M. $n = 4, \ell = 0, m_\ell = 0, m_s = -\frac{1}{2}; n = 3, \ell = 2, m_\ell = -2, m_s = +\frac{1}{2}$

N. $5s^2\ 4d^2$ O. $n = 5, \ell = 2, m_\ell = +2, m_s = -\frac{1}{2}; n = 6, \ell = 1, m_\ell = -1, m_s = +\frac{1}{2}$

P. $5d^1\ 4f^1$ Q. $4d^5$ R. $3d^3$

S. $4p^4$ T. $n = 5, \ell = 0, m_\ell = 0, m_s = +\frac{1}{2}; n = 5, \ell = 0, m_\ell = 0, m_s = -\frac{1}{2}$

(3) Give the electron configuration of each of the following elements with both the abbreviated *spdf* notation and also an abbreviated orbital diagram.

			U. N	V. S
W. Cl	X. Si	Y. Cu	Z. Na	Γ. Cs
Δ. Ce	Θ. Sc	Λ. O	Π. Fe	Σ. Sr

9. (1) Give the general valence electron configuration for each of the following groups of elements. Use n to designate the principal quantum number; use x as a superscript if the number of electrons in a subshell can have different values. (For example, for the inner transition elements, one would write: $ns^2(n-1)d^1(n-2)f^x$)

A. Transition element	B. Alkaline earths	C. Halogens
D. Group 4A elements	E. Group 3A elements	F. Group 6A elements
G. Alkali metals	H. Noble gases	I. Chalcogens
J. Group 3B elements	K. Group 4B elements	L. Group 5A elements

(2) Determine the valence electron configuration of each of the following atoms by referring only to the periodic table. Omit the principal quantum number (thus, N would be s^2p^3). Also omit d electrons (they are not valence electrons).

		M. P	N. Si	O. Ne	P. Bi
Q. Rb	R. Se	S. In	T. I	U. Cs	V. Xe
W. Ba	X. Ga	Y. Po	Z. Pb	Γ. Cl	Δ. Be

(3) Refer only to the periodic table, and write the electron configuration of each element that follows. Use the abbreviated *spdf* notation, in which Mn is [Ar] $4s^2 3d^5$. Use the extended Hund's rule where necessary. Check your predictions with the actual configurations *in the Appendix of the text.*

Θ. Cu	Λ. Mo	Ξ. Au	Π. Eu	Σ. La	Υ. Os
Φ. Fe	Ψ. V	Ω. Nb	Ø. Y		

Further Questions. (1) Use the Aufbau principle, the Pauli exclusion principle, and Hund's rule of maximum multiplicity to write a reasonable electron configuration for an atom of each element.

A. K	B. C	C. Mn	D. Sc	E. Cr	F. Au
G. Si	H. Zn	I. Mg	J. F	K. P	L. Fe
M. U	N. Ne	O. Pr	P. Ag	Q. Al	R. Sn
S. Ge	T. Mo	U. Ga	V. Ca	W. I	X. W
Y. B	Z. Cu				

(2) Compare these to the known configurations in the *text.* Which are different? Why?

QUIZZES (20 minutes each) Choose the best answer for each question.

Quiz A

1. The statement that one cannot simultaneously measure the speed and the position of the electron is the (a) Heisenberg uncertainty principle; (b) Pauli exclusion principle; (c) rule of maximum multiplicity; (e) none of these.

2. An experiment or effect that demonstrates that light is a particle is the (a) sun's spectrum; (b) destructive interference of light; (c) photoelectric effect; (d) electron microscope; (e) none of these.

3. The shape of a *p* orbital is (a) spherical; (b) similar to a figure 8; (c) similar to a four-leaf clover; (d) conical; (e) none of these.

4. Which series of subshells is arranged in order of increasing energy for multielectron atoms? (a) 6*s*, 4*f*, 5*d*, 6*p*; (b) 4*f*, 6*s*, 5*d*, 6*p*; (c) 5*d*, 4*f*, 6*s*, 6*p*; (d) 4*f*, 5*d*, 6*s*, 6*p*; (e) none of these.

5. The light with highest energy among those following is (a) television waves; (b) infrared (heat) radiation; (c) ultraviolet waves; (d) microwaves; (e) radio waves.

6. The radiation that follows that has the highest frequency has a wavelength of (a) 300 cm; (b) 200 nm; (c) 8.2 m; (d) 1.00 nm; (e) 7.31 μm.

7. A certain radiation has a frequency of $6.7 \times 10^{14} \text{s}^{-1}$. What is its wavelength in nanometers?

   (a) $(6.63 \times 10^{-34})(6.7 \times 10^{14})$; (b) $(3.0 \times 10^{8})/[(6.7 \times 10^{14})(10^{9})]$; (c) $(3.0 \times 10^{8})(10^{9})/(6.7 \times 10^{14})$; (d) $(6.7 \times 10^{14})(10^{7})/(3.0 \times 10^{8})$; (e) none of these.

8. The energy in joules of a photon of wavelength 1.23×10^{-5} m is

   (a) $(6.63 \times 10^{-34})(3.00 \times 10^{8})/(1.23 \times 10^{-5})$;
   (b) $(6.63 \times 10^{-34})(3.00 \times 10^{8})/(1.23 \times 10^{-5})$; (c) $(3.00 \times 10^{8})/(1.23 \times 10^{-5})$; (d) $(1.23 \times 10^{-5})/(6.63 \times 10^{-34})$; (e) none of these.

9. The number of unpaired electrons in the outermost subshell of a Mg atom is (a) 0; (b) 1; (c) 2; (d) 3; (e) none of these.

10. $1s^2\ 2s^2\ 2p^6\ 3s^2\ 3p^6\ 4s^2\ 3d^3$ is the ground-state electron configuration of (a) chromium; (b) vanadium; (c) scandium; (d) niobium; (e) none of these.

11. A sodium atom must gain or lose how many electrons to achieve an inert gas electron configuration? (a) gain 2; (b) gain 1; (c) lose 1; (d) lose 2; (e) none of these.

12. [Ar] $4s^2\ 3d^1$ is the ground-state electron configuration for (a) titanium; (b) zirconium; (c) vanadium; (d) calcium; (e) none of these.

Quiz B

1. What is an electron? (a) a wave; (b) a particle; (c) either, depending on how it is observed; (d) neither; (e) none of these.

2. The principle that is based on electrons attempting to be as far apart as possible is (a) the Bohr theory; (b) the Heisenberg principle; (c) exclusion principle; (d) Hund's rule; (e) none of these.

3. Which two orbitals are both located between the axes of a coordinate system, and not along the axes? (a) d_{xy}, d_{z^2}; (b) d_{yz}, p_x; (c) d_{xz}, p_y; (d) $d_{x^2-y^2}, p_z$; (e) none of these.

4. All of the terms that follow are the names of quantum numbers *except* (a) principal; (b) magnetic; (c) spin; (d) valence; (e) no choice is correct.

5. The radiation with the longest wavelength among those given is (a) infrared; (b) microwave; (c) radiowave; (d) X-ray; (e) ultraviolet.

6. The radiation with the highest energy has a frequency of (a) 31.2 Hz; (b) 71.3 MHz; (c) 4.12 kHz; (d) 3.00×10^{10} Hz; (e) 296 kHz.

7. Radiation with a frequency of 3×10^{15} Hz has a wavelength of (a) 10 nm; (b) 100 nm; (c) 1000 nm; (d) 10,000 nm; (e) none of these.

8. Light with a frequency of $4.5 \times 10^{10}\ \text{s}^{-1}$ has an energy in joules given by (a) $(6.63 \times 10^{-34})(3.0 \times 10^{8})/(4.5 \times 10^{10})$; (b) $(3.0 \times 10^{8})(6.63 \times 10^{-34})$; (c) $(4.5 \times 10^{10})(6.63 \times 10^{-34})$; (d) $(6.63 \times 10^{-34})/(4.5 \times 10^{10})$; (e) none of these.

9. The number of unpaired electrons in the outermost subshell of a carbon atom is (a) 0; (b) 1; (c) 2; (d) 3; (e) none of these.

10. The four quantum numbers that could identify the third $3p$ electron in sulfur are (a) $n = 3,\ \ell = 0, m_\ell = +1, m_s = +\frac{1}{2}$; (b) $n = 2,\ \ell = 2, m_\ell = -1, m_s = +\frac{1}{2}$; (c) $n = 3,\ \ell = 2, m_\ell = +1, m_s = -\frac{1}{2}$; (d) $n = 3,\ \ell = 1, m_\ell = -1, m_s = +\frac{1}{2}$; (e) none of these.

11. [Kr] $5s^2\ 4d^{10}\ 5p^5$ is the electron configuration of (a) Br; (b) I; (c) At; (d) Te; (e) none of these.

12. Which of the following represents the electron configuration of the element having the atomic number 17? (a) $1s^2\ 2p^8\ 3d^7$; (b) $1s^2\ 2s^8\ 3p^7$; (c) $1s^2\ 2p^2\ 2d^6\ 3f^7$; (d) $1s^2\ 2s^2\ 2p^6\ 3s^2\ 3p^5$; (e) none of these.

Quiz C

1. That the electron configuration of nitrogen is $1s^2\,2s^2\,2p_x^1\,2p_y^1\,2p_z^1$ rather than $1s^2\,2s^2\,2p_x^2\,2p_y^1$ is postulated by the (a) Bohr-Sommerfeld theory; (b) the Pauli exclusion principle; (c) the Heisenberg uncertainty principle; (d) Hund's rule of maximum multiplicity; (e) none of these.

2. In the equation $1/\lambda = R(1/2^2 - 1/n^2)$, R is known as the (a) ideal gas constant; (b) Boltzmann constant; (c) Rydberg constant; (d) Balmer constant; (e) none of these.

3. The shape of most *d* orbitals is (a) spherical; (b) a figure 8; (c) a figure 8 with a donut; (d) a four-leaf clover; (e) none of these.

4. An electron in a 4*f* orbital has principal (*n*) and orbital (ℓ) quantum numbers, respectively, of (a) 3 and 4; (b) 4 and 4; (c) 3 and 3; (d) 4 and 3; (e) none of these.

5. Among those that follow, the unit of frequency is (a) cm^{-1}; (b) s^{-1}; (c) Hz^{-1}; (d) pm; (e) none of these.

6. Of the following, the radiation that has the highest energy has a wavelength of (a) 560 nm; (b) 56.0 cm; (c) 5.60 μm; (d) 300 nm; (e) 56.0 nm.

7. The frequency of a microwave with a wavelength of 0.750 cm is (a) $0.750/3.00\times10^{10}$; (b) $06.63/10^{-27}/0.750$; (c) $3.00\times10^{10}/0.750$; (d) $0.750/6.63\times10^{-27}$; (e) none of these.

8. Radiation with a frequency of $3.3\times10^{15}\,\text{s}^{-1}$ has what energy in joules per photon? (a) 2.0×10^{-49}; (b) 6.0×10^{-19}; (c) 3.0×10^{18}; (d) 2.2×10^{-18}; (e) none of these.

9. The number of unpaired electrons in the outermost subshell of a Cl atom is (a) 0; (b) 1; (c) 2; (d) 3; (e) none of these.

10. The four quantum numbers that could identify the *second 2 s* electron in the nitrogen atom are (a) $0, m_\ell = +1, m_s = +\frac{1}{2}$; (b) $n = 2, \ell = 1, m_\ell = 0, m_s = -\frac{1}{2}$; (c) $n = 2, \ell = 0, m_\ell = 0, m_s = -\frac{1}{2}$; (d) $n = 3, \ell = 0, m_\ell = 0, m_s = +\frac{1}{2}$; (e) none of these.

11. A Mg atom must gain or lose how many electrons to achieve an inert gas electron configuration? (a) lose 1; (b) lose 2; (c) lose 3; (d) gain 1; (e) none of these.

12. The ground-state electron configuration of iron is (a) $[\text{Ar}]4s^2\,3d^6$; (b) $[\text{Xe}]6s^2\,4f^{14}\,3d^7$; (c) $[\text{Ar}]3d^8$; (d) $[\text{Ar}]4s^1 3d^7$; (e) none of these.

Quiz D

1. In Bohr's atomic theory, when an electron moves from one energy level to another energy level more distant from the nucleus of the same atom (a) energy is emitted; (b) energy is absorbed; (c) no change in energy occurs; (d) light is given off; (e) none of these.

2. We employ several rules to determine electron configurations. Which do we not use? (a) Aufbau principle; (b) Heisenberg principle; (c) Pauli principle; (d) Hund's rule; (e) none, we use all of these.

3. Which two orbitals are both located between the axes of a coordinate system and not along the axes? (a) $p_z, d_{x^2-y^2}$; (b) d_{z^2}, d_{xy}; (c) d_{xy}, p_z; (d) d_{xz}, p_x; (e) none of these.

4. When an electron has a principal quantum number of $1\,(n=1)$ the orbital quantum number l must be (a) $+1$; (b) 0; (c) -1; (d) without restriction; (e) none of these.

5. In spectroscopy, λ represents (a) frequency; (b) energy; (c) speed; (d) wavelength; (e) none of these.

6. The light with the longest wavelength has which of the following frequencies? (a) 3.00×10^{13} Hz; (b) 4.12×10^{5} Hz; (c) 8.50×10^{20} Hz; (d) 9.12×10^{12} Hz; (e) 3.12×10^{9} Hz.

7. A radio wave has a frequency of 5.0 kHz. Its wavelength in meters is (a) $(3.0\times10^{10})/5.0$; (b) $(3.0\times10^{10})/(5.0\times10^{3})$; (c) $(3.0\times10^{8})/(5.0\times10^{3})$; (d) $(3.0\times10^{8})/5.0$; (e) none of these.

8. Radiation with a wavelength of 500.0 nm has an energy in joules of (a) $(6.63\times10^{-34})(3.00\times10^{8})/(5.00\times10^{-7})$; (b) $(6.63\times10^{-34})(5.00\times10^{-5})/(3.00\times10^{8})$; (c) $(6.63\times10^{-34})(3.00\times10^{8})/500.$; (d) $(3.00\times10^{8})((500.)/(6.63\times10^{-34})$; (e) none of these.

9. Unpaired electrons are found in the ground-state atoms of (a) Ca; (b) Ne; (c) Mg; (d) P; (e) none of these.

10. The four quantum numbers of the *last* electron of a calcium atom could be (a) $n=4, \ell=1, m_\ell=0, m_s=+\frac{1}{2}$; (b) $n=3, \ell=0, m_\ell=1, m_s=-\frac{1}{2}$; (c) $n=4, \ell=1, m_\ell=0, m_s=-\frac{1}{2}$; (d) $n=4, \ell=0, m_\ell=0, m_s=-\frac{1}{2}$; (e) none of these.

11. The number of electrons that a P atom must acquire to achieve a noble gas electron configuration is (a) 1; (b) 2; (c) 3; (d) 4; (e) none of these.

12. The ground-state electron configuration of silicon is (a) $1s^2\,2s^2\,2p^6\,3s^2\,3p_x^2$; (b) $1s^2\,2s^2\,2p^6\,3s^2\,3p_x^1, 3p_y^1$; (c) $1s^2\,2s^2\,2p^2$; (d) $1s^2\,2s^2\,2p^6\,3s^2\,3d^2$; (e) none of these.

Sample Test (20 minutes)

1. An FM station broadcasts with a 3.28 m wavelength. What is the energy (in kJ) per mole of photons emitted by this radio station?

2. Sketch the outline of each of the following orbitals: (a) $3p_x$; (b) $3d_{xy}$; (c) $4s$.

3. Give the full *spdf* electron configuration for each of these atoms: (a) P; (b) Mn; (c) Si

4. Give the abbreviated orbital diagram of the electron configurations of each of the following: (a) Co; (b) Al; (c) O.

9 THE PERIODIC TABLE AND SOME ATOMIC PROPERTIES

CHAPTER OBJECTIVES

1. Illustrate the periodic law with graphs of selected properties of the elements as a function of atomic number.

The periodic law states that the physical and chemical properties of the elements vary periodically with atomic number. This is shown in the graphs of atomic volume, atomic radius, ionization energy, and melting point (Figure 9-1) against atomic number. The same general pattern of peaks and valleys occurs in each graph. Although peaks or valleys do not occur for elements in the same family on each graph, there is one peak and one valley within each period of the periodic table for each property that is plotted. Similar periodic variation is evident in a periodic table of atomic properties, such as the data presented in Table 9-1.

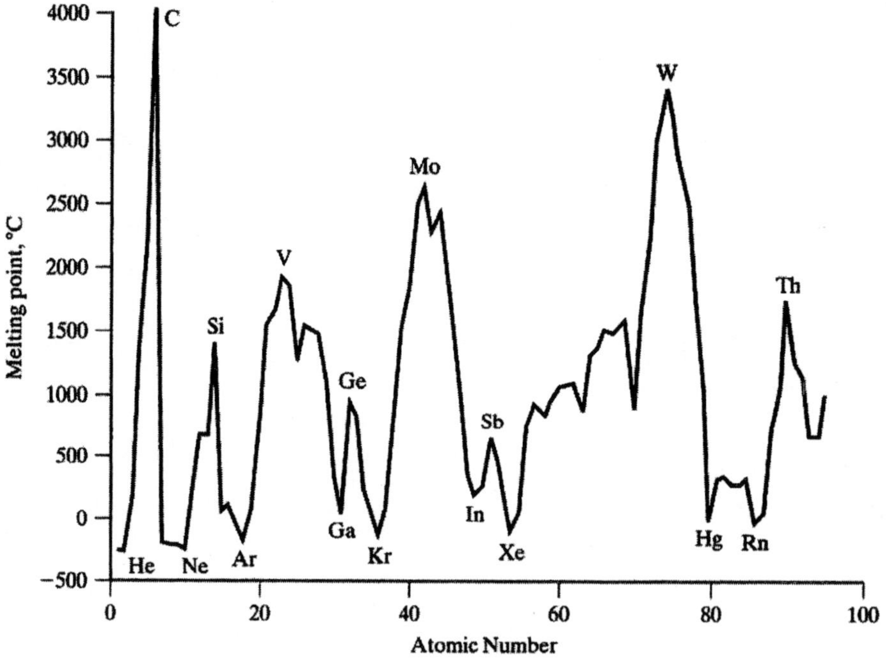

FIGURE 9-1 Melting Points of Elements vs. Atomic Number

2. Use the terms *periods, groups, families, representative elements*, and *transition elements* to describe individual elements and groupings of elements in the periodic table.

A *period* is a (horizontal) row in the periodic table. There are seven periods. A *group* or a *family* is a (vertical) column in the periodic table. Elements of the same family are called *cogeners*. For example, He, Ne, Ar, Kr, Xe, and Rn are cogeners. Several families have special names. Family 1A is the *alkali metals*. Family 2A is the *alkaline earth metals*. Family 7A is the *halogens* (meaning "salt-formers"). Family 8A is the *noble gases* (formerly the rare gases, or the inert gases). Family 6A is the *chalcogens* (meaning "chalk-formers"). Family 1B is the *coinage metals*. Fe, Co, and Ni form the *iron triad*. Ru, Rh, Pd, Os, Ir, and Pt are the *noble metals*. The *B* families (the "waist" of the periodic table) are the *transition metals*. or transition elements. Ce through Lu are the *lanthanide metals*, or the rare earth metals. Th through Lr are the *actinides*. The lanthanides and actinides together are the *inner transition elements*. The *A* families are the *main group elements*, or *representative elements*.

3. Describe metals, nonmetals, metalloids, and noble gases in several ways, and locate them in the periodic table.

Noble gases are the elements in Group 8A. Nonmetals are in the triangular block in the upper-right corner of the A groups H, C, N, O, F, P, S, Cl, Se, Br, and I. Metalloids are in the two diagonal rows just below and to the left of the nonmetals B, Si, Ge, As, Sb, Te, Po, and At. Metals are the remaining elements.

Physically, pure metals generally are malleable and ductile, melt at moderate temperatures, and conduct heat and electricity well. These properties may change substantially when two metals are alloyed, or melted together.

Nonmetals tend to be insulators and form small molecules that condense poorly into solids; hence the solids often melt easily (except carbon with its high melting point). Chemically, metals lose electrons and form cations, while nonmetals gain electrons and form anions. Metallic oxides form basic solutions with water, while nonmetallic oxides form acidic aqueous solutions. Metals rarely combine chemically with each other. Nonmetals combine chemically to form definite compounds (molecular compounds) with nonmetallic properties. When metals combine with nonmetals, ionic compounds are formed. These compounds have high melting points; the melt conducts electricity, while the solid does not.

Metalloids have physical and chemical properties that are between those of metals and those of nonmetals. For example, the aqueous solutions of their oxides often are *amphoteric*, meaning that they can react with either acids or bases.

*** 4. State the factors that influence atomic size; distinguish among covalent, ionic, metallic, and van der Waals radii; and describe the general trends in atomic size that occur within families and groups.**

Atoms are not small, hard balls, but indefinite spheres. Defining where the electron cloud of an atom ends is like trying to measure the extent of the metropolitan area of a city. (Such a nonbonded interaction defines the van der Waals radius.) But with both cities and atoms, we can determine the center. For atoms, one-half the distance between two centers (one-half the internuclear distance) is called the atomic size. This idea is complicated by the fact that not all atoms are normally bound together in the same way. Some are bound by covalent bonds in molecules, some are attracted to each other in ionic crystals, and some are held in metallic crystals by the force of a "sea" of electrons acting as "glue" between cations. (The distance between nuclei in such a metallic crystal equals the metallic radius.) Fortunately, it is possible to form molecules of nearly every element (except the noble gases) in which two like atoms are held together by a single covalent bond. The covalent radius of these molecules often is called the atomic radius. Several family and periodic trends in atomic size follow. (Data is in Table 10-1.)

1. Atomic size regularly increases from top to bottom in (that is, "down") a family (see Figure 9-2). More electrons with the same outer configuration are present around the nucleus.

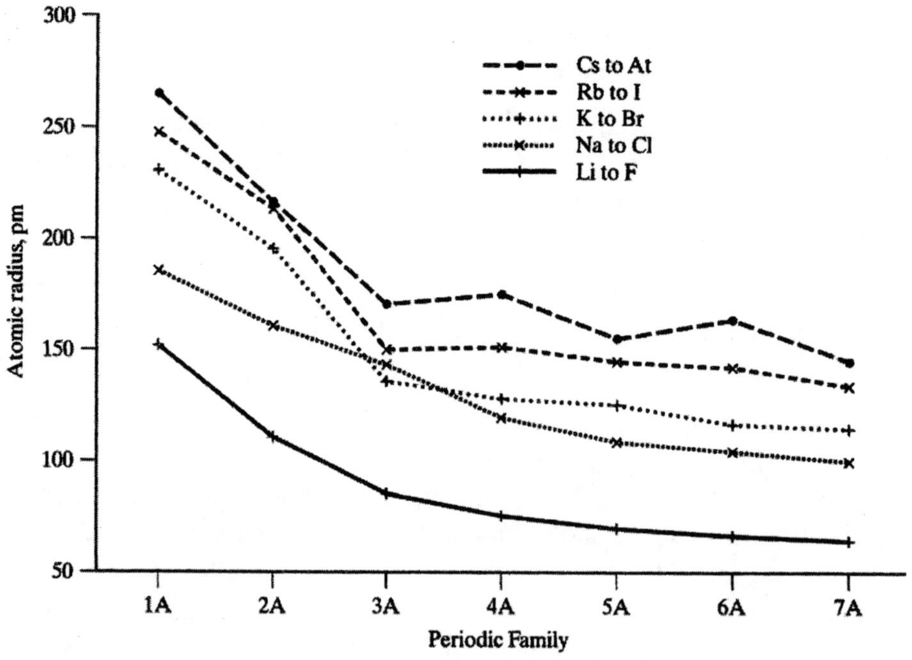

FIGURE 9-2 Atomic Radii of the Representative Elements

TABLE 9-1 Metallic, Covalent, and Ionic Radii; Ionization Energies and Electron Affinities

1A	2A	3B	4B	5B	6B	7B	--	8B	--	1B	2B	3A	4A
H					**Sym**			Outlined for metalloids					
r 37					R	MetCov pm		Covalent radii are underlined. Metallic radii are not.					
r					R	Ionic pm		Each ionic radius is followed by the ion's charge.					
154 – 1													
I 1312					I	Ion.En.		kJ/mol					
E – 72.					E	-ElnAff		kJ/mol Estimated values are followed by *					
Li	**Be**											***B***	**C**
r 152	111											R86	<u>77</u>
r 76 + 1	45 + 2											r27 + 3	260 – 4
r												r	16 + 4
I 520	899											I801	1086
E – 59.6	+241*											E-27	– 122.3
Na	**Mg**											**Al**	**Si**
r 186	160											r143	118
r 102 + 1	72 + 2											r54 + 3	<u>271 – 4</u>
r												r	41 + 4
I 496	738											I578	786
E – 52.9	+ 232*											E – 44	– 133 6
K	**Ca**	**Sc**	**Ti**	**V**	**Cr**	**Mn**	**Fe**	**Co**	**Ni**	**Cu**	**Zn**	**Ga**	**Ge**
r 232	197	162	147	134	128	127	126	125	124	128	134	135	128
r 138 + 1	100 + 2	75 + 3	86 + 2	79 + 2	80 + 2	83 + 2	78 + 2	75 + 2	69 + 2	74 + 2	62 + 3	<u>272 – 4</u>	58 + 3
									77 + 1				
r			61 + 4	64 + 3	62 + 3	54 + 4	65+3	61+3	60+3	73+1		120+2	73+2
I 419	590	631	658	650	653	717	759	758	737	745	906	579	762
E-48.4	+156*		–19	–48	–64		–24	–67	–111	–118.3	0	–29	–116
Rb	**Sr**	**Y**	**Zr**	**Nb**	**Mo**	**Tc**	**Ru**	**Rh**	**Pd**	**Ag**	**Cd**	**In**	**Sn**
r 248	215	180	160	146	139	136	134	134	137	144	149	167	151
r 152+1	118+2	90+	72+4	68+4	65+4	65+4	68+3	67+3	86+2	115+1	95+2	140+1	118+2
r				64+5	59+6	60+5	62+4	60+4	62+4			80+3	69+4
I 403	549	616	660	664	685	702	711	720	805	731	868	558	709
E–46.9	+167*	0	–48	–96	–97	–68	–106		–58	–125.7		–29	–107.3
Cs	**Ba**	**La**	**Hf**	**Ta**	**W**	**Re**	**Os**	**Ir**	**Pt**	**Au**	**Hg**	**Tl**	**Pb**
r 265	217	183	159	146	139	137	134	136	139	144	151	170	175
r 167+1	136+2	103.	71+4	64+5	66+4	63+4	63+4	68+3	80+2	137+1	119+1	150+1	119+2

r				72+3	60+6	55+6	55+6	63+4	63+4	85+3	102+2	89+3	78+4
I 376	503	538	680	761	770	760	840	880	870	890	1007	589	716
E–45 5	+52*	–48		–77	–58	–14	–106	–154	–205 3	–222.8	–48	–19.2	–106
Fr	**Ra**	**Ac**	**Ku**	**Ha**	**Sg**	**Ns**	**Hs**	**MtUunUuuUub**					
r 270	220	188											
r 180+1	140+2	111.											
I 375*	509	670											

	Ce	**Pr**	**Nd**	**Pm**	**Sm**	**Eu**	**Gd**	**Tb**	**Dy**	**Ho**	**Er**
R	182	182	182	183	180	208	180	177	178	176	176
R	102+3	99+3	98+3	97+3	97+3	95+3	94+3	92+3	91+3	90+3	89+3
R	87+4	85+4				117+2			107+2		
I	528	523	530	535	543	547	592	564	572	581	589
	Th	**Pa**	**U**	**Np**	**Pu**	**Am**	**Cm**	**Bk**	**Cf**	**Es**	**Fm**
R	179	163	156	155	159	173	174	----	186	186	
R	108+3	104+3	103+3	101+3	100+3	98+3	97+3	98+3	95+3	98+3	97+3
R	95+4	90+4	89+4	87+4	86+4	89+4	85+4	87+4	82+4 82 + 4		
I	671	568	587	597	585	579	581	601	608	619	627

Source for atomic & ionic radii, electron affinities and ionization energies: Lange's Handbook of Chemistry, 1

2. Atomic size gradually decreases from left to right in (that is, "across") a period of representative elements, as shown in Figure 9-2. The explanation is that, within a family, all electrons are being added to the same shell. It is helpful to think of these shells as similar to the layers of an onion. As each electron is added, the shell becomes more populated. But at the same time, protons are being added to the nucleus, making it more positively charged. This means the nucleus attracts the electrons more strongly. Therefore, the shell is pulled closer to the nucleus.

3. Across a transition series, size gradually—but somewhat irregularly—decreases then increases at the end of the series (see Figure 9-3). Within a transition series, electrons are added to an inner (*d*) shell. This inner shell gradually will decrease in size, but since these *d* electrons are between the nucleus and the outer electrons, they "shield" the outer electrons from the nuclear charge. Thus, the outer shell should increase in size. The two trends oppose each other, with the shielding effect initially being less important, but predominating at the end of the transition series. A consequence of the atomic radius being smallest in Family 8B—the iron triad (Fe, Co, Ni) and the noble metals (Ru, Rh, Pd, Os, Ir, Pt)—is that these metals are more dense than those in the rest of the periodic table. In fact, osmium (22.61 g/cm^3) and iridium (22.65g/cm^3) are the densest substances known.

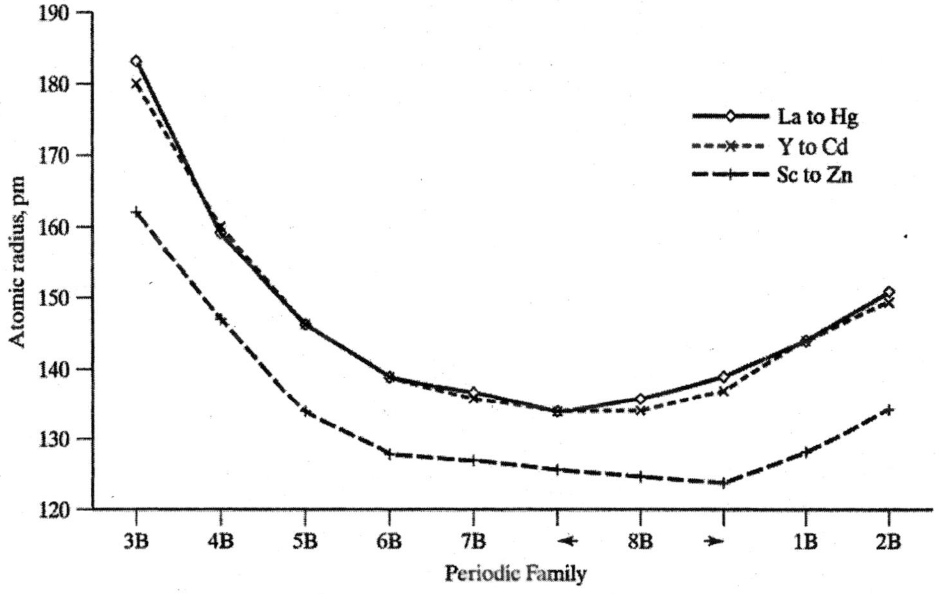

FIGURE 9-3 Atomic Radii of the Transition Elements

4. Across the lanthanides, atomic radii gradually—but irregularly—decrease, and the radii of the +3 cations steadily decrease (see Figure 9-4). This steady decrease in size, called the *lanthanide contraction*, is the result of the addition of electrons to a deep inner subshell (the 4*f* subshell). Note carefully the relative sizes of ions and atoms in Figure 9-4; the ions are about half the size of the atoms. The lanthanide contraction helps to explain the similarity of chemical and physical properties between the transition elements of the fifth (Y to Cd) and sixth (La to Hg) periods.

Because of the lanthanide contraction, atomic sizes are almost the same in these two transition periods (see Figure 9-3). Since the outer-electron configurations are the same, the properties are very similar.

5. The sizes of isoelectronic ions decrease as the positive charge on the ions increase (see Figure 9-5). The explanation is that the larger nuclear charge exerts more pull on each electron and, thus, pulls these electrons closer to the nucleus.

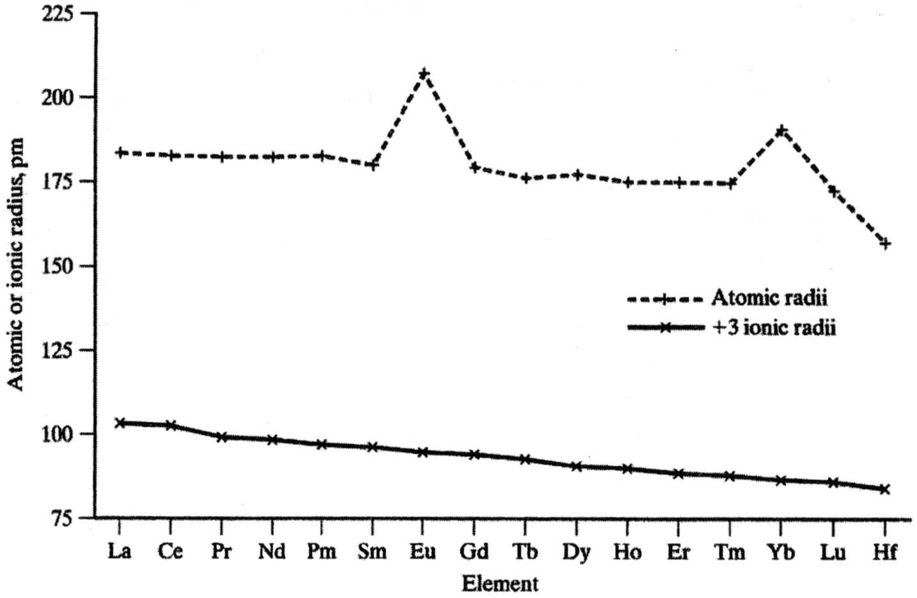

FIGURE 9-4 Atomic and Ionic Radii of the Lanthanide Elements

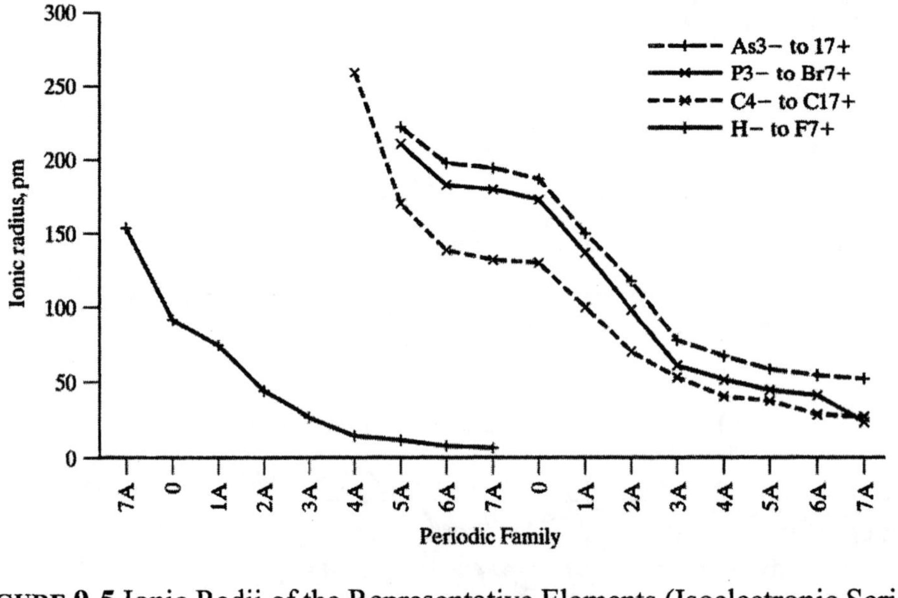

FIGURE 9-5 Ionic Radii of the Representative Elements (Isoelectronic Series)

*** 5. Define first, second, etc., ionization energies; describe the factors that affect the magnitude of these ionization energies; and relate ionization energies to the location of elements in the periodic table.**

Ionization energy (or ionization potential, IP) is the energy required by the process $M(g) \longrightarrow M^+(g) + e^-(g)$. The ease of losing an electron is a measure of the ability of an element to act as a metal. Ionization energies are measured in kJ/mol, or in electron volts per atom (eV/atom), where 1 eV/atom = 96.49 kJ/mol.

1. Ionization energy decreases down a family of representative elements (see Figure 9-6). As atoms get larger, the outermost electrons are farther from the nucleus and, hence, are less strongly held.

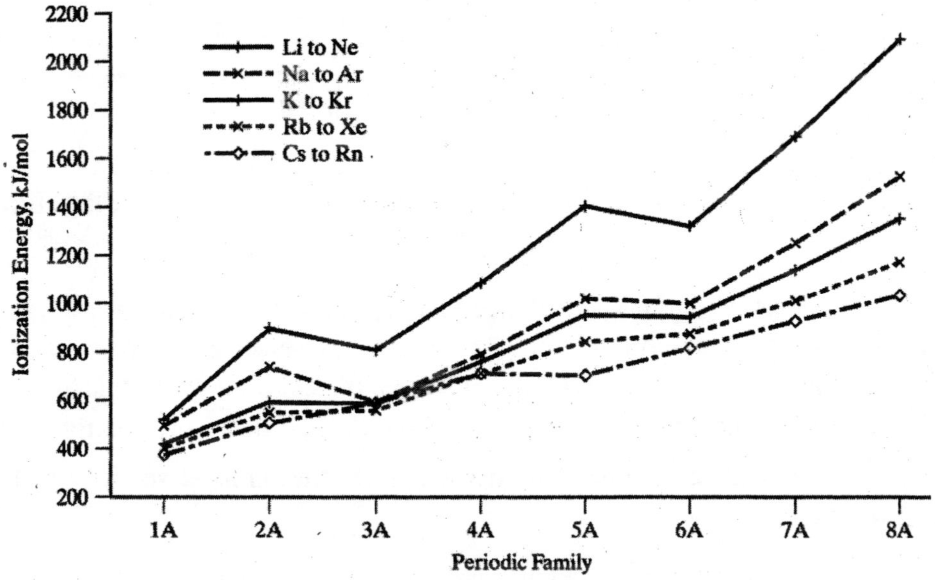

FIGURE 9-6 Ionization Energies (kJ/mol) of the Representative Elements

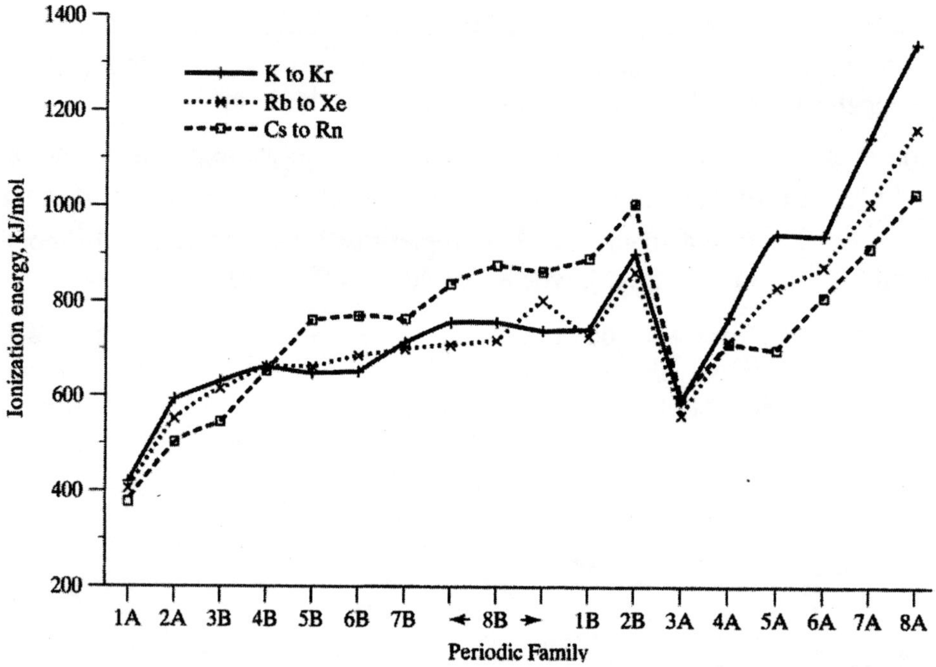

FIGURE 9-7 Ionization Energies (kJ/mol) for the Elements of the Fourth, Fifth, and Sixth Periods.

2. Ionization energy gradually, but irregularly, increases across a representative period (see Figure 9-6). As atomic size decreases, electrons are more strongly attracted to the nucleus. Note that the trend is irregular in that it is harder than expected to remove electrons from s^3 and p^3 configurations and relatively easy to ionize p^1 and p^4 configurations. This demonstrates the unusual stability of full and half-full subshells.

3. Ionization energy gradually, but irregularly, increases across a transition period (see Figure 9-7). Again, as atomic size decreases, removing the outer electrons becomes more difficult.

4. There is no noticeable trend in ionization potentials of the lanthanides.

5. The second ionization energy is greater than the first ionization energy, the third is greater than the second, and so forth. This is expected, because in the process accompanying the second ionization energy, an electron is being removed from a cation that already has a charge of $+1$, rather than from a neutral atom.

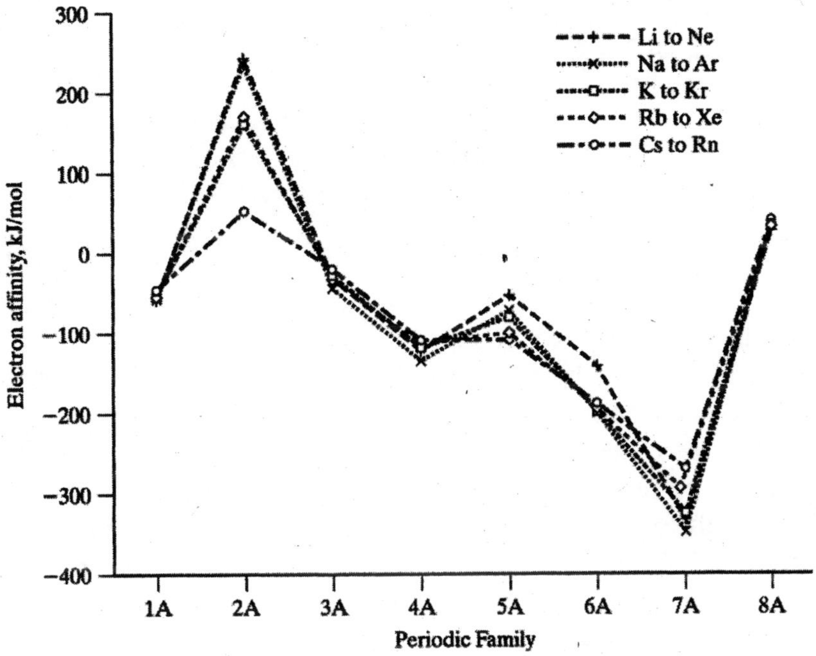

FIGURE 9-8 Electron Affinities: $X(g) + e^- \longrightarrow X^-(g)$

*** 6. Define electron affinity, cite the factors influencing its magnitude, and describe the variation of electron affinity within periods and groups of the periodic table.**

Electron affinity is the energy required by the process $X(g) + e^- \longrightarrow X^-(g)$. This is an indication of how nonmetallic an element is. Two trends are evident in Figure 9-8; electron affinity increases down a group and decreases gradually, but irregularly, across a period.

The increase in electron affinity down a group is explained by each atom being larger than the atom above it in the group. This means that the added electron is further away from the nucleus than in the previous smaller (and lighter) atom. With a larger distance between the negatively charged electron and the positively charged nucleus, the force of attraction is relatively smaller. Thus, electron affinity increases—becomes more positive, indicating a less exothermic reaction—as one proceeds down a group.

The trend toward smaller (more exothermic) electron affinities across a period also is size-related. Recall that atoms become smaller across a period (Figure 9-2). The added electron, thus, is closer to the nucleus and should be more strongly attracted. This results in a more exothermic electron affinity. The irregularities are due to adding an electron to a full subshell (an s^2 configuration) or a half-filled subshell (p^3). Because these two electron configurations are relatively stable, adding an electron to them is not very favorable energetically. In addition, adding an electron to an s^1, a p^2, or a p^5 electron will result in the creation of a full or half-filled subshell (s^2, p^3, or p^6, respectively). Creation of such a stable electron configuration is energetically favored—an exothermic process.

Although the first electron affinity (EA) often is negative (an exothermic process), the second EA usually is positive. Adding an electron to an ion that has a charge of -1 is a process that we expect to require energy.

*** 7. Relate the metallic character of an element to its position in the periodic table.**

We have seen that the ease of losing an electron is a measure of an element's metallic character. This ease is indicated by a small ionization potential, which decreases from right to left in a period and from top to bottom in a family. Similarly, the eagerness of gaining an electron is a measure of an element's nonmetallic character. This eagerness is measured by an exothermic (negative) electron affinity, which becomes more negative from left to right in a period and from bottom to top in a family. We realize that a good metal also is a poor nonmetal; that is, the two properties are opposites. Thus, we can combine the trends in ionization energy and electron affinity into metallic/nonmetallic character. Metallic character increases from right to left in a period and from top down to bottom in a family. Nonmetallic character increases from left to right in a period and from bottom up to top in a family.

*** 8. Relate the magnetic properties of an atom or ion to its electron configuration.**

An ion's electron configuration is determined from that of its atom. The electrons that are removed to form an ion are those of *highest principal quantum number*. Thus, when an iron atom ($[Ar]3d^6\,4s^2$) forms an iron(II) ion, the two electrons lost are those of the 4*s* subshell, resulting in the electron configuration $[Ar]\;3d^6\,4s^0$. The resulting ion obeys Hund's rule. When all electrons in the atom or ion are paired, the atom or ion is repelled by a magnetic field and is, thus, diamagnetic. If there are unpaired electrons, the atom or ion is paramagnetic and is attracted into a magnetic field. The degree of attraction depends on the number of unpaired electrons in each atom or ion. Paramagnetic character is an important way of verifying electron configurations.

9. Use the periodic law, the periodic table, and trends in atomic properties to make predictions about the physical and chemical behavior of various elements.

In predicting metallic and nonmetallic character, ionization energies and electron affinities indicate that, when metallic atoms chemically combine with nonmetallic atoms, the metallic atoms will form cations and the nonmetallic atoms will form anions. For example, when sodium combines with chlorine, sodium cations and chloride anions—the constituents of the compound sodium chloride—are formed. Lewis theory, studied in Chapter 11, enables us to predict that a sodium cation should have a charge of $+1$ and the chloride anion should have a charge of -1. But notice in Table 3-2 that alkali metal atoms (Group 1A) form cations with charges of $+1$, alkaline earth metal atoms (Group 2A) form cations with charges of $+2$, chalcogen atoms (Group 6A) form anions with charges of -2, and halogen atoms (Group 7A) form anions with charges of -1.

Physical and chemical properties of members of a family often are simply averages of the two adjacent elements, above and below.

EXAMPLE 9-1 Given data for calcium and barium, estimate the melting point, density, and atomic radius of strontium. For calcium: mp = 850 °C, density = 1.54 g/cm^3, and atomic radius = 197 pm. For barium, mp = 725 °C, density = 3.15 g/cm^3, and atomic radius = 217 pm.

We expect the properties of strontium to be averages of those of calcium and barium.

$$\text{melting point} = \frac{850\,°\text{C} + 725\,°\text{C}}{2} = 788\,°\text{C}$$

$$\text{density} = \frac{1.54\ \text{g/cm}^3 + 3.15\ \text{g/cm}^3}{2} = 2.35\ \text{g/cm}^3$$

$$\text{atomic radius} = \frac{197\ \text{pm} + 217\ \text{pm}}{2} = 207\ \text{pm}$$

Properties of strontium: mp = 757 °C, density = 2.6 g/cm^3, atomic radius = 215 pm

Predictions such as these can be refined by combining trends across a period with the family trends.

SELF-ASSESSMENT EXERCISES

73. The element is Sb. This element has some similarity to (c), another metalloid. Elements also have similarity to those above and below, in this case (b) Bi.

74. (a) K Atomic radii decrease from left to right in the periodic table.

78. (d) Se^{2-} will have the electron configuration [Ar]$3d^{10}4s^24p^6$. This configuration is isoelectronic with Kr. Sr^{2+} will lose the $5s^2$ electrons to have the Kr electron configuration.

85. (a) False. *s* electrons are more effective because of the greater probability of being close to the nucleus.

(b) Z_{eff} is the nuclear charge. The screening will be higher for a *s* electron, so Z_{eff} will be lower.

(c) True. Z is the nuclear charge without screening.

(d) True. $l = 1$ are *p* electrons, which penetrate better than $l = 2$, *d* electrons.

(e) The smaller atoms will have more effective screening and Z_{eff} will be less.

86. (a) The 1*s* orbital is closer to the nucleus so the Z_{eff} is higher.

(b) The *s* orbitals have a probability of being closer to the nucleus and, therefore, have greater penetration. Z_{eff} is higher.

(c) True. The *s* electron is less shielded.

(d) False. The *p* electrons have a probability farther from the nucleus and are, therefore, less shielded

DRILL PROBLEMS

1. Rewrite each of the lists that follow in order of increasing size (smallest to largest) of the species:

A. Ca, Sr, Mg, Be
B. Te, Po, O, S
C. Ar, He, Rn, Kr
D. Sn, Ge, Si, Pb
E. Ga, In, B, Al
F. B, Li, O, C
G. Bi, Cs, Tl, Ba
H. Sn, In, Te, Rb
I. K, Br, Ge, As
J. Mg, Al, Cl, P
K. K, V, Ti, Fe
L. Ba, Os, Hf, La
M. $Ce^{3+}, W^{3+}, TM^{3+}, Eu^{3+}$
N. Ac, W, U, Os
O. $P^{3-}, Ca^{2+}, Ar, Si^{4-}$
P. $O^{2-}, Na^{+}, C^{4-}, Ne$
Q. $Sc^{3+}, Cl^{-}, K^{+}, S^{2-}$
R. $Br^{-}, Ga^{5-}, Sr^{2+}, Y^{3+}$
S. Si, Ge, O, Ne
T. $Mg^{2+}, F^{-}, Be^{2+}, C^{4-}$
U. K, Al, Na, N
V. N, Na^{+}, Li, Ne
W. Tl, Ga, Ba, Ge

2. Rewrite each list that follows in order of increasing value (smallest to largest) of the first ionization energy:

A. Rb, K, Na, Cs
B. N, As, Sb, P
C. Al, B, Tl, In
D. Be, Sr, Mg, Ca
E. Na, Cl, Al, Ar
F. Ca, Ge, Kr, As
G. Rb, Te, In, I
H. N, Ne, Li, B
I. F, Se, Ba, Ga
J. Sr, P, Sb, Ne
K. Cs, Cl, As, Sn
L. Al, In, Rb, F

3. Rewrite each of the following lists in order of electron affinity: smallest or most negative, to largest:

A. Br, Cl, F, I
B. N, P, Bi, Sb
C. Na, Rb, Li, K
D. K, Se, Ge, Br
E. I, Sb, In, Rb
F. F, P, In, Cs
G. Sb, Se, Cl, Tl
H. Al, B, In, Tl
I. As, Cl, In, Sr

4. Rewrite each list in order of increasing metallic character: most nonmetallic to most metallic:

A. Rb, K, Na, Cs
B. Cs, F, In, P
C. As, Br, Ca, K
D. C, N, O, F
E. Al, Cl, Mg, Na
F. Ce, Fe, Ge, Se
G. Be, Ca, Mg, Sr
H. B, Be, C, Li
I. Al, Ga, Rb, Sr

5. Rewrite each of the following lists of atoms and ions in order of increasing value of the paramagnetism of each species. When two species have the same value, write them together in parentheses, such as (He, Ne).

A. Ce^{3+}, K^+, C, P B. He, Cs^+, Sc, S C. B, Si, P, S

D. K, Ca, Sc, Ti E. V, Cr, Mn, Cu F. Zn, Ca, As, Ti

G. Na, Na^+, S, S^- H. Li, Be, B, C I. $Fe^{2+}, Fe^{3+}, Cu^+, Cu^{2+}$

QUIZZES (20 minutes each) Choose the best answer for each question.

QUIZ A

1. The electrons lost when $Fe \longrightarrow Fe^{2+}$ are (a) 4*f*; (b) 3*d*; (c) 4*s*; (d) 3*p*; (e) none of these.

2. A metalloid in the same periodic group as P is (a) N; (b) S; (c) Al; (d) As; (e) none of these.

3. An atom has the general ground-state *valence* electron configuration of ns^2np^5, where *n* is the principal quantum number. The element is (a) an alkali metal; (b) an alkaline earth metal; (c) a noble gas; (d) a halogen; (e) none of these.

4. If an element has some properties that are metallic and others that are nonmetallic, it is said to be (a) inert; (b) isoelectronic; (c) a metalloid; (d) an alloy; (e) none of these.

5. From left to right in a period of the periodic table, nonmetallic character (a) decreases; (b) increases; (c) remains the same; (d) shows irregular changes; (e) none of these.

6. In which of the following is there a consistent decrease in atomic radius as the atomic number increases? (a) halogens; (b) representative elements; (c) transition elements; (d) lanthanides; (e) none of these.

7. Of the following, the one with the smallest ionization potential is (a) Mg; (b) Na; (c) K; (d) Ca; (e) Cs.

8. An atom with an atomic radius greater than that of S is (a) O; (b) Cl; (c) Ca; (d) Li; (e) none of these.

QUIZ B

1. The number of 3*d* electrons in a Co^{3+} ion is (a) 2; (b) 4; (c) 6; (d) 7; (e) none of these.

2. A nonmetal of group 5A is (a) Sb; (b) As; (c) N; (d) Bi; (e) none of these.

3. Which of these ions is the smallest? (a) O^{2-}; (b) F^-; (c) Na^+; (d) Mg^{2+}; (e) Al^{3+}.

4. The transition elements are all metallic because of their (a) *s* electrons; (b) lack of *s* electrons; (c) atomic size; (d) *d* electrons; (e) none of these.

5. Of the following elements, which possesses the lowest (most negative) electron affinity? (a) As; (b) O; (c) S; (d) Se; (e) Te.

6. Atoms in which either *s* or *p* electrons are being added are called (a) lanthanides; (b) metalloids; (c) representative elements; (d) transition elements; (e) none of these.

7. Going across a representative period from left to right, the radii of atoms (a) decrease; (b) increase; (c) remain constant; (d) show irregular patterns; (e) none of these.

8. The more metallic an element, the lower is its (a) ionization potential; (b) number of electrons; (c) electron affinity; (d) penetration of electrons.

Quiz C

1. The atom that is more nonmetallic than O is (a) H; (b) N; (c) S; (d) Rb; (e) none of these.

2. An atom with an atomic radius greater than that of Cs is (a) Ba; (b) Rb; (c) Sr; (d) Br; (e) none of these.

3. An element that is an example of a metalloid is (a) S; (b) Zn; (c) Ge; (d) Re; (e) none of these.

4. The most common ion of an atom with the electron configuration $1s^2 2s^2 2p^5$ would have the charge (a) $+1$; (b) -1; (c) $+2$; (d) 0; (e) none of these.

5. In the periodic table, the vertical (up and down) columns are called (a) periods; (b) transitions; (c) families; (d) metalloids; (e) none of these.

6. A series of atoms or ions that have the same electron configuration is said to be (a) degenerate; (b) isoelectronic; (c) amphoteric; (d) isotopic; (e) none of these.

7. A nonmetal will have a smaller _____ when compared to a metal of the same period. (a) atomic radius; (b) ionic radius; (c) ionization potential; (d) atomic weight; (e) none of these.

8. Usually in the periodic table, electron affinity *increases* (a) top $\rightarrow$ bottom and right $\rightarrow$ left; (b) bottom $\rightarrow$ top and right $\rightarrow$ left; (c) top $\rightarrow$ bottom and left $\rightarrow$ right; (d) bottom $\rightarrow$ top and left $\rightarrow$ right; (e) none of these.

Quiz D

1. The species with a radius less than that of Ne is (a) Mg^{2+}; (b) F^-; (c) O^{2}; (d) K^+; (e) none of these.

2. The representative elements are those that (a) are in the B groups; (b) fill *s* and *p* orbitals only; (c) fill *d* orbitals; (d) are metallic only; (e) none of these.

3. Which of the following is *not* considered a metalloid? (a) Si; (b) As; (c) Ge; (d) Br; (e) Sb.

4. In general, the ionization potentials of elements decrease as one proceeds in the periodic table (a) bottom → top and right → left; (b) top → bottom and right → left; (c) bottom → top and left → right; (d) top → bottom and left → right; (e) none of these.

5. A metal will have a larger _____ when compared to a nonmetal of the same period. (a) ionization potential; (b) electron affinity; (c) number of valence electrons; (d) atomic number; (e) none of these.

6. The electron lost when ionization occurs is the one with (a) highest principal quantum number; (b) lowest principal quantum number; (c) outer-shell electron with highest orbital quantum number; (d) highest orbital quantum number; (e) none of these.

7. An element that has a large radius and a small ionization potential is likely to be a(n) (a) nonmetal; (b) metalloid; (c) metal; (d) inert gas; (e) none of these.

8. Of the following, which element has the largest ionization potential? (a) Mg; (b) Ca; (c) Sr; (d) Ba; (e) Ra.

SAMPLE TEST (15 minutes)

P_4	15	S_8	16	Cl_2	17
$3s^2 3p^3$	30.97	$3s^2 3p^4$	32.06	$3s^2 3p^5$	35.45
3 upe	106	2 upe	102	1 upe	99
17.0	−74	15.5	−200.4	18.7	−348.7
11.0	0.741	10.4	0.732	13.0	0.485
As_4	33	Selenium	Z	Br_2	35
$4s^2 4p^3$	74.92	Config.	At.wt.	$4s^2 4p^5$	79.91
3 upe	120	No. upe	rad.	1 upe	114
13.1	−77	V	EA	23.5	−324.5
10.0	0.34	I.P.	sp ht	11.8	0.29
Sb_4	51	Te(metal)	52	I_2	53
$5s^2 5p^3$	121.75	$5s^2 5p^4$	127.60	$5s^2 5p^5$	126.90
3 upe	140	2 upe	136	1 upe	133
18.4	−101	20.5	−183	25.7	−295.3
8.6	0.21	9.0	0.20	10.5	0.22

Suppose that the element selenium is yet to be discovered. Use the information in the partial periodic table above, along with the principles of periodicity, to predict the following properties of selenium.

1. Number of atoms per molecule.
2. Atomic number (Z).
3. Valence electron configuration (Config.).
4. Atomic weight (At.wt.).
5. Number of unpaired electrons in the isolated atom (No. upe)

6. Atomic radius, in picometers (rad.).
7. Atomic volume in mL/mol (*V*).
8. Electron Affinity (*E A*).
9. First ionization potential in electron volts/atom (I.P.).
10. Specific heat in joules g^{-1} °C^{-1} (sp ht).

10 CHEMICAL BONDING I: BASIC CONCEPTS

CHAPTER OBJECTIVES

1. State the basic assumptions of the Lewis theory.

Lewis theory assumes that, in compounds, an atom strives for as many paired electrons as possible, and then the electron configuration of the nearest noble gas. This second part is known as the *octet rule* because all noble gases except helium have eight valence electrons. An octet may be obtained by losing, gaining, or sharing electrons. When electrons are lost, a cation is formed. When they are gained, an anion is formed. Cations and anions combine to form ionic compounds. When electrons are shared, a covalent bond is formed.

***2. Relate the Lewis symbol for an element to its position in the periodic table.**

The valence electron configurations of all elements in a representative family (or main group) are identical except for the value of the principal quantum number. The Lewis symbol is the symbol of the element surrounded by dots—one dot for each valence electron. See Table 10-1 where X is the symbol of the element. Note the following points.

1. The number of valence electrons of a representative element equals the family number.

2. There are four "sides" (top, bottom, left, and right) to a symbol. Each side can be thought of as representing one valence orbital (s, p_x, p_y, p_z).

3. No given "side" represents any particular orbital. The following are correct Lewis symbols of oxygen.

   ·Ö· ·Ȯ: :Ȯ· :Ȯ: ·Ȯ: :Ȯ·

4. Pairs of electrons can be represented by a line rather than two dots. Either Ba: or Ba| can represent barium.

TABLE 10-1 Lewis Symbols for Representative Elements

Group or Family	Valence Electron Configuration	Lewis Structures: General	Examples	Ions
1A	s^1	X·	Na·	X^+
2A	s^2	X: *or* ·X·	Ba: ·Be·	X^{2+}
3A	s^2p^1	:X· *or* ·Ẋ·	·B: ·Ȧl·	X^{3+}
4A	s^2p^2	:Ẋ·	·C̈· ·Ṡi·	$[|\overline{X}|]^{4-}$ *or* X^{4-}

5*A*	s^2p^3
6A	s^2p^4
7A	s^2p^5
8A	s^2p^6

***3. Write Lewis structures for simple ionic compounds.**

When electrons are gained or lost, the cations and anions shown in the last column of Table 10-1 are formed. Notice that anion Lewis symbols are surrounded by square brackets to separate the electron dots from the charge of the ion. For example, the Lewis structure of calcium fluoride is depicted by Ca^{2+} and two $\left[\,|\overline{\underline{F}}|\,\right]^{-}$.

Square brackets also are used around the Lewis structures of polyatomic ions, such as NH_4^+, for clarity.

TABLE 10-2 Electronegativities of Some Selected Elements

H																
2.1																
Li	Be											B	C	N	O	F
1.0	1.5											2.0	2.5	3.0	3.5	4.0
Na	Mg											Al	Si	P	S	Cl
0.9	1.2											1.5	1.8	2.1	2.5	3.0
K	Ca	Sc	Ti	V	Cr	Mn	Fe	Co	Ni	Cu	Zn	Ga	Ge	As	Se	Br
0.8	1.0	1.3	1.5	1.6	1.6	1.5	1.8	1.8	1.8	1.9	1.6	1.6	1.8	2.0	2.4	2.8
Rb	Sr	Y	Zr	Nb	Mo	Tc	Ru	Rh	Pd	Ag	Cd	In	Sn	Sb	Te	I
0.8	1.0	1.2	1.4	1.6	1.8	1.9	2.2	2.2	2.2	1.9	1.7	1.7	1.8	1.9	2.1	2.5
Cs	Ba	La*	Hf	Ta	W	Re	Os	Ir	Pt	Au	Hg	Tl	Pb	Bi	Po	At
0.8	0.9	1.1	1.3	1.5	2.4	1.9	2.2	2.2	2.2	2.4	1.9	1.8	1.8	1.9	2.0	2.2
Fr	Ra	Ac†														
0.7	0.9	1.1														

***4. Define electronegativity, and use its value to assess the relative metallic/nonmetallic character of an element.**

Electronegativity is a combination of ionization potential and electron affinity—one number expressing how strongly an atom attracts electrons. Thus, electronegativity is high for a nonmetallic element and low for a metallic element. Electronegativities are most often computed with a method devised by Linus Pauling (1901–1994); these are known as Pauling electronegativities. We see two trends: electronegativity increases

* Lanthanides: 1.1 – 1.3

† Actinides: 1.3 – 1.5

across a period and decreases down a family (see Table 10-2). These trends are a result of the trends in ionization energies and electron affinities. Generally speaking, metals have electronegativity values below 2.0 and nonmetals have values above 2.0. Metalloids have electronegativity values of about 2.0.

5. Describe the relationship between electronegativity difference and the percent ionic character of a bond.

The larger the electronegativity difference between two bonded atoms, the less covalent and the more ionic is the bond between them. An electronegativity difference of 0.5, as in S—Cl, produces about 6% ionic character. One of 1.0 (N—F) is about 22% ionic, one of 1.5 (C—F) about 43%, one of 2.0 (B—F) about 63%, one of 2.5 (Be—F) about 79%, and one of 3.0 (Li—F) is about 89% ionic. Often, in fact, it is possible to write both a covalent and an ionic Lewis structure for a molecule or an ion. The hydrogen halides (HF, HCl, HBr, and HI) are common examples. Others are drawn in Figure 10-2. The larger the electronegativity difference between the atoms, the greater the contribution of the ionic resonance structure. If the molecule is to take part in a reaction where it ionizes, the existence of an ionic resonance form indicates that the molecule should ionize more readily.

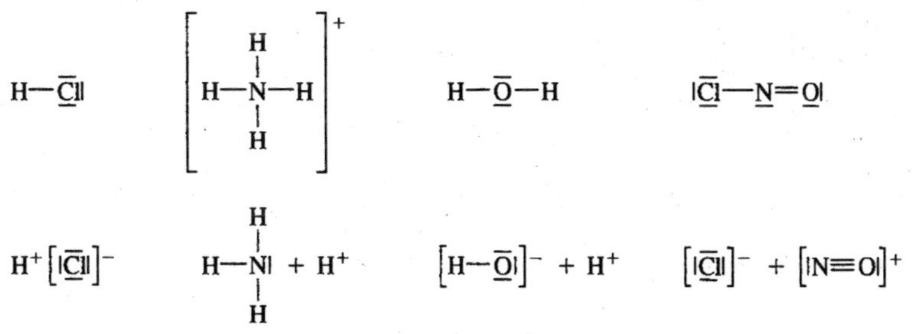

FIGURE 10-1 Some Ionic-Covalent Resonance Forms

***6. Use the basic rules of Lewis theory to propose a plausible skeleton structure for a molecule and assign valence electrons to this structure.**

Too often, students believe that they *must* start with the Lewis symbols for the elements in order to draw Lewis structures. The guidelines that follow avoid that difficulty:

1. Count the total number of valence electrons in the species, remembering to add electrons for anions and subtract them for cations. The number of valence electrons of an element equals its family number.

2. Divide the total number of valence electrons by 2 to obtain the number of electron pairs.

3. Arrange the atomic symbols in the correct molecular skeleton. In some cases, this will be given to you. If it is not, the central atom often is written first. Usually, the

least electronegative atom is the central atom (although H is almost never the central atom). Furthermore, realize that most molecules are compact clusters rather than atoms strung out in a line. You may have to choose among several skeletons; use the octet rule and the formal charge on each atom to decide.

4. Place one pair of electrons between each pair of adjacent atoms. Then place the remaining electron pairs so that each atom, except H, has an octet of electrons, if possible. (H just gets a pair of electrons, not an octet.) It is probably easier to complete the octets of terminal atoms (those on the outside) before central atoms.

5. If an atom lacks an octet of electrons, consider double and triple bonds between that atom and an adjacent atom. But remember that the halogens (F, Cl, Br, and I) rarely form double bonds and H never does.

6. A central atom may have an expanded octet (Objective 10-10), but only if it is from the third period or higher.

7. As a first guess, try the following number of bonds around an atom. One bond: H, F, Cl, Br, I. Two bonds: O, S, Se, Te, Be. Three bonds: N, P, As, Sb. Four bonds: C, Si, Ge.

EXAMPLE 10-1 Draw the Lewis structure of HClO.

(1) There are $1(\mathrm{H}) + 7(\mathrm{Cl}) + 6\,(\mathrm{O}) = 14$ valence electrons

(2) *or* $14/2 = 7$ valence electron pairs.

(3) Possible skeletons: H — Cl — O, H — O — Cl, O — H — Cl

(4) *and* $\mathrm{H-\overline{\underline{Cl}}-\overline{\underline{O}}|\quad H-\overline{\underline{O}}-\overline{\underline{Cl}}|\quad |\overline{\underline{O}}-H-\overline{\underline{Cl}}|}$

The third structure shows that we cannot have H as a central atom. H can have only one pair of electrons, not two. Based on Guideline 7, $\mathrm{H-\overline{\underline{O}}-\overline{\underline{Cl}}|}$ is correct. We shall see later (Example 10-3) that formal charges make the same prediction.

EXAMPLE 10-2 Draw the Lewis structure of N_2O.

(1 & 2) $5(\mathrm{N}) + 5\,(\mathrm{N}) + 6(\mathrm{O}) = 16$ valence electrons = 8 pairs.

(3) Two possible skeletons: N—N—O *and* N—O—N

(4) $\mathrm{|\overline{\underline{N}}-N-\overline{\underline{O}}|}$ and $\mathrm{|\overline{\underline{N}}-O-\overline{\underline{N}}|}$ are incorrect, since the central atom has no octet of electrons.

(5) There are many possibilities with multiple bonds. But based on Guideline 7, none of these structures is correct. We need another criterion to decide between them. That criterion is formal charge (see Example 10-4).

N as central atom : $\mathrm{|\underline{N}=N=\underline{O}|\quad |N\equiv N-\overline{\underline{O}}|\quad |\overline{\underline{N}}-N\equiv O|}$

O as central atom : $\mathrm{|\underline{N}=O=\underline{N}|\quad |N\equiv O-\overline{\underline{N}}|}$

*** 7. Compute the formal charge on each atom in a Lewis structure and use formal charges to determine which of several Lewis structures is the most plausible.**

Formal charge is simply the difference between the number of valence electrons an atom brings to a Lewis structure (its original number of valence electrons) and the number it possesses within the structure. To determine the number of valence electrons an atom possesses, assign to it all of its unshared electrons (also called *lone pair electrons*) and half of its bonding electrons. The assumption is that bonding electrons are equally shared between the atoms they join. Thus, to determine the formal charge (FC) of an atom, we can use either Equation [1] or Equation [2].

$$\text{FC} = (\text{group number}) - (\text{number of unshared electrons}) - (\text{number of shared electrons} \div 2)$$
[1]

Unshared electrons pairs are called *lone pairs* (l.p.) and shared electron pairs are called *bonding pairs* (b.p.).

$$\text{FC} = (\text{group number}) - (2 \times \text{number of lone pairs}) - (\text{number of bonding pairs}) \quad [2]$$

The most plausible structure is the one in which the formal charges are minimized, or the one in which the sum of the absolute values of the formal charges is the smallest. We want all formal charges to be as close to zero as possible. If formal charges remain in the species, the positive formal charges should be on the least electronegative atoms and the negative formal charges on the most electronegative atoms. Of course, there will always be some nonzero formal charges in an ion, since the sum of formal charges equals the charge on the ion (or zero for a molecule). Examples 10-3 and 10-4 illustrate the selection of the most plausible Lewis structure with the aid of formal charge.

EXAMPLE 10-3 Which of the structures of Example 10-1 is the most plausible?

The calculations (based on Equation [2] above) are shown at right, beneath the symbol of each atom.

Lewis structure	H	$—\overline{\underline{\mathrm{Cl}}}—$	$\overline{\underline{\mathrm{O}}}\vert$	H	$—\overline{\underline{\mathrm{O}}}—$	$\overline{\underline{\mathrm{Cl}}}\vert$
group number	1	7	6	1	6	7
− 2 × lone pairs	−0	−4	−6	−0	−4	−6
− bond paris	−1	−2	−1	−1	−2	−1
formal charge	0	+1	−1	0	0	0

Thus, the formal charges make the same prediction ($\mathrm{H—\overline{\underline{O}}—\overline{\underline{Cl}}|}$ as does Guideline 7 of Objective 10-7.

EXAMPLE 10-4 Which of the structures of Example 10-2 for N_2O is the most plausible?

Lewis structure	$\mathrm{N=N=\underline{O}|}$	$\mathrm{|N\equiv N—\overline{\underline{O}}|}$	$\mathrm{|\overline{\underline{N}}—N\equiv O|}$	$\mathrm{|\underline{N}=O=N|}$	$\mathrm{|N\equiv O—\overline{\underline{N}}|}$
Group number	5 5 6	5 5 6	5 6 5	5 6 5	5 6 5

−2× lone pairs	-4 −0 −4	−2 −0 −6	−6 −0 −2	−4 −0 −4	−2 −0 −6
− bond pairs	-2 −4 −2	−3 −4 −1	−1 −4 −3	−2 −4 −2	−3 −4 −1
Formal charge	-1 +1 0	0 +1 −1	−2 +1 +1	−1 +2 −1	0 +2 −2

$|\underline{N}=N=\underline{O}|$ and $|N\equiv N-\overline{\underline{O}}|$ have the minimum total formal charge. Of these two, $|N\equiv N-\overline{\underline{O}}|$ is the more plausible, as it has the negative formal charge (-1) on the more electronegative atom, oxygen.

*** 8. Recognize situations when resonance occurs, and draw plausible resonance structures.**

Some compounds cannot be represented with a single Lewis structure. For these species, we write several equivalent structures, called *resonance structures*. For example, ozone (O_3) has 18 valence electrons, or nine pairs. Two structures can be written:

$$|\overline{\underline{O}}-\underline{O}=\underline{O}| \quad \text{and} \quad |\underline{O}=\underline{O}-\overline{\underline{O}}|$$

These two resonance structures have the following features, that are true for *most* resonance structures:

1. The number of bonds of each type (single, double, and triple) is the same in each resonance structure.
2. The molecular skeleton is the same in each structure.
3. The formal charge distribution is the same in each structure. Here, the central atom has a formal charge of $+1$, one terminal atom has a formal charge of 0 (the double-bonded one), and the other oxygen has a formal charge of -1.
4. The atoms are symmetrically arranged (one O on each side of a central O), but the electrons have an unsymmetrical arrangement (a double bond and two lone pairs on one side, compared to a single bond and three lone pairs on the other).

All that is involved in creating one resonance structure from another is shifting a few pairs of electrons. Many students have the mistaken impression that the resonance hybrid (the molecule) "hops back and forth" between its various resonance forms. This is no more true than believing that a tangelo is sometimes an orange and other times a tangerine (or that you are sometimes your mother and at other times your father). There is nothing wrong with a resonance hybrid such as ozone (or a hybrid like a tangelo). Instead, the difficulty is with the Lewis theory itself; a single structure is inadequate. In the same way, a photograph is inadequate in fully representing a three-dimensional object, such as a coffee cup; several photographs from different angles are needed. So, also, a resonance hybrid is depicted by several Lewis structures.

*** 9. Draw Lewis structures for odd-electron and electron-deficient structures.**

When the total number of valence electrons is odd, there is no way to pair all the electrons.

EXAMPLE 10-5 Draw the Lewis structure for ClO_2. (Cl is the central atom.)

$7(Cl) + 2 \times 6(O) = 19$ valence electrons *or* $9\frac{1}{2}$ pairs. Possible Lewis structures and formal charges are:

Lewis structure	$|\overline{O}-\dot{C}-\overline{O}|$			$\cdot\overline{O}-\overline{C}l-\overline{O}|$		
group number	6	7	6	6	7	6
$-2 \times$ lone pairs	-6	-3	-6	-5	-4	-6
$-$ bond pairs	-1	-2	-1	-1	-2	-1
formal charge	-1	$+2$	-1	0	$+1$	-1

Thus, the resonance hybrid $(\cdot\overline{O}-\overline{C}l-\overline{O}| \longleftrightarrow |\overline{O}-\overline{C}l-\overline{O}\cdot)$ is the best choice, based on formal charge. (Note that unpaired electrons seldom are used as bonding electrons in Lewis structures.)

Electron-deficient compounds occur when an atom (other than hydrogen) is surrounded by fewer than eight valence electrons. The central atom is said to have an incomplete octet. Almost always, the central atom in an electron-deficient compound is Be, B, or Al.

EXAMPLE 10-6 Draw the Lewis structure of BF_3.

$3(B) + 3 \times 7(F) = 24$ valence electrons *or* 12 pairs

$$|\overline{F}-B-\overline{F}| \quad or \quad |\overline{F}-B=\overline{F} \quad \longleftrightarrow \quad |\overline{F}-B-\overline{F}| \quad \longleftrightarrow \quad \overline{F}=B-\overline{F}|$$
$$\underset{|\overline{F}|}{|} \qquad \underset{|\overline{F}|}{|} \qquad \underset{|\overline{F}|}{\parallel} \qquad \underset{|\overline{F}|}{|}$$

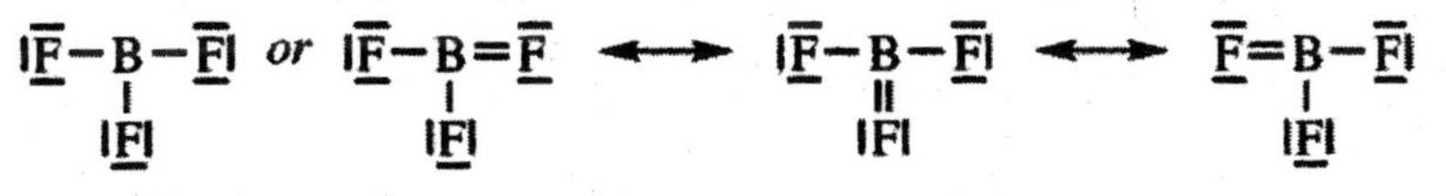

The structure at the left is electron deficient in that B is surrounded by only six electrons. Note that the formal charge on each atom in this structure is zero. In the resonance hybrid, there is no electron deficiency and the octet rule is obeyed. But B has a formal charge of -1, and the double-bonded F has a formal charge of $+1$. The true picture probably is a resonance hybrid of all four structures.

*** 10. State which elements can have expanded octets, and be able to draw Lewis structures with expanded octets.**

For elements of the first and second periods, it is very difficult to place more than an octet around the atom mainly because it requires too much energy (see Figure 9-7 for the relative energy spacing between the $2p$ and $3s$ orbitals), but also because the central atom is very small (see Figure 9-2 for relative sizes of atoms), leaving little room for more than four pairs of electrons. This is not the case for elements of the third period and higher; these elements can have expanded octets in Lewis structures. The purpose of an expanded octet is to bond all atoms to the central atom.

EXAMPLE 10-7 Draw the Lewis structure of ICl_3. (I is the central atom.)

$7(I) + 3 \times 7(Cl) = 28$ valence electrons or 14 pairs

	I	each Cl
group number	7	7
$-2 \times$ lone pairs	-4	-6
$-$ bond pairs	$\underline{-3}$	$\underline{-1}$
formal charge	0	0

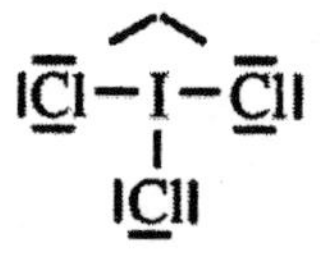

A common question is whether it is more important to follow the octet rule or to minimize formal charge. The best advice is to obey the octet rule and not to expand an atom's octet in an attempt to minimize formal charge. Consider, for example, the chlorite ion, ClO_2^-, with 20 valence electrons or 10 valence electron pairs. A structure in which all atoms have an octet, $[|\overline{\underline{O}}—\overline{\underline{Cl}}—\overline{\underline{O}}|]^-$, has -1 formal charge on each O and $+1$ formal charge on Cl. The structure (actually one of two resonance forms) with minimal formal charges, $[\overline{\underline{O}}=\overline{\underline{Cl}}—\overline{\underline{O}}|]^-$, has five valence electron pairs around Cl, zero formal charge on both Cl and the double-bonded oxygen, and -1 formal charge on the single-bonded oxygen. The structure that obeys the octet rule is considered the more plausible Lewis structure.

*** 11. Use bond distances to help in writing Lewis structures and bond energies to compute the enthalpy change of a reaction.**

Bond order (single, double, triple) lets us predict relative bond length and strength.

$$\text{bond strength: single} < \text{double} < \text{triple}$$

$$\text{bond length: single} > \text{double} > \text{triple}$$

This information can help us decide between various Lewis structures. In Example 10-6, we indicated the B—F bond to have some double-bond character. The bond length in BF_3 is somewhat shorter than an average B—F bond, confirming our prediction. Bond energies can be used to calculate an approximate $\Delta H°$ value for a reaction involving reactants and products that all are gases. A fairly complete set of bond energies and lengths is given in Table 10-3.

EXAMPLE 10-8 Calculate the approximate enthalpy change for the uncatalyzed combustion of ammonia.

$$4NH_3(g) + 3\,O_2(g) \longrightarrow 2N_2(g) + 6\,H_2O(g)$$

First, we draw the Lewis structures of all species:

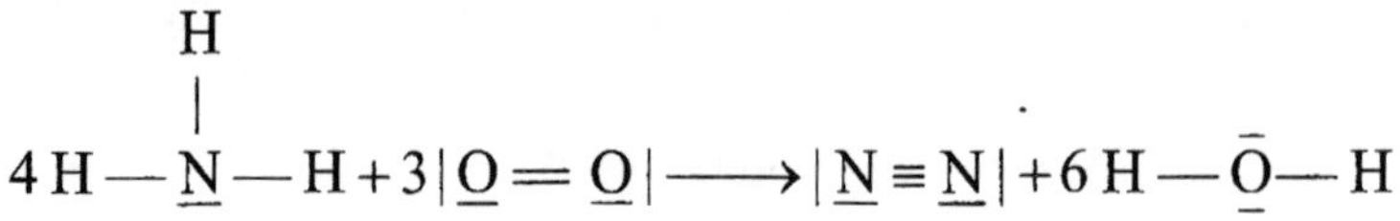

Then, we determine how many bonds of each type are broken or formed and the energy required to break the bonds and produced by forming them.

TABLE 10-3 Bond Energies (kJ/mol of bonds broken) and Bond Lengths (pm, in parentheses)

Bond	I	Br	Cl	S	P	Si	F	O	N	C	H	
H	297	364	431	368	318	293	565	464	389	414	436	
	(161)	(141)	(127)	(132)	(142)	(148)	(92)	(96)	(100)	(110)	(74)	
C	238	280	339	259	264	289	490	360	305	347		
	(214)	(194)	(178)	(182)	(156)	(187)	(138)	(143)	(147)	(154)		
N	176	276	201	205	209	326	264	222	163			
		(214)	(180)	(166)	(149)	(174)	(136)	(136)	(145)			P = O
O	201	235	205	268	360	368	184	142				P ≡ P
	(183)	(165)	(148)	(145)	(145)	(153)	(142)	(145)				B—B
F	268	234	255	285	490	540	159					
	(176)	(163)	(159)	(159)	(156)	(156)	(143)					B—H
Si	213	289	360	226	213	176						
	(246)	(217)	(201)	(214)		(230)						
P	213	272	318	230	172							
	(252)	(223)	(200)	(208)	(223)							
S	180	213	276	264								427
		(227)	(159)	(205)								
Cl	209	219	243								418	485
	(232)	(218)	(199)								(123)	
Br	180	193				946	N			498	615	331
		(228)				(110)				(121)	(120)	(143)
I	151				837	891	C		611	736	615	531
	(266)				(120)	(116)			(134)	(120)	(128)	(171)
					C	N	≡		C	O	N	S

Bonds broken	*Bonds formed*
$12\ N—H = 12 \times 398 = 4668\ kJ/mol$	$2\ N \equiv N = 2 \times 946 = 1892\ kJ/mol$
$3\ O = O = 3 \times 498 = \underline{1494\ kJ/mol}$	$12\ O—H = 12 \times 464 = \underline{5568\ kJ/mol}$
Total energy required $= 6162\ kJ/mol$	Total energy produced $= 7460\ kJ/mol$

Energy produced has a negative sign and energy required has a positive sign.

$$\Delta H^\circ = \textit{Net energy} = +6162\ kg/mol - 7460\ kJ/mol = -1298\ kJ/mol$$

This value of ΔH° is approximate. The actual value, computed from ΔH°_f values, is -1266.25 kJ/mol. Because ΔH° values computed from bond enthalpies often have an error of up to 10%, they can be used only for approximations.

*** 12. Predict the electron-pair geometry and molecular shape of a molecule or ion with VSEPR theory. Know that single bonds act similarly to multiple bonds to determine molecular shape.**

Lewis theory does not predict molecular shape. The valence-shell electron-pair repulsion (VSEPR) theory accounts for molecular shape. The basis of the theory is that pairs of electrons in the valence shell of an atom try to get as far apart as they can and still remain attached to the atom. VSEPR theory uses the following terms: AX_mE_n: the formula of a molecule or ion. *A* represents the *central atom*. *X* represents each of the *ligands*, or atoms or groups of atoms bound to the central atom. *E* represents each of the *lone pairs*, or unshared pairs of electrons on the central atom. *m* is the number of ligands or bonding pairs. (A double bond or a triple bond is counted as one bonding pair.) *n* is the number of lone pairs. (A lone electron, such as that of N in NO_2 is counted as a lone pair.) $m+n$ is the total number of electron pairs (also called electron groups). This total determines the electron-pair geometry of the molecule or ion, as given in Table 10-4. The dashed lines in each sketch represent electron pairs pointing back, away from you, and the wedges represent electron pairs pointing forward, toward you. You may have trouble seeing these electron-pair geometries. Most beginners find a set of models helpful. After you have built a couple dozen models of molecules and sketched the results, you will be able to more easily visualize molecular shapes.

Although the total number of electron pairs determines the electron-pair geometry, we only can see the ligands—we see only the *effect* of the lone pairs. The possible VSEPR structures are given in Table 10-4 and drawn in Figure 10-2. AX_4E is an irregular tetrahedron rather than a trigonal pyramid because of the differences in the space occupied by types of electron pairs. Sizes decrease in the order given in Expression [3], and repulsions between different types of electron pairs decrease in the order given in Expression [4].

TABLE 10-4 Electron-Pair Geometries

No. lone pairs	2	3	4	5	6
Electron-pair geometry	Linear	equilateral triangle	Tetrahedron	trigonal bipyramid	octahedron
Sketch	—A—				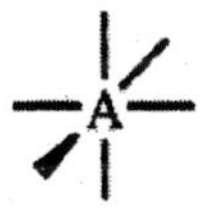
Bond angles	180°	120°	109.5°	120° (equatorial) 90° (axial)	90°

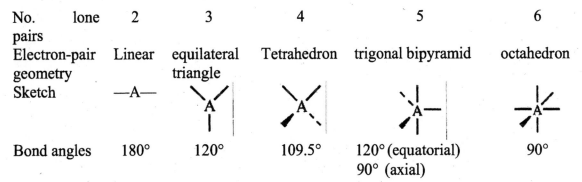

Sizes: lone pair > lone electron > triple bond > double bond > single bond [3]

Repulsions:
lone pair — lone pair > lone pair — bond pair > bond pair — bond pair [4]

Since there is more room around the equator of a trigonal bipyramid than at its axes, the lone pair in AX_4E occupies the equatorial position as shown in Figure 10-3(a). You should make a model of each shape shown in Figure 10-3 so that you have a clear idea of each geometry.

The application of VSEPR theory requires that you first draw the Lewis structure of the molecule or ion. Then, for each central atom, count the number of ligands and the number of lone pairs. From this information and Table 10-4 (which you should have memorized), name the molecular shape and draw a sketch of the molecule. The correct molecular shape is predicted, even if you have not drawn the most plausible Lewis structure. Consider the three possibilities for N_2O of Example 10-4. All have two bonded groups and no lone pairs on the central nitrogen atom, and all predict a linear structure. On the other hand, ICl_3 of Example 10-7 has three bond pairs and two lone pairs. Its electron pair geometry is trigonal pyramid and the molecule is T-shaped.

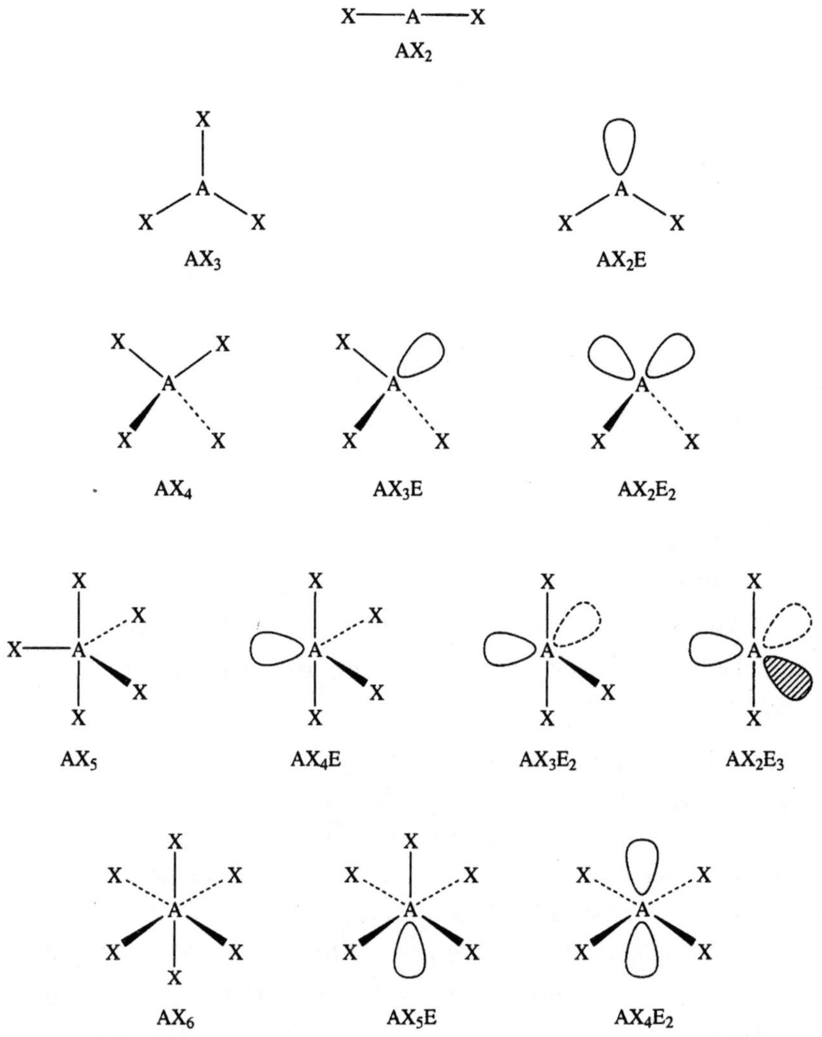

FIGURE 10-2 Shapes of the Formulas of Table 10-4.

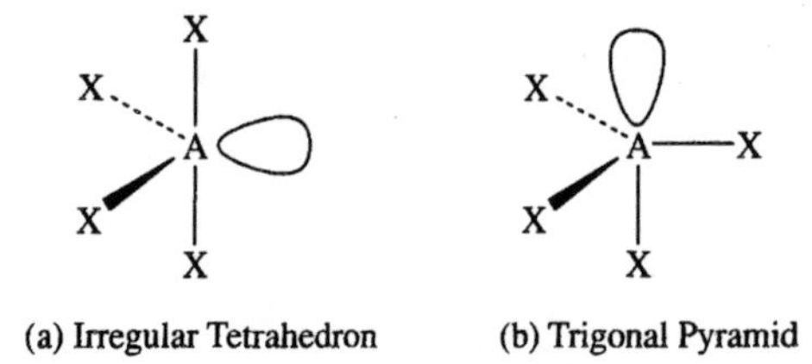

FIGURE 10-3 Possible Shapes of AX_4E Species.

TABLE 10-5 Summary of VSEPR Possibilities

		No. of Electron Pairs				
Formula	Example	Bonding	Lone	Total	Molecular shape	Electron-pair geometry
AX_2	$BeCl_2$	2	0	2	linear	linear
AX_3	BF_3	3	0	3	triangular	triangular
AX_2E	$SnCl_2$	2	1	3	bent	triangular
AX_4	CF_4	4	0	4	tetrahedron	tetrahedron
AX_3E	NF_3	3	1	4	triangular pyramid	tetrahedron
AX_2E_2	OF_2	2	2	4	bent	tetrahedron
AX_5	PCl_5	5	0	5	trigonal bipyramid	trigonal bipyramid
AX_4E	SF_4	4	1	5	seesaw	trigonal bipyramid
AX_3E_2	ClF_2	3	2	5	T-shaped	trigonal bipyramid
AX_2E_3	XeF_2	2	3	5	linear	trigonal bipyramid
AX_6	SF_6	6	0	6	octahedron	octahedron
AX_5E	BrF_5	5	1	6	square pyramid	octahedron
AX_4E_2	XeF_4	4	2	6	square planar	octahedron

*** 13. Use electronegativities to determine if a bond is polar, and use bond polarities and molecular shape to predict whether a molecule has a dipole moment.**

In a pure covalent bond, the bonding pair of electrons is shared equally between the two atoms. In a polar bond, the bonding pair is shifted toward the more electronegative atom. Thus, CO_2 and H_2O both possess polar bonds. A bond dipole is drawn as an arrow with a cross at the base $(+\!\rightarrow)$. The dipole points away from the more electropositive atom. VSEPR theory predicts that CO_2 is linear and H_2O is bent. Since the two bond dipoles in CO_2 are equal and point in opposite directions, they cancel. CO_2 has no net dipole moment. But in the H_2O molecule, the bond dipoles do not point in exactly opposite directions and, therefore, they do not cancel. H_2O has a dipole moment.

Self-Assessment Exercises

118. (b) CN^- has a triple bond. There are five valence electrons for N and four for carbon, plus one for the negative charge for a total of 10. With three pairs (a triple bond) between C and N and one lone pair on each C and N, both atoms have an octet. (a) NO_3^- has a double bond between one O and N and has resonance structures, but does not have a triple bond. (c) CO_2 has six electrons from each O and four from carbon. The structure will have double bonds between each O and C. (d) $AlCl_3$ is an ionic compound.

119. (c) Zero. Write the Lewis structure O-N-O with two unshared pairs on each oxygen and a double bond between O and N. The charge for O is then calculated from:

valence electrons in free atom - # lone pair electrons -1/2 (# of bond pair electrons)

There are six valence electrons – 4 lone pair electrons – 4 bond electrons/2 = 0

120. (a) SO_2 has an unshared pair of electrons, and will be trigonal planar for electron geometry and bent for molecular geometry.

121. (a) SO_3 is nonpolar. The geometry is trigonal planar and the dipoles will cancel.

125. (a) SO_2 will be bent or angular. The Lewis dot structure will have a double bond between S and O and a single bond for the other S-O. There is one unshared pair of electrons on S, giving it a bent geometry. The SO_2 structure is a resonance structure.

(b) SO_3^{2-}.There are 24 + 2 electrons (6 for each atom and the -2 charge). Distribute eight around each atom giving S one unshared pair and an AX_3E geometry, trigonal pyramidal.

(c) SO_4^{2-}. The lone pair described above is used in a bond to the fourth O. All of the O and the S have eight electrons. The geometry is tetrahedral.

DRILL PROBLEMS

1. Give the Lewis structure for each of the following atoms. You may use a periodic table for assistance.

A. He	B. Li	C. Ne	D. In	E. I	F. Sb
G. Sn	H. Sr	I. S	J. F	K. C	L. Ca
M. K	N. Al	O. P	P. Se		

2. Give the Lewis structure of each of the following ionic species.

A. LiH	B. Na_2S	C. $CaBr_2$	D. SrO	E. Sc_2S_3	F. Sc_2S_3
G. NaF	H. TiO_2	I. MgS	J. $LaCl_3$	K. K_2Se	L. Mg_2C
M. $CeCl_4$	N. AlN	O. Cs_2Te	P. Rb_3P		

3. Rewrite each of the following lists in order of increasing value of electronegativity.

A. N, F, O, C	B. K, Br, As, Ca	C. F, Cl, I, Br
D. Rb, Cs, K, Li	E. Rb, In, Ru, Se	F. Li, Na, F, B
G. Cs, F, Cl, As	H. Si, Ge, O, N	I. Ba, Bi, Cs, Tl

4. Give the Lewis structure of each of the following covalently bonded species.

A. CH_4	B. H_2CO	C. CH_3OH	D. N_2H_4	E. OF_2	F. NF_3
G. SiO_4^{4-}	H. NOCl	I. $GeCl_4$	J. C_2H_4	K. $SiCl_4$	L. Cl_2CO
M. $HClO_2$	N. HOCN	O. HFO	P. HNO_2	Q. H_2O_2	R. C_2H_2
S. AsF_3	T. BrO_4^-	U. PO_4^{3-}	V. $SeBr_2$	W. PO_3^{3-}	X. BF_4^-
Y. OF^-	Z. ClO_3^-	Γ. HO_2^-	Δ. CN^-	Θ. HI	Λ. CCl_2H_2
Π. H_2CO_2	Σ. C_2H_6	Φ. O_2^{2-}			

5. Give at least two Lewis structures for each of the following species, evaluate the formal charge on each atom of each structure, and predict which structure is the most plausible.

A. HOCN	B. HCN	C. $HClO_3$	D. CO	E. ONCl	F. CNS^-
G. HCNS	H. BCl_3	I. HON	J. $BeCl_2$	K. $OSeF_2$	L. CN_2^{2-}
M. HNO_2	N. $POCl_3$	O. ClCN	P. HN_3	Q. H_2CN_2	R. NO_2Cl

6. Give at least two resonance structures for each of the following resonance hybrids.

A. HCO_{2^-}	B. NO_{2^-}	C. N_3^-	D. SO_2	E. CO_3^{2-}	F. BO_3^{3-}
G. NO_{3^-}	H. PO_{3^-}	I. C_2O_3	J. SO_3	K. $C_2O_4^{2-}$	L. O_3

7. Give the most plausible Lewis structure for each of the following species.

A. NO	B. NO_2	C. $SnBr_2$	D. ClO_2	E. PbI_2	F. $BeCl_2$
G. PO_3	H. BeH_2	I. NCN^-	J. BN	K. BI_3	L. CH_3
M. SO_2^+	N. BCl_2	O. O_{2^-}			

8. Give the most likely Lewis structure for each of the following species.

A. BrF_3	B. $SbCl_5^{2-}$	C. BrO_3^-	D. I_{3^-}	E. PF_3Cl_2	F. $PbCl_2$
G. H_2SO_3	H. $SbCl_5$	I. ICl_{4^-}	J. $SnCl_6{}^{2-}$	K. IF_5	L. $SO_3{}^{2-}$
M. XeF_4	N. SeF_6	O. ArF_2	P. SCl_4	Q. ClF_3	R. $PO_4{}^{3-}$
S. IS_2-	T. XeF_2	U. XeO_3			

9. Use the bond enthalpies given in Table 10-3 to predict the value of $\Delta H°$ for each of the following reactions.

A. $CS_2(g) + 3\ O_2(g) \longrightarrow CO_2(g) + 2\ SO_2\ (g)$

B. $CH_4\ (g) + 2\ O_2\ (g) \longrightarrow CO_2\ (g) + 2\ H_2O\ (g)$

C. $H_2CO(g) + O_2(g) \longrightarrow CO_2(g) + H_2O\ (g)$

D. $PCl_3\ (g) + F_2\ (g) \longrightarrow PCl_3F_2(g)$

E. $I_2(g) + 3Cl_2(g) \longrightarrow 2\,ICl_3\ (g)$

F. $N_2\ (g) + 3H_2\ (g) \longrightarrow 2\,NH_3\ (g)$

G $2N_2H_2(g) + O_2(g) \longrightarrow 2\,N_2(g) + 2\,H_2O(g)$

H. $4\,HCN\,(g)\,5\,O_2\ (g) \longrightarrow 2\,H_2O\,(g) + 2\,N_2\ (g) + 4\,CO_2\ (g)$

I. $CO_2\ (g) + 2\,Cl_2\ (g) \longrightarrow Cl_2CO\,(g) + Cl_2O\,(g)$

J. $SiH_4\ (g) + 4\,F_2\ (g) \longrightarrow SiF_4\ (g) + 4\,HF\,(g)$

K. $CO_2\ (g) + 2\ H_2\ (g) \longrightarrow CH_4O_2\ (g)$

L. $C_2H_6O\,(g) + 3\,O_2(g) \longrightarrow 2\,CO_2\ (g) + 3\,H_2O\,(g)$

M. $PCl_5\ (g) + 4\,H_2O\,(g) \longrightarrow H_3PO_4\ (g) + 5\,HCl\,(g)$ (Assume P = O is 710 kJ/mol)

N. $PCl_5\ (g) + H_2O\,(g) \longrightarrow POCl_3\ (g) + 2\,HCl\,(g)$

O. $HCNS\,(g) + H_2\ (g) \longrightarrow HCNO\,(g) + H_2S\,(g)$

P. $C_3H_8(g) + 5\,O_2\ (g) \longrightarrow 3\,CO_2\ (g) + 4\,H_2O\,(g)$

10. Give the electron-pair geometry, the molecular shape, and a perspective sketch (use solid, dashed, and wedged lines) of each of the following species.

A. BF_3	B. $GeCl_4$	C. HO_2^-	D. BrF_3	E. $SbCl_5$	F. $PbBr_2$
G. H_2CO	H. $BeCl_2$	I. KrF_4	J. ArF_2	K. SCl_4	L. NO_2^-
M. NF_3	N. ClF_3	O. IF_5	P. PF_5	Q. ClO_4^-	R. AsF_3
S. XeF_4	T. BCl_3	U. H_2S	V. $POCl_3$	W. SF_4	X. CO_3^{2-}
Y. NO_3^-	Z. N_2O	Γ. BF_4^-			

11. Which of the molecules in the question for Objective 10-12 has a dipole moment, and which way does the dipole point in each case?

QUIZZES (20 minutes each) Choose the best answer to each question.

QUIZ A

1. Which of the following should be paramagnetic? (a) HCl; (b) CO_2; (c) PCl_3; (d) ClO_2; (e) none of these.

2. In which compound do two ions composed of nonmetal atoms combine to form an ionic compound? (a) NH_4NO_3; (b) $Al_2(SO_4)_3$; (c) Na_2SO_3; (d) $AlCl_3$; (e) none of these.

3. Which of the following molecules does *not* obey the octet rule? (a)HCN (b) PF_3; (c) CS_2; (d) NO; (e) none of these.

4. Which molecule has no polar bonds? (a) H_2CO; (b) CCl_4; (c) OF_2; (d) N_2O; (e) none of these.

5. The incorrect Lewis structure of the following is (a) $\underline{\overline{O}}{=}C{=}\underline{\overline{O}}$; (b) $\mathrm{H{-}\overset{\overset{H}{|}}{B}{-}H}$; (c) $|C{=}\overline{O}$; (d) $H{-}\underline{\overline{O}}{-}\underline{\overline{Cl}}|$; (e) none of these.

6. Of the following molecules, which contains the shortest bond distance? (a) Cl_2; (b) N_2; (c) O_2; (d) NH_3; (e) CO_2.

7. Which of the following species is expected to have tetrahedral electron pair geometry? (a) AsF_4-; (b) SF_4; (c) $SiCl_4$; (d) XeF_4; (e) none of these.

8. According to VSEPR theory, the shape of the ICl_4- ion is (a) pyramidal; (b) linear; (c) tetrahedral; (d) square planar; (e) none of these.

9. Which of the following species has a tetrahedral shape? (a) ICl_4-; (b) SF_4; (c) SiF_4; (d) XeF_4; (e) none of these.

10. From the bond enthalpies I—Cl (209), H—H (435), H—Cl (341), and H—I (297) in kJ/mol, compute ΔH° in kJ/mol for the reaction $ICl_3\,(g) + 2\,H_2\,(g) \longrightarrow HI\,(g) + 3\,HCl\,(g)$. (a) -93; (b) +264; (c) -523; (d) -84; (e) none of these within 2%.

QUIZ B

1. Which of the following should be paramagnetic? (a) $NO-$; (b) NO; (c) N_2O; (d) NO_2-; (e) none of these.

2. Which of the following exhibits ionic bonding? (a) BF_3; (b) AsH_3; (c) SO_3; (d) BaO; (e) none of these.

3. When a molecule should be represented by more than one electronic structure, it is best described in terms of (a) VSEPR theory; (b) hybridization; (c) resonance; (d) multiple covalent bonding; (e) none of these.

4. Which of the following molecules has polar bonds, but is itself not polar? (a) CH_3Cl; (b) H_2O; (c) SCl_2; (d) Cl_2; (e) none of these.

5. The incorrect Lewis structure of the following is (a) $|C{\equiv}O|$; (b) $\underline{\overline{S}}{=}\underline{\overline{O}}$; (c) $H{-}\underline{\overline{O}}{-}\underline{\overline{O}}{-}H$; (d) $|\underline{\overline{O}}{-}\overline{N}{=}\overline{O}|$; (e) none of these.

6. Which of the following seldom forms multiple bonds? (a) S; (b) C; (c) Cl; (d) O; (e) none of these.

7. Which of the following would be expected to be trigonal planar? (a) $H_2C{=}O$; (b) H_3O^+; (c) NH_3; (d) CH_3; (e) none of these.

8. The electron-pair geometry of H_2O is (a) tetrahedral; (b) trigonal planar; (c) bent; (d) linear; (e) none of these.

9. Which of the following has a triangular planar shape? (a) BF_3; (b) PF_3; (c) IF_3; (d) NH_3; (e) none of these.

10. From the bond enthalpies C=O (707), O=O (498), H—O (464), and C—H (414) in kJ/mol, compute ΔH° in kJ/mol for the reaction $CH_4\,(g) + 2\,O_2\,(g) \longrightarrow CO_2\,(g) + 2\,H_2O(g)$. (a) +618; (b) -519; (c) -618; (d) +259; (e) none of these within 2%.

QUIZ C

1. Which of the following is not paramagnetic? (a) NO; (b) CO; (c) NO_2; (d) BC; (e) none of these.

2. Which of the following exhibits covalent bonding? (a) NaF; (b) ICl; (c) K_2O; (d) SrI_2; (e) none of these.

3. Which of the following molecules is adequately represented by a single Lewis structure? (a) O_3; (b) NOCl; (c) SO_2; (d) N_2O; (e) none of these.

4. Which of the following does *not* have a molecular dipole moment? (a) HCl; (b) CO; (c) NCl_3; (d) BCl_3; (e) none of these.

5. As indicated by its Lewis structure, which of the following species would probably not exist as a stable molecule? (a) CH_3O; (b) CH_2O; (c) CH_2O_2; (d) C_2H_2; (e) none of these.

6. Which of the following atoms often is present in compounds that are electron deficient? (a) F; (b) O; (c) H; (d) B; (e) none of these.

7. The electron-pair geometry of $NO_2{}^+$ is (a) triangular; (b) bent; (c) tetrahedral; (d) octahedral; (e) none of these.

8. The molecular shape of PF_5 is (a) octahedral; (b) trigonal bipyramid; (c) irregular tetrahedron; (d) square pyramid; (e) none of these.

9. Which of these has a square planar shape? (a) $ICl_4{}^-$; (b) SF_4; (c) $NH_4{}^+$; (d) CH_2F_2; (e) none of these.

10. From the bond enthalpies C — H (414), H — Cl (431), C — Cl (331), and Cl — Cl (243) in kJ/mol, compute ΔH° in kJ/mol for the reaction $CH_4(g) + 4\,Cl_2\,(g) \longrightarrow CCl_4(g) + 4\,HCl(g)$. (a) $+904$; (b) $+418$; (c) -252; (d) -105; (e) none of these.

QUIZ D

1. Which of these molecules is not paramagnetic? (a) NCl; (b) NO_2; (c) BrCl; (d) O_2; (e) none of these.

2. The_____ the electronegativity difference between two atoms in a bond, the greater the _____ of the bond. (a) less, polarity; (b) greater, ionic character; (c) less, ionic character; (d) greater, covalent character; (e) none of these.

3. The most likely arrangement of atoms in S_2Cl_2 is (a) $S—S—Cl—Cl$; (b) $S—Cl—S—Cl$; (c) $S—Cl—Cl—S$; (d) $Cl—S—S—Cl$; (e) none of these.

4. Of the following, which has the largest dipole moment? (a) CH_4; (b) CH_3Cl; (c) BCl_3, (d) CO_2; (e) none of these.

5. As indicated by Lewis structures, which of the following species could probably not exist as a stable molecule? (a) NH_3; (b) N_2H_2; (c) N_2H_4; (d) N_2H_6; (e) none of these.

6. Which of the following contains the bond of largest energy? (a) H_2; (b) Cl_2; (c) CO_2; (d) N_2; (e) H_2O.

7. According to VSEPR theory, the molecular shape of the I_3^- ion is (a) trigonal planar; (b) linear; (c) trigonal bipyramid; (d) bent; (e) none of these.

8. Which of the following electron pairs exhibits the largest degree of repulsion? (a) lone pair-lone pair; (b) bond pair-bond pair; (c) lone pair-bond pair; (d) lone pair-unpaired; (e) bond pair-unpaired.

9. Which of the following species is T-shaped? (a) BF_3; (b) PCl_3; (c) CH_3^-; (d) ClF_3; (e) none of these.

10. From the bond enthalpies $H—O$ (464), $Cl—Cl$ (243), $Cl—O$ (205), and $H—Cl$ (431) in kJ/mol, compute $\Delta H°$ in kJ/mol for the reaction $H_2O(g) + Cl_2(g) \longrightarrow HOCl(g) + HCl(g)$. (a) -1100; (b) $+393$; (c) -393; (d) $+71$; (e) none of these within 2%.

Sample Test

1. Draw the best Lewis structure of each of the following species and indicate the formal charge of each atom of the structure. If the species is a resonance hybrid, be sure to draw all of the resonance forms.

   A. PO_3^- B. NO_2^+ C. CH_3Cl D. IF_3

2. For each of the following compounds, give the names of the electron-pair geometry and the molecular shape. Sketch the molecule and indicate on the sketch the direction of the dipole moment of the molecule, if any. For sketches, use the "wedge and dash" convention.

   A. SiF_4 B. NF_3 C. SF_4 D. IF_5

3. Given the bond enthalpies H — O (464), H — C (414), C — C (347), C — O (360), C = O (707), and O = O (498) in kJ/mol, determine which of the following compounds produces the most energy in kJ/g when it burns completely in oxygen: $CH_4(g), CH_3OH(g), H_2CO(g), HCOOH(g)$. Is there any relationship between the oxidation state of carbon and the heat of combustion (in kJ/g or kJ/mol)?

11 CHEMICAL BONDING II: ADDITIONAL ASPECTS

CHAPTER OBJECTIVES

1. Explain the fundamental basis of valence bond theory; and explain why hybrid orbitals are often used to describe bonding in molecules rather than pure atomic orbitals.

Valence bond theory states that a bond forms when two atomic orbitals overlap. (Actually orbitals interpenetrate, but the term *overlap* is commonly used.) Each bond contains, at most, two electrons with their spins paired. A high degree of overlap means that a strong bond is formed. Atomic orbitals with directional character (those that "point" in a certain direction) overlap better than those with no directional character. For example, *p* orbitals overlap better than do *s* orbitals. We say they overlap more efficiently or there is a greater degree of overlap.

An attempt to explain the bonding in the CH_4 molecule provides an example of the need for hybridization. The promoted, or excited, configuration of C, given in Expression [1], permits the formation of four C — H bonds.

C: [He] [↿] [↿][↿][↿] [1]
2*s* 2*p*

Each bond would result from the overlap of a half-filled orbital on carbon (2*s* or 2*p*) with a half-filled orbital on hydrogen (1*s*). These bonds disagree in two ways with the experimental results for CH_4: (1) although the four bonds in CH_4 are the same, the bonds we have just formed differ from each other (three 1*s*-2*p* bonds and one 1*s*-2*s* bond); (2) the just-predicted bond angles are 90°, compared to observed bond angles of 109.5°. But we can combine the four pure atomic orbitals to produce four equivalent hybrid orbitals. (Notice that hybridization is used *after* we know the bond angles, which VSEPR theory often can predict.)

C: [He] [↿][↿][↿][↿] [2]
sp^3

***2. Write hybridization schemes for the formation of sp, sp^2, sp^3, sp^3d, and sp^3d^2 hybrid orbitals, and predict the geometrical shapes of molecules in terms of the pure and hybrid orbitals used in bonding.**

In the process of hybridization, pure atomic orbitals are combined to produce hybrid orbitals that have the correct geometry for the resulting molecule. The hybridization of a species can be predicted by starting with the Lewis structure. There are as many hybrid orbitals as there are total electron pairs on the central atom. Recall that the total number of electron pairs is found by adding the number of lone pairs to the number of ligands (bonding pairs).

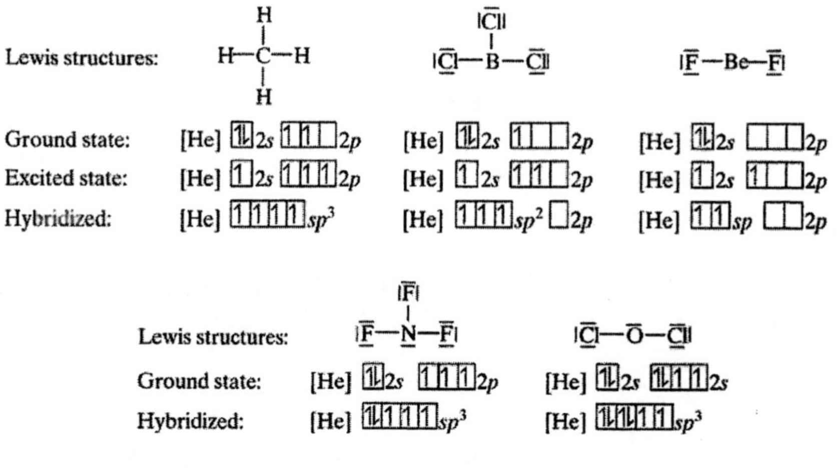

FIGURE 11-1 Bonding in CH_4, BCl_3, BeF_2 NF_3, **and** Cl_2O

The pure atomic orbitals are used (taken for use in the hybridization) in order of increasing energy. Thus, if the total number of electron pairs is two, an *s* orbital and one *p* orbital are combined or hybridized to form two *sp* hybrid orbitals. If there are three electron pairs total, an *s* orbital and two *p* orbitals are hybridized to form three sp^2 hybrid orbitals. And if there are four electron pairs total, an *s* orbital and three *p* orbitals are hybridized to form four sp^3 hybrid orbitals. (Notice that the symbol for a hybrid orbital—sp, sp^2, or sp^3—indicates which pure atomic orbitals have been combined to form it.) When hybrid orbitals are formed, those that are to hold lone pairs are full, while those that are to form bonds (or to hold bond pairs) usually have one electron. The exception is coordinate covalent bonds: formed from a full orbital (an electron pair) on one atom and an empty orbital on another. We can recognize coordinate covalent bonds because the atoms involved in them have nonzero formal charges in Lewis structures. Lewis structures, along with central atom ground state, excited state, and hybridized electron configurations for CH_4, BCl_3, BeF_2, NF_3, and Cl_2O are shown in Figure 11-1.

The properties of hybrid orbitals depend on the properties of the pure atomic orbitals that were combined to form them. Consider direction. A *p* orbital points along a line. Thus, if an *s* orbital is combined with a *p* orbital, the two *sp* hybrids formed must also point along a line. In similar fashion, sp^2 orbitals are confined to a plane (they must be flat), since any two *p* orbitals lie in a plane. And four sp^3 orbitals are three-dimensional.

When the number of ligands and bond pairs totals five or six, *d* orbitals must be included in the hybridization scheme, since there are only a total of four *s* and *p* orbitals in each shell. The Lewis structures and orbital diagrams for SF_4, ClF_3, XeF_4, and

Cl_2O in Figure 11-2 illustrate the inclusion of *d* orbitals in hybridization schemes. Notice that the technique used is similar to that of Figure 11-1, except for including *d* orbitals.

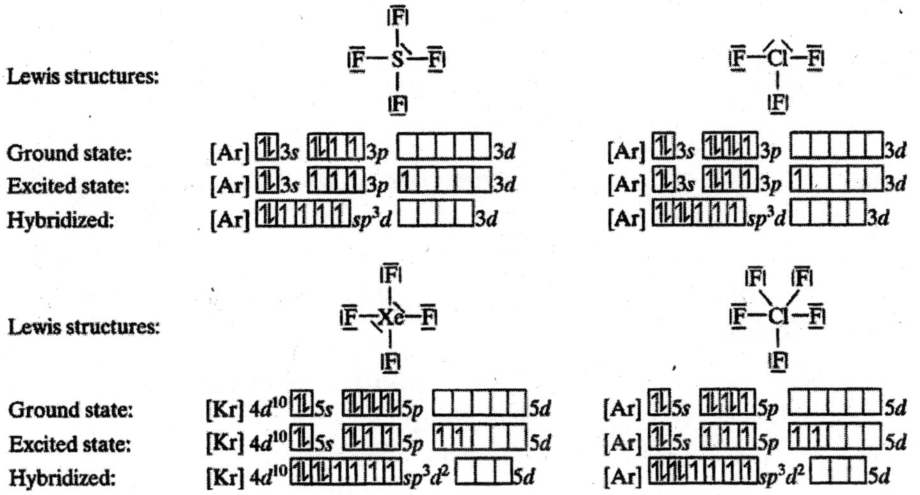

FIGURE 11-2 Bonding in SF_4, ClF_3, ClF_5, and XeF_4

*** 3. Use the relationship between VSEPR theory and valence bond theory to predict molecular geometries and bonding schemes.**

Table 11-1 is quite similar to Table 10-4, the electron-pair geometries of VSEPR theory. The bonding in species that contain only single bonds can be predicted with the following seven steps.

TABLE 11-1 Geometry of Hybrid Orbitals

Number of ligands and lone pairs	Hybridization	Geometry of hybrid	Bond angles
2	*sp*	linear	180°
3	sp^2	triangular	120°
4	sp^3	tetrahedral	109.5°
5	sp^3d	trigonal bipyramidal	120° (equatorial) 90° (axial)
6	sp^3d^2	octahedral	90°

(1) Draw the most plausible Lewis structure for the species.

(2) For each central atom, count the total number of bonded atoms or groups (ligands) and lone pairs.

(3) Use VSEPR theory to determine the electron-pair geometry and the molecular shape.

(4) Use the information in Table 11-2 to determine the hybridization of the central atom.

(5) Write the ground-state orbital diagram of the central atom (subtracting or adding electrons to account for the charge, if any, of the species) and the orbital diagram for the hybridization found in Step 4.

(6) Write the ground-state orbital diagram for each ligand. You may modify orbital diagrams if a coordinate covalent bond is to be formed. (Usually this just involves adding or subtracting electrons to account for formal charge: a N atom with a +1 formal charge will have four valence electrons.) Terminal atoms need not be hybridized.

(7) Overlap two half-filled orbitals—one from the ligand and one from the central atom—for each bond. (For each coordinate covalent bond, overlap one full orbital with one empty orbital.) Describe the result (i.e., the orbitals of each atom that overlap to form each bond in the molecule).

EXAMPLE 11-1 Describe the bonding in OF_2.

We follow the steps outlined above.

(1) $|\overline{\underline{F}}-\overline{\underline{O}}-\overline{\underline{F}}|$

(2) 2 ligands + 2 lone pairs = 4 electron pairs.

(3) tetrahedral electron-pair geometry; bent molecular shape.

(4) sp^3 hybridization on O.

(5) ground state O: hybridized O:

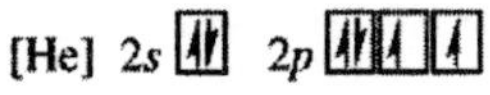

(6) ground state F:

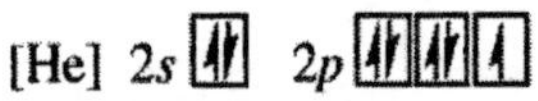

(7) Oxygen is bound to each fluorine by the overlap of a half-filled sp^3 orbital on O with a half-filled $2p$ orbital on F. The two lone pairs of oxygen are in sp^3 hybrid orbitals. Writing the two fluorines as F_a and F_b to tell them apart, the bonding scheme can be abbreviated as $O(sp^3)^1 — F_a(2p)^1$ $O(sp^3)^1 — F_b(2p)^1$. (The superscripts outside the parentheses indicate the number of electrons in each orbital before overlap.)

EXAMPLE 11-2 Describe the bonding in ClF_3.

(1)

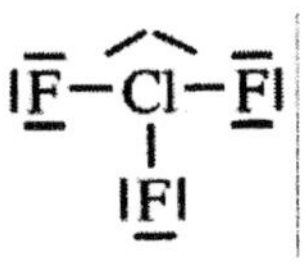

(2) 3 ligands + 2 lone pairs = 5 electron pairs.

(3) trigonal bipyramid electron-pair geometry; T-shaped molecular geometry.

(4) sp^3d hybridization

(5) ground state Cl:

[Ne] 3*s* [↑↓] 3*p* [↑↓][↑↓][↑] 3*d* [ ][ ][ ][ ][ ]

sp^3d – hybridized Cl:

[Ne] sp^3d [↑↓][↑↓][↑][↑][↑] 3*d* [ ][ ][ ][ ]

(6) ground state F:

[He] 2*s* [↑↓] 2*p* [↑↓][↑↓][↑]

(7) The two lone pairs of chlorine are in sp^3d orbitals. Chlorine is bonded to each fluorine by the overlap of a half-filled sp^3d hybrid orbital on Cl with a half-filled 2*p* orbital on F. Abbreviated, the bonding scheme is:

$$\text{Cl}(sp^3d)^1 — \text{F}_a(2p)^1 \qquad \text{Cl}(sp^3d)^1 — \text{F}_b(2p)^1 \qquad \text{Cl}(sp^3d)^1 — \text{F}_c(2p)^1$$

*** 4. Describe multiple bonds between second period elements in terms of the overlap of *sp*, sp^2 and pure 2*p* orbitals to form σ bonds, and the side-to-side overlap of *p* orbitals to form π bonds.**

A multiple bond consists of a σ bond and one or two π bonds. A π bond is formed by the side-to-side overlap of two *p* orbitals. σ bonds are formed by almost all types of overlap: *s-s*, *s-p*, *p-p* (end-to-end), *s*-hybrid, *p*-hybrid, and hybrid-hybrid. σ bonds lie along the *internuclear axis* (the line joining the two atoms), whereas the two lobes of a π bond lie on either side of the axis. To achieve side-to-side overlap, the *p* orbitals must, of course, be aligned. This means that the molecule is not free to rotate about a multiple bond as it is about a single bond. A π bond is weaker than a σ bond because its overlap is not as efficient. To account for π bonding, we add one more rule to the list first given in Objective 11-3.

8. Overlap one half-filled *p* orbital on a ligand with one half-filled *p* orbital on the central atom to form each π bond.

EXAMPLE 11-3 Describe the bonding in ClNO.

(1) $|\overline{\underline{\text{Cl}}} — \underline{\text{N}} = \underline{\text{O}}|$

(2) 2 ligands + 1 lone pair = 3 electron pairs.

(3) triangular electron-pair geometry; bent molecular shape.

(4) sp^2 hybridization.

(5) ground state N:

[He] $2s$ [$\uparrow\downarrow$] $2p$ [$\uparrow$][$\uparrow$][$\uparrow$]

sp^2 - hybidized N :

[He] sp^2 [$\uparrow\downarrow$][$\uparrow$][$\uparrow$] $2p$ [$\uparrow$]

(6) ground state Cl:

[Ne] $3s$ [$\uparrow\downarrow$] $3p$ [$\uparrow\downarrow$][$\uparrow\downarrow$][$\uparrow$]

ground state O:

[He] $2s$ [$\uparrow\downarrow$] $2p$ [$\uparrow\downarrow$][$\uparrow$][$\uparrow$]

(7) Sigma bonding:
$N(sp^2)^2$: lone pair. $\mathrm{N}(sp^2)^1 — \mathrm{Cl}(3p)^1$ $\mathrm{N}(sp^2)^1 — \mathrm{O}(2p_y)^1$

(8) Pi bonding: $\mathrm{N}(2p_z)^1 — \mathrm{O}(2p_z)^1$

The *y* and *z* subscripts on the $2p$ orbitals of oxygen simply remind us not to use the same $2p$ orbital twice.

5. Propose plausible bonding schemes from Lewis structures or from experimental information about molecules (that is, bond lengths, bond angles, and so on).

To this point, we have used theoretical models (Lewis theory and VSEPR theory) to predict molecular structure. However, when experimental information disagrees with theory, we discard the theoretical model. H_2S is an example of the conflict between theory and experiment. VSEPR theory predicts a tetrahedral electron-pair geometry.

$$\mathrm{H — \overline{\underline{S}} — H} \qquad [3]$$

This leads to sp^3 hybridization and a predicted H—S—H bond angle of 109.5°. The actual bond angle is 92°. It is very unlikely that the two lone pairs on S exert such a strong influence that they "close down" the bond angle from 109.5° to 92°. It is much more likely that hybridization does not occur and that the bonding is between half-filled $3p$ orbitals on S and half-filled $1s$ orbitals on H.

Do not become confused by this example of a conflict between theory and experiment. In practically all of the molecules and ions that you will encounter, VSEPR theory will correctly predict the shape of the species. The information in Table 11-1 then can be used to specify the hybridization on each central atom.

6. Explain the fundamental basis of molecular orbital theory.

In molecular orbital theory, the atom centers are put in place and the electrons are "poured" around them. The electrons find the orbitals of lowest energy. These *molecular orbitals* extend over at least two atoms; they are not restricted to one atom as are atomic orbitals. Electrons in a bonding orbital spend most of their time between two nuclei and act rather like a "glue" holding the atoms together. Electrons in an antibonding orbital spend most of their time outside the region between the two nuclei. In this case, the nuclear charges are not screened from each other by the negative charge of the electrons and, hence, they repel each other.

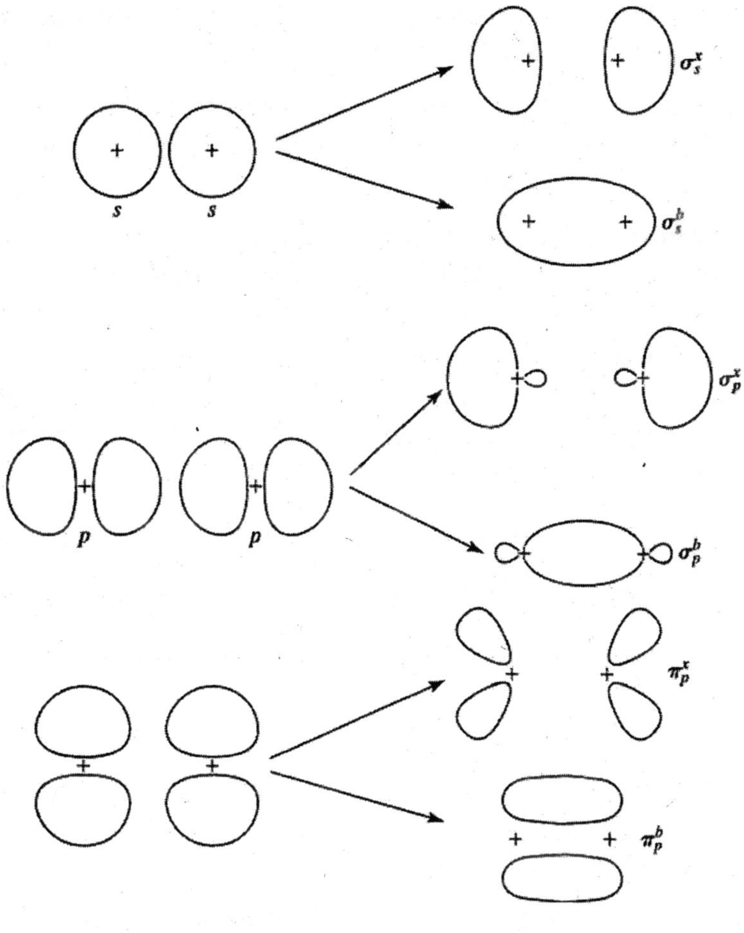

FIGURE 11-3 Molecular Orbitals Produced from *s* and *p* Atomic Orbitals

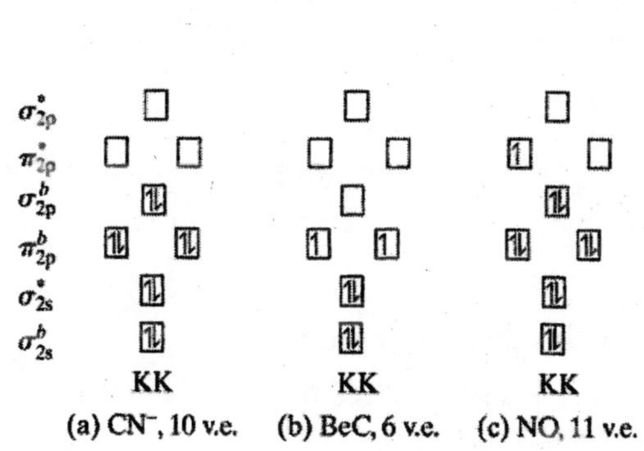

FIGURE 11-4 Molecular Orbital Energy Level Diagrams for CN^-, BeC, and NO.
(v.e. = valence electrons)

7. Know that two atomic orbitals combine to form a bonding and an antibonding molecular orbital. Sketch these molecular orbitals.

The bonding and antibonding molecular orbitals produced by the combination of two *s* atomic orbitals and of two *p* atomic orbitals are drawn in Figure 11-3. The asterisk (*) is used for an antibonding orbital and the ^{b} superscript (or sometimes no superscript) is used for a bonding orbital. The subscript (*s* or *p*) indicates the two atomic orbitals that were combined to form the molecular orbital.

*** 8. Assign probable electron configurations, determine bond orders, and predict magnetic properties of the diatomic molecules and ions of the first and second period elements.**

It is necessary to memorize the energy order of the molecular orbitals formed between two second period elements. This order is depicted in the molecular orbital energy level diagrams of Figure 11-4, where it is applied to CN^-, BeC, and NO. Notice that the molecular orbitals are filled with valence electrons from lowest to highest energy. Electrons are not paired until all orbitals of the same energy have one electron. In this regard, filling molecular energy level diagrams obeys Hund's rule. The abbreviation KK stands for σ_{1s}^b.

Molecular orbital theory defines bond order as follows:

Bond order = (number of bonding electrons - number of antibonding electrons) ÷ 2 [4]

Lewis theory and valence bond theory concepts equivalent to bond orders of 1, 2, and 3 are given in Table 11-2.

TABLE 11-2 Concepts Equivalent to Bond Order

Bond order	Lewis theory	Valence bond theory
1	single bond	one σ bond
2	double bond	one σ and one π bond
3	triple bond	one σ and two π bonds

EXAMPLE 11-4 Use the data of Figure 11-4 to determine the bond order of (a) CN^-, (b) BeC, and (c) NO.

(a) CN^- : bond order $= (8-2) \div 2 = 3$ (b) BeC: bond order $= (4-2) \div 2 = 1$

(c) NO : bond order $= (8-3) \div 2 = 2.5$

A molecule with unpaired electrons is *paramagnetic*, and one in which all electrons are paired is *diamagnetic*. Molecular orbital theory enables us to predict the number of unpaired electrons.

EXAMPLE 11-5 Use the data of Figure 11-4 to predict the number of unpaired electrons and the magnetic properties of (a) CN^-, (b) BeC, (c) NO.

(a) CN^- : all electrons are paired, diamagnetic (b) BeC: two unpaired electrons, paramagnetic

(c) NO: one unpaired electron, paramagnetic.

When molecular orbital energy level diagrams are written on one line, they are called molecular orbital diagrams. These are more compact, and are shown in Figure 11-5.

		σ_{2s}^{b}	σ_{2s}^{*}	π_{2p}^{b}	σ_{2p}^{b}	π_{2p}^{*}	σ_{2p}^{*}
CN^{-}	KK	⇅	⇅	⇅⇅	⇅	□□	□
BeC	KK	⇅	⇅	↑↑	□	□□	□
NO	KK	⇅	⇅	⇅⇅	⇅	↑□	□

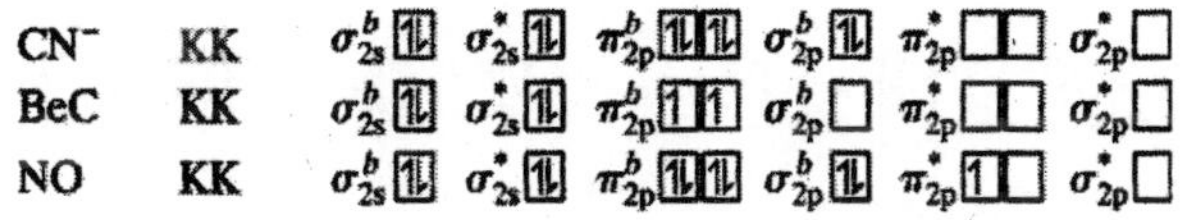

FIGURE 11-5 Molecular Orbital Diagrams for CN^{-}, BeC, and NO

9. **Describe the bonding in the benzene molecule (C_6H_6) through Lewis structures, valence bond theory, and molecular orbital theory.**

The benzene molecule is a specific example of a resonance hybrid, which we consider in general in Objectives 11-10 and 11-11. The delocalized p bonding consists of three filled bonding π orbitals (two electrons in each) and three empty antibonding π orbitals. Thus, the π bond order equals $(6-0) \div 2 = 3$, exactly as predicted by Lewis theory. The difference is that the π molecular orbitals are not localized between two carbon atoms but are spread out over the entire molecule.

*** 10.** **Explain why a single valence structure is inadequate to explain the bonding in a resonance hybrid.**

We considered resonance hybrids in Objective 10-8. Here we extend these ideas by first considering the nitrite ion (NO_2^{-}) by the procedure outlined in Objective 11-3. We write the two oxygens as O and Ø to distinguish them.

(1) $[|\overline{O}\!-\!N\!=\!\overline{\emptyset}]^{-} \leftrightarrow [\overline{O}\!=\!N\!-\!\overline{\emptyset}|]^{-}$

resonance form a resonance form b

(2) 2 ligands + 1 lone pair = 3 electron pairs.

(3) triangular electron-pair geometry; bent (or angular) molecular shape.

(4) sp^2 hybridization on N.

(5) ground state N:

[He] $2s$ ⇅ $2p$ ⇅↑↑

sp^2 - hybridized N :

[He] sp^2 ⇅⇅↑ $2p_y$ ↑

(6) ground state O:

[He] $2s$ ⇅ $2p$ ⇅↑↑

modified O:

[He] 2s [↑↓] 2p [↑↓][↑↓][ ]

The modified O is for a coordinate covalent bond; this is the single-bonded

O: [He] 2s [↑↓] $2p_x$ [↑↓] $2p_y$ [↑↓] $2p_z$ [ ]   O: [He] 2s [↑↓] $2p_x$ [↑↓] $2p_y$ [↑] $2p_z$ [↑]

Ø: [He] 2s [↑↓] $2p_x$ [↑↓] $2p_y$ [↑] $2p_z$ [↑]   Ø: [He] 2s [↑↓] $2p_x$ [↑↓] $2p_y$ [↑↓] $2p_z$ [ ]

oxygen in each resonance form.

(7) Sigma bonding descriptions:

resonance form *a*: $N(sp^2)^2 — O(2p_z)^0$ $N(sp^2)^1 — Ø(2p_z)^1$

resonance form *b*: $N(sp^2)^1 — O(2p_z)^1$ $N(sp^2)^2 — Ø(2p_z)^0$

Let us draw the 2*p* orbitals that remain, indicating the electrons in each with arrows. These are $2p_y$ orbitals. The 2*p* orbitals of the two resonance forms are drawn in Figure 11-6. In resonance form *a*, the π bond forms between N and Ø; in resonance form *b*, the π bond forms between N and O. It seems that valence bond theory, strictly applied, still does not explain resonance structures. The problem is transferred from "lines between atoms" to pictures of *p* orbitals.

11. Explain what is meant by a delocalized molecular orbital and how such orbitals are used to describe resonance hybrids.

Although it does not provide a complete explanation, Figure 11-6 does help us understand resonance hybrids. It reveals the very small difference between the two resonance structures for NO_2^- : the pairing of *p* electrons. Perhaps we should revise our concept of π bonding. Consider the following.

(1) The electron is not a classical particle, but has properties that require us to speak only of the probability of finding it in a small region. How can we demand that an electron remain in the small region of a π bond?

(2) The kinetic energy of a particle is $\text{K.E.} = mv^2/2 = p^2/2m$, where $p = mv$. The de Broglie wavelength of a particle is $\lambda = h/p$. Thus, $p = h/\lambda$ and $\text{K.E.} = h^2/2m\lambda^2$. The longer the wavelength of a particle, the lower its kinetic energy. The maximum wavelength of a particle is about the same size as the region in which it is confined. Thus, allowing electrons to move between more than two atoms should produce a molecule of lower energy: one that is more stable. And we have observed that a resonance hybrid is more stable than expected (that is, more stable than one resonance structure by itself).

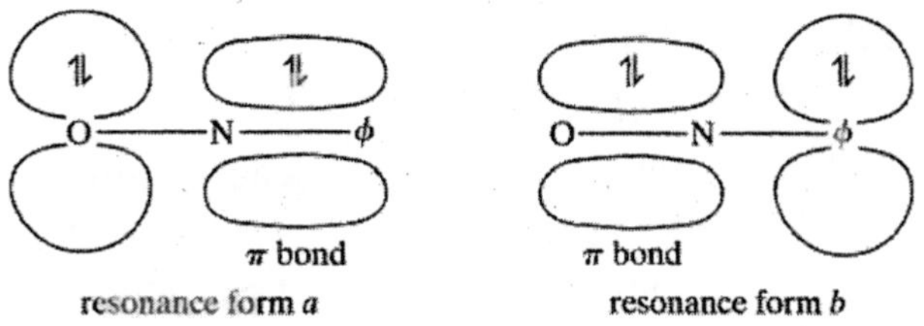

FIGURE 11-6 Valence Bond Pictures of the Two Resonance Forms of NO$_2^-$

Antibonding π molecular orbital	π^*	☐
Nonbonding π molecular orbital	π^n	⇅
Bonding π molecular orbital	π^b	⇅

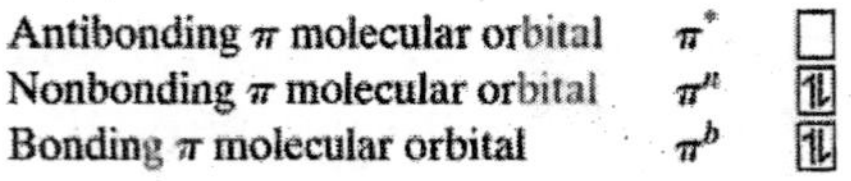

FIGURE 11-7 Molecular Orbital Energy Level Diagram for the π Electrons of NO$_2^-$

The extra stability possessed by a molecule due to resonance is known as the *resonance stabilization energy*. The stability occurs because the electrons are not restricted to one region; they are *delocalized*. For NO$_2^-$, we would expect three π molecular orbitals extending over the entire molecule. There are three molecular orbitals because three $2p$ atomic orbitals contribute to the π bonding: one orbital of each atom of the molecule ($2p_{z,O}, 2p_{z,N}$, and $2p_{z,\varnothing}$). When molecular orbitals form, they usually divide into the same number of bonding as antibonding orbitals. If the original number of molecular orbitals is an odd number (as is the case in NO$_2^-$), there usually is a nonbonding molecular orbital. Hence, these three molecular orbitals in NO$_2^-$ consist of one bonding, one nonbonding, and one antibonding orbital. Since there are four π electrons in this ion, there are two π electrons in the lowest energy bonding MO and two in the nonbonding MO. The antibonding molecular orbital is empty. This yields a π bond order of one, as depicted in the molecular orbital energy level diagram of Figure 11-7 for the π molecular orbitals of nitrite ion.

12. State the special physical properties of metals and describe how the electron sea and band theories account for them.

The special physical properties of metals are that they (1) conduct electricity, (2) conduct heat, (3) are malleable and ductile, and (4) have luster. A metal's ability to conduct electricity (its *conductance*) decreases when the metal is heated or when it contains impurities. Both the electron sea and the band theories account for the ability to conduct heat and electricity and the ease of deformation by considering the valence electrons as relatively free to move throughout the crystal. The electrons conduct electricity and their vibrations conduct heat. Bonds between the centers of metal atoms break and reform readily, accounting for the ease of deformation. The electron sea theory gives these electrons very little order. The band theory requires that they be in molecular orbitals that extend over the entire crystal.

Luster is explained by the band theory as an electron absorbing light and going to a higher energy orbital. The light is reemitted when the electron drops back down to a lower orbital. Because there are so many closely spaced orbitals, all wavelengths of light are absorbed and reemitted. Lower conductance at high temperatures is explained as being due to a disruption in the molecular orbitals because the atoms are vibrating (rather like walking across a stream on rocks during an earthquake). This also explains why small amounts of impurities lower a metal's conductance (the rocks are at quite different levels).

13. State how conductors, semiconductors, and insulators differ and explain how band theory explains these differences. Understand the meaning of *n*-type and *p*-type semiconductors and the impurities used to form each.

Conductors readily conduct an electrical current. Insulators do not conduct unless a very high voltage is applied to them. Semiconductors conduct a current when only a moderate voltage is applied. The band theory explains these differences by noting that, in insulators, there is a wide energy gap (the band gap) between the full valence (or bonding) band and the empty conduction band. The band gap is small in semiconductors, and there is no band gap (or a very small one) in conductors. A large band gap means that electrons must have a large voltage applied to promote them into the conduction band.

Insulators can be made into semiconductors by "doping" them with impurities. In one type, each impurity atom contains more valence electrons than does each atom of the insulator. Thus, due to the impurity, there is a set of filled or partly filled orbitals within the band gap, as shown in Figure 11-8a. A moderate voltage excites these impurity electrons into the conduction band, where they conduct a current. Thus, a negative particle (the electron) conducts the current in *n*-type semiconductors.

In *p*-type semiconductors, shown in Figure 11-8b, each impurity atom contains fewer valence electrons than does each atom of the insulator. Hence, empty impurity orbitals are present in the band gap. A moderate voltage excites electrons from the full valence band into these empty orbitals, leaving "holes" behind in the valence band. These holes move in the opposite direction of electron flow. It seems as if positive holes carry the current.

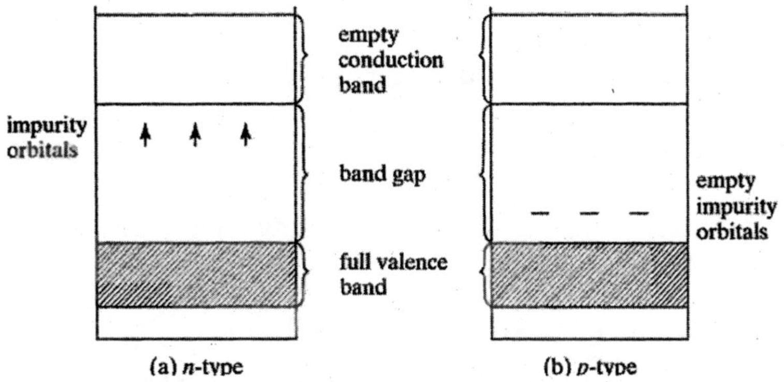

FIGURE 11-8 *n*-Type and *p*-Type Semiconductors

SELF-ASSESSMENT EXERCISES

86. (c) SO_2. The molecular geometry will be trigonal planar corresponding to sp^2 hybridization. PCl_5 will be sp^3d hybridized, N_2 is *sp* hybridized with π bond, and He_2 will be linear with *s* orbital overlap.

87. H_2Se has one unshared pair of electrons, giving it a bent molecular geometry. The trigonal planar electron geometry will give it a 120^0 bond angle, but an unshared pair takes up more room than a bonding pair, making the angle less than 120^0. The angle will not be as small as 109^0, as is found in a tetrahedron.

88. (a) I_3^-. The Lewis structure has an expanded octet, with three unshared pairs of electrons around the central I. This gives an AX_2E_3 geometry, which is sp^3d hybridized. PCl_3 has trigonal pyramid geometry (AXE) and sp^3 hybridization. NO_3^- is trigonal planar, sp^2 hybridization, and H_2Se is tetrahedral with two unshared pairs and sp^3 hybridization.

89. (b) Li_2 is correct. There are two electrons that fill the σ_{2s} orbital. The bond order is (2 bonding – 0 antibonding)/2 = 1.

H_2^+ has one electron that fills the σ_{1s} orbital. The bond order is ½. He_2 has four electrons: 2 σ_{1s} bonding and 2 σ_{1s}*. The bond order = (2 – 2)/2 = 0. H_2^- has three electrons: two σ_{1s} and one σ_{1s}*. Bond order = (2 – 1)/2 = ½ .

90. The Lewis structure for XeF_2 has two unshared pairs of electrons and three lone pairs around Xe, giving it a hybridization of sp^3d and five equivalent orbitals.

91. Delocalized molecular orbitals extend over three or more atoms, and resonance structures can be written. Therefore, the correct choice is (d) CO_3^{2-}.

92. (a) Li(s) is correct because, with one valence electron, it has only a partially filled conductance band. Br_2 has a filled valence band, and Ge and Si are semiconductors.

94. (a) False. Pi bonds require unshared *p* electrons, and hydrogen does not have *p* orbitals.

(b) True. Hydrogen forms sigma bonds by *s* orbital overlap.

(c) C-C bonds can only be σ bonds, or they can be σ and π bonds. π bonds are present in multiple bonds.

(d) False. Same reason as above.

DRILL PROBLEMS

1. The orbitals of the central atom are hybridized in each of the following species. Write the hybridization scheme for each species and give the electron-pair geometry and molecular shape of each one.

(1) Species with no *d* orbitals in the hybridization.

A. BeH_2 B. NH_4^+ C. $SnCl_2$ D. H_3O^+ E. PI_3 F. NH_2^-
G. SI_2 H. OF_2 I. BCl_3 J. CCl_2 K. CF_2Cl_2 L. BF_4^-
M. CH_3^- N. SiH_4 O. CH_3^+ P. PCl_2^-

(2) Species with an expanded octet.

Q. SeF_2Cl_2 R. TeF_6 S. ClF_3 T. PCl_5 U. BrF_3 V. ArF_2
W. IF_5 X. ICl_2^- Y. $SbCl_5$ Z. I_3^- Γ. PF_3Cl_2 Δ. BrF_5
Θ. KrF_4 Λ. $TeCl_4$ Ξ. $SbCl_5^{2-}$ Π. SF_4 Σ. ICl_4^- γ. ICl_3
Φ. SeF_6 Ψ. XeF_4 Ω. $SnCl_6^{2-}$

2. Describe the bonding in each of the following species.

A. BeH_2 B. CH_3 C. KrF_4 D. BF_4^- E. $SnCl_2$ F. NH_4^+
G. ICl_4^- H. CH_3^+ I. SI_2 J. SeF_4 K. XeF_2 L. CCl_2
M. NH_2^- N. BrF_3 O. IF_5 P. ClO_2^- Q. BCl_3 R. $SbCl_5$
S. SeF_6 T. H_3O^+

3. Describe the bonding in each of the following species. The central atoms are underlined.

A. $N\underline{C}N^{2-}$ B. $H_2\underline{C}S$ C. $Cl_2\underline{CC}H_2$ D. $Cl_2\underline{C}S$ E. $S\underline{C}N^-$ F. $N\underline{N}O$
G. CO_2 H. $H\underline{OC}N$ I. $H\underline{N}O$ J. $H\underline{C}N$ K. $H\underline{CC}H$ L. $\underline{N}O_2^+$

M. $\underline{C}OS$ N. $F\underline{P}O$ O. $H\underline{C}O\underline{O}H$

4. Write the simplified molecular orbital diagram of each of the following species. Based on that electron configuration, determine the number of (1) bonding and (2) antibonding electrons, (3) the bond order, the number of (4) σ and (5) π electrons, and (6) the number of unpaired electrons.

A. NO^+ B. O_2^{2-} C. N_2^- D. CN^+ E. BeO F. BN
G. Be_2^+ H. CN I. NeF J. F_2^- K. O_2^- L. NO^-
M. LiB

5. Give a complete description of the σ and π bonding in each of the following species. The central atom is underlined in each case.

A. $\underline{N}O_2$ B. $\underline{S}O_3$ C. $\underline{B}Cl_3$ D. $\underline{O}_3$ E. $H\underline{C}O_2^-$ F. $O\underline{C}F^-$
G. $\underline{C}O_3^{2-}$ H. $\underline{B}F_3$

QUIZZES (20 minutes each) Choose the best answer for each question.

QUIZ A

1. The hybridization of N in NCl_2F is (a) *sp*; (b) sp^2; (c) sp^3; (d) not hybridized; (e) none of these.

2. The hybridization of P in PCl_5 is (a) not hybridized; (b) sp^3d^2; (c) sp^3d; (d) sp^3; (e) none of these.

3. Which of the following exhibits sp^2 hybridization? (a) $BeCl_2$; (b) BCl_3; (c) CCl_4; (d) NCl_3; (e) none of these.

4. Which of the following molecules has the shortest C—O bond length? (a) CO; (b) CO_2; (c) CH_3OH; (d) H_2CO; (e) none of these.

5. A bond that has negligible electron density on the internuclear axis is a (a) single bond; (b) pi bond; (c) polar bond; (d) sigma bond; (e) none of these.

6. According to valence bond theory, bond strength is related to the degree of (a) orbital overlap; (b) covalent character; (c) resonance; (d) hybridization; (e) none of these.

7. A theory that can be used to explain why all bond distances and angles in a molecule such as methane are the same is (a) bond resonance; (b) delocalization of electrons; (c) bond polarities; (d) electronegativity; (e) bond hybridization.

8. Which of the following is predicted by molecular orbital theory to be paramagnetic? (a) CN^-; (b) CN^+; (c) NO^+; (d) LiB^{2+}; (e) none of these.

QUIZ B

1. In the acetylene molecule C_2H_2, the hybridization of each C atom is (a) *sp*; (b) sp^2; (c) sp^3; (d) d^2sp^3; (e) none of these.

2. Which of the following molecules displays sp^2 hybridization? (a) OF_2; (b) $F_2C=O$; (c) BeH_2; (d) NH_3; (e) none of these.

3. The hybridization of S in SF_4 is (a) not hybridized; (b) sp^3d; (c) sp^3d^2; (d) sp^3; (e) none of these.

4. The orientation in space (the geometry) of *sp*-hybridized orbitals is (a) perpendicular; (b) linear; (c) trigonal planar; (d) bent; (e) none of these.

5. If a double bond results from the sharing of four electrons, valence bond theory states that it consists of (a) 2 σ bonds; (b) 1 σ and 1 π bond; (c) 2 π bonds; (d) 1 π bond; (e) none of these.

6. Of the following, the molecule that has the shortest bond is (a) F_2; (b) N_2; (c) O_2; (d) HCl; (e) CO_2.

7. According to simple molecular orbital theory, the peroxide ion (O_2^{2-}) has (a) no unpaired electrons; (b) a bond order of 3/2; (c) its highest energy electrons in σ^* orbitals; (d) no *s* electrons; (e) none of these.

8. The double covalent bond in ethylene (C_2H_4) (a) includes a lone pair of electrons on one of the carbon atoms; (b) is a sigma bond; (c) consists of one sigma and one π bond; (d) is free to rotate because of delocalized electrons; (e) has a low electron density.

QUIZ C

1. Which of the following molecules contains an sp^2-sp^2 sigma bond? (a) C_2H_4, (b) C_2H_6; (c) C_2H_2; (d) HCN; (e) CH_3OH.

2. The hybridization of central N in N_2O is (a) not hybridized; (b) *sp*; (c) sp^2; (d) sp^3; (e) none of these.

3. The hybridization of Xe in XeF_4 is (a) not hybridized; (b) sp^3; (c) dsp^3; (d) sp^3d^2; (e) none of these.

4. If the wave functions describing the one 2*s* and two of the 2*p* orbitals are combined, the identical orbitals obtained form bond angles of (a) $90°$; (b) $120°$; (c) $180°$; (d) $109.5°$; (e) none of these.

5. A π bond is (a) formed from two *s* orbitals; (b) formed from two *p* orbitals; (c) formed from an *s* and a *p* orbital; (d) stronger than a sigma bond; (e) none of these.

6. Generally, short bonds are (a) between highly electronegative atoms; (b) strong bonds; (c) polar bonds; (d) ionic bonds; (e) none of these.

7. The concept of an antibonding orbital is unique to the (a) valence bond theory; (b) theory of bond hybridization; (c) concept of resonance; (d) molecular orbital theory; (e) electron-pair repulsion theory.

8. A diatomic molecule has 15 electrons (including KK). Its bond order is (a) 2.5; (b) 2.0; (c) 1.5; (d) 1.0; (e) none of these.

QUIZ D

1. Which of the following exhibits *sp* hybridization? (a) CO_2; (b) CCl_4; (c) H_2S; (d) NH_3; (e) none of these.

2. The hybridization of Cl in the perchlorate anion is (a) sp^3; (b) sp^2; (c) dsp^3; (d) not hybridized; (e) none of these.

3. The hybridization of Br in BrF_3 is (a) not hybridized; (b) sp^3d^2; (c) sp^3d; (d) sp^3; (e) none of these.

4. Which of the following does not use hybridized orbitals in its bonding? (a) CH_4; (b) CO_2; (c) H_2O; (d) HCl; (e) none of these.

5. A σ bond (a) cannot be formed by two *p* orbitals; (b) is weaker than a π bond; (c) has high electron probability between nuclei; (d) can only be formed from *s* orbitals; (e) none of these.

6. The shorter the internuclear distance between two bonded atoms, the (a) less stable the bond; (b) less multiple bonding character there is; (c) more energy is needed to break the bond; (d) more resonance there is; (e) none of these.

7. A π bond is (a) stronger than a σ bond; (b) concentrated along the bond axis; (c) a concept of bond hybridization theory only; (d) formed by the side-to-side overlap of *p* orbitals; (e) none of these.

8. According to molecular orbital theory, NO has (not counting KK) (a) a bond order of 3/2; (b) four antibonding electrons; (c) six σ electrons; (d) three π electrons; (e) none of these.

SAMPLE TEST (25 minutes)

1. Consider a molecule with the Lewis structure given below. For the central atoms C^a, C^b, and O, list the following:

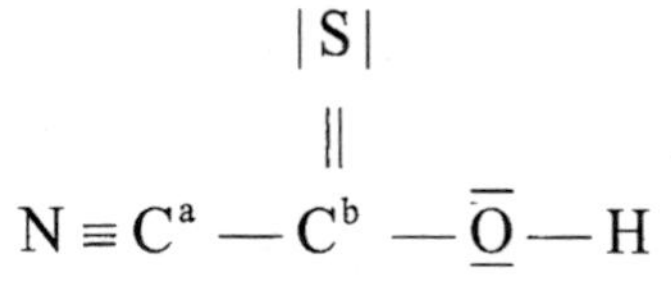

A. number of ligands.

B. number of lone pairs.

C. electron-pair geometry.

D. molecular shape.

E. orbitals in which the lone pairs (if any) are located.

F. hybridization (if any).

Now, describe the bonding in the molecule by specifying which orbitals on each atom overlap to form the bonds in the molecule. Describe each bond separately.

G. σ bonds.

H. π bonds.

What are the (ideal) values of the following bond angles?

I. $N—C^a—C^b$. J. $C^a—C^b—S$. K. $C^a—C^b—O$. L. $C^b—O—H$.

2. Use the principles of molecular orbital theory to answer the following:

A. What is the bond order of NO?

B. Which of the ions, NO^+ or O_2^+, is more stable? Why?

C. How many unpaired electrons are there in Ne_2^+, OF^+, and CO?

12 LIQUIDS, SOLIDS, AND INTERMOLECULAR FORCES

CHAPTER OBJECTIVES

1. Explain surface tension and describe several phenomena based on it.

Energy is given off when atoms are attracted to each other. Forces of attraction exist between molecules, and even between individual unbonded atoms. A molecule will have low energy if it is attracted to many other molecules. It then is more stable than a free molecule. A molecule in the body of a liquid is surrounded by and, therefore, attracted to more molecules than is a molecule on the surface. We say that the surface molecule is less stable. Thus, a liquid becomes more stable by decreasing its surface area, resisting attempts to increase its surface area and forming drops. When forces to another substance (adhesive forces) are stronger than forces within the liquid (cohesive forces), the liquid will wet that surface. Such a liquid also will rise in a capillary tube in a vain attempt to decrease the extra surface created by the liquid rising to meet the wall of the tube.

*** 2. Describe vaporization, including its enthalpy change, ΔH_{vap}; use the latter in calculations.**

Adding energy to a liquid causes some of the molecules to leave the surface of the liquid; they vaporize. The energy needed to vaporize a mole of the liquid is the molar enthalpy of vaporization (ΔH_{vap}). Vaporization always is an endothermic process—one that requires energy; thus, ΔH_{vap} is always positive. Molar enthalpies of vaporization of some common compounds are given in Table 12-1.

EXAMPLE 12-1 226.2 kJ of heat is added to 150 g of liquid water at 75.0 °C. What is the final temperature and how much water remains as liquid? Use values from Table 12-1 as needed.

First, we determine the heat needed to raise the temperature of the water to 100.0 °C.

heat = (amount in moles) (molar heat capacity)(temperature change)

$$= \left(150.\text{ g} \times \frac{\text{mol H}_2\text{O}}{18.0\text{ g}}\right) \times 75.3\text{ J mol}^{-1}\text{ K}^{-1} \times (100.0\,°\text{C} - 75.0\,°\text{C})$$

$$= 15.7 \times 10^3\text{ J} = 15.7\text{ kJ}$$

TABLE 12-1 Heat Capacities and Molar Heats of Vaporization of Some Common Compounds

Formula	Name	ΔH_{vap}, kJ/mol	Heat capacities, Liquid	$J\ mol^{-1}\ K^{-1}$ Vapor	Normal b.p., °C	M g/mol
NH_3	ammonia	31.2	76.1	44.4	−33.35	17.03
SO_2	sulfur dioxide	26.8	85.2	39.9	−10.0	64.06
HCN	hydrogen cyanide	30.7	70.7	36.0	26.0	27.03
$(C_2H_5)_2O$	diethyl ether	29.1	172.	108.	34.6	74.12
CS_2	carbon disulfide	28.4	76.1	45.6	46.3	76.13
$(CH_3)_2CO$	acetone	31.0	126.	75.3	56.2	58.08
Br_2	bromine	33.9	75.7	36.1	58.78	159.81
$CHCl_3$	chloroform	31.4	115.	65.7	61.2	119.38
CH_3OH	methanol	39.2	81.6	43.9	64.96	32.04
CCl_4	carbon tetrachloride	34.4	133.	83.3	76.8	153.82
C_2H_5OH	ethanol	40.5	113.	65.7	78.5	46.07
H_2O	water	44.0	75.3	36.	100.0	18.02
HCOOH	formic acid	41.4	98.7	45.2	100.7	46.03
CH_3COOH	acetic acid	39.7	123.	66.9	118.5	60.05

The heat remaining $(226.2\text{ kJ} - 15.7\text{ kJ} = 210.5\text{ kJ})$ is used to vaporize the water. We now determine the amount of water that vaporizes.

$$n_{\text{vap}} = 210.5\text{ kJ} \times \frac{\text{mol H}_2\text{O}}{44.0\text{ kJ}} = 4.78\text{ mol H}_2\text{O} \times \frac{18.0\text{ g}}{\text{mol H}_2\text{O}} = 86.0\text{ g H}_2\text{O}$$

Then, we determine the mass of liquid water remaining, m_1.

$$m_1 = 150.\text{ g liquid} - 86.0\text{ g vapor} = 64\text{ g liquid water remaining.}$$

3. Define *viscosity* and describe how intermolecular forces affect the viscosity.

Viscosity is the resistance of a liquid to flow. Strong intermolecular forces prevent liquid molecules from easily moving with respect to neighboring molecules; therefore, the liquid will have a high viscosity. Weak intermolecular forces will lead to low viscosity.

4. Explain what occurs when a liquid vaporizes in a closed container (i.e., the dynamic equilibrium between a liquid and its vapor).

When a liquid is placed in an open container, eventually it all vaporizes. If the container is closed, the vapor molecules cannot escape. The concentration of molecules in the vapor increases. It becomes more likely that some of these vapor molecules will condense. Soon molecules condense from the vapor just as rapidly as they vaporize; the vapor is saturated. From the ideal gas law $(PV = nRT \text{ or } P = RT \times n/V)$, we see that the vapor concentration

(n/V) is proportional to its pressure. The pressure of the saturated vapor is known as the liquid's *vapor pressure*. When the vapor is saturated, the two processes of vaporization and condensation occur at the same speed.

$$\text{rate of vaporizatin} = \text{rate of condensation} \qquad [1]$$

The rate of vaporization depends on three factors: temperature, the surface area of the liquid, and the forces between molecules. Increasing temperature causes liquid to vaporize more rapidly. Vaporization also occurs more rapidly from a large surface area than from a small one. Increasing the forces between molecules slows down the rate of vaporization. The rate of condensation depends on the pressure of the vapor and on the surface area. High pressure and large surface area both increase the rate of condensation. High pressure means there are more molecules present in the vapor that can condense. A large surface area gives the condensing molecules a larger area to hit and condense on.

***5. Determine vapor pressure experimentally, estimate values from tables or graphs, and predict whether vapor and/or liquid is present under specific conditions.**

In the transpiration method, a known volume of an inert gas is bubbled through a weighed liquid. Both liquid and gas are at the same temperature. The gas that leaves the liquid is saturated. The liquid then is reweighed. The loss in liquid mass divided by the liquid's mole weight is the amount in moles of liquid (n_1) that has vaporized.

$$n_1 = \frac{\text{liquid mass before bubbling} - \text{liquid mass after bubbling}}{\text{mole weight of liquid}} \qquad [2]$$

The final mixture of gases is assumed to occupy the same volume as the inert gas does initially; only the pressure changes. Because both the temperature and the volume of the gas are known, the vapor pressure of the liquid (P_l) can be determined from the ideal gas law.

$$P_l = n_l\,RT/V \qquad [3]$$

EXAMPLE 12-2 4.87 L of He is bubbled through isopropanol $(CH_3CHOHCH_3)$ at 312.7 K. 1.50 g of liquid vaporizes. What is the vapor pressure of isopropanol (60.0 g/mol) at this temperature?

$$P_l = \frac{\left(1.50\,\text{g} \times \dfrac{1\,\text{mol}}{60.0\,\text{g}}\right)\left(0.0821\dfrac{\text{L atm}}{\text{mol K}}\right)(312.7\,\text{K})}{4.87\,\text{L}} = 0.132\,\text{atm}$$

Determining the amount of liquid and vapor present in a container of known volume and temperature involves applying the ideal gas law to determine the moles of vapor present. We use the vapor pressure of the liquid as the pressure and the known volume and temperature. Thus, we can determine the amount in moles of vapor. Any substance left over must be liquid and is assumed to be of negligible volume.

EXAMPLE 12-3 4.00 g of isopropanol is placed in a 10.0-L container at 312.7 K. What mass of liquid is present after equilibrium is achieved?

First, determine the mass of isopropanol present as vapor, using a 0.132-atm vapor pressure from Example 12-2.

$$m_{\text{vap}} = \frac{PVM}{RT} = \frac{(0.132\text{ atm})(10.0\text{ L})(60.0\text{ g/mol})}{(0.0821\text{ L atm mol}^{-1}\text{ K}^{-1})(312.7\text{ K})} = 3.08\text{ g vapor}$$

Then, the mass of liquid is determined by difference.

$$m_{\text{liq}} = 4.00\text{ g total} - 3.08\text{ g vapor} = 0.98\text{ g liquid}$$

6. Describe the significance of the critical point.

Above the critical temperature, a gas will not condense to a liquid no matter how much pressure is exerted. It will get denser and may become as dense as a liquid, but it will never show a meniscus. Similarly, above the critical pressure, the familiar processes of vaporization and condensation do not occur no matter how the temperature changes. The critical temperature represents a very high kinetic energy of the molecules. Their speeds are so large that short-range intermolecular forces are not strong enough to hold molecules together.

***7. Explain how vapor pressure and temperature are related and the meaning of boiling. Also, be able to use the Clausius-Clapeyron equation.**

As the temperature increases, the vaporization rate increases as well as the vapor pressure. Vapor pressure is often tabulated or graphed against pressure. When the vapor pressure of the liquid reaches atmospheric pressure, bubbles of vapor form within the liquid and the liquid boils. This occurs at the boiling point, called the normal boiling point if the external pressure is 1.000 atm.

The Clausius-Clapeyron equation, Equation [4], is important for two reasons. First, it permits us to predict the vapor pressure of a liquid at several different temperatures. Second, it has the same form as two other important equations: the Arrhenius equation, Equation [5], which predicts the rate constant of a reaction at different temperatures (see Objective 15-12); and the van't Hoff equation, Equation [6], which predicts the equilibrium constant of a chemical reaction at different temperatures (see Objective 20-13).

$$\ln\frac{P_2}{P_1} = \frac{-\Delta H_{\text{vap}}}{R}\left(\frac{1}{T_2} - \frac{1}{T_1}\right) = 2.303\log\frac{P_2}{P_1} = \frac{+\Delta H_{\text{vap}}}{R}\left(\frac{1}{T_1} - \frac{1}{T_2}\right) \quad [4]$$

$$\ln\frac{k_2}{k_1} = \frac{-E_a}{R}\left(\frac{1}{T_2} - \frac{1}{T_1}\right) \quad [5]$$

$$\ln\frac{K_2}{K_1} = \frac{-\Delta H_{\text{rxn}}}{R}\left(\frac{1}{T_2} - \frac{1}{T_1}\right) \quad [6]$$

Thus, once we know how to use the Clausius-Clapeyron equation, we also will know how to use the other two. (Equations [5] and [6] are given just to show you the similarity of form. You do not need to learn them now.) There are five variables in the Clausius-Clapeyron equation: ΔH_{vap}, P_2, P_1, T_2, and T_1.

EXAMPLE 12-4 The vapor pressure of isopropyl alcohol is 100 mmHg at 39.5 °C and 400 mmHg at 67.8 °C. What is ΔH_{vap} of isopropyl alcohol?

Remember that temperature *must* be in kelvins.

$P_1 = 100\text{ mmHg}, T_1 = 312.7\text{ K}, P_2 = 400\text{ mmHg}, T_2 = 341.0\text{ K}$

$$\ln\frac{P_2}{P_1} = \frac{-\Delta H_{vap}}{R}\left(\frac{1}{T_2}-\frac{1}{T_1}\right) = \ln\frac{400\text{ mmHg}}{100\text{ mmHg}} = \frac{-\Delta H_{vap}}{8.3145\text{ J mol}^{-1}\text{ K}^{-1}}\left(\frac{1}{341.0\text{ K}}-\frac{1}{312.7\text{ K}}\right)$$

$\Delta H_{vap} = 43{,}430\text{ J/mol}$

EXAMPLE 12-5 What is the normal boiling point of isopropyl alcohol?

$P_1 = 100\text{ mmHg}, \text{T}_1 = 312.7\text{ K}, P_2 = 760\text{ mmHg}, \text{T}_2 = ?\text{K}, \Delta H_{vap} = 43430\text{ J/mol}$ from Example 12-4.

$$\ln\frac{760\text{ mmHg}}{100\text{ mmHg}} = \frac{-43430\text{ J mol}^{-1}}{8.3145\text{ J mol}^{-1}\text{ K}^{-1}}\left(\frac{1}{T_2}-\frac{1}{312.7\text{ K}}\right)$$

$T_2 = 355.9\text{ K} = 82.7\text{ °C}$

*8. Understand terms that apply to phase changes of solids.

The melting point and the freezing point are the same temperature. Melting also is called *fusion* and, thus, the molar heat absorbed on melting is called the heat of *fusion* (ΔH_{fus}). *Solidification* (freezing), the reverse of fusion, has an associated molar heat of solidification, ΔH_{sol}. ΔH_{sol} is negative since solidification is exothermic.

$$\Delta H_{sol} = -\Delta H_{fus} \qquad [7]$$

When a solid *sublimes*, it vaporizes without first becoming a liquid. The process of sublimation is the reverse of *deposition* (in which a gas becomes a solid). The molar enthalpy of sublimation, ΔH_{sub}, is positive; that is, sublimation is an endothermic process.

$$\Delta H_{sub} = -\Delta H_{dep} \qquad [8]$$

Sublimation also accomplishes the same phase change as melting and vaporization combined.

$$\Delta H_{sub} = \Delta H_{fus} + \Delta H_{vap} \qquad [9]$$

$$\Delta H_{dep} = \Delta H_{sol} + \Delta H_{cond} \qquad [10]$$

*9. Interpret simple phase diagrams and use phase diagrams to predict changes that occur as a substance is heated, cooled, or undergoes a change in pressure.

A phase diagram summarizes the changes in phase of a given substance. The liquid-vapor curve is determined by the Clausius-Clapeyron equation. The points on the sublimation curve (the curve between solid and vapor) also come from the Clausius-Clapeyron equation, if ΔH_{sub} is used for ΔH_{vap}. The slope of the fusion line is given by Equation [11].

$$\frac{\Delta P}{\Delta T} = \frac{\Delta H_{fus}}{T\Delta V_{fus}} \qquad [11]$$

where ΔV_{fus} is the volume increase of 1 mole on melting. The line between solid and liquid is straight enough to draw with a ruler. This line—the fusion curve—usually slopes upward and to the right (it has a positive slope) since ΔV_{fus} generally is positive. (Of course, ΔH_{fus} always is positive; fusion is an endothermic process.) In the case of water, however, ΔV_{fus} is negative (ice floats on water), and the fusion curve slopes upward and to the left (it has a negative slope). The point where the vapor pressure, sublimation, and fusion curves meet is the triple point where solid, liquid, and vapor exist together. You should remember that by convention, pressure is plotted vertically and temperature is plotted horizontally.

Example 12-6. It has been suggested that the pressure exerted on the blade of an ice skate is sufficient to melt the ice beneath the blade and make it easier for the blade to move over the ice. Using a phase diagram, predict whether this is a likely explanation. How much pressure would need to be exerted to lower the melting temperature 1 °C. (The decrease in volume is 0.0771 L/mol.)

Solution: Since solid water is less dense than liquid water, the line between solid-liquid slopes to the left; therefore, an increase in pressure will lower the melting point. However, the magnitude of the effect is small.

$$\frac{\Delta P}{\Delta T} = \frac{\Delta H_{Fusion}}{T\Delta V_{Fusion}} = \frac{P_2 - P_1}{-1\,K} = \frac{-285.8\,kJ/mol}{273\,K \times 0.0771\,L/mol}$$

Converting kJ to L atm:

$$\frac{P_2 - P_1}{-1\,K} = \frac{-285.8\,kJ/mol}{273\,K \times 0.0771\,L/mol} \times \frac{1\,L\,atm}{101.3\,J}$$

$$P_2 - P_1 = 134\,atm$$

***10. Describe the difference between intra- and intermolecular forces, distinguish the different types of forces between molecules, and explain how these forces influence molecular properties.**

Intermolecular forces are those that exist between molecules, whereas intramolecular forces hold a molecule together. Strong intermolecular forces make a substance difficult to break apart, so it has relatively high melting and boiling points, heats of vaporization and fusion, and surface tension. Intermolecular forces are also called van der Waals forces and are of three types, of which hydrogen bonds (considered in Objective 12-10) are the strongest.

The second type, and the weakest, are dipole-dipole attractions. The negative end of one dipole is attracted to the positive end of another. The more polar the molecules involved, the stronger their forces. Dipole-dipole forces are relatively weak, ranging from 0.01 to 0.2 kJ/mol.

The third type of force occurs between an instantaneous dipole in a normally nonpolar molecule and the dipole that the instantaneous dipole creates or induces in another nonpolar molecule. These forces are called *London forces*, or *exchange forces*. Temporary dipoles are more easily created in large molecules, especially those having atoms of high atomic number. In these atoms, the outer electrons are far from the nucleus and not held very firmly. Thus, large atoms are said to be very polarizable. This means that London forces are strong for molecules of high atomic weight. They range from 0.4 to 10 kJ/mol.

EXAMPLE 12-7 Rank Cl_2, OF_2, HF, and Ar in order of increasing boiling point and explain your reasoning.

Only HF is capable of hydrogen bonding; it should be the highest boiling point even though it has the lowest molar mass. Cl_2 has the highest molar mass of the remaining three; it should have the next–to-highest boiling point. Ar is nonpolar and should have the lowest boiling point. Actual boiling points are in parentheses in the following ranking.

HF (+19.52 °C) > Cl_2 (−34.05 °C) > OF_2 (−145.3 °C) > Ar(−185.9 °C)

11. State conditions that lead to hydrogen bonds, explain how hydrogen bonds differ from other types of intermolecular forces, and describe the effect of hydrogen bonds on physical properties.

Hydrogen bonds are formed when a hydrogen atom is bound to a F, O, or N atom and attracted to another F, O, or N atom. A Cl or S atom may substitute for either F, O, or N atom, but the resulting hydrogen bond is quite weak. One explanation for H bonds is that the high electronegativity of F, O, or N pulls so much of the bond pair of electrons toward it that the hydrogen nucleus is almost a bare proton. Thus, a very intense dipole is created. It is doubtful that this is the full explanation, for the following reasons:

(1) The hydrogen bond is very strong, ranging from 15 to 40 kJ/mol. Strengths of other intermolecular forces rarely exceed 10 kJ/mol.

(2) It is common to see a fixed limit on the number of hydrogen bonds formed. For example, two hydrogen bonds bond to oxygen in water.

(3) Hydrogen bonds have a reasonably well-defined length: 2.74 Å between oxygens in ice. Dipole-dipole attractions normally show a range of lengths.

(4) The X — H···Y line is a straight one, with X—H often pointing directly at a lone pair on Y. (Both X and Y represent F, O, or N.)

The required distance, direction, and number of hydrogen bonds make the structure of ice more open than that of liquid water. Ice is less dense than liquid water; it floats on water. The strength of the hydrogen bond results in unusually high melting and boiling points and heats of vaporization for substances comprised of small molecules that are hydrogen bonded.

An increase of 134 atm is, therefore, required.

Example 12-8. Carbon tetrachloride (CCl_4) has a lower enthalpy of vaporization than water. Explain.

Water will have hydrogen bonds, which will make the molecules more difficult to separate and, hence, vaporize.

12. Describe aspects of network covalent bonding (i.e., the conditions that lead to it, the properties of solids in which it occurs, and some examples of these solids).

A network covalent solid is held together by covalent bonds. In that respect, a crystal of such a solid is a single molecule. Covalent bonds range from 150 to 900 kJ/mol (as shown in Table 11-3). Network covalent solids are formed between the nonmetals and the metalloids of Families 3A, 4A, 5A, and 6A. (Recall that metals are held together by metallic bonds, as described in Objective 12-12.) It is very difficult to break the strong covalent bonds that hold network covalent solids together. These solids have very high melting points (2000 °C is not uncommon) and heats of fusion. They are very hard and make good abrasives. But they are brittle; once the bonds are broken they do not readily reform.

13. Predict relative lattice energies of ionic compounds and relate their magnitudes to the physical properties of those compounds (melting point, molar enthalpy of vaporization, and solubility in water).

The lattice energy of an ionic compound is the energy of the reaction in which gaseous cations and gaseous anions unite to form 1 mole of the solid compound. Since this reaction releases energy, lattice energies are invariably exothermic (that is, they have a negative sign). Lattice energies can be determined from experimental data with the Born-Fajans-Haber cycle, as explained in Objective 10-14. As also explained there, lattice energies can be calculated from information obtained from the crystal structure of the compound. The calculations reveal that the lattice energy depends on terms that look like $Z_A Z_C/(r_A + r_C)$, where Z_A and Z_C are the anion and cation charge (such as $-1, -2, +1$, and +3), and r_A and r_C are the anion and cation radii. Thus, a compound that contains small, highly charged ions will have a large lattice energy.

A large lattice energy means that the ions are firmly bonded into the crystal. Hence, the crystal will be difficult to disrupt and will have a high melting point and a large heat of fusion. Because the ions also must be separated when the compound dissolves in water, we also would expect a large lattice energy to predict a low solubility in water. But there are some exceptions, because solubility also depends on the attractions between ions and water molecules. These ion-molecule attractions also are strong for small, highly charged molecules. Thus, although it may require a great deal of energy to separate the ions, most of that energy (perhaps more) may be released when the ions are hydrated (surrounded by water molecules).

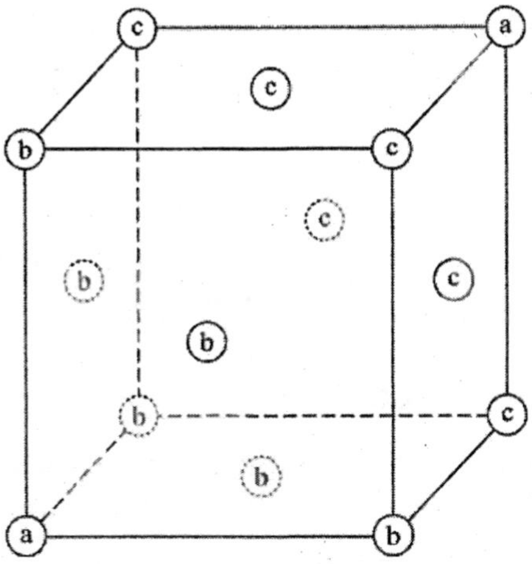

FIGURE 12-1 Close-Packing, Face-Centered Cube

14. Describe the three packing models for spheres and how X-rays determine crystal structures.

The three common ways to arrange atoms in solid metals are cubic close packed (also known as *abc*), hexagonal close packed (*aba*), and body-centered cubic. It helps a great deal to make models to see these structures. You can use Styrofoam balls touching each other and held together with glue or toothpicks. To see that the *abc* packing really is face-centered cubic, find three atoms on three different layers that touch each other and lie along a straight line. This is the diagonal of a face (corner, face center, corner; such as atoms *b* in Figure 12-1). From this start you should be able to see the rest of the face-centered cube. In Figure 12-1, the atoms of a face-centered cube are labeled with the letters of the layer (*a, b,* or *c*) each is in.

$$n\lambda = 2d \sin\theta \qquad [12]$$

The Bragg equation, Equation [12], relates the spacing between layers of atoms (d) to the angle (θ) at which X-rays of a certain wavelength (λ) are diffracted from the crystal. There are two types of layers easily seen in Figure 12-1: the layers parallel to each face and the *b* or *c* atom layers. Yet there are many other types of layers as well. These are hard to see in three dimensions unless you have a very good model. But you can see an arrangement of many rows in many orderly two-dimensional patterns: seat backs in a theater or tombstones in a military cemetery. X-rays bounce off planes of atoms somewhat as light is reflected from a mirror. A plane of atoms is a good "mirror" for X-rays if it contains many electrons. Thus, if there are many atoms in a plane or if the atoms have high atomic numbers, the reflected X-ray beam will be strong. The spacings between layers and the strengths of the reflected X-ray beams are analyzed by computer to determine the structure of the crystal.

15. Explain what a unit cell is, use its dimensions in calculations, and analyze unit cells to determine crystal coordination numbers and chemical formulas.

The smallest part of a crystal that will produce the entire crystal by simple repetition is called the *unit cell.* You can see two-dimensional unit cells in a wallpaper pattern (Figure 12-2). The dashed lines (in the upper left corner of the figure) outline a "formula unit."

Four of these formula units are needed in a unit cell since we cannot turn a unit cell around to repeat the pattern. The solid lines outline a unit cell. (Four unit cells are outlined.) Calculations involving unit cells are straightforward if you have a clear picture of which atoms are touching. In a face-centered cubic unit cell, the atoms touch along the face diagonal. In a body-centered cubic unit cell, the atoms touch along the body diagonal. And in a simple cubic unit cell, the atoms touch along the edge. These ideas, along with a knowledge of geometry, allow one to use unit cell dimensions to calculate atomic radii. In addition, one can readily establish crystal coordination numbers in the same fashion, since one knows and can count all of the ions of opposite charge touching a given ion.

In order to determine densities and chemical formulas, one should know how many unit cells "share" atoms on the outside. For a cubic unit cell, a face atom is shared by two unit cells, an edge atom by four, and a corner atom by eight.

EXAMPLE 12-9 Calcium sulfide crystals are a face-centered cubic sulfide anion lattice with calcium cations in the octahedral holes: at the center of each edge of the unit cell as well as at the body center of the unit cell's cube. The unit cell length is 569 pm. What is the density of CaS in g/cm^3?

The unit cell has 12 edges; an ion on an edge is shared among four unit cells.

FIGURE 12-2 Wallpaper Unit Cells and Formula Units

$$\frac{Ca^{2+}\ \text{ions}}{\text{unit cell}} = \left(\frac{12\ \text{edges}}{1\,\text{unit cell}} \times \frac{\tfrac{1}{4}\ Ca^{2+}\ \text{ion}}{1\ \text{edge}}\right) + \frac{1\ \text{body center ion}}{1\ \text{unit cell}} = \frac{4\ Ca^{2+}\ \text{ions}}{1\ \text{unit cell}}$$

There are six faces of the unit cell; an ion on a face is shared between two unit cells. There are eight corners of the unit cell; a corner ion is shared among eight unit cells.

$$\frac{S^{2-}\ \text{ions}}{\text{unit cell}} = \left(\frac{6\ \text{faces}}{1\,\text{unit cell}} \times \frac{\tfrac{1}{2}\ S^{2-}\ \text{ion}}{1\ \text{face}}\right) + \left(\frac{8\ \text{corners}}{1\ \text{unit cell}} \times \frac{\tfrac{1}{8}\ S^{2-}\ \text{ion}}{1\ \text{corner}}\right) = \frac{4\ S^{2-}\ \text{ions}}{1\ \text{unit cell}}$$

$$\text{unit cell volume} = (569\ \text{pm})^3 = 1.84 \times 10^8\ \text{pm}^3$$

$$\text{density} \times \frac{4\ Ca^{2+}\ \text{ions} + 4\ S^{2-}\ \text{ions}}{1.84 \times 10^{8}\ \text{pm}^3} \times \frac{1\ \text{CaS f.u}}{Ca^{2+}\ \text{ion} + S^{2-}\ \text{ion}} \times \frac{1\ \text{mol CaS}}{6.022 \times 10^{23}\ \text{CaS f.u.}} \times \frac{72.14\ \text{g CaS}}{1\ \text{mol CaS}}$$

$$\times \left(\frac{10^{12}\ \text{pm}}{1\ \text{m}} \times \frac{1\ \text{m}}{10^{2}\ \text{cm}} \right)^3 = 2.60\ \text{g/cm}^3$$

16. Use the Born-Fajans-Haber cycle to calculate lattice energies of ionic compounds from thermochemical, atomic, and molecular data.

The Born-Fajans-Haber cycle divides the formation reaction [5] of an ionic solid into several steps, as shown in Figure 12-3 for $MgBr_2$.

$$Mg(s) + Br_2(l) \longrightarrow MgBr_2(s)$$
$$\Delta H_f^\circ = -517.6\ \text{kJ/mol}$$

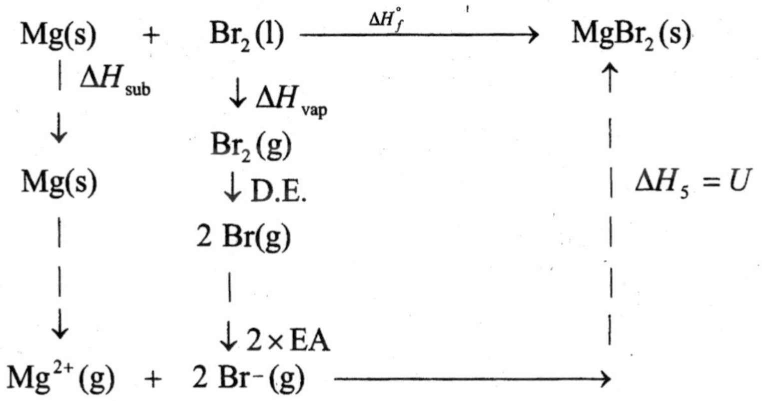

FIGURE 12-3 Born-Fajans-Haber Cycle for $MgBr_2$.

The energy of each step is found as follows.

(1) The metallic solid is melted and then vaporized (or it is sublimed), and then the heat of sublimation is measured: $\Delta H_1 = \Delta H_{sub}$

$Mg(s) \longrightarrow Mg(g) \qquad \Delta H_1 = \Delta H_{sub} = +150.2\ \text{kJ/mol}$

(2) The metallic vapor is ionized to produce cations of the correct positive charge and the ionization potentials (IP) are measured: $\Delta H_2 = IP_2 + IP_2$ for Mg

$Mg(g) \longrightarrow Mg^{2+}(g) + 2\ e^-$

$\Delta H_2 = 737.7\ \text{kJ/mol} + 1451\ \text{kJ/mol} = +2189\ \text{kJ/mol}$

(3) The nonmetal is vaporized, if necessary, and the heat of vaporization is measured: $\Delta H_3 = \Delta H_{vap}$ (or $= \Delta H\ \Delta H_{sub}$ if the nonmetal is a solid).

$Br_2(l) \longrightarrow Br_2(g) \qquad \Delta H_3 = +30.7\ \text{kJ/mol}$

(4) The nonmetal molecules are dissociated, or broken apart into atoms, and the dissociation energy (D.E.) is measured: $\Delta H_4 = \text{D.E.}$

$$Br_2(g) \longrightarrow 2\ Br(g) \qquad \Delta H_4 = +193.9\ kJ/mol$$

(5) Electrons are added to the nonmetal atoms to produce anions of the correct negative charge and the electron affinity (EA) is measured: $\Delta H_5 = \text{electron affinity} \times 2$ (for 2 Br).

$$2\ Br(g) + 2\ e^- \longrightarrow 2\ Br^-(g) \qquad \Delta H_5 = -1355\ kJ/mol \times 2 = -2710\ kJ/mol$$

(6) The gaseous ions are combined and the lattice energy is calculated: $\Delta H_6 = U = \text{lattice energy}$.

$$Mg^{2+}(g) + 2\ Br^-(g) \longrightarrow MgBr(s) \qquad \Delta H_6 = ?\ kJ/mol$$

We have created two pathways for making $MgBr_2(s)$ from the elements: the formation reaction and Steps (1) through (6). The energy for each pathway must be the same.

$$\Delta H_f^\circ = \Delta H_1 + \Delta H_2 + \Delta H_3 + \Delta H_4 + \Delta H_5 + \Delta H_6$$

$$= \Delta H_{sub} + (IP_1 + IP_2) + \Delta H_{vap} + D.E. + (2 \times EA) + U \qquad [12]$$

$$-517.6\ kJ/mol = +150.2\ kJ/mol + (2189\ kJ/mol) + 30.7\ kJ/mol$$
$$+ 193.9\ kJ/mol - 2710\ kJ/mol + U$$

or $U = -37.1\ kJ/mol$

Electron affinities are imprecise because they are hard to measure. If the lattice energy were known by some other technique, then the Born-Fajans-Haber cycle could be used to determine the electron affinity. In fact, the lattice energy of an ionic crystal can be *calculated* and the results agree excellently with those of the Born-Fajans-Haber cycle. The only data needed are the type of crystal structure and the ionic charges and radii. All of this data comes from X-ray analysis of crystals (see Objectives 12-12 through 12–14).

SELF-ASSESSMENT EXCERCISES

116. (c) Three of these, HF, CH_3OH, and N_2H_4. Hydrogen must be bonded to O, N, or F to form hydrogen bonds. It can form weak hydrogen bonds when bound to Cl; but, in $CHCl_3$, the H is bonded to carbon, not Cl.

117. (b) Eight. The coordination number is the number of atoms a given atom is in contact with.

119. (d) At 1 atm, the liquid will vaporize to gas. (f) The gaseous state will need to be cooled. A phase diagram of carbon dioxide will show these two statements to be true.

120. (a) $C_{10}H_{22}$, because larger hydrocarbon molecules have higher boiling points.

(b) $(CH_3)O$ will have greater induced dipole forces.

(c) CH_3CH_2OH will have stronger H bonds.

121. Ar will have a lower boiling point because it does not have dipole or other intermolecular forces.

122. The expected order is Ne< C_3H_8 < CH_3CH_2OH < $CH_2OHCHOHCH_2OH$ < KI < MgO < K_2OS_4. The order is based on (1) size, smaller molecules melt more easily and (2) H bonds will attract molecules and ionic bonds. Induced dipoles for KI, MgO and K_2SO_4 are very strong.

125. (a) Liquid and gas will be present. The temperature is above the melting point so no solid will be present. The liquid will vaporize until the liquid and gas are in equilibrium.

(b) There are 1.59 g/ml x 2.00 L x 10^3 ml/L x 1 mol/153.6 g = 20.7 mol.

Using $\ln\dfrac{P_2}{P_1} = \dfrac{\Delta H_{vap}}{R}\left(\dfrac{1}{T_1} - \dfrac{1}{T_2}\right)$, where P_2 = 1 atm, T_2 = 350 K, P_1 = 0.110 atm, and T_1 = 298 K. Solving gives us ΔH_{vap} = 36.8 kJ mol^{-1}. The heat required is 20.7 mol x 36.8 kJ mol^{-1} = 762 kJ.

126. (a) The length of the unit cell is the length of the side of the cube. The diagonal of the square is 4 x radius. Using $c^2 = a^2 + b^2$ for a right triangle, and the fact that a = b for a square, $c^2 = (4r)^2 = 2\,a^2$, $a = (8\,r^2)^{1/2}$ and a = 2.83 r , or 362 pm.

(b) The volume = $a^3 = (362)^3 = 4.75 \times 10^7$ pm^3.

(c) The number of atoms will be: 6 faces (1/2 atom/ face) + 8 corners (1/8 atom/corner) = 4 Cu atoms.

(d) The volume of 4 Cu atoms is 4 x 4/3 $\pi r^3 = 3.51 \times 10^7$ pm. The % occupied

= vol Cu/vol of cell x 100% = $\dfrac{3.52\times10^7}{4.75\times10^7}$ x 100% = 73.9%.

DRILL PROBLEMS

1. In each line below, the specified mass (m, g) of the indicated compound is heated from the initial temperature (t_i, °C) to the final temperature (t_f, °C) with the given amount of heat (q, kJ). Occasionally, liquid and vapor are present together at the boiling point, in which case the mass of the liquid is given: (m_l, g). Fill in the blanks in the table below. Data is included in Table 12-1.

Compound	m, g	t_i, °C	t_f, °C	q, kJ	m_l, g
A. methanol	200.	50.0	_____	184.7	_____
B. ethanol	204	20.2	100.0	_____	0.00
C. sulfur dioxide	186	−51.0	−10.0	_____	46
D. hydrogen cyanide	_____	−14.0	36.5	20.20	0.00
E. acetic acid	400.	50.0	_____	204	_____
F. carbon disulfide	300.	25.0	90.	_____	0.00
G. diethyl ether	55.0	25.0	_____	1.43	_____

H. carbon tetrachloride	_____	25.0	76.8	16.4	14
I. chloroform	314	0.00	_____	14.3	_____
J. formic acid	87.0	41.0	126.0	_____	0.00
K. acetone	435	8.2	56.2	_____	100.
L. bromine	124	20.40	_____	_____	24
M. ammonia	100.	−52.6	_____	8.17	_____

2. (1) In each of the following transpiration experiments, a volume of noble gas (V) is passed through a liquid at temperature t. m grams of liquid becomes vapor. Determine the vapor pressure in mmHg of each liquid.

Liquid	V, L	t, °C	m, g	*Liquid*	V, L	t, °C	m, g
A. acetone	214	−32.0	7.44	B. formic acid	86.7	20.0	7.64
C. diethyl ether	9.52	−10.0	4.94	D. chloroform	15.2	0.0	4.79
E. bromine	10.7	8.0	9.82	F. ethanol	83.1	−5.0	1.83
G. methanol	98.3	−43.0	0.241				

(2) Use the data of Figure 12-4 to help determine how much liquid and how much vapor of each substance is present in each container that has volume V and is at temperature t. (M = molar mass)

Substance	m, g	V, L	t, °C
H. diethyl ether	25.0	10.0	20
I. benzene	10.0	5.2	40
J. water	2.14	11.6	80
K. aniline	5.79	8.2	120
L. toluene	14.2	7.3	100
M. benzene	12.0	4.9	60
N. diethyl ether	12.0	6.1	30

Substance	M g/mol
benzene	78.11
toluene	92.14
aniline	93.12

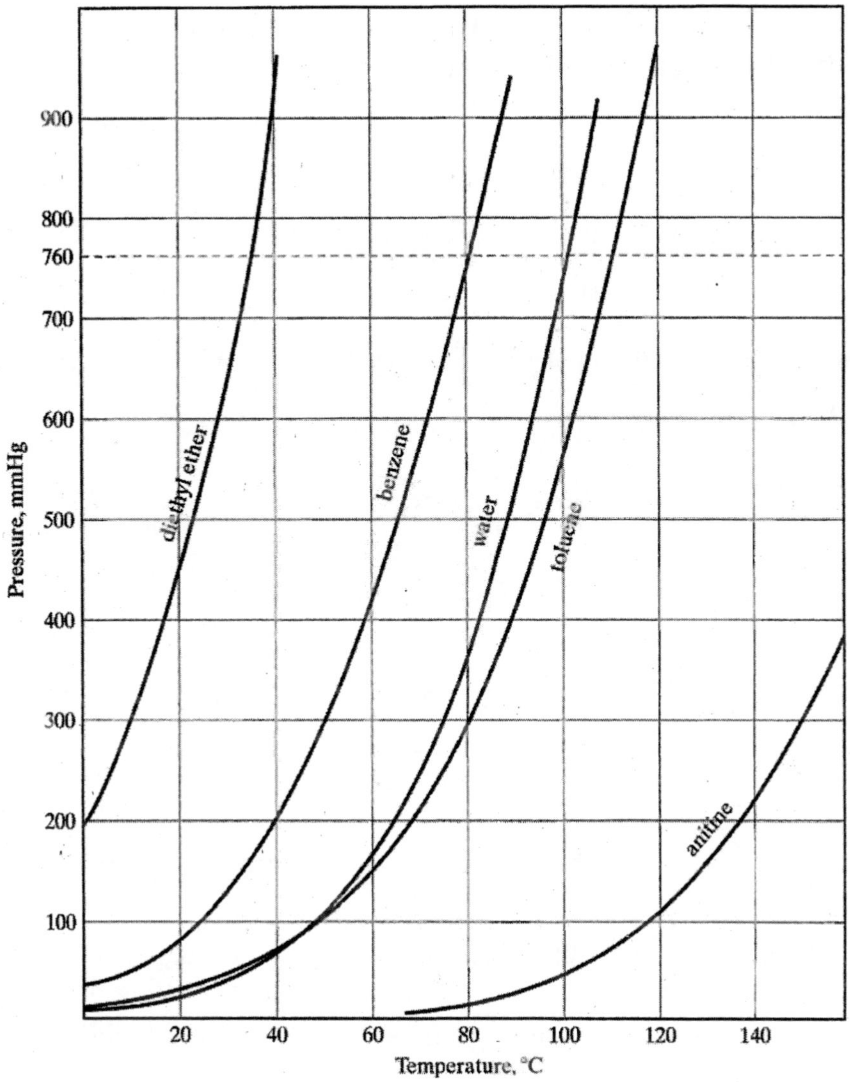

FIGURE 12-4 Vapor Pressure vs. Temperature

3. (1) The vapor pressure of several liquids at two temperatures follow. Determine the molar enthalpy change of vaporization of each liquid and its normal boiling point.

Liquid	P, mmHg	t_1, °C	P_2, mmHg	t_2, °C
A. BrF_5	10.0	−41.9	400.	+25.7
B. $COCl_2$	40.0	−50.3	100.	−35.6
C. H_2O_2	1.00	15.3	40.0	77.0
D. Si_2Cl_6	10.0	38.8	100	85.4
E. SO_2Cl_2	40.0	−1.0	100	17.8
F. SiH_3F	10.0	−141.2	400	−106.8
G. Rb	1.00	297	400	620

(2) Use the data of Table 12-1 to first determine the vapor pressure at the given temperature, and then the temperature needed to attain the given vapor pressure for each of the following liquids.

H. methanol, $t = 24.2$ °C, $P = 15$ mmHg

I. bromine, $t = 75.8$ °, $= P = 250$ mmHg

J. acetone, $t = 15.0$ °C, $P = 30$ mmHg

K. $CHCl_3$, $t = 37.0$ °C, $P = 300$ mmHg

L. SO_2, $t = -50.0$ °C, $P = 500$ mmHg

M. HCOOH, $t = 25.0$ °C, $P = 50$ mmHg

N. NH_3, $t = -50.0$ °C, $P = 180$ mmHg

O. ethanol, $t = 50.0$ °C, $P = 200$ mmHg

4. Fill in the blanks in the table below. All enthalpies are in kJ/mol.

Substance	ΔH_{fus}	ΔH_{vap}	ΔH_{sub}	ΔH_{sol}	ΔH_{cond}	HH_{dep}
A. H_2O	6.01	44.0	_____	_____	_____	_____
B. In	_____	_____	_____	_____	−232.8	−243.5
C. LiI	_____	176.9	_____	_____	_____	−204.1
D. NaCl	_____	181.8	229.0	_____	_____	_____
E. LiBr	20.1	_____	_____	_____	−158.6	_____
F. HCOOH	_____	41.4	60.4	_____	_____	_____
G. BeO	_____	_____	725.1	−56.6	_____	_____

5. A. Given the following data for hydrogen, roughly sketch its phase diagram and label all regions. Solid vapor pressure at 10 K: 0.001 atm; normal melting point: 14.01 K; normal boiling point: 20.38 K; triple point: 13.97 K, 0.07 atm; critical point: 33.3 K, 12.8 atm.

B. Does H_2 freeze at 9.0 atm pressure? If so, what is the freezing point? If not, why not?

C. Does H_2 boil at 14.0 atm pressure? If so, what is the boiling point? If not, why not?

D. Does solid H_2 float on the liquids? Why or why not?

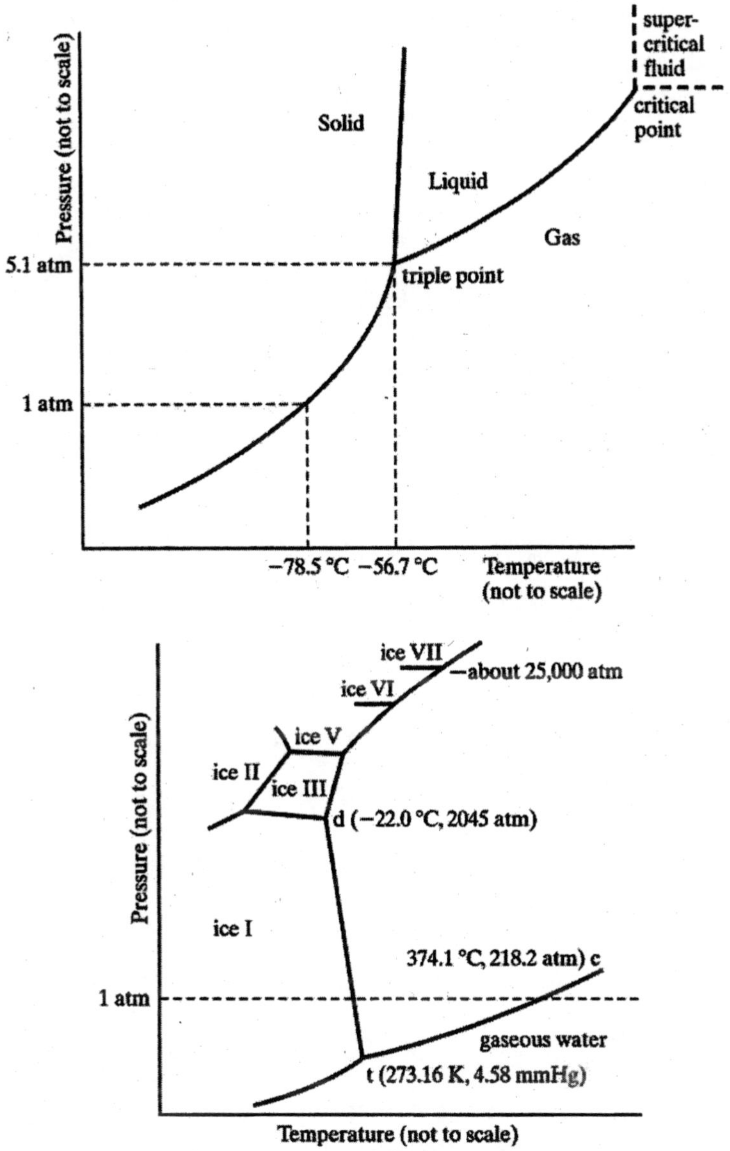

FIGURE 12-5 Phase Diagrams of Carbon Dioxide (left) and Water (right)

E. For hypothetical compound W, the vapor pressures of liquid and solid are as follows:

t, °C (solid)	−10	10	30	50	
P, atm (solid)	0.10	0.20	0.30	0.40	
t, °C (liquid)	50	150	200	250	300
P, atm (liquid)	0.40	0.64	0.76	0.88	1.00

The melting point of the solid varies with pressure as follows:

P, atm	0.40	1.00	1.60	2.20
t, °C	50	49.5	49.0	48.5

Above $t = 300\ °C$ and 1.0 atm, liquid and vapor are indistinguishable. Sketch the phase diagram of compound W and label all points and regions.

F. Does compound W freeze at 2.00 atm pressure? If so, what is the freezing point? If not, why not?

G. Does compound W boil at 0.80 atm pressure? If so, estimate the boiling point. If not, why not?

H. Does solid compound W float on the liquid? Why or why not?

Refer to the phase diagrams of CO_2 and H_2O above and the two phase diagrams you have just drawn in questions A. and E. to describe the phase changes that occur during each of the following changes in pressure or temperature. Give the approximate temperature and pressure at which each phase change occurs.

I.	H_2O	$P_1 = 2\ \text{mmHg}$ to $P_2 = 100\ \text{mmHg}$ at $T = 273.155\ \text{K}$ (fixed)
J.	H_2O	$T_1 = 253\ \text{K}$ to $T_2 = 423\ \text{K}$ at $P = 3.00\ \text{atm}$ (fixed)
K.	CO_2	$T_1 = 194.7\ \text{K}$ to $T_2 = 473.2\ \text{K}$ at $P = 2.00\ \text{atm}$ (fixed)
L.	CO_2	$T_1 = 173.2\ \text{K}$ to $T_2 = 473.2\ \text{K}$ at $P = 6.40\ \text{atm}$ (fixed)
M.	CO_2	$T_1 = 173.2\ \text{K}$ to $T_2 = 555.1\ \text{K}$ at $P = 80.0\ \text{atm}$ (fixed)
N.	H_2	$T_1 = 5.0\ \text{K}$ to $T_2 = 40.1\ \text{K}$ at $P = 1.00\ \text{atm}$ (fixed)
O.	H_2	$P_1 = 0.001\ \text{atm}$ to $P_2 = 10.0\ \text{atm}$ at $T = 14.0\ \text{K}$ (fixed)
P.	W	$P_1 = 0.10\ \text{atm}$ to $P_2 = 2.00\ \text{atm}$ at $T = 322.4\ \text{K}$ (fixed)
Q.	W	$T_1 = 273.0\ \text{K}$ to $T_2 = 573.2\ \text{K}$ at $P = 0.80\ \text{atm}$ (fixed)

6. For each of the following groups of substances, state which has the largest and which has the smallest value of the indicated property.

A. Boiling point: Xe, HI, H_2Te

B. Heat of vaporization: PF_3, NF_3, BCl_3

C. Vapor pressure at $-25\ °C$: HCl, HBr, HI

D. Vapor pressure at $-10\ °C$: CH_4, SiH_4, GeH_4

E. Boiling point: AlF_3, PF_3, ClF_3

F. Heat of vaporization: $C_3H_8OH, C_2H_5OH, CH_3OH$

G. Boiling point: NH_3, H_2O, HF,

H. Heat of vaporization: Cl_2, Br_2, I_2

I. Vapor pressure at $-10\ °C$; CH_4, CF_4, CCl_4

J. Boiling point: CCl_3F, GeF_4, Br_2

7. For each of the following groups of compounds, state which has the largest and which has the smallest value of the indicated property. (Remember that lattice energies are exothermic!)

A. Lattice energy: MgS, SrS, CaS

B. Melting point: NaH, NaBr, NaCl

C. Solubility (M): $CaCl_2, CaI_2, SrI_2$

D. Melting point: MgS, Na_2S, Na_3P

E. Solubility (M): $CaF_2, CaO, CaCl_2$

F. Lattice energy: $MgCO_3, Na_2CO_3,$

G. Melting point: LiF, LiCl, LiI

H. Lattice energy: K_2S, Na_2S, MgS, MgS

I. Solubility (M): $Pb(NO_3)_2, Mg(NO_3)_2, Ca(NO_3)_2$

J. Melting point: RbCl, $InCl_3, SrCl_2$

QUIZZES (20 minutes each) Choose the best answer to each question.

QUIZ A

1. The property of a liquid that does *not* depend on intermolecular forces is (a) surface tension; (b) boiling point; (c) vapor pressure; (d) heat of vaporization; (e) none of these.

2. 31.4 g of a solid requires 1.42 kJ of heat to melt. The solid has a mole weight of 154 g/mol. What is its heat of fusion in kJ/mol? (a) 31.4(1.42/154); (b) 154(1.42/31.4); (c) 31.4(1.42/154); (d) $1.42(1.54 \times 31.4)$; (e) none of these.

3. In the phase diagram to the right, point *A* is called the (a) triple point; (b) critical point; (c) melting point; (d) boiling point; (e) none of these.

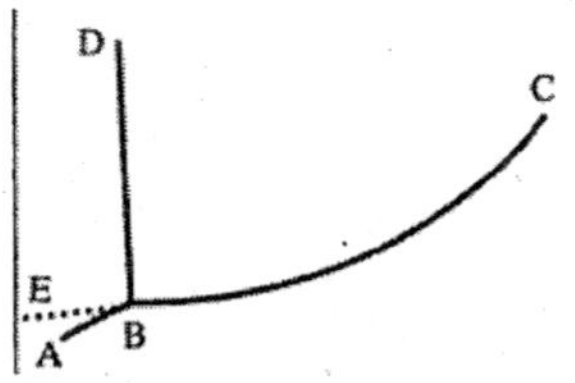

4. A liquid is in equilibrium with its vapor. If some of the vapor escapes, what is the immediate result? (a) vaporization rate decreases; (b) condensation rate decreases; (c) vaporization rate increases; (d) condensation rate increases; (e) none of these.

5. The process in which a solid is transformed into a vapor is called (a) vaporization; (b) sublimation; (c) gasification; (d) condensation; (e) none of these.

6. The vapor pressure of trichloroethene is given at four different temperatures. Which temperature is the normal boiling point? (a) 40 torr at 40.1 °C; (b) 100 torr at 61.3 °C; (c) 400 torr at 100.0 °C; (d) 760 torr at 120.8 °C; (e) none of these.

7. For anions of the same size, which type of hole is the smallest? (a) tetrahedral; (b) octahedral; (c) hexagonal; (d) cubic; (e) cannot be determined.

8. 3.42 g of a liquid of mole weight 100.0 g/mol vaporizes when 42.0 L of inert gas at 300 K is bubbled through. The vapor pressure of the liquid in mmHg is (a) $(3.42/100.0)(0.0821 \times 760)$; (b) $(3.42/100.0)(8.314 \times 760)(300/42.0)$; (c) $(3.42 \times 0.0821)(300/42.0)$; (d) insufficient information is given; (e) none of these.

QUIZ B

1. *A* has a higher normal boiling point than *B*. This could be due to any of the following factors except the higher (a) mole weight of *A*; (b) vapor pressure of *A*; (c) intermolecular attractions of *A*; (d) vapor pressure of *B*; (e) none of these.

2. The liquid with the highest vapor pressure at $-50\,°C$ is which of the following? (a) Xe; (b) Ne; (c) Cl_2; (d) F_2; (e) CCl_4.

3. The process in which a gas is transformed into a solid is called (a) vaporization; (b) condensation; (c) solidification; (d) deposition; (e) none of these.

4. In the phase diagram to the right, point *H* represents the (a) melting point; (b) normal melting point; (c) boiling point; (d) normal boiling point; (e) critical point.

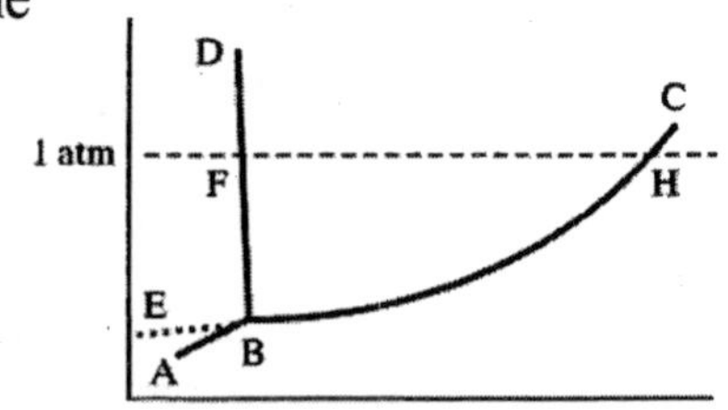

5. A liquid is in equilibrium with its vapor. A thin layer of a nonvolatile oil is poured on top of the liquid. What will now occur? (a) vaporization rate decreases; (b) condensation rate decreases; (c) vaporization rate increases; (d) condensation rate increases; (e) none of these.

6. The temperature at which a solid sublimes at 760 torr is called the (a) boiling point; (b) normal boiling point; (c) melting point; (d) normal melting point; (e) none of these.

7. A face-centered cubic cell of a metallic element contains all or part of how many different atoms? (a) 2; (b) 6; (c) 8; (d) 14; (e) none of these.

8. 105 kJ of heat vaporizes what mass of liquid if its mole weight is 86.0 g/mol and its molar heat of vaporization is 55.2 kJ/mol? (a) (105/55.2)/86.0; (b) 86.0(105/55.2); (c) 55.2(105/86.0); (d) (86.0/55.2)/105; (e) none of these.

QUIZ C

1. The factor that has the largest effect on vapor pressure is (a) liquid surface area; (b) molecular dipole moment; (c) presence of hydrogen bonding; (d) liquid mole weight; (e) volume available for the vapor.

2. Which intermolecular force of the following is the strongest? (a) London; (b) hydrogen bonding; (c) dipole-dipole; (d) exchange; (e) all have the same strength.

3. In the phase diagram to the right, point *C* is called

the (a) critical point; (b) triple point; (c) melting point; (d) boiling point; (e) none of these.

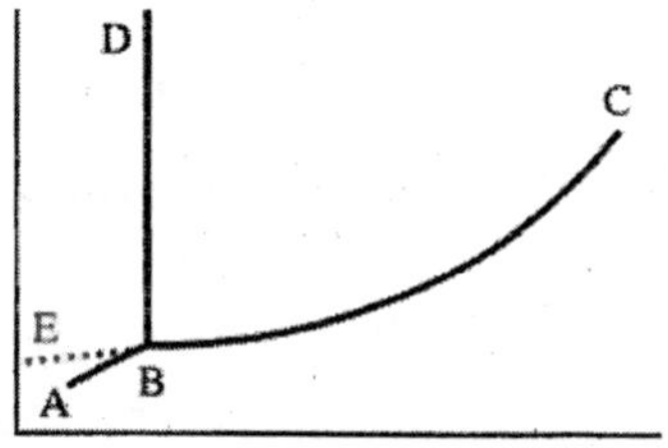

4. At the boiling point, the vapor pressure of a liquid is (a) 760 mmHg; (b) steadily decreasing; (c) constantly fluctuating; (d) atmospheric pressure; (e) none of these.

5. A liquid is in equilibrium with its vapor. Some vapor is allowed to escape. After equilibrium has been reestablished at the same temperature, it is true that the (a) vaporization rate is greater; (b) vaporization rate is less; (c) condensation rate is greater; (d) condensation rate is less; (e) none of these.

6. The process in which a liquid is changed into a gas is called (a) vaporization; (b) sublimation; (c) gasification; (d) condensation; (e) none of these.

7. 6.12 g of a solid with a mole weight of 79.0 g/mol is melted with just 3.76 kJ of heat. The molar heat of fusion of the solid in kJ/mol is (a) 79.0(3.76/6.12); (b) (3.76/6.12)/79.0; (c) 6.12(3.76/79.0); (d) $3.76(6.12 \times 79.0)$; (e) none of these.

8. Of the following, the solid that melts at the lowest temperature is (a) MgO; (b) NaCl; (c) CaS; (d) AlN; (e) Li_2O.

QUIZ D

1. The mole weight of a compound directly affects all of the following *except* (a) vapor pressure of liquid; (b) vapor density; (c) solid density; (d) vapor pressure of solid; (e) molar volume of vapor.

2. 4.25 kJ of heat is sufficient to melt 42 g of a solid of mole weight 210 g/mol. What is the heat of fusion of this solid in kJ/mol? (a) 42(4.25/210); (b) 210(4.25/42); (c) (4.25/210)/42; (d) $4.25(210 \times 42)$; (e) none of these.

3. When the vapor pressure of a liquid equals atmospheric pressure, the temperature of the liquid equals (a) $100\,°C$; (b) the boiling point; (c) the normal boiling point; (d) the vaporization point; (e) none of these.

4. The heat of sublimation equals the negative of the heat of (a) condensation; (b) vaporization; (c) solidification; (d) fusion; (e) none of these.

5. A liquid is in equilibrium with its vapor. Suddenly the pressure on the vapor is doubled. What will now occur? (a) vaporization rate increases; (b) condensation rate decreases; (c) condensation rate increases; (d) vaporization rate decreases; (e) none of these.

6. The highest melting solid of the following is (a) Mg_3N_2; (b) Ca_3P_2; (c) CaO; (d) NaCl; (e) $BaSO_4$.

7. Of the following, the liquid with the highest vapor pressure at 232 K is (a) H_2O; (b) NH_3; (c) CH_3OH; (d) CCl_3F; (e) CF_4.

SAMPLE TEST (20 minutes)

1. The normal boiling point of acetone is 56.2 °C and the molar heat of vaporization is 32.0 kJ/mol. At what temperature will acetone boil under a pressure of 50.0 mmHg?

2. Sketch a face-centered cubic cell.

3. Describe the processes that lead to the formation of exchange forces within molecules.

4. You have the following ions available: cations: $Na^+, K^+, Ca^{2+}, Mg^{2+}$; anions: $F^-, Br^-, O^{2-}, S^{2-}$. Which cation and which anion would you expect to combine to form the highest melting compound? Carefully explain your choice.

13 SOLUTIONS AND THEIR PHYSICAL PROPERTIES

CHAPTER OBJECTIVES

1. Explain how the relative forces between molecules predict whether an ideal solution, a nonideal solution, or a heterogeneous mixture will form.

(1) In an ideal solution, the two components (*A* and *B*) of the solution act as if each were pure. The volume of the solution is the sum of the volumes of the two components. No heat is absorbed or given off when the solution forms. (The heat of solution is zero, $\Delta H_{\text{soln}} = 0$). When the forces between like molecules ($A \leftrightarrow A$ and $B \leftrightarrow B$) are about the same as the forces between unlike molecules ($A \leftrightarrow B$), the solution formed is ideal.

> *But* the "like" forces ($A \leftrightarrow A$ and $B \leftrightarrow B$) may *not* be the same as the "unlike" forces ($A \leftrightarrow B$).

(2) When the "unlike" forces ($A \leftrightarrow B$) are stronger, molecules in the solution are more strongly attracted to each other than when they are pure. The two components readily mix to form a solution and the extra energy is given off as heat: $\Delta H_{\text{soln}} < 0$. Likewise, we expect the volume of the solution to be less than the sum of the component volumes, since the stronger forces in the solution should draw the molecules (and ions) closer together.

(3) When the "unlike" forces ($A \leftrightarrow B$) are weaker than the "like" ones ($A \leftrightarrow A$ and $B \leftrightarrow B$), the molecules in the solution are attracted to each other less strongly than when pure. Energy must be added during the solution process: $\Delta H_{\text{soln}} > 0$. And the solution's volume typically is larger than the sum of the components' volumes.

(4) When the "like" forces are a great deal larger than the "unlike" forces, a solution does not form; a heterogeneous mixture is produced. The attractions between unlike molecules ($A \leftrightarrow B$) do not produce enough energy to overcome the much stronger attractions between like molecules ($A \leftrightarrow A$ and $B \leftrightarrow B$). Sometimes raising the temperature provides sufficient energy to overcome this energy deficit. As we know, this is the case with oil and water; it is easier to wash dishes—even greasy ones—in hot water.

*2. Be able to use percent concentration units: volume %, mass %, and mass/vol %.

Many different concentration units exist because of the many different ways in which solutions are used. Although chemists usually are concerned with amounts (moles) of substances, others often are interested in masses or volumes. Most concentration units express the quantity of solute in a certain quantity of solution. Then we can measure out solution and know how much solute we have. All types of concentration calculations will be much easier if you use the following general technique:

(1) Write the definition of the concentration you are using. For example, the definition of mass % is:

$$\text{mass}\,\% = \frac{\text{mass of solute}}{\text{mass of solution}} \times 100$$

(2) Determine the quantity of solute given in the statement of the problem and use this as the numerator.

(3) Determine the quantity of either solution or solvent (as required by the definition), and use this as the denominator.

(4) Use the conversion factor method to separately change the units of numerator and denominator to agree with those given in the definition.

EXAMPLE 13-1 A solution of ethylene glycol in water contains 1.73 gal. of ethylene glycol in 20.0 gal. of solution. The densities are: solution, 1.0119 g/mL; ethylene glycol, 1.1108 g/mL; and water, 1.0000 g/mL. The molar masses are: ethylene glycol, 62.07 g/mol; and water, 18.02 g/mol. From this information, compute the (a) volume %, (b) mass/vol %, and (c) mass % of this solution.

(a) $$\text{volume}\,\% = \frac{1.73\text{ gal. solute}}{20.0\text{ gal. solution}} \times 100 = 8.65\text{ volume}\,\%$$

(b) $$\text{mass / volume}\,\% = \frac{1.73\text{ gal. solute} \times \dfrac{4\text{ qt.}}{1\text{ gal.}} \times \dfrac{946\text{ mL}}{1\text{ qt.}} \times \dfrac{1.1108\text{ g}}{1\text{ mL}}}{20.0\text{ gal. solution} \times \dfrac{4\text{ qt.}}{1\text{ gal.}} \times \dfrac{946\text{ mL}}{1\text{ qt.}}} \times 100 = 9.61\%\text{ (m/v)}$$

(c) $$\text{mass}\,\% = \frac{1.73\text{ gal. solute} \times \dfrac{4\text{ qt.}}{1\text{ gal.}} \times \dfrac{946\text{ mL}}{1\text{ qt.}} \times \dfrac{1.1108\text{ g}}{1\text{ mL}}}{20.0\text{ gal. solution} \times \dfrac{4\text{ qt.}}{1\text{ gal.}} \times \dfrac{946\text{ mL}}{1\text{ qt.}} \times \dfrac{1.0119\text{ g}}{1\text{ mL}}} \times 100 = 9.50\text{ mass}\,\%\text{(m/m)}$$

***3. Know and be able to use the definitions of *molarity* and *molality*.**

Two concentration units frequently used by chemists are molarity and molality.

$$\text{molality }(m) = \frac{\text{moles of solute}}{\text{kilogram of solvent}} \qquad [1]$$

$$\text{molarity}(M) = \frac{\text{moles of solute}}{\text{liter of solution}} \qquad [2]$$

Molality is convenient for freezing-point depression and boiling-point elevation measurements because it does not change with temperature. On the other hand, *molarity does* change with temperature, since solution volume generally expands as temperature increases. Molarity is valuable when one wishes to react two solutes. To obtain a certain amount of solute, in moles, one merely measures out the appropriate volume of solution.

EXAMPLE 13-2 What are the (a) molarity and (b) molality of the aqueous ethylene glycol solution of Example 13-1?

$$\text{(a) molarity} = \frac{1.73\ \text{gal.solute} \times \dfrac{4\ \text{qt.}}{1\ \text{gal.}} \times \dfrac{946\ \text{mL}}{1\ \text{qt.}} \times \dfrac{1.108\ \text{g}}{1\ \text{mL}} \times \dfrac{1\ \text{mol}}{62.07\ \text{g}}}{20.0\ \text{gal.solution} \times \dfrac{4\ \text{qt.}}{1\ \text{gal.}} \times \dfrac{946\ \text{mL}}{1\ \text{qt.}} \times \dfrac{1\ \text{L}}{1000\ \text{mL}}} = \frac{117\ \text{mol solute}}{75.7\ \text{L solution}}$$

$$= 1.55\ \text{M}$$

(b) To determine molality, we need the mass of the solvent.

$$\text{solution mass} = 75.7\ \text{L} \times \frac{1000\ \text{mL}}{\text{L}} \times \frac{1.0119\ \text{g}}{\text{mL}} \times \frac{\text{kg}}{1000\ \text{g}} = 76.6\ \text{kg solution}$$

$$or\ \text{solution mass} = 75.7\ \text{L} \times \frac{1.0119\ \text{kg}}{\text{L}} = 76.6\ \text{kg solution}$$

$$\text{solute mass} = 117\ \text{mol solute} \times \frac{62/07\ \text{g}}{\text{mol}} \times \frac{\text{kg}}{1000\ \text{g}} = 7.27\ \text{kg solute}$$

$$\text{solvent mass} = \text{solution mass} - \text{solute mass} = 76.6\ \text{kg solution} - 7.27\ \text{kg solute}$$
$$= 69.3\ \text{kg solvent}$$

$$\text{molality} = \frac{11.7\ \text{mol solute}}{69.3\ \text{kg solvent}} = 1.69\ m$$

Notice in Part (a) that the beginning units of solvent and solution volumes need not be the same, since both volumes are eventually converted to mL. Furthermore, as in this case, if the calculation is set up with the conversion factor method, duplicate operations [such as 4 qt./gal. in Part (a)] stand out. Finally, be careful not to confuse concentration units such as molarity with amounts of material (moles). Thinking of moles per liter as being the same as moles is similar to saying that the population of Australia is 5 persons, when actually there are 5 persons per square mile in this country of 16 million persons. Unfortunately, this is a very common mistake among beginning chemists.

***4. Know and be able to use the definitions of mole fraction and mole percent.**

Mole fraction is defined as the amount (in moles) of solute divided by the total amount of solution. If n_A is the symbol for the amount of substance A, then

$$\chi_A = \frac{n_A}{n_A + n_B + \ldots} = \text{mole fraction of A in solution} \quad [3]$$

We need to determine the amount in moles of each individual component and sum these amounts. We cannot directly use the total mass of the solution to determine the total amount in moles present in the solution.

EXAMPLE 13-3 What is the mole fraction of ethylene glycol in the solution of Example 13-2?

$$n_{\text{EtGly}} = 117 \text{ mol} \qquad n_{H_2O} = 69.3 \text{ kg solvent} \times \frac{1000 \text{ g}}{1 \text{ kg}} \times \frac{1 \text{ mol } H_2O}{18.02 \text{ g } H_2O} = 3846 \text{ mol } H_2O$$

$$\chi_{\text{EtGly}} = \frac{n_{\text{EtGly}}}{n_{\text{EtGly}} + n_{H_2O}} = \frac{117 \text{ mol}}{117 \text{ mol} + 3846 \text{ mol}} = 0.0296$$

The mole percent is 100 times the mole fraction: mol % EtGly $= 0.0296 \times 100\% = 2.96\%$

5. Distinguish among unsaturated, saturated, and supersaturated solutions and describe how a solute can be purified by recrystallization.

An unsaturated solution can dissolve more solute. A saturated solution holds the maximum amount of solute that it can at equilibrium. A supersaturated solution holds more than it can at equilibrium. Solute dissolves when it is added to an unsaturated solution. If solute is added to a saturated solution, the amount of undissolved solute will not change. If solute is added to a supersaturated solution, the amount of undissolved solute will increase.

To purify a solute by recrystallization, we lower the temperature of the solution containing that solute until the solute precipitates. Then we separate the crystallized solute from the saturated solution and dissolve it in pure, fresh solvent. This process is repeated until pure solute is obtained.

***6. Apply Henry's law to calculations of gas solubility as a function of pressure.**

Henry's law states that the concentration (c) of gas dissolved in solution increases as the gas pressure (P) increases.

$$c = kP \qquad [4]$$

There is no general agreement on the units to be used for P or for c. Thus, we must be careful to use units correctly when solving Henry's law problems. Henry's law data are given in Table 13-1.

TABLE 13-1 Gas Solubilities per 100 mL H_2O Under 1.00 atm Gas Pressure

Gas	*t* °C	*Gas Volume*, mL at STP	*t*, °C	*Mass of Gas Dissolved*, mg
Ar	0	5.6	50	5.36
O_2	0	4.89	50	3.51
N_2	0	2.33	40	1.77
Kr	0	11.0		17.5
H_2	0	2.14	50	0.170
O_3	0	49	0	105
NO	0	7.34	60	3.17r
CH_4	17	3.50	17	2.50

EXAMPLE 13-4 At 0°C, 49 mL of ozone (O_3), measured at STP, dissolves in 100. mL of water under a pressure of 1.00 atm. What mass in grams of ozone dissolves in 57.0 mL of water under a pressure of 1750 mmHg?

First, we compute the Henry's law constant for ozone in water at 0 °C by applying Equation [4]:

$$\frac{49\text{ mL at STP}}{100\text{ mL }H_2O} \times \frac{\text{mol }O_3}{22414\text{ mL at STP}} \times \frac{48.0\text{ g }O_3}{100.\text{ mL }H_2O} = k \times 1.00\text{ atm} \quad k = \frac{0.105\text{ g }O_3}{100.\text{ mL }H_2O \cdot \text{atm}}$$

Then, we compute the concentration at 1750 mmHg:

$$c = \frac{0.105\text{ g }O_3}{100.\text{ mL }H_2O \cdot \text{atm}} \times 1750\text{ mmHg} \times \frac{1\text{ atm}}{760\text{ mmHg}} = \frac{0.242\text{ g }O_3}{100.\text{ mL }H_2O}$$

Finally, we compute the mass of O_3 in 57.0 mL of H_2O:

$$57.0\text{ mL }H_2O\frac{0.242\text{ g }O_3}{100.\text{ mL }H_2O} = 0.138\text{ g }O_3$$

7. Describe the properties of solutions that are colligative properties.

Colligative properties depend on the *number* of solute particles (molecules or ions) in a given quantity of solution.

(1) *Vapor pressure lowering* obeys Raoult's law.

$$P_A = \chi_A P_A^\circ \qquad [5]$$

P_A = the vapor pressure of substance A above a solution in which χ_A is the mole fraction of A in solution. P_A° is the vapor pressure of pure A.

(2) *Boiling-point elevation.*

$$\Delta T_b = K_b m \qquad [6]$$

m is the molality of the solution, ΔT_b is the increase in the boiling point, and K_b is the boiling-point elevation constant. K_b depends only on the solvent.

(3) *Freezing-point depression.*

$$\Delta T_f = -K_f m \qquad [7]$$

m is the molality of the solution, ΔT_f is the decrease in the freezing point, and K_f is the freezing-point depression constant. K_f depends only on the solvent.

(4) *Osmotic pressure.*

$$\pi = cRT \qquad [8]$$

c is the molarity of the solution, T is its absolute temperature, and R is the ideal gas constant.

*8. Apply Raoult's law. Describe the applications and the limitations of the law.

Raoult's law, Equation [5], is true only for ideal solutions (see Objective 13-1, Type 1). The two components of an ideal solution often are quite similar chemically. For dilute *real* solutions, Raoult's law often is valid for the solvent but not the solute.

EXAMPLE 13-5 At 25 °C the vapor pressures of pure benzene and pure toluene are 95.1 mmHg and 28.4 mmHg, respectively. The total vapor pressure above a solution of these two liquids is 50.0 mmHg. What is the mole fraction of benzene in this solution?

Let subscript $_B$ represent benzene and $_T$, toluene. Raoult's law gives the vapor pressures:

$$P_T = \chi_T P_T^\circ = \chi_T(28.4 \text{ mmHg}) \qquad P_B = \chi_B P_B^\circ = \chi_B(95.1 \text{ mmHg})$$

Dalton's law gives the total vapor pressure:

$$P = P_1 + P_2 = 50.0 \text{ mmHg} = \chi_B(95.1 \text{ mmHg}) + \chi_T(28.4 \text{ mmHg})$$

Note that $\chi_B + \chi_T = 1$ or $\chi_T = 1 - \chi_B$. *THUS*

$$P = 50.0 \text{ mmHg} = (1-\chi_B)(28.4 \text{ mmHg}) + \chi_B(95.1 \text{ mmHg})$$

$$P = 28.4 \text{ mmHg} - \chi_B(28.4 \text{ mmHg}) + \chi_B(95.1 \text{ mmHg}) \quad \text{OR} \quad \chi_B = (50.0-28.4)/(95.1-28.4) = 0.324$$

The partial pressure of benzene is given by

$$P_B = \chi_B P^\circ_B = (0.324)(95.1 \text{ mmHg}) = 30.8 \text{ mmHg}$$

Then, the mole fraction of benzene in the *vapor* is

$$y_B = 30.8 \text{ mmHg} / 50.0 \text{ mmHg} = 0.616$$

$\chi_B = 0.324$ is the mole fraction of benzene in the liquid and $y_B = 0.616$ is its mole fraction in the vapor. The vapor is richer in the more volatile component—the one of higher vapor pressure. A plot of vapor pressure against mole fraction is called a *vapor pressure diagram*.

9. Describe how solution components can be separated by fractional distillation.

In fractional distillation, two components of a mixture are separated by repeatedly vaporizing and condensing portions of the mixture. Notice, from the answer to Example 13-5, that the vapor is richer or more concentrated in the more volatile component (the one with the higher vapor pressure). If this vapor is condensed and the resulting liquid is partially vaporized, the vapor that is produced also will be richer, yet in the more volatile component. This process can be extended through many cycles of partial vaporization followed by condensation. Eventually, the condensed vapor will be the pure liquid of the more volatile component. But what of the liquid that does not vaporize? Since the more volatile component has vaporized, the unvaporized liquid becomes more concentrated in the less volatile component. Given a sufficient number of cycles, the liquid left behind is pure; it is the less volatile component.

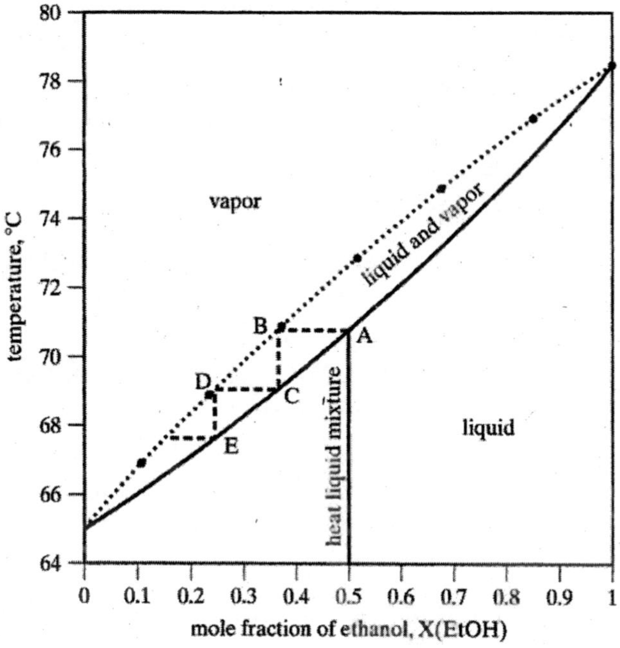

FIGURE 13-1 Boiling-Point Diagram for the Methanol-Ethanol System

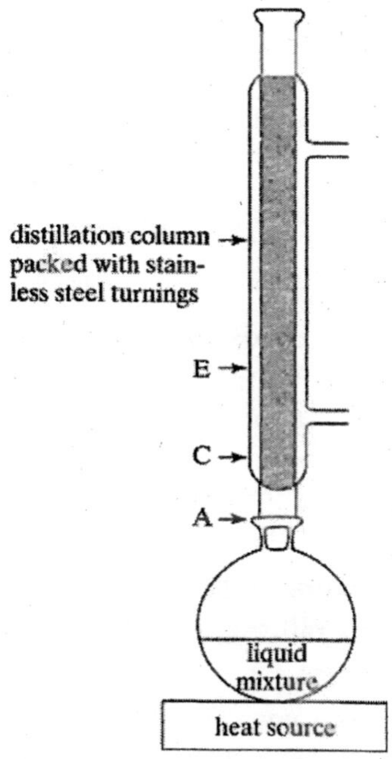

FIGURE 13-2 Fractional Distillation Apparatus

As an example of this process, consider Figure 13-1, a boiling-point diagram for the ethanol-methanol system. Below the solid curved line, the entire system is liquid. Consider what occurs when a mixture with $\chi_{\text{ethanol}} = 0.50$ is heated (solid vertical line). At $71.5\,^{\circ}\text{C}$ (point *A*), vapor appears with $y_{\text{ethanol}} = 0.37$. Note that the vapor (point *B*) is less concentrated in the less volatile component, ethanol. (The liquid and the vapor with which it is in equilibrium are connected by a dashed horizontal line.) This vapor condenses if the temperature is lowered to $69\,^{\circ}\text{C}$ (point *C*). The vapor (*D*) in equilibrium with this condensate has $\chi_{\text{ethanol}} = 0.25$. An apparatus to achieve this separation is pictured in Figure 13-2. The mixture is heated in the round-bottomed flask at the bottom. The resulting vapors rise into the vertical column, which is loosely packed with an inert material, such as stainless steel turnings. The packing is cooler the farther it is from the heat source; thus, the vapor begins to condense. Points *A*, *C*, and *E* at the start of heating are indicated approximately on the column. Of course, the concentrated liquid that does not vaporize drips back down the column until it encounters a high enough temperature to vaporize. As time continues, the temperature of the flask must continually increase because its contents are becoming richer in the less volatile, higher boiling component. (The more volatile one has vaporized.)

10. Explain how liquid-vapor equilibrium differs in nonideal and ideal solutions.

A solution that is not ideal does not obey Raoult's law. This can occur in two ways. If the vapor pressure of the solution is *less* than that predicted by Raoult's law, the solution is said to show a negative deviation from Raoult's law. This is due to the "unlike" forces ($A \leftrightarrow B$; see Objective 13-1) being stronger than the "like" forces and the molecules being more strongly held in the solution than they were in the pure components ($A \leftrightarrow A$ and $B \leftrightarrow B$).

If a solution's vapor pressure is *greater* than that predicted by Raoult's law, the solution is said to show a positive deviation from Raoult's law. In this case, the "unlike" forces are weaker than the "like" forces and the molecules are not held in solution as strongly as they were in the pure components.

***11. Explain how vapor pressure lowering leads to boiling-point elevation and also to freezing-point depression. Use Equations [6] and [7] for computing ΔT_{f} and ΔT_{b}.**

According to Raoult's law (Equation [5]), the vapor pressure of a component depends on its concentration. If the other component is nonvolatile, the vapor pressure of the solution will decrease as the concentration of the nonvolatile substance increases. Since the vapor pressure of the solution is lower at all temperatures, the liquid-vapor curve on a phase diagram, such as Figure 13-3, is lowered by the addition of a nonvolatile solute. In a similar fashion, if the solute does not freeze out with the solvent (and it seldom does), the temperature at which the solution begins to freeze is lower than the freezing point of pure solvent. The constants K_{b} and K_{f} in Equations [6] and [7] depend only on the solvent.

EXAMPLE 13-6 0.202 g of naphthalene (128 g/mol) lowers the freezing point of 10.453 g of cyclohexane by $3.08\,^{\circ}\text{C}$. 0.164 g of a solid unknown depressed the freezing point of 12.011 g of cyclohexane by 1.28 °C. What is the molar mass of the unknown?

First, compute the molality of the naphthalene solution and then K_f for cyclohexane:

$$\text{molality} = \frac{0.202 \text{ g naphthalene} \times 1 \text{ mol}/128 \text{ g}}{10.453 \text{ g cyclohexane} \times 1 \text{ kg}/100 \text{ g}} = 0.151\ m$$

$$K_f = \frac{-\Delta T_f}{m} = \frac{-(-3.08\ ^\circ\text{C})}{0.151\ m} = 20.4\ ^\circ\text{C}/m$$

Then, compute the molality of the unknown solution, which we use to determine the amount of solute and its molar mass:

$$m = \frac{\Delta T_f}{K_f} = \frac{1.28\ ^\circ\text{C}}{20.4\ ^\circ\text{C}/m} = 0.0627\ m$$

$$\text{amount solute} = 12.011 \text{ g solvent} \times \frac{1 \text{ kg}}{1000 \text{ g}} \times \frac{0.0627 \text{ mol solute}}{1 \text{ kg solvent}} = 7.54 \times 10^{-4} \text{ mol}$$

$$\text{molar mass} = \frac{0.164 \text{ g solute}}{7.54 \times 10^{-4} \text{ mol solute}} = 218 \text{ g/mol}$$

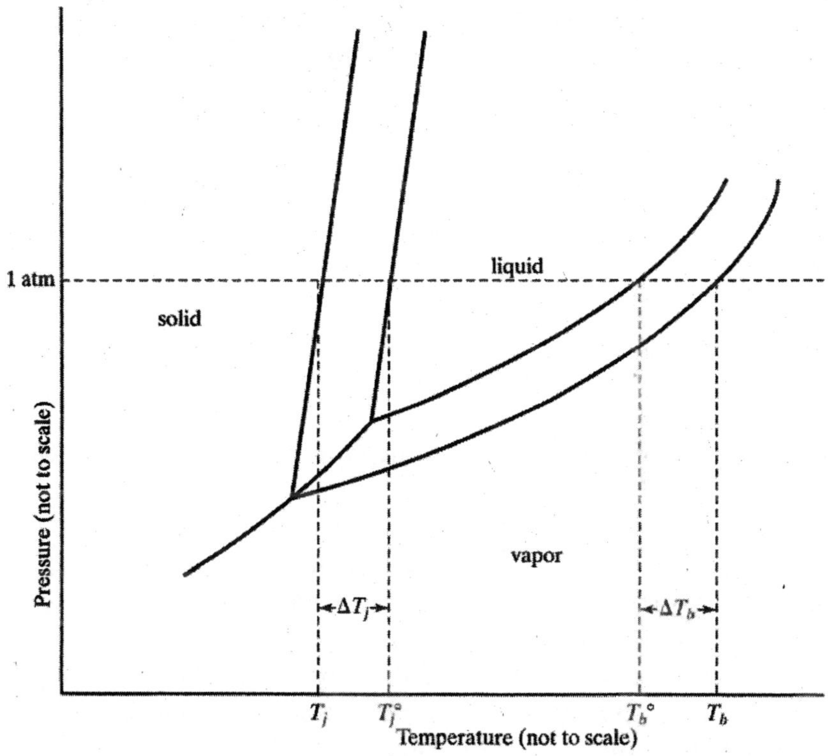

FIGURE 13-3 Vapor Pressure Lowering by a Non-Volatile Solute

*** 12. Describe the process of osmosis and use the law of osmotic pressure (Equation [8]).**

Osmosis occurs when two solutions of different concentrations are separated by semipermeable membrane. If the membrane were not present, the solutions would mix, resulting in a solution of intermediate concentration. But only the solvent can flow through the membrane. The direction of flow is from the dilute to the concentrated solution, making it less concentrated. The pressure that we exert on the concentrated solution to halt the flow of solvent is the *osmotic pressure*. If we let c_c indicate the concentration of the concentrated solution and c_d that of the diluted one, Equation [8] becomes:

$$\pi = (c_c - c_d)\,\mathrm{RT} \qquad [9]$$

EXAMPLE 13-7 An isotonic sugar solution has a concentration of 0.308 M. What osmotic pressure develops when a living cell (which basically contains such an isotonic solution) is placed in a 0.104 M sugar solution at 37 °C?

$$\pi = (0.308 - 0.104)\mathrm{M} \times 0.0821\,\mathrm{L\,atm\,mol^{-1}\,K^{-1}} \times 310\,\mathrm{K}$$
$$= 5.19\,\mathrm{atm}\ (= 76.3\,\mathrm{lb/in.^2})$$

*** 13. Describe how the theory of electrolytic dissociation explains the behavior of aqueous solutions of strong, weak, and nonelectrolytes, including ionic concentrations and differences in the values of colligative properties from the value computed from Equations [6] through [9].**

Colligative property data indicates that Equations [6] through [9] should be modified as follows:

boiling-point elevation:	$\Delta T_b = iK_b m$	[10]
freezing-point depression:	$\Delta T_f = -iK_f m$	[11]
osmotic pressure	$\pi = icRT$	[12]
	$\pi = i(c_c - c_d)RT$	[13]

The value of *i*, the van't Hoff factor, depends on the type of solute and somewhat on its concentration, but not on which colligative property is measured. When the electrical conductance of aqueous solutions is measured, the solutes can be grouped into three types. *Nonelectrolytes* produce aqueous solutions with a conductance equal to that of water. The van't Hoff factor equals 1. *Weak electrolytes* produce aqueous solutions with a slightly greater conductance than that of pure water. The van't Hoff factor is slightly larger than 1. Both the molar conductance and the van't Hoff factor increase as the solution becomes more dilute. *Strong electrolytes* produce aqueous solutions with a conductance much greater than that of pure water. The molar conductance of a strong electrolyte increases slightly as its solution becomes more dilute. The van't Hoff factor of its aqueous solution is a bit less than 2, 3, 4, or some whole number larger than 1, and increases slightly as the solution becomes more dilute. The molar conductances and van't Hoff factors for a strong and a weak electrolyte are given in Table 13-2.

Svante Arrhenius pointed out the similarity of the conductance trend on dilution to that of the van't Hoff factor. He used this similarity to argue that ions must be present in solutions of electrolytes at all times, even when they are not conducting a current. Dilute solutions have higher van't Hoff factors and molar conductances because more ions are present, according to Arrhenius. Arrhenius believed that the larger amount of water present in dilute solutions of electrolytes causes a larger fraction of solute to dissociate into ions. Presently we believe that strong electrolytes dissociate *completely* into ions in aqueous solution. All ionic compounds are strong electrolytes. In addition, a few covalent compounds also are strong electrolytes, namely the strong acids HCl, HBr, HI, HNO_3, H_2SO_4, $HClO_4$, and some others. (See Objective 13-14.)

TABLE 13-2 Variation of Molar Conductance (Λ, cm^2 mho/mol) and van't Hoff Factor (i) with Concentration for a Strong Electrolyte (HCl) and a Weak Electrolyte ($HC_2H_3O_2$)

Concentration,	*HCl*		$HC_2H_3O_2$	
mol/L	Λ	i	Λ	i
0.1	391	1.89	5.2	1.01
0.05	399	1.90	7.4	1.02
0.01	412	1.94	16.3	1.04
0.005	415	1.95	23.9	1.06
0.001	421	1.98	49.2	1.12
0.0005	423	2.00	67.7	1.17
0.0001	425	2.00	128	1.33

EXAMPLE 13-8 50.0 mL of 1.20 M NaCl is mixed with 30.0 mL of 0.800 M $CaCl_2$. What is the concentration of each ion in the final solution?

final solution volume $(V_f) = 50.0\text{ mL} + 30.0\text{ mL} = 80.0\text{ mL}$

First, compute the concentration of NaCl in the final solution:

$$50.0\ mL \times \frac{1.20\text{mmol NaCl}}{\text{mL soln}} = 60.0\text{ mmol NaCl}$$

$$[\text{NaCl}]_f = \frac{60.0\text{ mmolNaCl}}{80.0\text{ mLsoln}} = 0.750\text{ M}$$

The final $[CaCl_2]_f$ is computed in a different but equivalent way:

$$[CaCl_2]_f V_f = [CaCl_2]_i V_i \qquad [CaCl_2]_f (80.0\text{mL}) = (0.800\text{ M})(30.0\text{ mL})$$

$$[CaCl_2]_f = \frac{(0.800\text{ M})(30.0\text{ mL})}{80.0\text{ mL}} = 0.300\text{ M}$$

The only source of Ca^{2+} ions is $CaCl_2$ Thus, $[Ca^{2+}] = [CaCl_2] = 0.300\text{ M}$. Likewise, the only source of Na^+ ions is

NaCl; hence, $[\text{Na}^+] = [\text{NaCl}] = 0.750\ \text{M}$. But Cl^- ions are present in both solutes.

$$[\text{Cl}^-] = [\text{Cl}^- \text{ (from CaCl}_2)] + [\text{Cl}^- \text{ (from NaCl)}]$$

$$= 0.300\ \text{M} \times \frac{2\ \text{mol Cl}^-}{\text{mol CaCl}_2} + 0.750\ \text{M} \times \frac{1\ \text{mol Cl}^-}{\text{mol NaCl}}$$

$$= 0.600\ \text{M} + 0.750\ \text{M} = 1.350\ \text{M}$$

14. Describe how interionic attractions in solution require modifications to Arrhenius's theory.

We stated in Objective 13-13 that all strong electrolytes dissociate into ions or ionize completely in aqueous solution. (These ions are surrounded by water molecules. We say they are hydrated.) Why, then, are the van't Hoff factors not precisely equal to integers and why do the van't Hoff factors decrease as the solution becomes more concentrated? The explanation given by Peter Debye and Erich Hückel is that positive and negative ions cluster around each other in aqueous solution. These clusters are more likely to form in concentrated solutions because the ions are closer together. In concentrated solutions, therefore, there are more clusters, fewer particles (since a cluster of ions behaves in solution as if it were one particle), and the van't Hoff factor will be smaller than in dilute solutions.

TABLE 13-3 Some Common Types of Colloids

Dispersed phase	Dispersing medium	Colloid type	Examples
solid	liquid	sol	clay sols, colloidal gold
liquid	liquid	emulsion	oil in water, milk, mayonnaise
gas	liquid	foam	soap and detergent suds, whipped cream
solid	gas	aerosol	(cigarette) smoke, dust-laden air
liquid	gas	aerosol	fog, mist (as in aerosol products)
solid	solid	solid sol	ruby glass, certain natural and synthetic gems
liquid	solid	solid emulsion	opal, pearl
gas	solid	solid foam	pumice, lava, volcanic ash

14a. Describe some of the properties of colloids, how colloidal dispersions differ from solutions and heterogeneous mixtures, and classify colloidal mixtures.

Colloidal mixtures, or *colloidal dispersions*, appear uniform—that is, like solutions—to the naked eye, but often they are cloudy or foggy rather than clear. Colloidal dispersions often can be coagulated or separated by adding a solution of a strong electrolyte, such as NaCl, NaOH, or HCl. This partly neutralizes the positive or negative charges that are present on the surface of each colloidal particle and that keeps it separated from its neighbors. The various types of colloidal dispersions are summarized in Table 13-3.

SELF-ASSESSMENT EXERCISES

109. (b) For dilute solutions, molarity (moles/L of solution) and molality (mass/L of solvent) will be nearly the same. The small amount of solute will not increase the volume very much. This is not true for more concentrated solutions.

110. Ideal solutions have intermolecular forces of the same type and strength. Answer (a) will not be ideal because water has H bonds; the same is true for (b), where ethanol will have H bonds. For answer (c), the molecules do not have similar forces because, again, water has H bonds. Answer (d) is a mixture of hydrocarbons and is likely to be ideal.

111. Generally, solubility will increase with pressure and decrease with temperature. Therefore, Answer (a) is correct.

112. Answer (d) will have the lowest freezing point because $MgCl_2$ will ionize to give three ions and the freezing-point depression depends on the number of ions. It will have an activity of about 3 x (0.0080) = 0.0240.

115. (a) The concentration of Na^+ is

$$\frac{0.92g\ NaCl}{100g} \times \frac{1\ mol}{22.99\ g} \times \frac{1.005g}{mL} \times \frac{1000\ mL}{L} = 0.402\ M$$

(b) The molarity of ions will be twice 0.402 = 0.804 M.

(c) The osmotic pressure is $\Pi = M \times RT = 0.804 \times 0.0826\ L\ atm\ mol^{-1}\ K^{-1} \times 310\ K$

$$= 20.4\ atm$$

(d) Without knowing the volume of the NaCl to convert molarity to molality, we can estimate using molarity: $\Delta T_f = K_f c = 1.858 \times 0.804 = 1.49$. $T_f = -1.49\ ^0C$.

116. (a) 1 L of solution = 1159 g. The solution is 62% glycerol, so 0.62 x 1159 = 718.6 g in 1 L. 718.6 g L^{-1}/ 92.0 g mol^{-1} = 7.81 M.

(b) For water, 1159 g x 0.38 = 440.4 g in 1 L. Dividing by GAM, 440.4/18.0 = 24.47 M

(c) Molality of H_2O. 24.47 moles/L of solution and 1 L of solution is 62.0% glycerol (solvent). 24.47/0.620 = 39.5 m.

(d) Mole fraction of glycerol. There are 7.81 moles glycerol and 24.47 moles of water in each liter of solution. Mole fraction = 7.81/(24.47 + 7.81) = 0.242.

(e) Mole % of water = 24.47/(24.47 + 7.81) x 100% = 78.8%.

DRILL PROBLEMS

1. The aqueous solution described in each of the following lines is at 20°C. At 20°C, the density of water (the solvent) is 0.99823 g/mL. All solutes are liquids. Fill in the blanks of each line (the first line is an example).

	Solute Formula	*Volume, ML* *Soln.*	*Solute*	*Mass, g* *Solvent*	*Soln.*	*Density, g/mL* *Soln*	*Solute*	*mass %*	*mass vol %*	*volume*
Ex.	CH_3COOH	88.39	12.01	77.40	90.00	1.0182	1.0491	14.00	14.26	13.59
A.	CH_3COOH	___	___	___	31.16	1.0385	1.0491	30.00	___	___
B.	CH_3COCH_3	50.444	3.16	47.50	___	___	0.7908	___	___	___
C.	CH_3COCH_3	40.00	___	36.48	___	0.9967	0.7908	___	___	
D.	C_2H_5OH	134.51	___	___	___	0.8921	0.7893	___	53.53	___
E.	C_2H_5OH	___	___	___	6.87	0.9820	0.7893	10.00	___	___
F.	$(CH_2OH)_2$	59.441	4.87	54.60	___	___	1.1088	___	___	___
G.	$(CH_2OH)_2$	120.00	___	115.64	___	1.0038	1.1088	___	___	3.63
H.	HCOOH	137.99	___	___	___	1.0870	1.220	___	34.78	___
I.	HCOOH	___	___	___	38.11	1.0587	1.220	24.00	___	___
J.	$C_3H_5(OH)_3$	95.479	15.06	81.00	___	___	1.2613	___	___	___
K.	$C_3H_5(OH)_3$	75.00	___	43.72	___	1.1209	1.2613	___	___	42.65
L.	CH_3OH	83.673	___	___	___	0.8366	0.7914	___	70.27	___
M	CH_3OH	___	___	___	62.56	0.8762	0.7914	68.00	___	___

2. Fill in the blanks of each line in the following table. All of this data is for actual aqueous solutions.

	Solute *Mol. wt. or Formula*	*Number of moles*	*Solution* *Volume,* mL	*Density* g/mL	*Molarity,* M	*Molality,* *m*	*Mass %*
A.	$HC_2H_3O_2$	___	250.0	1.0250	3.414	___	___

B.	___	___	500.0	1.020	1.334	___	7.00
C.	$BaCl_2$	___	275.0	___	0.888	___	16.00
D.	HCl	1.270	400.0	1.0544	___	___	___
E.	___	1.245	___	1.056	2.490	___	10.00
F.	$Pb(NO_3)_2$	0.221	___	1.3509	1.104	___	___
G.	___	0.289	300.0	1.023	___	___	4.00
H.	$KHCO_3$	___	200.0	___	1.064	___	10.00
I.	KBr	___	750.0	1.107	___	___	14.00
J.	___	___	340.0	1.0402	0.907	___	6.50
K.	$NaHCO_3$	___	270.0	1.0408	___	___	6.00
L.	K_2CrO_4	0.200	___	1.172	1.207	___	___
M.	$NaNO_3$	___	750.0	___	5.043	___	34.00
N.	H_2SO_4	___	400.0	1.681	13.026	___	___
O.	NaOH	6.00	___	1.4299	___	___	40.00
P.	___	___	300.0	1.1972	5.326	___	26.00
Q.	H_3PO_4	2.047	___	1.2536	5.117	___	___
R.	CsCl	___	240.0	1.107	___	___	13.00

3. Find solute mole fraction and solvent mole percent in each solution of Objective 13-3's Drill Problems.

4. First use the data of Table 13-1 to determine the molality of each gas at each temperature under a pressure of 1.00 atm. The mass of 100.0 mL of water at each of the temperatures follows:

t, °C	0	17	40	50	60
mass, kg	0.1000	0.0999	0.0992	0.0988	0.0983

Then, compute the pressure or the molality of the gas in each solution described below.

A. Ar, $t = 0°C$, $P = 14.2$ atm

B. CH_4, $t = 17°C$, concn. = 0.0120 m

C. O_2, $t = 50°C$, $P = 160.$ mmHg

D. N_2, $t = 40°C$, concn. = 0.000314 m

E. NO, $t = 60°C$, $P = 1000.$ mmHg

F. O_3, $t = 0°C$, concn. = 0.100 m

G. Kr, $t = 50°C$, $P = 1000.$ mmHg

H. O_2, $t = 0°C$, concn. = 0.0100 m

I. H_2, $t = 50°C$, $P = 100.$ mmHg

J. H_2, $t = 50°C$, concn. = 0.0100 m

5. In each of the following ideal solutions, two liquids (A and B) are mixed together. The vapor pressures of the pure liquids are P°_{A} and P°_{B}, and their mole fractions are χ_A and χ_B. The vapor pressure of the solution is P. Fill in the blanks of the table below.

	P°_{A}, mHg	χ_A	P°_{B}, mmHg	χ_B	P, mmHg
A.	100.	0.75	50.	___	___
B.	180.	___	60.	___	125.
C.	___	0.25	20.	___	100.
D.	140.	0.30	___	___	220.

E.	____	____	160.	0.14	130.
F.	190.	____	270.	0.35	____
G.	120.	____	80.	____	85.
H.	220.	____	70.	____	100.

6. In each line that follows, the given mass of solute (in grams) is dissolved in the specified mass of solvent. The solvent has the boiling-point elevation constant K_b and the freezing-point depression constant K_f. t_b and t_f are the boiling and freezing points of the pure solvent. t_b' and t_f' are the boiling and freezing points of the solution. m is the molality of the solution, and M is the mole weight of the solute in grams/mole. Fill in the blanks of the following table.

	Solvent					*Solute*			*Solution*	
	mass, g	t_f, °C	t_b, °C	K_b, °C/*m*	K_f, °C/*m*	*mass*, g	*M* g/mol	*m*	t_b' °C	t_f' °C
A.	800.	16.62	117.9	3.07	3.90	17.2	____	0.201	118.52	____
B.	640.	5.51	80.09	2.53	____	19.4	140	____	____	4.45
C.	750.	5.68	210.81	5.24	____	____	126	0.244	____	3.97
D.	320.	43.02	181.75	____	7.40	7.62	____	0.284	182.76	____
E.	500.	0.00	100.00	0.512	1.86	12.7	78.2	____	____	−0.60
F.	600.	6.55	80.74	2.79	19.6	____	64.1	____	81.77	____

7. In the table that follows, each line represents a solution with osmotic pressure π (mmHg), concentration C (mol/L), and temperature t (°C). V is the volume of the solution (L), and M is the mole weight of the solute (g/mol). Fill in the blanks of each line.

Solution				*Solute*	
π, mmHg	C, mol/L	t (°C).	*V*,L	*mass*, g	M, g/mol
A. 740.	____	25	0.250	____	75.2
B. 143	____	37	0.375	1.42	____
C. ____	____	18	0.800	3.14	146
D. 1476	____	20.	0.750	____	128
E. 476	____	23	0.240	7.51	____
F. ____	____	15	0.150	0.904	187
G. 1793	____	24	0.400	____	207
H. 500.	____	19	0.500	1.37	____
I. ____	____	30.	0.650	3.27	209
J. 473	____	23	0.450	____	150.
K. 760.	____	20.8	0.120	0.816	____
L. ____	____	26.8	0.350	2.06	314
M. 946	____	16.4	0.100	____	200.
N. 640.	____	19.9	0.175	0.943	____
O. ____	____	29.0	0.180	0.0843	187

8. Solution *A* is mixed with Solution *B*. Compute the concentration of each ion in the final solution.

	Solution A			Solution B		
Vol.		*Solute*	*Molarity*	*Volume*	*Solute*	*Molarity*
A.	100. mL	$LiNO_3$	0.241	240. mL	$Ca(NO_3)_2$	0.618
B.	200. mL	H_2SO_4	0.314	450. mL	HCl	0.114
C.	400. mL	NaBr	0.186	200. mL	$AlBr_3$	0.916
D.	400. mL	Na_3PO_4	0.126	150. mL	NaCl	0.224
E.	620. mL	HNO_3	1.07	180. mL	$NaNO_3$	2.03
F.	250. mL	$CaCl_2$	0.143	150. mL	$MgCl_2$	0.321
G.	150. mL	Na_2SO_4	0.187	350. mL	$Al_2(SO_4)_3$	0.103
H.	100. mL	Na_2SO_4	0.946	900. mL	Na_3PO_4	0.204
I.	176 mL	NaBr	0.700	224 mL	NaCl	0.600
J.	894 mL	LiCl	0.600	106 mL	Li_2SO_4	0.800

QUIZZES (20 minutes each) Choose the best answer for each question.

QUIZ A

1. Moles of solute per mole of solvent is a definition of (a) mole fraction; (b) molarity; (c) molality; (d) solubility; (e) none of these.

2. A solution in which a relatively large amount of solute has been dissolved is called (a) unsaturated; (b) saturated; (c) supersaturated; (d) dilute; (e) none of these.

3. The concentration unit used in Raoult's law calculations is (a) mole fraction; (b) percent by weight; (c) molarity; (d) molality; (e) none of these.

4. 8.72 mL of a 14.0 M solution is diluted to 50.0 mL. What is the molarity of the resulting solution? (a) 14.0(8.72/50.0); (b) (14.0/8.72)/50.0; (c) 14.0(50.0/8.72); (d) 14.0(50.0)(8.72); (e) none of these.

5. 141 mL of 0.175 M solution of silver nitrate contains what solute mass (g)? (a) (141/1000)(0.175)(227.8); (b) (141)(0.175)(153.9); (c) (141/169.9)(0.175); (d) (141/1000)(0.175)(169.9); (e) none of these.

6. The normal boiling point of hexane is 68.7 °C. What is the mole fraction of nonvolatile solute in a hexane solution that has a vapor pressure of 600 mmHg at this temperature? (a) 68.7/600; (b) 600/760; (c) 160/600; (d) 160/760; (e) none of these.

7. 0.23 mol of solute depresses the freezing point of 267 g of solvent by 6.15 °C. What is the value of K_f for this solvent? (a) 6.15/(0.23/267); (b)

$6.5/(0.23 \times 0.267)$; (c) 6.15(0.23/0.267); (d) (6.15)(0.23)(0.267); (e) none of these.

8. 8.40 g (0.20 mol) NaF dissolves in 153 g (8.50 mol) of water. What is the mole fraction of NaF in the solution? (a) $0.20/(8.50+0.20)$; (b) 0.20/(153/1000); (c) $[8.40/(153+8.40)](100)$; (d) (8.40/153)(100); (e) none of these.

QUIZ B

1. Moles of solute per liter of solution is a definition of (a) mole fraction; (b) molarity; (c) molality; (d) solubility; (e) none of these.

2. A solution that has the capacity to dissolve more solute is called (a) unsaturated; (b) saturated; (c) supersaturated; (d) concentrated; (e) none of these.

3. Which of the following can be most directly linked to the lowering of vapor pressure by a nonvolatile solute? (a) osmotic pressure; (b) freezing-point depression; (c) boiling-point elevation; (d) solubility; (e) melting-point depression.

4. 52.5 mL of a solution was diluted to 6.25 L. It then had a concentration of 0.0316 M. What was the molarity of the original concentrated solution? (a) 0.0316(52.5/6.25); (b) 0.0316(6.25/52.5); (c) 0.0316(52.5/6250); (d) 0.0316(52.5)(6.25); (e) none of these.

5. At 34.9 °C, ethanol has a vapor pressure of 100 mmHg. What is the mole fraction of nonvolatile solute present in an ethanol solution that has a vapor pressure of 92 mmHg at this temperature? (a) 34.9/100; (b) 92/34.9; (c) 100/92; (d) 92/100; (e) none of these.

6. 87.5 L of a 3.17 M solution of hydrochloric acid contains how many grams of solute? (a) (87.5/1000)(3.17)(36.5); (b) (87.5)(3.17)(52.5); (c) (87.5)(3.17)(36.5); (d) (87.5/1000)(3.17)(68.5); (e) none of these.

7. 17.4 L of a 0.623 M starch solution displays what osmotic pressure in atmospheres at 75 °C? (a) (17.4)(0.623)(0.0821)(348); (b) (0.623)(0.0821)(75); (c) (17.4)(0.623)(75); (d) (17.4)(0.623/0.0821)(348); (e) none of these.

8. 23.4 g (0.40 mol) NaCl dissolves in 108 g (6.0 mol) H_2O. The solution's molality is (a) $0.40/(0.40+6.0)$; (b) 0.40/(108/1000); (c) $[23.4/23.4+108)](100)$; (d) (23.4/108)(100); (e) none of these.

QUIZ C

1. Moles of solute per liter of solvent is a definition of (a) mole fraction; (b) molality; (c) molarity; (d) solubility; (e) none of these.

2. The following always form solutions (if they do not react chemically): (a) a gas in a liquid; (b) a solid in a liquid; (c) two liquids; (d) two gases; (e) none of these.

3. Which of the following is not a colligative property? (a) freezing-point depression; (b) boiling-point elevation; (c) osmotic pressure; (d) solubility; (e) none of these.

4. How many mL of 6.25 M solution can be produced from 57.5 mL of 16.4 M solution? (a) (57.5)(6.25)(16.4); (b) 57.5(6.25/16.4); (c) 57.5(16.4/6.25); (d) 16.4(6.25/57.5); (e) none of these.

5. 81.2 g of cesium chloride (168.4 g/mol) is enough solute for how many mL of 4.00 M solution? (a) (81.2/168.4)/4.00; (b) (81.2/168.4)(4.00/1000); (c) (81.2/4.00)(168.4)(1000); (d) (81.2)(4.00)(168.4/1000); (e) none of these.

6. The normal boiling point of acetic acid is 118.1°C. What is the vapor pressure in mmHg of a solution in which there is 0.35 mol of nonvolatile solute for every mole of acetic acid at this temperature? (a) 760(0.35/1.35); (b) 760(0.35/1.00); (c) 118.1/760; (d) 760/118.1; (e) none of these.

7. 0.25 mol of a nonvolatile, nonionic solute dissolved in 300 g of water $(K_f = 1.86)$ will lower the freezing point how many degrees? (a) (1.86)(0.25)(1000/300); (b)(1.86/0.25)(300/1000); (c) (1.86)(0.25)(300/1000); (d) (0.25/1.86)(1000/300); (e) none of these.

8. 23.4 g (0.30 mol) of calcium fluoride dissolves in 189 g (10.5 mol) of water. What is the weight percent of calcium fluoride in this solution? (a) $0.30/(0.30+10.5)$; (b) 0.30/(189/1000); (c) $[23.4/(189+23.4)](100)$; (d) (23.4/189)(100); (e) none of these.

QUIZ D

1. Moles of solute per mole of solution is a definition of (a) percent by weight; (b) mole fraction; (c) molality; (d) molarity; (e) none of these.

2. Which are all homogeneous mixtures? (a) solvents; (b) solutes; (c) solutions; (d) fluids; (e) none of these.

3. Colligative properties are similar in that they all (a) were discovered in college laboratories; (b) are due to solvent molecules linked together; (c) have no harmful side effects; (d) depend on the number of solute particles in solution; (e) none of these.

4. What volume (in liters) of 0.0325 M solution can be produced from 41.6 mL of 0.742 M solution? (a) 41.6(0.742/0.0325); (b) 41.6(0.0325/0.742); (c) (41.6/1000)(0.0325/0.742); (d) (1000/41.6)(0.742/0.0325); (e) none of these.

5. 864 mL of solution contains 143.9 g of sodium phosphate (164.0 g/mol). What is the molarity of the solution? (a) (164.0/143.9)(864/1000); (b) (143.9/164.0)(864/1000); (c) (143.9/164.0)(1000/864); (d) (164.0/143.9)(1000/864); (e) none of these.

6. 841 mL of a 0.625 M sugar solution displays what osmotic pressure in atmospheres at 100 °C? (a) (0.625)(0.0821)(100); (b) (0.625)(0.841)(0.0821)(100); (c) (0.625)(0.841)(0.0821)(373); (d) (0.625)(0.0821)(373); (e) none of these.

7. 0.87 mol of solute depresses the freezing point of 742 g of solvent by 0.43 °C. What is K_f for this solvent? (a) (0.43)(0.87/742); (b) 0.43(0.87)(0.742); (c) 0.43/(0.87/742); (d) 0.43/(0.87/0.742); (e) none of these.

8. 15.9 g (0.15 mol) of sodium carbonate dissolves in 81 g (4.50 mol) of water. What is the weight percent of sodium carbonate in this solution? (a) $[15.9/(15.9+81)](100)$; (b) (15.9/81)(100); (c) 0.15/(81/1000); (d) $0.15/(0.15+4.50)$; (e) none of these.

SAMPLE TEST (20 minutes)

1. A solution contains 750.0 g of ethanol (46.0 g/mol) and 85.0 g of sucrose (180.0 g/mol). The volume of the solution is 810.0 mL. Determine the values of

   A. the density of the solution.

   B. the percent of sucrose in the solution C. the mole fraction of sucrose

   D. the molality of the solution E. the molarity of the solution

2. What volume of ethylene glycol ($C_2H_6O_2$, 62. g/mol, density $= 1.12$ g/mL) must be added to 20.0 L of water (H_2O, 18.0 g/mol, density $= 1.00$ g/mL) to produce a solution that freezes at 14.0 °F? The freezing-point depression constant of water is 1.86 °C/m.

14 CHEMICAL KINETICS

CHAPTER OBJECTIVES

*** 1. Describe how the rate of a reaction is related to the rate of disappearance of a reactant or formation of a product.**

When a reaction goes faster, the reactants are used up more rapidly and the products form more rapidly. But how much more rapidly? Let us consider an example: the reaction of $C_2H_4Br_2$ with KI:

$$C_2H_4Br_2 + 3\,KI \longrightarrow C_2H_4 + 2\,KBr + KI_3 \qquad [1]$$

We can express the rate of a reaction as the rate of change of concentration of a reactant ($\Delta[\text{reactant}]/\Delta t$) or of a product ($\Delta[\text{product}]/\Delta t$). But we need to be more specific, for at a given reaction rate, KI is used up three times as fast as $C_2H_4Br_2$, and KBr is produced twice as fast as either C_2H_4 or KI_3. By convention, the reaction rate is defined as in Equation [2]:

$$\text{rate of reaction} = \frac{-\Delta[\text{reactant}]}{\nu_r \Delta t} = \frac{+\Delta[\text{product}]}{\nu_p \Delta t}, \qquad [2]$$

where ν_r is the stoichiometric coefficient of the reactant and ν_p is the stoichiometric coefficient of the product. The negative sign is placed before the change in the concentration of the reactant because its change always will be negative and we want the reaction rate to be positive. Thus, for the reaction between $C_2H_4Br_2$ and KI, we have

$$\text{rate of reaction} = \frac{-\Delta[C_2H_4Br_2]}{\Delta t} = \frac{-\Delta[KI]}{3\Delta t} = \frac{+\Delta[C_2H_4]}{\Delta t} = \frac{+\Delta[KBr]}{2\Delta t} = \frac{+\Delta[KI_3]}{\Delta t} \qquad [3]$$

2. Explain how to obtain the data needed for a kinetic study from the results of a simple chemical analysis.

There are many methods of determining concentrations relatively rapidly, but two of the simplest are *colorimetric* (by measuring the light absorbed) and *titrimetric*. In Equation [1], only one species, KI_3 (actually the triiodide ion, I_3^-), absorbs visible light. Thus, the concentration of KI_3 can be followed by recording the absorbance of visible light by the reaction mixture, since concentration and absorbance are directly related.

$$CH_3COOC_2H_5 + H_2O \longrightarrow CH_3COOH + C_2H_5OH \qquad [4]$$

In the hydrolysis of ethyl acetate ($CH_3COOC_2H_5$), Equation [4], neither the products nor the reactants absorb visible light. But the concentration of one of the products, acetic acid (CH_3COOH), can be determined by titrating the reaction mixture. Thus, in either case, concentration vs. time data can be obtained readily.

*** 3. Establish the exact rate of a chemical reaction from the slope of a tangent line to the concentration vs. time graph. Also, explain how to determine the initial rate.**

Although concentration vs. time data can be tabulated, they often are presented graphically. Such a graph, known as a concentration vs. time curve, is presented in Figure 14-1 for the decomposition of HI(g) at 600 K.

$$2\,HI(g) \longrightarrow H_2(g) + I_2(g) \qquad [5]$$

To determine the instantaneous rate from a concentration vs. time curve, a tangent line is drawn to the curve at the time for which one wishes to know the rate. This straight line should just touch the curve at the desired time and should "follow" the curve as far before and after this time as possible. The slope of this line then is determined from two widely separated points on the line.

$$\text{slope} = \frac{\Delta[\text{graphed species}]}{\Delta t} = \frac{\text{concentration at time } t_1 - \text{concentration at time } t_2}{t_1 - t_2} \qquad [6]$$

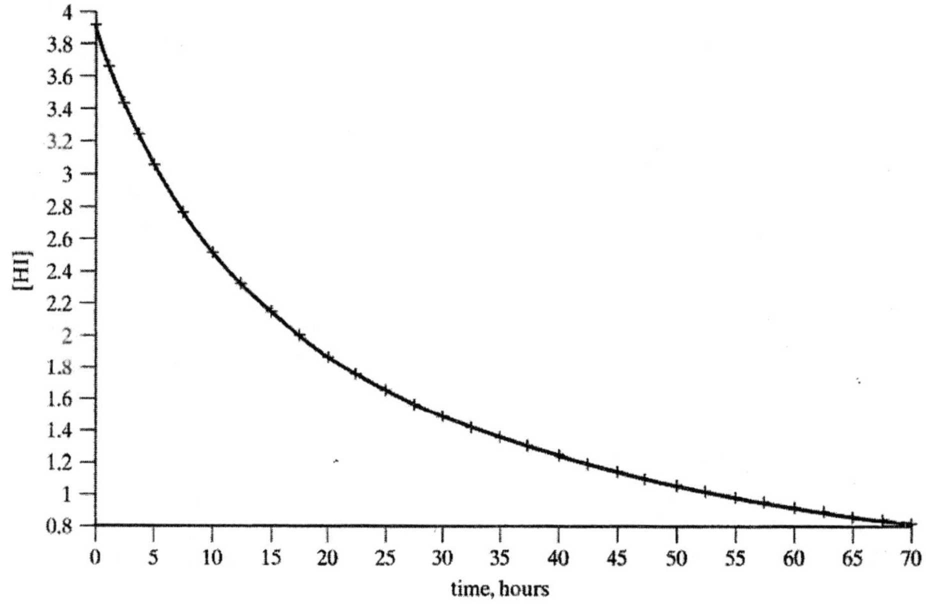

FIGURE 14-1 Concentration vs. Time Data for $2\,HI(g) \longrightarrow H_2(g) + I_2(g)$ at 600 K.

EXAMPLE 14-1 Use the data of Figure 14-1 to determine the rate of the decomposition of HI at 15 hours after the start of the reaction.

First, determine (with the dashed vertical line) the point on the curve that represents 15 hours from the start of the reaction. Then, draw the tangent to the curve at that point. (The tangent is drawn until it intersects the axes to make the slope easier to determine.) [HI] falls from 3.13 M to 0.76 M in the time interval from 0.00 hours to 36.0 hours. From this information, the slope of the tangent line can be determined.

$$\text{slope} = \frac{\Delta[\text{HI}]}{\Delta t} = \frac{3.13\,\text{M} - 0.76\text{M}}{0.00\,\text{hr} - 36.0\,\text{hr}} = -0.0658\,\text{mol L}^{-1}\,\text{hr}^{-1}$$

This slope represents the rate of change of [HI]. Thus, the rate of decomposition of HI is $0.0658\,\text{mol L}^{-1}\,\text{hr}^{-1}$.

We can also determine an instantaneous rate by calculation. We need just two data points: the concentrations at two times very close to, but on either side of, the time at which the rate is to be determined (along with the corresponding times). We compute the rate in the same way as the slope of the tangent line.

The initial rate of a reaction is a valuable source of information about that reaction because it is the rate of the forward reaction only. It is not influenced by the reverse reaction, in which the products react to form the reactants, or by any other reactions that involve the products. This is simply because the products have not reached sufficient concentrations to react at appreciable rates. To determine the instantaneous value of the initial rate, the tangent line is drawn at the point where time (t) = 0. For slow reactions (ones that are complete in minutes, rather than in seconds or even shorter), the initial rate is obtained by determining the change in the concentration of a product or a reactant during the first few seconds.

4. State the meaning of reaction order; use the rate law to determine the order of a reaction.

The law of mass action states that the rate of a reaction depends on the concentrations of its reactants. The reaction rate of Equation [1] is given by its rate law, which in general form is Equation [7].

$$\text{rate of reation} = k[\text{C}_2\text{H}_4\text{Br}_2]^x[\text{KI}]^y \qquad [7]$$

In the rate law, x and y are determined by experiment. In this case, $x = 1$ and $y = 1$. Thus, this reaction is first order in $[\text{C}_2\text{H}_4\text{Br}_2]$ (since $x = 1$) and first order in $[\text{KI}]$ (since $y = 1$). The reaction is said to be second order overall (since $x + y = 2$). We also say that this is a second-order reaction. The overall order is the sum of the orders of the various components.

*** 5. Apply the method of initial rates to determine the rate law for a reaction.**

We reviewed the determination of initial rates by two methods in Objective 14-3. Another general method is the study of clock reactions: those that produce a color change after a short period of time. The time period can be varied by changing the concentrations of the reactants. Consider the general reaction (Equation [8]), in which one of the products (species C) reacts to produce a color change (Equation [9]).

$$a\text{A} + b\text{B} \longrightarrow c\text{C} + d\text{D} \qquad [8]$$

$$\text{C} + \text{indicator} \longrightarrow \text{colored product} \qquad [9]$$

That same product, *C*, also reacts rapidly with yet another substance, *E*, as in Equation [10]. Because of this reaction, *C* cannot produce a color until all of *E* is consumed.

$$\mathrm{C + E \longrightarrow other\ products} \qquad [10]$$

A typical clock reaction uses I_2 as substance *C*, starch as the indicator, and $S_2O_3^{\ 2-}$ as substance *E*. For this example, Equations [8–10] become Equations [11–13]

$$2\,I^- + H_2O_2 + 2H+ \longrightarrow I_2 + 2\,H_2O \qquad [11]$$

$$I_2 + \mathrm{starch} \longrightarrow \mathrm{dark\ blue\ color} \qquad [12]$$

$$I_2 + S_2O_3^{\ 2-} \longrightarrow 2\,I^- + 2\,S_4O_6^{\ 2-} \qquad [13]$$

The reaction shown in Equation [12] does not occur until $S_2O_3^{\ 2-}$ is used up. Such a small $[S_2O_3^{\ 2-}]$ is present compared to $[I^-]$ and $[H_2O_2]$ that the changes in those latter two concentrations are negligible. Data from a typical series of experiments are given in Table 14-1, where the time is the period elapsed before the dark blue color appears.

TABLE 14-1 Initial Rates for $2\,I^- + H_2O_2 + 2\,H^+ \longrightarrow I_2 + 2\,H_2O$

	Initial concentration, M			
	$[I^-]$	$[H_2O_2]$	*Time,* s	*Rate,* $\mathrm{mol\,L^{-1}\,s^{-1}}$
1.	0.015	0.015	91	1.37×10^{-5}
2.	0.030	0.015	46	2.72×10^{-5}
3.	0.045	0.015	29	4.31×10^{-5}
4.	0.045	0.025	18	6.94×10^{-5}
5.	0.045	0.033	12	9.62×10^{-5}

$[S_2O_3^{\ 2-}] = 0.0025\,\mathrm{M}$ and $[H^+] = 1.7\times10^{-5}$ M throughout.

The rate of the reaction in Equation [11] is given by equation [14].

$$\mathrm{rate} = -\frac{\Delta[H_2O_2]}{\Delta t} = -\frac{\Delta[I^-]}{2\Delta t} = \frac{\Delta[S_2O_3^{\ 2-}]}{2\Delta t} \qquad [14]$$

The last equality of Equation [14] is established by the fact that the stoichiometry of the reactions in Equations [11] and [13] show that the number of moles of I^- reacted equals the number of moles of $S_2O_3^{\ 2-}$ consumed. Since all the $S_2O_3^{\ 2-}$ is used up when the color appears, $\Delta[S_2O_3^{\ 2-}] = 0.025\,\mathrm{M}$ and Δt is the elapsed time from Table 14-1. This data now can be used to determine the order of the reaction.

EXAMPLE 14-2 Use the data in Table 14-1 to determine the order of the reaction shown in Equation [11].

$\mathrm{rate} = k[I^-]^x[H_2O_2]^y[H^+]^z$

Note that we do not know the values of x and y. They cannot be found from the balanced chemical equation. In addition, in this case we cannot determine z since $[H^+]$ is constant throughout the experiment. The order of reaction with respect to a component cannot be determined unless the concentration of that component changes. Thus, we are dealing with the simplified rate law in Equation [15].

$$\text{rate} = k'[\text{I}^-]^x[\text{H}_2\text{O}_2]^y \qquad [15]$$

Any two of the first three lines of data in Table 14-1 can be used to determine x. (They cannot be used to determine y because $[H_2O_2]$ is the same in all three cases.) Using the data in Lines 1 and 2 with Equation [15], we have:

1.37×10^{-5} mol L^{-1} s^{-1} $= k'(0.015\text{ M})^x(0.015\text{ M})^y$ *AND*

2.72×10^{-5} mol L^{-1} s^{-1} $= k'(0.030\text{ M})^x(0.015\text{ M})^y$

Division of these two equations, one by the other, gives us $0.504 = (0.500)^x$. The other terms cancel. Taking the logarithm of both sides yields $\log 0.504 = x \log 0.500$ *OR* $x = 0.298/0.301 = 0.990$. In like fashion, we use Lines 5 and 3 in Table 14-1 to determine y. Substituting in Equation [15], we have:

9.62×10^{-5} mol L^{-1} s^{-1} $= k'(0.045\text{ M})^x(0.033\text{ M})^y$ *AND*

4.31×10^{-5} mol L^{-1} s^{-1} $= k'(0.045\text{ M})\text{s}^x(0.015\text{ M})^y$

The result is $2.23 = (2.20)^y$ *or* $y = 1.017$. Normally we round off reaction orders: $x \approx 1$ and $y \approx 1$.

This reaction is first order in each of $[\text{I}^-]$ and $[H_2O_2]$, and second order overall.

*** 6. Use the rate law and rate data to calculate a rate constant (*k*), or use the rate law and rate constant to calculate rate data.**

After we know the reaction order, we often can calculate the rate constant, and then the reaction rate under various circumstances. In Example 14-2, we determined, for the reaction in Equation [11]:

$$\text{rate} = k'[\text{I}^-]^1[\text{H}_2\text{O}_2]^1 \qquad [16]$$

EXAMPLE 14-3 Find the value of the specific rate constant (k')and write the rate law for the reaction shown in Equation [11].

We use Line 3 of Table 14-1:

$$4.31 \times 10^{-5}\text{ mol L}^{-1}\text{ s}^{-1} = k'\ (0.045\text{ M})\ (0.015\text{ M})$$

$k' = 0.639$ L mol^{-1} s^{-1} $\qquad$ rate $= (0.639\text{ L mol}^{-1}\text{ s}^{-1})\ [\text{I}^-]^1[\text{H}_2\text{O}_2]^1$

EXAMPLE 14-4 What is the rate of the reaction in Equation [10] when $[\text{I}^-] = 0.020$ M and $[H_2O_2] = 0.080$ M?

$$\text{rate} = (0.0639\text{ L mol}^{-1}\text{ s}^{-1})(0.020\text{M})(0.080\text{ M}) = 1.02 \times 10^{-4}\text{ mol L}^{-1}\text{ s}^{-1}$$

*** 7. Apply the integrated rate law. Establish through rate data, equations, and graphs, whether a reaction is zero order, first order, or second order.**

We shall consider the general reaction $A \longrightarrow$ products for which rate $= -\Delta[A]/\Delta t$.

Zero order: The rate law is $-\Delta[A]/\Delta t = k[A]^0 = k[A]^0 = k$, because $[A]^0 = 1$. The concentration of A at time t is calculated with the integrated rate law.

$$[A]_t = [A]_0 - kt \qquad [17]$$

If we plot $[A]_t$ against time, the result is a straight line with a slope $= -k$. It is helpful to remember that the rate of a zero-order reaction is constant.

First order: The rate law is $-\Delta[A]/\Delta t = k[A]$. The integrated rate law allows us to determine $[A]_t$, the concentration of A at time t.

$$\ln[A]_t - \ln[A]_0 = \ln([A]_t/[A]_0) = -kt \qquad [18]$$

If we plot $\ln[A]_t$ against time, the result is a straight line of slope $= -k$. Also, it helps to remember that the half-life of a first-order reaction is constant. (See Objective 14-8.)

Second order: The rate law is $-\Delta[A]/\Delta t = k[A]^2$. The value of [A] at time t is given by the integrated rate law.

$$1/[A]_t = 1/[A]_0 + kt \qquad [19]$$

If we plot $1/[A]_t$ against time, the result is a straight line of slope $= k$.

Of course, for all three orders of reaction, the rate *constant* is constant. Three sets of data are given in Table 14-2, and each set is plotted in the three ways (see Figure 14-2). Notice from Figure 14-2 that the reaction involving *A* is first order, that involving *B* is zero order, and that involving *C* is second order.

EXAMPLE 14-5 Of the reactions for which data is given in Table 14-2, one is zero order, one is first order, and one is second order. Without plotting data, determine the order of each.

Because [B] decreases by 0.25 M every 25 seconds, the rate is constant for the reaction $B \longrightarrow$ products. Thus, this reaction is zero order.

[A] drops to one-half of its initial value in slightly less than 75 seconds. In another 75 seconds (at $t = 150$ s), [A] drops again by a factor of about one half. Finally, it seems as though [A] drops yet again by a factor of one-half in another 75 seconds (at $t = 225$ s). Because this reaction has a constant half-life, it must be first order.

By the process of elimination, the second-order reaction is $C \longrightarrow$ products. We could also establish this by substituting several values of $1/[A]_t$ into Equation [19] and noting that the resulting value of k is constant.

EXAMPLE 14-6 Use the graphs in Figure 14-2 to determine the rate constant for each reaction in Table 14-2.

Zero order: The negative of the slope of the straight-line graph ([B] vs. *t*) for the reaction B$\longrightarrow$ products equals the rate constant.

$$k = -\text{slope} = -\frac{[B]_2 - [B]_1}{t_2 - t_1} = -\frac{0.00 - 1.00}{100 - 0} = 1.00 \times 10^{-2} \text{ mol L}^{-1} \text{ s}^{-1}$$

First order: The negative of the slope of the straight-line graph (ln[A] *vs. t*) for the reaction A$\longrightarrow$ products equals the rate constant.

$$k = -\text{slope} = -\frac{\ln[A]_2 - \ln[A]_1}{t_2 - t_1} = -\frac{-2.526 + 0.755}{250 - 75} = +1.01 \times 10^{-2} \text{ s}^{-1}$$

Second order: The slope of the straight-line graph (1/[C] vs. t) for the reaction C$\longrightarrow$ products equals the rate constant.

$$k = -\text{slope} = \frac{1/[C]_2 - 1/[C]_1}{t_2 - t_1} = \frac{3.45 - 1.75}{250 - 75} = 9.71 \times 10^{-3} \text{ L mol}^{-1} \text{ s}^{-1}$$

TABLE 14-2 Concentration-Time Data for the Reactions: A$\longrightarrow$ products

Time,s	[A]	log[A]	1/[A]
0	1.00	0.000	1.00
25	0.78	−0.248	1.28
50	0.61	−0.494	1.64
75	0.47	−0.755	2.13
100	0.37	−0.994	2.70
150	0.22	−1.514	4.55
200	0.14	−1.966	7.14
250	0.08	−2.526	12.5

B$\longrightarrow$ products

Time,s	[B]	log[B]	1/[B]
0	1.00	0.000	1.00
25	0.75	−0.288	1.33
50	0.50	−0.693	2.00
75	0.25	−1.386	4.00
100	0.00	_____	_____
150			
200			

C$\longrightarrow$ products

Time,s	[C]	log[C]	1/[C]
0	1.00	0.000	1.00
25	0.80	−0.223	1.25
50	0.67	−0.400	1.49
75	0.57	−0.562	1.75
100	0.50	−0.693	2.00

150	0.40	−0.916	2.50
200	0.33	−1.109	3.03
250	0.29	−1.238	3.45

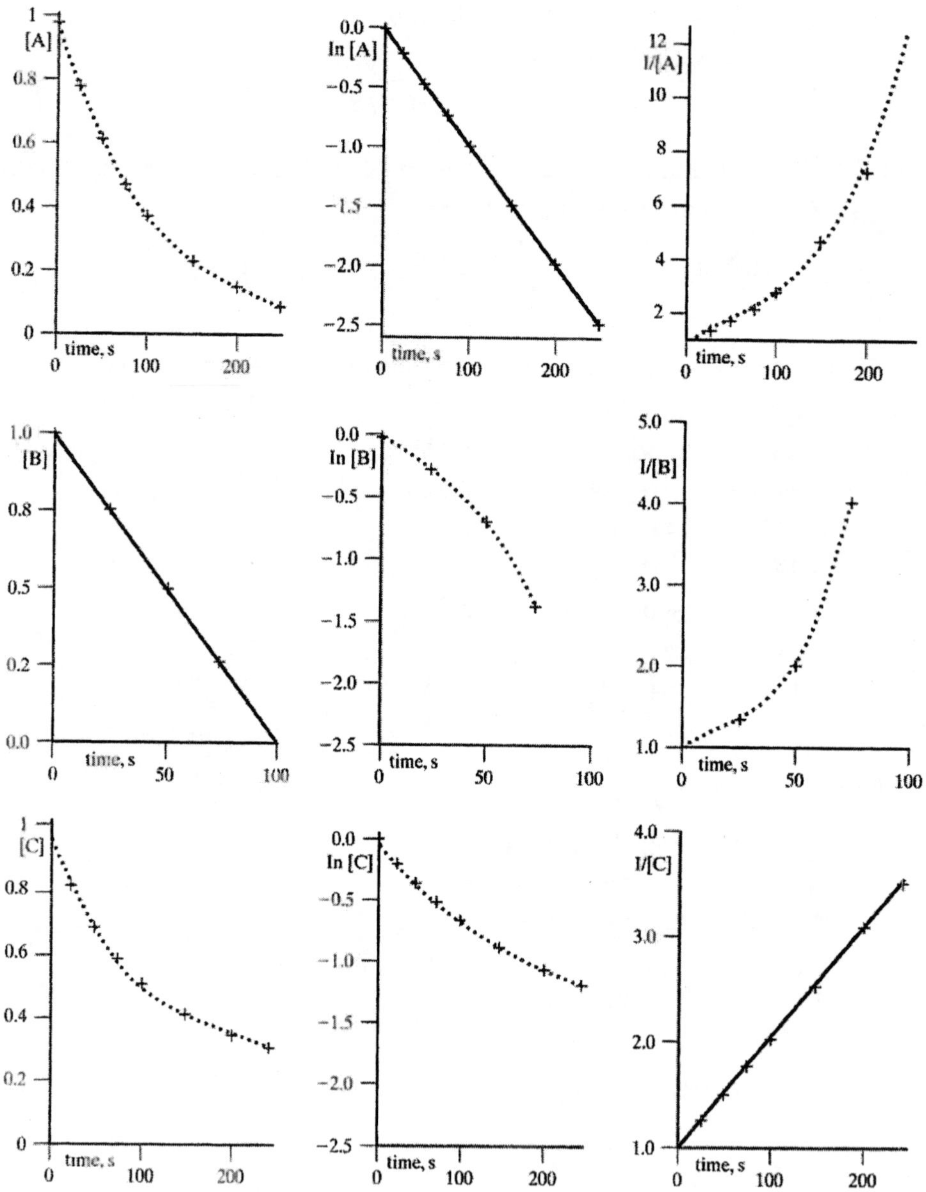

FIGURE 14-2 Graphs of the Data of Table 14-2

*** 8. Determine the half-life of a reaction that is zero order, first order, or second order.**

The half-life of a reaction is the time needed for half of the reactant to be consumed. For a zero-order reaction, the half-life is $t_{1/2} = [A]_0/2k$. For a first-order reaction, the half-life is $t_{1/2} = 1/k[A]_0$. And for a second-order reaction, the half-life is $t_{1/2} = 1/k[A]_0$.

EXAMPLE 14-7 Determine the initial value of the half-life for each of the reactions of Table 14-2.

$$\textit{Zero order: } t_{1/2} = \frac{[B]_0}{2k} = \frac{1.00\,\text{mol L}^{-1}}{2\times 1.00\times 10^{-2}\ \text{mol L}^{-1}\ \text{s}^{-1}} = 50.0\ \text{s}$$

$$\textit{First order: } t_{1/2} = \frac{0.693}{k} = \frac{0.693}{1.01\times 10^{-2}\ \text{s}^{-1}} = 50.0\ \text{s}$$

$$\textit{Second order: } t_{1/2} = \frac{1}{k[A]_0} = \frac{1}{9.71\times 10^{-3}\ \text{L mol}^{-1}\ \text{s}^{-1}\times 1.00\,\text{mol L}^{-1}} = 103\ \text{s}$$

9. Describe the collision theory of reactions, stating the factors that affect collision frequency and those that lead to favorable collisions.

Collision theory states that all reactions are the result of collisions between molecules. For a gaseous reaction, the number of collisions in a liter each second is the collision number, denoted Z_{AB}.

$$Z_{AB} = \text{constant}\,[A][B]\,k_{AB}\,T^{1/2} \qquad [20]$$

k_{AB} is a constant that depends on the properties of the *A* and *B* molecules. The factors are about the same for reactions in aqueous solution. Collision theory also requires that the collisions be good ones. That is, the molecules must be oriented correctly in order to react. This means that not all collisions lead to reactions.

10. Explain the concept of activation energy.

In addition to being oriented correctly, each collision must possess enough energy for the molecules to react. Molecules repel each other because of the electron clouds that surround them. The collision energy must be large enough to overcome these repulsive forces. The needed energy is called the *activation energy.* If a correctly oriented collision has at least this much energy, a reaction occurs between the colliding molecules. Activation energy can be thought of as an energy barrier that must be overcome before reaction can occur. Think for a moment of an automobile on a mountain road. The car can easily be driven off the road and down the mountainside. If a guardrail is present, however, the car must have a high energy (similar to an activation energy) in order to break through the barrier and roll down the mountainside.

11. Show how transition-state theory extends theoretical explanations of chemical kinetics.

Collision theory fails to explain many of the details of even simple reactions. Transition-state theory explains the rate of a reaction by considering the intermediate or activated complex that forms as the reactants are transformed into products. Transition-state theory considers the bonding in the activated complex, the orientation of reactants needed for it to form, the energy released when bonds are formed and required to break bonds, and the products that result. A reaction profile depicts the path from reactants through activated complex to products. The reaction profile for the decomposition of $H_2O_2(aq)$, both catalyzed and uncatalyzed, is shown in Figure 14-3. The reaction profile also shows that any exothermic reaction would occur instantaneously if there were no activation energy.

*** 12. Use the Arrhenius equation in calculations involving rate constants, temperatures, and activation energies.**

The Arrhenius equation (Equation [21]) relates the rate constants (k_2 and k_2) of a chemical reaction at two different Kelvin temperatures (T_1 and T_2) to the activation energy (E_a):

$$\ln\frac{k_1}{k_2} = 2.303\log\frac{k_1}{k_2} = -\frac{E_a}{R}\left(\frac{1}{T_1}-\frac{1}{T_2}\right) \qquad [21]$$

The form of the Arrhenius equation is the same as that of the Clausius-Clapeyron equation, discussed in Objective 12-7. If you have forgotten how to use this type of equation, you should review that objective.

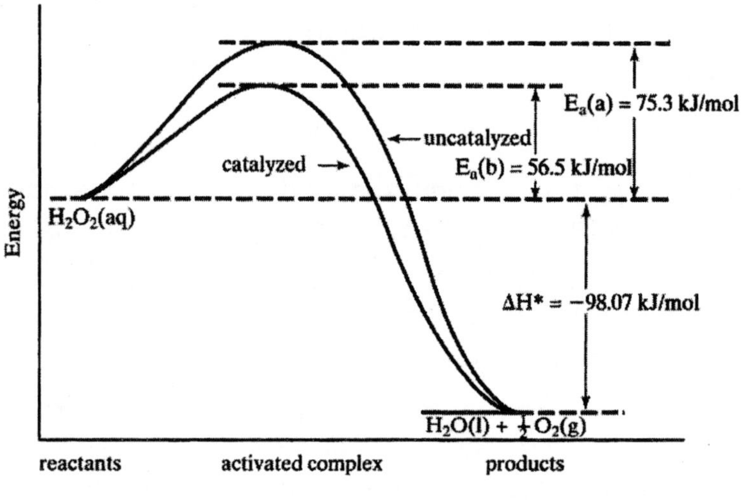

FIGURE 14-3 Reaction Profiles of Catalyzed Reactions.

(a) Uncatalyzed: $2\,H_2O_2(aq) \longrightarrow 2\,H_2O(l) + O_2(g)$

(b) Catalyzed: $H_2O_2(aq) + I^-(aq) \longrightarrow H_2O(l) + IO^-(aq)$

$$IO^-(aq) + H_2O_2(aq) \longrightarrow H_2O(l) + O_2(g) + I^-(aq)$$

EXAMPLE 14-8 At 25 °C, $k = 7.3\times10^{-4}\ s^{-122}$ for the uncatalyzed decomposition of H_2O_2. What is the value of the rate constant at 125 °C? By what factor does the rate constant of the catalyzed reaction increase over the same temperature range, from 25 °C to 125 °C? Use data from Figure 14-3.

For the uncatalyzed reaction, we know all terms in the Arrhenius equation except the rate constant at 125 °C, k_2 :

$$\ln\frac{k_1}{k_2} = \ln\frac{7.3\times10^{-4}\ \mathrm{s}^{-1}}{k_2} = -\frac{E_a}{R}\left(\frac{1}{T_1}-\frac{1}{T_2}\right)$$

$$= -\frac{75.3\ \times10^{3}\ \mathrm{J/mol}}{8.3145\ \mathrm{J\ mol^{-1}\ K^{-1}}}\left(\frac{1}{298\ \mathrm{K}}-\frac{1}{398\ \mathrm{K}}\right) = -7.64$$

$$\frac{7.3\times10^{-4}\ \mathrm{s}^{-1}}{k_2} = \mathrm{e}^{-7.64} = 4.8\times10^{-4} \qquad k_2 = \frac{7.3\times10^{-4}\ \mathrm{s}^{-1}}{4.8\times10^{-4}} = 1.5\ \mathrm{s}^{-1}$$

The rate constant has increased by a factor equal to $\dfrac{1.5\ \mathrm{s}^{-1}}{7.3\times10^{-4}\ \mathrm{s}^{-1}} = 2.1\times10^{3}$.

To determine the factor of increase for the catalyzed reaction, we simply solve for the ratio k_2 / k_1 :

$$\ln\frac{k_2}{k_1} = -\frac{56.5\times10^{3}\ \mathrm{J/mol}}{8.3145\ \mathrm{J/mol\ K}}\left(\frac{1}{398}-\frac{1}{298}\right) = 5.73 \qquad \frac{k_2}{k_1} = \mathrm{e}^{+5.73} = 3.1\times10^{2}$$

Because its energy demands are not as great, the catalyzed reaction is less affected by an increase in temperature than the uncatalyzed reaction.

13. Describe a reaction mechanism, and distinguish between elementary processes and a net chemical reaction.

A reaction mechanism is a series of elementary steps that result in the overall reaction. For example, the catalyzed decomposition of $N_2O(g)$ (Equation [29]) has the three reactions in Equation [28] as its mechanism. The mechanism of a reaction must meet several criteria. First, each of the elementary steps must involve, at most, three molecules as reactants. That is, the steps may be unimolecular, bimolecular, or (in extremely rare cases) termolecular. (This is because there is a very small chance that three molecules will collide at the same time.) Second, the elementary steps must sum to the overall reaction. Finally, we must be able to derive the rate law from the mechanism. In addition, the chances that the mechanism is correct are greatly enhanced if we can detect some of the reactive intermediates of the mechanism. For example, mechanism shown in Equation [28] is more plausible if we can detect Cl and ClO in the reaction mixture.

14. Derive the rate law from a simple mechanism with the concepts of steady-state condition and rate-determining step.

An example of a reaction with two mechanisms is Equation [26], where $X_2 = H_2, O_2, Cl_2$, or Br_2. (We shall use H_2 as a specific example in this objective.)

$$2\ \mathrm{NO(g)} + \mathrm{X_2(g)} \longrightarrow \mathrm{N_2O(g)} + \mathrm{X_2O(g)} \qquad [22]$$

An example of a reaction with one mechanism is shown in Equation [27].

$$2\,NO(g) \underset{K_2}{\overset{K_1}{\rightleftharpoons}} N_2O_2(g) \quad \text{(rapid)}$$
$$N_2O_2(g) + H_2(g) \xrightarrow{k_3} N_2O(g) + H_2O(g) \quad \text{(slow)} \qquad [23]$$

As shown in [23], the rate of reaction is found from the slow step, or the rate-determining step:

$$\text{rate} = \frac{\Delta[N_2O]}{\Delta t} = -\frac{\Delta[N_2O_2]}{\Delta t} = k_3[N_2O_2][H_2]$$

Since N_2O_2 does not appear in the overall equation, we are unable to determine $[N_2O_2]$ experimentally. Thus, we replace $[N_2O_2]$ in the rate law by assuming that the rapid first step is proceeding as quickly forward as reverse. Therefore, $[N_2O_2]$ remains virtually constant during the course of the reaction, a condition known as the "steady-state condition.", forward rate = reverse rate, $k_1[NO]^2 = k_2[N_2O_2]$, or $[N_2O_2] = k_1[NO]^2/k_2$.

$$\text{rate} = k_3\frac{k_1[NO]^2}{k_2}[H_2] = \frac{k_3k_1}{k_1} = [NO]^2[H_2] \qquad [24]$$

This rate law, determined from the mechanism, is in agreement with the experimentally-determined rate law. The other mechanism of the reaction shown in [22] is equations shown in [25]. This mechanism gives the same rate law as mechanism [23]. (One way to distinguish between these mechanisms is to detect N_2O_2 or $NOH_2(g)$ in the reaction mixture.)

$$NO(g) + H_2(g) \underset{K_2}{\overset{K_1}{\rightleftharpoons}} NOH_2(g) \quad \text{(rapid)}$$
$$NOH_2(g) + NO(g) \xrightarrow{k_3} N_2O(g) + H_2O(g) \quad \text{(slow)} \qquad [25]$$

***15. Describe the role of a catalyst and explain the difference between homogeneous and heterogeneous catalysis.**

A *catalyst* is a substance that alters the rate of a chemical reaction while not appearing as a product or a reactant in the overall reaction. Some catalysts speed up reactions (*positive catalysts*, or simply *catalysts*), and some slow reactions down (*negative catalysts*, or more commonly *inhibitors*). Heterogeneous catalysts are in a different physical state of matter than the reaction mixture (a solid catalyst in contact with a liquid or gas, for instance). All of the catalytic activity occurs at the surface of the solid (*surface catalysis*). Heterogeneous catalysts can be counteracted by poisons that bond strongly to the surface (as lead poisons the catalytic converter in an automobile). In some cases, the catalyst determines the products of the reaction. An example is the reaction between CO(g) and :

$$CO(g) + 3\,H_2(g) \xrightarrow{Ni} CH_4(g) + H_2O(g) \qquad [26]$$

$$CO(g) + 2\,H_2(g) \xrightarrow{ZnO/Cr_2O_3} CH_3OH \qquad [27]$$

Homogeneous catalysts are in solution with the reaction mixture. For example, the decomposition of $N_2O(g)$, Equation [29], is catalyzed by $Cl_2(g)$, as the mechanism in [28] shows:

$$\text{Mechanism:}\quad \left.\begin{array}{r} Cl_2(g) \longrightarrow 2\,Cl(g) \\ 2\times[N_2O(g) + Cl(g) \longrightarrow N_2(g) + ClO(g)] \\ 2\,ClO(g) \longrightarrow Cl_2(g) + O_2(g) \end{array}\right\} \qquad [28]$$

$$\text{Overall:}\quad 2\,N_2O(g) \longrightarrow 2\,N_2(g) + O_2(g) \qquad [29]$$

The decomposition of ozone is catalyzed by $N_2O_5(g)$:

$$\text{Mechanism:}\quad 2\,N_2O_5(g) \longrightarrow 2\,N_2O_4(g) + O_2(g)$$

$$2\times[O_3(g) + N_2O_4(g) \longrightarrow O_2(g) + N_2O_5(g)]$$

$$\text{Overall:}\quad 2\,O_3(g) \longrightarrow 3\,O_2(g)$$

Catalysts enhance the reaction rate by providing a reaction pathway that has a lower activation energy.

SELF ASSESSMENT EXCERCISES

100. (a) The units of k will be $mol^{-1}s^{-1}$ so that the rate will be moles/s. (b) This is a second-order reaction, so $t_{1/2}$ is not constant. (c) Correct. k is a rate constant and is independent of concentration. (d) Two A will disappear for each C formed. The rate of appearance of C is ½ that of the disappearance of A.

101. (a) is not true because, at two half-lives, the reactant will be ¼ the initial concentration. (b) True. (c) False. The first half-life consumes ½ the reactant; the second consumes ½ of what was left after the first half-life–¼ the initial concentration. (d) The change in concentration is a log relationship, not a linear relationship. (e) True. The half-life is the time for ½ of the reactant to be consumed. (f) False. See the answer for (c).

102. (a) 0.020 mol $L^{-1}min^{-1}$. Using the relationship for a first-order reaction, $t_{1/2} = 0.693/k$ and $k = 0.0498\ min^{-1}$; the rate = k[A] = $0.0498\ min^{-1}$ x 0.40 M = 0.020 mol L^{-1} min^{-1}.

103. (a) False. The rate = $k[A]^2$, so doubling [A] will increase the rate four-fold. (b) False. See previous answer . (d) True.

105. (c) True. Answer (a) gives the reaction for a first-order process. (b) The [C] will be twice [A]. Answer (d) is the expression for a first-order reaction.

106. Rate of reaction = k[A]. To find the rate constant (k), we can use the slope of $\ln[A]_t$ vs. time or at t = 100 s. $[A]_o$ has decreased by half, so $t_{1/2}$ = 100 s. k = .693/100s = 0.00693 s^{-1}. The rate is rate = 0.00693 s^{-1}x 0.44 M = 3.05 x 10^{-3} M s^{-1}.

107. (a) A first-order reaction will take two half-lives. (b) A zero-order reaction will take 1 ½ half-lives or 45 min. The decrease of [A] with time is linear.

109. (a) The initial rate of reaction is (1.204-1.180 M)/1.0 min = 0.024 M min^{-1}. (b) The rate = $k[A]^2$ for a second-order reaction. The initial concentration is twice that for Experiment 1; the rate will be 2^2, or 4 times. Rate = 0.096 M min^{-1} at 1 min. [A] = 2.408 – 0.096 = 2.312 M. (c) For a first-order reaction, the rate = k[A], so the rate will be the same. Rate = 0.024 m min^{-1}, change in [A] = (0.024 M min^{-1}) (30 min) = 0.72 M, and [A] = 2.408 – 0.72 = 1.688 M.

110. (a) It will be consumed as quickly as it is formed, and the reaction will depend only on the first step. (b) In this case, the rate-limiting step is the second step. The rate of reaction = $k_3[A][B_2]$. However, B_2 is an intermediate. The first reaction step is an equilibrium, so the rate forward = rate reverse and $k_1[B]^2 = k_2[B_2]$, or $K = \frac{[B]^2}{[B_2]}$. Rearranging and substituting into the rate equation, rate = $k_3[A][B]^2/K$, which is a third-order reaction and not consistent with the observed rate law.

DRILL PROBLEMS

1. Give the expression that relates the rate of each reaction in the drill problems of Objective 14-5 (Problem 3) to the rate of change of concentration of each reactant and each product in turn.

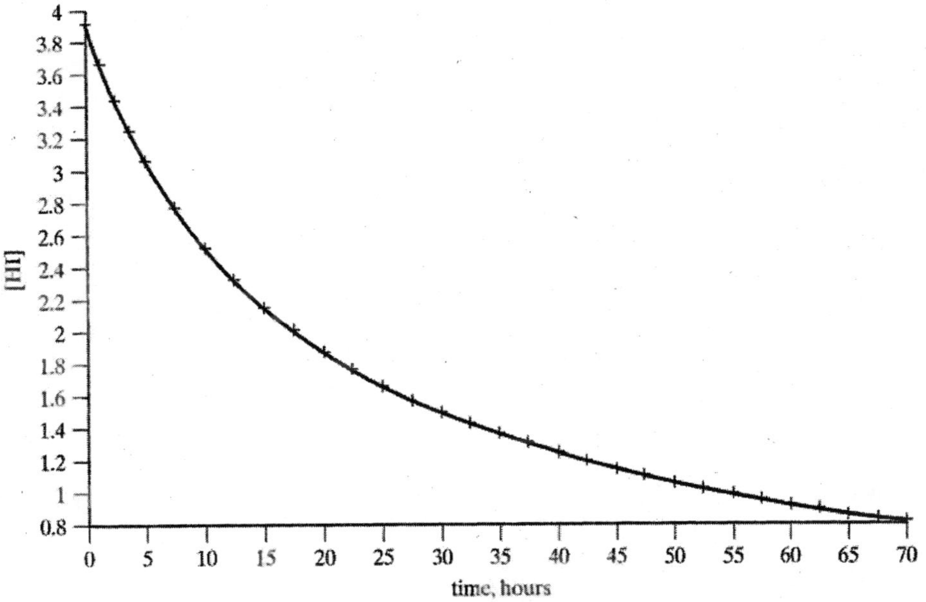

FIGURE 14-4 Concentration vs. Time Data for $C_2H_5Cl \longrightarrow C_2H_4 + HCl$ at 700 K.

2. Concentration *vs.* time curves for the reactions in Equations [5] and [30] are drawn in Figures 14-1 and 14-4.

$$2\,HI(g) \longrightarrow H_2(g) + I_2(g) \text{ at } 600\text{ K} \qquad [5]$$

$$C_2H_5Cl \longrightarrow C_2H_4 + HCl \text{ at } 700\text{ K} \qquad [30]$$

(1) Determine the exact rate of the indicated reaction at each of the following times:

For [30]: A. 0 h B. 400 h C. 800 h D. 80 h E. 500 h F. 200 h G. 1000 h

For [5]: H. 5 h I. 10 h J. 20 h K. 60 h L. 30 h M. 40 h N. 35 h

(2) Determine the time at which the indicated reactant has each of the following concentrations and the exact rate at this time:

For $[C_2H_5Cl]$ (reaction [30]): O. 0.90 M P. 0.45 M Q. 0.30 M R. 0.18 M

For [HI] (reaction [6]): S. 3.60 M T. 1.80 M U. 0.90 M V. 1.20 M

3. The initial rate of each of the following reactions at various concentrations of the reactant(s) is given in the following lines. Determine the rate law for each reaction (but do not determine a value for *k*) and state the overall order of the reaction.

A. $2\ N_2O_5(g) \longrightarrow 4\ NO_2(g) + O_2(g)$

$[N_2O_5]$, M	0.170	0.340	0.680
$-\Delta[N_2O_5]/\Delta t$, mol $L^{-1}\ h^{-1}$	0.050	0.100	0.200

B. $CH_3CHO(g) \longrightarrow CH_4(g) + CO(g)$

$[CH_3CHO]$, M	0.050	0.100	0.200
$-\Delta[CH_3CHO]/\Delta t$, mol $L^{-1}\ s^{-1}$	2.25×10^{-7}	9.00×10^{-7}	3.60×10^{-6}

C. $CO(g) + NO_2(g) \longrightarrow CO_2(g) + NO(g)$

[CO], M	0.10	0.10	0.20
$[NO_2]$, M	0.10	0.20	0.10
$\Delta[CO_2]/\Delta t$, mol $L^{-1}\ s^{-1}$	0.012	0.024	0.024

D. $2\ NO(g) + 2\ H_2(g) \longrightarrow N_2(g) + 2\ H_2O(l)$

$[H_2]$, M	0.37	0.37	0.74
[NO], M	0.60	1.20	1.20
$\Delta[N_2]/\Delta t$, mol $L^{-1}\ h^{-1}$	0.18	0.72	1.44

E. $NOCl(g) \longrightarrow 2\ NO(g) + Cl_2(g)$

[NOCl], M	0.30	0.60	0.90
$\Delta[Cl_2]/\Delta t$, mol $L^{-1}\ s^{-1}$	3.60×10^{-9}	1.44×10^{-8}	3.24×10^{-8}

F. $F_2(g) + 2\ ClO_2(g) \longrightarrow 2\ FClO_2(g)$

$[F_2]$, M	0.10	0.10	0.20
$[ClO_2]$, M	0.010	0.010	0.010
$\Delta[FClO_2]/\Delta t$, mol $L^{-1}\ s^{-1}$	0.0024	0.0048	0.0048

G. $F^- + CH_3Cl \longrightarrow CH_3F + Cl^-$

$[CH_3Cl]$, M	1.2×10^{-6}	2.4×10^{-6}	1.2×10^{-6}	2.4×10^{-6}
$[F^-]$, M	1.2×10^{-6}	1.2×10^{-6}	2.4×10^{-6}	2.4×10^{-6}
$-\Delta[F^-]/\Delta t$, mol $L^{-1}\ s^{-1}$	0.69	1.38	1.38	2.7

H. $BrO_3^-(aq) + 5\ Br^-(aq) + 6\ H^+(aq) \longrightarrow 3\ Br_2(l) + 3\ H_2O(l)$

$[BrO_3^-]$, M	0.0333	0.0333	0.0667	0.0333
$[Br^-]$, M	0.0667	0.133	0.0667	0.0667
$[H^+]$, M	0.100	0.100	0.100	0.200
t, s	42.6	21.5	21.1	10.7

(*t* = time for 0.0020 M BrO_3^- to be consumed. Solution volume is 1.00 L)

I. $NH_4^+(aq) + NO_2^-(aq) \longrightarrow N_2(g) + 2\ H_2O(l)$

$[NH_4^+]$, M	0.0092	0.0092	0.0488	0.0249
$[NO_2^-]$, M	0.098	0.049	0.196	0.196
$\Delta[N_2]/\Delta t$, mol $L^{-1}\ s^{-1}$	3.49×10^{-7}	1.75×10^{-7}	3.70×10^{-6}	1.88×10^{-6}

J. $H_2PO_2^-(aq) + OH^-(aq) \longrightarrow HPO_3^{2-}(aq) + H_2(g)$ (Assume that the solution volume is 1.00 L.)

$[H_2PO_2^-]$, M	0.50	0.25	0.25
$[OH^-]$, M	1.28	1.28	3.84
V_{H2}, cm^3 produced at STP	19.5	25.0	135.0
t, s	10	50	30

K. $2\ NO(g) + H_2(g) \longrightarrow N_2O(g) + H_2O(g)$ (at 1000 K)

$P_i(NO)$, atm	0.150	0.075	0.150
$P_i(H_2)$, atm	0.400	0.400	0.200
$-\Delta P_{total}/\Delta t$, atm/min	0.020	0.005	0.010

4. For each of the reactions of the drill problems of Objective 14-5, first determine the specific rate constant, making sure to express it with the correct units. Then, determine the rate(s) of the reaction when the concentrations are those given below:

A. $[N_2O_5] = 0.200$ M; $[N_2O_5] = 1.00$ M.

B. $[CH_3CHO] = 0.150$ M; $[CH_3CHO] = 0.337$ M

C. $[CO] = 0.50$ M and $[NO_2] = 0.50$ M; $[CO] = 1.20$ M; and $[NO_2] = 0.60$ M.

D. $[H_2] = 0.12$ M and $[NO] = 0.21$ M; $[H_2] = 0.43$ M and $[NO] = 0.24$ M.

E. $[NOCl] = 0.50$ M; $[NOCl] = 0.23$ M.

F. $[F_2] = 0.13$ M and $[ClO_2] = 0.27$ M; $[F_2] = 0.29$ M and $[ClO_2] = 0.11$ M.

G. $[CH_3Cl] = 9.4 \times 10^{-7}$ M and $[F^-] = 8.7 \times 10^{-7}$ M; $[CH_3Cl] = 4.3 \times 10^{-7}$ M and

$[F^-] = 10.2 \times 10^{-7}$ M.

H. $[BrO_3^-] = 0.100$ M, $[Br^-] = 0.243$ M, and $[H^+] = 0.342$ M;

$[BrO_3^-] = 0.076$ M, $[Br^-] = 0.111$ M; and $[H^+] = 0.114$ M.

I. $[NH_4^+] = 0.0100$ M and $[NO_2^-] = 0.100$ M; $[NH_4^+] = 0.0046$ M and

$[NO_2^-] = 0.0043$ M.

J. $[H_2PO_2^-] = 1.04$ M and $[OH^-] = 0.29$ M; $[H_2PO_2^-] = 0.47$ M and $[OH^-] = 0.50$ M.

K. $P_i(NO) = 0.147$ atm and $P_i(H_2) = 0.089$ atm; $P_i(NO) = 1.06$ atm and $P_i(H_2) = 0.104$ atm.

5. Concentration-time data for various reactions is given below. Determine the order of each reaction for the reactant whose concentration varies. Then determine the value—with the correct units—of the specific rate constant. (mM $= 10^{-3}$ mol/L, μM $= 10^{-6}$ mol/L)

A. $2\ NH_3(g) \longrightarrow N_2(g) + 3\ H_2(g)$

$[NH_3]$, M	2.000	1.67	1.43	1.11	0.91	0.77	0.67
t, s	0	50.	100.	200.	300.	400.	500.

B. $CH_3COCIN—C_6H_5(A) \longrightarrow CH_3COCN — C_6H_4{}^+Cl^-$

[A], mM	24.5	18.1	13.3	9.7	7.1	5.2
t, min	0	15	30	45	60	75

C. $2\ AsH_3(g) \longrightarrow 2\ As(s) + 3\ H_2(g)$

$[AsH_3]$, mM	21.6	16.4	15.1	12.6	10.5	8.01	7.32
T, h	0	3.00	4.00	6.00	8.00	11.00	12.00

D. $NH_4CNO \longrightarrow (NH_2)_2CO$

$[NH_4CNO]$, M	0.400	0.320	0.253	0.207	0.167	0.125
t, min	0	10.0	25.0	40.0	60.0	95.0

E. $2\ C_4H_6 \longrightarrow C_8H_{12}$

$[C_4H_6]$, M	0.2000	0.1500	0.1200	0.1000	0.0857	0.0750	0.0667
t, min	0	500	1000	1500	2000	2500	3000

F. Chinnamylidene chloride(A) + $C_2H_5OH \longrightarrow$ products (at 23 °C, in C_2H_5OH.)

[A], μM	21.1	19.8	17.6	13.3	11.6	9.56	7.43
t, min	0	10	31	67	100	133	178

G. $(CH_3)_3CBr + H_2O \longrightarrow (CH_3)_3COH + HBr$

$[(CH_3)_3CBr]$, mM	103.9	77.6	63.9	52.9	38.0	27.0	20.7
t, min	0	6.20	10.0	13.5	18.3	26.0	30.8

H. $CH_3COOCH_3 + H_2O \longrightarrow CH_3COOH + CH_3OH$

$[CH_3OH]$, mM	0	25.16	49.60	73.24	96.21	119.0	140.4
t, min	0	15	30	45	60	75	90

($[CH_3COOCH_3] = 0.850$ M and $[H_2O] = 51.0$ M initially.)

I. $CH_3Br + I^- \longrightarrow CH_3I + Br^-$

$[I^-]$, M	1.00	0.902	0.813	0.662	0.539	0.439	0.357	0.291

t, min	0	10	20	40	60	80	100	120

J. $BzOOPh(aq) + H_2O \longrightarrow BzOOH(aq) + PhOH(aq)$ $[H_2O] = 55.5$ M

[BzOOPh], μM	33.88	27.83	22.90	18.92	15.38	12.63	10.36
t, min	0	10	20	30	40	50	60

6. Determine the half-life of each reaction of Objective 14-7. Use the initial concentration of reactant, if needed.

7. The rate constants—or the rates at the same concentrations of reactants—of various reactions at different temperatures are given in the following lines. Determine the activation energy in kJ/mol of each reaction. Also, determine the rate constant, or the rate, at the specified temperature.

A. $2\ NO_2(g) \longrightarrow 2\ NO(g) + O_2(g)$ $k = 0.775\ s^{-1}$ at 313 K, $k = 4.02\ s^{-1}$ at 378 K

$E_a =$ ________ kJ/mol At 298 K, $k =$ ________ s^{-1}

B. $CO + NO_2 \longrightarrow CO_2 + NO$ $k = 0.220\ M^{-1}\,s^{-1}$ at 650 K, $k = 23.0\ M^{-1}\,s^{-1}$ at 800 K

$E_a =$ ________ kJ/mol At 300 K, $k =$ ________ $M^{-1}\,s^{-1}$

C. $(CH_2COOH)_2CO \longrightarrow (CH_3)_2C{=}O + 2\ CO_2$ $k = 0.0548\ s^{-1}$ at 60 °C $k = 2.46 \times 10^{-5}$ at 0 °C

$E_a =$ ________ kJ/mol At 100 K, $k =$ ________ s^{-1}

D. $(CH_2)_3 \longrightarrow CH_2CHCH_3$ $E_a =$ ________ kJ/mol At 1000 K, $k =$ ________ s^{-1}

t, °C	470	485	500	510	530
k, s^{-1}	1.10×10^{-4}	2.61×10^{-4}	5.70×10^{-4}	10.21×10^{-4}	2.86×10^{-3}

E. $CH_3I + C_2H_5ONa \longrightarrow CH_3OC_2H_5 + NaI$

t, °C	0	6	12	18	24
k, $M^{-1}\,s^{-1}$	5.60×10^{-5}	1.18×10^{-4}	2.45×10^{-4}	4.88×10^{-4}	1.00×10^{-3}

$E_a =$ ________ kJ/mol At 100 °C, $k =$ ________ $M^{-1}\,s^{-1}$

8. (1) For the reaction $CO + NO_2 \longrightarrow CO_2 + NO$, the rate law under certain conditions is rate $= k[NO_2]^2$. Which of the following mechanisms is the most plausible for this reaction? Find the rate law for each one.

A. $CO + NO_2 \longrightarrow CO_2 + NO$ B. $2\ NO_2 \rightleftharpoons N_2O_4$(fast)

C. $2\ NO_2 \longrightarrow N_2 + 2\ O_2$ (slow)
$2\ CO + O_2 \longrightarrow 2\ CO_2$ (fast)
$N_2 + O_2 \longrightarrow 2\ NO$ (fast)

D. $N_2O_4 + 2\ CO \longrightarrow 2\ CO_2 + 2\ NO$ (slow)
$2\ NO_2 \longrightarrow NO_3 + NO$ (slow)
$NO_3 + CO \longrightarrow NO_2 + CO_2$ (fast)

(2) For the reaction $CO(g) + Cl_2(g) \longrightarrow COCl_2(g)$, under certain conditions, the rate law is rate $= k[CO][Cl_2]^{3/2}$. Which of the following is the most plausible mechanism? Find the rate law for each one.

E. $\frac{1}{2}\ Cl_2 \rightleftharpoons Cl$ (fast)
$Cl + Cl_2 \rightleftharpoons Cl_3$ (fast)
$Cl_3 + CO \longrightarrow COCl_2 + Cl$ (slow)
$Cl \rightleftharpoons CO \longrightarrow \frac{1}{2}\ Cl_2$ (fast)

F. $\frac{1}{2}\ Cl_2 \rightleftharpoons Cl$ (fast)
$Cl + CO \rightleftharpoons COCl$ (fast)
$COCl + Cl_2 \longrightarrow COCl_2 + Cl$ (slow)
$Cl \rightleftharpoons \frac{1}{2}\ Cl_2$ (fast)

G. $Cl_2 + CO \longrightarrow CCl_2 + O$ (slow)
$O + Cl_2 \longrightarrow Cl_2O$ (fast)
$Cl_2O + CCl_2 \longrightarrow COCl_2 + Cl_2$ (fast)

H $Cl_2 + CO \rightleftharpoons N_2O_2$ (fast)
$COCl + Cl_2 \longrightarrow COCl_2 + Cl$ (slow)
$Cl + Cl \longrightarrow Cl_2$ (fast)

(3) For the reaction $H_2 + Br_2 \longrightarrow 2$, under certain conditions, the rate law is rate $= k[H_2][Br_2]^{1/2}$. Which of the following is the most plausible mechanism? Find the rate law for each mechanism.

I. $H_2 + Br_2 \longrightarrow 2\ HBr$

J. $H_2 \rightleftharpoons 2\ H$ (fast)
$H + Br_2 \longrightarrow HBr + Br$ (slow)
$Br + H_2 \longrightarrow HBr + H$ (fast)

K. $Br_2 \rightleftharpoons 2\ Br$ (fast)
$Br + H_2 \longrightarrow HBr + H$ (slow)
$H + Br_2 \longrightarrow HBr + Br$ (fast)

(4) Which of the following mechanisms is the most plausible for $2\ NO(g) + O_2(g) \longrightarrow 2\ NO_2(g)$ if the rate law is rate $= k[NO]^2[O_2]$? Find the rate law for each mechanism.

L. $2\ NO + O_2 \longrightarrow 2\ NO_2$

M. $2\ NO \rightleftharpoons N_2O_2$ (fast)
$N_2O_2 + O_2 \rightleftharpoons 2\ NO_2$ (slow)

N. $2\ NO \rightleftharpoons N_2 + O_2$ (fast)
$N_2 + 2\ O_2 \rightleftharpoons 2\ NO_2$ (fast)

(5) Which mechanism is the most plausible for $C_4H_9Br + OH^- \longrightarrow C_4H_9OH + Br^-$ if the rate law is rate $= k[C_4H_9Br]$? Find the rate law for each mechanism.

O. $C_4H_9Br + OH^- \longrightarrow C_4H_8Br^- + H_2O$ (fast)
$C_4H_8Br^- \longrightarrow C_4H_8 + Br^-$ (slow)
$C_4H_8 + H_2O \longrightarrow C_4H_9OH$ (fast)

P. $C_4H_9Br \longrightarrow C_4H_9^+ + Br^-$ (slow)
$C_4H_9^+ + OH^- \longrightarrow C_4H_9OH$ (fast)

Q. $C_4H_9Br + OH^- \longrightarrow C_4H_9OH + Br^-$

(6) R. Show that the following three-step mechanism is plausible for $Cl_2 + CHCl_3 \longrightarrow CCl_4 + HCl$. Give the rate law that results.

$$Cl_2 \rightleftharpoons 2\,Cl\,(slow)$$

$$Cl + CHCl_3 \longrightarrow CCl_4 + H \;\; (slow)$$

$$H + Cl \longrightarrow HCl \;\; (fast)$$

(7) S. Is the following mechanism plausible for the NO-catalyzed decomposition of O_3?

$$2\,O_3(g) \xrightarrow{NO} 3\,O_2(g)$$

What rate law is obtained from this mechanism?

$$NO + O_3 \longrightarrow NO_2 + O_2 \;\; (slow)$$

$$NO_2 + O_3 \longrightarrow NO + 2\,O_2 \;\; (fast)$$

QUIZZES (20 minutes each) Choose the best answer to each question.

QUIZ A

1. For $4\,OH + H_2S \longrightarrow SO_2 + 2\,H_2O + H_2$, initial rate data is as follows:

[OH], M	13×10^{-12}	39×10^{-12}	39×10^{-12}	
[H_2S], M	21×10^{-12}	21×10^{-12}	42×10^{-12}	The rate law is
Rate, M s^{-1}	1.4×10^{-6}	4.2×10^{-6}	8.4×10^{-6}	$k[OH]^x[H_2S]^y$

Thus, (a) $x=2, y=2$; (b) $x=2, y=1$; (c) $x=1, y=2$; (d) $x=1, y=1$; (e) none of these.

2. The reaction $2\,NO_2Cl \longrightarrow 2\,NO_2 + Cl_2$ is first order. The rate equation is thus: rate = (a) $k[Cl_2]$; (b) $k[NO_2]^2[Cl_2]/[NO_2Cl]^2$; (c) $k[NO_2Cl]^2$; (d) $k[NO_2Cl]$; (e) none of these.

3. For a given reaction, a reactant that has an initial concentration of 0.75 M is completely gone in 30 min. The reaction is, thus, _____ order with a rate constant of _____. (a) first, 0.0462 min^{-1}; (b) zero, 0.025 M min^{-1}; (c) first, 0.025 min^{-1}; (d) zero, 0.050 M min^{-1}; (e) none of these.

4. What is the first-order rate constant when the half-life is 57 seconds? (a) 0.018 s^{-1}; (b) 0.024 s^{-1}; (c) 0.009 s^{-1}; (d) 0.012 s^{-1}; (e) none of these.

5. For the reaction rate $= k[A][B]^2$, which of the following will change the value of k the most? (a) halving [B]; (b) doubling [A]; (c) removing product; (d) increasing both [A] and [B]; (e) none of these.

6. The rate of a specific chemical reaction is independent of the concentrations of the reactants. Thus, the reaction is (a) first order; (b) second order; (c) exothermic; (d) catalyzed; (e) none of these.

7. For the reaction $A \longrightarrow$ products, time-concentration data is as follows:

time, min	0	2	4	6	8
[A], M	1.00	0.80	0.60	0.40	0.20

This reaction is (a) catalyzed; (b) zero order; (c) first order; (d) second order; (e) none of these.

8. In the Haber-Bosch process, nickel metal is used to increase the rate of $N_2(g) + 3\ H_2(g) \longrightarrow 2\ NH_3(g)$. The nickel is called (a) an inhibitor; (b) a negative catalyst; (c) a homogeneous catalyst; (d) a heterogeneous catalyst; (e) none of these.

QUIZ B

1. For the reaction $2\ HgCl_2 + C_2O_4^{2-} \longrightarrow$ products, initial rate data is as follows:

$[HgCl_2]$, M	0.0836	0.0836	0.0418
$[C_2O_4^{2-}]$, M	0.202	0.404	0.404
Init. Rate, M h^{-1}	0.26	1.04	0.52

The rate law is $\text{rate} = [HgCl_2]^x[C_2O_4^{2-}]^y$

Thus, (a) $x = 1, y = 1$; (b) $x = 2, y = 2$; (c) $x = 2, y = 2$; (d) $x = 1, y = 2$; (e) none of these.

2. For the reaction $2\ NO_2 + F_2 \longrightarrow 2\ NO_2F$, the rate equation is $\text{rate} = k[NO_2][F_2]$. The order of the reaction is, thus, (a) zero; (b) third; (c) first; (d) second order in $[NO_2]$; (e) none of these.

3. A certain reaction is 1st order in [A]; 2nd order in [B]. When $[A] = 3.0$ M and $[B] = 2.0$ M, rate $= 0.144$ M s^{-1}. What is the rate constant of this reaction? (a) 0.008; (b) 0.024; (c) 0.004; (d) 0.018; (e) none of these.

4. The rate constant for a first-order reaction is $0.0396\ \text{min}^{-1}$. Its half-life is (a) 175 min.; (b) 17.5 min.; (c) 25.3 min.; (d) 35.0 min.; (e) none of these.

5. Which of the following does *not* determine the rate of a reaction? (a) value of $\Delta H°$; (b) activation energy; (c) presence of a catalyst; (d) temperature of reactants; (e) none of these.

6. The units of the specific rate constant for a zero-order reaction are (a) s^{-1}; (b) $M^{-1}\ s^{-1}$; (c) $M^{-2}\ s^{-1}$; (d) s; (e) none of these.

7. Which plot is typical of data for a first-order reaction?

(a) 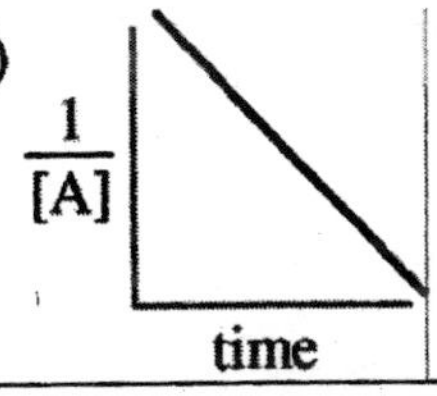

(b) 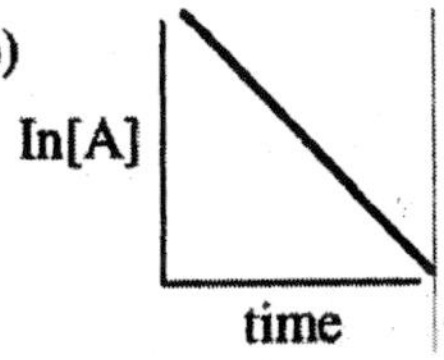

(c) 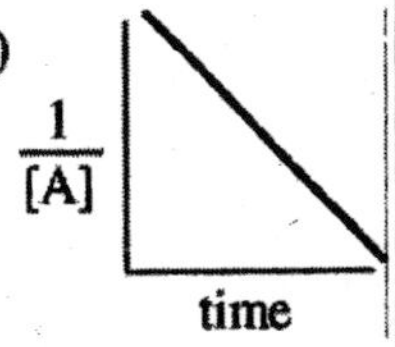

(d) 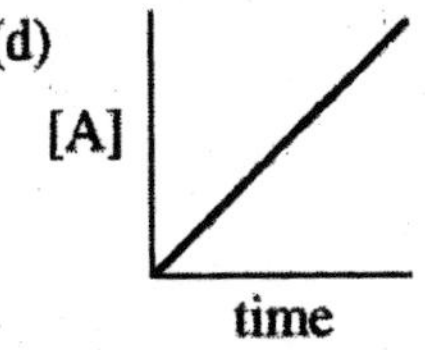

(e) none of these.

8. Which part of an elementary reaction must be exothermic? (a) reactants $\longrightarrow$ products; (b) products $\longrightarrow$ reactants; (c) reactants $\longrightarrow$ activated complex; (d) products $\longrightarrow$ activated complex; (e) activated complex $\longrightarrow$ products.

QUIZ C

1. A combination of reactants that exists for an instant at the time of collision between reactants is called a(n) (a) catalyst; (b) activated complex; (c) reaction product; (d) rate-determining step; (e) none of these.

2. If the half-life of a reaction depends on the concentration of the reactant, then the reaction cannot be _____ order. (a) zero; (b) first; (c) second; (d) third; (e) none of these.

3. A reaction has the rate law $\text{rate} = k[A][B]^2$. Which of the following will cause the rate to increase the most? (a) tripling [B]; (b) doubling [B]; (c) doubling [A]; (d) quadrupling [A]; (e) lowering temperature.

4. A reaction is first order. If its initial rate is $0.0200\ \text{M s}^{-1}$, and 25.0 days later its rate is $6.25 \times 10^{-4}\ \text{Ms}^{-1}$, then its half-life is (a) 25.0 days; (b) 12.5 days; (c) 50.0 days; (d) 5.0 days; (e) none of these.

5. For $2\ NO + O_2 \longrightarrow 2\ NO_2$, initial rate data is as follows:

[NO], M	0.010	0.010	0.030	The rate law is
$[O_2]$, M	0.010	0.020	0.020	$\text{rate} = k[NO]^x[O_2]^y$
Rate, M s^{-1}	2.5×10^{-9}	5.0×10^{-9}	45.0×10^{-9}	

Thus, (a) $x = 1, y = 1$; (b) $x = 2, y = 1$; (c) $x = 1,\ y = 2$; (d) $x = 2, y = 2$; (e) none of these.

6. The correct units of the specific rate constant for a first-order reaction could be (a) $M^{-2}\ s^{-1}$; (b) s; (c) $M^{-1}\ s^{-1}$; (d) s^{-1}; (e) none of these.

7. For the reaction A $\longrightarrow$ products, time-concentration data is as follows:

time, min	0	1	3	4
[A], M	1.00	0.50	0.25	0.20

The order of this reaction is (a) zero; (b) first; (c) second; (d) cannot be determined; (e) none of these.

8. Which of the following can lower the activation energy of a reaction? (a) adding reactants; (b) lowering the temperature; (c) removing products; (d) adding a catalyst; (e) raising temperature.

QUIZ D

1. For $C_2H_4Br_2 + 3\,KI \longrightarrow C_2H_4 + 2\,KBr + KI_3$, initial rate data is as follows:

$[C_2H_4Br_2]$, M	0.500	0.500	1.500	
[KI], M	1.80	7.20	1.80	The rate law is
Rate, $M\,min^{-1}$	0.269	1.08	0.807	rate $= k[C_2H_4Br_2]^x[KI]^y$

Thus, (a) $x = 1, y = 1$; (b) $x = 2, y = 1$; (c) $x = 1, y = 2$; (d) $x = 0, y = 0$; (e) none of these.

2. The time required for the initial amount of reactant in a first-order reaction to decrease by a factor of 2 is known as the (a) reaction time; (b) activation time; (c) half-life; (d) reaction rate; (e) none of these.

3. If a reaction has a rate equation of rate $= k[A]^2[B]$, then it is (a) second order in [B]; (b) first order in [A]; (c) second order overall; (d) spontaneous; (e) none of these.

4. A first-order reaction has a $5.48 \times 10^{-2}\ s^{-1}$ rate constant. How long is required for three-fourths of the initial concentration of reactant to be used up? (a) 5.25 s; (b) 25.3 s; (c) 12.6 s; (d) 0.19 s; (e) none of these.

5. The effect of a catalyst in a chemical reaction is to change the (a) heat evolved; (b) work done; (c) yield of product; (d) activation energy; (e) none of these.

6. The units of a first-order rate constant could be (a) s^{-1}; (b) M^{-1}; (c) $M^{-1}\,s^{-1}$; (d) M s; (e) none of these.

7. Which plot is typical of data for a second-order reaction?

(a)
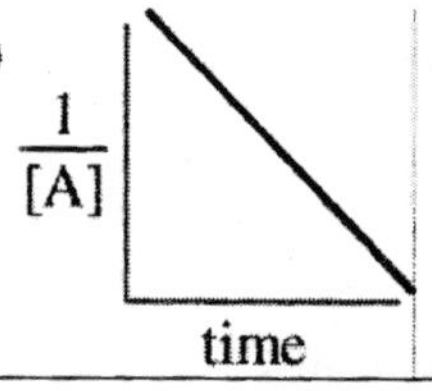

(b)
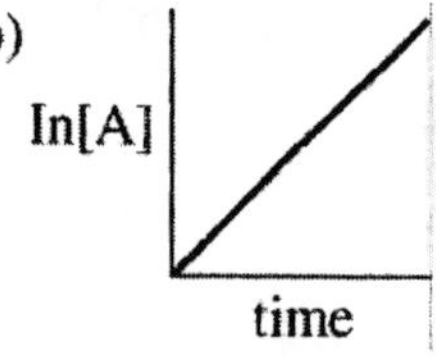

(c)
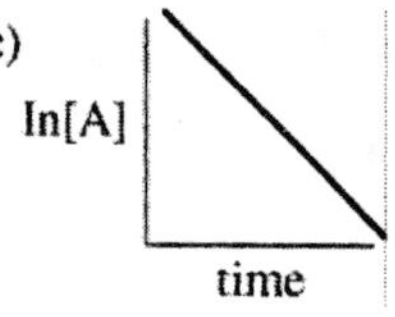

(d)
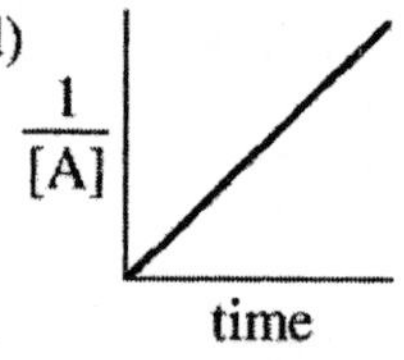

(e)none of these.

8. A reaction has rate $= k[A][B]^2$. Which of the following change the rate constant, k, the least? (a) adding homogeneous catalyst; (b) adding inhibitor; (c) raising temperature $10\,°C$; (d) raising temperature $20\,°C$; (e) increasing [A].

SAMPLE TEST

1. A chemical reaction has a rate constant of $7.00 \times 10^{-9}\ h^{-1}$ at 400 K and $1.00 \times 10^{-4} h^{-1}$ at 500 K. What is the activation energy in kJ/mol for this reaction?

2. Benzene diazonium chloride ($C_6H_5N \equiv N^+Cl^-$) decomposes to nitrogen gas and other products. The concentration of benzene diazonium chloride (B) in isoamyl

alcohol is measured at various times (the results follow). What is the order of the decomposition reaction and what is the value of the specific-rate constant?

[B], mM	3.116	2.822	2.542	2.291	2.064	1.859	1.674
t, min	0	40	80	120	160	200	240
[B], mM	1.505	1.358	1.223				
t, min	280	320	360				

What is the rate of this reaction when [B] = 0.00504 M?

15 PRINCIPLES OF CHEMICAL EQUILIBRIUM

CHAPTER OBJECTIVES

1. Describe the condition of equilibrium in a reversible reaction.

Most chemical reactions do not "go to completion," as we assumed in Chapter 4 and in our studies of most reactions since then. Rather, the final reaction mixture contains some unreacted product. In this way, chemical reactions are similar to many common processes. Consider the following "reaction."

$$\text{woman} + \text{man} \longrightarrow \text{married couple} \qquad [1]$$

Even though Equation [1] is written as a one-way process, we know that it occurs both ways. So do chemical reactions. In addition, we can envision reaching a final state from different starting conditions. For example, if everyone in a large city were to marry on January 1, we would expect some to be unmarried (through divorce or death, for example) at the end of the year. In similar manner, if all were single at the beginning of the year, we would expect many to be married by December 31. In fact, we expect the percent of married persons in each of these cases to be essentially the same. The final situation is known as a *dynamic equilibrium*—dynamic because both the forward (marriage) and the reverse ("unmarriage") reactions are occurring constantly.

Much the same thing occurs for a chemical reaction. Both the forward and the reverse reactions occur continuously, producing a dynamic equilibrium. We can start with either pure reactants, pure products, or at some intermediate state and reach the same equilibrium condition. Data for three experiments for the reaction in Equation [2] are given

$$2\,NO_2(g) \rightleftharpoons N_2O_4(g) \text{ at } 352.6\text{ K} \qquad [2]$$

in Table 15-1. The reaction takes place in a 50.0-L container, except for Trial 3, which carried out in a 20.0-L container.

TABLE 15-1 Equilibrium Data for $2\,NO_2(g) \rightleftharpoons N_2O_4(g)$ at 352.6 K

		Amount mol		*Concentration, M*	
Trial	*Substance*	*Initial*	*Final*	*Initial*	*Final*
1	NO_2	1.000	0.8175	0.0200	0.01635
	N_2O_4	0.000	0.0912	0.000	0.00182
2	NO_2	0.000	0.4457	0.000	0.00891
	N_2O_4	0.250	0.0271	0.00500	0.000542
3	NO_2	0.160	0.210	0.00800	0.01050
	N_2O_4	0.040	0.0150	0.00200	0.000750

*** 2. Describe how equilibrium concentrations are established experimentally.**

Often, an equilibrium mixture can be "frozen" by quickly cooling it or by adding an excess of a reagent that reacts completely with a reactant or a product. After the mixture has been frozen, its composition can be determined by separating the components and identifying them. This procedure works for the mixture resulting from the reaction in Equation [3].

$$2\,HI(g) \rightleftharpoons H_2(g) + I_2(g) \qquad [3]$$

But Reaction [2] cannot be "frozen" readily because it is quite rapid. Thus, this reaction must be studied under the conditions of a *dynamic* equilibrium, 352.6 K and a 50.0-L vessel. $NO_2(g)$ is red-brown, while $N_2O_4(g)$ is colorless. The intensity of the color and, thus, the value of NO_2 can be determined visually or with a colorimeter. If the value of $[NO_2]$ is combined with the known initial amounts of NO_2 and N_2O_4, we can determine the final amounts in moles and the molar concentrations of both species.

EXAMPLE 15-1 0.160 mol $NO_2(g)$ and 0.040 mol $N_2O_4(g)$ are placed in a 20.0-L container. At equilibrium (for Reaction [2]) at 352.6 K, $[NO_2] = 0.01050$ M. What are the equilibrium amounts of all substances?

The equilibrium amount of NO_2 in the 20.0-L container is

$$\text{mol } n_{NO_2} = 0.01050\,\text{M} \times 20.0\,\text{L} = 0.210\ \text{mol}$$

NO_2 0.210 mol − 0.160 mol = 0.050 mol NO_2 is produced, and then we have

Equation:	$2\,NO_2(g)$	$\rightleftharpoons$	$N_2O_4(g)$
Initial:	0.160 mol		0.040 mol
Changes:	+0.050 mol		−0.025 mol
Equil.:	0.210 mol		0.015 mol

Carefully observe the setup for Example 15-1. The three lines "Initial," "Changes," and "Equil." organize the data of the problem. The values on the "Changes" line is related by stoichiometry. All these values are related as are the stoichiometric coefficients (two to one in this case), and the values for products have opposite signs from those for reactants. Further, notice that "Inititial" + "Changes" = Equilibrium for each species.

*** 3. Write the equilibrium constant expression in terms of concentrations K_c for a reaction, and use the value of K_c and the concentrations of all species but one to determine the equilibrium concentration of that species.**

The equilibrium constant is written as the ratio of concentrations (product concentrations over those of reactants), each raised to the power of its stoichiometric coefficient in the balanced chemical equation. The equilibrium constant expressions for all reactions are written in the same way (see Objectives 15-4 and 15-5). Finding one equilibrium concentration given all the others and the value of K_c is a matter of rearranging the expression to solve for the necessary concentration.

EXAMPLE 15-2 What is $[N_2O_4]$ at equilibrium if for the reaction in Equation [3]?

$$2\,NO_2(g) \rightleftharpoons N_2O_4(g) \qquad K_c = 6.81 \text{ at } 352.6\text{ K} \qquad [3]$$

$$K_c = \frac{[N_2O_4]}{[NO_2]^2} \text{ OR } [N_2O_4] = K_c[NO_2]^2 \text{ THUS } [N_2O_4] = (6.81)(0.100)^2 = 0.681\text{ M}$$

*** 4. Derive *K* values for situations where chemical equations are reversed, multiplied through by constant coefficients or added together.**

Reaction [4] is the *reverse* of Reaction [2]. $N_2O_4(g) \rightleftharpoons 2\,NO_2(g)$ [4]

For Reaction [4], the equilibrium constant expression is $K_4 = [NO_2]^2/[N_2O_4]$. K_4 for Reaction [4] is merely the inverse of K_2 for Reaction [2]. Thus, $K_4 = 1/K_2 = 1/6.81 = 0.147$. In similar fashion, Reaction [4] can be *multiplied by one-half* to obtain Reaction [5].

$$\tfrac{1}{2}N_2O_4(g) \rightleftharpoons NO_2(g) \qquad [5]$$

The equilibrium constant expression for Reaction [5] is

$K_5 = [NO_2]/[N_2O_4]^{1/2} = K_4^{1/2} = \sqrt{K_4} = \sqrt{0.147} = 0.383.$

In addition, consider Reaction [6].

$$2\,NO_2(g) \rightleftharpoons 2\,NO(g) + O_2(g) \qquad [6]$$

For this reaction, $K_6 = [NO]^2[O_2]/[NO_2]^2 = 5.01 \times 10^5$. If we *add* Reactions [4] and [6], we obtain Reaction [7].

$$N_2O_4(g) \rightleftharpoons 2\,NO(g) + O_2(g) \qquad [7]$$

The equilibrium constant expression for Reaction [7], K_7, is the product of the equilibrium constant expressions for Reactions [4] and [6].

$$K_4 \times K_6 = \frac{[NO_2]^2}{[N_2O_4]} \times \frac{[NO]^2[O_2]}{[NO_2]^2} = 0.147 \times 5.01 \times 10^5 = 7.36 \times 10^6 = K_7 = \frac{[NO]^2[O_2]}{[N_2O_4]}$$

Thus, when we modify reactions, we can follow the following rules to determine values of equilibrium constants.

(1) When a reaction is reversed, the new equilibrium constant is the inverse of the old equilibrium constant.

(2) When a reaction is multiplied through by a constant coefficient, the new equilibrium constant is the old *K* raised to the power of the coefficient.

(3) When two reactions are added together, the equilibrium constant of the new reaction is the product of the equilibrium constants of the original two reactions.

5. Assess the relative importance of the forward and reverse reactions from the magnitude of an equilibrium constant.

A large equilibrium constant ($\approx > 1000$) indicates that the products predominate over the reactants in the equilibrium mixture. Thus, the forward reaction occurs more readily than does the reverse reaction. A small equilibrium constant ($\approx < 0.001$), on the other hand, indicates that products do not readily form from the reactants. In fact, the opposite is true: the reverse reaction occurs more readily than the forward reaction.

*** 6. Write an equilibrium constant expression in terms of partial pressures of gases (K_p), and relate the value of K_p to the corresponding value of K_c.**

K_p has the same form as does K_c, except partial pressures are used rather than concentrations. These K_p expressions are used in reactions that involve gases only. Although K_p and K_c may not be equal, they both are constant for a given reaction, as long as the temperature is fixed. K_c and K_p are related through the ideal gas equation—$PV = nRT$ or $P = (n/V)RT$. Notice that (n/V) is a molar concentration. Thus, for Reaction [2] we have

$$K_p = \frac{P_{N_2O_4}}{(P_{NO_2})^2} = \frac{[N_2O_4]RT}{([NO_2]RT)^2} = \frac{[N_2O_4]}{[NO_2]^2} \times \frac{1}{RT} = \frac{K_c}{RT}$$

(Note that a different expression of K_p will result from another reaction, specifically one in which the stoichiometric coefficients for gases differ.) The value of R that must be used is 0.082057 L atm $mol^{-1}K^{-1}$ and the pressure is in atmospheres. Of course, the temperature is expressed in kelvins.

EXAMPLE 15-3 What is K_p for Reaction [2]?

$2\,NO_2(g) \rightleftharpoons N_2O_4(g) \qquad K_c = 6.81$ at 352.6 K

The relationship is given above. $K_p = \dfrac{K_c}{RT} = \dfrac{6.81}{(0.082057)(352.6)} = 0.235$

Just as K_c has no units, neither does K_p. By convention, partial pressures always are expressed in atmospheres. It is important to recognize that K_p and K_c are valid only at a specific temperature. Merely multiplying or dividing by RT will not produce the value of the equilibrium constant at a different temperature. Obtaining that value is the subject of Objective 15-13.

*** 7. Know that the concentrations of pure liquids and solids are omitted from equilibrium constant expressions.**

Many experiments have investigated both the rates of chemical reactions and the position of equilibrium. In all of them, it has been found that the amount of pure liquid or pure solid present has no effect as long as *some* of the liquid or solid is present. Thus, the concentrations of pure liquids and solids are omitted from both rate laws and equilibrium constant expressions. Usually, this means that the solvent in a liquid solution also is omitted. But in concentrated solutions, when the solvent is a reactant or a product, its concentration must be included. Consider the hydrolysis of ethyl acetate, Reaction [8].

$$CH_3COOC_2H_5 + H_2O \rightleftharpoons CH_3COOH + C_2H_5OH \qquad [8]$$

In dilute solution, the solvent—H_2O in this instance—behaves as if it were pure and $K_c = [CH_3COOH][C_2H_5OH]/[CH_3COOC_2H_5]$. In contrast, in concentrated solution, the $[H_2O]$ varies significantly during the course of the reaction and

$K_c = [CH_3COOH][C_2H_5OH]/[CH_3COOC_2H_5][H_2O]$

*** 8. Calculate a numerical value of an equilibrium constant if equilibrium conditions are given.**

We now can determine the value of an equilibrium constant. Often it helps to write the concentrations of reactants and products beneath their formulas in the balanced chemical equation. It is very useful to write three lines: "Initial Concentrations," "Changes of Concentrations," and "Equilibrium Concentrations."

EXAMPLE 15-4 The reaction $H_2(g) + CO_2(g) \rightleftharpoons H_2O(g) + CO(g)$ reaches equilibrium at 500 K. Initially, $[H_2]_i = 0.0420\,M$ and $[CO_2]_i = 0.0160\,M$. At equilibrium, there is 0.194 mol CO_2 in the 14.0-L container. What is the value of K_c?

The equilibrium concentration of CO_2 is $[CO_2]_e = 0.194\,mol/14.0\,L = 0.0139\,M$. Thus, the concentration of CO_2 that has reacted is $[CO_2]_r = [CO_2]_i - [CO_2]_e = 0.0160\,M = 0.0139\,M = 0.0021\,M$.

Equation:	$H_2(g)$ +	$CO_2(g) \rightleftharpoons$	$H_2O(g)+$	$CO(g)$
Initial:	0.0420 M	0.0160 M	0.0000 M	0.0000 M
Changes:		−0.0021 M		
Equil.:		0.0139 M		

Since 1 mole of $H_2(g)$ *reacts* for every mole of $CO_2(g)$ reacting, and since 1 mole each of $H_2O(g)$ and $CO(g)$ are *produced* for each mole of $CO_2(g)$ that reacts, we have

$[H_2]_r = [CO_2]_r = +0.0021\,M$ *and* $[H_2]_p = [CO_2]_p = -0.0021\,M$

These values are related by the reaction's stoichiometry.

Equation:	$H_2(g)+$	$CO_2(g) \rightleftharpoons$	$H_2O(g)+$	$CO(g)$
Initial:	0.0420 M	0.0160 M	0.0000 M	0.0000 M
Changes:	−0.0021 M	−0.0021 M	+0.0021 M	+0.0021 M
Equil.:	0.0399 M	0.0139 M	0.0021 M	0.0021 M

Notice in every case $[\,]_e = [\,]_i + [\,]_r$. Finally, notice that we can add and subtract concentrations in the same way we would combine amounts in moles *as long as the species involved are in the same solution* (which they should be in an equilibrium system). Then, the value of K_c is computed.

$$K_c = \frac{[CO][H_2O]}{[CO_2][H_2]} = \frac{(0.0021)(0.0021)}{(0.0139)(0.0399)} = 8.0 \times 10^{-3}$$

*** 9. Predict the direction in which a reaction proceeds toward equilibrium by comparing the reaction quotient, Q_c, to K_c.**

The reaction quotient expression is written in the same way as the equilibrium constant expression. But, in the reaction quotient, the values of the concentrations and partial pressures are not necessarily those at equilibrium. The reaction quotient is valuable in predicting the direction in which the reaction will shift or move in reaching equilibrium. There are three possibilities.

If $Q_c = K$, the reaction is at equilibrium

If $Q_c > K$, the reaction will shift left

[9]

If $Q_c < K$, the reaction will shift right

"Shift right" means that more products will be produced and reactants will be consumed. This is the case in Trial 1 of Table 15-1.

$$Q_{c,1} = \frac{[N_2O_4]_{1i}}{[NO_2]_{1i}^2} = \frac{0.000}{(0.0200)^2} = 0.00 < 6.81 = K_c$$

"Shift left" means that reactants are produced at the expense of products. This is the case in Trial 2 of Table 15-1.

$$Q_{c,2} = \frac{[N_2O_4]_{2i}}{[NO_2]_{2i}^2} = \frac{0.00500}{(0.000)^2} = \infty > 6.81 = K_c$$

The same procedure is used in evaluating Q_p and comparing it with K_p, except partial pressures in atmospheres are used rather than molarities.

*** 10. Use the ideal gas law and Dalton's law of partial pressures in working with K_p expressions.**

We can add and subtract partial pressures as we did molarities. One advantage is that the total pressure is readily measured and Dalton's law of partial pressures can be used.

EXAMPLE 15-5 A container is filled with $CH_3OH(g)$ at 350 K, and equilibrium is achieved for the reaction $CO(g) + 2\,H_2(g) \rightleftharpoons CH_3OH(g)$. At equilibrium, the partial pressure of $CH_3OH(g)$ is 1.000 atm and the total pressure is 1.420 atm. What is the value of K_p?

We can use the setup that was first introduced in Objective 15-2.

Equation:	$CO(g)$	+ $2\,H_2(g)$	$\rightleftharpoons$	$CH_3OH(g)$
Initial:	0 atm	0 atm		y atm
Changes:	$+x$ atm	$+2x$ atm		$-x$ atm
Equil.:	x atm	$2x$ atm		$y-x\text{ atm}=1.000\text{ atm}$

$$P_{\text{total}} = P\{CO\} + P\{H_2\} + P\{CH_3OH\} = 1.420\text{ atm}$$
$$= x\text{ atm} + 2x\text{ atm} + 1.000\text{ atm} = (1.000 + 3x)\text{ atm}$$

$$x = P\{CO\} = (1.420 - 1.000)/3 = 0.140\text{ atm} \quad K_p = \frac{P\{CH_3OH\}}{P\{CO\}P\{H_2\}^2} = \frac{1.000}{(0.140)(0.280)^2} = 91.1$$

$$P\{H_2\} = 2 \times 0.140\text{ atm} = 0.280\text{ atm}$$

*** 11. Make qualitative predictions of how equilibrium conditions change when an equilibrium mixture is disturbed. That is, apply Le Châtelier's principle.**

Le Châtelier's principle states that, when a stress is applied to an equilibrium system, the system will react ("shift") in a direction that will relieve the stress. One way to account for the change in the equilibrium position brought about by varying concentration is to consider the effect that such variation would have on the rates of the forward and reverse reactions.

Chemical equilibria proceed toward products ("shift to the right") when the forward reaction is speeded up relative to the reverse reaction. For example, if the reactant concentrations are increased, the forward reaction speeds up. Or, if the concentrations of the products are decreased, the reverse reaction slows down. In each case, the system eventually reaches a new equilibrium. In similar fashion, conditions that cause the reverse reaction's rate to increase relative to the forward rate makes the reaction proceed toward reactants ("shift to the left").

In the special case of reactions involving gaseous species, concentrations can be changed by varying the partial pressures within the system. These pressures can be increased by squeezing the system into a smaller volume. If the pressure is increased on a system at equilibrium, the system tries to make its volume as small as possible. Thus, it proceeds toward the side where the sum of the stoichiometric coefficients of gaseous species is smaller.

Changing the temperature of a system at equilibrium causes the reaction to shift right or left depending on whether the reaction is endothermic or exothermic. An easy way to remember the direction of the shift is to think of heat as a product (in an exothermic reaction) or as a reactant (in an endothermic reaction) of the reaction.

$$\text{Exothermic: reactants} \longrightarrow \text{products} + \text{heat}$$

$$\text{Endothermic: reactants} + \text{heat} \longrightarrow \text{products} \qquad [10]$$

Raising the temperature "increases the concentration of heat" and the exothermic reaction proceeds toward reactants. Of course, heat is not a substance but a form of energy, and relations [10] are misleading. However, they are a convenient memory aid. All of these changes in conditions and their effects are summarized in Table 15-2.

TABLE 15-2 Effects of Changing Conditions on Equilibrium Systems

Changed condition	*Type of reaction*	*Shift direction*
Increase in reactant concentration	General	Right
Increase in product concentration	General	Left
Increase pressure *or*	$n_{g,R} < n_{g,P}$	Left
decrease volume	$n_{g,R} > n_{g,P}$	Right
Decrease pressure *or*	$n_{g,R} < n_{g,P}$	Right
increase volume	$n_{g,R} > n_{g,P}$	Left
Increase temperature	Exothermic	Left
	Endothermic	Right
Decrease temperature	Exothermic	Right
	Endothermic	Left

$n_{g,R}$ = sum of stoichiometric coefficients of gaseous reactants

$n_{g,P}$ = sum of stoichiometric coefficients of gaseous products

*** 12. Calculate the final equilibrium condition in a reversible reaction from a given set of initial conditions.**

Once the value of the equilibrium constant is known, it can be used to predict the final concentrations of reactants and products. An easy technique to help keep track of the calculation was first used in Objective 15-8. The equation for the reaction is written out. Below the formula of each species, we write the initial concentration or partial pressure with the changes in these values immediately below, and the equilibrium values, in turn, below them. In this type of problem, one or several concentrations are not known. If possible, we need to express all concentrations or partial pressures in terms of one unknown.

EXAMPLE 15-6 For $H_2(g) + CO_2(g) \rightleftharpoons H_2O(g) + CO(g)$, $K_c = 8.0 \times 10^{-3}$ at 500 K. What are the equilibrium concentrations of all reagents if initial concentrations are $[H_2] = 0.0200$ M, $[CO_2] = 0.014$ M and $[CO] = 0.0100$ M?

Some $H_2O(g)$ must be produced to attain equilibrium. Let $[H_2O]_{eq} = x$.

Equation:	$H_2(g)$ +	$CO_2(g)$ $\rightleftharpoons$	$H_2O(g)$ +	$CO(g)$
Initial:	0.0200 M	0.0144 M	0.0000 M	0.0100 M
Changes:	$-x$ M	$-x$ M	$+x$ M	$+x$ M
Equil.:	$(0.0200 - x)$ M	$(0.0144 - x)$ M	x M	$(0.0100 + x)$ M

$$K_c \frac{[H_2O]_{eq}[CO]_{eq}}{[H_2]_{eq}[CO_2]_{eq}} = \frac{(x)(0.0100 + x)}{(0.0200 - x)(0.0144 - x)} = 8.0 \times 10^{-3} = \frac{0.0100x + x^2}{2.88 \times 10^{-4} - 0.0344x + x^2}$$

or $0.992\,x^2 + 0.01028\,x - 2.304 \times 10^{-6} = 0$.

We use the quadratic equation to find the solution:

$$x = \frac{-b \pm \sqrt{b^2 - 4ac}}{2a} = \frac{-0.01028 \pm \sqrt{(0.01028)^2 - (4)(0.992)(-2.304 \times 10^{-6})}}{2(0.992)}$$

$$= -5.181 \times 10^{-3} \pm 5.401 \times 10^{-3} = 2.20 \times 10^{-4}\,\mathrm{M}$$

We choose the positive root rather than the negative root, because a negative concentration (-0.0010582 M) has no physical meaning. The equilibrium concentrations follow:

$$[H_2] = 0.0200 - x = 0.0200 - 0.00022 = 0.0198\,\mathrm{M}$$

$$[CO_2] = 0.0144 - x = 0.0144 - 0.00022 = 0.0142\,\mathrm{M}$$

$$[H_2O] = \qquad x = 0.00022\,\mathrm{M}$$

$$[CO] = 0.0100 + x = 0.0100 + 0.00022 = 0.0102\,\mathrm{M}$$

To check our results, we substitute these equilibrium values into the K_c expression:

$$K_c = \frac{[H_2O][CO]}{[H_2][CO_2]} = \frac{(0.00022)(0.0102)}{(0.0198)(0.0142)} = 8.0 \times 10^{-3}$$

EXAMPLE 15-7 For $Br_2(g) + Cl_2(g) \rightleftharpoons 2\,BrCl(g)$, $K_p = 7.21$ at 298 K. What is the equilibrium pressure of BrCl(g) if the`initial pressure of $Br_2(g)$ is 1.000 atm and that of Cl_2 is 2.000 atm?

Let x be the decrease in the pressure of $Br_2(g)$. Then x also is the decrease in the pressure of $Cl_2(g)$ and $2x$ is the increase in the pressure of BrCl(g), as summarized below.

Equation:	$Br_2(g) + Cl_2(g)$	$\rightleftharpoons$	$2\,BrCl(g)$
Initial:	1.000 atm	2.000 atm	0.000 atm
Changes:	$-x$ atm	$-x$ atm	$+2x$ atm
Equil.:	$(1.000 - x)$ atm	$(2.000 - x)$ atm	$2x$ atm

$$K_p = \frac{P^2_{BrCl}}{P_{Br2}P_{Cl2}} = \frac{(2x)^2}{(1.000 - x)(2.000 - x)} = 7.21 \text{ or } 3.21x^2 - 21.630x + 14.420 = 0$$

$$x = \frac{21.630 \pm \sqrt{(21.630)^2 - 4(3.21)(14.420)}}{2(3.21)} = 3.369 \pm 2.619 = 0.750\,\text{atm}$$

$P_{Br2} = 1.000 - x = 1.000 - 0.750 = 0.250$ atm

$P_{Cl2} = 2.000 - x = 2.000 - 0.750 = 1.250$ atm

$P_{BrCl} = 2x \quad = 2 \times 0.750 \quad = 1.500$ atm

$$K_p = \frac{P_{BrCl}^{\ 2}}{P_{Br2}P_{Cl2}} = \frac{(1.500)^2}{(0.250)(1.250)} = 7.20$$

In this case, the positive root is not used because it would produce negative pressures of the reactants. Notice also that the line labeled "Changes" has x multiplied by the stoichiometric coefficients of each species, and that the signs on this line are different for products than for reactants.

13. Calculate the new equilibrium concentrations or partial pressures after an equilibrium system has adapted to changed conditions.

Changes in concentrations or pressures cause no change in the value of the equilibrium constant. (Of course, that's why it's called a constant.) But individual pressures and concentrations may indeed change.

EXAMPLE 15-8 The volume of Trial 3 in Table 15-1 is expanded from 20.0 L to 50.0 L. In which direction will the reaction proceed and what will be the final concentrations once equilibrium is reestablished?

The "initial" concentrations after the expansion are determined by multiplying the 20.0-L concentrations by $20.0\text{ L}/50.00\text{ L} = 0.400$. The decrease in pressure that occurs causes the reaction to proceed toward reactants ("shift left").

Equation:	$2\,NO_2(g)$ $\rightleftharpoons$	$N_2O_4(g)$
Initial equil.:	0.01050 M	0.000750 M
Volume to 50.0 L:	0.00420 M	0.000300 M
Changes:	$+2x$ M	$-x$ M
Final equil.:	$(0.00420 + 2x)$ M	$(0.000300 - x)$ M

The line labeled "Volume to 50.0 L" gives the concentrations at the instant that the volume is increased to 50.0 L, but before the final equilibrium is achieved.

$$K_c = \frac{[N_2O_4]}{[NO_2]^2} = \frac{(0.000300 - x)}{(0.00420 + 2x)^2} = 6.81$$

The quadratic equation gives $x = -0.02046 \pm 0.02062 = 0.00016\text{ M}$. The positive root is chosen since the negative root gives $x = -0.04104\text{ M}$. This produces $[NO_2] = -0.07788\text{ M}$. This root is discarded because negative concentrations have no physical meaning. Thus, $[NO_2] = 0.00452\text{ M}$ and $[N_2O_4] = 0.000140\text{ M}$.

Notice that $[NO_2]$ has increased and $[N_2O_4]$ has decreased, as predicted. But have we decided the outcome of the calculation by *assuming* that $[NO_2]$ would increase? Suppose we had assumed that $[NO_2]$ decreased by $2y$ and $[N_2O_4]$ increased by y. Then we would have $[NO_2]_e = 0.00420 - 2y$ and $[N_2O_4]_e = 0.000300 + y$

$$K_c = \frac{[N_2O_4]}{[NO_2]^2} = \frac{(0.000300 + y)}{(0.00420 - 2y)^2} = 6.81 \quad \text{or} \quad y = -0.00016\ \text{M}$$

(The positive root gives the physically meaningless result of a negative $[NO_2]_e$.) We obtain the same result as before. $[NO_2]_e = 0.00420 - 2(-0.00016) = 0.00452\ \text{M}$ and

$[N_2O_4]_e = 0.000300 + (-0.00016) = 0.00014\ \text{M}$

When a temperature change occurs, the *value* of the equilibrium constant changes. Its new value is determined by the van't Hoff equation, as outlined in Objective 19-13.

SELF-ASSESSMENT EXERCISES

87. (c) Less than 2 moles. If the reaction went to completion, 2 mol would be produced from the 1 mol of I_2. Because the reaction is an equilibrium, less than 2 mol will be produced.

88. The equilibrium expression is $\frac{[SO_3]^2}{[SO_2][O_2]} = 100$. If $[SO_3] = [SO_2]$, then

$[O_2] = 0.010$ M.

89. Because all of the concentrations will initially decrease, to reestablish equilibrium the amount of $[SO_2]$ and $[Cl_2]$ will have to increase.

90. Answer (c) is correct. Reversing the reaction will give the inverse of K_c. As long as the equation is balanced, multiplying the coefficients of the reaction will have no effect on K_c.

91. For this reaction, $\Delta n = 1$. Therefore, $K_p = K_c(RT)$, so answer (b) is correct—K_c is greater than K_p.

92. Because the number of moles remains the same, $K_p = K_c$. A catalyst increases the rate but does not change the equilibrium. Increasing temperature will have no effect. Similarly, changing the volume will change all of the concentrations. Therefore answer (d) is correct.

93. $K_c = \dfrac{[C]^2}{[A]^2[B]} = \dfrac{[0.43]^2}{[0.55]^2[0.33]} = 1.85\ M^{-1}$

94. (a) Adding O_2 will increase products.

(b) Removing HCl will shift the equilibrium to the left and decrease product.

(c) The pressure of all will decrease. There are a greater number of moles of reactants and, therefore, the reaction will shift to the left.

(d) There will be no change. The rate will change, but not the position of equilibrium.

(e) $K_p = K_c(RT)^{-1}$, so increasing T will decrease K_p, and the amount of products will decrease.

95. (a) $K_c = \dfrac{[SO_3]^2}{[SO_2][O_2]}$. The $[O_2] = 1/K_c = 0.028$ M and

the number of moles = 0.028 $molL^{-1}$ x 2.05 L = 0.0577 mol

(b) $[SO_3] = 2 \times [SO_2]$ and $K_c = \dfrac{[2SO_2]^2}{[SO_2][O_2]} = 4/[O_2]$. $[O_2] = (0.028)(4) = 0.112\ mol\ L^{-1}$

number of moles = 0.230 mol.

DRILL PROBLEMS

1. The initial amount of each substance in each of the following reactions is given below, as well as the chemical formula of that substance. The equilibrium amount of one substance then is given, to the right, along with the volume of the container. Determine the equilibrium concentrations of all substances. Each line represents a different situation.

$CH_4(g)$ +	$CCl_4(g) \rightleftharpoons$	$2CH_2Cl_2(g)$	at 298 K	
A. 0.100 mol	0.640 mol	0.220 mol	$n_{CH4} = 0.1502$ mol,	$V = 5{,}000$ L
B. 0.200 mol	0.200 mol	0.320 mol	$n_{CCl4} = 0.3118$ mol,	$V = 8.00$ L
C. 0.650 mol	0.320 mol	0.320 mol	$n_{CH2Cl2} = 0.1569$ mol,	$V = 4.00$ L
D. 0.3400 mol	0.3400 mol	0.5100 mol	$n_{CH2Cl2} = 0.1593$ mol,	$V = 8.50$ L
$PCl_5(g) \rightleftharpoons$	$PCl_3(g)$ +	$Cl_2(g)$	at 600 K	
E. 0.100 mol	0.100 mol	0.100 mol	$n_{Cl2} = 0.1848$ mol,	$V = 4.00$ L
F. 0.300 mol	0.204 mol	0.153 mol	$n_{PCl5} = 0.0897$ mol,	$V = 3.00$ L
G. 0.144 mol	0.276 mol	0.072 mol	$n_{PCl3} = 0.408$ mol,	$V = 12.0$ L

$HCO(g) \rightleftharpoons$	$H_2(g) +$	$CO(g)$	at 773 K	
H. 2.50 mol	0.00 mol	0.00 mol	$n_{H2} = 1.00$ mol,	$V = 2.50$ L
I. 0.00 mol	3.10 mol	2.00 mol	$n_{CO} = 0.500$ mol,	$V = 2.00$ L
J. 0.900 mol	0.000 mol	0.600 mol	$n_{HCHO} = 0.500$ mol,	$V = 3.00$ L
$2\,NO(g) +$	$Br_2(g) \rightleftharpoons$	$2\,NOBr(g)$	at 350 K	
K. 0.000 mol	0.400 mol	1.000 mol	$n_{NO} = 0.200$ mol,	$V = 0.200$ L
L. 1.50 mol	4.60 mol	0.00 mol	$n_{NOBr} = 1.20$ mol,	$V = 1.60$ L
M. 0.00 mol	8.75 mol	2.00 mol	$n_{NO} = 0.500$ mol,	$V = 6.40$ L
$3\,H_2(g) +$	$N_2(g) \rightleftharpoons$	$2\,NH_3(g)$	at 473 K	
N. 0.5800 mol	0.1930 mol	0.0000 mol	$n_{N2} = 0.0097$ mol,	$V = 0.525$ L
O. 0.0000 mol	0.0117 mol	0.5134 mol	$n_{H2} = 0.200$ mol,	$V = 0.287$ L

2. (1) Write the equilibrium constant expression in terms of concentrations for each of the following reactions. (Some numerical values of K_c given will be used in the following problems.)

A. $2\,NOBr(g) \rightleftharpoons 2\,NO(g) + Br_2(g)$

B. $4\,HCl(g) + O_2(g) \rightleftharpoons 2\,H_2O(g) + 2\,Cl_2(g)$ $K_c = 1.49 \times 10^3$ at 320 K

C. $N_2(g) + O_2(g) \rightleftharpoons 2\,NO(g)$ $K_c = 6.37 \times 10^{-16}$ at 298 K

D. $CH_4(g) + CCl_4(g) \rightleftharpoons 2\,CH_2Cl_2(g)$

E. $PCl_5(g) \rightleftharpoons PCl_3(g) + Cl_2(g)$

F. $HCHO(g) \rightleftharpoons H_2(g) + CO(g)$

G. $N_2(g + 3\,H_2(g) \rightleftharpoons 2\,NH_3(g)$

H. $2\,H_2S(g) + 3\,O_2(g) \rightleftharpoons 2\,H_2O(g) + 2\,SO_2(g)$ $K_c = 1.74 \times 10^{45}$ at 1000 K

I. $2\,NH_3(g) + 6\,F_2(g) \rightleftharpoons 2\,NF_3(g) + 6\,HF(g)$

(2) For each of the following reactions, the value of K_c and the equilibrium concentrations of all reagents but one are given. Determine the value of the omitted concentration.

$CH_3ONO(g) +$	$HCl(g) \rightleftharpoons$	$CH_3OH(g) +$	$NOCl(g)$ $K_c = 0.242$ at 323 K
J. 0.500 M	0.274 M	1.26 M	_____ M
K. _____ M	4.96 M	1.14 M	1.49 M
$CH_4(g) +$	$CCl_4(g) \rightleftharpoons$	$2\,CH_2Cl_2(g)$	$K_c = 0.0956$ at 298 K
L. 1.124 M	1.124 M	____ M	
M. 0.104 M	____ M	0.256 M	

$2\,NOBr(g) \rightleftharpoons 2\,NO(g) + Br_2(g)$ $K_c = 1.40 \times 10^5$ at 350 K

N. 0.275 M	0.124 M	____M
O. ____M	0.173 M	0.822 M

$N_2(g) + 3\,H_2(g) \rightleftharpoons 2\,NH_3(g) \quad K_c = 1.40\times10^5$ at 473 K

P. ____M	0.163 M	0.970 M
Q. 0.00100 M	0.0220 M	____M

$Br_2(g) + Cl_2(g) \rightleftharpoons 2\,BrCl(g) \quad K_c = 6.30$ at 350 K

R. 0.146 M	____M	0.227 M
S. 0.082 M	0.077 M	____M

$PCl_3(g) + Cl_2(g) \rightleftharpoons PCl_5(g) \quad K_c = 1.784$ at 600 K

T. 0.149 M	____M	0.206 M
U. 0.200 M	0.111 M	____M

3. Based on your answers to (and the data of) the drill problems for Objectives 15-2 and 15-3, write the equilibrium constant expression and determine its value for each of the following equations.

A. $H_2(g) + CO(g) \rightleftharpoons HCHO(g)$

B. $PCl_3(g) + Cl_2(g) \rightleftharpoons PCl_5(g)$

C. $2\,NH_3(g) \rightleftharpoons N_2(g) + 3\,H_2(g)$

D. $2\,BrCl(g) \rightleftharpoons Br_2(g) + Cl_2(g)$

E. $2\,CH_2Cl_2(g) \rightleftharpoons CH_4(g) + CCl_4(g)$

F. $CH_2Cl_2(g) \rightleftharpoons \frac{1}{2}CH_4(g) + \frac{1}{2}CCl_4(g)$

G. $NO(g) + \frac{1}{2}Br_2(g) \rightleftharpoons NOBr(g)$

H. $\frac{1}{2}N_2(g) + \frac{3}{2}H_2(g) \rightleftharpoons NH_3(g)$

I. $\frac{1}{2}Br_2(g) + \frac{1}{2}Cl_2(g) \rightleftharpoons BrCl(g)$

J. $NO(g) \rightleftharpoons \frac{1}{2}N_2(g) + \frac{1}{2}O_2(g)$

K. $H_2O(g) + Cl_2(g) \rightleftharpoons 2\,HCl(g) + \frac{1}{2}O_2(g)$

L. $H_2S(g) + \frac{3}{2}O_2(g) \rightleftharpoons H_2O(g) + SO_2(g)$

4. The value of K_p or K_c is given for each following reaction. Write the K_p expression for each reaction and find the value of the one K_p or K_c that is not given.

A. $PCl_3(g) + Cl_2(g) \rightleftharpoons PCl_5(g) \quad K_c = 1.784$ at 600 K

B. $Br_2(g) + Cl_2(g) \rightleftharpoons 2\,BrCl(g) \quad K_p = 6.30$ at 350 K

C. $2\,H_2S(g) + 3\,O_2(g) \rightleftharpoons 2\,H_2O(g) + 2\,SO_2(g) \quad K_c = 1.74\times10^{45}$ at 1000 K

D. $N_2(g) + 3\,H_2(g) \rightleftharpoons 2\,NH_3(g) \quad K_c = 1.40\times10^5$ at 473 K

E. $N_2O_4(g) \rightleftharpoons 2\,NO_2(g) \quad K_p = 0.113$ at 298 K

F. $2\,NOBr(g) \rightleftharpoons 2\,NO(g) + Br_2(g) \quad K_c = 0.1563$ at 350 K

G. $CO(g) + 2\,H_2(g) \rightleftharpoons CH_3OH(g) \quad K_p = 2.05\times10^4$ at 298 K

H. $CH_4(g) + CCl_4(g) \rightleftharpoons 2\,CH_2Cl_2(g) \quad K_c = 0.0956$ at 298 K

I. $2\,NOCl(g) \rightleftharpoons 2\,NO(g) + Cl_2(g) \quad K_p = 0.0658$ at 298 K

J. $4\,HCl(g) + O_2(g) \rightleftharpoons 2\,H_2O(g) + 2\,Cl_2(g) \quad K_c = 1.49\times10^3$ at 320 K

K. $2\,H_2S(g) \rightleftharpoons 2\,H_2(g) + S_2(g)$ $\quad K_p = 3.16\times10^{-3}$ at 830 K

L. $N_2(g) + O_2(g) \rightleftharpoons 2\,NO(g)$ $\quad K_c = 6.37\times10^{-6}$ at 298 K

5. Write the equilibrium constant expression in terms of concentrations K_c for each of these equations.

A. $CO_2(g) + 2\,NH_3(g) \rightleftharpoons CO(NH_2)_2(s) + H_2O(l)$

B. $S(s) + 2\,CO(g) \rightleftharpoons SO_2(g) + 2\,C(s)$

C. $NH_4Cl(s) \rightleftharpoons NH_3(g) + HCl(g)$

D. $NH_3(g) + CH_3COOH(l) \rightleftharpoons NH_2CH_2COOH(s) + H_2(g)$

E. $SO_2Cl_2(l) \rightleftharpoons SO_2(g) + Cl_2(g)$

F. hydrogen(g) + oxygen(g) $\rightleftharpoons$ water(l)

G. ammonium nitrate(s) $\rightleftharpoons$ dinitrogen oxide(g) + water(g)

H. hydrogen sulfide(aq) + sodium hydroxide(aq) $\rightleftharpoons$
sodiumisulfide(aq) + water(l)

I. phosphorus trichloride(g) + water(l) $\rightleftharpoons$
phosphorous acid(aq) + hydrochloric acid(aq)

J. $4\,HNO_3(aq) + 3Ag(s) \rightleftharpoons 3\,AgNO_3(aq) + NO(g) + 2\,H_2O(l)$

K. $S(s) + 6\,HNO_3(aq) \rightleftharpoons H_2SO_4(aq) + 6\,NO_2(g)$

L. calcium carbonate(s) $\rightleftharpoons$ calcium oxide(s) + carbon dioxide(g)

M. $Mg_3N_2(s) + 6\,H_2O(l) \rightleftharpoons 3\,Mg(OH)_2(s) + 2NH_3(g)$

6. (1) Use the data from the drill problems for Objective 15-2 to determine the value of the equilibrium constant K_c for each of the following reactions.

A.
$CH_4(g) + CCl_4(g) \rightleftharpoons$
$2\,CH_2Cl_2(g)$ at 298 K

B. $PCl_5(g) \rightleftharpoons PCl_3(g) + Cl_2(g)$ at 600 K

C. $HCHO(g) \rightleftharpoons H_2(g) + CO(g)$ at 773 K

D.
$2\,NO(g) + Br_2(g) \rightleftharpoons 2\,NOBr(g)$ at 350 K

E..
$3H_2(g) + N_2(g) \rightleftharpoons 2\,NH_3(g)$ at 473 K

(2) The equilibrium pressure of each species is written below its formula. Evaluate K_p in each case.

F. $2\,HI(g) \rightleftharpoons H_2(g) + I_2(g)$ at 298 K
0.400 atm 0.0136 atm 0.0136 atm

G. $CO(g) + 2\,H_2(g) \rightleftharpoons CH_3OH(g)$ at 298 K
0.300 atm 0.300 atm 554 atm

H. $SO_2Cl_2(l) \rightleftharpoons SO_2(g) + Cl_2(g)$
0.00442 atm 0.00442 atm 0.00442 atm

I. $H_2(g) + CO_2(g) \rightleftharpoons H_2O(g) + CO(g)$ at 350 K
1.00 atm 1.00 atm 0.0110 atm 0.0110 atm

7. The initial pressure or concentration is given below the formula of each species in each of the following equations. Evaluate Q_P or Q_c; compare the value obtained with that given for K_c or ; and predict whether the reaction will shift left, shift right, or is at equilibrium.

$4\,HCl(g) + O_2(g) \rightleftharpoons 2H_2O(g) + 2Cl_2(g)$ $K_c = 1.49\times10^3$ at 320 K
A. 0.100 M 0.0200 M 0.122 M 0.200 M
B. 0.100 M 0.103 M 0.409 M 0.010 M

$CH_3ONO(g) + HCl(g) \rightleftharpoons CH_3OH(g) + NOCl(g)$ $K_c = 0.242$ at 323 K
C. 0.104 M 0.246 M 0.115 M 0.0941 M
D. 0.346 M 0.102 M 0.212 M 0.198 M

$CH_4(g) + CCl_4(g) \rightleftharpoons 2\,CH_2CL_2(g)$ $K_c = 0.0956$ at 298 K
E. 0.1060 M 0.02075 M 0.01450 M
F. 0.207 M 0.0156 M 0.0142 M

$2\,NOBr(g) \rightleftharpoons 2\,NO(g) + Br_2(g)$ $K_c = 0.1563$ at 350 K
G. 0.396 M 0.214 M 0.289 M
H. 0.204 M 0.311 M 0.149 M

$3\,H_2(g) + N_2(g) \rightleftharpoons 2\,NH_3(g)$ $K_c = 1.40\times10^5$ at 473 K
I. 0.0163 M 0.214 M 0.270 M
J. 0.163 M 0.0214 M 0.0270 M

$PCl_3(g) + Cl_2(g) \rightleftharpoons PCl_5(g)$ $K_C = 1.784$ at 600 K
K. 0.100 M 0.100 M 0.173 M
L. 0.204 M 0.306 M 0.104 M

$N_2O_4(g) \rightleftharpoons 2\,NO_2(g)$ $K_p = 0.113$ atm at 298 K
M. 0.260 atm 0.152 atm
N. 0.115 atm 0.214 atm

$Br_2(g) + Cl_2(g) \rightleftharpoons 2\,BrCl(g)$ $K_p = 6.30$ at 350 K
O. 0.216 atm 0.216 atm 0.304 atm
P. 0.143 atm 0.157 atm 0.424 atm

$PCl_5(g) \rightleftharpoons PCl_3(g) + Cl_2(g)$ $K_p = 0.675$ at 500 K
Q. 0.165 atm 0.243 atm 0.114 atm
R. 0.0464 atm 0.302 atm 0.147 atm

$H_2(g) + CO_2(g) \rightleftharpoons H_2O(g) + CO(g)$ $K_p = 0.00831$ at 500 K
S. 0.206 atm 0.206 atm 0.114 atm 0.114 atm

T. 0.814 atm 0.514 atm 0.0621 atm 0.0214 atm

$SO_2Cl_2(l) \rightleftharpoons SO_2(g) + Cl_2(g)$ $K_p = 2.69$ at 360 K

U. 1.76 atm 0.423 atm

V. 1.70 atm 1.70 atm

8. Evaluate for each of the following reactions. The pressure written below a species is its equilibrium pressure. *P* is the total equilibrium pressure.

A. $N_2O_4(g) \rightleftharpoons 2\,NO_2(g)$ at 298 K
1.58 atm $P = 2.00$ atm

B. $2\,HI(g) \rightleftharpoons H_2(g) + I_2(g)$ at 320 K
0.464 atm $P = 1.000$ atm and $P_{H2} = P_{I2}$

C. $CO(g) + 2\,H_2(g) \rightleftharpoons CH_3OH(g)$ at 350 K
0.254 atm 0.254 atm $P = 2.000$ atm

D. $Br_2(g) + Cl_2(g) \rightleftharpoons 2\,BrCl(g)$ at 298 K
$P_{Br2} = P_{Cl2}$ 0.543 atm $P = 2.00$ atm

E. $SO_2Cl_2(l) \rightleftharpoons SO_2(g) + Cl_2(g)$E at 360 K
0.277 atm $P = 10.00$ atm

F. $H_2(g) + CO_2(g) \rightleftharpoons H_2O(g) + CO(g)$ at 500 K
0.916 atm 0.916 $P_{H2O} = P_{CO}$ $P = 2.000$ atm

G. $PCl_5(g) \rightleftharpoons PCl_3(g) + Cl_2(g)$ at 500 K
0.551 0.551 atm $P = 1.552$ atm

9. Each of the following reacting systems initially is at equilibrium. The indicated changes are made individually on this equilibrium system. Predict whether the system will shift right (R), shift left (L), or be unchanged (U) in regaining equilibrium. Write the appropriate letter in each blank.

$HCHO(g) \rightleftharpoons CO(g) + H_2(g)$ $\Delta H° = +5.36$ kJ/mol

A. increase temperature _____ B. add 1.00 mol CO(g)____

C. remove 0.50 mol $H_2(g)$____ D. decrease volume _____

$NH_3(g) + CH_3COOH(l) \rightleftharpoons NH_2CH_2COOH(s) + H_2(g)$ $\Delta H° = -6.90$ kJ/mol

E. add 1.00 mol $CH_3COOH(l)$____ F. increase pressure____

G. increase temperature ____ H. remove 0.010 mol $H_2(g)$____

$2\,NOCl(g) \rightleftharpoons 2\,NO(g) + Cl_2(g)$ $\Delta H° = +71.38$ kJ/mol

I. increase temperature ____ J. add 1.00 mol $Cl_2(g)$____

K. increase volume ____ L. increase pressure _____

$S(s) + 2\,CO(g) \rightleftharpoons SO_2(g) + 2\,C(s)$ $\Delta H° = -75.81$ kJ/mol

M. increase pressure ____ N. decrease temperature ____

O. remove 0.0024 mol S(s) ____ P. add 0.0024 mol $SO_2(g)$____

10. Use the data from the drill problems for Objective 15-9. Determine the equilibrium concentrations or partial pressures for each part except A, B, G, H, I, and J.

In the two additional problems that follow, the initial concentration is given below the formula of each species.

$HCHO(g)$	$\rightleftharpoons$	$H_2(g)$	+	$CO(g)$	$K_c = 0.267$ at 773 K
W. 0.200 M		0.100 M		0.100 M	
X. 0.400 M		0.000 M		0.000 M	

11. For eight of the equilibrium mixtures described in the drill problems for Objective 15-3, the total volume (V) of the equilibrium system is varied (by expansion, ×, or compression, ÷) as specified below. Determine the direction in which the reaction shifts and the new equilibrium value of the indicated concentration.

J. $V \times 2, [HCl]$ K. $V \div 4, [NOCl]$ L. $V \times 5$, $[CH_4]$ M. $V \div 3, [CH_2Cl_2]$

R. $V \times 3, [Br_2]$ S. $V \div 5$ T. $V \times 4, [Cl_2]$ U. $V \div 3$, $[PCl_3]$

QUIZZES (15 minutes each) For each quiz, answer the first question. Then choose the best answer for each of the remaining questions.

QUIZ A

1. Write the equilibrium constant expression K_c for the reaction
   sodium sulfite(aq) + chloric acid(aq) $\rightleftharpoons$
   sodium chlorite(aq) + sulfur dioxide(g) + water(l)

2. For the reaction $PCl_3(g) + ICl(g) \rightleftharpoons PCl_4I(g)$, initially there are 1.20 mol PCl_3, 0.48 mol ICl, and 1.26 mol PCl_4I. If there is 1.06 mol PCl_3 present at equilibrium, how many moles of PCl_4I are then present? (a) 1.33; (b) 1.54; (c) 1.68; (d) 1.44; (e) none of these.

3. For $I_2(s) + 4\,Cl_2(g) \rightleftharpoons 2\,ICl(g), K_c = [ICl]^2/[Cl_2]$. At equilibrium there is 0.300 mol Cl_2, 0.100 mol I_2, and 2.0 mol ICl in a 5.0-L container. What is the value of K_c? (a) 0.075; (b) 2.7; (c) 0.0075; (d) 133; (e) none of these.

4. $CH_4(g) + 4\,Cl_2(g) \rightleftharpoons CCl4(l) + 4\,HCl(g)$ $\Delta H° = -397$ kJ/mol. The equilibrium represented by this equation is displaced to the *right* if (a) the temperature is raised; (b) the pressure is lowered; (c) some carbon tetrachloride is removed; (d) some hydrogen chloride is removed; (e) none of these.

5. $N_2(g) + O_2(g) \rightleftharpoons 2\,NO(g)$ $\Delta H° = +21$ kJ/mol. The equilibrium represented by this equation is displaced to the *right* if (a) the temperature is raised; (b) the pressure is lowered; (c) some nitrogen is removed; (d) some nitrogen monoxide is added; (e) none of these.

6. Which factor influences the value of the equilibrium constant for a reversible reaction? (a) addition of a catalyst; (b) raising the temperature; (c) removing product; (d) removing reactant; (e) raising the pressure.

7. For the reaction $SbCl_5(g) \rightleftharpoons SbCl_3(g) + Cl_2(g)$, K_c equals (a) K_c; (b) RT/K_p; (c) $K_P(RT)$; (d) K_P/RT; (e) none of these.

QUIZ B

1. Write the equilibrium constant expression K_p for the reaction $SCl_4(l) + O_2(g) \rightleftharpoons SO_2(g) + 2\,Cl_2(g)$.

2. 10.0 mol NO and 26.0 mol O_2 are placed in an empty container. At equilibrium for >, there is 8.8 mol NO_2 present. How many moles of O_2 are present at equilibrium? (a) 17.2; (b) 8.4; (c) 21.6; (d) 4.4; (e) none of these.

3. For $CO_2(g) + H_2(g) \rightleftharpoons CO(g) + H_2O(g)$, $K_c = [CO][H_2O]/[CO_2][H_2]$. If there are 1.43 mol each of CO and H_2O, 0.572 mol H_2, and 4.572 mol CO_2 in a 4.00-L container at equilibrium, what is the value of K_c? (a) 0.547; (b) 1.27; (c) 0.782; (d) 0.137; (e) none of these.

4. $N_2H_6CO_2(s) \rightleftharpoons 2\,NH_3(g) + CO_2(g)\ \Delta H° = +33\ kJ/mol$. The equilibrium represented by this equation is displaced to the *left* if (a) the pressure is lowered; (b) the temperature is lowered; (c) some $N_2H_6CO_2$ is removed; (d) some ammonia is removed; (e) none of these.

5. In the equilibrium $2\,O_3(g) \rightleftharpoons 3\,O_2(g)$, one can conclude that (a) high temperature favors ozone formation; (b) low pressure favors ozone formation; (c) pressure does not affect the equilibrium position; (d) ozone decomposes rapidly into dioxygen; (e) none of these.

6. Which of these cannot shift the position of a reversible reaction at equilibrium? (a) concentration change; (b) pressure change; (c) temperature increase; (d) addition of reactant; (e) addition of heterogeneous catalyst.

7. For the reaction $CS_2(l) + 3\,O_2(g) \rightleftharpoons CO_2(g) + 2\ SO_2(g)$, K_c equals (a) K_p; (b) RT/K_P; (c) $K_p(RT)$; (d) K_p/RT; (e) none of these

QUIZ C

1. Write the equilibrium constant expression K_c for the reaction $3\,Cu(s) + 8\,HNO_3(aq) \rightleftharpoons 3\,Cu(NO_3)_2(aq) + 2\,NO(g) + 4\,H_2O(l)$.

2. 0.300 mol NO, 0.200 mol Cl_2, and 0.500 mol NOCl are placed in a 25.0-L vessel. At equilibrium for $2\,NO(g) + Cl_2(g) \rightleftharpoons 2\,NOCl(g)$, there is 0.600 mol NOCl present. What amount in moles of Cl_2 is present at equilibrium? (a) 0.200; (b) 0.050; (c) 0.100; (d) 0.150; (e) none of these.

3. For the formation of NO, $K_c = [NO_2]^2/[NO]^2[O_2]$. At equilibrium in a 2.50-L container, there are 3.00 mol NO, 4.00 mol O_2, and 22.0 mol NO_2. The value of K_c is (a) 13.4; (b) 33.6; (c) 5.38; (d) 0.0116; (e) none of these.

4. $CO_2(g) + 2\,HCl(g) \rightleftharpoons Cl_2CO(g) + H_2O(g)$. The position of equilibrium of this exothermic reaction shifts to the *left* if (a) catalyst is removed; (b) pressure is increased; (c) carbon dioxide is removed; (d) temperature is lowered; (e) none of these.

5. All of the following may shift the position of a reaction at equilibrium except (a) concentration increase; (b) pressure increase; (c) temperature increase; (d) addition of product; (e) addition of a homogeneous catalyst.

6. In a reaction at equilibrium involving only gases, a change in pressure of the reaction mixture shifts the position of equilibrium only when (a) heat is absorbed by the reaction proceeding to the right; (b) the gases are impure; (c) the collision rate increases; (d) the reaction is exothermic as written; (e) none of these.

7. For the reaction $2\,NO(g) \rightleftharpoons N_2O_4(g)$, K_p equals (a) K_c; (b) RT/K_c; (c) $K_c(RT)$; (d) K_c/RT; (e) none of these.

QUIZ D

1. Write the equilibrium constant expression K_p for the reaction $CaCO_3(s) + SO_2(g) \rightleftharpoons CaSO_3(s) + CO_2(g)$.

2. In the reaction $2\,N_2O(g) + N_2H_4(g) \rightleftharpoons 3\,N_2(g) + 2\,H_2O(s)$, 0.100 mol N_2O and 0.250 mol N_2H_4 are placed in a 10.0-L container. If there is 0.060 mol N_2O present at equilibrium, how many moles of N_2 are then present? (a) 0.090; (b) 0.040; (c) 0.060; (d) 0.020; (e) none of these.

3. For the reaction of PBr_3, with $Br_2(g)$, $K_c = [PBr_5]/[PBr_3][Br_2]$. At equilibrium in an 8.00-L vessel there are 1.60 mol PBr_3, 0.800 mol Br_2, and 3.20 mol PBr_5. What is the value of K_c? (a) 2.50; (b) 20.0; (c) 0.313; (d) 0.400; (e) none of these.

4. Consider the equilibrium reaction $N_2(g) + O_2(g) \rightleftharpoons 2\,NO(g)\ \Delta H° = +192\ kJ/mol$. At 2000 K, the value of the equilibrium constant is 5.0×10^{-4}. At 2500 K, the value of the equilibrium constant (a) is greater than 5.0×10^{-4}; (b) is less than 5.0×10^{-4}; (c) is 5.0×10^{-4}; (d) depends on [NO]; (e) depends on $[O_2]$.

5. $S_2Cl_2(l) + CCl_4(g) \rightleftharpoons CS_2(g) + 3\,Cl_2(g)\ \Delta H° = +84.31\ kJ/mol$. The equilibrium represented by this equation is displaced to the right if (a) some carbon tetrachloride is removed; (b) the temperature is lowered; (c) the pressure is raised; (d) some carbon disulfide is removed; (e) none of these.

6. For the reaction $2\,N_2O(g) + N_2H_4(g) \rightleftharpoons 3\,N_2(g) + 2\,H_2O(g)$, K_p equals (a) K_c (b) RT/K_c; (c) $K_c(RT)^2$; (d) K_c/RT; (e) none of these.

7. The system $2\,H_2O(g) \rightleftharpoons 2\,H_2(g) + O_2(g)$ is at equilibrium when (a) $[H_2]/[O_2] = 2$; (b) $[H_2O]$ decreases to half of its initial value; (c) $[H_2O]/([H_2]+[O_2]) = 2/3$; (d) H_2O molecules no longer decompose; (e) the total pressure remains constant.

SAMPLE TEST (20 minutes)

1. At 85 °C, $K_p = 1.19$ for $PCl_5(g) \rightleftharpoons PCl_3(g) + Cl_2(g)$. What is the partial pressure of $PCl_5(g)$ at equilibrium if the initial pressures are 2.00 atm PCl_3, 1.00 atm Cl_2, and 0.00 atm PCl_5?

2. At 773 K, for the reaction $HCHO(g) \rightleftharpoons H_2(g) + CO(g)$, there are 0.900 mol HCHO(g), 0.000 mol $H_2(g)$, and 0.600 mol CO(g) present initially in a 3.00-L container. At equilibrium, there is 0.500 mol HCHO(g) present in this same container. If the container that holds this equilibrium mixture is now compressed to 1.00 L, what will $[H_2]$ be after equilibrium is reestablished?

3. Predict whether each of the following actions will cause the equilibrium system to shift left, shift right, or remain unchanged.

$AgCl(s) \rightleftharpoons Ag^+(aq) + Cl^+(aq)$	
A. Add solid $AgNO_3$	B. Add solid AgCl
$2\,NO(g) + O_2(g) \rightleftharpoons 2\,NO_2(g)$	$\Delta H°_{1000} = -116.9$ kJ/mol
C. Increase $[O_2]$	D. Add NO(g)
E. Increase total pressure by compression	F. Raise temperature
$CaCO_3(s) \rightleftharpoons CaO(s) + CO_2(g)$	(endothermic)
G. Remove $CO_2(g)$	H. Decrease total pressure by expansion
I. Add CaO(s)	J. Add catalyst

16 ACIDS AND BASES

CHAPTER OBJECTIVES

1. Describe the similarities and differences among the Arrhenius, Brønsted-Lowry, and Lewis theories of acids and bases.

Arrhenius acid-base theory defines an acid as a substance that produces hydrogen ions (H^+) when it ionizes in water. A base produces hydroxide ions (OH^-). All *strong* acids and bases obey the Arrhenius definition. (Recall that the common strong acids and bases are listed in Table 5-1). The ionization of a typical acid is shown in Equation [1] and a base in Equation [2].

$$HCl \longrightarrow H^+(aq) + Cl^-(aq) \qquad [1]$$

$$NaOH \longrightarrow Na^+(aq) + OH^-(aq) \qquad [2]$$

The Arrhenius theory does not explain the behavior of some common weak bases in aqueous solution or reactions that seem quite like acid-base reactions but occur in other solvents. The Brønsted-Lowry theory explains most of these cases. In terms of this theory, an acid is a *proton donor* and a base is a *proton acceptor*. Typical reactions are shown in Equations [3] and [4], in which a proton (H^+) is donated and accepted.

$$H_2O(l) + HCN(aq) \longrightarrow CN^-(aq) + H_3O^+(aq) \qquad [3]$$

$$NH_3(aq) + H_2O(l) \longrightarrow OH^-(aq) + NH_4^+(aq) \qquad [4]$$

In each equation, the base (proton acceptor) appears first on each side, and the acid (proton donor) appears second. The term *conjugate* is used to describe species that differ by only a proton (H^+). Consider Equation [3]. $CN^-(aq)$ is the conjugate base of HCN(aq), HCN(aq) is the conjugate acid of $CN^-(aq)$, *and* HCN(aq) and $CN^-(aq)$ are a conjugate acid-base pair. (The term *conjugate* is used in a similar way to the word *twin*: Tom is Jim's twin, Jim is Tom's twin, *and* Tom and Jim are twins.) All Arrhenius acids and bases also are Brønsted-Lowry acids and bases.

There remain some acid-base reactions that do not involve proton transfer. For example, Equation [5] is an acid-base reaction; thus, the reaction in [6] between the two anhydrides CaO(s) and $SO_3(g)$ should be also.

$$Ca(OH)_2(aq) + H_2SO_4(aq) \longrightarrow CaSO_4(aq) + 2\,H_2O(l) \qquad [5]$$

$$CaO(s) + SO_3(g) \longrightarrow CaSO_4(s) \qquad [6]$$

Yet Equation [6] cannot be interpreted by either the Arrhenius or the Brønsted-Lowry theory, since there are no protons (H^+ ions) present. (*Anhydride* means literally "no water." SO_3 is the anhydride of H_2SO_4 because H_2SO_4 can be made by adding H_2O to SO_3. Note that *hydroxide* was derived from "hydrated oxide." Thus, calcium oxide, CaO, is the anhydride of calcium hydroxide.) Reactions like [6] are explained by Lewis acid-base theory, which defines an acid as an electron-pair acceptor (SO_3 in Equation [6]) and a base as an electron-pair donor (the oxide ion of CaO in Equation [6]). Lewis acid-base reactions result in the formation of a covalent bond between the acid and the base. The donated (and accepted) pair of electrons forms the bond—a coordinate covalent bond. The acid-base reaction is more easily seen once we draw Lewis structures.

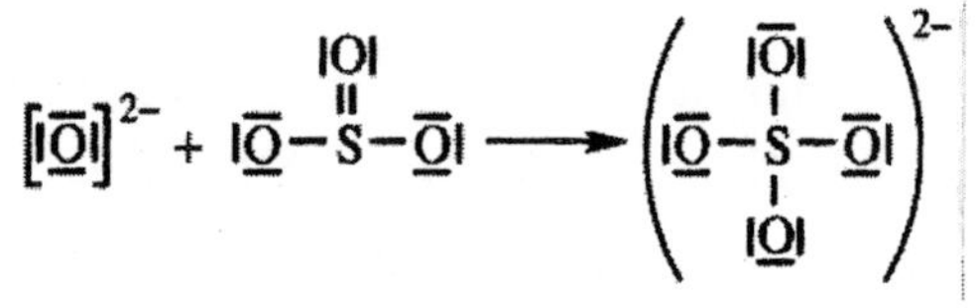

$$\left[\overline{\underline{O}}\right]^{2-} + \ |\overline{\underline{O}} - \overset{\displaystyle |\overline{O}|}{\underset{}{\overset{\displaystyle \|}{S}}} - \overline{\underline{O}}| \longrightarrow \left(\ |\overline{\underline{O}} - \overset{\displaystyle |\overline{O}|}{\underset{\displaystyle |\underline{O}|}{\overset{\displaystyle |}{\underset{\displaystyle |}{S}}}} - \overline{\underline{O}}|\ \right)^{2-} \qquad [7]$$

*** 2. Identify Brønsted-Lowry conjugate acids and bases; write equations of acid-base reactions.**

We discussed methods of identifying acids and bases in Objective 5-1. With regard to the Brønsted-Lowry theory, there always are four species in a Brønsted-Lowry equation: a base and an acid as reactants and the conjugate base and acid as products. Equation [4] is typical.

$$NH_3(aq) + H_2O(l) \rightleftharpoons OH^-(aq) + NH_4^+(aq) \qquad [4]$$

In the forward reaction, $NH_3(aq)$ is a base because it accepts a proton to become $NH_4^+(aq)$. $H_2O(l)$ is an acid because it donates a proton to $NH_3(aq)$. If the reverse reaction is considered, then $OH^-(aq)$ is a base because it accepts a proton to become $H_2O(l)$. $H_2O(l)$ and $OH^-(aq)$ are another conjugate acid-base pair. In the reverse reaction, $NH_4^+(aq)$ is an acid because it donates a proton to $OH^-(aq)$ and becomes $NH_3(aq)$. $NH_4^+(aq)$ and $NH_3(aq)$ are the other conjugate acid-base pair in this reaction. Writing acid-base reactions is as easy as it seems. Do not try so hard that you make it difficult. All that happens is that H^+ is being transferred: the conjugate acid loses H and its charge becomes one unit less positive; the conjugate base gains H and its charge becomes one unit more positive. An amphiprotic species can act as either a proton donor (an acid) or as a proton acceptor (a base). Water is amphiprotic: H_2O is an acid in Equation [4] and a base in Equation [3].

*** 3. Identify Lewis acid-base reactions and write equations for acid-base reactions that involve them.**

To be able to identify Lewis acids and bases, you must be able to draw Lewis structures. (Review Objectives 10-2, 10-3, and 10-6 through 10-10, if necessary.) A Lewis base *must* have at least one lone pair of electrons; this pair will be donated. A Lewis acid will either be an electron-deficient species or one that can have an "expanded octet." When dealing with an ionic compound (such as CaO in Equation [6]), consider each ion separately. In the case of CaO, O^{2-} is the Lewis base. Sometimes O^{2-} is called the *primary* Lewis base and CaO is called the *secondary* Lewis base. Also, review the last paragraph of Objective 17-1.

4. Explain what *self-ionization* (or *autoionization*) is, and describe the nature of the proton in aqueous solution.

Autoionization is the reaction in which a compound, that may be thought of as covalent, produces cations and anions. There are many examples.

$$2\,NH_3 \rightleftharpoons NH_4^+ + NH_2^- \qquad [8]$$

$$2\,SO_2 \rightleftharpoons SO^{2+} + SO_3^{2-} \qquad [9]$$

$$2\,HC_2H_3O_2 \rightleftharpoons H_2C_2H_3O_2^+ + C_2H_3O_2^- \qquad [10]$$

$$N_2O_4 \rightleftharpoons NO^+ + NO_3^- \qquad [11]$$

$$COCl_2 \rightleftharpoons COCl^+ + Cl^- \qquad [12]$$

$$2\,H_2O \rightleftharpoons H_3O^+(aq) + OH^-(aq) \qquad [13]$$

The equilibrium constant for reaction in [13] is given the symbol K_w and termed the *ion product* of water.

$$K_w = [H_3O^+][OH^-] = 1.0 \times 10^{-14} \text{ at } 25\,°C \qquad [14]$$

A free proton cannot exist in aqueous solution. It is associated strongly with one water molecule (H_3O^+) and weakly with three others ($H_3O^+, 3\,H_2O$, *or* $H_9O_4^+$). The hydroxide ion (OH^-) is similarly associated with three water molecules ($OH^-, 3\,H_2O$, *or* $H_7O_4^-$). But most chemists write the aqueous hydrogen ion as H_3O^+, often called the *hydronium ion*, and the aqueous hydroxide ion as OH^-.

*** 5. Calculate ion concentrations in solutions of strong electrolytes, and relate $[H_3O^+]$ and $[OH^-]$ through K_w.**

Strong electrolytes ionize completely in water. This includes strong acids and bases and all salts. When 0.0200 moles of NaCl dissolve to form a liter of aqueous solution, that solution contains sodium ions and chloride ions with $[Na^+] = 0.0200\,M$ and $[Cl^-] = 0.0200\,M$. Likewise, when 1.5×10^{-4} mol $CaCl_2$ forms a liter of solution, $[Ca^{2+}] = 1.5 \times 10^{-4}\,M$ and $[Cl^-] = 3.0 \times 10^{-4}\,M$. In similar fashion, a 0.0200 M NaOH solution has $[Na^+] = 0.0200\,M$ and $[OH^-] = 0.0200\,M$. Furthermore, Equation [14] relates $[OH^-]$ and $[H_3O^+]$. Since we know $[OH^-] = 0.0200\,M$, we can compute $[H_3O^+]$: $[H_3O^+](0.0200\,M) = 1.0 \times 10^{14}$ *or* $[H_3O^+] = 5.00 \times 10^{-11}\,M$

EXAMPLE 16-1 What is $[H_3O^+]$ in a 1.50×10^{-4} M solution?

$$1.50\times10^{-4}\ \text{M Ca(OH)}_2 \times \frac{2\ \text{mol OH}^-}{1\ \text{mol Ca(OH)}_2} = 3.00\times10^{-4}\ \text{M} = [\text{OH}]^-$$

$$[H_3O^+][OH^-] = [H_3O^+](3.00\times10^{-4}\ \text{M}) = 1.0\times10^{-14} \quad \textit{or} \quad [H_3O^+] = 3.33\times10^{-9}\ \text{M}$$

*** 6. Given a value of any one of $[H_3O^+]$, $[OH^-]$, pH, and pOH, be able to compute values of the other three.**

Four expressions can be used to express the acidity of a solution as $[H_3O^+]$, $[OH^-]$, pH, and pOH. They are: Equation [14], which relates $[H_3O^+]$ and $[OH^-]$; the definitions of pH and pOH, Equations [15] and [16]; and the relationship between pH and pOH at $25\,^\circ\text{C}$, Equation [17].

$$K_w = [H_3O^+][OH^-] = 1.0\times10^{-14}\ \text{at}\ 25\,^\circ\text{C} \qquad [14]$$

$$\text{pH} = -\log\,[H_3O^+] \qquad [15]$$

$$\text{pOH} = -\log\,[OH^-] \qquad [16]$$

$$\text{pH} + \text{pOH} = 14.00 \qquad [17]$$

EXAMPLE 16-2 A solution with $[OH^-] = 2.50\times10^{-5}$ M has what pH, pOH, and $[H_3O^+]$?

pOH is most readily determined:

$$\text{pOH} = -\log[OH^-] = -\log(2.50\times10^{-5}) = 4.602.$$

Then we determine pH:

$$\text{pH} + \text{pOH} = 14.00 = \text{pH} + 4.602 \quad \textit{or} \quad \text{pH} = 9.40$$

Finally, we can find $[H_3O^+]$ in two ways:

$$[H_3O^+] = 10^{-\text{pH}} = 10^{-9.40} = 4.0\times10^{-10}\ \text{M}$$

or

$$[H_3O^+][OH^-] = 1.0\times10^{-14} = [H_3O^+](2.50\times10^{-5}\ \text{M})$$

$$[H_3O^+] = 4.00\times10^{-10}\ \text{M}$$

*** 7. Identify a weak acid or base, write a chemical equation to represent its ionization, and set up its ionization constant expression.**

A common acid or base that is not strong (Table 5-1) must be weak. It's as simple as that. (Recall Objective 5-1 for identifying weak acids and bases. Also, remember that *un*common strong acids and bases will be clearly identified as such.) *Weak* means not completely ionized. Some of the dissolved acid or base is present in molecular form. The ionization equations for all weak acids in water involve the acid donating a proton to a water molecule, forming H_3O^+ and the anion of the acid.

$$HCN(aq) + H_2O(l) \rightleftharpoons H_3O^+(aq) + CN^-(aq) \qquad [18]$$

$$HF(aq) + H_2O(l) \rightleftharpoons H_3O^+(aq) + F^-(aq) \qquad [19]$$

The ionization constant expression is written in the same way that *all* equilibrium constant expressions are written: concentrations of products in the numerator and those of reactants in the denominator. The concentration of the solvent (water) does not appear in the expression.

$$K_a = \frac{[H_3O^+][CN^-]}{[HCN]} \qquad K_a = \frac{[H_3O^+][F^-]}{[HF]} \qquad [20]$$

Ionization equations of weak bases in water involve the base accepting a proton from H_2O, producing OH^- and the cation of the base (the protonated original base). Note that many weak bases are derived from ammonia, NH_3.

$$NH_3(aq) + H_2O(l) \rightleftharpoons OH^-(aq) + NH_4{}^+(aq) \qquad [21]$$

$$N_2H_4(aq) + H_2O(l) \rightleftharpoons OH^-(aq) + N_2H_5{}^+(aq) \qquad [22]$$

Again, the ionization constant expressions are written in the usual manner.

$$K_b = \frac{[NH_4{}^+][OH^-]}{[NH_3]} \qquad K_b = \frac{[N_2H_5{}^+][OH^-]}{[N_2H_4]} \qquad [23]$$

The ionization constant of a weak acid usually is symbolized K_a, and called the weak acid ionization constant; sometimes the more general symbol K_i (ionization constant) is used. A similar substitution may be made for K_b.

*** 8. Calculate one of K_a, $[H_3O^+]$, or the molarity of a weak acid, given the other two (and perform similar calculations for a weak base). Know how to simplify these calculations by making suitable approximations.**

The ionization equation for a weak acid is

$$HA(aq) + H_2O(l) \rightleftharpoons H_3O^+(aq) + A^-(aq) \qquad [24]$$

For Equation [24], the ionization constant expression is

$$K_a = \frac{[H_3O^+][CN^-]}{[HCN]} \qquad K_a = \frac{[H_3O^+][F^-]}{[HF]} \qquad [25]$$

If HA is the only solute, then it is the only source of $A^-(aq)$ ions.

$$[H_3O^+] = [A^-] \qquad [26]$$

We use these expressions in three types of problem:

(1) *Determining* K_a *from* $[H^+]$ *and [HA].* A common way to determine K_a experimentally is to measure the pH of a solution of known initial acid concentration. The initial concentration of acid (or base) sometimes is referred to as the *formality* since it is based on how the solution is made up or *formulated.*

EXAMPLE 16-3 A 0.10 M solution of HN_3 has pH = 2.86. What is K_a for this acid?

We first write the ionization equation and the ionization constant expression:

$$HN_3(aq) + H_2O(aq) \rightleftharpoons H_3O^+(aq) + N_3^-(aq) \quad K_a = \frac{[H_3O^+][N_3^-]}{[HN_3]}$$

Then, we determine $[H_3O^+] = 10^{-pH} = 10^{-2.68} = 1.4 \times 10^{-3}$ M.

Next, we write the concentration of each species below its formula: that before ionization (Initial = initially), the change in concentration needed to reach equilibrium (Changes), and that at equilibrium (Equil). (Recall that this is the technique we first employed in Objective 15-12 to set up other equilibrium problems.)

Equation:	$HN_3(aq)$	+ $H_2O(l)$	$\rightleftharpoons$	$H_3O^+(aq)$	$+N_3^-(aq)$
Initial:	0.101 M				
Changes:	-0.0014 M			+0.0014 M	+0.0014 M
Equil:	0.100 M			0.0014 M	0.0014 M

Then use the equilibrium concentrations to compute K_a.

$$K_a = \frac{[H_3O^+][N_3^-]}{[HN_3]} = \frac{(0.0014)(0.0014)}{0.100} = 2.0 \times 10^{-5}$$

(2) *Determining* $[H_3O^+]$ *from* K_a *and [HA].* K_a and the initial acid concentration are given and $[H_3O^+]$ is to be computed. The setup is like that of Example 16-3, with $[H_3O^+] = x$.

EXAMPLE 16-4 What is $[H_3O^+]$ in a solution with $[HF] = 0.500$ M? $K_a = 6.6 \times 10^{-4}$.

Again, the setup is based on the ionization equation. A benefit of using the same type of setup each time is that you can concentrate on the new and different aspects of the problem and spend little of your effort on routine matters.

Equation:	HF(aq)	$+H_2O(l)$	$\rightleftharpoons$	$H_3O^+(aq)$	$+F^-(aq)$
Initial:	0.500 M				
Changes:	$-x$ M			$+x$ M	$+x$ M

Equil: $(0.500 - x)\,\text{M}$ $\quad x\,\text{M}$ $\quad x\,\text{M}$

Then substitute into the K_a

expression: $K_a = \dfrac{[H_3O^+][F^-]}{[HF]} = 6.6\times10^{-4} = \dfrac{(x)(x)}{0.500 - x}$

There are two ways to solve this equation. The first uses the quadratic equation (Objective 14-8). The second uses an approximation. With a small equilibrium constant, as is the case for most weak acids and weak bases, we expect very little product. Thus, we assume $x \ll 0.500$ *or* $0.500 - x \approx 0.500$. This gives

$$6.6\times10^{-4} \approx \frac{x^2}{0.500} \quad \text{or} \quad x = 1.8\times10^{-2}\ \text{M} = 0.018\ \text{M}$$

0.018 M may not seem small compared to 0.500 M. Let us substitute into the original expression for K_a:

$$K_a = \frac{(0.018)(0.018)}{0.500 - 0.018} = 6.7\times10^{-4}$$

We see that $[H_3O^+] = 0.018$ M indeed satisfies the original equation.

A convenient rule of thumb is that x is small compared to the original acid concentration. [HA], if x is less than 5% of [HA]. Of course, you can only be certain that x is small enough by substituting into the original equation. The approximation is worth trying if the acid concentration is at least 100 times larger than K_a. This is a cautious guideline and can be ignored if you are willing to learn a simple technique, as is illustrated in Example 16-5.

EXAMPLE 16-5 What is $[H_3O^+]$ in a $HClO_2$ solution? $K_a = 1.2\times10^{-2}$ for $HClO_2$.

Equation: $HClO_2(aq) + H_2O(l) \rightleftharpoons H_3O^+(aq) + ClO_2^-(aq)$

$$K_a = \frac{[H_3O^+][ClO_2^-]}{[HClO_2]} = \frac{(x)(x)}{0.50 - x}$$

Initial:	0.50 M		
Changes:	$-x$ M	$+x$ M	$+x$ M
Equil:	$(0.50 - x)$ M	x M	x M

$$= 1.2\times10^{-2} \approx \frac{x^2}{0.50}$$

We have disregarded the guideline cited above and assumed $x \ll 0.50$:

$$x^2 = (1.2\times10^{-2})(0.50) \quad \text{or} \quad x = 0.077\,\text{M}$$

Obviously 0.077 M is not small compared to 0.50 M. However, we can use it as a first guess:

$$\frac{(x)(x)}{0.50 - 0.077} \approx \frac{x^2}{0.42} \approx 1.2\times10^{-2} \quad \text{or} \quad x = 0.071\ \text{M}$$

Repeat the process: $\dfrac{(x)(x)}{0.50 - 0.071} \approx \dfrac{x^2}{0.43} \approx 1.2\times10^{-2}$ or $x = 0.072$ M

These last two values (0.071 M and 0.072 M) agree, for an answer of $[H_3O^+] = 0.072$ M.

This technique is known as the method of successive approximations. It is quite time-saving (especially with an electronic calculator) and much easier than solving the quadratic equation. Note that when two successive approximations agree to two significant figures (the same precision as the K_a value) you should stop and use the last value. (See the first part of Example 16-8 for another illustration of the method.)

(3) *Determining [HA] from* K_a *and* $[H^+]$. This last type of calculation determines the concentration of a given weak acid needed to produce a solution of a certain pH.

EXAMPLE 16-6 What HNO_2 ($K_a = 7.2 \times 10^{-4}$) molarity is needed to produce a solution with $pH = 2.20$?

We first find $[H_3O^+] = 10^{-pH} = 10^{-2.20} = 6.3 \times 10^{-3}$ M.

Equation:	$HNO_2(aq) + H_2O(l)$ $\rightleftharpoons$	$H_3O^+(aq)$ +	$NO_2^-(aq)$
Initial:	x M		
Changes:	-0.0063 M	$+0.0063$ M	$+0.0063$ M
Equil:	$(x - 0.0063)$ M	0.0063 M	0.0063 M

$$K_a = \frac{(0.0063)^2}{x - 0.0063} = 7.2 \times 10^{-4} \qquad \text{OR} \quad x = 0.061 \text{ M} = HNO_2 \text{ molarity}$$

Often students can solve problems for weak acids but not for weak bases. Weak base problems are easier if you remember that $[OH^-]$ is as important for bases as $[H_3O^+]$ is for acids. A possible source of mistakes is that problems often furnish and request pH values, whereas pOH values are more "natural" for solutions of bases. Just remember to convert with Equation [17]: $pOH = 14.00 - pH$.

EXAMPLE 16-7 What is the pH of a 0.750 M $C_2H_5NH_2$ solution? $K_b = 4.4 \times 10^{-4}$ for $C_2H_5NH_2$.

Equation:	$C_2H_5NH_2(aq) + H_2O(l)$ $\rightleftharpoons$	$C_2H_5NH_3^+(aq)$ +	$OH^-(aq)$
Initial:	0.750 M		
Changes:	$-x$ M	$+x$ M	$+x$ M
Equil:	$(0.750 - x)$ M	x M	x M

$$K_b = \frac{[C_2H_5NH_3^+][OH^-]}{[C_2H_5NH_2]} = 4.4 \times 10^{-4} = \frac{(x)(x)}{0.750 - x} \approx \frac{x^2}{0.750}$$

We have assumed $x << 0.750\,\text{M}$. This gives $x = 1.8 \times 10^{-2}\,\text{M} = 0.018\,\text{M}$.

Substitution produces $\frac{x^2}{0.750 - x} = \frac{(0.018)^2}{0.750 - 0.018} = 4.4 \times 10^{-4}$. This agrees with the value of K_b. Thus, $x = [\text{OH}^-] = 0.018\,\text{M}$.

$\text{pOH} = -\log(0.018) = 174$ AND $\text{pH} = 14.00 - \text{pOH} = 14.00 - 1.74 = 12.26$

9. Describe the ionization of a polyprotic acid in aqueous solution and calculate the concentrations of the different species present in such a solution.

Polyprotic acids ionize in several steps, one step for each ionizable hydrogen in the acid.

$$H_2C_2O_4(aq) + H_2O(l) \rightleftharpoons H_3O^+(aq) + HC_2O_4^-(aq) \qquad K_{a1} = 5.4 \times 10^{-2}$$

$$HC_2O_4^-(aq) + H_2O(l) \rightleftharpoons H_3O^+(aq) + C_2O_4^{2-}(aq) \qquad K_{a2} = 5.4 \times 10^{-5}$$

Example 16-8 What is the concentration of each species present in a 1.00 M oxalic acid solution?

First, we solve the first ionization, that of $H_2C_2O_4(aq)$:

Equation:	$H_2C_2O_4(aq)$	+	$H_2O(l)$	$\rightleftharpoons$	$H_3O+(aq)$	+	$HC_2O_4^-(aq)$
Initial:	1.00 M						
Changes:	$-x$ M				$+x$ M		$+x$ M
Equil:	$(1.00 - x)$ M				x M		x M

$K_{a1} = \frac{[H_3O^+][HC_2O_4^-]}{[H_2C_2O_4]} = 5.4 \times 10^{-2} = \frac{(x)(x)}{1.00 - x} \approx \frac{x^2}{1.00}$. We assumed $x << 1.00$ and found $x = 0.232\,\text{M}$.

Clearly, our assumption is invalid. Successive approximations yield a final answer:

$\frac{x^2}{1.00 - 0.232} = 5.4 \times 10^{-2}$ $\qquad$ $\frac{x^2}{1.00 - 0.204} = 5.4 \times 10^{-2}$ $\qquad$ $\frac{x^2}{1.00 - 0.207} = 5.4 \times 10^{-2}$

or $x = 0.204\,\text{M}$ or $x = 0.207\,\text{M}$ or $x = 0.207\,\text{M}$

Then, the second ionization, that of HC_2O_4- (aq), is solved:

Equation:	$HC_2O_4-(aq)$ + $H_2O(l)$	$\rightleftharpoons$	$H_3O^+(aq)$	+	$C_2O_4^{2-}(aq)$
Initial:	0.207 M		0.207 M		
Changes:	$-y$ M		$+y$ M		$+y$ M
Equil:	$(0.207 - y)$ M		$(0.207 + y)$ M		y M

$$K_{a2} = \frac{[H_3O^+][C_2O_4^{2-}]}{[HC_2O_4^-]} = 5.4 \times 10^{-5} = \frac{(y)(0.207 + y)}{0.207 - y} \approx \frac{y(0.207)}{0.207}$$

We have assumed $y << 0.207$. This produces $y = 5.4 \times 10^{-5}$ M $= [C_2O_4^{2-}]$. Clearly our assumption is true. Notice that the small $[H_3O^+]$ produced in the second ionization does not significantly alter the $[H_3O^+]$ produced in the first step. Notice, in addition, that $[C_2O_4^{2-}] = K_{a2}$, a fact that is generally true for solutions of polyprotic acids. Then the remaining concentrations are obtained. $[C_2O_4^{2-}] = y = 5.4 \times 10^{-5}$ M

$[H_3O^+] = 0.207 + y = 0.207$ M $[HC_2O_4^-] = 0.207 - y$
$= 0.207$ M $[H_2C_2O_4] = 1.00 - x = 0.793$ M

***10. Predict which ions hydrolyze and whether salt solutions are acidic, basic, or neutral.**

All salts ionize completely when they dissolve in water (see Objectives 5-1 and 16-5). The anion of a strong acid does not react with water (Equation [27]). However, the anion of a weak acid accepts a proton from a water molecule to produce an acid molecule and a hydroxide ion (Equation [28]).

$$NO_3^-(aq) + H_2O(l) \longrightarrow \text{no reaction} \qquad [27]$$

$$NO_2^-(aq) + H_2O(l) \rightleftharpoons HNO_2(aq) + OH^-(aq) \qquad [28]$$

This process is frequently called *hydrolysis*. Notice that NO_2^-, the anion of a weak acid, is a base (proton acceptor). In fact, there is little difference in overall form between Equations [28] and [29]:

$$NH_3(aq) + H_2O(l) \rightleftharpoons NH_4^+(aq) + OH^-(aq) \qquad [29]$$

Cations of strong bases do not react with water (Equation [30]) other than to become hydrated. Yet the cation of a weak base hydrolyzes, forming the base and H_3O^+ (Equation [31]). Thus the cations of weak bases are proton donors (acids).

$$Na^+(aq) + H_2O(l) \longrightarrow \text{no reaction} \qquad [30]$$

$$CH_3NH_3^+(aq) + H_2O(l) \rightleftharpoons CH_3NH_2(aq) + H_3O^+(aq) \qquad [31]$$

Salts (which contain a cation and an anion) can, therefore, be of four types:

1. *Cation of a strong base and anion of a strong acid.* Neither ion hydrolyzes (see Reactions [27] and [30]), and the resulting solution is neutral with $H = 7.00$.

2. *Cation of a strong base and anion of a weak acid.* The cation does not hydrolyze ([30]), but the anion does (as in [28]). The OH^- thus formed produces a basic solution with $pH > 7.00$.

3. *Cation of a weak base and anion of a strong acid.* The anion does not hydrolyze ([27]), but the cation does (as in [31]). The resulting solution is acidic with $pH < 7.00$.

4. *Cation of a weak base and anion of a weak acid.* Both the cation and the anion hydrolyze. The one which does so most extensively determines the acidity of the solution.

A convenient memory aid is that the relative strengths of the acid that supplied the anion and the base that supplied the cation indicate the acidity of the salt solution. If the base is stronger, the solution is basic. If the acid is stronger, the solution is acidic.

11. Calculate values of K_a for cations and K_b for anions from ionization constants of their conjugates and K_w of water; calculate the pH values of salt solutions in which hydrolysis occurs.

Hydrolysis problems are set up in the same way as weak acid and weak base problems (see Objective 16-8). The sole difference is that the ionization constant (sometimes called the hydrolysis constant, K_h), normally is not tabulated but must be computed. For the anion of a weak acid, where $K_{a,HA}$ is the ionization constant of the weak acid, the ionization constant of the anion (K_b) is obtained as follows.

$$K_w = K_{a,HA} K_b \qquad or \qquad K_b = K_w / K_{a,HA} \qquad [32]$$

For the cation of a weak base, where $K_{b,B}$ is the ionization constant of the weak base, the ionization constant of the cation (K_a) is obtained as follows.

$$K_w = K_{b,B} K_a K_a \qquad or \qquad K_a = K_w / K_{b,B} \qquad [33]$$

Another way to state these two relationships is to realize that

$$K_a = \frac{[H_3O^+][\text{conjugate base}]}{[\text{weak acid}]} \qquad and \qquad K_b = \frac{[OH^-][\text{conjugate acid}]}{[\text{weak base}]}.$$

EXAMPLE 16-9 Determine the pH of 0.125 M sodium benzoate solution. $K_a = 6.2 \times 10^{-5}$ for benzoic acid.

$$NaC_7H_5O_2 \longrightarrow Na^+(aq) + C_7H_5O_2^-(aq) \qquad Na^+(aq) \text{ does not hydrolyze}$$

but benzoate anion does.

Equation:	$C_7H_5O_2^-(aq) + H_2O(l) \rightleftharpoons$	$OH^-(aq)$ +	$HC_7H_5O_2(aq)$
Initial:	0.125 M		
Changes:	$-x$ M	$+x$ M	$+x$ M
Equil:	$(0.125 - x)$ M	x M	x M

$$K_b = \frac{[OH^-][HC_7H_5O_2]}{[C_7H_5O_2^-]} = \frac{(x)(x)}{0.125 - x} = \frac{K_w}{K_a} = \frac{1.0 \times 10^{-14}}{6.3 \times 10^{-5}} \approx \frac{x^2}{0.125} \qquad \text{or}$$

$x = 4.5 \times 10^{-6}$ M $= [OH^-]$

We have assumed that $x << 0.125\text{ M}$, a valid assumption, since 4.5×10^{-6} M clearly is much smaller than 0.125 M. We obtain $\text{pOH} = 5.35$ *or* $\text{pH} = 14.00 - \text{pOH} = 8.65$.

Notice the very strong similarity between this calculation and that of Example 16-7.

*** 12. Use the relative strengths of Brønsted-Lowry acids and bases to predict the direction of acid-base reactions.**

The products of the reaction of an acid with a base are the conjugate acid and the conjugate base.

$$HC_2H_3O_2 + CH_3NH_2 \rightleftharpoons CH_3NH_3^+ + C_2H_3O_2^- \qquad [34]$$

Do products or reactants predominate in this mixture? Remember that the stronger acid is the better proton donor. When the acid is a very good proton donor, there will be very little of it present in solution. Instead, mainly its conjugate base will be present. Since $HC_2H_3O_2$ ($K_a = 1.8 \times 10^{-5}$) is a stronger acid than $CH_3NH_3^+$ ($K_a = K_w / K_b = 2 \times 10^{-9}$), reaction [34] lies to the right. Notice also that CH_3NH_2 ($K_b = 5 \times 10^{-4}$) is a stronger base than $C_2H_3O_2^-$ ($K_b = K_w / K_a = 5.6 \times 10^{-10}$). Hence the equilibrium lies to the side of the weaker acid and the weaker base. The stronger acid has done the better job of donating protons and the stronger base has done the better job of accepting them. Their weaker conjugates predominate in solution.

*** 12a. Predict whether certain oxides and hydroxo compounds are acidic, basic, or neutral.**

Oxides can be grouped into three types: acidic, basic, and amphoteric. Acidic oxides form an acidic solution when they dissolve in water; basic oxides form an alkaline (a basic) solution. Solutions of amphoteric oxides act as bases when treated with strong acid, and as acids when treated with strong base. Acidic oxides are those of the nonmetals Cl, Br, I, S, Se, N, P, As, C, Si, and B. Basic oxides are those of the metals Li, Na, K, Rb, Cs, Mg, Ca, Sr, Ba, Tl, and Bi. Amphoteric oxides are those of the metalloids Be, Al, Ga, Ge, Sn, Pb, and Sb. For transition elements, acidic oxides have the element in a high oxidation state (+6 or higher), such as CrO_3 and Mn_2O_7. The common amphoteric oxides of transition metals are ZnO, CdO, and Cr_2O_3. Other transition metal oxides are basic.

The acidic character of hydroxo compounds parallels that of the oxides. Metal hydroxides are bases while nonmetal "hydroxides" are more commonly known as *oxoacids*. In both cases, we are considering the group of atoms E—O—H. In bases, the E—O bond cleaves more readily than the O—H bond. In acids, the situation is reversed. The electronegativity of E determines whether the hydroxide is acidic or basic. Breakage of the O—H bond (and acidic character) is favored if E is highly electronegative (a nonmetallic atom or an atom of a high oxidation state). Breakage of the E—O bond (and basic character) is favored if E is electropositive.

Example 16-10. Indicate whether you expect each of the following solutions to be acidic, basic, or neutral:

A. KNO_3 (aq) B. CH_3CH_2COONa C. CH_4

Solution: A. KNO_3 is the salt of a strong acid. A strong acid is completely dissociated in solution. NO_3^- does not hydrolyze. Therefore, the solution will be neutral. B. CH_3CH_2COONa is the salt of a weak acid. CH_3CH_2COONa will hydrolyze in solution to form $CH_3CH_2COO^-$, which will react with H_3O^+ from water, reducing the concentration and causing the resulting solution to be basic. C. CH_4 is an organic hydrocarbon and does not ionize. The solution will be neutral.

*** 13. Define what an *oxoacid* is and predict the relative strengths of oxoacids from their molecular structures.**

An *oxoacid* is one in which the ionizable proton is bound to oxygen, as in Structure [35].

$$\text{rest of the molecule} — E — O — H \qquad [35]$$

When the O—H bond breaks readily, the oxoacid is a strong one. The O—H bond will break readily if atom E is highly electronegative. Thus, for the hypohalous acids, $HOI < HOBr < HOCl$.. However, the electronegativities of elements can be altered by the atoms in the "rest of the molecule" bonded to E. In general, the following factors increase the electronegativity of E (and increase the oxoacid strength):

(1) *Strongly electronegative atoms bonded to E* ($HCOOH < ClCOOH < FCOOH$). Highly electronegative atoms bonded to atoms next to E also increase the strength of an acid , although their effect is less than that of atoms directly bonded to E.

(2) *High positive oxidation state and formal charge of E.* This also means that, for the oxoacids of a specific element, acid strength increases with the number of terminal oxygens in the acid molecule. Thus, for the chlorine oxoacids, $HClO < HClO_2 < HClO_3 < HClO_4$, which is the same direction in which chlorine's oxidation state increases.

SELF-ASSESSMENT EXERCISES

103. An amphiprotic ion can act as both an acid and a base. (a) HCO_3^- is amphiprotic because it can gain an H^+ and act as a base or lose H^+ forming CO_3^{-2} to act as an acid.

104. (c) Greater than the pH in 0.010 M HBr. (a) Acetic acid is a weaker acid than nitric acid, so the pH will be higher (lower H_3O^+). (b) HI is a stronger acid, so the pH of acetic acid will be higher; the same is true for (c).

105. Answer (d) is correct. CH_3NH_2 will act as a base, but will not be fully ionized; therefore, (a) and (b) are incorrect. Since it acts as a base, the pH will be basic; therefore, (d) pH < 13 is correct.

106. (a) H_2O. The reaction will proceed farthest when the product is a weak base so that H^+ is not re-ionized.

107. (b) 0.11 M H_2SO_4 will fully ionize the first proton to give 0.01 M $[H_3O^+]$.

The second ionization: $HSO_4^- + H_2O \leftrightarrow H_3O^+ + SO_4^{-2}$

from 1st ionization:	0.10	0.10	
change	-x	+x	+x

Assume x is much smaller than 0.10 M, $K_a = \frac{0.10 \times X}{0.10} \approx x = 1.1 \times 10^{-2}$

108. (b) $[SO_3^{-2}] = 6.3 \times 10^{-8}$ M. Use the first ionization to determine $[H_3O^+]$ and $[HSO_3^-]$. Using the assumption that x is small compared to 0.10 M, $K_a = x^2/(0.10) = 1.3 \times 10^{-2}$ and $x = 0.036$ M. Using this value for $[HSO_3^-]$ and solving for the second ionization as above gives $[SO_3^{-2}] = 6.3 \times 10^{-8}$ M.

109. This is a mixture of a strong acid and a strong base. First calculate the mole of each to determine how much is neutralized, assuming complete ionization.

$0.02480 \text{ L} \times 0.248 \text{ mol L}^{-1} = 6.15 \times 10^{-3} \text{ mol H}^+$

$0.01540 \text{ L} \times 0.394 \text{ mol L}^{-1} = 6.07 \times 10^{-3} \text{ mol OH}^-$

Subtracting gives an excess of 8.0×10^{-5} mol H^+. To obtain the pH, use the ionization of water.

110. For a pH of 3.25, $-\log [H_3O^+] = 3.25 = 5.62 \times 10^{-4}$. K_a for acetic acid is 1.8×10^{-5}. The concentration of acetic acid to achieve the desired pH can be found from

$K_a = \frac{[H_3O^+][C_2H_3O_2^-]}{[HC_2H_3O_2]} = 1.8 \times 10^{-5}$ and $[H_3O^+] = [C_2H_3O_2^-] = 5.62 \times 10^{-4}$. Substituting into the K_a expression and solving for $[HC_2H_3O_2] = 1.74 \times 10^{-4}$. The acetic acid concentration is the sum of the ionized and nonionized, which is $5.62 \times 10^{-4} + 1.74 \times 10^{-4} = 7.34 \times 10^{-4}$. For 12.5 L, $12.5 \times 7.34 \times 10^{-4} \text{ mol L}^{-1} = 9.17 \times 10^{-3}$ mol.

$9.17 \times 10^{-3} \text{ mol} \times 60.0 \text{ g mol}^{-1} = 0.550$ g. D= m/V and V = 1.50 L of conc. acetic acid.

111. Ionization will yield Na^+, $C_2H_2ClO_2H$, and OH^-

$$K_b = \frac{[C_2H_2ClO_2H][OH^-]}{[C_2H_2ClO_2^-]} = \frac{K_w}{K_a} = \frac{1.0\times10^{-14}}{1.4\times10^{-3}} = 7.1 \text{ x } 10^{-11}$$

	$C_2H_2ClO_2Na + H_2O \leftrightarrow$	$C_2H_2ClO_2H$	$+ OH^-$
initial	2.05 M		
change	-x	+x	+x
final	2.05 – x	x	x

$x^2/2.05 = 7.1 \text{ x } 10^{-11}$ and $x = 1.20 \text{ x } 10^{-6} = [OH^-]$, pOH = 5.92

pH = 14 – pOH = 8.08.

113. The solution is basic. pH + pOH = 14 and pH = 5 pOH. pH + pH/5 = 14 = 1.2 pH and pH = 11.7.

DRILL PROBLEMS (Assume throughout that H_2SO_4 is a diprotic strong acid.) Ionization constant values are in Table 16-1 at the end of this section.

1. (1) Label the Brønsted-Lowry acids and bases in these equations. Make sure to label both conjugate pairs.

A. $H_2S + CO_3^{2-} \rightleftharpoons HCO_3^- + HS^-$

B. $H_3O^+ + HS^- \rightleftharpoons H_2S + H_2O$

C. $HS^- + OH^- \rightleftharpoons H_2O + S^{2-}$

D. $H_2O + NH_2^- \rightleftharpoons NH_3 + OH^-$

E. $HCO_3^- + OH^- \rightleftharpoons H_2O + H_2O + CO_3^{2-}$

F. $HCN + OH^- \rightleftharpoons H_2O + CN^-$

G. $NH_4^+ + OH^- \rightleftharpoons H_2O + NH_3$

H. $HSO_4^- + C_2H_3O_2^- \rightleftharpoons HC_2H_3O_2 + SO_4^{2-}$

I. $H_3PO_4 + CN^- \rightleftharpoons HCN + H_2PO_4^-$

J. $HSO_4^- + CN^- \rightleftharpoons HCN + SO_4^{2-}$

(2) Complete the following equations. (The acid is first in each.) Label each conjugate acid-base pair.

K. $HCl + NH_3 \rightleftharpoons$

L. $H_3O^+ + H_2PO_4^- \rightleftharpoons$

M. $HC_2H_3O_2 + HS^- \rightleftharpoons$

N. $NH_3 + CH_3NH_2 \rightleftharpoons$

O. $CH_3NH_3 + OH^- \rightleftharpoons$

P. $HNO_2 + NH_{3^-} \rightleftharpoons$

2. Draw the Lewis structure for each reactant in these reactions; indicate which is the acid and which is the base.

A. $H_2O + SO_2 \longrightarrow H_2SO_3$

B. $BeF_2 + 2\,F^- \longrightarrow BeF_4^{2-}$

C. $NH_3 + HCl \longrightarrow NH_4Cl$

D. $CO_2 + OH^- \longrightarrow HCO_3^-$

E. $Ag^+ + 2\,NH_3 \longrightarrow BF_4^-$

F. $O^2 + H_2O \longrightarrow 2\,OH^-$

G. $BF_3 + F^- \longrightarrow BF_4^-$

H. $HCl + H_2O$ \$\$\$ $H_3O^+ + Cl^-$

I. $H_2O + BF_3 \longrightarrow H_2OBF_3$

J. $S^{2-} + SO_3 \longrightarrow S_2O_3^{2-}$

K. $H^+ + CO_3^{2-} \longrightarrow HCO_3^-$

L. $H_2O + CO_2 \longrightarrow H_2CO_3$

M. $BF_3 + CH_3OH \longrightarrow CH_3OH \cdot BF_3$

3. Fill in the blanks in the table that follows. The first line is an example.

Solute	*Concentration*		*Acidity*			
Formula	M	g/L	$[H^+]$	$[OH^-]$	pH	pOH
HCl	0.020	0.73	0.020	5.0×10^{-13}	1.70	12.30
A. NaOH	0.040	____	____	____	____	____
B. H_2SO_4	____	____	0.00125	____	____	____
C. $Ca(OH)_2$	____	____	____	0.0333	____	____
D. RbOH	____	____	____	0.0375	____	____
E. $HClO_4$	____	3.84	____	____	____	____
F. $Ba(OH)_2$	____	0.513	____	____	____	____
G. HNO_3	0.0125	____	____	____	____	____
H. KOH	____	____	____	0.045	____	____
I. $Sr(OH)_2$	____	____	____	____	9.13	____
J. CsOH	____	____	____	____	____	4.65
K. HBr	____	____	____	____	3.15	____
L. HI	____	____	____	____	____	12.75

4. For each weak acid or base, give the equation for ionization in water and the ionization constant expression.

A. HOBr B. $(CH_3)_2NH$ C. $HC_6H_7O_2$ D. N_2H_4 E. $HC_3H_5O_3$

F. $(CH_3)_3N$ G. $HC_3H_5O_2$ H. $HC_2HCl_2O_2$ I. H_3BO_3 J. $HC_2Cl_3O_2$

5. (1) The initial concentration (or formality) of each acid or base is given below, along with an indication of the resulting solution's acidity. Compute the ionization constant of the acid or base in each case.

A. $HOBr, 0.250\ M, pH = 4.640$ B. $(CH_3)_2NH, 0.400\ M, [OH^-] = 0.0165\ M$
C. $HC_6H_7O_2, 0.300\ M, pOH = 11.687$ D. $N_2H_4, 0.0100\ M, pH = 9.996$
E. $HC_3H_5O_3, 1.20\ M, [H+] = 0.0313$ F. $(CH_3)_3N, 0.140\ M, pH = 11.503$
G. $HC_3H_5O_2, 0.0200\ M, pH = 3.282$ H. $HC_2HCl_2O_2, 2.20\ M, [H^+] = 0.254\ M$
I. $H_3BO_3, 0.0400\ M, pOH = 9.683$ J. $HC_2Cl_3O_2, 1.50\ M, [H^+] = 0.466\ M$

(2) Determine the pH of each of the following solutions. Use the data of Table 16-1.

K. $0.250\ M\ HC_2H_3O_2$ L. $1.20\ M\ HC_7H_5O_2$ M. $0.820\ M\ HCHO_2$ N. $0.400\ M\ HNO_2$
O. $0.230\ M\ HF$ P. $1.00\ M\ HClO_2$ Q. $1.20\ M\ NH_3$ R. $2.60\ M\ CH_3NH_3$
S. $5.10\ M\ C_2H_5NH_2$ T. $0.510\ M\ HN_3$ U. $0.510\ M\ HONH_2$ V. $0.150\ M\ C_6H_5NH_2$

6. Determine the pH, the final concentration of undissociated acid, and the concentration of all anions in the following solutions. Use the data of Table 16-1.

A. $0.00460\ M\ H_2CO_3$ B. $0.0100\ M\ H_2Se$ C. $1.20\ M\ H_2C_2O_4$
D. $0.750\ M\ H_3PO_4$ E. $0.900\ M\ H_3PO_3$ F. $1.00\ M\ H_2SO_3$

7. For each of the following salts, write the reaction for the salt dissolving in water. Based on the ions present, predict whether the solution is acidic, basic, or neutral. Finally, write any hydrolysis reactions that occur.

A. $NaCl$ B. NH_4Cl C. N_2H_5Br D. KNO_3 E. $C_2H_5NH_3NO_3$
F. Na_2CO_3 G. $C_6H_5NH_3I$ H. $Ca(ClO)_2$ I. $RbClO_2$ J. $(HONH_3)_2SO_4$
K. SrF_2 L. $CsC_2H_3O_2$ M. NH_4CN N. $HONH_3F$ O. $CH_3NH_3ClO_2$

8. (1) 1.00 mol of each of the salts in parts A through L of the drill problems for Objective 16-10 is used to form 1.00 L of aqueous solution. Determine the pH of each solution.

(2) Determine the pH of each of the following solutions.

M. 3.00 mmol sodium acetate in 15.0 mL soln
N. 5.00 mmol calcium benzoate in 50.0 mL soln
O. 12.5 mmol potassium formate in 12.5 mL soln
P. 3.20 mmol barium nitrite in 16.0 mL soln
Q. 1.79 mmol rubidium fluoride in 20.0 mL
R. 12.1 mmol cesium chlorite in 10.0 mL

soln

soln

S. 6.10 mmol ammonium sulfate in 25.0 mL soln

T. 5.37 mmol methylammonium sulfate, 20. mL soln

U. 26.4 mmol ethylammonium nitrate in 60.0 mL soln

V. 177.44 mmol hydrazine hydrochloride (N_2H_5Cl) in 180.0 mL soln ($K_b = 9.8 \times 10^{-7}$)

W. 2.28 mmol aniline bromide in 250.0 mL soln

X. 11.0 mmol ammonium iodide in 80.0 mL soln

9 Equilibrium in each reaction in the two groups of drill problems for Objective 16-2 lies to the right. Rank all the conjugate acid-base pairs in those drill problems in order of decreasing acid strength (strongest to weakest). Use the data of Table 16-1 only if necessary. Then complete the following equations and use your ranking to predict whether each equilibrium lies to the left or to the right.

A. $H_2O + N_3^- \rightleftharpoons$ B. $HNO_2 + OH^- \rightleftharpoons$ C. $H_3O^+ + CN^- \rightleftharpoons$

D. $NH_3 + CN^- \rightleftharpoons$ E. $HCN + H_2PO_4^- \rightleftharpoons$ F. $H_3PO_4 + NH_2^- \rightleftharpoons$

G. $H_3PO_4 + OH^- \rightleftharpoons$ H. $H_3O^+ + NH_2^- \rightleftharpoons$ I. $HCO_3^- + C_2H_3O_2^- \rightleftharpoons$

J. $HSO_4^- + HS^- \rightleftharpoons$ K. $H_2S + OH^- \rightleftharpoons$ L. $H_2O + C_2H_3O_2^- \rightleftharpoons$

M. $HCO_3^- + SO_4^{2-} \rightleftharpoons$ N. $HSO_4^- + OH^- \rightleftharpoons$

10. (1) Classify each as acidic, basic, or amphoteric.

A. CaO B. CO_2 C. PbO

D. SiO_2 E. SO_2 F. ClO_2 G. BeO H. NO_5 I. SnO

J. BaO K. Al_2O_3 L. Tl_2O M. Na_2O N. P_2O_5 O. MgO

(2) Complete and balance each of the following reactions. ("No reaction" may be the correct answer.)

P. $CaO + SiO_2 \longrightarrow$ Q. $Al_2O_3 + SO_2 \longrightarrow$ R. $NO + CO_2 \longrightarrow$

T. $BaO + SO_2 \longrightarrow$ U. $PbO + P_2O_5 \longrightarrow$ V. $Na_2O + Al_2O_3 \longrightarrow$

S. $NaO + CaO \longrightarrow$

11. Predict which acid of each of the following pairs is the stronger, and briefly explain why.

A. H_3BO_3 *or* H_2CO_3 B. H_2SO_4 *or* H_2SO_3 C. H_3PO_4 *or* H_3AsO_4

D. $HCLO_3$ *or* $HClO_4$ E. HBrO *or* HClO F. HNO_2 *or* HNO_3

G. H_2SO_4 *or* H_2SeO_4 H. CF_3OH *or* CH_3OH I. FC_6H_4OH *or* ClC_6H_4OH

J. $HBrO_3$ *or* H_2SeO_3 K. H_3PO_4 *or* $HClO_4$ L. H_2CO_3 *or* H_4SiO_4

TABLE 16–1 Ionization Constants for Some Weak Acids and Weak Bases in Water at 25 °C‡

Name	Ionization Equilibrium Equation	K_i	pK_i
acetic	$HC_2H_3O_2 + H_2O \rightleftharpoons H_3O^+ + C_2H_3O_2^-$	1.8×10^{-5}	4.74
acrylic	$HC_3H_3O_2 + H_2O \rightleftharpoons H_3O^+ + C_3H_3O_2^-$	5.5×10^{-5}	4.26
aluminum ion	$[Al(H_2O)_6]^{3+} + H_2O \rightleftharpoons$	9.8×10^{-6}	5.01
	$[Al(H_2O)_5OH]^{2+} + H_3O^+$		
arsenic	$H_3AsO_4 + H_2O \rightleftharpoons H_3O^+ + H_2AsO_4^-$	6.0×10^{-3}	2.22
	$H_2AsO_4- + H_2O \rightleftharpoons H_3O^+ + HAsO_4^{2-}$	1.0×10^{-7}	7.00
	$HAsO_4^{2-} + H_2O \rightleftharpoons H_3O^+ AsO_4^{3-}$	3.2×10^{-12}	11.50
arsenous	$HAsO_2 + H_2O \rightleftharpoons H_3O^+ + AsO_2^-$	6.6×10^{-10}	9.18
benzoic	$HC_7H_5O_2 + H_2O \rightleftharpoons H_3O + C_7H_5O_2^-$	6.3×10^{-5}	4.20
bromoacetic	$HC_2H_2BrO_2 + H_2O \rightleftharpoons H_3O^+ C_2H_2BrO_2^-$	1.3×10^{-3}	2.90
butyric	$HC_4H_7O_2 + H_2O \rightleftharpoons H_3O^+ C_4H_7O_2^-$	1.5×10^{-5}	4.83
carbonic	ς$H_2CO_3 + H_2O \rightleftharpoons H_3O^+ + HCO_3^-$	4.4×10^{-7}	6.36
	$HCO_3^- + H_2O \rightleftharpoons H_3O^+ + CO_3^{2-}$	4.7×10^{-11}	10.33
chloroacetic	$HC_2H_2ClO_2 + H_2O \rightleftharpoons H_3O^+ + C_2H_2ClO_2^-$	1.4×10^{-3}	2.85
chlorous	$HClO_2 + H_2O \rightleftharpoons H_3O^+ + ClO_2^-$	1.1×10^{-2}	1.92
chromium(III) ion	$[Cr(H_2O)_6]^{3+} + H_2O \rightleftharpoons$	1×10^{-4}	4.0
	$[Cr(H_2O)_5OH]^{2+} + H_3O^+$		
cinnamic	$HC_9H_7O_2 + H_2O \rightleftharpoons H_3O^+ + C_9H_7O_2^-$	3.6×10^{-5}	4.44
citric	$H_3C_6H_5O_7 + H_2O \rightleftharpoons H_3O^+ + H_2C_6H_5O_7^-$	7.4×10^{-4}	3.13
	$H_2C_6H_5O_7^- + H_2O \rightleftharpoons H_3O^+ + HC_6H_5O_7^{2-}$	1.7×10^{-5}	4.76
	$HC_6H_5O_7^{2-} + H_2O \rightleftharpoons H_3O^+ + C_6H_5O_7^3$	4.0×10^{-7}	6.40
cyanic	$HOCN + H_2O \rightleftharpoons H_3O^+ + OCN^-$	3.5×10^{-4}	3.46

‡ Although some of these K and pK values could be expressed with an additional significant figure, the circumstances of a calculation do not warrant this

ςH_2CO_3 cannot be isolated. It is in equilibrium with $CO_2(aq)$. The value for K_{a1} is actually that for $CO_2(aq) + H_2O \rightleftharpoons H_3O^+ + HCO_3^-$.

cyanuric	$HC_3H_2N_3O_3 + H_2O \rightleftharpoons H_3O^+ + C_3H_2N_3O_3^-$	1.7×10^{-7}	6.78
dichloroacetic	$HC_2HCl_2O_2 + H_2O \rightleftharpoons H_3O^+ + C_2HCl_2O_2^-$	5.5×10^{-2}	1.26
fluoroacetic	$HC_2H_2FO_2 + H_2O \rightleftharpoons H_3O^+ + C_2H_2FO_2^-$	2.6×10^{-3}	2.59
formic	$HCHO_2 + H_2O \rightleftharpoons H_3O^+ + CHO_2^-$	1.8×10^{-4}	3.74
germanic	$H_2GeO_3 + H_2O \rightleftharpoons H_3O^+ + HGeO_3^-$	1×10^{-9}	9.0
	$HGeO_3^- + H_2O \rightleftharpoons H_3O^+ + GeO_3^{2-}$	4×10^{-13}	12.4
hydrazoic	$HN_3 + H_2O \rightleftharpoons H_3O^+ + N_3^-$	1.9×10^{-5}	4.72
hydrocyanic	$HCN + H_2O \rightleftharpoons H_3O^+ + CN^-$	6.2×10^{-10}	9.21
hydrofluoric	$HF + H_2O \rightleftharpoons H_3O^+ + F^-$	6.6×10^{-4}	3.18
hydrogen peroxide	$H_2O_2 + H_2O \rightleftharpoons H_3O^+ + HO_2^-$	2.2×10^{-12}	11.65
hydroselenic	$H_2Se + H_2O \rightleftharpoons H_3O^+ + HSe^-$	1.3×10^{-4}	3.89
	$HSe^- + H_2O \rightleftharpoons H_3O^+ + Se^{2-}$	1×10^{-11}	11.0
hydrosulfuric	$H_2S + H_2O \rightleftharpoons H_3O^+ + HS^-$	1.0×10^{-7}	7.00
	$HS^- + H_2O \rightleftharpoons H_3O^+ + S^{2-}$	1×10^{-19}	19.0
hydrotelluric	$H_2Te + H_2O \rightleftharpoons H_3O^+ + HTe^-$	2.3×10^{-3}	2.64
	$HTe^- + H_2O \rightleftharpoons H_3O^+ + Te^{2-}$	1.6×10^{-11}	10.80
hypobromous	$HOBr + H_2O \rightleftharpoons H_3O^+ + OBr^-$	2.5×10^{-9}	8.62
hypochlorous	$HOCl + H_2O \rightleftharpoons H_3O^+ + OCl^-$	2.9×10^{-8}	7.54
hypoiodous	$HOI + H_2O \rightleftharpoons H_3O^+ + OI^-$	2.3×10^{-11}	10.64
hyponitrous	$HON{=}NOH + H_2O \rightleftharpoons H_3O^+ + HON{=}NO^+$	8.9×10^{-8}	7.05
	$HON{=}NO^+ + H_2O \rightleftharpoons H_3O^+ + {}^+ON{=}NO^+$	4×10^{-12}	11.4
hypophosphorous	$HPH_2O_2 + H_2O \rightleftharpoons H_3O^+ + PH_2O_2^-$	5.9×10^{-2}	1.23
iodic	$HIO_3 + H_2O \rightleftharpoons H_3O^+ + IO_3^-$	1.6×10^{-1}	0.80
iodoacetic	$HC_2H_2IO_2 + H_2O \rightleftharpoons H_3O^+ + C_2H_2IO_2^-$	6.7×10^{-4}	3.18
iron(II) ion	$[Fe(H_2O)_6]^{2+} + H_2O \rightleftharpoons [Fe(H_2O)_5OH]^+ + H_3O^+$	1.8×10^{-7}	6.74
iron(III) ion	$[Fe(H_2O)_6]^{3+} + H_2O \rightleftharpoons [Fe(H_2O)_5OH]^{2+} + H_3O^+$	1.5×10^{-3}	2.83
	$[Fe(H_2O)_5OH]^{2+} + H_2O \rightleftharpoons [Fe(H_2O)_4(OH)]^+ + H_3O^+$	2.6×10^{-5}	4.59

malonic	$H_2C_3H_2O_4 + H_2O \rightleftharpoons H_3O^+ + HC_3H_2O_4^-$	1.5×10^{-3}	2.82
	$HC_3H_2O_4^- + H_2O \rightleftharpoons H_3O^+ + C_3H_2O_4^{2-}$	2.0×10^{-6}	5.70
nitrous	$HNO_2 + H_2O \rightleftharpoons H_3O^+ + NO_2^-$	7.2×10^{-4}	3.14
oxalic	$H_2C_2O_4 + H_2O \rightleftharpoons H_3O^+ + HC_2O_4^-$	5.4×10^{-2}	1.27
	$HC_2O_4^- + H_2O \rightleftharpoons H_3O^+ + C_2O_4^{2-}$	5.3×10^{-5}	4.28
Peroxomono sulfuric	$H_2SO_5 + H_2O \rightleftharpoons H_3O^+ + HSO_5^-$	1×10^{-1}	1.0
	$HSO_5^- + H_2O \rightleftharpoons H_3O^+ + SO_5^{2-}$	5×10^{-10}	9.3
phenol	$HOC_6H_5 + H_2O \rightleftharpoons H_3O^+ + C_6H_5O^-$	1.0×10^{-10}	10.00
phenylacetic	$HC_8H_7O_2 + H_2O \rightleftharpoons H_3O^+ + C_8H_7O_2^-$	4.9×10^{-5}	4.31
phosphoric	$H_3PO_4 + H_2O \rightleftharpoons H_3O^+ + H_2PO_4^-$	7.1×10^{-3}	2.15
	$H_2PO_4^- + H_2O \rightleftharpoons H_3O^+ + HPO_4^{2-}$	6.3×10^{-8}	7.20
	$HPO_4^{2-} + H_2O \rightleftharpoons H_3O^+ + PO_4^{3-}$	4.2×10^{-13}	12.38
phosphorous	$H_3PO_3 + H_2O \rightleftharpoons H_3O^+ + H_2PO_3^-$	3.7×10^{-2}	1.43
	$H_2PO_3^- + H_2O \rightleftharpoons H_3O^+ + HPO_3^{2-}$	2.1×10^{-7}	6.68
propionic	$HC_3H_5O_2 + H_2O \rightleftharpoons H_3O^+ + C_3H_5O_2^-$	1.3×10^{-5}	4.87
pyrophosphoric	$H_4P_2O_7 + H_2O \rightleftharpoons H_3O^+ + H_3P_2O_7^-$	3.0×10^{-2}	1.52
	$H_3P_2O_7^- + H_2O \rightleftharpoons H_3O^+ + H_2P_4O_7^{2-}$	4.4×10^{-3}	2.36
	$H_2P_4O_7^{2-} + H_2O \rightleftharpoons H_3O^+ + HP_4O_7^{3-}$	2.5×10^{-7}	6.60
	$HP_4O_7^{3-} + H_2O \rightleftharpoons H_3O^+ + P_4O_7^{4-}$	5.6×10^{-10}	9.25
selenic	$H_2SeO_4 + H_2O \rightleftharpoons H_3O^+ + HSeO_4^-$	1×10^{3}	-3
	$HSeO_4^- + H_2O \rightleftharpoons H_3O^+ + SeO_4^{2-}$	2.2×10^{-2}	1.66
selenous	$H_2SeO_3 + H_2O \rightleftharpoons H_3O^+ + HSO_3^-$	2.3×10^{-3}	2.64
	$HSeO_3^- + H_2O \rightleftharpoons H_3O^+ + SO_3^{2-}$	5.4×10^{-9}	8.27
succinic	$H_2C_4H_4O_4 + H_2O \rightleftharpoons H_3O^+ + HC_4H_4O_4^-$	6.2×10^{-5}	4.21
	$HC_4H_4O_4^- + H_2O \rightleftharpoons H_3O^+ + C_4H_4O_4^{2}$	2.3×10^{-6}	5.64
sulfuric	$H_2SO_4 + H_2O \rightleftharpoons H_3O^+ + HSO_3^-$	$\approx 1\times10^{3}$	≈ -3
	$HSO_4^- + H_2O \rightleftharpoons H_3O^+ + SO_4^{2-}$	1.1×10^{-2}	1.96
sulfurous	$^{\#}H_2SO_3 + H_2O \rightleftharpoons H_3O^+ + HSO_3^-$	1.3×10^{-2}	1.89

H_2SO_3 is a hypothetical, nonisolable species produced in the reaction $SO_2(aq) + H_2O \rightleftharpoons H_2SO_3(aq)$. Many chemists prefer to write $SO_2.H_2O$.

	$HSO_3^- + H_2O \rightleftharpoons H_3O^+ + SO_3^{2-}$	6.2×10^{-8}	7.21
thiophenol	$HSC_6H_5 + H_2O \rightleftharpoons H_3O^+ + C_6H_5S^-$	3.2×10^{-7}	6.49
trichloroacetic	$HC_2Cl_3O_2 + H_2O \rightleftharpoons H_3O^+ + C_2Cl_3O_2^-$	3.0×10^{-1}	0.52
vanadic	$H_3VO_4 + H_2O \rightleftharpoons H_3O^+ + H_2VO_4^-$	1.7×10^{-4}	3.78
	$H_2VO_4^- + H_2O \rightleftharpoons H_3O^+ + HVO_4^{2-}$	2×10^{-8}	7.7
	$HVO_4^{2-} + H_2O \rightleftharpoons H_3O^+ + VO_4^{3-}$	1×10^{-13}	13.0
zinc ion	$[Zn(H_2O)_6]^{2+} + H_2O \rightleftharpoons$ $[Zn(H_2O)_5OH]^+ + H_3O^+$	1.1×10^{-9}	8.96
Bases			
allylamine	$C_3H_5NH_2 + H_2O \rightleftharpoons C_3H_5NH_3^+ + OH^-$	4.9×10^{-5}	4.31
ammonia	$NH_3 + H_2O \rightleftharpoons NH_4^+\ OH^-$	1.8×10^{-5}	4.74
aniline	$C_6H_5NH_2 + H_2O \rightleftharpoons C_6H_5NH_3 + OH^-$	7.4×10^{-10}	9.13
codeine	$C_{18}H_{21}O_3N + H_2O \rightleftharpoons C_{18}H_{21}O_3NH^+\ OH^-$	8.9×10^{-7}	6.05
diethylamine	$(C_2H_5)_2NH + H_2O \rightleftharpoons$ $(C_2H_5)_2NH_2^+ + OH^-$	6.9×10^{-5}	4.16
dimethylamine	$(CH_3)_2NH + H_2O \rightleftharpoons (CH_3)_2NH_2^+ + OH^-$	5.9×10^{-4}	3.23
ethylamine	$C_2H_5NH_2 + H_2O \rightleftharpoons C_2H_5NH_3^+ + OH^-$	4.3×10^{-4}	3.37
hydrazine	$H_2NNH_2 + H_2O \rightleftharpoons H_2NNH_3^+ + OH^-$	8.5×10^{-7}	6.07
	$H_2NNH_3^+ + H_2O \rightleftharpoons H_3NNH_3^+ + OH^-$	8.9×10^{-16}	15.05
hydroxylamine	$HONH_2 + H_2O \rightleftharpoons HONH_3^+ + OH^-$	9.1×10^{-9}	8.04
isoquinoline	$C_9H_7N + H_2O \rightleftharpoons C_9H_7NH^+ + OH^-$	2.5×10^{-9}	8.60
methylamine	$CH_3NH_2 + H_2O \rightleftharpoons CH_3NH_3^+ + OH^-$	4.2×10^{-4}	3.38
morphine	$C_{17}H_{19}O_3N + H_2O \rightleftharpoons$ $C_{17}H_{19}O_3NH^+ + OH^-$	7.4×10^{-7}	6.13
piperidine	$C_5H_{11}N + H_2O \rightleftharpoons C_5H_{11}NH^+ + OH^-$	1.3×10^{-3}	2.88
pyridine	$C_5H_5N + H_2O \rightleftharpoons C_5H_5NH^+ + OH^-$	1.5×10^{-9}	8.82
quinoline	$C_9H_7N + H_2O \rightleftharpoons C_9H_7NH^+ + OH^-$	6.3×10^{-10}	9.20
triethanolamine	$C_6H_{15}O_3N + H_2O \rightleftharpoons C_6H_{15}O_3NH^+ + OH^-$	5.8×10^{-7}	6.24
triethylamine	$(C_2H_5)_3N + H_2O \rightleftharpoons (C_2H_5)_3NH^+ + OH^-$	5.2×10^{-4}	3.28
trimethylamine	$(CH_3)_3N + H_2O \rightleftharpoons (CH_3)_3NH^+ + OH^-$	6.3×10^{-5}	4.20

QUIZZES (20 minutes each) Choose the best answer for each question.

QUIZ A

1. In the reaction $BF_3 + NH_3 \longrightarrow F_3B:NH_3, BF_3$ accepts an electron pair and acts as (a) an Arrhenius base; (b) a Brønsted acid; (c) a Lewis acid; (d) a Lewis base; (e) none of these.

2. Which of the following five statements about Brønsted-Lowry acids and bases is true? (a) an acid and its conjugate base react to form a salt and water; (b) the acid H_2O is its own conjugate base; (c) the conjugate base of a strong acid is a strong base; (d) the conjugate base of a very weak acid is a strong base; (e) a base and its conjugate acid react to form a neutral solution.

3. Nitrous acid, HNO_2, is 0.37% ionized in 3.0 M solution. For HNO_2 K_a = (a) $(3.0)(0.37)^2/99.63$; (b) $(3.0)(0.0037)^2/0.9963$; (c) $(0.0037)^2/0.9963$; (d) $[0.9963/(0.0037)^2]/3.0$; (e) none of these.

4. A 2.4 M solution of a weak base has $[OH^-] = 0.0034$ M. What is K_b for this base? (a) 0.0034; (b) 0.0034/2.4; (c) $(2.4)^2/0.0034$; (d) $(0.0034/2.4)^2$; (e) none of these.

5. 7200. mL of solution contains 0.216 mmol of potassium hydroxide. What is the pH of this solution? (a) 10.143; (b) 4.523; (c) 3.857; (d) 9.477; (e) none of these.

6. $pOH = 3.14$ is equivalent to (a) $pH = 11.86$; (b) $[H^+] = 1.4 \times 10^{-10}$; (c) $[OH^-] = 7.3 \times 10^{-4}$; (d) acidic solution; (e) none of these.

7. $CO_2(g)$ acts as an acid in the reaction $CaO(s) + CO_2(g) \longrightarrow CaCO_3(s)$ because it (a) turns blue litmus red; (b) reacts with a metal; (c) is a proton donor; (d) is an electron-pair acceptor; (e) none of these.

8. In order of decreasing strength: $OH^- > ClO^- > NO_2^-$. Which of the following reactions lies to the right? (a) $NO_2^- + HClO \rightleftharpoons HNO_2 + ClO^-$; (b) $H_2O + HClO \rightleftharpoons H_3O^+ + ClO^-$; (c) $OH^- + HNO_2 \rightleftharpoons H_2O + NO_2^-$; (d) $H_2O + ClO^- \rightleftharpoons OH^- + HClO$; (e) none of these.

9. Which of these is the strongest acid? (a) H_2SO_4; (b) H_2SO_3; (c) H_2SeO_4; (d) $HONH_2$; (e) H_2O.

QUIZ B

1. In the reversible reaction $HCO_3^-(aq) + OH^-(aq) \rightleftharpoons CO_3^{2-}(aq) + H_2O$, the Brønsted acids are (a) HCO_3^- and CO_3^{2-}; (b) HCO_3^- and H_2O; (c) OH^- and H_2O; (d) OH^- and CO_3^{2-}; (e) none of these.

2. *Proton donor* is an abbreviated definition of a (a) Brønsted-Lowry acid; (b) Brønsted-Lowry base; (c) Lewis acid; (d) Lewis base; (e) none of these.

3. Propionic acid, $HC_3H_5O_2$, is 0.42% ionized in 0.80 M solution. For this acid, K_a = (a) $0.80(0.0042)^2/0.9958$; (b) $(0.80)(0.42)^2/99.58$; (c) $(0.0042)^2/0.9958$; (d) $(0.9958/0.80)/(0.0042)^2$; (e) none of these.

4. In a 2.00 M solution of HNO_2, $[H^+] = 0.030$ M. What is K_a for this acid? (a) 2.0/0.030; (b) 0.30/2.0; (c) $(0.030)^2/2.00$; (d) $2.00/(0.030)^2$; (e) none of these.

5. 250. mL of solution contains 0.132 mmol of perchloric acid. What is the pH of this solution? (a) 3.880; (b) 2.398; (c) 10.120; (d) 3.277; (e) none of these.

6. $[OH^-] = 3.0 \times 10^{-10}$ M is equivalent to (a) pH = 4.522; (b) pOH = 10.478; (c) 3.33×10^{-4} M = $[H_3O^+]$; (d) acidic solution; (e) none of these.

7. The following are all polyprotic acids *except* (a) H_2SO_4; (b) H_3BO_3; (c) $HClO_4$; (d) H_3PO_3; (e) $H_2C_2O_4$.

8. In order of decreasing strength: HF > $HC_2H_3O_2$ > HCN. Which of the following equilibria lies to the right? (a) $HCN + C_2H_3O_2^- \rightleftharpoons CN^- + HC_2H_3O_2$; (b) $HF + CN^- \rightleftharpoons HCN + F^-$; (c) $HCN + H_2O \rightleftharpoons H_3O^+ + CN^-$; (d) $HC_2H_3O_2 + F^- \rightleftharpoons C_2H_3O_2^- + HF$; (e) none of these.

9. Which of these is the strongest acid? (a) $HC_2H_3O_2$; (b) $HC_2HF_2O_2$; (c) H_2O; (d) NH_3; (e) $HC_2H_2FO_2$.

QUIZ C

1. In the reaction $HCl(g) + NaOH(s) \longrightarrow NaCl(s) + H_2O(g)$, HCl(g) acts as an acid because it (a) turns blue litmus red; (b) reacts with a metal; (c) is an electron-pair acceptor; (d) is a proton donor; (e) none of these.

2. A Lewis base is defined as a (a) proton donor; (b) proton acceptor; (c) hydrogen ion donor; (d) hydrogen ion acceptor; (e) none of these.

3. Acetic acid, $HC_2H_3O_2$, is 0.11% ionized in 1.5 M solution. What is K_a for this acid? (a) $(0.11)^2/99.89$; (b) $(0.0011)^2/0.9989$; (c) $99.89/(0.11)^2$; (d) $(0.9989/1.5)/(0.0011)^2$; (e) none of these.

4. A 0.30 M weak acid solution has $[H^+] = 6.8 \times 10^{-5}$. The value of K_a for this acid is (a) 1.5×10^{-8}; (b) 6.8×10^{-5}; (c) 4.6×10^{-9}; (d) 2.3×10^{-4}; (e) none of these.

5. 500.0 mL of solution contains 1.50 mmol of nitric acid. What is the pH of this solution? (a) 2.824; (b) 2.699; (c) 2.523; (d) 11.477; (e) none of these.

6. $pH = 4.25$ is equivalent to (a) $pOH = 9.75$; (b) $[H^+] = 5.61 \times 10^{-4}$ M; (c) $[OH^-] = 5.61 \times 10^{-4}$ M; (d) an alkaline solution; (e) none of these.

7. Which pair does not represent an acid and its anhydride? (a) H_3PO_4, P_4O_{10}; (b) H_2CO_3, CO_2; (c) $HClO_2$, Cl_2O; (d) HNO_3, N_2O_5; (e) H_3BO_3, B_2O_3.

8. The equilibria $HClO + CN^- \rightleftharpoons ClO^- + HCN$ *and* $HC_2H_3O_2 + ClO^- \rightleftharpoons C_2H_3O_2^- + HClO$ lie to the right. Which list in order of decreasing strength is correct? (a) $C_2H_3O_2^- > ClO^- > CN^-$; (b) $C_2H_3O_2^- > CN^- > ClO^-$; (c) $HC_2H_3O_2 > HCN > HClO$; (d) $HCN > HClO > HC_2H_3O_2$; (e) none of these.

9. Which of these is the strongest acid? (a) HBrO; (b) HIO_2; (c) $HClO_2$; (d) $HClO_2$; (e) $HClO_3$.

QUIZ D

1. The conjugate acid of HPO_4^{2-} is (a) PO_4^{3-}; (b) $H_2PO_4^-$; (c) H_3PO_4; (d) H_3O^+; (e) P_2O_5.

2. Proton acceptor is an abbreviated definition of a (a) Brønsted-Lowry base; (b) Brønsted-Lowry acid; (c) Lewis base; (d) Lewis acid; (e) none of these.

3. Formic acid, $HCHO_2$, is 0.915% ionized in 2.5 M solution. What is K_a for this acid? (a) $(0.99085/2.5)/(0.00915)^2$; (b) $(0.00915)^2/0.99085$; (c) $(2.5)(0.00915)^2/0.99085$; (d) $(2.5)(0.915)^2/99.085$; (e) none of these.

4. In a 2.00 M solution of weak base, $[OH^-] = 0.040$ M. What is the value of K_b for this base? (a) 0.02; (b) 8×10^{-4}; (c) 4×10^{-4}; (d) 1.6×10^{-3}; (e) none of these.

5. 1200. mL of solution contains 0.0720 mmol of sulfuric acid. What is the pH of this solution? (a) 3.921; (b) 4.222; (c) 9.699; (d) 1.143; (e) none of these.

6. $[H^+] = 2.5 \times 10^{-4}$ M is equivalent to (a) $pH = 4.40$; (b) $pOH = 11.60$; (c) $[OH^-] = 4.0 \times 10^{-10}$; (d) acidic solution; (e) none of these.

7. P_4O_6 is the anhydride of (a) HPO_3; (b) $H_4P_2O_7$; (c) H_3PO_4; (d) H_3PO_3; (e) none of these.

8. The equilibria $OH^- + HClO \rightleftharpoons H_2O + ClO^-$ *and* $ClO^- + HNO_2 \rightleftharpoons HClO + NO_2^-$ both lie to the right. Which list in order of decreasing strength is correct? (a) $HClO > HNO_2 > H_2O$; (b) $ClO^- > NO_2^- > OH^-$; (c) $NO_2^- > ClO^- > OH^-$; (d) $HNO_2 > HClO > H_2O$; (e) none of these.

9. Which of these is the strongest acid? (a) H_3AlO_3; (b) H_2SiO_3; (c) H_3PO_3; (d) H_2SO_3; (e) $HClO_3$.

SAMPLE TEST

1. Benzoic acid (C_6H_5COOH, 122.0 g/mol) has $K_a = 6.3 \times 10^{-5}$. What is the pH of a solution that was formulated by dissolving 3.050 g of benzoic acid in enough water to make 0.250 L of solution?

2. Classify the following acid-base reactions in terms of the acid-base concept that they represent. In each case, choose the most elementary concept that is applicable. In order of increasing complexity, the acid-base concepts are Arrhenius, Brønsted-Lowry, and Lewis. Indicate the acid(s) and the base(s) in each equation.

A. $HF + N_2H_4 \rightleftharpoons F^- + N_2H_5^+$

B. $O^{2-} + H_2O \rightleftharpoons 2\ OH^-$

C. $SOI_2 + BaSO_3 \rightleftharpoons Ba^{2+} + 2\ I^- + 2\ SO_2$

D. $HgCl_3^- + Cl^- \rightleftharpoons HgCl_4^{2-}$

3. All four of these equilibria lie to the right:

$N_2H_5^+ + CH_3NH_2 \rightleftharpoons N_2H_4 + CH_3NH_3$

$H_2SO_3 + F^- \rightleftharpoons HSO_3^- + HF$ $CH_3NH_3^+ + OH^- \rightleftharpoons CH_3NH_2 + H_2O$

$HF + N_2H_4 \rightleftharpoons F^- + N_2H_5^-$

A. Rank all of the acids involved in order of decreasing acid strength.

B. Rank all of the bases involved in order of decreasing base strength.

C. State whether each of the following two equilibria lies primarily to the right or to the left.

(i) $HF + OH^- \rightleftharpoons F^- + H_2O$ (ii) $CH_3NH_3^+ + HSO_3^- \rightleftharpoons CH_3NH_2 + H_2SO_3$

4. 3.00 mol of calcium chlorite is dissolved in enough water to produce 2.50 L of solution. $K_a = 2.9 \times 10^{-8}$ for HClO and $K_a = 1.1 \times 10^{-2}$ for $HClO_2$. Compute the pH of the solution produced.

17 ADDITIONAL ASPECTS OF ACID-BASE EQUILIBRIA

CHAPTER OBJECTIVES

*** 1. Describe the effect of common ions on the ionization of weak acids and bases, and calculate the concentrations of all species present in solutions of weak acids or bases and their common ions.**

The ionization of a weak acid or base is a dynamic equilibrium, governed by Le Châtelier's principle. If we increase the concentration of one of the products, the equilibrium will shift left. Consider the ionization of a weak acid. We use the techniques of Chapter 16 to compute $[H_3O^+]$ and $[F^-]$ in an aqueous solution of HF.

EXAMPLE 17-1 What are $[F^-]$ and $[H_3O^+]$ in 0.500 M HF? From Table 16-1, $K_a = 6.6\times10^{-5}$.

Equation:	$HF(aq)$	$+\ H_2O(l) \rightleftharpoons$	$H_3O^+(aq)$	$+\ F^-(aq)$
Initial:	0.500 M			
Changes:	$-x$ M		$+x$ M	$+x$ M
Equil:	$(0.500-x)$ M		x M	x M

$$K_a = \frac{[H_3O^+][F^-]}{[HF]} = 6.6\times10^{-4} = \frac{(x)(x)}{0.500-x} \approx \frac{x^2}{0.500}$$

We assumed $x << 0.500\text{ M}$. Then, $x = 0.018\text{ M} = [H_3O^+] = [F^-]$. x indeed is small (less than 5%) compared to 0.500 M. The percent ionization of HF is now computed.
$$\%\text{ ionization} = \frac{0.018\text{ M}}{0.500\text{ M}}\times100 = 3.6\%$$

If we add H_3O^+ or F^- to the solution, the equilibrium will shift to the left, and the percent ionization of HF will decrease. Example 17-2 shows what occurs if we add 0.100 mol H_3O^+ (for example, 0.100 mol HCl) to each liter of the solution of Example 17-1.

EXAMPLE 17-2 What are $[H_3O^+]$ and $[F^-]$ in a solution that has $[HF]_i = 0.500\text{ M}$ and $[HCl]_i = 0.100\text{ M}$?

Equation:	$HF(aq)$	$+\ H_2O(l) \rightleftharpoons$	$H_3O^+(aq)$	$+$	$F^-(aq)$
Initial:	0.500 M		0.100 M		
Changes:	$-x$ M		$+x$ M		$+x$ M
Equil:	$(0.500-x)$ M		$(0.100+x)$ M		x M

$$K_a = \frac{[H_3O^+][F^-]}{[HF]} = 6.6\times10^{-4} = \frac{(0.100+x)\,(x)}{0.500-x} \approx \frac{(0.100+x)\,(x)}{0.500}$$

We have assumed $x << 0.100\text{ M}$.

This gives $x = (0.500)(6.6\times10^{-4})/0.100 = 3.3\times10^{-3} = [F^-]$. Thus, x is small (less than 5%, actually 3.3%) compared to 0.100.

$$\%\text{ ionization} = \frac{0.0033\text{ M}}{0.500\text{ M}}\times100 = 0.66\% \qquad [H_3O^+]_{\text{total}} = 0.103\text{ M}$$

This 0.66% is less than the 3.6% of Example 17-2, as we predicted. The presence of one of the product ions (H_3O^+) has suppressed the ionization of the weak acid, as predicted by Le Châtelier's principle. This effect on a reaction that has ions as products is known as the *common ion effect.* Adding either one of the product ions to the solution suppresses ionization. Let us see what effect the addition of F^- has. Consider that if we add less F^- than we did H_3O^+ in Example 17-2, the suppression of ionization should be smaller. This means that the percent ionization we calculate should be larger than in Example 17-2 but smaller than in Example 17-1.

EXAMPLE 17-3 What are $[H_3O^+]$ and $[F^-]$ in a solution with $[HF]_i = 0.500\text{ M}$ and $[F^-]_i = 0.0400\text{ M}$?

Equation:	$HF(aq)$	$+\ H_2O(l)$	$\rightleftharpoons H_3O^+(aq)$	$+\ F^-(aq)$
Initial:	0.500 M			0.0400 M
Changes:	$-x$ M		$+x$ M	$+x$ M
Equil:	$(0.500-x)$ M		x M	$(0.0400+x)$ M

$$K_a = \frac{[H_3O^+][F^-]}{[HF]} = 6.6\times10^{-4} = \frac{(x)(0.0400+x)}{0.500-x} \approx \frac{(0.0400)(x)}{0.500}$$

We assumed $x \ll 0.0400$. This gives $x = 8.25\times10^{-3}$, which is *not* small compared to 0.0400. We use the technique of successive approximations (Example 16-5, Objective 16-8).

$$\frac{(x)(0.0400+x)}{0.500-x} = \frac{(x)(0.0400+0.0083)}{0.500-0.0083} = \frac{0.0483\,x}{0.492} \approx 6.6\times10^{-4}$$

Then, $x = (6.6\times10^{-4})(0.492)/0.0483 = 0.00672\text{ M}$. We use another cycle of approximation.

$$\frac{(x)(0.0400+x)}{0.500-x} = \frac{(x)(0.0400+0.0067)}{0.500-0.0067} = \frac{0.0467\,x}{0.493} \approx 6.6\times10^{-4}$$

Then, $x = (6.6\times10^{-4})(0.493)/0.0467 = 6.98\times10^{-3}\text{ M}$ *or* 0.00698 M. The last two values (0.00672 and 0.00698) are in fair agreement. The quadratic equation gives 0.0069 M, as does one more cycle of successive approximations. Thus, $[H_3O^+] = 0.0069\text{ M}$ and $[F^-] = 0.0400 + 0.0069 = 0.0469\text{ M}$.

$$\%\text{ ionization} = \frac{0.0069\text{ M}}{0.500\text{ M}}\times100 = 1.4\%$$

This is more than the 0.66% of Example 17-2, as we predicted. But it still is less than the 3.6% of Example 17-1, where HF was the only solute. We now work the same types of problems for a solution of a weak base [trimethylamine, $(CH_3)_3N$, $K_b = 6.3\times10^{-5}$].

EXAMPLE 17-4 What are $[OH^-], [(CH_3)_3\ NH^+]$, and the percent ionization in solutions with the following initial concentrations? (a) $[(CH_3)_3N]_i = 0.400\ M$; (b) $[(CH_3)_3N]_i = 0.400\ M$ and $[OH^-]_i = 0.0900\ M$; (c) $[(CH_3)_3N]_i = 0.400\ M$ and $[(CH_3)_3NH^+]_i = 0.0500\ M$.

(a) Equation:	$(CH_3)_3N(aq) + H_2O(l)$	$\rightleftharpoons$	$(CH_3)_3NH^+(aq)$	$+ OH^-(aq)$
Initial:	0.400 M			
Changes:	$-x$ M		$+x$ M	$+x$ M
Equil:	$(0.400 - x)$ M		x M	x M

$$K_b = \frac{[(CH_3)_3NH^+][OH^-]}{[(CH_3)_3N]} = 6.3\times10^{-5} = \frac{(x)(x)}{0.400 - x} \approx \frac{x^2}{0.400}$$

We assumed $x << 0.40$ M.

Then, $x = 5.0\times10^{-3}\ M = [(CH_3)_3NH^+] << 0.400\ M$. $\quad \%\ \text{ionization} = \frac{5.0\times10^{-3}}{0.400}\times100 = 1.3\%$

(b) Equation:	$(CH_3)_3N(aq) + H_2O(l)$	$\rightleftharpoons$	$(CH_3)_3NH^+(aq)$	$+ OH^-(aq)$
Initial:	0.400 M			0.0900 M
Changes:	$-x$ M		$+x$ M	$+x$ M
Equil:	$(0.400 - x)$ M		x M	$(0.0900 + x)$ M

$$K_b = \frac{[(CH_3)_3NH^+][OH^-]}{[(CH_3)_3N]} = 6.3\times10^{-5} = \frac{(x)(0.0900 + x)}{0.400 - x} \approx \frac{(0.0900)(x)}{0.400}$$

We assume $x << 0.0900$ M. Then, $x = 2.8\times10^{-4}\ M = [(CH_3)_3NH^+] << 0.0900\ M$;

$[OH^-] = 0.0900 + 0.0003 = 0.0903\ M$. $\quad \%\ \text{ionization} = \frac{2.8\times10^{-4}}{0.400}\times100 = 0.070\%$

(c) Equation:	$(CH_3)_3N(aq) + H_2O(l)$	$\rightleftharpoons$	$(CH_3)_3NH^+(aq)$	$+ OH^-(aq)$
Initial:	0.400 M		0.0500 M	
Changes:	$-x$ M		$+x$ M	$+x$ M
Equil:	$(0.400 - x)$ M		$(0.0500 + x)$ M	X M

$$K_b = \frac{[(CH_3)_3NH^+][OH^-]}{[(CH_3)_3N]} = 6.3\times10^{-5} = \frac{(0.0500 + x)(x)}{0.400 - x} \approx \frac{(0.0500)(x)}{0.400}$$

We assumed $x << 0.500$ M. Then, $x = 5.9\times10^{-4}\ M = [OH^-] << 0.0500\ M$;

$[(CH_3)_3NH^+] = 0.0500 + 0.0006 = 0.0506\ M$. $\quad \%\ \text{ionization} = \frac{5.9\times10^{-4}}{0.400}\times100 = 0.15\%$

For a particular situation, you should be able to identify the source of the common ion based on the following information. H_3O^+ or OH^- comes from a strong acid or base (see Table 5-1 in Objective 5-1). The anion of a weak acid comes from a salt in which the cation is that of a strong base. For example,

NaF will add F^- ion to solution. The cation of a weak base comes from a salt in which the anion is that of a strong acid. For instance, $(CH_3)_3NHCl$ will add $(CH_3)_3NH^+$ to solution.

2. Explain why the pH of water changes markedly when a small amount of H_3O^+ or OH^- is added, and why the pH of a buffer does not change very much with a similar addition.

A buffer can be either (a) a solution of a weak acid and about the same amount of the weak acid's anion (which, of course, is a base) *or* (b) a solution of a weak base and about the same amount of the weak base's cation (an acid). The solution of Example 17-4(c) is a buffer. It contains a weak base, 0.400 M $(CH)_3N$, and its cation, 0.0500 M $(CH_3)_3NH^+$. The solution of Example 17-3 contains a weak acid, 0.500 M HF, and its anion, 0.0400 M F^-. However, most chemists would not think of this latter solution as a buffer because the concentrations of the two species should differ by less than a factor of 10.

Thus, for a buffer that contains a weak acid, $0.10 \le \dfrac{[\text{acid's anion}]}{[\text{acid}]} = \dfrac{[A^-]}{[HA]} \le 10.0$ [2]

And, for a buffer that contains a weak base, $0.10 \le \dfrac{[\text{base's cation}]}{[\text{base}]} = \dfrac{[BH^+]}{[B]} \le 10.0$ [3]

Equations [2] and [3] can be combined in a single expression.

$$0.10 \le \frac{[\text{conjugate acid}]}{[\text{conjugate base}]} \le 10.0 \qquad [4]$$

We shall often refer to Expression [2], [3], or [4] as the *buffer condition.* In Example 17-3, $[F^-]_i/[HF]_i = 0.0400\text{ M}/0.500\text{ M} = 0.0800$, and we would not consider the solution to be a buffer.

A buffer keeps the pH of a solution constant by reacting with small amounts of added H_3O^+ or OH^- to remove them from solution. Consider first the buffer of a weak acid, a solution with $[HF]_i = 1.00\text{ M}$ and $[F^-]_i = 0.500\text{ M}$. A small amount of added H_3O^+ will react with the weak acid's anion to produce the weak acid

$$H_3O^+(aq) + HF(aq) \longrightarrow H_2O + F^-(aq) \qquad [5]$$

A small amount of added OH^- will react with the weak acid to produce its anion,

$$OH^-(aq) + HF(aq) \longrightarrow H_2O + F^-(aq) \qquad [6]$$

In both Reactions [5] and [6], one of the fluorine-containing species, either acid or anion, is consumed and the other is produced. The only effects of adding H_3O^+ or OH^- are to produce H_2O and change the ratio of $[F^-]/[HF]$. Also, notice in both cases that a strong acid (H_3O^+) or a strong base (OH^-) is consumed and a weak acid (HF) or a weak base (F^-) is formed in its place. This is the major function of a buffer: to consume strong acid or strong base by reaction with weak base or acid. Thus, both a weak acid and a weak base must be present in a buffer.

A buffer that contains a weak base operates on the same principle, but a bit differently from a weak acid buffer. Consider a buffer of 0.400 M CH_3NH_2 and 1.00 M $CH_3NH_3^+$. A small amount of added H_3O^+ reacts with the weak base and produces its cation,

$$H_3O^+(aq) + CH_3NH_2(aq) \longrightarrow CH_3NH_3^+(aq) + H_2O \qquad [7]$$

A small amount of added OH^- reacts with the weak base cation to produce the weak base

$$OH^-(aq) + CH_3NH_3^+(aq) \longrightarrow CH_3NH_2(aq) + H_2O \qquad [8]$$

Thus, the sole effects of adding H_3O^+ or OH^- are to produce H_2O and to change the $[CH_3NH_3^+]/[CH_3NH_2]$ ratio. It now is clear why pure water is not a buffer. There are no species present to react with added H_3O^+ or OH^-.

Finally, some thoughts on naming buffer solutions: As Equation [4] reveals, every buffer solution contains both a weak acid and its conjugate base, a weak base. We will, however, find it useful to distinguish between two types of buffer solutions. The first of these consists of an uncharged weak acid (such as acetic acid, $HC_2H_3O_2$) and its anion ($C_2H_3O_2^-$) as the conjugate base. We shall often call this a *weak acid buffer*. The second type of buffer solution consists of an uncharged weak base (such as ammonia, NH_3) and its cation (NH_4^+) as the conjugate acid. We often shall call this a *weak base buffer*.

3. Describe how buffer solutions can be prepared.

There are three ways of making a buffer that contains an uncharged weak acid, that is, a *weak acid buffer*.

(1) Place nearly equal quantities of a weak acid and the salt of its anion in solution. 0.600 mol $HC_2H_3O_2$ plus 0.400 mol $KC_2H_3O_2$ in solution is an acetic acid-acetate buffer.

(2) Partly neutralize the weak acid with a strong base. 1.000 mol $HC_2H_3O_2$ plus 0.400 mol KOH in solution yields the same acetic acid-acetate buffer as in Part 1.

$$HC_2H_3O_2 + KOH \longrightarrow C_2H_3O_2^- + K^+ + H_2O$$

	$HC_2H_3O_2$	KOH	$C_2H_3O_2^-$	K^+
before rxn:	1.000 mol	0.400 mol		
after rxn:	0.600 mol	0 mol	0.400 mol	0.400 mol

(3) Partly neutralize the salt of a weak acid's anion with strong acid. 1.000 mol $KC_2H_3O_2$ (a source of $C_2H_3O_2^-$) plus 0.600 mol HCl (a source of H_3O^+) in solution also produce a buffer. This buffer has more $K^+(aq)$ and $Cl^-(aq)$ ions (0.600 mol more of each) than did the previous two. However, these two ions are spectator ions and do not influence the pH of the solution. Specifically, neither $K^+(aq)$ nor $Cl^-(aq)$ hydrolyzes.

$$C_2H_3O_2^- + H_3O^+ \longrightarrow HC_2H_3O_2 + H_2O$$

before rxn:	1.000 mol	0.600 mol	
after rxn:	0.400 mol	0 mol	0.600 mol

A buffer that contains an uncharged weak base (a weak base buffer) also can be made in three ways.

(1) The weak base and about the same amount of its salt (such as, 0.700 mol NH_3 + 0.300 mol NH_4Cl).

(2) The weak base partly titrated with a strong acid (1.000 mol NH_3 + 0.300 mol HCl).

(3) A salt of the weak base's cation partly titrated with a strong base (1.000 mol NH_4Cl + 700 mol NaOH).

4. Know the limitations of the basic equations used to determine the pH of buffer solutions (specifically the Henderson-Hasselbalch equation) by understanding their derivations.

Let us carefully analyze the calculation of the pH of a buffer solution.

EXAMPLE 17-5 What are the $[H_3O^+]$ and pH of a solution with $[HF]_\iota = 1.00$ M and $[F^-]_\iota =$ M?

Equation:	$HF(aq) + H_2O(l) \rightleftharpoons$	$H_3O^+(aq)$ +	$F^-(aq)$
Initial:	1.00 M		0.500 M
Changes:	$-x$ M	$+x$ M	$+x$ M
Equil:	$(1.00E - x)$ M	x M	$(0.500 + x)$ M

$$K_a = \frac{[H_3O^+][F^-]}{[HF]} = 6.6 \times 10^{-4} = \frac{(x)(0.500 + x)}{1.00 - x} \approx \frac{(0.500)(x)}{1.00}$$

We assumed that $x \ll 0.500$ M and find $x = 1.3 \times 10^{-3}$ M $\ll 0.500$ M, pH = 2.89.

$$\text{The exact expression for } K_a \text{ is } K_a = \frac{x([F^-]_\iota + x)}{([HF]_\iota - x)} \qquad [9]$$

The concentrations of HF and F^- are initial ones before dissociation occurs.

$$\text{If we assume that } [F^-]_\iota >> x << [HF]_\iota, \text{ we obtain } K_a = \frac{x([F^-]_\iota)}{([HF]_i)} \qquad [10]$$

Equation [10] is considerably easier to solve that the exact expression. It can be used if $[HF]_i$ and $[F^-]_i$ differ by less than a factor of 10. This is the same condition that we imposed on a buffer, Expression [2]. Recognizing that $x = [H_3O^+]$, we obtain Equation [11] by taking the negative logarithm of Equation [10].

$$pK_a = pH - \log \frac{[F^-]_\iota}{[HF]_\iota} \quad \text{or} \quad pH = pK_a + \log \frac{[F^-]_i}{[HF]_\iota} \qquad [11]$$

If Expression [2] (the buffer condition for a weak acid buffer, $0.10 \le \frac{[A^-]}{[HA]} \le 10.0$) is true, then for a buffer that contains a weak acid (where HA = weak weak acid and A^- = its anion),

$$pH = pK_a + \log \frac{[A^-]}{[HA]_i} \qquad [12]$$

If Expression [3] (the buffer condition for a weak base buffer, $0.10 \le \frac{[BH^+]}{[B]} \le 10.0$) is true, then for a buffer that contains a weak base (where $B =$ weak base and $BH^+ =$ its cation),

$$pOH = pK_b + \log \frac{[BH^+]}{[B]_i} \qquad [13]$$

Equations [12] and [13] are two forms of the Henderson-Hasselbalch equation. Note very carefully that they *require* that the buffer condition—Expressions [2], [3], or [4]—be satisfied. The benefit of using the Henderson-Hasselbalch equation is that all of the work of the derivation has been done, and the calculation is easier. The disadvantage is that you may forget to check to make sure that the buffer condition is satisfied.

*** 5. Calculate the pH of a buffer solution from concentrations of the buffer components and a value of K_a or K, and describe how to prepare a buffer that has a specific pH.**

This objective illustrates the application of Equations [12] and [13].

EXAMPLE 17-6 A solution has $[HCOOH]_i = 0.800$ M and $[Ca(HCOO)_2]_i = 0.230$ M. What is its pH?

We first compute a value of $[HCOO^-]_i$, the initial concentration of formate anion.

$$[HCOO^-]_i = \frac{0.230 \text{ mol } Ca(HCOO)_2}{1 \text{ L}} \times \frac{2 \text{ mol } HCOO^-}{\text{mol } Ca(HCOO)_2} = 0.460 \text{ M}$$

The ratio $[HCOO^-]_i/[HCOOH]_i = (0.460 \text{ M}/0.800 \text{ M}) = 0.575$ and satisfies Expression [2], the weak acid buffer condition. From Table 17-1, for formic acid, $K_a = 1.8 \times 10^{-4}$ and $pK_a = 3.74$.

$$pH = pK_a + \log \frac{[HCOO^-]_i}{[HCOOH]_i} = 3.74 + \log 0.575 = 3.50$$

EXAMPLE 17-7 What is the pH of a solution with $[C_2H_5NH_2]_i = 0.750$ M and $[C_2H_5NH_3]_i = 0.250$ M? The ratio $[C_2H_5NH_3^+]_i/[C_2H_5NH_2]_i = (0.250 \text{ M}/0.750 \text{ M}) = 0.333$ and satisfies Expression [3], the weak base buffer condition. From Table 17-1 for ethylamine, $K_b = 4.4 \times 10^{-4}$ and $pK_b = 3.37$.

Buffer preparation involves first choosing an appropriate acid or base. Equation [12] in conjunction with Condition [2] ($\log 10.0 = +1.0$ and $\log 0.10 = -1.0$) limit the pK_a of the acid to within one unit of the desired pH for a weak acid buffer. In like fashion, Equation [13] in conjunction with Condition [3] limits the pK_a of the base to within one unit of the desired pOH for a weak base buffer. After the weak acid or base is selected, obtaining the desired pH or pOH requires that the correct ratio of either $[A^-]_i/[HA]_i$ or $[BH^+]_i/[B]_i$ be obtained by adjusting the concentrations of the species involved.

*** 6. Determine the changes in pH of buffer solutions resulting from adding acids or bases.**

In Objective 2, we considered the effect of adding a small amount of strong acid (H_3O^+) or strong base (OH^-) to two specific buffers. We now consider the effect of addition to two general buffers—first a weak acid buffer. When a small amount of H_2O^+ is added to this weak acid buffer, some of the weak acid's anion reacts.

$$H_3O^+(aq) + A^-(aq) \longrightarrow H_2O + HA(aq) \qquad [5a]$$

When a small amount of OH^- is added to a weak acid buffer, some of the weak acid reacts.

$$OH^-(aq) + HA(aq) \longrightarrow H_2O + A^-(aq) \qquad [6a]$$

When a small amount of H_3O^+ is added to a weak base buffer, some of the weak base reacts.

$$H_3O^+(aq) + B \longrightarrow H_2O + BH^+(aq) \qquad [7a]$$

The situation is only slightly different with a weak base buffer. When a small amount of OH^- is added to a weak base buffer, some of the weak base's cation reacts.

$$OH^-(aq) + BH^+(aq) \longrightarrow H_2O + B(aq) \qquad [8a]$$

In each case, we first determine the component of the buffer that reacts with the added reagent. The buffer's weak base reacts with strong acid; its weak acid reacts with strong base. Reaction continues until all of either acid or base is consumed. Be careful not to interpret the adjective *weak* as meaning "ineffective." One mole of weak acid reacts with 1 mole of strong base to neutralize it just as effectively as 1 mole of strong acid reacts with 1 mole of that same strong base. *Weak* just means that the acid is not completely ionized. As H_3O^+ from the acid is consumed by the strong base, the undissociated acid produces more H_3O^+ for further reaction until all the acid (or the base) is consumed.

In summary, the effects of these small additions are to produce water and to change the ratio $[A^-]/[HA]$ for a weak acid buffer or the ratio $[BH+]/[B]$ for a weak base buffer. Thus, to calculate the new pH when a small amount of strong acid or base is added, we first compute the concentrations of the components of the buffer present after the appropriate reaction ([5a], [6a], [7a], or [8a]) occurs. Then we compute the ratio $[A^-]/[HA]$ or $[BH+]/[B]$, make sure it satisfies the appropriate buffer condition ([2] or [3]), and finally determine pH or pOH with Equation [12] or [13].

EXAMPLE 17-8. 0.050 mol NaOH is added to 1.00 L of the buffer of Example 17-7. What is the pH?

Equation:	$C_2H_5NH_2(aq) + H_2O \longrightarrow$	$C_2H_5NH_3^+$	$+ OH^-$
Initial:	0.750 M	0.250 M	0.050 M
after rxn [8a]:	0.800 M	0.200 M	

$[C_2H_5NH_3^+]_1/[C_2H_5NH_2]_1 = (0.200\ M/0.800\ M) = 0.250$, which satisfies Condition [3]. From Equation [13], we determine the pOH:

$$\mathrm{pOH} = \mathrm{pK_b} + \log\frac{[\mathrm{C_2H_5NH_3}^+]_i}{[\mathrm{C_2H_5NH_2}]_i} = 3.37 + \log 0.250 = 2.77$$

$$\mathrm{pH} = 14.00 - \mathrm{pOH} = 14.00 - 2.77 = 11.23$$

This is an increase of 0.12 pH unit (from 11.11 to 11.23). If NaOH had been added to pure water, the pH would have gone from pH = 7.00 to pH = 12.70, an increase of 5.70 pH units.

*** 7. Define and compute values for *buffer range* and *buffer capacity*.**

Buffer range is the range of pH values over which a buffer satisfies Condition [2] or [3]. We have

$$\text{pH range of an acid buffer} = \mathrm{p}K_a \pm 1.00 \qquad \text{pOH range of a base buffer} = \mathrm{p}K_b \pm 1.00 \quad [14]$$

The capacity of a buffer is the number of moles of strong acid (H_3O^+) or strong base (OH^-) that the buffer can absorb before Condition [2] or [3] becomes invalid. The capacity of a buffer for OH^- may differ from its capacity for H_3O^+. Also, buffer capacity depends on the total volume of buffer solution; 10 L of solution has five times the buffer capacity of 2.0 L. Another factor is the concentrations of the two components of the buffer. A buffer with $[\mathrm{HF}]_i = 1.00\,\mathrm{M}$ and $[\mathrm{HF}]_i = 0.500\,\mathrm{M}$ has twice the capacity as one with $[\mathrm{HF}]_i = 0.500\,\mathrm{M}$ and $[\mathrm{F}^-]_i = 0.250\,\mathrm{M}$.

EXAMPLE 17-9 What is the capacity toward (a) strong acid and (b) strong base of 2.00 L of buffer with $[\mathrm{HF}]_i = 0.400$ M and $[\mathrm{F}^-]_i = 0.700$ M? (Note that $[\mathrm{HF}]_i + [\mathrm{F}^-]_i = 1.100\,\mathrm{M}$ throughout this problem since no fluorine-containing species are added. For simplicity, we assume that the volume of the buffer solution also does not change.)

(a) The weak base in the buffer reacts with all of the added strong acid.

Equation:	$F^-(aq)$ +	$H_3O^+(aq)$ $\longrightarrow$	$HF(aq)$ + H_2O
Initial:	0.700 M	x M	0.400 M
Changes:	$-x$ M	$-x$ M	$+x$ M
Equil:	$(0.700 - x)$	M	$(0.400 + x)$ M

The buffer is exhausted when enough acid is added so that $[\mathrm{F}^-]_i/[\mathrm{HF}]_i = 0.10.0 = (0.700 + x)/(0.400 - x)$. This gives $x = 0.600$. The buffer capacity toward H_3O^+ is $0.600\,\mathrm{M}\,\mathrm{H_3O^+} \times 2.00\,\mathrm{L\ solution} = 1.20\,\mathrm{mol}\,\mathrm{H_3O^+}$.

(b) The weak acid in the buffer reacts with all of the added strong base.

Equation:	$HF(aq)$ +	$OH^-(aq)$ $\longrightarrow$	$F^-(aq)$ + H_2O
Initial:	0.400 M	y M	0.700 M
Changes:	$-y$ M	$-y$ M	$+y$ M
Equil:	$(0.400 - y)$	M	$(0.700 + y)$ M

The buffer is exhausted when enough base is added so that $[\mathrm{F}^-]_i/[\mathrm{HF}]_i = 10.0 = (0.700 + y)/(0.400 - y)$. This gives $y = 0.300$. The buffer capacity toward OH^- is $0.300\,\mathrm{mol\,OH^-/L} \times 2.00\,\mathrm{L\ soln} = 0.600\,\mathrm{mol\,OH^-}$.

7a. Describe how the blood buffer system works.

Blood contains at least three buffers, the most important being a $H_2CO_3 - HCO_3^-$ buffer.

$$H_2CO_3(aq) + H_2O \rightleftharpoons H_3O^+(aq) + HCO_3^-(aq) \qquad [15]$$

To maintain $pH = 7.4$, the ratio $[HCO_3^-]/[H_2CO_3] - 20$. Thus, there is a large amount of $HCO_3^-(aq)$ present to absorb excess acidity. Excess base reacts with H_2CO_3, which is replenished from CO_2 in the lungs.

$$CO_2(g) + H_2O \rightleftharpoons H_2CO_3(aq) \qquad [16]$$

8. Explain how an acid-base indicator works to determine the equivalence point in a titration.

An indicator is a weak acid, HIn, which has a different color than its anion, In^-. The pH color change range of an indicator spans two pH units, from one unit below pK_{In} of the indicator to one pH unit above. In the titration procedure, the material to be titrated is placed in a flask, indicator is added, and titrant of known concentration is added from a buret until the indicator changes color— at the *endpoint*. The *equivalence point* occurs when chemically equivalent amounts of acid and base are present in solution. If the indicator has been chosen correctly (see objective 17-11), the endpoint is the same as the equivalence point.

*** 9. Calculate pH values and plot the titration curve of a strong acid with a strong base, or a strong base with a strong acid.**

The common features of plotting titration curves are most easily seen by first considering the titration of a strong acid with a strong base, where the complications of a small ionization constant are absent. Consider the titration of 25.00 mL of 0.112 M NaOH with 0.100 M HCl solution. We use strong acid as titrant to provide an example that is different from that of the text. We have divided the procedure into seven steps.

(1) Write the titration reaction, [17], as a net ionic equation, [18].

$$NaOH(aq) + HCl(aq) \longrightarrow NaCl(aq) + H_2O \qquad [17]$$

$$OH^-(aq) + H_3O^+(aq) \longrightarrow 2\,H_2O \qquad [18]$$

(2) Determine the amount of substance being titrated. (We use $0.112\,M = 0.112\,mol/L = 0.112\,mmol/mL$.)

$$25.00\,mL\,NaOH\,soln \times \frac{0.112\,mmol\,NaOH}{mL\,soln} = 2.80\,mL\,HCl\,soln$$

(3) Find the volume of titrant needed to reach the equivalence point.

$$2.80\,mmol\,NaOH \times \frac{1\,mmol\,HCl}{1\,mmol\,NaOH} \times \frac{1.00\,mL\,HCl\,soln}{0.100\,mmol\,HCl} = 28.0\,mL\,HCl\,soln$$

This volume is known also as the *100% titration point*. Now consider that we want to draw a curve like Figure 17-1 by calculating as few points as possible. A good titration curve can be drawn from the pH at 0%, 10%, 50%, 90%, 100%, and 110% titration. For each point (except

0%) we need (a) the volume of titrant added, V_t; (b) the total volume of solution, V_s; (c) the amount (in moles) of each species in solution; and (d) the concentration of each species.

(4) Determine the pH at 0% titration.

$[OH^-] = 2.80\ \text{mmol}/25.00\ \text{mL} = 0.112 = [Na^+]$ $pOH = -\log(0.112) = 0.951$ *or* $pH = 13.049$

(5) Determine the pH at 100% titration ($V_t = 28.00$ mL and $V_s = 53.00$ mL).

We find the number of moles of each species:

$$OH^-(aq) + H_3O(aq) \longrightarrow 2\,H_2O(l)$$

before rxn:	2.80 mmol	2.80 mmol
after rxn:	0 mmol	0 mmol

Thus, the solution is neutral and pH = 7.000. At the endpoint of a *strong* acid and *strong* base titration, pH = 7.000.

(6) Determine the pH at 110% titration ($V_t = 30.00$ mL titrant and $V_s = 55.80$ mL solution total).

$$OH^-(aq) + H_3O(aq) \longrightarrow 2\,H_2O(l)$$

before rxn:	2.80 mmol	3.08 mmol
after rxn:	0 mmol	0.28 mmol

Since neither Na^+ nor Cl^- hydrolyzes, they do not affect the pH of the solution and we need not calculate their concentrations:

$[H_3O^+] = (0.28\ \text{mmol}/55.8\ \text{mL}) = 5.0 \times 10^{-3}$ *or* $pH = 2.30$

The pH after the equivalence point is based on the total solution volume and the excess amount (mol or mmol) of titrant added. This is true of *every* titration in which the titrant is a strong acid or strong base.

(7) Determine the pH between 0% and 100% titration.

50% titration ($V_t = 14.00$ mL and $V_s = 39.00$ mL).

$$OH^-(aq) + H_3O^+(aq) \rightarrow 2\,H_2O(l) \qquad [OH^-] = \frac{1.40\ \text{mmol}}{39.0\ \text{mL}} = 0.0359\ \text{M}$$

before: 2.80 mmol 1.40 mmol
after: 1.40 mmol 0 mmol $pOH = 1.444$ and $pH = 12.556$

10% titration ($V_t = 2.80$ mL and $V_s = 27.80$ mL).

$$OH^-(aq) + H_3O^+(aq) \rightarrow 2\,H_2O(l) \qquad [OH^-] = \frac{2.52\ \text{mmol}}{27.8\ \text{mL}} = 0.0906\ \text{M}$$

before: 2.80 mmol 0.28 mmol
after: 2.52 mmol 0 mmol $pOH = 1.444$ and $pH = 12.556$

90% titration ($V_t = 25.20$ mL and $V_s = 50.20$ mL).

$$OH^-(aq) \quad + \quad H_3O^+(aq) \rightarrow 2H_2O(l) \qquad [OH^-] = \frac{0.28 \text{ mmol}}{50.2 \text{ mL}} = 0.00558 \text{ M}$$

before: 2.80 mmol   2.52 mmol

after: 0.28 mmol   0 mmol $pOH = 2.253$ and $pH = 11.747$

*** 10. Calculate pH values and plot the titration curve of a weak acid with a strong base or of a weak base with a strong acid.**

We now consider the titration of 25.00 mL of 0.112 M $(CH_3)_3N$ ($K_b = 7.4 \times 10^{-5}$) with 0.100 M HCl. To better see the differences between titrating strong base and weak base, the concentrations of acid and base and the volume of base to be titrated are the same as Objective 17-9. We follow the procedure of that objective:

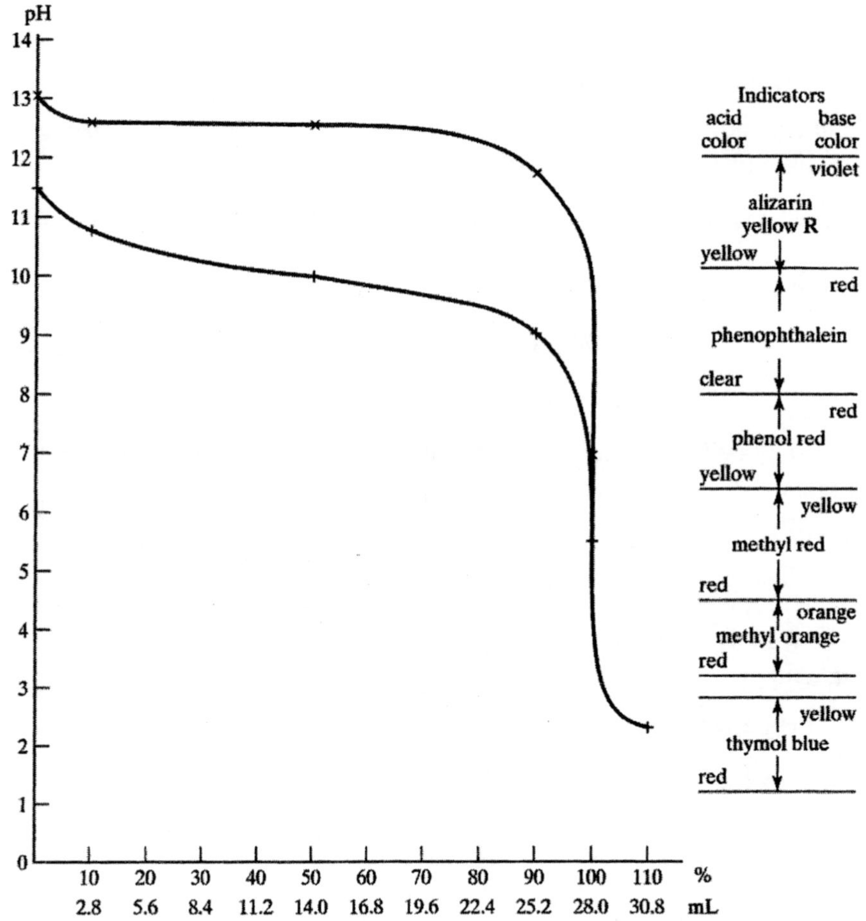

FIGURE 17-1 Titration Curves: Titrating 25.00 mL of 0.112 M NaOH (×) or $(CH_3)_3N$ (+) with 0.100 M HCl.

(1) The titration reaction:

$(CH_3)_3N(aq) + HCl(aq) \longrightarrow (CH_3)_3NHCl(aq)$ whole equation

$(CH_3)_3N(aq) + H_3O^+(aq) \longrightarrow (CH_3)_3NH^+(aq) + H_2O$ net ionic equation

(2) The amount of $(CH_3)_3N$ being titrated:

$$25.00\ \text{mL}\ (CH_3)_3N\ \text{soln} \times \frac{0.112\ \text{mmol}\ (CH_3)_3N}{\text{mL soln}} = 2.80\ \text{mmol}\ (CH_3)_3N$$

(3) The volume of titrant needed:

$$2.80\ \text{mmol}\ (CH_3)_3N \times \frac{\text{mmol HCl}}{\text{mmol}\ (CH_3)_3N} \times \frac{1.00\ \text{mL HCl soln}}{0.100\ \text{mmol HCl}} = 2.80\ \text{mL HCl soln}$$

(4) pH at 0% titration. This is a weak base problem (see Objective 17-8, if necessary).

Equation:	$(CH_3)_3N(aq) + H_2O(l) \rightleftharpoons$	$(CH_3)_3NH^+(aq) +$	$OH^-(aq)$
Initial:	0.112 M		
Changes:	$-x$ M	$+x$ M	$+x$ M
Equil:	$(-0.112 - x)$ M	x M	x M

$$K_b = \frac{[(CH_3)_3NH^+][OH^-]}{[(CH_3)_3N]} = 7.4 \times 10^{-5} = \frac{(x)(x)}{0.112 - x} \approx \frac{x^2}{0.112} \text{(assuming } x << 0.112)$$

$$x = 2.9 \times 10^{-3}\ \text{M} = [OH^-] << 0.112 \quad \text{pOH} = 2.54 \quad \text{AND} \quad \text{pH} = 11.46$$

(5) pH at 100% titration ($V_t = 28.0$ mL titrant and $V_s = 53.0$ mL solution total).

$$(CH_3)_3N(aq) + H_3O^+(aq) \longrightarrow (CH_3)_3NH^+(aq) + H_2O$$

before:	2.80 mmol	2.80 mmol	
after:	0 mmol	0 mmol	2.80 mmol

Now we have the hydrolysis of a weak base's cation (see Objective 17-11).

$[(CH_3)_3NH^+]_i = 2.80\ \text{mmol}/53.0\ \text{mL} = 0.0528\ \text{M}$

Equation:	$(CH_3)_3NH^+(aq) + H_2O(l) \rightleftharpoons$	$(CH_3)_3N(aq) +$	$H_3O^+(aq)$
Initial:	0.0528 M		
Changes:	$-x$ M	$+x$ M	$+x$ M
Equil:	$(0.0528 - x)$ M	x M	x M

Note carefully: pH = 7.000 at the equivalence point *only* for strong acid-strong base titrations.

(6) pH at 110% titration: This is the same calculation and result as Step (6) of Objective 17-9. pH = 2.30.

(7) pH at points between 0% and 100% titration:

50% titration ($V_t = 14.00$ mL and $V_s = 39.00$ mL 39.00 mL)

$$(CH_3)_3N(aq) + H_3O^+(aq) \longrightarrow (CH_3)_3NH^+(aq) + H_2O$$

before:	2.80 mmol	1.40 mmol	
after:	1.40 mmol	0 mmol	1.40 mmol

$$[(CH_3)_3N] = 1.40\ \text{mmol}/39.00\ \text{mL} = [(CH_3)_3NH^+]$$

This is a buffer solution; we can use Equation [13]:

$$\text{pOH} = \text{p}K_b + \log\frac{[(CH_3)_3NH^+]_i}{[(CH_3)_3N]_i} = 4.13 + \log\frac{1.40\ \text{mmol}/39.0\ \text{mL}}{1.40\ \text{mmol}/39.0\ \text{mL}}$$

Notice that the two volumes are the same; thus, they cancel. This is particularly helpful in titration calculations. Note also that the amounts in millimoles of weak base and its conjugate acid are the same at the 50% point:

$$\text{pOH} = \text{p}K_b + \log 1.00 = \text{p}K_b = 4.13 \quad \textit{and} \quad \text{pH} = 9.87$$

The 50% titration point is the *half-equivalence point*. When a weak acid is titrated with a strong base, $\text{pH} = \text{p}K_a$ at the 50% titration point.

10% titration ($V_t = 2.80\ \text{mL}$ and $V_s = 27.80\ \text{mL}$)

$$(CH_3)_3N(aq) + H_3O^+(aq) \longrightarrow (CH_3)_3NH^+(aq) + H_2O$$

before:	2.80 mmol	0.28 mmol	
after:	2.52 mmol	0 mmol	0.28 mmol

Again, we have a buffer; thus, we use Equation [12].

$$\text{pOH} = 4.13 + \log\frac{0.28\ \text{mmol}}{2.52\ \text{mmol}} = 4.13 - 0.95 = 3.18 \quad \text{and} \quad \text{pH} = 10.82$$

We used the amounts in mmol of weak base and its cation in the ratio, since they are in the same volume. A further simplification is possible. Consider that 10% titration means that the fraction titrated = 0.10 and the fraction untitrated = 0.90. Notice that, in this case, 0.28 mmol = (0.10)(2.80 mmol) and 2.52 mmol = (0.90)(2.80 mmol), where 2.80 mmol is the amount of $(CH_3)_3N$ being titrated (Step 2). Thus,

$$\text{pOH} = 4.13 + \log\frac{(0.10)(2.80\ \text{mmol})}{(0.90)(2.80\ \text{mmol})} = 4.13 + \log\frac{0.10}{0.90}$$

In general, for the titration of a weak base by a strong acid,

$$\text{pOH} = \text{p}K_b + \log\frac{\text{fraction of base titrated}}{\text{fraction of base untitrated}} \qquad [19]$$

For the titration of a weak acid by a strong base

$$\text{pH} = \text{p}K_a + \log \frac{\text{fraction of acid titrated}}{\text{fraction of acid untitrated}} \qquad [20]$$

Equations [19] and [20] are based on Equations [14] and [13], which in turn depend on Conditions [3] and [2]. Thus, Equations [19] and [20] are restricted to the region from 9% titrated to 91% titrated—the *buffer region.*

90% titrated: We use Equation [19] directly.

$$\text{pOH} = \text{p}K_b + \log(0.90/0.10) = 4.13 + 0.95 = 5.08 \quad \text{AND} \quad \text{pH} = 8.92$$

*** 11. Plot titration curves. Use those curves to determine the initial pH, buffer region, the pH of the equivalence point, as well as to select an appropriate indicator.**

The results of Objectives 17-9 and 17-10 are collected in Table 17-1 and plotted in Figure 17-1, with the NaOH results as x's and the $(CH_3)_3N$ results as +'s. The *abscissa* (the horizontal axis) can be labeled either "percent titration" or "volume of titrant." Notice the gradual change in pH for the titration of the weak base in the buffer region.

TABLE 17-1 Strong Acid Titration of a Strong and a Weak Base

Percent Titration	Volume of titrant, mL	pH for the titration of NaOH	pH for the titration of $(CH_3)_3N$
0	0.00	13.049	11.43
10	2.80	12.597	10.75
50	14.00	12.556	9.80
90	25.20	11.747	8.85
100	28.00	7.000	5.54
110	30.80	2.30	2.30

To choose an indicator, we consider the equivalence point. (Note the indicators given at the right in Figure 17-1.) For the NaOH-HCl titration, we could choose phenophthalein, which is red at $\text{pH} = 10$ and colorless at $\text{pH} = 8$, but the color change from very faint red to just colorless is a bit hard to see. A better indicator is phenol red: red at $\text{pH} = 8$ and yellow at $\text{pH} = 6.4$. Phenol red would not work for the $(CH_3)_3N-HCl$ titration because it would be starting to change color at about 96% titration. A better indicator would be methyl red: yellow at $\text{pH} = 6.2$ and red at $\text{pH} = 4.5$. It might seem that this indicator would be suitable for the NaOH-HCl titration. But we prefer that the equivalence point fall within the pH color change range of the indicator.

*** 12. Calculate the pH of certain solutions of salts of polyprotic acids.**

Polyprotic acids produce three distinct kinds of chemical species in aqueous solution: (1) the acid itself [such as $H_2CO_3(aq)$], (2) the anion of the partly neutralized acid [such as $HCO_3^-(aq)$], and (3) the anion of the totally neutralized acid [such as $CO_3^{2-}(aq)$]. Calculation of the pH of a solution of the original acid was discussed in Objective 16-9. The pH of a solution of the anion of the totally neutralized acid is a hydrolysis problem (see Objective 16-11). Calculation of the pH of a solution of the anion of the partly neutralized acid[$HCO_3^-(aq)$] is not a straightforward task, because two reactions are occurring: ionization and hydrolysis.

$$HCO_3^-(aq) + H_2O \rightleftharpoons CO_3^{2-}(aq) + H_3O^+(aq) \qquad [21]$$

$$HCO_3^-(aq) + H_2O \rightleftharpoons H_2CO_3(aq) + OH^-(aq) \qquad [22]$$

For Equation [21], $K_a = K_{a2}$, and for Reaction [22], $K_b = K_w / K_{a1}$. The pH of a solution that contains significant hydrogen carbonate ion (at a concentration of $[HCO_3^-]_i > 0.10$ M) is the average of pK_{a1} and pK_{a2}.

$$pH = \frac{1}{2}(pK_{a1} + pK_{a2}) \qquad [23]$$

Note that the ionization constants used in Equation [23] are those of two species: the anion placed in solution (HCO_3^-) and the species that results from hydrolysis (H_2CO_3) of that anion. Solutions of these anions (of partly neutralized polyprotic acids) are important for a reason other than the uniqueness of the calculation of their pH's. These ions are both anions of weak acids and "cations" of weak bases. Thus, their solutions can act as buffers, reacting with both strong acid and with strong base (Equations [24] and [25]).

$$H_3O^+(aq) + HCO_3^-(aq) \longrightarrow H_2CO_3(aq) + H_2O \qquad [24]$$

$$OH^-(aq) + HCO_3^-(aq) \longrightarrow CO_3^{2-}(aq) + H_2O \qquad [25]$$

SELF-ASSESSMENT EXERCISES

96. (a) $K_a = \dfrac{[C_7H_5O_2^-][H_3O^+]}{[HC_7H_5O_2]} = 6.5 \times 10^{-5}$

$$= \frac{x^2}{0.0100 - x} = 6.5 \times 10^{-5}$$

Using the quadratic equation and solving for x, x = 5.39×10^{-3} M.

pH = -log 5.39×10^{-3} = 2.27.

(b) The $[H_3O^+] = 5.39 \times 10^{-3}$ M. By dilution, the concentration will be reduced to 4.31×10^{-3} M.

4.31×10^{-3} mol $L^{-1} \times 0.3125$ L = 1.35×10^{-4} moles.

$Ba(OH)_2$ contains 0.010 mol $L^{-1} \times 2$ mol $[OH^-]$/mol $Ba(OH)_2 \times 0.00625$ L =

1.25×10^{-4} mol to neutralize H_3O^+, $[H_3O^+] = 1.0 \times 10^{-5}$ mol in 0.03125 L

= 3.2×10^{-6} M and pH = 5.50.

(c) At the equivalence point, 100 % titration, $V_t = 25.0$ mL, $V_s = 12.5$ mL

	$HC_7H_5O_2$	$+ OH^- \leftrightarrow$	$C_7H_5O_2^-$	$+ H_2O$
initial	0.25 mmol	0.25 mmol	0	
final	0	0	0.25 mmol	

The problem is now one of a conjugate base of a weak acid, $[C_7H_5O_2^-] = 0.25$ mmol/37.5 mL

$= 6.7 \times 10^{-3}$ M.

	$C_7H_5O_2^-$	$+ H_2O$	$\leftrightarrow HC_7H_5O_2$	$+ OH^-$
initial	6.7×10^{-3}		0	0
change	-x		+x	+x

And $K_b = K_w/K_a = 1.0 \times 10^{-14}/6.5 \times 10^{-5} = x^2/\ 6.7 \times 10^{-3}$

$X = 1.01 \times 10^{-4}$, pOH = 4.00 and pH = 10.0.

(d) Calculate the number of moles:

0.01 M $HC_7H_5O_2$ x 0.025 L = 2.5×10^{-4} mol $[H_3O^+]$

0.01 M $Ba(OH)_2$ x 0.015 L x 2 mol $[OH^-]/Ba(OH)_2 = 3.0 \times 10^{-4}$ $[OH^-]$

After neutralization, there is an excess of 0.5×10^{-4} mol $[OH^-]$.

The total volume is 40 mL and 5×10^{-5} mol/0.040 L = 1.25×10^{-3} M $[OH^-]$.

pOH = 2.90 and pH = 14 – 2.90 = 11.1.

97. To repress ionization, add something that increases ions on the right-hand side of the reaction or remove ions from the left-hand side (c) $NaCHO_2$ will repress ionization.

98. (d) $NaHCO_3$ because HCO_3^- will react with some of the $[H^+]$ on the right-hand side, forcing the reaction to the right.

99. (b) Raising the pH will increase the $[OH^-]$, which will react with NH_4^+ to increase NH_3. Adding acid will increase NH_4^+.

100. (a) KOH is a base, so it will raise the pH. NH_3 - NH_4Cl is a buffer, so the effect will not be very large.

DRILL PROBLEMS

1. Determine the pH of each of the following solutions. First, when only the uncharged weak acid or weak base is present, and then when both substances are present. Initial concentrations are given in each case.

A. $[HC_2H_3O_2] = 0.250\ M, [NaC_2H_3O_2] = 0.120M$
B. $[HC_2H_3O_2] = 0.250\ M, [HCl] = 0.140\ M$
C. $[HC_7H_5O_2] = 1.20\ M, [Ca(C_7H_5O_2)_2] = 0.13\ M$
D. $[HCHO_2] = 0.820\ M, [HNO_3] = 0.240\ M$
E. $[HNO_2] = 0.400\ M, [Ba(NO_2)_2] = 0.0480\ M$
F. $[HCHO_2] = 0.620\ M, [KCHO_2] = 0.340\ M$
G. $[HClO_2] = 1.00\ M, [CsClO_2] = 0.138\ M$
H. $[NH_3] = 1.20\ M, [KOH] = 0.320\ M$
I. $[CH_3NH_2] = 0.260\ M, [CH_3NH_3Cl] = 0.130M$
J. $[C_2H_5NH_2] = 0.150\ M, [NaOH] = 0.216\ M$
K. $[(CH_3)_3N] = 0.460\ M, [(CH_3)_3NHI] = 0.819\ M$
L. $[NH_3] = 0.510\ M, [(NH_4)_2SO_4] = 0.774\ M$
M. $[CH_3NH_2] = 0.104\ M, [Ca(OH)_2] = 0.0102\ M$
N. $[C_2H_5NH_2] = 0.630M, [C_2H_5NH_3NO_3] = 1.89\ M$

2 The buffer solutions in each following part are made up as described in Objective 17-3. Determine the initial concentrations of the weak acid and its anion, or of the weak base and its cation. Use these concentrations to determine the pH of each buffer. (The concentrations given below are formalities; that is, how the solution was made up, not the concentrations at equilibrium.)

A. $[HC_2H_3O_2] = 1.450M, [NaC_2H_3O_2] = 0.270M$
B. $[HC_7H_5O_2 = 1.300\ M, [KOH] = 0.640\ M$
C $[NaCHO_2] = 0.815\ M, [HBr] = 0.105\ M$
D. $[NH_3] = 0.750\ M, [NH_4I] = 0.250\ M$
E. $[C_2H_5NH_2] = 2.31\ M, [HNO_3] = 1.02\ M$
F. $[CH_3NH_3Cl] = 0.555\ M, [CsOH] = 0.114\ M$
G. $[HF] = 1.296\ M, [NaF] = 1.045\ M$
H. $[HC_2H_2ClO_2] = 2.93\ M, [Mg(OH)_2] = 1.07\ M$
I. $[Ca(NO_2)_2] = 0.926\ M, [HClO_4] = 1.302\ M$
J. $[CH_3NH_2] = 0.990M, [CH_3NH_3ClO_4] = 0.443\ M$
K. $[(CH_3)_3N] = 1.236\ M, [HI] = 0.314\ M$
L. $[(NH_4)_2SO_4] = 0.1140\ M, [KOH] = 0.0731\ M$
M. $[HCHO_2] = 1.946M, [Mg(CHO_2)_2] = 0.104M$
N $[HC_2H_3O_2] = 0.0561\ M, [Ca(OH)_2] = 0.0112\ M$
O. $[NaC_7H_5O_2] = 0.743\ M, [H_2SO_4] = 0.256\ M$
P. $[(CH_3)_3NHBr] = 0.831\ M, [KOH] = 0.218\ M$
Q. $[CH_3NH_2] = 1.214\ M, [CH_3NH_3NO_3] = 0.885\ M$
R. $[C_2H_5NH_2] = 1.149\ M, [HClO_4] = 1.001\ M$

3. The amount of strong acid and strong base given in each following part is added to the given volume of the corresponding buffer in the drill problems for Objective 17-5. Determine the final pH in each case (addition of strong acid, then addition of strong base) for each solution.

A. 0.0520 mol to 0.750 L B. 0.0130 mol to 0.800 L C. 0.00780 mol to 1.00 L

D. 0.0110 mol to 121 mL	E. 0.544 mol to 2.04 L	F. 0.123 mol to 5.91 L
G. 0.247 mol to 1.20 L	H. 1.32 mol to 11.4 L	I. 0.963 mol to 8.71 L
J. 0.0231 mol to 0.604 L	K. 0.150 mol to 0.890 L	L. 1.09 mol to 75.9 L
M. 0.0231 mol to 0.561 L	N. 0.196 mol to 103 L	O. 0.0931 mol to 1.52 L
P. 0.161 mol to 3.75 L	Q. 7.14 mol to 10.5 L	R. 0.100 mol to 1.00 L

4. For each of the buffers in the drill problems for Objective 17-5, determine the buffer range. Then, using the volume of each buffer given in the problems for Objective 17-6, determine the capacity of each buffer for both strong acid and strong base. Express each capacity in moles of strong acid, and moles of strong base.

5. Determine the pH at 0%, 10%, 50%, 90%, 100%, and 110% titration for each of the following titrations. (Objective 17-9: A through G; Objective 17-10: H through N.) Volumes are given in mL.

	Substance being titrated			*Titrant*	
	Volume	*Formula*	*Molarity*	*Formula*	*Molarity*
A.	25.00	HCl	0.245	KOH	0.200
B.	15.00	NaOH	0.250	HNO_3	0.100
C.	30.00	CsOH	2.50	HI	1.00
D.	10.00	$HClP_4$	0.120	RbOH	0.100
E.	20.00	HBr	0.250	KOH	0.100
F.	16.00	NaOH	0.624	H_2SO_4	0.100
G.	40.00	$Ca(OH)_2$	0.0100	HCl	0.0250
H.	25.00	HNO_2	0.245	KOH	0.200
I.	15.00	CH_3NH_2	0.250	HNO_3	0.100
J.	30.00	$C_2H_5NH_2$	2.50	HI	1.00
K.	10.00	HF	0.120	RbOH	0.100
L.	20.00	$HCHO_2$	0.250	KOH	0.100
M.	16.00	CH_3NH_2	0.624	H_2SO_4	0.100
N.	40.00	C_5H_5N	0.0200	HCl	0.0250

6. For each of the titrations in the drill problems of Objective 17-9 and 17-10, plot the titration curve. Indicate the equivalence point and, for the titration of a weak acid or weak base, the buffer region. Finally, choose a suitable indicator from those given in Figure 17-1.

7. Determine the pH of a solution of each of the following compounds. Use the data of Table 16-1. (The numbers in parentheses are the pK_a values for successive ionizations of the parent acid.)

A. $KHCO_3$ B. $Ca(HSe)_2$ C. NaH_2PO_3 D. KHC_2O_4

E. NaH_2AsO_3 (2.22, 6.98, 11.50) F. $Ca(HCrO_4)_2$ (-0.98, 6.50)

G. $Ca(HAsO_4)$ (2.22, 6.98, 11.50) H. $KHC_8H_4O_4$ (2.95, 5.41)

I. $NaHN_2O_2$ (7.05, 11.4) J. $CaHVO_4$ (3.78, 7.8, 13.0)

K. $NaHGeO_3$ (9.0, 12.4) L. $CaHC_9H_3O_6$ (2.52, 3.84, 5.20)

QUIZZES (20 minutes each) Choose the best answer for each question.

QUIZ A

1. What volume in mL of 2.0 M H_2SO_4 is needed to neutralize 7.8 g of $Al(OH)_3$ (78.0 g/mol)? $Al_2(SO_4)_3$ and water are the titration products. (a) 150; (b) 75; (c) 300; (d) 500; (e) none of these.

2. A weak base, B, has $K_b = 4.0 \times 10^{-5}$. If $[OH^-] = 2.0 \times 10^{-5}$ M, what must $[BH^+]/[B]$ equal? (a) 1.0; (b) 0.50; (c) 1.5; (d) 8.0; (e) none of these.

3. Which of the following mixtures, when added to 1.00 L of water, will produce a buffer solution? (a) 1 mol $HC_2H_3O_2$ and 0.5 mol HCl; (b) 1 mol NH_3 and 0.5 mol NaOH; (c) 1 mol NH_4Cl and 0.5 mol HCl; (d) 1 mol $HC_2H_3O_2$ and 0.5 mol NaOH; (e) none of these.

4. 20.0 mL of 0.200 M NaOH is added to 50.0 mL of 0.100 M $HC_2H_3O_2$ ($K_a = 1.8 \times 10^{-5}$). What is the pH of this solution? (a) 3.72; (b) 11.13; (c) 2.87; (d) 10.71; (e) 3.29.

5. The solution that is added from the buret during a titration is the (a) buffer; (b) titrant; (c) indicator; (d) base; (e) none of these

6. A weak acid and the sodium salt of its anion in solution in equimolar amounts has (a) $pH > 7$; (b) $pH < 7$; (c) $pH = 7$; (d) depends only on concentration; (e) none of these.

7. An indicator, HIn, is purple, and its anion, In^-, is yellow. The indicator changes color in the range pH 8-10. This indicator will be yellow in which of the following? (a) 0.010 M HCl(aq); (b) 0.10 M $NaC_2H_3O_2$ ($K_a = 1.8 \times 10^{-5}$ for $HC_2H_3O_2$); (c) pure H_2O; (d) 0.010 M NaOH; (e) none of these.

QUIZ B

1. What volume in mL of 1.50 M NaOH is needed to neutralize 45.0 mL of 0.500 M acetic acid? (a) 15.0; (b) 22.5; (c) 33.8; (d) 135; (e) none of these.

2. A weak acid has $K_a = 4.2 \times 10^{-3}$. If $[A^-] = 2.0$ M, what must be [HA] so that $[H^+] = 2.1 \times 10^{-3}$ M? (a) 0.500 M; (b) 2.0 M; (c) 1.0 M; (d) 1.5 M; (e) none of these.

3. The addition of NH_4Cl to a solution of NH_3 would (a) decrease total $[NH_4^+]$; (b) decrease $[NH_3]$; (c) decrease the K_i of NH_3; (d) increase $[OH^-]$ in the solution; (e) decrease the pH of the solution.

4. A strong base and the chloride salt of its cation in water has (a) $pH > 7$; (b) $pH < 7$; (c) $pH = 7$; (d) pH that depends on $[Cl^-]$; (e) none of these.

5. Phenophthalein (color-change range of 8–10) may be used for the titration of (a) a weak acid with a weak base; (b) a weak base with a strong acid; (c) a weak acid with a strong base; (d) any acid with any base; (e) any base with any acid.

6. 10.0 mL of 0.100 M $NH_3 = (K_b = 1.8\times10^{-5})$ is mixed with 5.00 mL of 0.200 M NH_4Cl. The resulting solution has (a) pH = 4.76; (b) $[NH_4^+]$ larger than in 0.100 M NH_3; (c) $[OH^-] = 3.6\times10^{-5}$ M; (d) $[H^+]$ of about 10^{-3} M; (e) none of these.

7. In the neutralization of 50.0 mL of 0.100 M B (a weak base with $K_b = 1.6\times10^{-7}$) with 0.100 M H_2SO_4, the most correct description of the solution at the midpoint of the titration (that is, half-neutralized) is a solution (a) whose volume is 75.0 mL and which contains significant concentrations of unionized B molecules and BH^+, OH^-, and HSO_4^- ions; (b) whose volume is 62.5 mL and which contains the same species listed in (a); (c) in which $[BH^+]$ is about twice [B]; (d) in which $[BH^+]$ is about 4×10^{-4} M; (e) none of these.

QUIZ C

1. What volume in mL of 1.7 M H_2SO_4 is needed to neutralize 68 mL of 2.5 M NaOH? Na_2SO_4 is produced. (a) 100; (b) 200; (c) 50; (d) 46; (e) none of these.

2. A weak acid has $K_a = 1.00\times10^{-3}$. If $[HA] = 1.00$ M, what must be $[A^-]$ for the pH to be 2.7? (a) 2.7 M; (b) 2.0 M; (c) 0.50 M; (d) 0.37 M; (e) none of these.

3. Which of the following would *not* be a good buffer pair? (a) potassium carbonate and potassium bicarbonate; (b) ammonium chloride and ammonia; (c) boric acid and sodium borate; (d) sodium chloride and sodium hydroxide; (e) all would be good.

4. A substance that displays little change in pH when small amounts of acid or base are added is called a(n) (a) buffer; (b) indicator; (c) salt; (d) titrant; (e) none of these.

5 Which condition characterizes the equivalence point of the titration of a weak acid with a strong base? (a) a slight excess of titrant is present; (b) pH = 7.00; (c) the indicator must change color; (d) pH = pK_a; (e) none of these.

6 Which factor governs the selection of an indicator? (a) volume of titrant; (b) pH at the equivalence point; (c) concentration of the titrant; (d) final solution volume; (e) none of these.

7. 25.0 mL of 0.200 M NaOH is added to 50.0 mL of 0.100 M $HC_2H_3O_2 (K_a = 1.74\times10^{-5})$. What is the pH of the final solution? (a) 5.21; (b) 8.79; (c) 11.12; (d) 2.88; (e) none of these.

QUIZ D

1. What volume in mL of 2.50 M H_3PO_4 is needed to neutralize 150.0 mL of 0.500 M KOH? Potassium phosphate is produced. (a) 30.0; (b) 10.0; (c) 90.0; (d) 750.0 ; (e) none of these.

2. A weak acid HA has $K_a = 1.4\times10^{-3}$. In order to produce a solution with $[H^+] = 5.6\times10^{-3}$, the ratio $[A^-]/[HA]$ must equal (a) 4.0; (b) 0.25; (c) 2.0; (d) 1.0; (e) none of these.

3. $HC_2H_3O_2$ will be ionized to the greatest degree in which of the following solutions? (a) 0.010 M $HC_2H_3O_2$ and 0.10 M HCl; (b) 0.10 M $HC_2H_3O_2$ and 0.10 M $KC_2H_3O_2$; (c) 0.0010 M $HC_2H_3O_2$ and 0.10 M $NaC_2H_3O_2$; (d) 0.010 M $HC_2H_3O_2$; (e) 0.0010 M $HC_2H_3O_2$.

4. The region of pH over which an indicator changes color is the (a) buffer region; (b) equivalence point; (c) over-titration region; (d) pH color change range; (e) none of these.

5. At the equivalence point of a titration (a) the indicator must change color; (b) $pH = pK_a$; (c) $pH = 7.00$; (d) equivalents of acid = equivalents of base; (e) none of these.

6. An indicator, HIn, is yellow and its anion, In^-, is green. The indicator changes color from $pH = 5.1$ to $pH = 7.1$. This indicator will be yellow in which of the following? (a) 0.01 M HCl(aq); (b) 0.10 M NH_4Cl ($K_b = 1.8 \times 10^{-5}$ for NH_3); (c) pure H_2O; (d) 0.10 M NaOH; (e) none of these.

7. 50.0 mL of 0.100 M NH_3, 10.0 mL of 0.100 M NH_4Cl, and 40.0 mL of 0.050 M HCl are mixed. The resulting solution (a) contains 5.0 mmol NH_3, 2.0 mmol HCl, and 0.1 mmol NH_4Cl; (b) is a buffer with $pH = pK_b$ for ammonia; (c) has $[Cl^-] - 0.30\ M$; (d) has $[NH_3] - 0.30\ M$; (e) none of these.

SAMPLE TEST (15 minutes)

1. 7.500 g of weak acid HA is added to sufficient distilled water to produce 500.0 mL of solution with $pH = 2.716$. This solution is titrated with NaOH(aq) solution. At the half-equivalence point, $pH = 4.602$. What is the molar mass of the acid?

2. The titration curve at right was obtained in a general chemistry laboratory with a pH meter. The titrant was either 0.100 M NaOH or 0.100 M HCl. The unknown being titrated weighed 0.400 g.

   A. Was the unknown a weak acid or a weak base?

   B. What is the molar mass of the unknown?

   C. What is the value of the ionization constant of the unknown?

   D. What would be the pH color change range of a suitable indicator for this titration? Choose a suitable indicator from those given in Figure 17-1.

18 Solubility and Complex Ion Equilibria

Chapter Objectives

*** 1. Write the solubility product expression K_{sp} for a slightly soluble ionic compound.**

The K_{sp} expression is written in the same way as an equilibrium constant expression. The solubility equilibrium has a solid ionic compound as reactant and its ions in solution as products. The concentration of solid does not appear in the equilibrium constant expression. Thus, the K_{sp} expression simply is a product of two ionic concentrations, raised to appropriate powers. (Sometimes the solubility product expression is called the *ion product constant.*) There are 12 different types of ionic compounds (Objective 3–11, Table 3–4). These produce six different forms of solubility product expressions, which are given in Table 18-1. Of course, these expressions are determined from the stoichiometry of the solubility reaction. (We make these determinations in Examples 18-1 and 18-2.)

Table 18-1 Types of Solubility Equilibria

Salt formula	Solubility equilibrium	Solubility constant expression: General	Solubility constant expression: In terms of s
MX	$MX(s) \rightleftharpoons M^{+}(aq) + X^{-}(aq)$	$[M^{+}][X^{-}]$	s^2
	or $M^{2+}(aq) + X^{2-}(aq)$	$[M^{2+}][X^{2-}]$	
	or $M^{3+}(aq) + X^{3-}(aq)$	$[M^{3+}][X^{3-}]$	
MX_2	$MX_2(s) \rightleftharpoons M^{2+}(aq) + 2\,X^{-}(aq)$	$[M^{2+}][X^{-}]^2$	$4s^3$
	or $M^{4+}(aq) + 2\,X^{2-}(aq)$	$[M^{4+}][X^{2-}]^2$	
M_2X	$M_2X(s) \rightleftharpoons 2\,M^{+}(aq) + X^{2-}(aq)$	$[M^{+}]^2[X^{2-}]$	
MX_3	$MX_3(s) \rightleftharpoons M^{3+}(aq) + 3\,X^{-}(aq)$	$[M^{3+}][X^{-}]^3$	$27s^4$
M_3X	$M_3X(s) \rightleftharpoons 3\,M^{+}(aq) + X^{3-}(aq)$	$[M^{+}]^3[X^{3-}]$	
M_2X_3	$M_2X_3(s) \rightleftharpoons 2\,M^{3+}(aq) + 3\,X^{2-}(aq)$	$[M^{3+}]^2[X^{2-}]^3$	$108s^5$
M_3X_2	$M_3X_2(s) \rightleftharpoons 3\,M^{2+}(aq) + 2\,X^{3-}(aq)$	$[M^{2+}]^3[X^{3-}]^2$	
MX_4	$MX_4(s) \rightleftharpoons M^{4+}(aq) + 4\,X^{-}(aq)$	$[M^{4+}][X^{-}]^4$	$256s^5$
M_3X_4	$M_3X_4(s) \rightleftharpoons 3\,M^{4+}(aq) + 4\,X^{3-}(aq)$	$[M^{4+}]^3[X^{3-}]^4$	$6912s^7$

K_{sp} values are not easy to determine because they can be based on a very small mass of compound in solution. Even a small error in this mass is magnified when the concentration is squared or cubed. For example, for AgCl, the solubility is 1.9 mg/L. This gives $K_{sp} = 1.8 \times 10^{-10}$. If the solubility were determined as 1.8 mg/L, then $K_{sp} = 1.6 \times 10^{-10}$. This difficulty in measurement is partly responsible for various values of K_{sp} cited by different sources. For AgCl, five sources give $1.56 \times 10^{-10}, 1.7 \times 10^{-10}, 1.78 \times 10^{-10}, 1.79 \times 10^{-10}$, and 1.8×10^{-10} for K_{sp}.

*** 2. Calculate K_{sp} from the solubility of an ionic compound or solubility from the value of K_{sp}.**

The right-hand column of Table 18-1 gives the expression for the solubility product constant of a sparingly soluble ionic compound (a salt). In each expression, *s* is the solubility of the salt in moles per liter. These expressions can only be used when the salt is the sole source of its ions in solution. They cannot be used when one of the ions of the salt already is present in solution, that is, in common ion problems.

EXAMPLE 18-1 What is the molar solubility of aluminum hydroxide? $K_{sp} = 1.3 \times 10^{-33}$ for $Al(OH)_3$.

The solubility equilibrium is

$$Al(OH)_3(s) \rightleftharpoons Al^{3+}(aq) + 3\,OH^-(aq) \qquad [1]$$

The solubility product constant expression for $Al(OH)_3(s)$ is then

$$K_{sp} = [Al^{3+}][OH^-]^3 \qquad [2]$$

Let *s* be the molar solubility of $Al(OH)_3$. Then $[Al^{3+}] = s$ and $[OH^-] = 3s$ since 3 moles of OH^- are in solution for each mole of $Al(OH)_3$ that dissolves.

$$[Al^{3+}] = s \text{ AND } [OH^-] = 3s \qquad [3]$$

Substitution of the expressions in [3] into Equation [2] produces Equation [4].

$$K_{sp} = (s)(3s)^3 = 27\,s^4 \qquad [4]$$

Notice that the forms of the solubility equilibrium ([1]) and of the solubility product expression ([2] and [4]) are the same as those in Table 18-1 for a salt with the general formula MX_3.

$K_{sp} = 1.3 \times 10^{-33} = 27\,s^4$ *or* $s^4 = (1.3 \times 10^{-33})/27$ This gives $s = 2.6 \times 10^{-9}$ M.

EXAMPLE 18-2 The solubility of $Ca_3(PO_4)_2$ is 0.000022 g/100 mL of solution at 25°C. What is the value of K_{sp} at this temperature?

The solubility equilibrium for $Ca_3(PO_4)_2$ is

$$Ca_3(PO_4)_3(s) \rightleftharpoons 3\,Ca^{2+}(aq) + 2\,PO_4^{3-}(aq) \qquad [5]$$

The K_{sp} expression is $K_{sp} = [Ca^{2+}]^3[PO_4^{3-}]^2$ [6]

Let s be the molar solubility of $Ca_3(PO_4)_2$. Remember that for every mole of $Ca_3(PO_4)_2$ that dissolves, there are in solution 3 moles of Ca^{2+} and 2 moles of PO_4^{3-}. Thus,

$$[Ca^{2+}] = 3s \quad and \quad [PO_4^{3-}] = 2s \qquad [7]$$

Substituting the expressions in [7] into Equation [6] gives Equation [8].

$$K_{sp} = (3s)^3(2s)^2 = 27s^3\,4s^2 = 108\,s^5 \qquad [8]$$

Notice that Equations [5], [6], and [8] have the same form as that given for a salt of general formula M_3X_2 in Table 18-1. The molar solubility of $Ca_3(PO_4)_2$ is computed.

$$s = \frac{0.000022\ g\,Ca_3(PO_4)_2}{100\ mL\,soln} \times \frac{1000\ mL}{L} \times \frac{1\ mol\ Ca_3(PO_4)_2}{310.3\ g} = 6.4 \times 10^{-7}\ M$$

With this value of s in Equation [8], we obtain a value of $K_{sp} = 108(6.4 \times 10^{-7})^5 = 1 \times 10^{-29}$

We used two significant figures throughout the calculation of Example 18-2 and then rounded to one significant figure in the final result. Carrying one or more additional significant figures through a calculation is an acceptable and generally followed practice. Just do not forget to round the result. Also notice that we used the term *salt* to mean ionic compound. Strictly a chemist would typically consider only a soluble ionic compound to be a salt. However, chemists freely use the phrase "sparingly soluble salt."

3. Calculate the effect of common ions on the aqueous solubilities of sparingly soluble salts.

Objective 18-2 deals with the solubility of a salt in solutions where it is the only solute. This means that both ions come from the salt. But if an ion in solution is the same as one of the ions of the salt, then less salt will dissolve than in pure water. Because the solution and the salt have one ion in common, this lowering of solubility is called the common ion effect. It can be explained with Le Châtelier's principle. Since some of the product (one of the ions) of the solubility equilibrium is present in solution, the equilibrium will lie to the left (toward undissolved solute) compared to where it would be if no common ion were present. The expressions in the right-hand column of Table 18-1 *cannot* be used in common ion problems.

EXAMPLE 18-3 What is the molar solubility of $Al(OH)_3$ in a solution in which $[Al(NO_3)_3] = 0.0200$ M?

The solubility equilibrium follows, with the equilibrium concentration written below the formula of each ion. The solubility of $Al(OH)_3$ is represented by s.

Equation:	$Al(OH)_3(s)$	$\rightleftharpoons$	$Al^{3+}(aq)$	+ $3\ OH^-(aq)$
Initial:			0.0200 M	
Changes:			$+s$	$+3s$
Equil:			$(0.0200 + s)$M	3s

$K_{sp} = [Al^{3+}][OH^-]^3 = (0.0200\ M + s)(3s)^3 = 1.3 \times 10^{-33}$ Assume that s is small compared to 0.200 M.

Then we have $(0.0200\text{M})(3s)^3 = 1.3\times10^{-33}$ AND $s = 1.3\times10^{-11}$ M

We see that s indeed is small with respect to 0.0200 M. The presence of the common ion, Al^{3+}, greatly reduces the solubility of $Al(OH)_3$ from 2.6×10^{-9} M (Example 18-1) to 1.3×10^{-11} M, about 200 times less soluble.

4. Describe how the presence of "uncommon" ions in solution or the formation of ion pairs in concentrated solutions increases the solubility of a sparingly soluble salt.

In a dilute solution, each ion is completely surrounded by water molecules and is not affected by the presence of the other ions because they are too far away. When the solution becomes more concentrated, the ions are closer together. Ions of opposite charge attract and those of like charge repel. Hence ions clump into ion pairs. An ion pair acts more like a molecule than like two separate ions. Therefore, more of the solute dissolves because more ions can be present in solution. The effective concentration (or the activity) becomes less than the formal concentration (the amount of solute that dissolved). One indication of this difference between activity (a) and formal concentration (c) is that the activity coefficient (γ) becomes steadily smaller as the solution becomes more concentrated.

$$\gamma = a / c \qquad [9]$$

Another indication of this difference is the decrease in the van't Hoff factor as the solution becomes concentrated, as shown by the data in Table 13-2 (Objective 13-13). The charge, not the identity, of ions influences the activity coefficient. Hence, a sodium ion (Na^+) and a silver ion (Ag^+) should have about the same effect.

*** 5 Determine if a salt will precipitate from solution based on the concentrations of its ions.**

The ion product expression has the same form as the solubility product expression, but the ionic concentrations are not necessarily those at equilibrium. The ion product (Q_{sp}) is related to the solubility product (K_{sp}) as shown in Expression [10] (recall Objective 15–9 for a discussion of reaction quotient [Q]).

If $Q_{sp} < K_{sp}$, more solute can dissolve.

If $Q_{sp} > K_{sp}$, precipitate forms. [10]

If $Q_{sp} = K_{sp}$, the solution is saturated.

EXAMPLE 18-4 150 mL of solution with $[Ca^{2+}] = 1.2\times10^{-3}\text{ M} = 0.0012$ M is mixed with 100 mL of solution with $[PO_4^{3-}] = 4.0\times10^{-3}\text{ M} = 0.0040$ M. $K_{sp} = 2.0\times10^{-29}$ for $Ca_3(PO_4)_2$. Should $Ca_3(PO_4)_2$ precipitate when the two solutions are mixed?

We first find the amount of each ion present:

Ca^{2+} amt. $= 150\text{ mL}\times 0.0012\text{ M} = 0.18\text{ mmol } Ca^{2+}$

PO_4^{3-} amt. $= 100\text{ mL}\times 0.0040\text{ M} = 0.040\text{ mmol } PO_4^{3-}$

Then, we find the concentration of each ion: Solution volume $= 150\text{ mL} + 100\text{ mL} = 250$ mL

$[Ca^{2+}] = 0.18 \text{ mmol}/250 \text{ mL} = 7.2 \times 10^{-4} \text{ M}$

$[PO_4^{3-}] = 0.040 \text{ mmol}/250 \text{ mL} = 1.6 \times 10^{-3} \text{ M}$

Q_{sp} is compared to K_{sp}: $Q_{sp} = (7.2 \times 10^{-4})^3 (1.6 \times 10^{-3})^2 = 9.56 \times 10^{-16} > 2.0 \times 10^{-29} = K_{sp}$

Since $Q_{sp} > K_{sp}$, precipitation should occur.

*** 6. Determine the concentration of ions remaining in solution after precipitation and predict whether precipitation will be complete.**

Determining whether precipitation is complete when the concentration of one ion remains constant involves using that constant concentration in the solubility product constant expression. This gives the maximum possible concentration of the other ion. Most chemists agree that 99.9% of the original amount of an ion present in a solution must precipitate in order for precipitation to be considered complete. That is, when c is the molar concentration of the ion being considered, precipitation is complete if

$$c_{\text{final}} \le 0.001\, c_{\text{initial}} \qquad [11]$$

EXAMPLE 18-5 A solution has $[I^-] = 1.3 \times 10^{-3}$ M. If $[Pb^{2+}]$ is kept equal to 2.0×10^{-2} M, will precipitation of I^- be complete? For PbI_2, $K_{sp} = 7.1 \times 10^{-9}$.

Obtain the maximum $[I^-]$ that can exist in this solution:

$$K_{sp} = [Pb^{2+}][I^-]^2 = 7.1 \times 10^{-9} = (2.0 \times 10^{-2})[I^-]^2$$

$$[I^-] = \sqrt{\frac{7.1 \times 10^{-9}}{2.0 \times 10^{-2}}} = 5.9 \times 10^{-5} \text{ M} \qquad \% \text{ left in solution} = \frac{5.9 \times 10^{-5} \text{ M}}{1.3 \times 10^{-3} \text{ M}} \times 100\% = 4.5\%$$

This is 4.5% of the original $[I^-]$. Thus, according to Condition [11], precipitation of I^- is not complete.

Recognize that the concentration of one ion will effectively remain constant if it is initially much more concentrated than the other. When the concentrations of both ions decrease, the calculation is somewhat more difficult. The easiest method is to assume that solute precipitates until all of one ion is consumed (its concentration reaches zero), and then solve a common ion problem.

EXAMPLE 18-6 What will be the final concentration of each ion in the solution of Example 18-4?

Equation:	$Ca_3(PO_4)_2(s) \rightleftharpoons$	$3Ca^{2+}(aq)$ +	$2\,PO_4^{3-}(aq)$
Initial:		7.2×10^{-4} M	1.6×10^{-3} M
Max. pptn:		0 M	$(1.6 \times 10^{-3} - 4.8 \times 10^{-4})$ M
Changes:		$+3s$	$+2s$
Equil:		$3s$	$(1.12 \times 10^{-3} + 2s)$ M

The decrease in $[PO_4^{3-}]$ comes from: $\Delta[PO_4^{3-}] = \dfrac{7.2 \times 10^{-4} \text{ mol } Ca^{2+}}{1 \text{ L}} \times \dfrac{2 \text{ mol } PO_4^{3-}}{3 \text{ mol } Ca^{2+}} = 4.8 \times 10^{-4}$ M.

The equilibrium expressions for ionic concentrations are then substituted into the K_{sp} expression Remember:

$K_{sp} = 2.0\times10^{-29}$ for $Ca_3(PO_4)_2$. $[Ca^{2+}]^3[PO_4^{\ 3-}]^2 = (3s)^3(1.12\times10^{-3} + 2s)^2 = 2.0\times10^{-29}$

Assume $2s \ll 1.12\times10^{-3}$ M. $(3s)^3(1.12\times10^{-3})^2 = 2.0\times10^{-29}$ or $s = 8.4\times10^{-9}$ M

We now need to evaluate our assumption that $2s \ll 1.12\times10^{-3}$. We determine whether the following equation is valid by substituting $s = 8.4\times10^{-9}$ M and evaluating both sides:

$$(3s)^3(1.12\times10^{-3})^2 = (3s)^3(1.12\times10^{-3} + 2s)^2$$

$$(3\times8.4\times10^{-9})^3(1.12\times10^{-3})^2 = (3\times8.4\times10^{-9})^3(1.12\times10^{-3} + 2\times8.4\times10^{-9})^2$$

$$2.0\times10^{-29} = 2.0\times10^{-29}$$

Thus, the assumption is valid. An easier way to evaluate the assumption is to make sure that the omitted quantity ($2s$ in this case) is less than 5% of the value to which it is added (1.12×10^{-3} M). In this problem, $2s = 1.7\times10^{-8}$ M $= 0.00015\%$ of 1.12×10^{-3} M; thus, the assumption is valid. The final concentrations now are computed:

$[PO_4^{\ 3-}] = (1.12\times10^{-3} + 2s)\text{ M} = 1.12\times10^{-3}\text{ M}$ $[Ca^{2+}] = 3s = 2.5\times10^{-8}$ M

Notice that the concentration of calcium ion has decreased to less than 0.1% (actually to 0.0034%) of its original value. Hence the precipitation of calcium ion is complete.

*** 7. Explain how fractional precipitation works and when it can be used.**

Fractional precipitation is the technique of separating one or more ions from solution in a precipitate, while leaving others behind in solution. To separate two cations, an anion is added to the solution. As the concentration of the anion increases, eventually the value of the ion product (Q_{sp}) for one of the salts exceeds the value of its solubility product constant (K_{sp}). The anion concentration is increased until the first cation has completely precipitated (99.9% of the original amount of cation has precipitated). It is hoped that other cations will not have begun to precipitate at this point. If they have, fractional precipitation is not possible, at least not using this anion as a precipitating agent.

EXAMPLE 18-7 A solution has $[Ba^{2+}] = 0.50$ M, $[Ag^+] = 3.0\times10^{-5}$ M, and $[Zn^{2+}] = 2.0\times10^{-5}$ M. Solid sodium oxalate is added gradually. What is the $[C_2O_4^{\ 2-}]$ when each cation (Ba^{2+}, Ag^+, Zn^{2+}) (a) just begins to precipitate? and (b) is completely precipitated? Which of the ions is completely separated from the others by fractional precipitation and which is not? K_{sp} values at 25 °C are 2.3×10^{-8} for BaC_2O_4, 2.7×10^{-8} for ZnC_2O_4, and 3.6×10^{-11} for $Ag_2C_2O_4$.

We first compute the $[C_2O_4^{\ 2-}]$ needed to just start precipitation of each cation. At this point $Q_{sp} = K_{sp}$.

BaC_2O_4: $K_{sp} = [Ba^{2+}][C_2O_4^{2-}] = 2.3\times10^{-4}[C_2O_4^{2-}] = 2.3\times10^{-8} = (0.50)[C_2O_4^{2-}]$

or $[C_2O_4^{2-}] = 4.6\times10^{-8}$ M

ZnC_2O_4: $K_{sp} = [Zn^{2+}][C_2O_4^{2-}] = 2.7\times10^{-9} = (2.0\times10^{-9})[C_2O_4^{2-}]$

or $[C_2O_4^{2-}] = 1.4\times10^{-3}$ M

$Ag_2C_2O_4$: $K_{sp} = [Ag^+]^2[C_2O_4^{2-}] = 3.6\times10^{-11} = (3.0\times10^{-5})^2[C_2O_4^{2-}]$

or $[C_2O_4^{2-}] = 0.040$ M

Next, we determine the $[C_2O_4^{2-}]$ when each of the ions has completely precipitated by substituting 0.001 (0.1%) of each original ion concentration into the K_{sp} expression.

BaC_2O_4: $2.3\times10^{-8} = 5.0\times10^{-4}[C_2O_4^{2-}]$ or $[C_2O_4^{2-}] = 4.6\times10^{-5}$ M

ZnC_2O_4: $2.7\times10^{-8} = 2.0\times10^{-8}[C_2O_4^{2-}]$ or $[C_2O_4^{2-}] = 1.4$ M

$Ag_2C_2O_4$: $3.6\times10^{-11} = (3.0\times10^{-8})^2[C_2O_4^{2-}]$ or $[C_2O_4^{2-}] = 4.0\times10^{4}$ M

Thus, as $[C_2O_4^{2-}]$ increases, Ba^{2+} is the first ion to precipitate at $[C_2O_4^{2-}] = 4.6\times10^{-8}$ M. The precipitation of Ba^{2+} is complete when $[C_2O_4^{2-}] = 4.6\times10^{-5}$ M. When $[C_2O_4^{2-}]$ rises to 1.4×10^{-3} M, ZnC_2O_4 begins to precipitate. Thus Ba^{2+} is completely separated. When $[C_2O_4^{2-}] = 1.4$ M, Zn^{2+} is completely precipitated, but Ag^+ begins to precipitate before this, when $[C_2O_4^{2-}] = 0.040$ M. Thus Ag^+ and Zn^{2+} are not separated by fractional precipitation with oxalate ion. Also, we cannot completely precipitate Ag^+; 4.0×10^{4} M is an impossible concentration to reach.

*** 8. Describe, through net ionic equations and calculations, the effect of pH on the precipitation and dissolving of certain substances.**

Many metal hydroxides are insoluble. (Recall the solubility rules in Objective 5–3.) $Fe(OH)_3$ is an example:

$$Fe(OH)_3(s) \rightleftharpoons Fe^{3+}(aq) + 3\,OH^-(aq) \quad K_{sp} = 4\times10^{-38} \qquad [12]$$

If we add strong acid to $Fe(OH)_3(s)$ and its saturated solution, $H_3O^+(aq)$ from the acid reacts with $OH^-(aq)$s from the $Fe(OH)_3$. The equilibrium of Equation [12] shifts right. $Fe(OH)_3$ will dissolve in a highly acidic solution. This is an acid-base neutralization.

$$Fe(OH)_3(s) + 3\,H_3O^+(aq) \rightleftharpoons Fe^{3+}(aq) + 6\,H_2O(aq) \qquad [13]$$

All acids will enhance the solubility of $Fe(OH)_3$, *including* the cation of a weak base.

$$Fe(OH)_3(s) + 3\,NH_4^+(aq) \rightleftharpoons Fe^{3+}(aq) + 3\,NH_3(aq) + 3\,H_2O \qquad [14]$$

In addition, an insoluble hydroxide can precipitate from a neutral solution because of hydrolysis of the cation.

$$Fe^{3+}(aq) + 6\,H_2O \rightleftharpoons Fe(OH)_3(s) + 3H_3O^+(aq) \qquad [15]$$

EXAMPLE 18-8 What is the solubility of $Fe(OH)_3, K_{sp} = 4 \times 10^{-38}$ in (a) pure water, and (b) a solution maintained at $pH = 2.00$?

(a) $K_{sp} = [Fe^{3+}][OH^-]^3 = (s)(3s)^3 = 4.0 \times 10^{-38}$ *or* $s = 2 \times 10^{-10}$ M

(b) $pH = 2.00$ is $pOH = 12.00$; thus $[OH^-] = 10^{-pOH} = 10^{-12.00} = 1.0 \times 10^{-2}$ M.

$4 \times 10^{-38} = [Fe^{3+}](1.0 \times 10^{-12})^3$ or $[Fe^{3+}] = 4 \times 10^{-2}$ M

Another group of compounds whose solubilities are affected by pH are those in which the anion is the anion of a weak acid, such as strontium fluoride.

$$SrF_2(s) \rightleftharpoons Sr^{2+}(aq) + 2\,F^-(aq) \qquad [16]$$

The addition of $H_3O^+(aq)$ removes $F^-(aq)$ from solution, and Equation [16] shifts to the right.

$$H_3O^+(aq) + F^-(aq) \rightleftharpoons HF(aq) + H_2O \qquad [17]$$

*** 9. Write equations showing the effect of complex ion formation on other equilibrium processes such as solubility equilibria.**

In the analysis of the ammonium sulfide group in qualitative analysis, Fe^{3+} is precipitated as the hydroxide (see Equation [12]). If NaCN(aq) is added, free $Fe^{3+}(aq)$ forms a complex ion $[Fe(CN)_6]^{3-}(aq)$.

$$Fe^{3+}(aq) + 6\,CN^-(aq) \rightleftharpoons [Fe(CN)_6]^{3-}(aq) \qquad K_f = 1 \times 10^{42} \qquad [18]$$

With the addition of cyanide ion to the solution, Equation [12] is displaced to the left and some of the $Fe(OH)_3(s)$ dissolves. Notice that the stronger Lewis base (CN^-) displaces the weaker one (OH^-) from the Lewis acid (Fe^{3+}). The relative strengths of Lewis bases is given by the spectrochemical series (Chapter 25, Expression [2]). Notice there that halide ligands generally are weak Lewis bases, nitrogen-containing ligands generally are strong, while ligands that contain oxygen are intermediate in strength.

*** 10. Use complex ion formation constants, K_f (from Table 18-4 at the end of the chapter), to compute the concentrations in solution of free ions, ligands, and complex ions.**

Although Pb^{2+} will precipitate as $PbCl_2(s)$, the addition of excess Cl^- will redissolve the solid because of the formation of $[PbCl_3^-](aq)$.

$$Pb^{2+}(aq) + 3\,Cl^-(aq) \rightleftharpoons [PbCl_3^-](aq) \qquad K_f = 5.0 \times 10^1 \qquad [19]$$

EXAMPLE 18-9 What is the $[Pb^{2+}]$ at equilibrium if a 0.0100 M $[Pb^{2+}]$ solution is made 0.400 M in Cl^-? The line labeled "Right" in the setup below gives the concentrations when all the $Pb^{2+}(aq)$ reacts. We then approach equilibrium from that point.

Equation:	$Pb^{2+}(aq) + 3\,Cl^-(aq)$	$\rightleftharpoons$	$[PbCl_3]^-(aq)$
Initial:	0.0100 M	0.400 M	
Right:	0 M	0.370 M	0.0100 M
Changes:	$+x$ M	$+3x$ M	$-x$M
Eqil:	x M	$(0.370+3x)$ M	$(0.0100-x)$ M

$$K_f = \frac{0.0100-x}{x(0.370-3x)^3} \approx \frac{0.0100-x}{x(0.370)^3} = 50 \quad \text{OR} \quad x = 0.00283$$

We have assumed $3x << 0.370$ M. This is not quite the case, so we go through two cycles of approximation to find $x = [Pb^{2+}] = 0.0030$ M, the concentration of free $Pb^{2+}(aq)$.

We used the method of successive approximations before (Objective 16-8, Example 16-5). This method is essential here since the exact equation is fourth order and has no simple solution.

*** 11. Use K_f values along with K_{sp} values to determine the solubilities of slightly soluble solutes in the presence of complexing ligands.**

The presence of a complexing ligand reduces the concentration of the free metal ion. This reduction may be enough that no precipitate forms even in the presence of a precipitating reagent. For example, $CdCO_3$ will precipitate when $Cd^{2+}(aq)$ and $CO_3^{2-}(aq)$ are present even at low concentrations (Equation [20]). But the addition of $NH_3(aq)$ lowers the concentration of free $Cd^{2+}(aq)$ (Equation [21]) so that precipitation may not occur.

$$CdCO_3(s) \rightleftharpoons Cd^{2+}(aq) + CO_3^{2-}(aq) \qquad K_{sp} = 5.2\times10^{-12} \qquad [20]$$

$$Cd^{2+}(aq) + 4\,NH_3(aq) \rightleftharpoons [Cd(NH_3)_4]^{2+} \qquad K_f = \frac{[Cd(NH_3)_4{}^{2+}]}{[Cd^{2+}][NH_3]^4} = 1.3\times10^7 \qquad [21]$$

EXAMPLE 18-10 One liter of solution has $[CO_3^{2-}] = 0.032$ M. (a) What is the maximum amount of $Cd(NO_3)_2$ that may be added without a precipitate forming? (b) If the solution also has $[NH_3] = 0.100$ M, what is the maximum amount of $Cd(NO_3)_2$ that may be added without $CdCO_3(s)$ precipitating?

(a) Substitution into the K_{sp} expression for Reaction [20] produces the maximum $[Cd^{2+}]$.

$$[Cd^{2+}][CO_3^{2-}] = [Cd^{2+}][0.032\text{ M}) = 5.2\times10^{-12} \qquad [Cd^{2+}] = 1.6\times10^{-10}\text{ M}$$

Thus, we could add only 1.6×10^{-10} mol $Cd(NO_3)_2$.

(b) We know $[Cd^{2+}] = 1.6\times10^{-10}$ M at equilibrium before $NH_3(aq)$ is added. This value will remain fixed because the $[CO_3^{2-}]$ does not change.

Equation:	$Cd^{2+}(aq)$ +	$4\,NH_3(aq)$	$\rightleftharpoons$	$[Cd(NH_3)_4]^{2+}(aq)$
Initial:	1.6×10^{-10} M	0.100 M		
Changes:		$-4x$ M		$+x$ M

Equil:	1.6×10^{-10} M	$(0.100-4x)$ M	x M

$$K_f = \frac{x}{(1.6\times10^{-10})(0.100-4x)^4} = 1.3\times10^7 \approx \frac{x}{(1.6\times10^{-10})(0.100)^4}$$

We assumed $4x << 0.100$ M.. This gives $x = 1.1\times10^{-4}$ M $= [Cd(NH_3)_4^{2+}]$. The assumption is valid. Practically all of the cadmium ion is present in the complex; thus, we can add 1.1×10^{-4} mol $Cd(NO_3)_2$ to 1 L of solution before precipitation occurs, 1.3×10^3 times more than in Part (a).

12. Calculate the solubilities of certain solutes in the presence of complexing ligands.

This is similar to the previous objective, except that the complexing reagent is added to a mixture containing the already formed precipitate. The problem is solved by combining the equilibrium constants of the precipitation reaction (Equation [22]) and the complex ion formation reaction (Equation [23]) to obtain an equilibrium constant for the overall reaction (Equation [24]). (Objective 16-4 discusses how the resultant equilibrium constant is calculated when two reactions are combined.)

$$ZnCO_3(s) \rightleftharpoons Zn^{2+}(aq) + CO_3^{2-}(aq) \qquad K_{sp} = 1.4\times10^{-11} \qquad [22]$$

$$Zn^2 + (aq) + 4\,CN^-(aq) \rightleftharpoons [Zn(CN)_4]^{2-}(aq) \qquad K_f = 5\times10^{16} \qquad [23]$$

$$ZnCO_3(s) + 4\,CN^-(aq) \rightleftharpoons [Zn(CN)_4]^{2-}(aq) + CO_3^{2-}(aq) \quad K = K_{sp}\times K_f = 7\times10^5 \qquad [24]$$

EXAMPLE 18-11 What is the molar solubility of $ZnCO_3$ in (a) pure water and (b) 0.800 M $NH_3(aq)$?

(a) The K_{sp} expression is solved to determine the molar solubility.

$$[Zn^{2+}][CO_3^{2-}] = 1.4\times10^{-11} = s^2 \qquad \text{OR} \qquad s = 3.7\times10^{-6}\text{ M}$$

(b) We solve this problem with the equilibrium constant of Equation [24], first forming as much complex as possible, since the large value of that equilibrium constant indicates the reaction lies to the right.

Equation:	$ZnCO_3(s)$ +	$4\,CN^-(aq)$	$\rightleftharpoons [Zn(CN)_4]^{2-}(aq)$ +	$CO_3^{2-}(aq)$
Initial:		0.800 M	0 M	
Form complex:		0 M	0.200 M	0.200 M
Changes:		$+4x$ M	$-x$ M	$-x$ M
Equil:		$+4x$ M	$(0.200-x)$ M	$(0.200-x)$ M

$$K = \frac{[[Zn(CN)_4]^{2-}][CO_3^{2-}]}{[CN]^4} = 7\times10^5 = \frac{(0.200-x)^2}{(4x)^4} \qquad \frac{0.200-x}{(4x)^2} = 8\times10^2 \approx \frac{0.200}{16\,x^2}$$

$x = \sqrt{\dfrac{0.200}{16\times8\times10^2}} = 4\times10^{-3}$ M. We have assumed $x << 0.200$ M, which we see is true.

The molar solubility of zinc(II) is $(0.200-x)$ M $= 0.196$ M.

Practically all of the zinc ion in solution is present as the complex ion.

13. Predict if metal sulfides will precipitate from saturated $H_2S(aq)$ solutions of known pH.

The most common situation, the dissolving of a metal sulfide (MS) in an *acidic* solution, can be represented by Equation [25], with an equilibrium constant denoted by K_{spa}.

$$MS(s) + 2\,H_3O \rightleftharpoons M^{2+}(aq) + H_2S(aq) + 2\,H_2O \quad K_{spa} = \frac{[M^{2+}][H_2S]}{[H_3O^+]^2} \qquad [25]$$

In an alkaline or a neutral solution, however, there are far fewer hydrogen ions present and the approproate equation is [26], with an equilibrium constant denoted by K_{sp}.

$$MS(s) + H_2O \rightleftharpoons M^{2+}(aq) + OH^-(aq) + HS^-(aq) \qquad K_{sp} = [M^{2+}][OH^-][HS^-] \qquad [26]$$

Numerical values for these equilibrium constants are presented in Table 18-2. In solving problems, one needs to be aware that a saturated H_2S solution has $[H_2S] = 0.10\,M$. In both acidic and basic solutions, we may be asked whether a precipitate will form; that is, whether Equation [25] or [26] lies to the left. In either case, we calculate a reaction quotient, Q_{spa} or Q_{sp}. If Q_{spa} K_{spa} or $Q_{sp} > K_{sp}$, a precipitate will indeed form.

TABLE 18-2 Corrected Solubility Product Constant Values for Metal Sulfides

Solid	K_{sp}	K_{spa}	Solid	K_{sp}	K_{spa}
HgS(red)	4×10^{-54}	4×10^{-33}	HgS(black)	2×10^{-53}	2×10^{-32}
Ag_2S	6×10^{-51}	6×10^{-30}	CuS	6×10^{-37}	6×10^{-16}
PbS	3×10^{-28}	3×10^{-7}	CdS	8×10^{-28}	8×10^{-7}
SnS	2×10^{-25}	2×10^{-4}	α-ZnS	2×10^{-25}	2×10^{-4}
β-ZnS	3×10^{-23}	3×10^{-2}	FeS	6×10^{-19}	6×10^{2}
MnS(green)	3×10^{-14}	3×10^{7}	MnS(pink)	3×10^{-11}	3×10^{10}

EXAMPLE 18-12 A solution is maintained at $pH = 4.00$ by an $HC_2H_3O_2 - NaC_2H_3O_2$ buffer. The solution contains Pb^{2+} and Mn^{2+}, each at a concentration of 0.0500 M. The solution is saturated with H_2S. Will either of the metal sulfides precipitate?

$pH = 4.00$ means $[H_3O^+] = 1.0\times10^{-4}$ M. In addition, $[H_2S] = 0.10\,M$ and $[M^{2+}] = 0.0500\,M$. Therefore, we can calculate the value of Q_{spa}.

$$Q_{spa} = \frac{[M^{2+}][H_2S]}{[H_3O^+]^2} = \frac{(0.0500\,M)(0.10\,M)}{(1.0\times10^{-4}\,M)^2} = 5.0\times10^{5}$$

This value of Q_{spa} is smaller than $K_{spa} = 3\times10^{-7}$ for PbS. PbS will not precipitate. But the calculated value of Q_{spa} is larger than $K_{spa} = 3\times10^{10}$ for MnS(pink) or $K_{spa} = 3\times10^{7}$ for MnS(green). MnS should precipitate.

SELF-ASSESSMENT EXERCISES

92. (a) $[I^-]$ will be twice $[Pb^{+2}]$. (b) and (c) are incorrect. $K_{sp} = [Pb^{+2}]\,[I^-]^2$; therefore, (d) is correct.

93. (a) Reduce $[Ba^{2+}]$. $K_{sp} = [Ba^{2+}][SO_4^{-2}]$ and sulfate ions will increase so $[Ba^{2+}]$ will decrease because K_{sp} is constant.

94. (a) pure water. Adding a common ion will reduce solubility, and since KNO_3 does not contain an ion that reacts with either species, it will have no effect.

95. (b) H_2SO_4. $PbSO_4$ is soluble. Copper sulfate is insoluble.

96. (a) and (b) will not yield precipitates. (c) and (d) will yield precipitates because an increase in $[NH^+]$ for (c) and formation of a complex ion with Cu for (d). (e) The $[OH^-]$ is increasing; therefore, $Al(OH)_3$ will precipitate.

98. (a) add acid. Sulfide ion is a strong base and very little exists in acidic solution. The precipitating agent is actually $[H_3O^+]$.

99. (a) $H_2C_2O_4$ is a weak acid, so it will be more soluble in basic solution. (b) $MgCO_3$, acid because HCO_3^- will be formed. (c) CdS will be more soluble in base because acid increases precipitation. (d) KCl will be independent of pH. (e) $NaNO_3$ will be independent because HNO_3 is a strong acid and NaOH is a strong base. (f) $Ca(OH)_2$ is more soluble in acid, which will react with OH^-.

100. NH_3 because Cu^{2+} reacts with NH_3 to from a common ion, which can be precipitated.

101. $K_{sp} = [Al^{3+}][OH^-]^3 = 1.3 \times 10^{-33}$

Using the Henderson-Hasselbach equation: $pH = pK_a + \log \dfrac{[A^-]}{[HA]}$

$pH = 4.74 + \log \dfrac{0.35}{0.45} = 4.63$ and $pOH = 9.37$ $[OH^-] = 4.27 \times 10^{-10}$

$Q = (0.275)(4.27 \times 10^{-10})^3 = 2.1 \times 10^{-31} > K_{sp}$ Precipitation will occur.

102. The source of Ag^+ will be from the complex ion. $K_f = \dfrac{[Ag(CN)_2]}{[Ag+][CN^-]^2}$

$$5.6 \times 10^{18} = \frac{0.012\ M}{[Ag^+](1.05)^2} \quad \text{and } [Ag^+] = 1.94 \times 10^{-21}\ M$$

$Q = (1.94 \times 10^{-21})(2.0\ M) = 3.88 \times 10^{-21} < K_{sp}$ Precipitation will not occur.

DRILL PROBLEMS

1. Write the solubility equilibrium and the solubility product expression for each of the following salts. The first is done as an example.

$Ag_2SO_4 : Ag_2SO_4(s) \rightleftharpoons 2\,Ag^+(aq) + SO_4^{2-}(aq)$ $\qquad K_{sp} = [Ag^+]^2[SO_4^{2-}]$

A. AgCN	B. SrF_2	C. TlBr	D. $Zn_3(AsO_4)_3$	E. Hg_2Br_2	F. Ag_3AsO_4
G. Li_2CO_3	H. PbC_2O_4	I. $La(IO_3)_3$	J. $Ce_2(C_2O_4)_3$	K. $CdSO_3$	L. $AlPO_4$
M. Al_2S_3	N. $Ca(IO_3)_2$	O. CoS	P. $FeAsO_4$		

2. Fill in the blanks in the following lines. The first line is completed as an example.

Compound	K_{sp}	*Molar solubility of* cation, M	anion, M	*Cmpd solubility, g/100 mL soln*
Ag_2SO_4	1.4×10^{-5}	0.030	0.015	0.47
A. AgCNS	____	1.08×10^{-6}	____	____
B. $SrCrO_4$	____	____	4.7×10^{-3}	____
C. $TlIO_3$	____	____	____	6.7
D. Hg_2CrO_4	2.0×10^{-9}	____	____	____
E. $AlPO_4$	____	7.9×10^{-10}	____	____
F. $Ba(BrO_3)_2$	____	____	1.9×10^{-2}	____
G. $Be(OH)_2$	____	____	____	1.5×10^{-5}
H. $Ca(IO_3)_2$	7.1×10^{-7}	____	____	____
I. Cu_2S	____	17×10^{-16}	____	____
J. Tl_2CrO_4	____	____	6.3×10^{-5}	____
K. $Ag_2C_2O_4$	____	____	____	0.62
L. BiI_3	8.1×10^{-19}	____	____	____
M. Ag_3PO_4	____	1.4×10^{-4}	____	____
N. $La(IO_3)_3$	____	____	2.1×10^{-3}	____
O. $Ca_3(PO_4)_2$	____	____	____	2.2×10^{-3}
P. Al_2S_3	2.0×10^{-7}	____	____	____

3. Use the data of Table 18-3 at the end of this chapter to predict the molar solubility of each of the following salts in a solution that contains the given concentration of one of its ions.

A. $BaSO_4; [SO_4^{2-}] = 4.3 \times 10^{-5}$ M
B. $AgI; [I^-] = 7.2 \times 10^{-6}$ M
C. $SrSO_3; [SO_3^{2-}] = 3.1 \times 10^{-2}$ M
D. $MgCO_3; [Mg^{2+}] = 1.7 \times 10^{-4}$ M
E. $Ag_2CrO_4; [Ag^+] = 1.3 \times 10^{-3}$ M
F. $Hg_2Cl_2; [Hg_2^{2+}] = 0.85$ M
G. $Fe(OH)_3; [Fe^{3+}] = 1.6 \times 10$ M
H. $PbI_2; [I^-] = 3.4$ M
I. $Li_3PO_4; [PO_4^{3-}] = 0.29$ M
J. $Mg_3(PO_4)_2; [PO_4^{3-}] = 0.87$ M

4. (1) In each part, determine the concentration of the second compound that is just large enough to cause precipitation from a solution of the given concentration. Use the data of Table 18-3 at the end of this chapter.

A. $[Cu(NO_3)_2] = 1.45 \times 10^{-3}$ M; Na_2CrO_4
B. $[AgNO_3] = 1.2 \times 10^{-6}$ M; NaBr
C. $[Sr(NO_3)_2] = 1.4 \times 10^{-3}$ M; Na_2CO_3
D. $[NiSO_4] = 4.7 \times 10^{-6}$ M; Na_2S
E. $[Pb(NO_3)_2] = 8.9 \times 10^{-3}$ M; $MgSO_4$
F. $[MgCl_2] = 3.4 \times 10^{-5}$ M; NaF
G. $[AgNO_3] = 5.0 \times 10^{-3}$ M; Na_2SO_4
H. $[Pb(NO_3)_2] = 1.6 \times 10^{-3}$ M; MgI_2
I. $[Al(NO_3)_3] = 1.7 \times 10^{-14}$ M; NaOH
J. $[MgSO_4] = 2.4 \times 10^{-8}$ M; Na_3PO_4

(2) Determine if precipitation will occur in each of the following solutions. Use the data of Table 18-3 at the end of this chapter. All concentrations are given in moles per liter of solution.

K. $[Co^{2+}] = 3.2 \times 10^{-5}; [CO_3^{2-}] = 1.6 \times 10^{-12}$
L. $[Pb^{2+}] = 0.0047; [SO_4^{2-}] = 7.0 \times 10^{-6}$
M. $[Ag^+] = 1.2 \times 10^{-5}; [I^-] = 9.2 \times 10^{-8}$
N. $[Ba^{2+}] = 0.064; [SO_4^{2-}] = 1.6 \times 10^{-5}$
O. $[Sr^{2+}] = 3.2 \times 10^{-5}; [CO_3^{2-}] = 7.2 \times 10^{-7}$
P. $[Pb^{2+}] = 8.6 \times 10^{-6}; [CrO_4^{2-}] = 0.0017$
Q. $[Ag^+] = 0.017; [CrO_4^{-2}] = 4.6 \times 10^{-5}$
R. $[Mg^{2+}] = 3.7 \times 10^{-6}; [F^-] = 0.0088$
S. $[Li^+] = 1.6; [PO_4^{3-}] = 0.0042$
T. $[Mg^{2+}] = 1.4 \times 10^{-5}; [PO_4^{3-}] = 7.9 \times 10^{-6}$

5. (1) For each of the compounds that precipitates in Part (2) of the drill problems for Objective 18-5, determine the final concentration of each ion after precipitation and then determine if precipitation of either the anion or the cation is complete.

(2) Do the same for each of the following solutions. In each of these cases, precipitation *does* occur. Again, all concentrations are given in moles per liter of solution.

A. $[Zn^{2+}] = 1.1 \times 10^{-3}; [CO_3^{2-}] = 9.5 \times 10^{-7}$
B. $[Hg_2^{2+}] = 3.4 \times 10^{-4}; [Cl^-] = 1.2 \times 10^{-7}$
C. $[Al^{3+}] = 1.6 \times 10^{-12}; [OH^-] = 2.7 \times 10^{-4}$
D. $[Pb^{2+}] = 1.0; [Cl^-] = 0.015$
E. $[Ag^+] = 1.7 \times 10^{-6}; [Br^-] = 3.3 \times 10^{-3}$
F. $[Ca^{2+}] = 2.7 \times 10^{-6}; [PO_4^{3-}] = 4.6 \times 10^{-15}$

6. Use the solubility rules in Objective 5-3 and the data of Table 18-3 at the end of this chapter to determine what anion or cation will form a precipitate with each member of the following pairs of ions. (In some cases, the precipitating ion's formula is given in parentheses.) Determine the concentration of precipitating agent needed to just cause precipitation of the given ions, and the concentration needed to

completely precipitate each ion. Finally, determine if the two ions can be separated by fractional precipitation or not. All concentrations are in moles per liter.

A. $[Cl^-] = 1.2\times10^{-6}; [Br^-] = 8.9\times10^{-4}$
B. $[Ba^{2+}] = 1.6\times10^{-5}; [Pb^{2+}] = 1.3\times10^{-5}$
C. $[Pb^{2+}] = 3.9\times10^{-4}; [Hg_2{}^{2+}] = 3.9\times10^{-4}$
D. $[Sr^{2+}] = 1.6\times10^{-8}; [Pb^{2+}] = 3.4\times10^{-5}\ (SO_4{}^{2-})$
E. $[OH^-] = 9.6\times10^{-4}; [F^-] = 0.13\ \ (Mg^{2+})$
F. $[Ag^+ = 4.6\times10^{-3}; [Pb^{2+}] = 0.023\ (CrO_4{}^{2-})$
G. $[Zn^{2+}] = 7.9\times10^{-14}; [Sn^{2+}] = 2.9\times10^{-12}$
H. $[Li^+] = 0.60; [Mg^{2+}] = 7.4\times10^{-4}$

7. Determine the solubility of each of the following compounds in a solution maintained at pH = 2.00, 7.00, and 10.00. Use the data of Table 18-3 at the end of this chapter where needed.

A. $Al(OH)_3$ B. $Cr(OH)_3$ C. $Mg(OH)_2$ D. $Be(OH)_2$ $K_{sp} = 1.6\times10^{-27}$
E. $Ba(OH)_2$ F. $Fe(OH)_3$ G. $Ca(OH)_2$ H. AgOH $K_{sp} = 2.0\times10^{-8}$

8. Write equations for the equilibrium reactions that occur in aqueous solutions that contain the following species. Does the formation of a complex ion increase or decrease the amount of precipitate?

A. $Ag^+, S_2O_3{}^{2-}, Cl^-, Na^+$
B. $Hg^{2+}, I^-, Ag^+, NO_3{}^-$
C. $Zn^{2+}, CN^-, S^{2-}, Na^+$
D. $Fe^{2+}, CN^-, OHG^-, K^+$
E. $Al^{3+}, F^-, Ca^{2+}, Li^+$
F. Cu^+, CN^-, Ag^+, Cl^-
G. Al^{3+}, F^-, OH^-, Na^+
H. $Co^{3+}, S^{2-}, NH_3, Cl^-$
I. $Cu^{2+}, NH_3, OH^-, Na^+$

9. Determine the concentrations of free metal ion and complex ion in each solution in which the initial metal ion and ligand concentrations (all in moles per liter) are given below. Neglect the acid and base reactions of the ligands and cations. Use the data of Table 18-4 at the end of this chapter as needed.

A. $[F^-] = 0.100; [Al^{3+}] = 0.002$
B. $[Cl^-] = 1.00, [Pb^{2+}] = 8.2\times10^{-5}$
C. $[S_2O_3{}^{2-}] = 0.310; [Ag^+] = 0.0067$
D. $[CN^-] = 0.850; [Zn^{2+}] = 5.0\times10^{-4}$
E. $[CN^-] = 0.850; [Cu^+] = 5.0\times10^{-4}$
F. $[CN^-] = 0.850; [Fe^{2+}] = 5.0\times10^{-4}$
G. $[CN^-] = 0.850; [Fe^{3+}] = 5.0\times10^{-4}$
H. $[NH_3] = 1.52; [Co^{3+}] = 1.4\times10^{-5}$
I. $[NH_3] = 1.52; [Ag^+] = 1.4\times10^{-5}$
J. $[I^-] = 1.00; [Hg^{2+}] = 2.4\times10^{-6}$
K. $[Cl^-] = 1.00; [Hg^{2+}] = 2.4\times10^{-6}$

10. Determine the concentration of free metal ion (after a complex forms, but before precipitation occurs) and state whether a precipitate will form or not. The initial concentration of each species is given. You may neglect the acid and base reactions between water and the ligand, and the cation.

A. $[Ag^+] = 0.0100\ M, [I^-] = 0.0100\ M, [NH_3] = 5.00\ M$
B. $[Ag^+] = 0.0100\ M, [Cl^-] = 0.0100\ M, [NH_3] = 5.00\ M$
C. $[Zn^{2+}] = [S^{2-}] = 0.0100\ M. [EDTA^{4-}] = 0.100\ M$
D. $[Ag^+] = 0.0200\ M, [Br^-] = 0.0100\ M, [S_2O_3{}^{2-}] = 0.250\ M$
E. $[Cu^{2+}] = 0.0400\ M, [OH^-] = 0.0200\ M, [NH_3] = 2.00\ M$

11. Each of the following paragraphs gives the results of the qualitative analysis for a different unknown. Use these results to state which ions are definitely present and absent and which are uncertain.

All unknowns contain ions from qualitative analysis groups 3 through 5. (See Figure 23-3. Flame tests are summarized in Table 8-2 of Objective 8-3.)

A. The unknown was treated with NH_4Cl-NH_3 buffer and no precipitated formed. Further treatment with $(NH_4)_2S$ produced a light-colored precipitate that was separated from the solution. When the solution was treated with $(NH_4)_2CO_3$, a white precipitate formed. The remaining solution gave an orange-yellow color to a flame. The white precipitate dissolved in HCl and gave a brick-red color to a flame. The light-colored precipitate dissolved completely in HCl(aq) and did not reform when NaOH was added.

B. An unknown was treated with NH_4Cl-NH_3 buffer and a brown-red precipitate formed. Addition of $(NH_4)_2S$ produced no further precipitate. The remaining solution gave no precipitate with $(NH_4)_2CO_3$, but gave a violet color to a flame. The brownish red precipitate dissolved in HCl(aq), but reformed when the solution was made basic. Treatment of this precipitate with HCl(aq), followed by addition of KSCN(aq), produced a dark-red color in the solution.

C. The unknown gave no precipitate on treatment with NH_4Cl-NH_3 buffer, but a black precipitate formed when $(NH_4)_2S$ was added. The resulting solution gave a white precipitate on treatment with $(NH_4)_2CO_3$. The solution above the white precipitate gave no positive flame tests. However, when the white precipitate was dissolved in HCl(aq), the solution gave a yellow-green color to a flame. The black precipitate did not dissolve in HCl(aq), but did in aqua regia. Addition of NH_3(aq) produced a faint blue solution, which turned red on the addition c dimethylglyoxime.

QUIZZES (20 minutes each) Choose the best answer for each question.

QUIZ A

1. A saturated solution of copper(I) bromide has a concentration of 7.0×10^{-5} M. What is the value of K_{sp} of this compound? (a) 4.9×10^{-9}; (b) 4.9×10^{-10}; (c) 7.0×10^{-5}; (d) 14.0×10^{-5}; (e) none of these.

2. For MX, the solubility product constant is 2.8×10^{-9}. What is the molar solubility of this compound? (a) 2.8×10^{-9} M; (b) $\sqrt{28}\times10^{-5}$ M; (c) $(2.8)^{1/3}\times10^{-3}$ M; (d) $(2.8/4)^{1/3}\times10^{-3}$ M; (e) none of these.

3. The value of the solubility product constant of $Mg(OH)_2$ is 9.0×10^{-12}. If a solution has $[Mg^{2+}]=0.010$ M, $[OH^-]$, the amount required to just start the precipitation of $Mg(OH)_2$ is (a) 9.0×10^{-10} M; (b) 1.5×10^{-7} M; (c) 3.0×10^{-7} M; (d) 1.5×10^{-5} M; (e) 3.0×10^{-5} M.

4. The value of K_{sp} for AgCl is 1.10×10^{-10}. AgCl(s) will precipitate from solution whenever (a) $[Ag^+]$ exceeds 1.10×10^{-10}; (b) a solution containing Ag^+ is added to a solution containing

Cl^-; (c) $[Ag^+]+[Cl^-]$ exceeds 1.10×10^{-10}; (d) $[Ag^+][Cl^-]$ exceeds 1.10×10^{-10}; (e) $\sqrt{[Ag^+][Cl^-]}$ exceeds 1.10×10^{-10}.

5. Which of the following has the highest molar solubility? (a) CuS, $K_{sp}=8\times10^{-27}$; (b) $Bi_2S_3, K_{sp}=1\times10^{-70}$; (c) $Ag_2S, K_{sp}=6\times10^{-51}$; (d) MnS, $K_{sp}=7\times10^{-16}$; (e) cannot be determined.

6. Saturated solutions of cesium iodide, mercury(I) acetate, and calcium chlorate are mixed together. What precipitates? (a) calcium acetate; (b) mercury(I) chlorate; (c) mercury(I) iodide; (d) cesium acetate; (e) nothing.

7. AgI has the highest solubility in which of the following solutions? (a) 0.100 M $Mg(NO_3)_2$; (b) 0.100 M 0.100 M $AgNO_3$; (c) 0.100 M NaI; (d) 0.100 M MgI_2; (e) 0.100 M NaCN.

8. Which would be the least effective in dissolving $CuSO_3$? (a) add $SO_2(aq)$; (b) add $Na_2S_2O_3(aq)$; (c) add $HC_2H_3O_2(aq)$; (d) add $H_2O(l)$; (e) all would be equally effective.

QUIZ B

1. A saturated solution of lithium carbonate has a concentration of 1.62×10^{-2} M. For this compound, K_{sp} = (a) $4(1.62\times10^{-2})^3$; (b) $(1.62\times10^{-2})^3$; (c) $(1.62\times10^{-2})^2$; (d) 1.62×10^{-2}; (e) none of these.

2. The solubility product constant of copper(II) sulfide is 9.0×10^{-36}. The molar solubility of this compound is (a) 1.7×10^{18} M; (b) 1.5×10^{-12} M; (c) 1.2×10^{-18} M; (d) 3.0×10^{-11} M; (e) none of these.

3. Saturated solutions of sodium sulfate, ammonium carbonate, and nickel(II) nitrate are mixed together. The precipitate that forms is (a) nickel(II) carbonate; (b) sodium carbonate; (c) nickel(II) sulfate; (d) ammonium sulfate; (e) nothing precipitates.

4. Which of the following has the highest molar solubility? (a) $BaSO_4, K_{sp}=1.1\times10^{-10}$; (b) AgCl, $K_{sp}=1.6\times10^{-10}$; (c) $Mg(OH)_2, K_{sp}=2\times10^{-11}$; (d) $Cr(OH)_3, K_{sp}=6.3\times10^{-11}$.

5. The value of K_{sp} for AgCl is 1.6×10^{-10}. What is the maximum $[MnCl_2]$ that can exist in solution before precipitation occurs if the solution has $[AgNO_3]=3.4\times10^{-4}$ M? (a) 4.7×10^{-5} M; (b) 2.4×10^{-5} M; (c) 4.7×10^{-7} M; (d) 2.4×10^{-7} M; (e) none of these.

6. An ion is said to be completely precipitated when it (a) has a concentration of 0.00 M; (b) has 0.1% of its initial concentration; (c) cannot be precipitated by any known method; (d) has a concentration of less than $1/(6.022\times10^{23})$ M; (e) none of these.

7. Which would be the least effective in dissolving $MnCO_3$? (a) add $Na_2C_2O_4(aq)$; (b) add NaCN(aq); (c) add $CO_2(aq)$; (d) add $H_2O(l)$; (e) all would be equally effective.

8. $CuCO_3$ is least soluble in which of the following solutions? (a) 0.100 M HCl; (b) 1.00 M $CuCl_2$; (c) 0.100 M NaCl; (d) 0.100 M NaCl; (e) 0.100 M $MnSO_4$.

QUIZ C

1. A saturated solution of barium chromate has a concentration of 1.55×10^{-5} M. For this compound, K_{sp} = (a) 1.55×10^{-5}; (b) $(2 \times 1.55 \times 10^{-5})^2$; (c) $2(1.55 \times 10^{-5})$; (d) $1.55^2 \times 10^{-10}$; (e) none of these.

2. The value of the solubility product constant of silver carbonate is 8.2×10^{-12}. What is the molar solubility of this compound? (a) 2.9×10^{-6}; (b) 1.3×10^{-4}; (c) 8.2×10^{-12}; (d) 9.1×10^{-7}; (e) none of these.

3. Saturated solutions of sodium phosphate, copper(II) chloride, and ammonium acetate are mixed together. The precipitate is (a) copper(II) acetate; (b) copper(II) phosphate; (c) sodium chloride; (d) ammonium phosphate; (e) nothing precipitates.

4. Which of the following has the highest molar solubility? (a) $MgF_2, K_{sp} = 3.7 \times 10^{-8}$; (b) $MgCO_3, K_{sp} = 3.5 \times 10^{-8}$; (c) $Mg_3(PO_4)_2$, $K_{sp} = 1 \times 10^{-25}$; (d) Li_3PO_4, $K_{sp} = 3.2 \times 10^{-9}$.;

5. A solution has $[F^-] = 1.4 \times 10^{-5}$ M. What is the maximum $[Ca^{2+}]$ that can be present before CaF_2 ($K_{sp} = 2.7 \times 10^{-11}$) will precipitate? (a) 1.4×10^{-5}; (b) 7.0×10^{-6}; (c) 3.5×10^{-6}; (d) 5.6×10^{-5}; (e) none of these.

6. In which liquid will $AgNO_3$ be the most soluble? (a) 0.01 M HCl; (b) pure water; (c) 0.01 M HNO_3; (d) cannot be determined; (e) equally soluble in the two solutions.

7. Which of the following would be the least effective in producing a solution from $CuSO_4$? (a) add H_2O; (b) add $H_2SO_4(aq)$; (c) add $CO_2(aq)$; (d) add $CuCl_2(aq)$; (e) all are equally effective.

8. In which of the following solutions is $PbCl_2$ the most soluble? (a) 0.100 M NaCl; (b) 0.100 M $Pb(NO_3)_2$; (c) 0.100 M $NaNO_3$; (d) 0.100 M $NaNO_3$; (e) 0.100 $MnSO_4$.

QUIZ D

1. A saturated solution of magnesium fluoride has a concentration of 1.17×10^{-3} M. For this compound, K_{sp} = (a) 1.17×10^{-3}; (b) $(1.17 \times 10^{-3})^2$; (c) $4(1.17 \times 10^{-3})^3$; (d) $(1.17 \times 10^{-3})^3$; (e) none of these.

2. The value of the solubility product constant of silver chloride is 1.6×10^{-10}. What is the molar solubility of this compound? (a) 7.5×10^{-4}; (b) 9.2×10^{-5}; (c) 1.3×10^{-5}; (d) 1.7×10^{-10}; (e) none of these.

3. Which of the following has the highest molar solubility? (a) $Al(OH)_3, K_{sp} = 1.3 \times 10^{-33}$; (b) $CoS, K_{sp} = 4.0 \times 10^{-21}$; (c) $CuS, K_{sp} = 6.3 \times 10^{-36}$; (d) $Ag_2S, K_{sp} = 6.3 \times 10^{-50}$;

4. Saturated solutions of barium chloride, calcium nitrate, and cobalt(III) iodide are mixed. The precipitate is (a) calcium iodide; (b) calcium chloride; (c) barium iodide; (d) cobalt(III) chloride; (e) none of these.

5. In which solution will $BaSO_4$ be the most soluble? (a) $BaSO_4$ is insoluble in everything; (b) pure water; (c) 0.010 M $NaNO_3$; (d) 0.010 M $Ba(NO_3)_2$; (e) cannot be determined.

6. A solution has $[Ag^+] = 1.3 \times 10^{-4}$ M. What is the minimum $[CO_3^{2-}]$ needed to precipitate Ag_2CO_3 ($K_{sp} = 8.1 \times 10^{-12}$)? (a) 1.2×10^{-4} M; (b) 9.6×10^{-4} M; (c) 4.8×10^{-4} M; (d) 3.1×10^{-8} M; (e) none of these.

7. Which constant is numerically equal to the inverse of the formation constant of $[AgCl_2]^-$? (a) step-wise dissociation constant; (b) dissociation constant; (c) solubility product constant; (d) step-wise formation constant; (e) ionization constant.

8. In which of the following solutions is ZnS the least soluble? (a) 1.00 M $NH_3(aq)$; (b) 1.00 M $Na_2S_2O_3$; (c) 1.00 M NaCl; (d) 1.00 M $Mg(NO_3)_2$; (e) 1.00 M Na_2CO_3.

SAMPLE TEST (20 minutes)

1. A holding pond has $[PO_4^{3-}] = 8.06 \times 10^{-4}$ M. It measures $100.\text{ m} \times 200.\text{ m} \times 8.00\text{ m}$. What mass of $Ca(NO_3)_2$ must be added to this pond to reduce $[PO_4^{3-}]$ to 1.00×10^{-12} M? $K_{sp} = 1.30 \times 10^{-32}$ for $Ca_3(PO_4)_3$.

2. An aqueous solution has $[CrO_4^{2-}] = 2.0 \times 10^{-3}$ M and $[Cl^-] = 1.0 \times 10^{-5}$ M. Concentrated $AgNO_3$ solution is added. Will AgCl ($K_{sp} = 1.6 \times 10^{-10}$) or Ag_2CrO_4 ($K_{sp} = 1.1 \times 10^{-12}$) precipitate first? Eventually, $[Ag^+]$ should increase enough to cause precipitation of the second compound. What is that remaining concentration of the first ion that precipitates when the second ion just begins to precipitate?

3. 350.0 mL of 0.200 M $AgNO_3(aq)$ is added to 250.0 mL of 0.240 M $Na_2SO_4(aq)$, and Ag_2SO_4 ($K_{sp} = 1.4 \times 10^{-5}$) precipitates. 400.0 mL of 0.500 M $Na_2S_2O_3(aq)$ is added to this mixture. What mass of $Ag_2SO_4(s)$ remains? $K_f = 1.7 \times 10^{13}$ for $[Ag(S_2O_3)_2]^{3-}$.

TABLE 18-3 Solubility Product Constants at 25°C

Solute	Solubility equilibrium	K_{sp}
aluminum hydroxide	$Al(OH)_3(s) \rightleftharpoons Al^{3+}(aq) + 3\,OH^-(aq)$	1.3×10^{-33}
aluminum phosphate	$AlPO_4(s) \rightleftharpoons Al^{3+}(aq) + PO_4^{3-}(aq)$	6.3×10^{-19}
barium carbonate	$BaCO_3(s) \rightleftharpoons Ba^{2+}(aq) + CO_3^{2-}(aq)$	5.1×10^{-9}
barium chromate	$BaCrO_4(s) \rightleftharpoons Ba^{2+}(aq) + CrO_4^{2-}(aq)$	1.2×10^{-10}
barium fluoride	$BaF_2(s) \rightleftharpoons Ba^{2+}(aq) + 2\,F^-(aq)$	1.0×10^{-6}
barium hydroxide	$Ba(OH)_2(s) \rightleftharpoons Ba^{2+}(aq) + 2\,OH^-(aq)$	5×10^{-3}
barium oxalate	$BaC_2O_4\cdot H_2O(s) \rightleftharpoons Ba^{2+}(aq) + C_2O_4^{2-}(aq)$	2.3×10^{-8}
barium sulfate	$BaSO_4(s) \rightleftharpoons Ba^{2+}(aq) + SO_4^{2-}(aq)$	1.1×10^{-10}
barium sulfite	$BaSO_3(s) \rightleftharpoons Ba^{2+}(aq) + SO_3^{2-}(aq)$	8×10^{-7}
barium thiosulfate	$BaS_2O_3(s) \rightleftharpoons Ba^{2+}(aq) + S_2O_3^{2-}(aq)$	1.6×10^{-5}
bismuthyl chloride	$BiOCl(s) \rightleftharpoons BiO^+(aq) + Cl^-(aq)$	1.8×10^{-31}
bismuthyl hydroxide	$BiOOH(s) \rightleftharpoons BiO^+(aq) + OH^-(aq)$	4×10^{-10}
bismuth(III) sulfide	$Bi_2S_3(s) \rightleftharpoons 2\,Bi^{3+}(aq) + 3\,S^{2-}(aq)$	1×10^{-97}
cadmium carbonate	$CdCO_3(s) \rightleftharpoons Cd^{2+}(aq) + CO_3^{2-}(aq)$	5.2×10^{-12}
cadmium hydroxide	$Cd(OH)_2(s) \rightleftharpoons Cd^{2+}(aq) + 2\,OH^-(aq)$	2.5×10^{-14}
cadmium sulfide	$CdS(s) + H_2O \rightleftharpoons Cd^{2+}(aq) + HS^-(aq) + OH^-(aq)$	8×10^{-28}
calcium carbonate	$CaCO_3(s) \rightleftharpoons Ca^{2+}(aq) + CO_3^{2-}(aq)$	2.8×10^{-9}
calcium chromate	$CaCrO_4(s) \rightleftharpoons Ca^{2+}(aq) + CrO_4^{2-}(aq)$	7.1×10^{-4}
calcium fluoride	$CdCO_3(s) \rightleftharpoons Cd^{2+}(aq) + CO_3^{2-}(aq)$	5.3×10^{-9}
calcium hydroxide	$Ca(OH)_2(s) \rightleftharpoons Ca^{2+}(aq) + 2\,OH^-(aq)$	5.5×10^{-6}

calcium oxalate	$CaC_2O_4 \cdot H_2(s) \rightleftharpoons Ca^{2+}(aq) + C_2O_4{}^{2-}(aq)$	4×10^{-9}
calcium hydrogen phosphate	$CaHPO_4(s) \rightleftharpoons Ca^{2+}(aq) + HPO_4{}^{2-}(aq)$	1×10^{-7}
calcium phosphate	$Ca_3(PO_4)_2(s) \rightleftharpoons 3\,Ca^{2+}(aq) + 2\,PO_4{}^{3-}(aq)$	2.0×10^{-29}
calcium sulfate	$CaSO_4(s) \rightleftharpoons Ca^{2+}(aq) + SO_4{}^{2-}(aq)$	9.1×10^{-6}
calcium sulfite	$CaSO_3(s) \rightleftharpoons Ca^{2+}(aq) + SO_3{}^{2-}(aq)$	6.8×10^{-8}
chromium(II) hydroxide	$Cr(OH)_2(s) \rightleftharpoons Cr^{2+}(aq) + 2\,OH^-(aq)$	2×10^{-16}
chromium(III) hydroxide	$Cr(OH)_3(s) \rightleftharpoons Cr^{3+}(aq) + 3\,OH^-(aq)$	6.3×10^{-31}
cobalt(II) carbonate	$CoCO_3(s) \rightleftharpoons Co^{2+}(aq) + CO_3{}^{2-}(aq)$	1.4×10^{-13}
cobalt(II) hydroxide	$Co(OH)_2(s) \rightleftharpoons Co^{2+}(aq) + 2\,OH^-(aq)$	1.6×10^{-15}
cobalt(III) hydroxide	$Co(HO)_3(s) \rightleftharpoons Co^{3+}(aq) + 3\,OH^-(aq)$	1.6×10^{-44}
cobalt(II) sulfide (α form)	$CoS(s) \rightleftharpoons Co^{2+}(aq) + S^{2-}(aq)$	4×10^{-21}
copper(I) chloride	$CuCl(s) \rightleftharpoons Cu^+(aq) + Cl^-(aq)$	1.2×10^{-6}
copper(I) cyanide	$CuCN(s) \rightleftharpoons Cu^+(aq) + CN^-(aq)$	3.2×10^{-20}
copper(I) iodide	$CuI(s) \rightleftharpoons Cu^+(aq) + I^-(aq)$	1.1×10^{-12}
copper(I) sulfide	$Cu_2S(s) \rightleftharpoons 2\,Cu^+(aq) + S^{2-}(aq)$	2.5×10^{-48}
copper(II) arsenate	$Cu_3(AsO_4)_2(s) \rightleftharpoons 3\,Cu^{2+}(aq) + 2\,AsO_4{}^{3-}(aq)$	7.6×10^{-36}
copper(II) carbonate	$CuCO_3(s) \rightleftharpoons Cu^{2+}(aq) + CO_3{}^{2-}(aq)$	1.4×10^{-10}
copper(II) chromate	$CuCrO_4(s) \rightleftharpoons Cu^{2+}(aq) + CrO_4{}^{2-}(aq)$	3.6×10^{-6}
copper(II) ferrocyanide	$Cu_2[Fe(CN)_6] \rightleftharpoons 2\,Cu^{2+}(aq) + [Fe(CN)_6]^{4-}$	1.3×10^{-16}
copper(II) hydroxide	$Cu(OH)_2(s) \rightleftharpoons Cu^{2+}(aq) + 2\,OH^-(aq)$	2.2×10^{-20}

copper(II) sulfide	$Cu_2S(s) \rightleftharpoons 2\,Cu^{+}(aq) + S^{2-}(aq)$	6×10^{-37}
iron(II) carbonate	$FeCO_3(s) \rightleftharpoons Fe^{2+}(aq) + CO_3{}^{2-}(aq)$	3.2×10^{-11}
iron(II) hydroxide	$Fe(OH)_2(s) \rightleftharpoons Fe^{2+}(aq) + 2\,OH^{-}(aq)$	8.0×10^{-16}
iron(II) sulfide	$FeS(s) + H_2O \rightleftharpoons Fe^{2+}(aq) + HS^{-}(aq) + OH^{-}(aq)$	6×10^{-19}
iron(III) arsenate	$FeAsO_4(s) \rightleftharpoons Fe^{3+}(aq) + AsO_4{}^{3-}(aq)$	5.7×10^{-21}
iron(III) ferrocyanide	$Fe_4[Fe(CN)_6]_3(s) \rightleftharpoons 4\,Fe^{3+}(aq) + 3\,[Fe(CN)_6]^{4-}(aq)$	3.3×10^{-41}
iron(III) hydroxide	$Fe(OH)_3(s) \rightleftharpoons Fe^{3+}(aq) + 3\,OH^{-}(aq)$	4×10^{-38}
iron(III) phosphate	$FePO_4(s) \rightleftharpoons Fe^{3+}(aq) + PO_4{}^{3-}(aq)$	1.3×10^{-22}
lead(II) arsenate	$Pb_3(AsO_4)_2(s) \rightleftharpoons 3\,Pb^{2+}(aq) + 2\,AsO_4{}^{3-}(aq)$	4.0×10^{-36}
lead(II) azide	$Pb(N_3)_2(s) \rightleftharpoons Pb^{2+}(aq) + 2\,N_3{}^{-}(aq)$	2.5×10^{-9}
lead(II) bromide	$PbBr_2(s) \rightleftharpoons Pb^{2+}(aq) + 2\,Br^{-}(aq)$	4.0×10^{-5}
lead(II) carbonate	$PbCO_3(s) \rightleftharpoons Pb^{2+}(aq) + CO_3{}^{2-}(aq)$	7.4×10^{-14}
lead(II) chloride	$PbCl_2(s) \rightleftharpoons Pb^{2+}(aq) + 2\,Cl^{-}(aq)$	1.6×10^{-5}
lead(II) chromate	$PbCrO_4(s) \rightleftharpoons Pb^{2+}(aq) + CrO_4{}^{2-}(aq)$	2.8×10^{-13}
lead(II) fluoride	$PbF_2(s)_2 \rightleftharpoons Pb^{2+}(aq) + 2\,F^{-}(aq)$	2.7×10^{-8}
lead(II) hydroxide	$Pb(OH)_2\,(s) \rightleftharpoons Pb^{2+}(aq) + 2\,OH^{-}(aq)$	1.2×10^{-15}
lead(II) iodide	$PbI_2(s) \rightleftharpoons Pb^{2+}(aq) + 2\,I^{-}(aq)$	7.1×10^{-9}
lead(II) sulfate	$PbSO_4(s) \rightleftharpoons Pb^{2+}(aq) + SO_4{}^{2-}(aq)$	1.6×10^{-8}
lead(II) sulfide	$PbI_2(s) \rightleftharpoons Pb^{2+}(aq) + 2\,I^{-}(aq)$	3×10^{-28}
lithium carbonate	$Li_2CO_3(s) \rightleftharpoons 2\,Li^{+}(aq) + CO_3{}^{2-}(aq)$	2.5×10^{-2}
lithium fluoride	$LiF(s) \rightleftharpoons Li^{+}(aq) + F^{-}(aq)$	3.8×10^{-3}
lithium phosphate	$Li_3PO_4(s) \rightleftharpoons 3\,Li^{+}(aq) + PO_4{}^{3-}(aq)$	3.2×10^{-9}

magnesium arsenate	$Mg_3(AsO_4) \rightleftharpoons 3\,Mg^{2+}\,(aq) + 2\,AsO_4{}^{3-}\,(aq)$	2.1×10^{-20}
magnesium carbonate	$MgCO_3\,(s) \rightleftharpoons Mg^{2+}\,(aq) + CO_3{}^{2-}\,(aq)$	3.5×10^{-8}
magnesium fluoride	$Mg(OH)_2\,(s) \rightleftharpoons Mg^{2+}\,(aq) + 2\;F-(aq)$	3.7×10^{-8}
magnesium hydroxide	$Mg(OH)_2\,(s) \rightleftharpoons Mg^{2+}\,(aq) + 2\,OH-(aq)$	1.8×10^{-11}
magnesium phosphate	$Mg_3(PO_4)_2\,(s) \rightleftharpoons 3\,Mg^{2+}\,(aq) + 2\,PO_4{}^{3-}\,(aq)$	1×10^{-25}
magnesium ammonium phosphate	$MgNH_4PO_4\,(s) \rightleftharpoons Mg^{2+}\,(aq) + NH_4{}^{+}\,(aq) + PO_4{}^{3-}\,(aq)$	2.5×10^{-13}
manganese(II) carbonate	$MnCO_3(s) \rightleftharpoons Mn^{2+}\,(aq) + CO_3{}^{2-}\,(aq)$	1.8×10^{-11}
manganese(II) hydroxide	$Mn(OH)_2\,(s) \rightleftharpoons Mn^{2+}\,(aq) + 2\,OH^-\,(aq)$	1.9×10^{-13}
manganese(II) sulfide	$MnS(s)\;Br_2\,(s) \rightleftharpoons Mn^{2+}\,(aq)\;HS^-\,(aq) + OH-$	3×10^{-14}
mercury(I) bromide	$Hg_2Br_2\,(s) \rightleftharpoons Hg_2{}^{2+}(aq) + 2\,Br^-\,(aq)$	5.6×10^{-23}
mercury(I) chloride	$Hg_2Cl_2 \rightleftharpoons Hg_2{}^{2+}(aq) + 2\;Cl^-\,(aq)$	1.3×10^{-18}
mercury(I) iodide	$Hg_2I_2 \rightleftharpoons Hg_2{}^{2+}\,(aq) + 2\,I^-\,(aq)$	4.5×10^{-29}
mercury(I) sulfate	$Hg_2SO_4\,(s) \rightleftharpoons Hg_2{}^{2+}(aq) + SO_4{}^{2-}(aq)$	7.4×10^{-7}
mercury(I) sulfide	$Hg_2S(s) \rightleftharpoons Hg_2{}^{2+}\,(aq) + S^{2-}\,(aq)$	1.0×10^{-47}
mercury(II) sulfide	$HgS(s) + H_2O \rightleftharpoons Hg^{2+}\,(aq) + HS-(aq) + OH-(aq)$	2×10^{-53}
nickel(II) carbonate	$NiCO_3\,(s) \rightleftharpoons Ni^{2+}\,(aq) + CO_3{}^{2-}\,(aq)$	6.6×10^{-9}
nickel(II) hydroxide	$Ni(OH)_2(s) \rightleftharpoons Ni^{2+}\,(aq) + 2\;OH^-\,(aq)$	2.0×10^{-15}
nickel(II) sulfide	$NiS(s) \rightleftharpoons Ni^{2+}\,(aq) + S^{2-}\,(aq)$	3×10^{-19}
scandium fluoride	$ScF_3(s) \rightleftharpoons Sc^{3+}\,(aq) + 3\;F-(aq)$	4.2×10^{-18}
scandium hydroxide	$Sc(OH)_3(s) \rightleftharpoons Sc^{3+}\,(aq) + 3\,OH^-\,(aq)$	8.0×10^{-31}
silver arsenate	$Ag_3AsO_4\,(s) \rightleftharpoons 3\;Ag^{+}\,(aq) + AsO_4{}^{3-}\,(aq)$	1.0×10^{-22}

silver azide	$AgN_3(s) \rightleftharpoons Ag^+(aq) + N_3-(aq)$	2.8×10^{-9}
silver bromide	$AgBr(s) \rightleftharpoons Ag^+(aq) + Br^-(aq)$	5.0×10^{-13}
silver carbonate	$Ag_2CO_3(s) \rightleftharpoons 2\,Ag^+(aq) + CO_3^{2-}(aq)$	8.1×10^{-12}
silver chloride	$AgCl(s) \rightleftharpoons Ag^+(aq) + Cl-(aq)$	1.8×10^{-10}
silver chromate	$Ag_2CrO_4(s) \rightleftharpoons 2\,Ag^+(aq) + CrO_4^{2-}(aq)$	1.1×10^{-12}
silver cyanide	$AgCN(s) \rightleftharpoons Ag^+(aq) + CN-(aq)$	1.2×10^{-16}
silver iodate	$AgIO_3(s) \rightleftharpoons Ag^+(aq) + IO_3-(aq)$	3.0×10^{-8}
silver iodide	$AgI(s) \rightleftharpoons Ag^+(aq) + I-(aq)$	8.5×10^{-17}
silver nitrite	$AgNO_2(s) \rightleftharpoons Ag^+(aq) + NO_2-(aq)$	6.0×10^{-4}
silver oxalate	$Ag_2C_2O_4(s) \rightleftharpoons 2\,Ag^+(aq) + C_2O_4^{2-}(aq)$	3.6×10^{-11}
silver sulfate	$Ag_2SO_4(s) \rightleftharpoons 2\,Ag^+(aq) + SO_4^{2-}(aq)$	1.4×10^{-5}
silver sulfide	$Ag_2S(s) + H_2O \rightleftharpoons 2\,Ag^+(aq) + HS^-(aq) + OH^-(aq)$	6×10^{-51}
silver sulfite	$Ag_2SO_3(s) \rightleftharpoons 2\,A^+(aq) + SO_3^{2-}(aq)$	1.5×10^{-14}
silver thiocyanate	$AgSCN(s) \rightleftharpoons Ag^+(aq) + SCN^-(aq)$	1.0×10^{-12}
strontium carbonate	$SrCO_3(s) \rightleftharpoons Sr^{2+}(aq) + CO_3^{2-}(aq)$	1.1×10^{-10}
strontium chromate	$SrCrO_4(s) \rightleftharpoons Sr^{2+}(aq) + CrO_4^{2-}(aq)$	2.2×10^{-5}
strontium fluoride	$SrF_2(s) \rightleftharpoons Sr^{2+}(aq) + 2\,F^-(aq)$	2.5×10^{-9}
strontium sulfate	$SrSO_4(s) \rightleftharpoons Sr^{2+}(aq) + SO_4^{2-}(aq)$	3.2×10^{-7}
strontium sulfite	$SrSO_3(s) \rightleftharpoons Sr^{2+}(aq) + SO_3^{2-}(aq)$	4×10^{-8}
thallium(I) bromide	$TlBr(s) \rightleftharpoons Tl^+(aq) + Br^-(aq)$	3.4×10^{-6}
thallium(I) chloride	$TlCl(s) \rightleftharpoons Tl^+(aq) + Cl^-(aq)$	1.7×10^{-4}
thallium(I) iodide	$TlI(s) \rightleftharpoons Tl^+(aq) + I^-(aq)$	6.5×10^{-8}
thallium(III)	$Tl(OH)_3(s) \rightleftharpoons Tl^{3+}(aq) + OH^-(aq)$	6.3×10^{-46}

hydroxide		
tin(II) hydroxide	$Sn(OH)_2(s) \rightleftharpoons Sn^{2+}(aq) + 2\,OH^-(aq)$	1.4×10^{-28}
tin(II) sulfide	$SnS(s) + H_2O \rightleftharpoons Sn^{2+}(aq) + HS^-(aq) + OH^-(aq)$	1×10^{-26}
zinc carbonate	$ZnCO_3(s) \rightleftharpoons Zn^{2+}(aq) + CO_3^{2-}(aq)$	1.4×10^{-11}
zinc hydroxide	$Zn(OH)_2(s) \rightleftharpoons Zn^{2+}(aq) + 2\,OH^-(aq)$	1.2×10^{-17}
zinc oxalate	$ZnC_2O_4(s) \rightleftharpoons Zn^{2+}(aq) + C_2O_4^{2-}(aq)$	2.7×10^{-17}
zinc phosphate	$Zn_3(PO_4)(s) \rightleftharpoons 3\,Zn^{2+}(aq) + 2\,PO_4^{3-}(aq)$	9.0×10^{-33}
zinc sulfide (α form)	$ZnS(s) + H_2O \rightleftharpoons ZnH^{2+}(aq) + HS^-(aq) + OH^-(aq)$	2×10^{-25}

TABLE 18-4 Formation Constants for Some Complex Ions

Complex ion	Equilibrium reaction	K_f
$[AgCl_2]^-$	$Ag^+ + 2\,Cl^- \rightleftharpoons [AgCl_2]^-$	1.1×10^5
$[Ag(CN)_2]^-$	$Ag^+ + 2\,CN^- \rightleftharpoons [Ag(CN)_2]^-$	5.6×10^{18}
$[AgI_2]^-$	$Ag^+ + 2\,I^- \rightleftharpoons [AgI_2]^-$	5.5×10^{11}
$[Ag(EDTA)]^{3-}$	$Ag^+ + EDTA^{4-} \rightleftharpoons [Ag(EDTA)]^{3-}$	2.1×10^7
$[Ag(en)_2]^+$	$Ag^+ + 2\,en \rightleftharpoons [Ag(en)_2]^+$	5.0×10^7
$[Ag(NH_3)_2]^+$	$Ag^+ + 2\,NH_3 \rightleftharpoons [Ag(NH_3)_2]^+$	1.6×10^7
$[Ag(py)_2]^+$	$Ag^+ + 2\,py \rightleftharpoons [Ag(py)_2]^+$	2.2×10^4
$[Ag(SCN)_4]^{3-}$	$Ag^+ + 4\,SCN^- \rightleftharpoons [Ag(SCN)_4]^{3-}$	1.2×10^{10}
$[Ag(S_2O_3)_2]^{3-}$	$Ag^+ + 4\,S_2O_3{}^{2-} \rightleftharpoons [Ag(S_2O_3)_2]^{3-}$	1.7×10^{13}
$[Al(C_2O_4)_3]^{3-}$	$Al^{3+} + 3\,C_2O_4^{2-} \rightleftharpoons [Al(C_2O_4)_3)^{3-}$	2×10^{16}
$[Al(EDTA)]^-$	$Al^{3+} + EDTA^{4-} \rightleftharpoons [Al(EDTA)]^-$	1.3×10^{16}
$[AlF_6]^{3-}$	$Al^{3+} + 6F^- \rightleftharpoons [AlF_6]^{3-}$	6.9×10^{19}
$[Al(OH)_4]^-$	$Al^{3+} + 4\,OH^- \rightleftharpoons [Al(OH)_4]^-$	1.1×10^{33}
$[CdCl_4]^{2-}$	$Cd^{2+} + 4Cl^- \rightleftharpoons [CdCl]^{2-}$	6.3×10^2
$[Cd(CN)_4]^{2-}$	$Cd^{2+} + 4\,CN^- \rightleftharpoons [Cd(CN)_4]^{2-}$	6.0×10^{18}
$[Cd(en)_3]^{2+}$	$Cd^{2+} + 3\,en \rightleftharpoons [Cd(en)_3]^{2+}$	1.2×10^{12}
$[Cd(NH_3)_4]^{2+}$	$Cd^{2+} + 4\,NH_3 \rightleftharpoons [Cd(NH_3)_4]^{2+}$	1.3×10^7
$[Cd(NH_3)_6]^{2+}$	$Cd^{2+} + 6\,NH_3 \rightleftharpoons [Cd(NH_3)_6]^{2+}$	1.4×10^5
$[Cd(py)_4]^{2+}$	$Cd^{2+} + 4\,py \rightleftharpoons [Cd(py)_4]^{2+}$	3.2×10^2
$[Co(C_2O_4)_3]^{4-}$	$CO^{2+} + 3\,C_2O_4^{2-} \rightleftharpoons [Co(C_2O_4)_3]^{4-}$	5×10^9
$[Co(EDTA)]^{2-}$	$Co^{2+} + EDTA^{4-} \rightleftharpoons [Co(EDTA)]^{2-}$	2.0×10^{16}
$[Co(en)_3]^{2+}$	$Co^{2+} + 3\,en \rightleftharpoons [Co(en)_3]^{2+}$	8.7×10^{13}
$[Co(NH_3)_6]^{2+}$	$Co^{2+} + 6\,NH_3 \rightleftharpoons [Co(NH_3)_6]^{2+}$	1.3×10^5
$[Co(SCN)_4]^{2=}$	$Co^{2+} + 4\,SCN^- \rightleftharpoons [Co(SCN)_4]^{2-}$	1.0×10^3
$[Co(C_2O_4)_3]^{3-}$	$Co^{3+} + 3\,C_2O_4^{2-} \rightleftharpoons [Co(C_2O_4)_3]^{3-}$	$\approx 1 \times 10^{20}$
$[Co(EDTA)]^-$	$Co^{3+} + EDTA^{4-} \rightleftharpoons [Co(EDTA)]^-$	1×10^{36}
$[Co(en)_3]^{3+}$	$Co^{3+} + 3\,en \rightleftharpoons [Co(en)_3]^{3+}$	4.9×10^{48}
$[Co(NH_3)_6]^{3+}$	$Co^{3+} + 6\,NH_3 \rightleftharpoons [Co(NH_3)_6]^{3+}$	4.5×10^{33}
$[Cr(EDTA)]^-$	$Co^{3+} + EDTA^{4-} \rightleftharpoons [Cr(EDTA)]^-$	1×10^{23}
$[Cr(OH)_4]^-$	$Cr^{3+} + 4\,OH^- \rightleftharpoons [Cr(OH)_4]^-$	8×10^{29}
$[CuCl_3]^{2-}$	$Cu^+ + 3\,Cl^- \rightleftharpoons [CuCl_3]^{2-}$	5×10^5
$[Cu(CN)_4]^{3-}$	$Cu^+ + 4\,CN^- \rightleftharpoons [Cu(CN)_4]^{3-}$	2.0×10^{30}

$[Cu(C_2O_4)_2]^{2-}$	$Cu^{2+} + 2\,C_2O_4^{2-} \rightleftharpoons [Cu(C_2O_4)_2]^{2-}$	3×10^{8}
$[Cu(EDTA)]^{2-}$	$Cu^{2+} + EDTA^{4-} \rightleftharpoons [Cu(EDTA)]^{2-}$	5×10^{18}
$[Cu(en)_2]^{2+}$	$Cu^{2+} + 2\,en \rightleftharpoons [Cu(en)_2]^{2+}$	1×10^{20}
$[Cu(NH_3)_4]^{2+}$	$Cu^{2+} + 4\,NH_3 \rightleftharpoons [Cu(NH_3)_4]^{2+}$	1.1×10^{13}
$[Cu(OH)_4]^{2-}$	$Cu^{2+} + 4\,OH^- \rightleftharpoons [Cu(OH)_4]^{2-}$	3×10^{18}
$[Cu(py)_6]^{2+}$	$Cu^{2+} + 6\,py \rightleftharpoons [Cu(p)_6]^{2+}$	2×10^{10}
$[Fe(CN)_6]^{4-}$	$Fe^{2+} + 6\,CN^- \rightleftharpoons [Fe(CN)_6]^{4-}$	$\approx 1\times10^{37}$
$[Fe(C_2O_4)_3]^{4-}$	$Fe^{2+} + 3\,C_2O_4{}^{2-} \rightleftharpoons [Fe(C_2O_4)_3]^{4-}$	1.7×10^{5}
$[Fe(EDTA)]^{2-}$	$Fe^{2+} + EDTA^{4-} \rightleftharpoons [Fe(EDTA)]^{2-}$	2.1×10^{14}
$[Fe(en)_3]^{2+}$	$Fe^{2+} + 3\,en \rightleftharpoons [Fe(en)_3]^{2+}$	5.0×10^{9}
$[Fe(CN)_6]^{3-}$	$Fe^{3+} + 6\,CN^- \rightleftharpoons [Fe(CN)_6]^{3-}$	$\approx 1\times10^{42}$
$[Fe(C_2O_4)_3]^{3-}$	$Fe^{3+} + 3\,C_2O_4{}^{2-} \rightleftharpoons [Fe(C_2O_4)_3]^{3-}$	2×10^{20}
$[Fe(EDTA)]^-$	$Fe^{3+} + EDTA^{4-} \rightleftharpoons [Fe(EDTA)]^-$	1.7×10^{24}
$[Fe(SCN)]^{2+}$	$Fe^{3+} + SCN^- \rightleftharpoons [Fe(SCN)]^{2+}$	8.9×10^{2}
$[HgBr_4]^{2-}$	$Hg^{2+} + 4\,Br^- \rightleftharpoons [HgBr_4]^{2-}$	1.0×10^{21}
$[HgCl_4]^{2-}$	$Hg^{2+} + 4\,Cl^- \rightleftharpoons [HgCl_4]^{2-}$	1.2×10^{15}
$[Hg(CN)_4]^{2-}$	$Hg^{2+} + 4\,CN^- \rightleftharpoons [Hg(CN)_4]^{2-}$	3×10^{41}
$[Hg(C_2O_4)_2]^{2-}$	$Hg^{2+} + 2\,C_2O_4{}^{2-} \rightleftharpoons [Hg(C_2O_4)_2]^{2-}$	9.5×10^{6}
$[Hg(EDTA)]^{2-}$	$Hg^{2+} + EDTA^{4-} \rightleftharpoons [Hg(EDTA)]^{2-}$	6.3×10^{21}
$[Hg(en)_2]^{2+}$	$Hg^{2+} + 2\,en \rightleftharpoons [Hg(en)_2]^{2+}$	2×10^{23}
$[HgI_4]^{2-}$	$Hg^{2+} + 4\,I^- \rightleftharpoons [HgI_4]^{2-}$	6.8×10^{29}
$[Ni(CN)_4]^{2-}$	$Ni^{2+} + 4\,CN^- \rightleftharpoons [Ni(CN)_4]^{2-}$	2×10^{31}
$[Ni(C_2O_4)_3]^{4-}$	$Ni^{2+} + 3\,C_2O_4{}^{2-} \rightleftharpoons [Ni(C_2O_4)_3]^{4-}$	3×10^{8}
$[Ni(EDTA)]^{2-}$	$Ni^{2+} + EDTA^{4-} \rightleftharpoons [Ni(EDTA)]^{2-}$	3.6×10^{18}
$[Ni(en)_3]^{2+}$	$Ni^{2+} + 3\,en \rightleftharpoons [Ni(en)_3]^{2+}$	2.1×10^{18}
$[Ni(NH_3)_6]^{2+}$	$Ni^{2+} + 6\,NH_3 \rightleftharpoons [Ni(NH_3)_6]^{2+}$	5.5×10^{8}
$[PbCl_3]^-$	$Pb^{2+} + 3\,Cl^- \rightleftharpoons [PbCl_3]^-$	2.4×10^{1}
$[Pb(C_2O_4)_2]^{2-}$	$Pb^{2+} + 2\,C_2O_4{}^{2-} \rightleftharpoons [Pb(C_2O_4)_2]^{2-}$	3.5×10^{6}
$[Pb(EDTA)]^{2-}$	$Pb^{2+} + EDTA^{4-} \rightleftharpoons [Pb(EDTA)]^{2-}$	2×10^{18}
$[Pb(OH)_3]^-$	$Pb^{2+} + 3\,OH^- \rightleftharpoons [Pb(OH)_3]^-$	3.8×10^{14}
$[Pb(S_2O_3)_3]^{4-}$	$Pb^{2+} + 3\,S_2O_3{}^{2-} \rightleftharpoons [Pb(S_2O_3)_3]^{4-}$	2.2×10^{6}
$[PbCl_4]^{2-}$	$Pb^{2+} + 4\,Cl^- \rightleftharpoons [PbCl_4]^{2-}$	4.0×10^{1}
$[PbI_4]^{2-}$	$Pb^{2+} + 4\,I^- \rightleftharpoons [PbI_4]^{2-}$	3.0×10^{4}
$[PtCl_4]^{2-}$	$Pt^{2} + 4\,Cl^- \rightleftharpoons [PtCl_4]^{2-}$	1×10^{16}

$[Pt(NH_3)_6]^{2+}$	$Pt^{2+} + 6\ NH_3 \rightleftharpoons [Pt(NH_3)_6]^{2+}$	2×10^{35}
$[Zn(CN)_4]^{2-}$	$Zn^{2+} + 4\ CN^- \rightleftharpoons [Zn(CN)_4]^{2-}$	1×10^{18}
$[Zn(C_2O_4)_3]^{4-}$	$Zn^{2+} + 3\ C_2O_4^{\ 2-} \rightleftharpoons [Zn(C_2O_4)_3]^{4-}$	1.4×10^{8}
$[Zn(EDTA)]^{2-}$	$Zn^{2+} + EDTA^{4-} \rightleftharpoons [Zn(EDTA)]^{2-}$	3×10^{16}
$[Zn(en)_3]^{2+}$	$Zn^{2+} + 3\ en \rightleftharpoons [Zn(en)_3]^{2+}$	1.3×10^{14}
$[Zn(NH_3)_4]^{2+}$	$Zn^{2+} + 4\ NH_3 \rightleftharpoons [Zn(NH_3)_4]^{2+}$	4.1×10^{8}
$[Zn(OH)_4]^{2-}$	$Zn^{2+} + 4\ OH^- \rightleftharpoons [Zn(OH)_4]^{2-}$	4.6×10^{17}
$[Zn(py)_4]^{2+}$	$Zn^{2+} + 4\ py \rightleftharpoons [Zn(py)_4]^{2+}$	8.5×10^{1}

19 SPONTANEOUS CHANGE: ENTROPY AND FREE ENERGY

CHAPTER OBJECTIVES

1. Explain the meaning of *spontaneous change* as it applies to chemical reactions.

A *spontaneous reaction* is one that eventually will occur when left to itself. A *nonspontaneous reaction* will not eventually occur. The distinction between spontaneity and speed is important. Simply because a reaction is spontaneous does not mean that it is rapid. Silver tarnishes spontaneously and diamond spontaneously changes to graphite, but neither process is particularly rapid. Even though *spontaneous* is in general use by scientists, perhaps *feasible* conveys the idea more clearly.

2. Explain why entropy is important and how it is related to the disorder of the system.

At first, the evolution of heat was thought to be a sign of a spontaneous change. But not all spontaneous reactions are exothermic. Vaporizing water is an example of a spontaneous endothermic process. (Water spontaneously vaporizes from your skin to cool you on a warm day.)

To arrive at a criterion for spontaneous change, let us consider everyday changes: diving into a pool, spilling milk, and an automobile collision. The reverse of each of these spontaneous changes is nonspontaneous. In the reverse of the automobile collision, the scattered parts fly off the roadway and reassemble into two undamaged cars that back away from each other. Nonspontaneous changes "look funny" because, in them, energy is not dissipated but rather assembled. When an automobile part strikes the ground, its energy of motion is scattered: some causing the ground to shake, some producing sound, and some scattering soil and pebbles. Reassembling all of this energy is extremely unlikely.

Thus, a nonspontaneous change does not occur because, in the reverse change—the spontaneous one—energy has been scattered. Another way to say this is that the entropy increases in a spontaneous process. Entropy is related to the amount of disorder present. The more disorder, the higher the entropy.

* 3. Predict whether entropy increases or decreases for certain processes.

Determining entropy change is easier if we think of entropy as a measure of disorder. When a solid melts, its orderly crystalline structure is broken apart, resulting in a liquid that is more disordered. Likewise, a gas is more disordered than the liquid because the gas molecules are not confined to the small volume of the liquid.

Entropy also increases when two substances are mixed together, since the molecules or atoms now are mixed up. Placing a solid solute in a liquid solution is accompanied by a large entropy increase because the regular crystal structure of the solid is broken apart and the ions or molecules of the solid are scattered throughout the solution. This entropy effect is large enough to cause ionic solids to dissolve, even when a great deal of energy must be supplied. (The solution process is very endothermic; for instance, the dissolving of NH_4NO_3 in water that is used for "instant cold packs.") On the other hand, entropy *decreases* when a gas dissolves in a liquid solution. Gases are very disordered, and confining them to the small volume of a liquid increases their order. Whenever the balanced chemical equation shows more moles of gaseous products than gaseous reactants, there will be an entropy increase. Thus, when $\Delta n_{gas} > 0$, $\Delta S > 0$.

4. Know the relationship between entropy, heat, and temperature.

The change in entropy, ΔS, is directly proportional to the heat, q, and inversely proportional to the Kelvin temperature, T.

$$\Delta S \propto \frac{q}{T}$$

This appears reasonable because, if a system absorbs a large amount of heat, more disorder will occur than if a small amount of heat is absorbed. The inverse relationship with temperature is less intuitive. If a system is at high temperature, it has relatively more disorder than at low temperature. Additional heat will create less disorder in the high-temperature system than if the same quantity of heat were added to the low-temperature system. This can be explained on a microscopic level by the number of energy states available.

Example 19-1. Predict whether the entropy will be greater for:

(a) 50 J of heat added to a block of ice or 50 J of heat added to the same mass of water.

(b) A spontaneous process or a nonspontaneous process.

(c) A particle in a box of length L or a particle in a box of length 2 L.

Solutions:

(a) The entropy change will be greater for the block of ice (entropy is inversely proportional to T).

(b) Cannot predict. Spontaneity is not determined by ΔS alone.

(c) The larger box will have more possible microstates.

5. Know the Boltzmann equation for entropy.

The more states a system can occupy, the greater the entropy. $S = k \ln W$.

Example 19-2. Four coins are placed heads up in a box. After the box is shaken, two coins are heads up. Determine the change in entropy for the system.

Solution: There is only one macrostate where all the coins are heads up. $S = k \ln W$ or $S = 2.3\ k \log W$, but $\log (1.0) = 0$; therefore, $S = 0$.

For the case where there are four coins, two of which are heads up, the number of macrostates is 6.

$$S = 2.3\,k \log 6$$
$$= 2.3\,(1.38 \times 10^{-23}\ J/K) \times \log 6$$
$$= 2.4 \times 10^{-23}\ J/K$$

and the entropy change is $\Delta S = 2.4 \times 10^{-23}\ J/K - 0 = +2.4 \times 10^{-23}\ J/K$. The change in entropy is positive as we would expect since the disorder of the system has increased.

6. Explain why entropy alone is not used to predict a spontaneous change and why free energy is needed.

It seems very easy to determine whether the entropy of the system increases ($\Delta S > 0$) or decreases ($\Delta S < 0$) for a given process. However, it is the *total* entropy change, that of the system *plus* that of the surroundings, that determines whether or not a process is spontaneous. Only if this total entropy change (the entropy change of the universe) is positive is the process spontaneous. For a spontaneous change,

$$\Delta S_{univ} = \Delta S_{syst} + \Delta S_{surr} > 0 \qquad [6]$$

ΔS_{surr} often is quite hard to compute. In addition, other measured properties (ΔH, T, P, volume, density, and so on) are properties of the *system*. We wish the criterion of spontaneous change to be a property of the system, and only the system. The free energy change is defined by

$$\Delta G_{system} = \Delta H_{system} - T\Delta S_{system} \qquad [7]$$

Free energy is *negative* for a spontaneous change: $\Delta G < 0$. In fact, we can show that $\Delta G = -T\Delta S_{univ}$. The defining equation for free energy is

$$G = H - TS \qquad [8]$$

Equation [7] is produced from Equation [8] only when the temperature is constant or when the initial and final temperatures of a process are the same. Another way to interpret Equation [7] is as a balance between two trends. First, the system tries to become disordered, making ΔS positive and, thus, $-T\Delta S$ negative. The second trend is that the system tries to reach the lowest possible energy, making ΔH negative.

*** 6a. Qualitatively predict whether reactions are spontaneous or nonspontaneous based on their ΔH and ΔS values.**

There are four possibilities in applying Equation [7]:

(1) An exothermic reaction ($\Delta H < 0$) where the order of the system decreases ($\Delta S > 0$ or $-T\Delta S < 0$) must be spontaneous.

(2) An exothermic reaction ($\Delta H < 0$) where the order of the system increases ($\Delta S < 0$ or $-T\Delta S < 0$) is spontaneous at low temperatures where the $-T\Delta S$ term is small and ΔH determines the sign of ΔG. As the temperature rises, $-T\Delta S$ gets larger

and the reaction becomes nonspontaneous. An example of such a process is the condensation of a gas to form a liquid.

(3) An endothermic reaction ($\Delta H > 0$) where the order of the system decreases ($\Delta S > 0$ or $-T\Delta S < 0$) is nonspontaneous at low temperatures where $-T\Delta S$ has a small effect. At high temperatures, the $T\Delta S$ term increases and eventually the reaction becomes spontaneous. The dissolving of most ionic salts and the dissociation of any substance into its atoms both fall into this category.

(4) An endothermic reaction ($\Delta H > 0$) where the order of the system increases ($\Delta S < 0$ or $-T\Delta S > 0$) is nonspontaneous at all temperatures.

*** 6b. Determine $\Delta G°$ from tabulated data, both tables of $\Delta G_f°$ and those of $\Delta H_f°$ and $S°$.**

$\Delta G°$ is the standard molar free energy change of a reaction. If $\Delta G°$ is negative, the reaction is spontaneous and the products in their standard states will be spontaneously produced from the reactants in their standard states. Consider Reaction [9].

$$2\ NO(g) + O_2(g) \longrightarrow 2\ NO_2(g) \qquad \Delta G° = -69.71\ \text{kJ/mol} \qquad [9]$$

69.71 kJ of free energy is produced when 2 mol $NO_2(g)$ at a total pressure of 1.00 atm is produced from 2 mol NO(g) and 1 mol $O_2(g)$, with each reactant having a partial pressure of 1.00 atm. In contrast, $\Delta G_f°$ is the standard molar free energy of formation. $\Delta G_f°$ is, thus, $\Delta G°$ for a reaction in which 1 mole of product is formed and in which the reactants are elements in their most stable forms. (That is what *formation* means, denoted by subscript $_f$.) Products and reactants are in their standard states, as the superscript ° indicates. (See Objective 19–10 for a definition of *standard state conditions*.) For example, Equation [10] is the formation reaction of $NO_2(g)$:

$$\frac{1}{2} N_2(g) + O_2(g) \longrightarrow NO_2(g) \qquad \Delta G_f° = 33.85\ \text{kJ/mol} \qquad [10]$$

$\Delta G°$ is determined from values of $\Delta G_f°$ in the same way that we determined $\Delta H°$ from values of $\Delta H_f°$ (see Objective 7–9),

$$\Delta G°_{rxn} = \sum_{\text{products}} \nu_{prod}\, \Delta G°_{f,prod} - \sum_{\text{reactants}} \nu_{reac}\, \Delta G°_{f,reac} \qquad [11]$$

where the v's are the stoichiometric coefficients in the balanced chemical equation. For example, for Reaction [12]

$$4\ NH_3(g) + 5\ O_2(g) \longrightarrow 4\ NO(g) + 6\ H_2O(l) \qquad [12]$$

$\Delta G°$ is computed as follows:

$$\begin{aligned}\Delta G°_{rxn} &= 4\ \Delta G_f°\ [NO(g)] + 6\ \Delta G_f°\ [H_2O(l)] - 4\ \Delta G_f°\ [NH_3(g)] - 5\ \Delta G_f°\ [O_2(g)] \\ &= [4(86.57) + 6(-237.2) - 4(-16.48) - 5(0.00)]\ \text{kJ/mol} = -1010.0\ \text{kJ/mol}\end{aligned}$$

Tabulated values of ΔG_f° are valid for only one temperature, often 298 K. We also can use the Gibbs-Helmholtz equation ([14]) to determine ΔG for a reaction. Because ΔH° and ΔS° vary very little with temperature, Equation [14] allows us to compute ΔG° at many temperatures.

$$\Delta G^\circ = \Delta H^\circ - T\Delta S^\circ \qquad [14]$$

ΔH° is computed from ΔH_f° data as described in Objective 7–10. ΔS° is computed from S° data in a similar manner:

$$\Delta S_{rxn}^\circ = \sum_{products} \nu_{prod}\, S_{prod}^\circ - \sum_{reactants} \nu_{reac}\, S_{reac}^\circ \qquad [15]$$

For Reaction [12], ΔS° is computed as follows:

$$\Delta S_{rxn}^\circ = 4\, S^\circ[NO(g)] + 6\, S^\circ[H_2O(l)] - 4\, S^\circ[NH_3(g)] - 5\, S^\circ[O_2(g)]$$
$$= [4(210.6) + 6(69.91) - 4\ (192.5) - 5\ (205.0)]\ \text{kJ/mol} = -\ 533.1\ \text{J mol}^{-1}\ \text{K}^{-1}$$

ΔH° is computed in a manner similar to that for ΔG°. One obtains $\Delta H^\circ = -1169.36$ kJ/mol. Thus, from the Gibbs-Helmholtz equation at $T = 298.15$ K,

$$\Delta G^\circ = -1169.36\ \text{kJ/mol} - 298.15\,\text{K} \times (-533.1\ \text{J mol}^{-1}\ \text{K}^{-1}) \times \text{KJ/1000 J}$$
$$= -1169.36\ \text{kJ/mol} + 158.91\ \text{kJ/mol} = -1010.41\ \text{kJ/mol}$$

Take care here. Values of ΔH_f° usually are tabulated with units of kJ, while values of S° are tabulated with J/K. Do not forget to convert between J and kJ before adding.

*** 7. Use $\Delta G^\circ = \Delta H^\circ - T\Delta S^\circ$, the Gibbs-Helmholtz equation, to determine ΔG° at various temperatures.**

In Objective 19-6b we learned how to determine ΔG° from values of ΔG_f° and from the Gibbs-Helmholtz equation ([14]). The Gibbs-Helmholtz equation can be used with ΔH° and ΔS° values for 298 K only when ΔH° and ΔS° do not change with temperature, which is a good assumption over small temperature ranges (of 100 degrees or less).

EXAMPLE 19-3 For Reaction [4]
$NH_3(g) + 5\ O_2(g) \rightleftharpoons 4\ NO(g) + 6\,H_2O(l)$, $\Delta H^\circ = -1169.36$ kJ/mol and $\Delta S^\circ = -532.97$ J mol^{-1} K^{-1}. Determine the value of ΔG° at 350 K.

$$\Delta G^\circ = -1169.36\ \frac{\text{kJ}}{\text{mol}} - \left(350\ \text{K} \times \frac{-533.1\ \text{J}}{\text{mol K}} \times \frac{1\ \text{kJ}}{1000\ \text{J}}\right) = -982.78\ \text{kJ/mol}$$

*** 8. Know that $\Delta G_{tr} = 0$ at equilibrium. For phase changes, use $\Delta S_{tr} = \Delta H_{tr} / T_{tr}$ = the molar entropy of transition.**

$\Delta G = 0$ at equilibrium, the point where a reaction changes from being spontaneous in the forward direction to being spontaneous in the reverse direction. We can reach equilibrium by varying the temperature (considered here), or by varying concentration (considered in Chapter 15). Recall the Gibbs-Helmholtz equation ($\Delta G = \Delta H - T\Delta S$). If ΔH and ΔS have the same sign (either both positive or both negative), the temperature can be varied so that ΔG changes sign. At the temperature where the sign change occurs, $\Delta G = 0$. For a phase change such as freezing or boiling, this temperature is the freezing point or the boiling point. For a general transition, it is the transition temperature, T_{tr}. At the normal transition point (where the external pressure is 1.000 atm),

$$\Delta G^\circ_{tr} = 0 = \Delta H^\circ_{tr} - T\Delta S^\circ_{tr} \quad \text{OR} \quad \Delta S^\circ_{tr} = \Delta H^\circ_{tr} / T_{tr} \qquad [16]$$

This provides a convenient way of determining ΔS°_{tr}, since ΔH°_{tr} can be measured readily.

EXAMPLE 19-4 Determine T_{tr} for the change S(monoclinic) $\rightleftharpoons$ S(rhombic) from the data that follows. mono: $\Delta H^\circ_f = 0.402$ kJ/mol, $S° = 32.97$ J mol^{-1} K^{-1} rhombic: $\Delta H^\circ_f = 0.000$ kJ/mol, $S° = 31.80$ mol^{-1} K^{-1}

For the transition reaction,

$\Delta H^\circ_{tr} = (0.000 - 0.402)\text{ kJ/mol} = -0.402\text{ kJ/mol}$

$\Delta S^\circ_{tr} = (31.80 - 32.97)\text{ J mol}^{-1}\text{ K}^{-1} = -1.09\text{ J mol}^{-1}\text{ K}^{-1}$

T_{tr} is computed from Equation [16].

$$-1.09\text{ J mol}^{-1}\text{ K}^{-1} = \frac{-402\text{ kJ/mol} \times 1000\text{ J/kJ}}{T_{tr}} \quad \text{OR} \quad T_{tr} = \frac{0.402 \times 1000}{1.09}\text{ K} = 368\text{ K}$$

*** 9. Write thermodynamic equilibrium constant expressions K_{eq} for reactions and relate these to K_p and K_c.**

*** 10. Compute values of K_{eq} from tabulated data and $\Delta G° = -RT \ln K_{eq}$**

The variation of the free energy change from standard conditions is given by Equation [17].

$$\Delta G = \Delta G° \; RT \ln Q \qquad [17]$$

Q is the reaction quotient, but it is somewhat more restricted than shown in Objectives 19-3 and 19-6. If the reactants or products are gases, we *must* use partial pressures in atmospheres, *not* concentrations, because Q expresses a change from standard state conditions.

Standard state conditions:

Gases: Ideal gas at 1.000 atm pressure. Liquids: Pure liquid under 1.000 atm total pressure.

Solutes: Ideal solution at 1.000 M concentration. Solids: Pure solid under 1.000 atm total pressure.

$$3\text{ Ag(s)} + 4\text{ H}^+\text{(aq)} + \text{NO}_3^-\text{(aq)} \rightleftharpoons 3\text{ Ag}^+\text{(aq)} + 2\text{ H}_2\text{O(l)} + \text{NO(g)}$$

$$Q = \frac{[\text{Ag}^+]^3 P\{\text{NO}\}}{[\text{H}^+]^4[\text{NO}_3^-]} \qquad [18]$$

The free energy of pure solids and liquids does not change very much in the normal range of pressures that can be produced in the laboratory (0 to 5 atm). Pure solids and liquids (as well as solvents in dilute solutions) are not included in the reaction quotient, Q. The free energy of a gas does depend strongly on pressure and, thus, we use the partial pressure in atmospheres in the reaction quotient. The free energy of solutes depends on their concentrations, which are used in the reaction quotient. Equation [17] is quite important at equilibrium, where it is transformed as follows.

$$\Delta G = 0 = \Delta G° + RT \ln Q_{eq} \text{ OR } -RT \ln Q_{eq} = \Delta G° = -RT \ln K_{eq} \quad [19]$$

We use K_{eq}, the thermodynamic equilibrium constant, to replace Q_{eq}. K_{eq} has the same restrictions as does Q_{eq}. For example, the partial pressures of gases must be used. For equilibrium [20], K_{eq} is given in [21].

$$CO_2(aq) \rightleftharpoons CO_2(g) \quad [20]$$

$$K_{eq} = \frac{P\{CO_2\}}{[CO_2]} \quad [21]$$

EXAMPLE 19-5 At 298 K, standard molar free energies of formation are CO_2 (g) = −394.38 kJ/mol and CO_2(aq) = −386.23 kJ/mol. What is $[CO_2]$ if the pressure of CO_2(g) above the solution is 2.000 atm?

$\Delta G° = -394.38 - (-386.23)$ kJ/mol $= -8.15$ kJ/mol $= -8150$

J/mol $= -(8.3145 \text{ J mol}^{-1} \text{ K}^{-1})(298\ K) \ln K$

$\ln K = 3.289$ OR $K = 26.8$ and $26.8\ K_{eq} = \frac{P(CO_2)}{[CO_2]} = \frac{2.000 \text{ atm}}{[CO_2(aq)]}$ or

$[CO_2(aq)] = 0.0745$ M

11. Explain how absolute entropies are determined with the third law of thermodynamics.

The third law states that the entropy of a pure perfect crystalline solid is zero at 0 K. All that is necessary to determine $S°$ at 298 K is to find the entropy change that occurs during heating from 0 K to 298 K. This has been done for several hundred substances. Those for 35 common elements in their most stable forms are compiled in Table 19-2 at the end of this chapter. With Equation [15], we can determine $S°$ for one substance in a reaction if we know $\Delta S°_{rxn}$ and have values of $S°$ for all the other reactants and products. In this way, we know the absolute entropies of most substances. We have no such absolute values of enthalpy (and, thus, no absolute values of free energy either). All enthalpies and free energies are expressed relative to the stable elements rather than as absolute values.

*** 12. Relate the equilibrium constant to the standard molar enthalpy of reaction, $\Delta H°$ and to Kelvin temperature, both graphically and algebraically.**

Note that there are two ways to compute the variation of the equilibrium constant with temperature. The first uses the van't Hoff equation, the subject of this objective. The second method uses the Gibbs-Helmholtz equation $(\Delta G° = \Delta H° - T\Delta S°)$, as described in Objective 19-8, followed by Equation [19] $(\Delta G° = -RT \ln K_{eq})$. In both cases, $\Delta H°$ and $\Delta S°$ are assumed not to vary with temperature. The van't Hoff equation, [22],

$$\ln \frac{K_1}{K_2} = \frac{-\Delta H°}{R}\left(\frac{1}{T_1} - \frac{1}{T_2}\right) \qquad [22]$$

is similar in form to both the Clausius-Clapeyron equation and the Arrhenius equation. The method used in solving these equations is described in Objective 13-5, which you should review, if necessary. One way to evaluate $\Delta H°$ is graphically, by plotting ln K against $1/T$. The slope of this graph equals $-\Delta H°/R$.

EXAMPLE 19-6 Consider the decomposition: $PCl_5(g) \rightleftharpoons PCl_3(g) + Cl_2(g)$. At 500 K, equilibrium partial pressures are $PCl_5(g) = 1.285$ atm, $PCl_3(g) = 0.715$ atm, and $Cl_2(g) = 1.215$ atm; $\Delta H° = 92.59$ kJ/mol. If the temperature is raised to 550 K, what is the partial pressure of each species present once equilibrium is reestablished?

The first step is to use the van't Hoff equation to compute the equilibrium constant K_2 at 550 K.

$$\ln \frac{K_1}{K_2} = \frac{-\Delta H°}{R}\left(\frac{1}{T_1} - \frac{1}{T_2}\right) = \frac{-92590 \text{ J/mol}}{8.314 \text{ J mol}^{-1} \text{ K}^{-1}}\left(\frac{1}{500} - \frac{1}{550}\right) = -2.0248.$$

Thus, $$\frac{K_1}{K_2} = e^{-2.0248} = 0.1320 \text{ (where } e = 2.718)$$ NOW

$$K_1 = \frac{P_{PCl3}\, P_{Cl2}}{P_{PCl5}} = \frac{(0.715)\,(1.215)}{1.285} = 0.676,$$

and $\dfrac{K_1}{K_2} = \dfrac{0.676}{K_2} = 0.1320$ or $K_2 = \dfrac{0.676}{0.1320} = 5.12$.

Equation:	$PCl_5(g)$	$\rightleftharpoons PCl_3(g)$	$+ Cl_2(g)$
Initial:	1.285 atm	0.715 atm	1.215 atm
Changes:	$-x$ atm	$+x$ atm	$+x$ atm
Equil:	$(1.285 - x)$ atm	$(0.715 \text{ atm} + x)$ atm	$(1.215 + x)$ atm

$$K_p = \frac{(0.715 + x)\,(1.215 + x)}{1.285 - x} = 5.12 \quad x = -3.525 \pm 4.259 \text{ atm} = 0.734 \text{ atm}$$

Partial pressures: $PCl_5(g) = 0.551$ atm, $PCl_3(g) = 1.449$ atm, $Cl_2(g) = 1.949$ atm.

SELF-ASSESSMENT EXERCISES

95. (c) This is one statement of the second law.

97. (a) Spontaneous at all temperatures. Because the moles of gas are increasing, ΔS will be positive. The term $-T\Delta S$ will always have negative values.

98. (b) The reaction is at equilibrium. ΔH can be equal to TΔS and ΔG will be zero while neither ΔH or ΔS are equal to zero.

99. (a) and (d) are correct. Because the moles of gas are equal on each side of the equation, a larger container will not affect the equilibrium. (d) is correct because $K_p = K_c(RT)^{\Delta n}$ and $\Delta n = 0$.

DRILL PROBLEMS

1. Determine q, w, and ΔE in each of the following cases.

Heat (of system)	*Work done*	*Heat (of system)*	*Work done*
A. 196 J evolved	123 J by system	B. 189 J evolved	247 J on system
C. 647 J absorbed	92.4 J on surr.	D. 1362 J absorbed	321 J by system
E. 142 J evolved	254 J by surr.	F. 802 J absorbed	897 J by surr.
G. −1432 J absorbed	937 J by surr.	H. 972 J evolved	432 J by system

2. Compute ΔH and ΔE in kJ/mol for each reaction. q_p or q_v is given in kJ/g of the first reactant. ($T = 298$ K)

A. $C_4H_9OH(l) + 6\,O_2(g) \longrightarrow 4\,CO_2(g) + 5\,H_2O(l)$ $q_P = -36.11$ kJ/g

B. $C_6H_5OH(s) + 7\,O_2(g) \longrightarrow 6\,CO_2(g) + 3\,H_2O(l)$ $q_v = -32.56$ kJ/g

C. $PCl_5(g) + H_2O(l) \longrightarrow POCl_3(g) + 2\,HCl(g)$ $q_p = -0.4410$ kJ/g

D. $2\,Cl_2(g) + 7\,O_2(g) \longrightarrow 2\,Cl_2O_7(g)$ $q_v = -3.864$ kJ/g

E. $H_2(g) + Br_2(g) \longrightarrow 2\,HBr(g)$ $q_p = -35.95$ kJ/g

F. $2\,B_5H_9(g) + 12\,O_2(g) \longrightarrow 5\,B_2O_3(s) + 9\,H_2O(l)$ $q_v = -71.137$ kJ/g

G. $SOCl_2(l) + H_2O(g) \longrightarrow SO_2(g) + 2\,HCl(g)$ $q_p = -2.842$ kJ/g

H. $CCl_4(g) + 2\,H_2O(g) \longrightarrow CO_2(g) + 4\,HCl(g)$ $q_v = 1.1526$ kJ/g

3. Predict whether ΔS is positive or negative for each of the following processes. (Do not *calculate* a value of ΔS.) In which case(s) is a prediction unclear? All substances are gases unless otherwise indicated.

A. $NH_3(g) + HCl(g) \longrightarrow NH_4Cl(s)$
B. $Li_2CO_3(s) \longrightarrow Li_2O(s) + CO_2(g)$
C. $NH_4HCO_3(s) \rightarrow NH_3 + H_2O(l) + CO_2$
D. $N_2O_4(g) \longrightarrow NO_2(g)$
E. $CaCO_3(s) \longrightarrow CaO(s) + CO_2(g)$
F. $CaCO_3(s) + SO_3 \longrightarrow CaSO_4(s) + CO_2$
G. $PCl_3(g) + Cl_2(g) \longrightarrow PCl_5(g)$
H. $2\,SO_2(g) + O_2(g) \longrightarrow 2\,SO_3(g)$
I. $Br_2(g) + Cl_2(g) \longrightarrow 2\,BrCl(g)$
J. $2\,H_2S + 3\,O_2 \longrightarrow 2\,H_2O(l) + 2\,SO_2(g)$
K. $N_2(g) + 3\,H_2(g) \longrightarrow 2\,NH_3(g)$
L. $CO(g) + 2\,H_2(g) \longrightarrow CH_3OH(g)$
M. $CH_4(g) + CCl_4(g) \longrightarrow 2\,CH_2Cl_2(g)$
N. $2\,NOCl(g) \longrightarrow 2\,NO(g) + Cl_2(g)$
O. $N_2(g) + O_2(g) \longrightarrow 2\,NO(g)$
P. $CO_2 + 2\,NH_3 \rightarrow CO(NH_2)_2(s) + H_2O(l)$
Q. $S(s) + 2\,CO(g) \longrightarrow SO_2(g) + 2\,C(gr)$
R. $SO_2Cl_2(g) \longrightarrow SO_2(g) + Cl_2(g)$
S. $CO_2(g) \longrightarrow CO_2(aq)$
T. $2\,O_3(g) \longrightarrow H_2O(g)$
U. $4\,HCl + O_2 \longrightarrow 2\,H_2O(g) + 2\,Cl_2$
V. $H_2 + CO_2 \longrightarrow H_2O(g) + CO$
W. $2\,HI(g) \longrightarrow H_2(g) + I_2(g)$
X. $HCHO(g) \longrightarrow H_2(g) + CO(g)$

4. Use the data of Table 19-1 (at the end of the chapter). State if the reaction is spontaneous (S) or nonspontaneous (N), and to which one of the categories (1, 2, 3, or 4) described in Objective 19-6a it belongs.

5. Determine the value of $\Delta G°$ and K for each of the reactions in the drill problems for Objective 19-6 at 200 K and 400 K. Assume that no phase changes occur.

6. Fill in the blanks in the following table. All values of ΔH_f° and ΔH_{tr}° are in kJ/mol, all values of $\Delta S°$ and ΔS_{tr}° are in J mol^{-1} K^{-1}, and all temperatures are in kelvins. Each line represents one transition, from substance A to substance B.

Substance A	ΔH_f°	S°	*Substance B*	ΔH_f°	S°	ΔH_{tr}°	ΔS_{tr}°	T_{tr}
A. $SbCl_3(g)$	−315	338	$SbCl_3(s)$	−382.2	186	____	____	____
B. $AsF_3(g)$	−314	327	$AsF_3(l)$	−336	____	____	____	228
C. BeO(c)	−610.9	14.1	BeO(g)	49.4	197.4	____	____	____
D. $BCl_3(l)$	−418.4	209	$BCl_3(g)$	−395	289.9	____	____	____
E. $Br_2(g)$	____	245.3	$Br_2(l)$	0.00	152	____	____	266
F. $Ca_3(PO_4)_2(\alpha)$	−4126.25	241	$Ca_3(PO_4)_2(\beta)$	−4138	236.9	____	____	____
G. $I_2(g)$	62.25	260.6	$I_2(s)$	0.00	116.7	____	____	____
H. Mn(a)	0.00	31.8	Mn(g)	1.55	32.3	____	____	____
I. HgO(red)	−90.71	____	HgO(yellow)	−90.21	73.2	____	____	400

7. (1) For each of the reactions in the drill problems for Objective 19-6, write the expression for the thermodynamic equilibrium constant and evaluate the constant at 298 K.

(2) Determine the value of $\Delta G°$ that corresponds to each of the following sets of equilibrium constant (K) and temperature (T).

	A. $K = 1$, $T = 298$ K	B. $K = 1$, $T = 500.$ K
C. $K = 10^{-3}$, $T = 298$ K	D. $K = 10^{-3}$, $T = 200.$ K	E. $K = 10^{-3}$, $T = 500.$ K
F. $K = 10^{3}$, $T = 200.$ K	G. $K = 10^{3}$, $T = 298$ K	H. $K = 10^{3}$, $T = 500.$ K

8. The value of $\Delta H°$ and/or the equilibrium constant at two different temperatures is given or requested for each of the following equations. Fill in each blank.

A. $CO(g) + 2\ H_2(g) \rightleftharpoons CH_3OH(g)$ $\quad \Delta H° =$ ____ kJ/mol

$K_p = 91.4$ at 350 K $\quad K_p = 2.05 \times 10^{-4}$ at 298 K

B. $HCHO(g) \rightleftharpoons H_2(g) + CO(g)$ $\quad \Delta H° = 5.36$ kJ/mol

$K_c =$ ____ at 450 K $\quad K_c = 0.267$ at 773 K

C. $4\ HCl(g) + O_2(g) \rightleftharpoons 2\ H_2O(g) + 2Cl_2(g)$ $\quad \Delta H° = -114.47$ kJ/mol

$K_c =$ ____ at 298 K $\quad K_c = 1.49 \times 10^{3}$ at 320 K

D. $2\ P$ (white) $+ 3\ H_2(g) \rightleftharpoons 2\ PH_3(g)$ $\quad \Delta H° =$ ____ kJ/mol

$K_p = 3.10 \times 10^{6}$ at 270 K $\quad K_p = 7.62 \times 10^{5}$ at 330 K

E. $SO_2Cl_2(l) \rightleftharpoons SO_2(g) + Cl_2(g)$ $\quad \Delta H° = 29.37$ kJ/mol

$K_p = 6.30$ at 350 K $\quad K_p =$ ____ 400 K $\quad K_p = 1.03$ at ____ K

F. $PCl_5(g) \rightleftharpoons PCl_3(g) + Cl_2(g)$ $\quad \Delta H° = 92.59$ kJ/mol

$K_p = 0.675$ at 500 K $\quad K_p =$ ____ at 60 K $\quad K_p = 1.000$ at ____ K

QUIZZES (20 minutes each) Choose the best answer for each question. Accept a numerical answer if it is within 5% of the correct value.

QUIZ A

1. $2\ NH_3(g) + \frac{3}{2}\ O_2(g) \longrightarrow N_2(g) + 3H_2O(g)$

   $S°$, J mol^{-1} K^{-1} 192.5 205.0 191.5 188.7. What is the value of $\Delta S°$ in J mol^{-1} K^{-1} for this reaction? (a) +65.1; (b) -65.1; (c) -17.3; (d) $+17.3$; (e) none of these.

2. $Cl_2O(g) + \frac{3}{2}\ O_2(g) \longrightarrow 2\ ClO(g)$

   $\Delta G_f°$, kJ/mol 97.9 0.00 123.4. What is the value of $\Delta G°$ in kJ/mol for this reaction? (a) $+25.5$; (b) $++148.9$; (c) $+221.3$; (d) -25.5; (e) none of these.

3. For the reaction $SO_3(g) \longrightarrow SO_2(g) + \frac{1}{2}O_2(g)$, $\Delta H° = 93.8$ kJ/mol and $\Delta S° = 326$ J mol^{-1} K^{-1}, $\Delta G°$ for this reaction at 27 °C in kJ/mol equals (a) 89.5; (b) 196; (c) 0.42; (d) 107.1; (e) none of these.

4. For $Cl_2(g) + F_2(g) \longrightarrow 2\ ClF(g)$, $\Delta H° = -11.3$ kJ/mol and $\Delta S° = 9.25$ J mol^{-1} K^{-1}. As temperature decreases for this reaction, (a) it eventually becomes nonspontaneous; (b) $\Delta G°$ decreases; (c) it reaches equilibrium; (d) it shifts left; (e) none of these.

5. The change in free energy of a reaction (a) $= \Delta H + T\Delta S$; (b) = heat; (c) predicts speed; (d) depends on the standard state chosen; (e) none of these.

6. For an endothermic reaction, (a) $\Delta H < 0$; (b) $\Delta H > 0$; (c) $\Delta G > 0$; (d) $\Delta G < 0$; (e) none of these.

7. Which of the following phrases is *not* used in the definition of the standard enthalpy of formation? (a) most stable form; (b) heat evolved at constant pressure; (c) elements in standard states; (d) work at constant volume; (e) none of these.

8. Which best expresses the increased degree of randomness associated with melting and sublimation, respectively?

(a) $\Delta S_{fus} = 4$ J mol^{-1} K^{-1}, $\Delta S_{sub} = 120$ J mol^{-1} K^{-1};
(b) $\Delta S_{fus} = 120$ J mol^{-1} K^{-1}, $\Delta S_{sub} = 4$ J mol^{-1} K^{-1}; (c) $\Delta S_{fus} = \Delta S_{sub} = 0$; (d) $\Delta S_{fus} = 4$ J mol^{-1} K^{-1}, $\Delta S_{sub} = -120$ J mol^{-1} K^{-1}; (e) $\Delta S_{fus} - \Delta S_{sub} = 0$.

QUIZ B

1. $C(graph) + 2H_2(g) \longrightarrow CH_4(g)$

$S°$, J mol^{-1} K^{-1} 5.69 130.58 186.19. What is the value of $\Delta S°$ in J mol^{-1} K^{-1} for this reaction? (a) +80.67; (b) +49.92; (c) −49.92; (d) +115.23; (e) none of these.

2. $B_2H_6(g) + 3Cl_2(g) \longrightarrow 2BCl_3(g) + 6HCl(g)$

$\Delta G_f°$, kJ/mol 86.6 0.00 −388.7 22.8. What is the value of $\Delta G°$ in kJ/mol for this reaction? (a) −291.6; (b) −279.3; (c) −727.2; (d) −452.5; (e) none of these.

3. For the reaction $Cl_2O(g) + \frac{3}{2}O_2(g) \longrightarrow 2\ ClO(g)$, $\Delta H° = +126.4$ kJ/mol and $\Delta S° = -74.9$ J mol^{-1} K^{-1}. $\Delta G°$ for this reaction at 377 °C in kJ/mol equals (a) +98.3; (b) +77.8; (c) +175.1; (d) +51.5; (e) none of these.

4. For $ICl_3(s) \longrightarrow ICl(g) + Cl_2(g)$, $\Delta H° = +105.9$ kJ/mol and $\Delta S° = +298.4$ J mol^{-1} K^{-1}. As temperature decreases for this reaction, (a) it eventually becomes spontaneous; (b) $\Delta G°$ decreases; (c) it shifts right; (d) equilibrium is eventually reached; (e) none of these.

5. "In a spontaneous process the entropy of the universe increases," is a statement of the (a) zeroth law; (b) third law; (c) first law; (d) second law; (e) none of these.

6. The free energy change for a particular process is $+136$ kJ. Thus, the process is (a) exothermic; (b) endothermic; (c) nonspontaneous; (d) spontaneous; (e) none of these.

7. Which of the following is *not* a state property? (a) heat; (b) enthalpy; (c) entropy; (d) free energy; (e) all are state properties.

8. The enthalpy of any reaction could be determined, at least approximately, from complete tables of each of the following kinds of data except (a) bond enthalpies; (b) heats of combustion; (c) ionization energies; (d) heats of formations; (e) all could be used.

QUIZ C

1. $CO_2(g) + 2\ HCl(g) \longrightarrow COCl_2(g) + H_2O(g)$

   ΔG_f°, kJ/mol -394.4 -95.3 -210.5 -228.6. What is the value of ΔG° in kJ/mol for this reaction? (a) -145.8; (b) -439.1; (c) $+56.6$; (d) $+146.0$; (e) none of these.

2. $C_3H_4(g) + 2\ H_2(g) \longrightarrow C_3H_8(g)$

   S°, J mol^{-1} K^{-1} 266.9 130.6 269.9. What is the value of ΔS° in J mol^{-1} K^{-1} for this reaction? (a) -127.6; (b) -258.2; (c) $+3.0$; (d) $+127.0$; (e) none of these.

3. For the reaction $B_2H_6(g) + 3\ Cl_2(g) \longrightarrow 2\ BCl_3(g) + 6\ HCl(g)$, $\Delta H^\circ = -1396.9$ kJ/mol and $\Delta S^\circ = 800.0$ J mol^{-1} K^{-1}. ΔG° for this reaction at 227 °C in kJ/mol equals (a) -996.8; (b) -2196.9; (c) -400.0; (d) -1578.5; (e) none of these.

4. For $2\ NO(g) + Cl_2(g) \longrightarrow 2\ NOCl(g)$, $\Delta H^\circ = -37.78$ kJ/mol and $\Delta S^\circ = -117.03$ J mol^{-1} K^{-1}. As temperature decreases for this reaction, (a) ΔH° increases; (b) it becomes more spontaneous; (c) it shifts right; (d) ΔS° increases; (e) none of these.

5. If $\Delta G = -10$ kJ for a reaction, then the reaction is said to be (a) spontaneous; (b) reversible; (c) endothermic; (d) exothermic; (e) none of these.

6. Thermodynamics *cannot* predict for a reaction its (a) speed; (b) position of equilibrium; (c) maximum amount of work available; (d) feasibility of the reaction; (e) it can predict them all.

7. $C(graphite) + O_2(g) \longrightarrow CO_2(g)$ $\Delta H^\circ = -393.5$ kJ/mol is not a (a) combustion reaction; (b) bond enthalpy; (c) formation reaction; (d) thermochemical equation; (e) it is all of these.

8. The energy available for useful work is (a) constant; (b) the free energy; (c) the entropy; (d) the internal energy; (e) the enthalpy.

QUIZ D

1. $2\ KClO_4(s) \longrightarrow 2\ KClO_3(s) + O_2(g)$

   $S°$, J mol^{-1} K^{-1} 151.0 143.0 205.0. What is the value of $\Delta S°$ in J mol^{-1} K^{-1} for this reaction? (a) $+196.9$; (b) -196.9; (c) $+188.9$; (d) -188.9; (e) none of these.

2. $4\ NH_3(g) + 7\ O_2(g) \longrightarrow 4\ NO_2(g) + 6\ H_2O(l)$

   ΔG_f°, kJ/mol 16.65 0.00 51.84 -237.19. What is the value of $\Delta G°$ in kJ/mol for this reaction? (a) -1282.40; (b) $+168.70$; (c) -168.70; (d) -210.97 (e) none of these.

3. For $NH_3(g) + HCl(g) \longrightarrow NH_4Cl(s), \Delta H° = -176.90$ kJ/mol and $\Delta S° = -284.6\ J\ mol^{-1}\ K^{-1}$. $\Delta G°$ for this reaction at 400. K in kJ/mol equals (a) $+63.05$; (b) -368.4; (c) $+290.7$; (d) -63.05; (e) none of these.

4. For $CaCO_3 \longrightarrow CaO(s) + CO_2(g), \Delta H = +178$ kJ/mol and $\Delta S° = +160.5\ J\ mol^{-1}\ K^{-1}$. As temperature decreases for this reaction, (a) it eventually becomes spontaneous; (b) $\Delta G°$ increases; (c) $\Delta S°$ increases; (d) it becomes nonspontaneous; (e) nothing happens.

5. Which material has the largest entropy? (a) salt; (b) powdered sugar; (c) pure water; (d) salt water; (e) cannot be determined.

6. A reaction is spontaneous if (a) $\Delta H < 0$; (b) $\Delta S < 0$; (c) $\Delta G < 0$; (d) $\Delta S > 0$; (e) none of these.

7. The state function most closely associated with the second law is (a) enthalpy; (b) entropy; (c) internal energy; (d) temperature; (e) none of these.

8. What must be true of an isothermal system? (a) constant temperature; (b) no heat exchanged; (c) no work done; (d) closed to mass transfer; (e) none of these.

SAMPLE TEST (20 minutes)

1. Write the thermodynamic equilibrium constant for each of the following reactions.

   A. $Mg_3N_2(s) + 6\ H_2O(l) \longrightarrow 3\ Mg(OH)_2(aq) + 2\ NH_3(aq)$

   B. $SO_2(g) + Cl_2(g) \longrightarrow SO_2Cl_2(g)$

   C. $CO(g) + 2\ H_2(g) \longrightarrow CH_3OH(l)$

   D. $Ag_2O(s) + H_2SO_4(aq) \longrightarrow H_2O(l) + Ag_2SO_4(aq)$

2. For reaction $PCl_5(g) \rightleftharpoons PCl_3(g) + Cl_2(g)$ at 298 K, $K_{eq} = 1.87 \times 10^{-7}, \Delta S° = 181.92\ J\ mol^{-1}\ K^{-1}$.

   A. Is this reaction spontaneous at 298 K? Explain why or why not.

B. Compute the value of $\Delta H°$ for this reaction.

C. Without performing any calculations, predict whether K_{eq} will increase or decrease when the temperature changes from 298 K to 200. K. Explain your prediction.

D. Compute $\Delta G°$ for this reaction at 46 °C. State assumptions that you make during this calculation.

3. Consider these four compounds.

Compound	CaO(s)	CO(g)	$CO_2(g)$	$COCl_2(g)$
$S°$, J mol^{-1} K^{-1}	40	198	214	289

Explain the increase in standard molar entropies in terms of disorder.

4. What is $\Delta H°$ for a reaction that has $K_p = 1.456$ at 273 K and $K_p = 14.2$ at 298 K?

TABLE 19-1 Standard Entropies of the Most Stable Forms of Some Elements at 298 K; $\Delta S°$ in J mol^{-1} K^{-1}

Element	$\Delta S°$	Element	$\Delta S°$	Element	$\Delta S°$	Element	$\Delta S°$	Element	$\Delta S°$
Al(s)	28.33	Ba(s)	62.8	Be(s)	9.50	Bi(s)	56.74	B(s)	5.86
Br_2(l)	152.2	Cd(s)	51.76	Ca(s)	41.42	C(gr.)	5.74	Cl_2(g)	223.0
Cr(s)	23.77	Co(s)	30.0	Cu(s)	33.15	F_2(g)	202.7	He(g)	126.0
H_2(g)	130.6	I_2(s)	116.1	Fe(s)	27.28	Pb(s)	64.81	Li(s)	29.12
Mg(s)	32.68	Mn(s)	32.01	Hg(l)	76.02	N_2(g)	191.5	O_2(g)	205.0
P(α)	41.09	K(s)	64.18	Si(s)	18.83	Ag(s)	42.55	Na(s)	51.21
S(rh.)	31.80	Sn(wh)	51.55	Ti(s)	30.63	U(s)	50.21	Zn(s)	41.63

In Tables 19-1 and 19-2, substances are at 1 atm pressure. For aqueous solutions, solutes are at unit activity (roughly 1M). Data for ions in aqueous solution are relative to values of *zero* for $\Delta H_f°$ $\Delta G_f°$, and $\Delta S°$ for H^+(aq).

TABLE 19-2 Thermodynamic Properties of Selected Substances $\Delta H_f°$ and $\Delta G_f°$ in kJ/mol; $\Delta S°$ in J mol^{-1} K^{-1}

Substance	$\Delta H_f°$	$\Delta G_f°$	$\Delta S°$
Al^{3+}(aq)	−531	−485	−321.7
$AlCl_3$(s)	−704.2	−628.8	110.7
Al_2Cl_6(g)	−1291	−122.0	490
AlF_3(s)	−1504	−1425	66.44
Al_2O_3(α,s)	−1676	−1582	50.92

$Al(OH)_3(s)$	−1276	----	----
$Al_2(SO_4)_3(s)$	−3441	−3100	239
$NH_3(g)$	−46.11	−16.48	192.3
$NH_3(aq)$	−80.29	−26.50	111.3
$NH_4^+(aq)$	−132.5	−79.31	113.4
$NH_4Br(aq)$	−270.8	−175.2	113
$NH_4Cl(s)$	−314.4	−202.9	94.6
$NH_4F(s)$	−464.0	−384.7	71.96
$NH_4HCO_3(s)$	−849.4	−666.0	120.9
$NH_4I(s)$	−201.4	−112.5	117
$NH_4NO_3(s)$	−365.6	−183.9	151.1
$NH_4NO_3(aq)$	−339.9	−190.6	859.8
$(NH_4)_2SO_4(s)$	−1181	−901.7	220.
$Ba^{2+}(aq)$	−537.6	−560.8	9.6
$BaCO_3(s)$	−1216	−1138	112
$BaCO_3(aq)$	−1214.7	−1088.6	−47.3
$BaCl_2(s)$	−858.6	−810.4	123.7
$BaF_2(s)$	−1207	−1157	96.36
$BaO(s)$	−553.5	−525.1	70.42
$Ba(OH)_2(s)$	−944.7	----	----
$Ba(OH)_2 \cdot 8H_2O$	−3342	−2793	427
$BaSO_4(s)$	−1473	−1362	132.2
$BeCl_2(\alpha,s)$	−490.4	−445.6	82.86
$BeF_2(\alpha,s)$	−1027	−979.4	53.35
$BeO(s)$	−609.6	−580.3	14.14
$BiCl_3(s)$	−379.1	−315.0	177.0
$Bi_2O_3(s)$	−573.9	−493.7	151.5
$BCl_3(l)$	−427.2	−387.4	206.3
$BF_3(g)$	−1137	−1120	254.0
$B_2H_6(g)$	35.6	86.6	232.0
$B_2O_3(s)$	−1273	−1194	53.97
$Br(g)$	111.9	82.42	174.9
$Br^-(aq)$	−121.6	−104.0	82.4
$Br_2(g)$	30.91	3.14	245.4
$HBr(g)$	−36.40	−53.43	198.6
$BrCl(g)$	14.64	−0.98	240.0
$BrF_3(g)$	−255.6	−229.4	292.4

$BrF_3(l)$	−300.8	−240.5	178.2
$Cd^{2+}(aq)$	−75.90	−77.61	−73.2
$CdCl_2(s)$	−391.5	−343.9	115.3
$CdO(s)$	−258.2	−228.4	54.8
$Ca^{2+}(aq)$	−542.8	−553.6	−53.1
$CaCO_3(s)$	−1207	−1129	92.9
$CaCl_2(s)$	−795.8	−748.1	104.6
$CaF_2(s)$	−1220	−1167	68.87
$CaH_2(s)$	−186.2	−147.2	42
$Ca(NO_3)_2(s)$	−938.4	−743.1	193.3
$Ca(OH)_2(s)$	−986.1	−898.5	83.39
$CaO(s)$	−635.1	−604.0	39.75
$Ca_3(PO_4)_2(s)$	−4121	−3885	236.0
$CaSO_4(s)$	−1434	−1322	106.7
$C(g)$	716.7	671.3	158.0
$C(diamond)$	1.90	2.90	2.38
$CO(g)$	−110.5	−137.2	197.6
$CO_2(g)$	−393.5	−394.4	213.6
$CO_2(aq)$	−413.8	−386.0	117.6
$CO_3^{2-}(aq)$	−677.1	−527.8	−56.9
$CH_4(g)$	−74.81	−50.75	186.2
$CH_3OH(g)$	−200.7	−162.0	239.7
$CH_3OH(l)$	−238.7	−166.3	126.8
$HCHO(g)$	−108.6	−102.5	218.7
$CH_2Cl_2(g)$	−92.47	−65.90	270.1
$CCl_4(g)$	−102.9	−60.62	309.7
$CCl_4(l)$	−135.4	−65.21	216.4
$COCl_2(g)$	−218.8	−204.6	283.4
$COS(g)$	−142.1	−169.3	231.5
$CS_2(l)$	89.70	65.27	151.3
$HCN(g)$	135.1	124.7	201.7
$CH_3NH_2(g)$	−22.97	32.09	243.4
$CO(NH_2)_2(s)$	−333.5	−197.4	104.6
$C_2H_2(g)$	226.7	209.2	200.8
$C_2H_4(g)$	52.26	68.12	219.5
$C_2H_6(g)$	−84.68	−32.89	229.5
$CH_3CHO(g)$	−166.2	−128.9	250.3
$CH_3CHO(l)$	−192.3	−128.1	160.2

$CH_3CH_2OH(g)$	−235.1	−168.6	282.6
$CH_3CH_2OH(l)$	−277.7	−174.8	160.7
$CH_3COOH(g)$	−432.3	−374.1	282.4
$CH_3COOH(l)$	−484.5	−389.9	159.8
$CH_3COOH(aq)$	−485.8	−396.5	178.7
$C_2N_2(g)$	309.0	297.4	241.8
$C_3H_8(g)$	−104.7	−23.6	270.2
$(CH_3)_2CO(g)$	−216.6	−153.1	294.9
$(CH_3)_2CO(l)$	−242.1	−155.8	200.6
$C_3O_2(g)$	−93.72	−109.8	276.4
$C_3O_2(l)$	−117.3	−105.0	181.1
$C_4H_{10}(g)$	−125.6	−17.2	310.1
$C_6H_6(g)$	82.6	129.7	269.2
$C_6H_6(l)$	49.0	124.4	173.4
$C_6H_{12}(g)$	−123.4	31.8	298.3
$C_6H_{12}(l)$	−156.4	26.7	204.4
$C_6H_5OH(s)$	−165.1	−50.4	144.0
$C_6H_5NH_2(g)$	86.86	166.7	319.2
$C_6H_5NH_2(l)$	31.3	149.2	191.4
$C_6H_5COOH(s)$	−385.2	−245.3	167.6
$C_{10}H_8(g)$	149	223.6	335.6
$C_{10}H_8(s)$	77.9	201.0	166.9
$Cl(g)$	121.7	105.7	165.1
$Cl^-(aq)$	−167.2	−131.2	56.5
$HCl(g)$	−92.31	−95.30	186.8
$HCl(aq)$	−167.2	−131.2	56.5
$ClF_3(g)$	−163.2	−123.0	281.5
$ClO_2(g)$	102.5	120.5	256.7
$Cl_2O(g)$	80.3	97.9	266.1
$[Cr(H_2O)_6]^{2+}$	−1999	----	----
$Cr_2O_3(s)$	−1140	−1058	81.2
$CrO_4^{2-}(aq)$	−881.2	−727.8	50.21
$Cl_2O_7^{2-}(aq)$	−1490	−1301	261.9
$CoO(s)$	−237.9	−214.2	52.97
$Co(OH)_2(pink)$	−539.7	−454.3	79
$Cu^{2+}(aq)$	64.77	65.49	−99.6

$CuCO_3 \cdot Cu(OH)_2$	−1051	−893.6	186.2
$CuO(s)$	−157.3	−129.7	42.63
$Cu(OH)_2(s)$	−449.8	----	----
$CuSO_4 \cdot 5H_2O$	−2280	−1880	300.4
$F(g)$	78.99	61.93	158.7
$F^-(aq)$	−332.6	−278.8	−13.8
$HF(g)$	−271.1	−273.2	173.7
$OF_2(g)$	24.7	41.9	247.3
$H(g)$	218.0	203.3	114.6
$H^+(aq)$	0	0	0
$H_2O(g)$	−241.8	−228.6	188.7
$H_2O_2(g)$	−285.8	−237.1	69.91
$H_2O_2(g)$	−136.3	−105.6	232.6
$H_2O_2(l)$	−187.8	−120.4	109.6
$I(g)$	106.8	70.28	180.7
$I^-(aq)$	−55.19	−51.57	111.3
$I_2(g)$	62.44	19.36	260.6
$IBr(g)$	40.84	3.72	258.7
$ICl(g)$	17.78	−5.44	247.4
$ICl(l)$	−23.89	−13.58	135.1
$HI(g)$	26.48	1.72	206.5
$Fe^{2+}(aq)$	−89.1	−78.90	−137.7
$Fe^{3+}(aq)$	−48.5	−4.7	−315.9
$FeCO_3(s)$	−740.6	−666.7	92.9
$FeCl_3(s)$	−399.5	−334.0	142.3
$FeO(s)$	−272	----	----
$Fe_2O_3(s)$	−842.2	−742.2	87.40
$Fe_2O_4(s)$	−1118	−1015	146.4
$Fe(OH)_3(s)$	−832.0	−696.5	106.7
$Pb^{2+}(aq)$	−1.7	−24.43	10.5
$PBI_2(s)$	−175.5	−173.6	174.9
$PbO_2(s)$	−277.4	−217.3	68.6
$PbSO_4(s)$	−919.9	−813.1	148.6
$Li^+(aq)$	−278.5	−293.3	13.4
$Li_2CO_3(s)$	−1215.9	−1132.1	90.37
$LiCl(s)$	−408.6	−384.4	59.33
$LiOH(s)$	−484.9	−439.0	42.80
$Li_2O(s)$	−597.9	−561.2	37.57
$LiNO_3(s)$	−483.1	−381.1	90.0

Mg^{2+}(aq)	−466.9	−454.8	−138.1
$MgCl_2$(s)	−641.3	−591.8	89.62
$MgCO_3$(s)	−1096	−1012	65.7
MgF_2(s)	−1123	−1070	57.24
MgO(s)	−601.7	−569.4	26.94
$Mg(OH)_2$(s)	−924.5	−833.5	63.18
$MgSO_4$(s)	−1285	−1171	91.6
Mn^{2+}(aq)	−220.8	−228.1	−73.6
MnO_2(s)	−520.0	−465.1	53.05
MnO_4^-(aq)	−541.4	−447.2	191.2
Hg(g)	61.32	31.82	175.0
HgO(s)	−90.83	−58.54	70.29
N(g)	472.7	455.6	153.2
NF_3(g)	−124.7	−83.2	260.6
N_2H_4(g)	95.40	159.3	238.4
N_2H_4(l)	50.63	149.3	121.2
NO(g)	90.25	86.55	210.7
N_2O(g)	82.05	104.2	219.7
NO_2(g)	33.18	51.29	240.0
N_2O_4(g)	9.16	97.82	304.2
N_2O_4(l)	−19.50	97.54	209.2
N_2O_5(g)	11.3	115.0	355.6
NOBr(g)	82.17	82.42	273.6
NOCl(g)	51.71	66.06	261.6
NO_3^-(aq)	−205.0	−108.7	146.4
HNO_3(l)	−174.1	−80.71	155.6
HNO_3(aq)	−207.4	−111.3	146.4
O(g)	249.2	231.7	160.9
OH^-(aq) −230.0	−157.2	−10.75	
O_3(g)	142.7	163.2	238.8
P(red)	−17.6	−12.1	22.80
P_4(g)	58.91	24.47	279.9
PCl_3(g)	−287.0	−267.8	311.7
PCl_5(g)	−374.9	−305.0	364.5
PH_3(g)	5.4	13.4	210.1
P_4O_{10}(s)	−2984	−2698	228.9
PO_4^{3-}(aq)	−1277	−1019	−222
K(g)	89.24	60.62	160.2

K(l)	2.28	0.26	71.46
$K^+(aq)$	−252.4	−283.3	102.5
KBr(s)	−393.8	−380.7	95.90
KCN(s)	−113.0	−101.9	128.5
KCl(s)	−436.7	−409.1	82.59
$KClO_3(s)$	−397.7	−296.3	143.1
$KClO_4(s)$	−432.8	−303.1	151.0
KF(s)	−567.3	−537.8	66.57
KI(s)	−327.9	−324.9	106.3
$KNO_3(s)$	−494.6	−394.9	133.1
KOH(s)	−424.8	−379.1	78.9
KOH(aq)	−482.4	−440.5	91.6
$K_2SO_4(s)$	−1438	−1321	175.6
$SiH_4(g)$	34.3	56.9	204.5
$Si_2H_6(g)$	80.3	127.2	272.6
SiO_2(quartz)	−910.9	−856.6	41.84
$Ag^+(aq)$	105.6	77.11	72.68
AgBr(s)	−100.4	−96.90	107.1
AgCl(s)	−127.1	−109.8	96.2
AgI(s)	−61.84	−66.19	115.5
$AgNO_3(aq)$	−124.4	−33.41	140.9
$Ag_2O(s)$	−31.05	−11.20	121.3
$Ag_2SO_4(s)$	−715.9	−618.4	200.4
Na(g)	107.3	76.79	153.6
Na(l)	2.41	0.50	57.86
$Na^+(aq)$	−240.1	−261.9	59.0
$Na_2(g)$	142.1	104.0	230.1
NaBr(s)	−361.1	−349.0	86.82
$Na_2CO_3(s)$	−1131	−1044	135.0
$NaHCO_3(s)$	950.8	−851.0	101.7
NaCl(s)	−411.2	−384.1	72.13
NaCl(aq)	−407.3	−393.1	115.5
$NaClO_3(s)$	−365.8	−262.3	123.4
$NaClO_4(s)$	−383.3	−254.9	142.3
NaF(s)	−573.6	−543.5	51.46
NaH(s)	−56.28	−33.46	40.02
NaI(s)	−287.8	−286.1	98.53
$NaNO_3(s)$	−467.9	−367.0	116.5
$NaNO_3(aq)$	−447.5	−373.2	205.4
$Na_2O_2(s)$	−510.9	−447.7	95.0
NaOH(s)	−425.6	−379.5	64.46

NaOH(aq)	−470.1	−419.2	48.1
$NaH_2PO_4(s)$	−1537	−1386	127.5
$Na_2HPO_4(s)$	−1748	−1608	150.5
$Na_3PO_4(s)$	−1917	−1789	173.8
$NaHSO_4(s)$	−1126	−992.8	113.0
$Na_2SO_4(s)$	−1387	−1270	149.6
$Na_2SO_4(aq)$	−1390	−1268	138.1
$Na_2SO_4 \cdot 10H_2O$	−4327	−3647	592.0
$Na_2S_2O_3(s)$	−1123	−1028	155
$S_8(g)$	102.3	49.66	430.9
$S_2Cl_2(g)$	−18.4	−31.8	331.4
$SF_6(g)$	−1209	−1105	291.7
$H_2S(g)$	−20.63	−33.56	205.7
$H_2SO_4(l)$	−814.0	−690.0	156.9
$H_2SO_4(aq)$	−909.3	−744.5	20.1
$SO_4^{2-}(aq)$	−909.3	−744.5	20.1
$S_2O_3^{2-}(aq)$	−648.5	−522.5	67
$SO_2(g)$	−296.8	−300.2	248.1
$SO_3(g)$	−395.7	−371.1	256.7
$SO_2Cl_2(g)$	−364.0	−320.0	311.8
$SO_2Cl_2(l)$	−394.1	----	----
Sn(gray)	−2.09	0.13	44.14
$SnCl_4(l)$	−511.3	−440.1	259
SnO(g)	−285.8	−256.9	56.5
$SnO_2(g)$	−580.7	−519.6	52.3
$TiCl_4(g)$	−763.2	−726.7	354.9
$TiCl_4(l)$	−804.2	−737.2	252.3
$TiO_2(s)$	−944.7	−889.5	50.33
$UF_6(g)$	−2147	−2064	377.8
$UF_6(s)$	−2197	−2069	227.6
$UO_2(s)$	−1085	−1032	77.03
$Zn^{2+}(aq)$	−153.9	−147.1	−112.1
ZnO(s)	−348.3	−318.3	43.64

20 ELECTROCHEMISTRY

CHAPTER OBJECTIVES

*** 1. Describe how a voltaic cell operates, using the concepts of electrodes, salt bridges, half-cell reactions, net cell reaction, and cell diagram.**

An oxidation-reduction reaction can be run in an electrochemical cell. If the reaction is spontaneous, a chemical change produces an electrical current. This is a *galvanic* or a *voltaic* cell. If the cell reaction is not spontaneous, electricity can produce a chemical change. This is an *electrolytic* cell (see Objective 20-8). To help understand the working of a galvanic cell, consider the following spontaneous reaction.

$$CuSO_4(aq) + Fe(s) \longrightarrow Cu(s) + FeSO_4(aq)$$

This reaction produces two half-equations:

$$Cu^{2+}(aq) + 2\,e^- \longrightarrow Cu(s) \qquad [1]$$

$$Fe(s) \longrightarrow Fe^{2+}(aq) + 2\,e^- \qquad [2]$$

If we physically separate these half-reactions, the electrons from Fe(s) have to travel to combine with the $Cu^{2+}(aq)$ ions. This movement of electrons is an electrical current.

The first step in physically separating the half-reactions is to create two half-cells: (1) an iron bar immersed in a 1.00 M solution of $FeSO_4(aq)$, and (2) a copper bar immersed in a 1.00 M $CuSO_4(aq)$ solution. If we connect the two electrodes (the iron and the copper bars) with a wire, as in Figure 20-1(a), no current flows. This is because, when electrons come from the iron bar, Fe^{2+} ions go into solution. Thus, the $FeSO_4$ solution becomes positively charged from the added Fe^{2+} ions. This positive charge attracts electrons and prevents current from flowing.

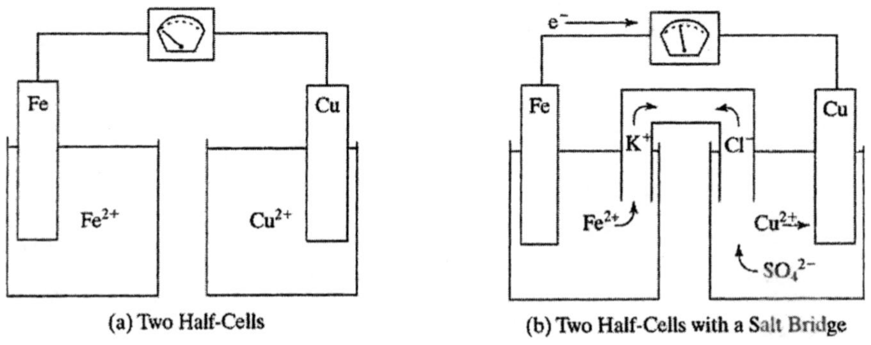

FIGURE 20-1 Construction of a Galvanic or Voltaic Cell

One way to make sure the solutions are uncharged is to allow them to mix together. But then the Cu^{2+} ions will move into the iron half-cell and react directly with the iron metal bar. If we place another solution between the $CuSO_4(aq)$ and $FeSO_4(aq)$ solutions (as in Figure 20-1b), the ions can move through that solution to keep both solutions uncharged. Yet the Cu^{2+} ions will not reach the iron bar. The solution in the middle is called a *salt bridge*. It is an aqueous solution of a strong electrolyte.

The *electrode* is the solid metal bar that is placed in the solution, although sometimes the entire half-cell is called the electrode. Oxidation occurs at the anode. Reduction occurs at the cathode. (To help you remember, note that both anode and oxidation begin with a vowel and both cathode and reduction begin with a consonant. Note also that anions move toward the anode and cations move toward the cathode.) The electrolyte is either (1) the solution containing ions, or (2) the solute of this solution. For instance, either 1.00 M $CuSO_4(aq)$, or simply $CuSO_4$, is the electrolyte.

A cell diagram is a more compact way of representing an electrochemical cell than is a sketch like Figure 20-1. In cell diagrams, the anode is on the left and the cathode is on the right. The electrolytes are in the center of the diagram and the electrodes are on the edges. A solid bar (I) indicates a boundary between two phases. The cell diagram for Figure 20-1(b) is $Fe|FeSO_4\ (1.00\ M)||CuSO_4\ (1.00\ M)|Cu$. The double bar represents the salt bridge. Spectator ions often are not included: $Fe|Fe^{2+}\ (1.00\ M)||Cu^{2+}\ (1.00\ M)|Cu$. Perhaps the easiest way to think of cell diagrams is to trace the path of the positive charge from the anode to the cathode through the cell. Simply write down every species in the order you encounter them.

Sometimes, neither the oxidized nor the reduced species in a half-cell can be made into a metal bar. Consider the oxidation-reduction reaction between the elements zinc and chlorine:

$$Zn(s) + Cl_2(g) \longrightarrow ZnCl_2(aq).$$

The half-equations for this reaction are

$$\text{anode: } Zn(s) \longrightarrow Zn^{2+}(aq) + 2\,e^-$$

$$\text{cathode: } Cl_2(g) + 2\,e^- \longrightarrow 2\,Cl^-(aq).$$

The best way to construct this cathode (which is a chlorine electrode) is to pass a stream of chlorine gas over a surface of finely divided platinum, called platinum black. The cell diagram is $Zn|Zn^{2+}(aq)||Cl^-(aq)|Cl_2(g)|Pt$.

2. Describe the standard hydrogen electrode and explain how other standard electrode potentials are related to it.

The standard hydrogen electrode (S.H.E.) is similar to the chlorine electrode described above. A platinum black electrode is immersed in a solution with $[H^+] = 1.00\ M$, and $H_2(g)$ at a pressure of 1.00 atm is bubbled past the electrode. The potential of this electrode (or half-cell) is set equal to 0.000 V. The potential of any other electrode is the voltage of a cell in which the other electrode is paired with the S.H.E. The sign of that potential is positive if the S.H.E is the anode, and negative if the S.H.E is the cathode.

*** 3. Use tabulated standard potentials, $E°$, to determine E°_{cell} for an oxidation-reduction reaction and predict whether the reaction is spontaneous.**

An oxidation-reduction reaction produces two half-equations. The potentials of both reactions are given in a table of standard reduction potentials, such as Table 20-1 at the end of this chapter.

$$5\,H_2O_2(aq) + 2\,KMnO_4(aq) + 6\,HCl(aq) \longrightarrow 2\,MnCl_2(aq) + 8\,H_2O + 5\,O_2(g) \qquad [3]$$

The two reduction half-equations for Equation [3] are

$$MnO_4^-(aq) + 8\,H^+(aq) + 5\,e^- \longrightarrow Mn^{2+}(aq) + 4\,H_2O \qquad E°\{MnO_4^- \mid Mn^{2+}\} = +1.51\text{ V} \quad [4]$$

$$O_2(g) + 2\,H^+(aq) + 2\,e^- \longrightarrow H_2O_2(aq) \qquad E°\{O_2 \mid H_2O_2\} = +0.682\text{ V} \qquad [5]$$

The balanced net ionic equation [3] is twice the half-equation [4], minus five times half-equation [5]. The multiplications ensure that reduction consumes as many electrons as oxidation produces. (Recall that balancing oxidation-reduction equations, by combining half-equations, was discussed in Objective 5–7.) The half-cell potentials are added in the same way as the half-equations, *except* that the potentials are *not* multiplied by integers.

$$E^{\circ}_{cell} = +1.51 - (+0.682) = +0.83\text{ V}$$

When the resulting potential of the cell is positive, as it is for Reaction [3], the reaction is spontaneous. A negative cell potential indicates a nonspontaneous reaction.

*** 4. Quantitatively and qualitatively predict the effect of varying conditions (concentrations and gas pressures) on value of E_{cell}.**

The Nernst equation ([6]) describes how cell potential depends on concentration and gas pressure.

$$E_{cell} = E°_{cell} - \frac{RT}{nF}\ln Q = E°_{cell} - \frac{0.05915}{n}\log Q \qquad \text{at } 25°\text{C} \qquad [6]$$

Increasing the concentration or gas pressure of the products decreases the cell voltage, making the reaction less spontaneous. Increasing the concentration or the gas pressure of the reactants increases the cell voltage. n is the number of electrons transferred in the balanced oxidation-reduction equation.

EXAMPLE 20-1 What is the potential of the cell $Cu \mid Cu(NO_3)_2(2.00\text{ M}) \parallel AgNO_3(0.100\text{ M}) \mid Ag$?

$$[Ag^+(aq) + e^- \longrightarrow Ag(s)] \times 2 \qquad E°\{Ag^+ \mid Ag\} = 0.800\text{ V}$$

$$Cu(s) \longrightarrow Cu^{2+}(aq) + 2\,e^- \qquad -E°\{Cu^{2+} \mid Cu\} = -(0.337\text{ V})$$

$$2\,Ag^+ + Cu \longrightarrow 2\,Ag + Cu^{2+} \qquad E° = +0.463\text{ V}$$

$n = 2$ for this reaction. The Nernst equation gives

$$E = +0.463\text{ V} - (0.05915/2)\log\frac{[Cu^{2+}]}{[Ag^+]^2} = 0.463\text{ V} - 0.02958$$

$$\log\frac{2.00}{(0.100)^2} = 0.463\text{ V} - 0.068\text{ V} = 0.395\text{ V}$$

*** 5. Know and be able to use the equations that relate $\Delta G°$, E°_{cell}, and K.**

A negative ΔG and a positive E_{cell} indicate a spontaneous reaction. However, ΔG is an extensive property, whereas E is intensive. The product of E and n, the number of moles of electrons transferred, is extensive.

$$\Delta G = -n\,FE \qquad [7]$$

F is the number of coulombs of charge per mole of electrons. $\hat{}c = 96{,}485$ C/mol = Faraday's constant. Now, if the reactants and products are in their standard states, we can compute the equilibrium constant of a reaction from its value of $E°$.

$$\Delta G° = -n\,FE° = -RT \ln K_{\text{eq}} \qquad [8]$$

$$nFE^{0} = RT\ln k_{eq} \qquad [9]$$

EXAMPLE 20-2 What are $\Delta G°$ and K_{eq} for $2\,\text{AgNO}_3\text{(aq)} + \text{Cu(s)} \longrightarrow \text{Cu(NO}_3)_2\text{(aq)} + 2\,\text{Ag(s)}$?

From Example 20-1, $E° = 0.463$ V. (Note also that 1 joule = 1 volt-coulomb).

$\Delta G° = -n\,FE° = -(2 \text{ mol e}^-)(96{,}485 \text{ C/mol})(0.463 \text{ V}) = -89.2\times10^3 \text{ J/mol} = -RT \ln K_{\text{eq}}$

$-89.2\times10^3 \text{ J/mol} = -(8.314 \text{ J mol}^{-1}\text{ K}^{-1})(298 \text{ K}) \ln K_{\text{eq}}$ and

$\ln K_{\text{eq}} = 36.0$ or $K_{\text{eq}} = 4.3\times10^{15}$

EXAMPLE 20-3 What is ΔG for $2\,\text{AgNO}_3\text{(aq)} + \text{Cu(s)} \longrightarrow \text{Cu(NO}_3)_2\text{(aq)} + 2\,\text{Ag(s)}$ when $[\text{AgNO}_3] = 0.100$ M and $[\text{Cu(NO}_3)_2] = 2.00$ M?

In Example 20-1 we found that $E = 0.395$ V with these same concentrations.

$\Delta G = -n\,FE = -(2 \text{ mol e}^-)(96{,}485 \text{ C/mol e}^-)(0.395 \text{ V}) = -76.0\times10 \text{ J/mol}$

6. Describe some common voltaic cells: the flashlight or dry cell, the lead storage battery, the silver-zinc cell, and the fuel cell.

A battery is a device in which several cells are linked together. For the dry cell, the cells of the lead storage battery, and the silver-zinc cell, you should know the answers to the following questions.

(1) What are the materials of the anode and of the cathode?

(2) What are the cell equation and the anode and cathode half-cell equations?

(3) What does the cell look like? Sketch the cell.

(4) What is the cell diagram?

(5) What are the standard cell potentials of each half-cell and of the full cell?

Fuel cells often are based on combustion reactions that are run so that an electrical current is generated. The standard cell potential is best calculated from ΔG by means of Equation [8].

7. Explain the corrosion of metals in electrochemical terms and describe methods of corrosion protection.

The corrosion of a metal is simply its oxidation in the presence of oxygen. Metals often are protected from corrosion by coating either with another metal (such as zinc or chromium) or with paint. Another means of protection is a sacrificial anode. An active metal (such as magnesium) is connected to the metal to be protected. The active metal oxidizes first; only when it is consumed will the protected metal corrode. Some metals protect themselves with an insoluble and firmly bound compound on their surface. Aluminum forms Al_2O_3, which gives the metal a slightly dull appearance, and copper forms blue-green $Cu_2(OH)_2CO_3$. When the insoluble compound is removed, the fresh surface soon corrodes.

8. Describe an electrolytic cell and how it differs from a voltaic cell.

In electrolytic cells, electricity causes a chemical change. The cell reaction usually is a decomposition. (*Electrolytic* literally means "breaking apart with electricity.") The cell reaction usually is nonspontaneous; electricity overcomes the nonspontaneity. The principal problem is keeping the products from reacting. For example, an important industrial process is the electrolysis of brine (NaCl solutions).

$$2\,NaCl(aq) + 2\,H_2O \longrightarrow 2\,NaOH(aq) + Cl_2(g) + H_2(g) \qquad [10]$$

$H_2(g)$ and $Cl_2(g)$ explode when mixed and exposed to light. Thus, the electrolysis cell is designed to keep these gases separate. Reaction [10] is nonspontaneous.

$$2\,Cl^-(aq) \longrightarrow Cl_2(g) + 2\,e^- \qquad E°\{Cl_2 \mid Cl^-\} = -1.358\text{ V}$$

$$2\,H_2O + 2\,e^- \longrightarrow H_2(g) + 2\,OH^-(aq) \qquad -E°\{H_2 \mid H_2O\} = -(0.828\text{ V})$$

$$2\,H_2O + 2\,Cl^- \longrightarrow Cl_2(g) + H_2(g) + 2\,OH^- \quad E° = -2.186\text{ V}$$

The applied voltage must be larger than $+2.186$ V to cause Reaction [10] to occur.

*** 9. Identify the possible half-reactions that might occur in an electrolysis and choose the pair that will occur based on the highest (least negative) cell potential.**

Determining the products of the electrolysis of a solution involves two steps. First, write all the half-reactions that might occur, along with their standard potentials, by referring to a table of standard electrode potentials (see Table 20-1 at the end of this chapter). Then select the one oxidation and the one reduction that sum to the largest (least negative) cell potential.

EXAMPLE 20-4 A solution with $[Cu(NO_3)_2] = 0.500$ M, $[ZnCl_2] = 0.500$ M, and $[FeBr_2] = 0.500$ M is electrolyzed between inert electrodes. What are the initial products at the anode and at the cathode, and what minimum voltage must be used?

There are eight possible half-equations, including those for the electrolysis of water:

$$Cu^{2+}(aq) + 2\,e^- \longrightarrow Cu(s) \qquad E°\{Cu^{2+}|Cu\} = +0.337\text{ V}$$

$$Zn^{2+}(aq) + 2\,e^- \longrightarrow Zn(s) \qquad E°\{Zn^{2+}|Zn\} = -0.763\text{ V}$$

$$Fe^{2+}(aq) + 2\,e^- \longrightarrow Fe(s) \qquad E°\{Fe^{2+}|Fe\} = -0.440\text{ V}$$

$$2\,H^+(aq) + 2\,e^- \longrightarrow H_2(g) \qquad E°\{H^+|H_2\} = 0.000\text{ V}$$

$$NO_3^-(aq) + 4\,H^+ + 3\,e^- \longrightarrow NO(g) + 2\,H_2O \qquad E°\{NO_3^-|NO\} = +0.956\text{ V}$$

$$2\,Br^-(aq) \longrightarrow Br_2(l) + 2\,e^- \qquad E°\{Br_2|Br^-\} = -1.065\text{ V}$$
$$2\,Cl^-(aq) \longrightarrow Cl_2(g) + 2\,e^- \qquad E°\{Cl_2|Cl^-\} = -1.358\text{ V}$$
$$2\,H_2O \longrightarrow O_2(g) + 4\,H^+(aq) + 4\,e^- \qquad E°\{O_2|H_2O\} = -1.229\text{ V}$$

We are tempted to pick the NO_3^- reduction until we realize that anions move away from the cathode where reduction occurs. The cation with the most positive half-cell potential is $Cu^{2+}(aq)$. $Br^-(aq)$ has the most positive oxidation potential.

$$Cu^{2+}(aq) + 2\,e^- \longrightarrow Cu(s) \qquad E°\{Cu^{2+}\,|\,Cu\} = +0.337\text{ V}$$
$$2\,Br^-(aq) \longrightarrow Br_2(l) + 2\,e^- \qquad -E°\{Br_2\,|\,Br^-\} = -(+1.065\text{ V})$$
$$Cu^{2+} + 2\,Br^- \longrightarrow Cu + Br_2 \qquad E° = -0.728\text{ V}$$

The Nernst equation now is used to compute E:

$$E = E° - \frac{0.05915}{n}\log\frac{1}{[Cu^{2+}][Br^-]^2} = -0.728\text{ V} - 0.02958\log\frac{1}{0.500(1.00)^2} = -0.737\text{ V}$$

*** 10. Use Faraday's laws to relate the quantity of chemical change produced by a given amount of electric charge.**

Voltage is the driving force of electrochemical change. The quantity of chemical change is measured by the quantity of charge produced or consumed. The rate of chemical change is measured by the current. If current and voltage seem similar, an analogy with water may help. Voltage is similar to water pressure and current is similar to rate of flow (perhaps measured in gallons/minute). Current is measured in amperes, where 1 ampere (A) = 1 coulomb/second = 1 C/s. Charge is measured in coulombs (C) or in Faradays (F), where 1 Faraday = 96,485 C = 1 mole of electrons. The conversion factor method uses these relationships.

EXAMPLE 20-5 A 20.0 A current is used to plate copper out of a $CuSO_4(aq)$ solution. How long must the current be applied in order to plate out 1.00 lb. of Cu?

The half-equation is that of the reduction of $Cu^{2+}(aq)$:

$$Cu^{2+}(aq) + 2\,e^- \longrightarrow Cu(s)$$

Then, we determine the time required:

$$1.00\text{ lb. Cu} \times \frac{454\text{ g Cu}}{1\text{ lb. Cu}} \times \frac{1\text{ mol Cu}}{63.5\text{ g Cu}} \times \frac{2\text{ mol e}^-}{1\text{ mol Cu}} \times \frac{1\text{ s}}{20.0\text{ C}} \times \frac{1\text{ hr}}{3600\text{ s}} = 19.2\text{ hr}$$

EXAMPLE 20-6 A 12.4 A current is used to electrolyze 5.20 L of a 3.00 M $AgNO_3(aq)$ solution. What is $[Ag^+]$ after 1.50 hr?

We first compute the amount of Ag plated out:

$$1.50\text{ hr} \times \frac{3600\text{ s}}{\text{hr}} \times 1\frac{12.4\text{ C}}{\text{s}} \times \frac{\text{mol e}^-}{96{,}485\text{ C}} \times \frac{\text{mol Ag}^+}{\text{mol e}^-} 0.964\text{ mol Ag}^+$$

The amount of Ag originally present is found: $5.20\,\text{L} \times \frac{3.00\,\text{mol AgNO}_3}{\text{L}} \times \frac{\text{mol Ag}^+}{\text{mol AgNO}_3} = 15.6\,\text{mol Ag}^+$

Then, we find the amount of Ag^+ left: $15.750\,\text{mol Ag}^+ - 0.694\,\text{mol Ag}^+ = 15.056\,\text{mol Ag}^+$

The last step is to compute the final $[Ag^+]$: $[\text{Ag}^+] = 15.056\,\text{mol Ag}^+ / 5.520\,\text{L} = 2.868\,\text{M}$

11. Describe the industrial electrolysis processes of electrorefining, electroplating, electrosynthesis, and the chlor-alkali process.

Electrorefining of metals involves the deposition of a metal at a cathode from a metal ion solution.

Electroplating, such as for silver-plated flatware, involves plating of one metal onto another by electrolysis. This is often done to protect the underlying metal or for decorative purposes.

Electrosynthesis is a method of producing substances through electrolysis reactions. This process can be used for purifying metal compounds and for some organic reactions. The chlor-alkali process is the industrial scale electrolysis of NaCl to produce chlorine and NaOH (aq).

SELF-ASSESSMENT EXERCISES

99. (a) True. (b) Flase. The salt bridge permits the flow of ions. (c) True. (d) False. The direction of the reaction will depend on the electrodes, but the anode is where oxidation occurs and the cathode is where reduction occurs. (e) True. (f). False. An electrolytic cell requires current to drive the reaction in a nonspontaneous direction. (g) True.

100. (a) False. A positive E^0 indicates the potential for reduction is greater than S.H.E. (b) True. (c) True. (d) False. In order to displace Zn from solution, Zn would need to be reduced and the Zn E^0 = -.0763. Adding oxidation of Hg will give an even more negative (nonspontaneous) number.

101. (d) $E_{cell} = 0.66\text{ V} - \frac{0.0592}{2}\log\frac{[\text{Zn}^{+2}]}{[\text{Pb}^{+2}]} = 0.66\text{V} - \frac{0.0592}{2}\log\frac{[0.10\text{ M}]}{[0.01\text{ M}]} = 0.63\text{ V}$

102. The reaction is spontaneous as written, so the reaction will occur. The reaction will proceed, but as the $[Ni^{2+}]$ is reduced, the reaction will eventually become nonspontaneous.

103. (c) SO_2. The reaction is $SO_4^{2-} + 4H^+ + 2\,e^- \rightarrow 2H_2O + SO_2(g)$ $E^0 = +0.17$ V

and the oxidation reaction is $H_2 \rightarrow 2\,H^+ + 2e^-$.

104. (d) 5.6 L. The charge required for Ag is 4.5 g x 1mol/26.98 x 3 mol e^-/1 mol Ag = 0.500 mol e^-. For H_2, $2H^+ + 2e^- \rightarrow H_2$, so 0.5 mol will produce 0.25 mol H_2 and 0.25 mol x 22.4 L mol^{-1} = 5.6 L.

106. Zn| Zn^{+2}|| Pt | H_2 | H^+. The two half reactions are

$Zn(s) \rightarrow Zn^{2+} + 2e^-$ (oxid., anode) and

$2H^+ + 2e^- \rightarrow H_2$ (reduction, cathode).

The balanced cell reaction is

$Zn(s) + 2H^+ + 2NO_3^- \rightarrow Zn^{2+} + H_2O + 2NO_3^-$ $\quad E^0_{cell} = +0.763$ V.

107. $E_{cell} = E^0_{cell} - \frac{0.0592}{2}\log\frac{[0.10\text{ M}]}{x}$

$0.108 = 0 - \frac{0.0592}{2}\log\frac{[0.10\text{ M}]}{x}$. Solving for x, x = 2.2 x 10^{-3} M = $[H^+]$

pH = 2.65

108. Because the concentrations are equal, $E_{cell} = E^0_{cell,}$ which is negative. ΔG for the reaction is $\Delta G = -nFE_{cell}$ and will be positive. The reaction is not spontaneous and the equilibrium will not have those concentrations. The net reaction will occur in the reverse direction.

DRILL PROBLEMS

1. Draw the voltaic cell for each equation that follows. Label the anode and the cathode. Indicate the direction of movement of all ions, including those of the salt bridge, $K^+(aq)$, and $Cl^-(aq)$. Draw the cell diagram. You may use either platinum or graphite as inert electrodes. (The equations are not balanced.)

A. $CuSO_4(aq) + H_2O_2(aq) \longrightarrow Cu(s) + O_2(g) + H_2SO_4(aq)$

B. $Mg(s) + AgNO_3(aq) \longrightarrow Ag(s) + Mg(NO_3)_2(aq)$

C. $Al(s) + H_2SO_4(aq) \longrightarrow Al_2(SO_4)_3(aq) + H_2(g)$

D. $Fe_2(SO_4)_3(aq) + Pb(s) \longrightarrow PbSO_4(s) + Fe(s)$

E. $NaI(aq) + F_2(g) \longrightarrow I_2(s) + NaF(aq)$

F. $H_2(g) + KClO(aq) \longrightarrow KCl(aq) + H_2O(l)$, in alkaline solution

G. $MnO_2(s) + HCl(aq) \longrightarrow MnCl_2(aq) + H_2O(l) + Cl_2(g)$

H. $I_2(s) + H_2O(l) + Br_2(l) \longrightarrow HBr(aq) + HIO_3(aq)$

I. $Ag(s) + HNO_3(aq) \longrightarrow AgNO_3(aq) + NO(g) + H_2O(l)$

J. $S(s) + H_2O(l) + Pb(NO_3)_2(aq) \longrightarrow Pb(s) + H_2SO_3(aq) + HNO_3(aq)$

2. For each reaction in the drill problems for Objective 20–1, along with each reaction that follows, use the data of Table 20-1 at the end of this chapter to determine E°_{cell} and state whether the standard cell reaction is spontaneous or nonspontaneous.

K. $O_2(g) + H_2O(l) \longrightarrow O_3(g) + H_2O_2(aq)$ L. $H_2S(g) + Br_2(l) \longrightarrow S(s) + HBr(aq)$
M. $SnSO_4(aq) + FeSO_4(aq) \longrightarrow Sn(s) + Fe_2(SO_4)_3(aq)$
N. $H_2SO_4(aq) + S(s) + H_2O(l) \longrightarrow H_2SO_3(aq)$ O. $I_2(s) + H_2O(l) \longrightarrow HI(aq) + HIO_3(aq)$

3. Determine the cell potential of each of the 10 cells in the drill problems for Objective 20–1, if all concentrations are 1.00 M and all gas pressures are 1.00 atm, except the following:

A. $[CuSO_4] = 4.00$ M, $P(O_2) = 0.210$ atm B. $[AgNO_3] = 0.150$ M, $[Mg(NO_3)_2] = 2.75$ M
C. $[H_2SO_4] = 6.00M$, $P(H_2) = 0.100$ atm D. $[PbSO_4] = 1.30 \times 10^{-5}$ M
E. $[NaF] = 2.50$ M, $[NaI] = 0.400$ M F. $P\{H_2\} = 6.40$ atm, $[KCl] = 1.50$ M
G. $[HCl] = 6.00$ M, $P(Cl_2) = 0.0120$ atm H. $[HBr] = 0.100$ M, $[HIO_3] = 0.100$ M
I. $[HNO_3] = 3.00$ M, $P(NO) = 4.00$ atm J. $[HNO_3] = 4.00$ M, $[H_2SO_3] = 0.250$ M

4. Determine values of ΔG° and K_{eq} for the 10 cells of the drill problems for Objective 20–1. Determine the value of ΔG for each of the cells in the drill problems for Objective 20–4.

5. For each of the following solutions, use the data of Table 20-1 at the end of the chapter to predict the products of electrolysis at the anode and at the cathode. Write the net ionic equation for the electrolysis reaction, then use the Nernst equation to predict the minimum voltage that must be imposed for electrolysis to occur. Disregard overpotentials. Assume that any products formed are in their standard states.

A. $[Na_2SO_4] = 1.00$ M, $[MgCl_2] = 1.50$ M B. $[CuCl_2] = 1.00$ M, $[NaNO_3] = 0.500$ M
C. $[HCl] = 6.20$ M, $[AlBr_3] = 2.00$ M D. $[KOH] = 1.25$ M, $[NaNO_3] = 2.00$ M
E. $[Cu(NO_3)] = 0.13$ M, $[SnCl_2] = 3.15$ M F. $[FeSO_4] = 1.25$ M, $[AlI_3] = 0.0240$ M

6. In these problems, assume that the ions in solution are not depleted during the course of the electrolysis.

A. A constant current of 10.0 A is passed through an electrolytic cell for 90.0 min. How many Faradays of charge are passed through the cell?

B. A solution of $Sn(NO_3)_2$ is electrolyzed by passing 85.0 A of current for 72.0 min. How many grams of tin plate out?

C. Two cells containing solutions of $AgNO_3$ and $CuSO_4$ are connected in series (that is, the current goes in the cathode of one and out its anode, and then passes into the cathode of the other cell and out its anode) and electrolyzed. The cathode in the $AgNO_3$ cell gains 1.078 g. What mass has the cathode in the other cell gained?

D. A solution of $Cu_3(PO_4)_2$ is electrolyzed by passing 17.2 A of current for 145 min. How many grams of Cu plate out?

E. A solution of $Hg_2(NO_3)_2$ is electrolyzed by passing 108 A of current for 95.0 min. How many grams of elemental mercury are produced?

F. A solution of copper(II) sulfate is electrolyzed. Assume that no gas evolution occurs at the cathode. For how long must a current of 1.93 A flow to deposit 64.0 g of copper?

G. A solution of $Au_2(SO_4)_3$ is electrolyzed by passing 104 A of current for 165 min. What mass of gold plates out?

H. What volume of oxygen (at STP) can be liberated during the passage of 5.00 F of electricity?

I. A solution of $Al(NO_3)_3$ is electrolyzed by passing 144 A of current for 102 min. What mass of aluminum plates out?

J. A solution of $Cr(NO_3)_3$ is electrolyzed with a current of 15.0 A of current for 36.2 min. What mass of chromium plates out?

K. A solution of $AuCl_3$ is electrolyzed by passing 85.0 A of current for 17.6 min. What mass of gold plates out?

L. What mass of silver can be deposited by the passage of a constant current of 5.00 A through a $AgNO_3$ solution for 2.00 hours?

QUIZZES (20 minutes each) Choose the best answer for each question. Since balancing oxidation-reduction equations is an important part of electrochemistry, some equation-balancing problems are included for review. (Balancing is discussed in Objective 5–8.)

QUIZ A

1. When the equation $HClO_3 + H_2O + I_2 \longrightarrow HIO_3 + HCl$ is balanced with the smallest integer coefficients, the coefficient of chloric acid is (a) 10; (b) 12; (c) 6; (d) 8; (e) none of these.

2. $Cr_2O_7^{2-} + I^- + H^+ \longrightarrow Cr^{3+} + I_2 + H_2O$. The correct coefficient for I^- in the balanced equation is (a) 2; (b) 3; (c) 6; (d) 14; (e) none of these.

3. When the equation $Br_2 + KOH + MnBr_2 \longrightarrow MnO_2 + KBr + H_2O$ is balanced with the smallest integer coefficients, the coefficient of H_2O is (a) 3; (b) 4; (c) 1; (d) 2; (e) none of these.

4. In the reaction $Br_2 + 2\,H_2O + SO_2 \longrightarrow 2\,HBr + H_2SO_4$, the mass of SO_2 oxidized by 1 F of electrons is (a) 64.1 g; (b) 128.2 g; (c) 21.4 g; (d) 32.1 g; (e) none of these.

5. The cell reaction and $E°$ of the spontaneous cell made from the Zn/Zn^{2+} (0.76 V) and $2\,Br^-/Br_2$ (−1.07 V) half-cells are (a) $Zn^{2+} + 2\,Br^- \longrightarrow Br_2 + Zn, +1.83$ V;

(b) $Br_2 + Zn \longrightarrow Zn^{2+} + 2\,Br^-$, +1.83 V; (c) $Br_2 + Zn \longrightarrow Zn^{2+} + 2\,Br^-$, +0.31 V; (d) $Zn^{2+} + 2\,Br^- \longrightarrow Br_2 + Zn$, +1.30 V; (e) none of these.

6. Consider the half-cell of Pb metal in contact with 1 M $Pb(NO_3)_2$, connected to the half-cell of Ag metal in contact with 1.0 M $AgNO_3$ in a complete circuit connected with an exterior wire (standard reduction potentials: Pb −0.13 V, Ag +0.80 V). If NaCl is added to the half-cell containing $AgNO_3$ and AgCl precipitates, (a) the voltage of the cell would become more positive; (b) the voltage of the cell would become less positive; (c) the voltage of the cell would not change; (d) the equilibrium constant of the cell reaction would change; (e) the ΔG of the cell reaction would remain the same.

7. The law that relates the amount of material oxidized or reduced to the amount of charge passed is due to (a) Volta; (b) Faraday; (c) Nernst; (d) Mendeleev; (e) Avogadro.

8. In the cell $Zn \mid Zn^{2+} \parallel Ni^{2+} \mid Ni$, the zinc is called the (a) electrolyte; (b) cathode; (c) anode; (d) oxidizing agent; (e) none of these.

9. What volume of H_2 (measured at STP) is produced by the electrolysis of a 2.0 M H_2SO_4 solution? (A current of 0.50 A is passed for 30.0 min.) (a) 418 mL; (b) 52 mL; (c) 209 mL; (d) 104 mL; (e) none of these.

10. Suppose that the cell $Sn^{2+} + Zn^\circ \longrightarrow Sn^\circ + Zn^{2+}$, $E^\circ = 0.537$ V will run until one of the reactants is completely gone. If the cell consists of 30.0 mL of 0.50 M Sn^{2+}, 30 mL of 0.50 M Zn^{2+}, 10.0 g Zn, and 18.2 g Sn, which one will run out first? (a) Sn^{2+} solution; (b) Zn^{2+} solution; (c) Sn strip; (d) Zn strip; (e) none of these; two reactants will run out simultaneously.

QUIZ B

1. When the half-equation $NO_3^- + H_2O \longrightarrow NO + OH^-$ is balanced with the smallest integer coefficients, the coefficient of OH^- is (a) 3; (b) 2; (c) 4; (d) 1; (e) none of these.

2. When the equation $KMnO_4 + KNO_2 + H_2O \longrightarrow MnO_2 + KNO_3 + KOH$ is balanced with the smallest integer coefficients, the coefficient of potassium nitrate is (a) 1; (b) 2; (c) 3; (d) 4; (e) none of these.

3. When the equation $HI + HNO_3 \longrightarrow H_2O + NO + I_2$ is balanced with the smallest integer coefficients, the sum of the coefficients of the products is (a) 5; (b) 7; (c) 6; (d) 8; (e) none of these.

4. In the reaction $2\,H_2O + 3\,KCN + 2\,KMnO_4 \longrightarrow 3\,KCNO + 2\,MnO_2 + 2\,KOH$, the mass of $KMnO_4$ reduced with 1 mole of electrons is (a) 158 g; (b) 52.7 g; (c) 79.0 g; (d) 316 g; (e) none of these.

5. The cell reaction and E° of the spontaneous cell made from the Ag/Ag^+ (−0.80 V) and Cl^-/Cl_2 (−1.36 V) half-cells are

(a) $Cl_2 + 2\,Ag \longrightarrow 2\,Cl^- + 2\,Ag^+$, 0.56 V;

(b) $Cl_2 + 2\,Ag \longrightarrow 2\,Cl^- + 2\,Ag^+, 1.12\,V;$

(c) $Cl_2 + 2\,Ag \longrightarrow 2\,Cl^- + 2\,Ag^+, 2.16\,V;$

(d) $2\,Ag^+ + 2\,Cl^- \longrightarrow 2\,Ag + Cl_2, 1.08\,V;$

(e) none of these.

6. In the cell $Fe^{2+} \mid Fe^{3+} \parallel Cu^{2+} \mid Cu$, which will increase the cell voltage the *most*? (a) halve $[Cu^{2+}]$; (b) halve $[Fe^{2+}]$; (c) double $[Cu^{2+}]$; (d) double $[Fe^{2+}]$; (e) cut Cu electrode in half.

7. If the voltage of an electrochemical cell is negative, the cell reaction is (a) nonspontaneous; (b) slow; (c) exothermic; (d) spontaneous; (e) none of these.

8. The equation that relates ion concentrations and gas pressures to cell voltages is called (a) Raoult's law; (b) Le Châtelier's principle; (c) Volta's theory; (d) Faraday's law; (e) none of these.

9. What mass of copper (63.54 g/mol) can be deposited by the passage of 9650 C of electricity through a $CuSO_4$ solution? (a) 12.7 g; (b) 6.35 g; (c) 3.18 g; (d) 7.98 g; (e) none of these.

10. A copper electrode weighs 35.42 g before the electrolysis of a copper(II) sulfate solution and 36.69 g after the electrolysis has run for 20.0 s. Compute the amperage of the current that is used. (a) 3860 A; (b) 157 A; (c) 193 A; (d) 259 A; (e) none of these.

QUIZ C

1. When the half-reaction $ClO_3{}^- + H^+ \longrightarrow Cl^- + H_2O$ is balanced with the smallest integer coefficients, the coefficient of H_2O is (a) 6; (b) 5; (c) 2; (d) 3; (e) none of these.

2. When the equation $Cs_2MnO_4 + H_2O \longrightarrow CsMnO_4 + CsOH + MnO_2$ is balanced with the smallest integer coefficients, the coefficient of MnO_2 is (a) 1; (b) 2; (c) 3; (d) 4; (e) none of these.

3. When the equation $HCl + H_2C_2O_4 + MnO_2 \longrightarrow MnCl_2 + CO_2 + H_2O$ is balanced with the smallest integer coefficients, the coefficient of CO_2 is (a) 1; (b) 2; (c) 3; (d) 4; (e) none of these.

4. In the reaction $HMnO_4 + 5\,AsH_3 + 8\,H_2SO_4 \longrightarrow 5\,H_3AsO_4 + 8\,MnSO_4 + 12\,H_2O$, the mass of AsH_3 oxidized by 1 mole of electrons is (a) 77.9 g; (b) 26.0 g; (c) 15.6 g; (d) 9.74 g; (e) none of these.

5. The cell reaction and $E°$ of the spontaneous cell made from Fe/Fe^{2+} (0.44 V) and Sn/Sn^{2+} (0.14 V) half-cells are (a) $Fe^{2+} + Sn \longrightarrow Sn^{2+} + Fe$, -0.30 V; (b) $Fe^{2+} + Sn \longrightarrow Sn^{2+} + Fe$, +0.30 V; (c) $Fe° + Sn^{2+} \longrightarrow Sn° + Fe^{2+}$, -0.30 V; (d) $Fe° + Sn^{2+} \longrightarrow Sn° + Fe^{2+}$, +0.58 V; (e) none of these.

6. For the cell $Cu|Cu^+||Al^{3+}|Al$, which will increase the cell voltage the most? (a) double $[Al^{3+}]$; (b) halve $[Cu^+]$; (c) double $[Cu^+]$; (d) halve $[Al^{3+}]$; (e) cut the Al electrode in half.

7. One mole of electrons has a charge of (a) 96,485 A; (b) 1.60×10^{-19} C; (c) 6.02×10^{23} A; (d) 96,485 C; (e) none of these.

8. In a galvanic cell, oxidation occurs at the (a) anode; (b) cathode; (c) salt bridge; (d) electrolyte; (e) none of these.

9. 1.00 L of a 1.500 M $Al_2(SO_4)_3$ solution is electrolyzed between platinum electrodes for 1.20 hr with a current of 62.0 A. What is $[Al^{3+}]$ at the end of the electrolysis? (a) 0.57 M; (b) 0.93 M; (c) 2.07 M; (d) not enough Al^{3+} present; (e) none of these.

10. A solution of CuBr is electrolyzed by passing 35.0 A of current for 17.0 min. What mass in grams of Cu (63.54 g/mol) plates out? (a) 26.5; (b) 53.1; (c) 11.8; (d) 23.5; (e) none of these.

QUIZ D

1. After balancing the ionic equation $MnO_4^- + Cl^- + H^+ \longrightarrow Mn^{2+} + Cl_2 + H_2O$, the sum of the simplest set of integer coefficients of the six substances would be (a) 21; (b) 24; (c) 38; (d) 43; (e) 51.

2. When the equation $ClO_2 + H_2O_2 + KOH \longrightarrow H_2O + KClO_2 + O_2$ is balanced with the smallest integer coefficients, the coefficient of H_2O is (a) 3; (b) 5; (c) 2; (d) 4; (e) none of these.

3. When the equation $HNO_2 \longrightarrow HNO_3 + NO + H_2O$ is balanced with the smallest integer coefficients, the sum of all four of the coefficients is (a) 7; (b) 6; (c) 9; (d) 10; (e) none of these.

4. In the reaction $2\ Bi(OH)_3 + 3\ Na_2SnO_2 \longrightarrow 2\ Bi + 3\ H_2O + 3\ Na_2SnO_3$, the mass of $Bi(OH)_3$ reduced with 1.00 Faraday of charge is (a) 86.7 g; (b) 260.0 g; (c) 130.0 g; (d) 173 g; (e) none of these.

5. The cell reaction and $E°$ of the spontaneous cell made from the Sn^{2+}/Sn^{4+} (–0.15 V) and $H_2S/S°$ (–0.14 V) half-cells are

(a) $Sn^{4+} + H_2S \longrightarrow 2\ H^+ + S° + Sn^{2+}$; 0.29 V;

(b) $2\ H^+ + S° + Sn^{2+} \longrightarrow Sn^{4+} + H_2S$, –0.29 V;

(c) $Sn^{4+} + H_2S \longrightarrow 2\ H^+ + S° + Sn^{2+}$, 0.01 V;

(d) $2\ H^+ + S° + Sn^{2+} \longrightarrow Sn^{4+} + H_2S$, –0.01 V; (e) none of these.

6. For the cell $Ag \mid Ag^+ \parallel Br^- \mid Br_2$, which of the following will increase the cell voltage the *most*? (a) double $[Ag^+]$; (b) triple $[Br^-]$; (c) cut the Ag electrode in half; (d) remove half of the $Br_2(l)$; (e) halve $[Ag^+]$.

7. Consider the half-cell of Pb metal in contact with 1.00 M $Pb(NO_3)_2$, connected to the half-cell of Ag metal in contact with 1.00 M $AgNO_3$ in a complete circuit connected with an external wire. (Standard reduction potentials: Pb = –0.13 V, Ag = +0.80 V.) As this cell operates, (a) the Pb electrode increases in mass; (b) electrons flow from the Pb electrode to the Ag electrode; (c) the Ag electrode is the anode of the cell; (d) NO_3^- ions migrate from the lead half-cell to the silver half-cell; (e) the Ag electrode dissolves slowly.

8. A solution of $CuCl_2$ is electrolyzed by passing 35.0 A for 17.0 min. What mass in grams of Cu (63.54 g/mol) plates out? (a) 24.9; (b) 49.8; (c) 47.0; (d) 23.5; (e) none of these.

9. The amount of substance that produces a mole of electrons is known as a(n) (a) Lewis base; (b) electron mole; (c) reducing agent; (d) oxidizer; (e) none of these.

10. For how many hours must a current of 14.0 A flow to produce 5.02 g Al (27.0 g/mol) from a solution of $Al(NO_3)_3$ (213 g/mol)? (a) 1.07; (b) 0.135; (c) 0.356; (d) 0.405; (e) none of these.

SAMPLE TEST (20 minutes)

1. The reaction $3\ Pb(NO_3)_2 + 2\ Al \longrightarrow 2\ Al(NO_3)_3 + 3\ Pb$ is used in a galvanic cell.

   A. Write the cell diagram, with a KCl salt bridge.

   B. From the following reduction couples, determine $E°$ of the cell.

   Pb^{2+}/Pb, $E° = -0.126$ V; Al^{3+}/Al, -1.66 V; Pb^{4+}/Pb^{2+} 1.455 V; NO_3^-/NO, 0.96 V

   C. Determine the value of E when $[Pb(NO_3)_2] = 4.00$ M and $[Al(NO_3)_3] = 0.0200$ M.

2. 50.0 mL of 0.100 M $CuSO_4(aq)$ is electrolyzed with a 0.965 A current for 1.00 min. When electrolysis is complete, what is (a) $[Cu^{2+}]$, and (b) pH, assuming pH = 7.00 initially?

3. Balance the following oxidation-reduction equations:

   A. $KMnO_4(aq) + H_2S(g) \longrightarrow MnO_2(s) + S(s) + KOH(aq) + H_2O(l)$

   B. $NiS(s) + KClO_3(aq) + HCl(aq) \longrightarrow NiCl_2(aq) + KCl(aq) + S(s) + H_2O(l)$

TABLE 20-1 Some Selected Standard Electrode Potentials

$E°$, V	Reduction half-reaction
	Acidic Solution
+0.800	$Ag^+(aq) + e^- \longrightarrow Ag(s)$
+1.98	$Ag^{2+}(aq) + e^- \longrightarrow Ag^+(aq)$
+0.071	$AgBr(s) + e^- \longrightarrow Ag(s) + Br^-(aq)$
+0.2223	$AgCl(s) + e^- \longrightarrow Ag(s) + Cl^-(aq)$
−0.152	$AgI(s) + e^- \longrightarrow Ag(s) + I^-(aq)$
−1.676	$Al^{3+}(aq) + 3\ e^- \longrightarrow Al(s)$
+1.36	$Au^{3+}(aq) + 2\ e^- \longrightarrow Au^+(aq)$
+1.52	$Au^{3+}(aq) + 3\ e^- \longrightarrow Au(s)$
+0.85	$[AuBr_4]^-(aq) + 3\ e^- \longrightarrow Au(s) + 4\ Br^-(aq)$
+1.002	$[AuCl_4]^-(aq) + 3\ e^- \longrightarrow Au(s) + 4\ Cl^-(aq)$
−2.92	$Ba^{2+}(aq) + 2\ e^- \longrightarrow Ba(s)$
+0.32	$BiO^+(aq) + 2\ H^+(aq) + 3\ e^- \longrightarrow Bi(s) + H_2O$
+1.065	$Br_2(l) + 2\ e^- \longrightarrow 2\ Br^-(aq)$

+1.478	$2\ BrO_3^-(aq) + 12\ H^+(aq) + 10\ e^- \longrightarrow Br_2(l) + 6\ H_2O$
−0.49	$2\ CO_2(g) + 2\ H^+(aq) + 2\ e^- \longrightarrow H_2C_2O_4(aq)$
+0.37	$C_2N_2(g) + 2\ H^+(aq) + 2\ e^- \longrightarrow 2\ HCN(aq)$
−2.84	$Ca^{2+}(aq) + 2\ e^- \longrightarrow Ca(s)$
−0.403	$Cd^{2+}(aq) + 2\ e^- \longrightarrow Cd(s)$
+1.358	$Cl_2(g) + 2\ e^- \longrightarrow 2\ Cl^-(aq)$
+1.175	$ClO_3^-(aq) + 2\ H^+(aq) + e^- \longrightarrow ClO_2(g) + H_2O$
+1.450	$ClO_3^-(aq) + 6\ H^+(aq) + 6\ e^- \longrightarrow Cl^-(aq) + 3\ H_2O$
+1.19	$ClO_4^-(aq) + 2\ H^+(aq) + 2\ e^- \longrightarrow ClO_3^-(aq) + H_2O$
−0.277	$Co^{2+}(aq) + 2\ e^- \longrightarrow Co(s)$
−0.90	$Cr^{2+}(aq) + 2\ e^- \longrightarrow Cr(s)$
+1.33	$Cr_2O_7^{2-}(aq) + 14\ H^+(aq) + 6\ e^- \longrightarrow 2\ Cr^{3+}(aq) + 7\ H_2O$
−0.424	$Cr^{3+}(aq) + e^- \longrightarrow Cr^{2+}(aq)$
−2.923	$Cs^+(aq) + e^- \longrightarrow Cs(s)$
+0.520	$Cu^+(aq) + e^- \longrightarrow Cu(s)$
+0.337	$Cu^{2+}(aq) + 2e^- \longrightarrow Cu(s)$
+0.159	$Cu^{2+}(aq) + e^- \longrightarrow Cu^+(aq)$
+0.86	$Cu^{2+}(aq) + I^-(aq) + e^- \longrightarrow CuI(s)$
+2.886	$F_2(g) + 2e^- \longrightarrow 2\ F^-(aq)$
−0.440	$Fe^{2+}(aq) + 2\ e^- \longrightarrow Fe(s)$
−0.036	$Fe^{3+}(aq) + 3\ e^+ \longrightarrow Fe(s)$
+0.771	$Fe^{3+}(aq) + e^- \longrightarrow Fe^{2+}(aq)$
+0.361	$[Fe(CN)_6]^{3-}(aq) + e^- \longrightarrow [Fe(CN)_6]^{4-}(aq)$
0	$2\ H^+(aq) + 2\ e^- \longrightarrow H_2(g)$
+1.763	$H_2O_2(aq) + 2\ H^+(aq) + 2\ e^- \longrightarrow 2\ H_2O$
+0.449	$H_2SO_3(aq) + 4\ H^+(aq) + 4\ e^- \longrightarrow S(s) + 3\ H_2O$
+0.581	$H_3AsO_4(aq) + 2\ H^+(aq) + 2\ e^- \longrightarrow HAsO_2(aq) + 2\ H_2O$
−0.276	$H_3PO_4(aq) + 2\ H^+(aq) + 2\ e^- \longrightarrow H_3PO_3(aq) + H_2O$
+0.240	$HAsO_2(aq) + 3\ H^+(aq) + 3\ e^- \longrightarrow As(s) + 2\ H_2O$
+0.854	$Hg^{2+}(aq) + 2\ e^- \longrightarrow Hg(l)$
+0.63	$2\ HgCl_2(aq) + 2\ e^- \longrightarrow Hg_2Cl_2(s) + 2\ Cl^-(aq)$
+0.2676	$Hg_2Cl_2(s) + 2\ e^- \longrightarrow 2\ Hg(l) + 2\ Cl^-(aq)$
+0.792	$Hg_2^{2+}(aq) + 2\ e^- \longrightarrow 2\ Hg(l)$
+0.535	$I_2(s) + 2\ e^- \longrightarrow 2\ I^-(aq)$
+1.20	$2\ IO_3^-(aq) + 12\ H^+(aq) + 10\ e^- \longrightarrow I_2(s) + 6\ H_2O$

-0.338	$In^{3+}(aq) + 3\ e^- \longrightarrow In(s)$
-2.924	$K^+(aq) + e^- \longrightarrow K(s)$
-2.38	$La^{3+}(aq) + 3\ e^- \longrightarrow La(s)$
-3.040	$Li^+(aq) + e^- \longrightarrow Li(s)$
-2.356	$Mg^{2+}(aq) + 2\ e^- \longrightarrow Mg(s)$
-1.18	$Mn^{2+}(aq) + 2\ e^- \longrightarrow Mn(s)$
$+1.23$	$MnO_2(g) + 4\ H^+(aq) + 2\ e^- \longrightarrow Mn^{2+}(aq) + 2\ H_2O$
$+1.70$	$MnO_4^-(aq) + 4\ H^+(aq) + 3\ e^- \longrightarrow MnO_2(s) + 2\ H_2O$
$+1.51$	$MnO_4^-(aq) + 8\ H^+(aq) + 5\ e^- \longrightarrow Mn^{2+}(aq) + 4\ H_2O$
$+0.56$	$MnO_4^-(aq) + e^- \longrightarrow MnO_4^{2-}(aq)$
-2.713	$Na^+(aq) + e^- \longrightarrow Na(s)$
-0.257	$Ni^{2+}(aq) + 2\ e^- \longrightarrow Ni(s)$
$+1.03$	$NO_2(g) + 2\ H^+(aq) + 2\ e^- \longrightarrow NO(g) + H_2O$
$+1.07$	$NO_2(g) + H^+(aq) + e^- \longrightarrow HNO_2(aq)$
$+0.956$	$NO_3^-(aq) + 4\ H^+(aq) + 3\ e^- \longrightarrow NO(g) + 2\ H_2O$
$+0.695$	$O_2(g) + 2\ H^+(aq) + 2\ e^- \longrightarrow H_2O_2(aq)$
$+1.229$	$O_2(g) + 4\ H^+(aq) + 4\ e^- \longrightarrow 2\ H_2O$
$+2.075$	$O_3(g) + 2\ H^+(aq) + 2\ e^- \longrightarrow O_2(g) + H_2O$
$+2.1$	$OF_2(g) + 2\ H^+(aq) + 4\ e^- \longrightarrow H_2O + 2\ F^-(aq)$
$+0.06$	$P(white) + 3\ H^+(aq) + 3\ e^- \longrightarrow PH_3(g)$
-0.125	$Pb^{2+}(aq) + 2\ e^- \longrightarrow Pb(s)$
$+0.28$	$PbO_2(s) + 2\ H^+(aq) + 2\ e^- \longrightarrow PbO(s) + H_2O$
$+1.455$	$PbO_2(s) + 4\ H^+(aq) + 2\ e^- \longrightarrow Pb^{2+}(aq) + 2\ H_2O$
$+1.69$	$PbO_2(s) + SO_4^{2-}(aq) + 4\ H^+(aq) \longrightarrow PbSO_4(s) + 2\ H_2O$
-0.356	$PbSO_4(s) + 2\ e^- \longrightarrow Pb(s) + SO_4^{2-}(aq)$
$+0.14$	$S(s) + 2\ H^+(aq) + 2\ e^- \longrightarrow H_2S(g)$
$+2.01$	$S_2O_8^{2-}(aq) + 2\ e^- \longrightarrow 2\ SO_4^{2-}(aq)$
$+0.08$	$S_4O_6^{2-}(aq) + 2\ e^- \longrightarrow 2\ S_2O_3^{2-}(aq)$
$+0.212$	$SbO^+(aq) + 2\ H^+(aq) + 3\ e^- \longrightarrow Sb(s)s + H_2O$
-0.137	$Sn^{2+}(aq) + 2\ e^- \longrightarrow Sn(s)$
$+0.154$	$Sn^{4+}(aq) + 2\ e^- \longrightarrow Sn^{2+}(aq)$
$+0.17$	$SO_4^{2-}(aq) + 4\ H^+(aq) + 2\ e^- \longrightarrow 2\ H_2O + SO_2(g)$
-2.89	$Sr^{2+}(aq) + 2\ e^- \longrightarrow Sr(s)$
-1.63	$Ti^{2+}(aq) + 2\ e^- \longrightarrow Ti(s)$
-1.66	$U^{3+}(aq) + 3\ e^- \longrightarrow U(s)$

−0.61	$U^{4+}(aq) + e^- \longrightarrow U^{3+}(aq)$
+0.33	$UO_2^{\ 2+}(aq) + e^- \longrightarrow UO_2^{\ +}(aq)$
−0.255	$V^{3+}(aq) + e^- \longrightarrow V^{2+}(aq)$
+0.337	$VO^{2+}(aq) + 2\,H^+(aq) + e^- \longrightarrow V^{3+}(aq) + H_2O$
+1.000	$VO^{2+}(aq) + 2\,H^+(aq) + e^- \longrightarrow VO^{2+}(aq) + H_2O$
−0.763	$Zn^{2+}(aq) + 2\,e^- \longrightarrow Zn(s)$
	Basic Solution
+0.342	$Ag_2O(s) + H_2O + 2\,e^- \longrightarrow 2\,Ag(s) + 2\,OH^-(aq)$
+0.604	$2\,AgO(s) + H_2O + 2\,e^- \longrightarrow Ag_2O(s) + 2\,OH^-(aq)$
−2.310	$Al(OH)_4^{\ -}(aq) + 3\,e^- \longrightarrow Al(s) + 4\,OH^-(aq)$
−1.21	$As(s) + 3\,H_2O + 3\,e^- \longrightarrow AsH_3(g) + 3\,OH^-(aq)$
−0.68	$AsO_2^{\ -}(aq) + 2\,H_2O + 3\,e^- \longrightarrow As(s) + 4\,OH^-(aq)$
−0.67	$AsO_4^{\ 3-}(aq) + 2\,H_2O + 2\,e^- \longrightarrow AsO_2^{\ -}(aq) + 4\,OH^-(aq)$
+0.584	$BrO_3^{\ -}(aq) + 3\,H_2O + 6\,e^- \longrightarrow Br^-(aq) + 6\,OH^-(aq)$
+0.766	$BrOH^-(aq) + H_2O + 2\,e^- \longrightarrow Br_2(l) + 2\,OH^-(aq)$
+0.455	$2\,BrO^-(aq) + 2\,H_2O + 2\,e^- \longrightarrow Br_2(l) + 4\,OH^-(aq)$
+0.622	$2\,BrO^-(aq) + 2\,H_2O + 6\,e^- \longrightarrow Br_2(l) + 6\,OH^-(aq)$
+0.17	$Co(OH)_3(s) + e^- \longrightarrow Co(OH)_2(s) + OH^-(aq)$
−0.12	$CrO_4^{\ 2-}(aq) + 2\,H_2O + 3\,e^- \longrightarrow CrO_2^{\ -}(aq) + 4\,OH^-(aq)$
−0.11	$CrO_4^{\ 2-}(aq) + 4\,H_2O + 3\,e^- \longrightarrow Cr(OH)_3(s) + 5\,OH^-$
−0.828	$2\,H_2O + 2\,e^- \longrightarrow H_2(g) + 2\,OH^-(aq)$
+0.88	$H_2O_2(aq) + 2\,e^- \longrightarrow 2\,OH^-(aq)$
−0.59	$HCHO(aq) + 2\,H_2O + 2\,e^- \longrightarrow CH_3OH(aq) + 2\,OH^-(aq)$
−0.54	$HPbO_2^{\ -}(aq) + H_2O + 2\,e^- \longrightarrow Pb(s) + 3\,OH^-(aq)$
−0.91	$HSnO_2^{\ -}(aq) + H_2O + 2\,e^- \longrightarrow Sn(s) + 3\,OH^-(aq)$
+0.42	$2\,IO^-(aq) + 2\,H_2O + 2\,e^- \longrightarrow I_2(s) + 4\,OH^-(aq)$
−2.687	$Mg(OH)_2(s) + 2\,e^- \longrightarrow Mg(s) + 2\,OH^-(aq)$
+0.1	$Mn(OH)_3(s) + e^- \longrightarrow Mn(OH)_2(s)$
+0.60	$MnO_4^{\ -}(aq) + 2\,H_2O + 3\,e^- \longrightarrow MnO_2(s) + 4\,OH^-(aq)$
+0.48	$Ni(OH)_3(s) + e^- \longrightarrow Ni(OH)_2(s) + OH^-(aq)$
+0.01	$NO_3^{\ -}(aq) + H_2O + 2\,e^- \longrightarrow NO_2^{\ -}(aq) + 2\,OH^-(aq)$
+0.401	$O_2(g) + 2\,H_2O + 4\,e^- \longrightarrow 4\,OH^-(aq)$
−0.076	$O_2(g) + H_2O + 2\,e^- \longrightarrow HO_2^{\ -}(aq) + OH^-(aq)$
+1.246	$O_3(g) + H_2O + 2\,e^- \longrightarrow O_2(g) + 2\,OH^-$
+0.890	$OCl^-(aq) + H_2O + 2\,e^- \longrightarrow Cl^- + 2\,OH^-$

−0.97	$OCN^{-}(aq) + H_2O + 2e^{-} \longrightarrow CN^{-}(aq) + 2OH^{-}(aq)$
−1.338	$Sb(s) + 3H_2O + 3e^{-} \longrightarrow SbH_3(g) + 3OH^{-}(aq)$
−0.66	$2SO_3^{2-}(aq) + 3H_2O + 4e^{-} \longrightarrow S(s) + 6OH^{-}(aq)$
−0.93	$SO_4^{2-}(aq) + H_2O + 2e^{-} \longrightarrow SO_3^{2-}(aq) + 2OH^{-}(aq)$
−1.246	$Zn(OH)_2(s) + 2e^{-} \longrightarrow Zn(s) + 2OH^{-}(aq)$

The electrode potentials in this table are arranged alphabetically by the first species in each half-equation. This can make them easier to use in some circumstances.

21 CHEMISTRY OF THE MAIN-GROUP ELEMENTS I: GROUPS 1, 2, 13 AND 14

CHAPTER OBJECTIVES

*** 1. Describe and explain physical properties of a number of representative elements in terms of their atomic properties and positions in the periodic table. Use periodic relationships to predict certain properties of the elements and their simple compounds (such as melting and boiling point).**

You should know the trends of physical and atomic properties and be able to predict relative values for different elements. These trends are summarized in Table 21-1. You should be able to apply them to all representative elements.

TABLE 21-1 Periodic Trends of Physical and Atomic Properties of Elements

Property	*Behavior down a group*	*Behavior across a period*
ionization potential	decrease	increase
atomic or cationic radius	increase	decrease
hydration energy of cation	decrease	increase
density of solid	increase	increase
electrical conductivity for metals	decrease	increase
melting points of metals	decrease	increase
melting points of nonmetals	increase	decrease

EXAMPLE 21-1 Arrange these elements in order of increasing melting point: Cl_2, Br_2, K, Mg, and Ca. Among the metals, we expect the melting point to increase from K to Ca to Mg. We know that both $Br_2(l)$ and $Cl_2(g)$ melt at lower temperatures than K. Thus, the expected order is: $Cl_2 < Br_2 < K < Ca < Mg$.

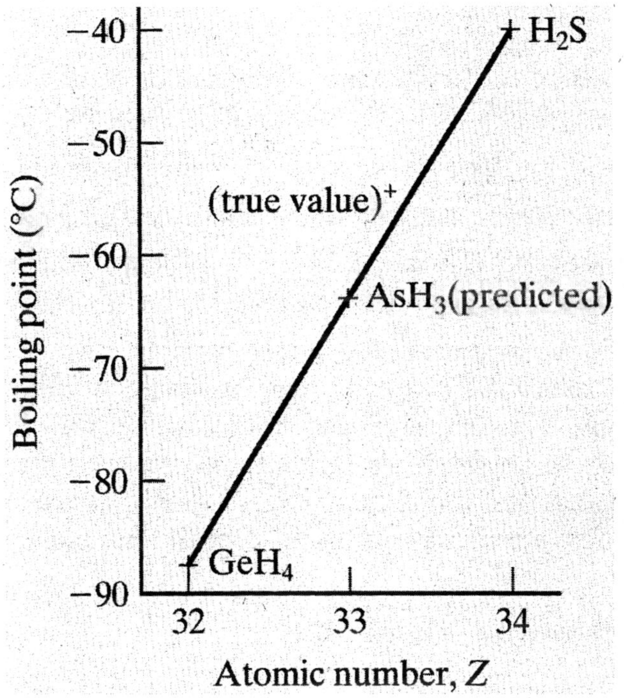

FIGURE 21-1a Boiling Point vs. Atomic number for GeH_4, AsH_3, and H_2Se.

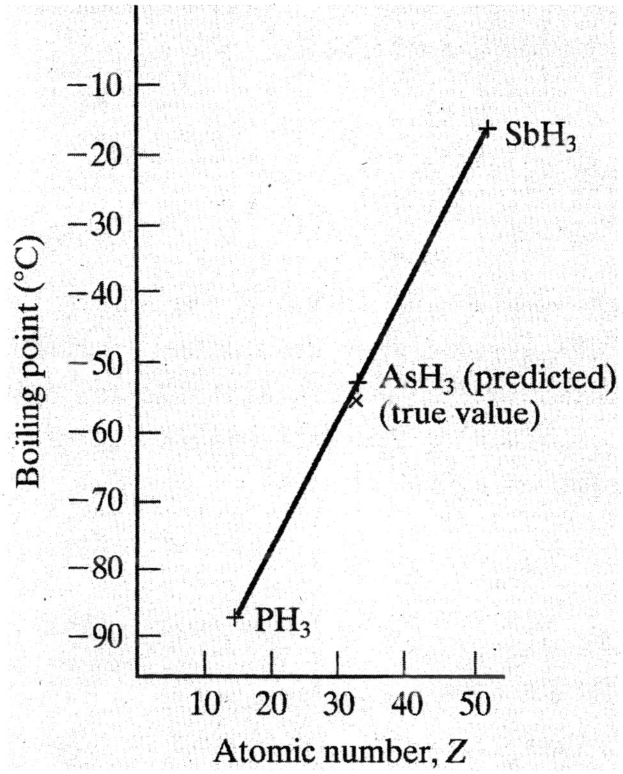

FIGURE 21-1b Boiling Point vs. Atomic Number for PH_3, AsH_3, and SbH_3.

If the values of a property are plotted against atomic number, a periodic trend often is evident. For example, the boiling points of some hydrides are $-87.7°C$ for PH_3, $-88.5°C$ for GeH_4, $-41.5°C$ for H_2Se, and $-17.1°C$ for SbH_3. Suppose we wish to predict the boiling point of AsH_3. We plot boiling point against the atomic number of the non-hydrogen atom, either within a period (Figure 21-1a) or within a family (Figure 21-1b). The predicted boiling point for AsH_3 of $-64°C$ from Figure 21-1a is not as good a prediction as the value of $-52°C$ from Figure 21-1b, because the true value is $-55°C$. In general, the family or group trend is a better predictor than the period trend, because elemental properties vary gradually within a family but often change abruptly between groups. You should master this predictive skill for representative elements. Other properties that can be predicted are density (and, hence, atomic volume), ionization potential, heats of vaporization and fusion, atomic size, and (of course) formulas of compounds. For numerical properties, one needs sufficient data for surrounding elements.

*** 2. Outline the methods that are used to produce Na, Mg, and Al.**

The raw material for the production of *sodium* is NaCl, which is either mined from underground deposits of rock salt or obtained from brine—NaCl(aq)—by evaporation. Sodium is produced by electrolysis of molten NaCl (mixed with $CaCl_2$ or Na_2CO_3 to lower the fusion point of the mixture below the boiling point of Na), Equation [1]. This reaction occurs in the Downs cell, especially designed to keep Na(1) separated from $Cl_2(g)$. Lithium and calcium are produced in similar fashion—Equations [1a] and [1b]—from chlorides that are mined.

$$2\,NaCl(l) \xrightarrow{\text{electrolysis}} 2\,Na(l) + Cl_2(g) \quad [1]$$

$$2\,LiCl\ (l,\ \text{with } KCl) \xrightarrow{\text{electrolysis}} 2\,Li(l) + Cl_2(g) \quad [1a]$$

$$CaCl_2\ (l,\ \text{with } CaF_2 \text{ or } KCl) \xrightarrow{\text{electrolysis}} Ca(l) + Cl_2(g) \quad [1b]$$

The source of *magnesium* is seawater, from which it is precipitated as $Mg(OH)_2$ by the addition of $Ca(OH)_2$, Equation [4]. (The $Ca(OH)_2$ is produced by first calcining limestone, Equation [2], and slaking the resulting lime, Equation [3]). Treatment of $Mg(OH)_2(s)$ with HCl(aq) produces an aqueous solution of $MgCl_2$, Equation [5]. The water is evaporated and the resulting $MgCl_2$ is electrolyzed to produce the metal, Equation [6].

$$CaCO_3(s) \xrightarrow{\Delta} CaO(s) + CO_2(g) \quad [2]$$

$$CaO(s) + H_2O \longrightarrow Ca(OH)_2(aq) \quad [3]$$

$$Mg^{2+}(aq) + Ca(OH)_2(aq) \longrightarrow Mg(OH)_2(s) + Ca^{2+}(aq) \quad [4]$$

$$Mg(OH)_2(s) + HCl(aq) \longrightarrow MgCl_2(aq) + 2\,H_2O \quad [5]$$

$$MgCl_2(l) \xrightarrow{\text{electrolysis}} Mg(l) + Cl_2(g) \quad [6]$$

The principal ore of *aluminum* is bauxite, which contains large quantities of SiO_2 and Fe_2O_3 as impurities. Treatment of bauxite with concentrated NaOH(aq) dissolves Al_2O_3 in the ore, Equation [7], and leaves Fe_2O_3 and SiO_2 behind. The solution is filtered and either diluted or treated with acid to precipitate $Al(OH)_3$, Equation [8]. Al_2O_3 is formed by heating the $Al(OH)_3(s)$, Equation [9]. Finally, Al_2O_3 is dissolved in molten cryolite (Na_3AlF_6) to lower its melting temperature and electrolyzed between carbon electrodes, Equation [10].

$$Al_2O_3(s) + 2\ OH^-(aq) + 3\ H_2O \longrightarrow 2\ [Al(OH)_4]^-(aq) \qquad [7]$$

$$Al(OH)_4^-(aq) + H_3O^+(aq) \longrightarrow Al(OH)_3(s) + 2H_2O \qquad [8]$$

$$2\ Al(OH)_3(s) \longrightarrow Al_2O_3(s) + 3\ H_2O(g) \qquad [9]$$

$$2\ Al_2O_3(l) + 3\ C(s) \xrightarrow{\text{electrolysis}} 4\ Al(l) + 3\ CO_2(g) \qquad [10]$$

3. Outline the Solvay process for the production of $NaHCO_3$, showing how substances are recycled in the process.

The raw materials for Na_2CO_3 in the Solvay process are NaCl (from rock salt or brine) and $CaCO_3$ (limestone). The limestone is calcined, Equation [2], to obtain $CO_2(g)$. The $CO_2(g)$ and $NH_3(g)$ are passed through an aqueous solution of NaCl, Equation [11]. The $NaHCO_3(s)$ that precipitates is heated to produce $Na_2CO_3(s)$, Equation [12]. (The CO_2 produced is recycled to yield more $NaHCO_3$.) NH_3 is recycled by treating the $NH_4Cl(aq)$ with strong base (Equation [13]), produced by slaking CaO(s), Equation [3]. The net reaction for the process is Equation [14] and the only by-product is $CaCl_2$. The Solvay process has been almost completely replaced in the United States by the mining and processing of trona, $Na_2CO_3 \cdot NaHCO_3 \cdot 2H_2O$.

$$NaCl(aq) + NH_3(g) + CO_2(g) + H_2O \longrightarrow NaHCO_3(aq) + NH_4Cl(aq) \qquad [11]$$

$$2\ NaHCO_3(s) \xrightarrow{\Delta} Na_2CO_3(s) + CO_2(g) + H_2O(g) \qquad [12]$$

$$2\ NH_4Cl(aq) + Ca(OH)_2(aq) \longrightarrow CaCl_2(aq) + 2\ NH_3(g) + 2\ H_2O \qquad [13]$$

$$2\ NaCl(aq) + CaCO_3(s) \longrightarrow Na_2CO_3(s) + CaCl_2(s) \qquad [14]$$

4. Outline the series of reactions whereby rainwater acquires temporary hardness, as well as how this hardness is removed by heating. Write chemical equations for reactions that can be used to soften temporary and permanent hard water.

When rain falls through the atmosphere, it absorbs $CO_2(g)$, forming a weakly acidic solution, Equation [15]. As this solution flows through underground limestone beds, some of the limestone dissolves, Equation [16]. Heating this solution drives off $CO_2(g)$, reversing equilibrium, Equation [15]. This in turn reverses equilibrium, Equation [16], and $CaCO_3(s)$ precipitates. Although the water is now soft, the $CaCO_3$ precipitate is undesirable if it occurs in the wrong place. For example, as $CaCO_3(s)$ builds up on the inside of a boiler, it acts as an insulator; more heat is then required to heat the water in the boiler.

$$CO_2(g) + H_2O \rightleftharpoons \text{“}H_2CO_3\text{”}(aq) \qquad [15]$$

$$CaCO_3(s) + \text{“}H_2CO_3\text{”}(aq) \rightleftharpoons Ca^{2+}(aq) + 2\,HCO_3^-(aq) \qquad [16]$$

Note that both a divalent (+2) cation ($M^{2+} = Ca^{2+}, Mg^{2+}$, or Fe^{2+}) and a divalent (–2) anion (CO_3^{2-}, SO_4^{2-}) must be present in solution in order for water to be considered hard. Permanent hard water differs from the soft variety in that its hardness cannot be removed by simple heating; the anion often is SO_4^{2-}. Chemical methods of removing hardness generally work by removing the cation, M^{2+}. The first of these is to precipitate $M(OH)_2$ by adding a strong base, Equation [17]. Slaked lime, $Ca(OH)_2$, often is used as a source of OH^- to treat municipal water supplies. Na_2CO_3 often is used in cleaning products to precipitate the cation, Equation [18]. In the second chemical method, a chelating agent, such as a polyphosphate ($P_2O_7^{4-}$) compound, is added to the water to sequester the hard-water cation as a soluble complex ion, Equation [19]. Finally, *zeolites* (naturally occurring silicate minerals) remove hard-water cations by the process of ion exchange. When water containing the divalent cations is passed through the channels in the zeolite (—*Z*—), the dipositive cations are attracted to its silicate structure and held there. The Na^+ ions originally present in the zeolite go into solution, Equation [20]. Thus, temporary hard water becomes a solution of $NaHCO_3$ (baking soda) on passage through a zeolite. The zeolite is "recharged" by passing a concentrated NaCl solution through it. The very high $[Na^+]$ forces the dipositive cations out of the zeolite by reversing equilibrium, Equation [20].

$$M^{2+}(aq) + 2\,OH^-(aq) \longrightarrow M(OH)_2(s) \quad M = Ca,\ Mg,\ \text{or}\ Fe \qquad [17]$$

$$Na_2CO_3 + H_2O + Ca(HCO_2)_2(aq) \longrightarrow 2\,Na^+(aq) + 2\,HCO_3^- + CaCO_3(s) \qquad [18]$$

$$M^{2+}(aq) + 2\,P_2O_7^{4-}(aq) \longrightarrow M(P_2O_7)_2^{6-}(aq) \qquad [19]$$

$$—Z(—Na^+)_2 + Ca^{2+}(aq) \rightleftharpoons —Z—Ca^{2+} + 2\,Na^+(aq) \qquad [20]$$

*** 5. Write equations to show the effect of H_2O; of dilute and concentrated HCl(aq), $H_2SO_4(aq)$, and $HNO_3(aq)$; and of NaOH(aq) on metals, such as those of Groups 1A and 2A (Al, Sn, Pb, Zn, Cd, and Hg).**

All alkali metals (Group 1A) and alkaline earth metals (2A), except for Be, readily react with water. The reactions of Na (Equation [21]) and Ca (Equation [22]) are typical. Mg requires steam for reaction, Equation [23]. Zn also reacts with steam in the same manner as Mg or, if the protective coating of basic zinc carbonate is removed, in the same manner as Ca. If its protective oxide coating is removed, Al also reacts with water (Equation [24]); otherwise, Al is unaffected. Cd, Hg, Bi, Sn, and Pb do not react with pure water. Pb dissolves slowly if $O_2(aq)$ is present.

$$2\,Na(s) + 2\,H_2O \longrightarrow 2\,NaOH(aq) + H_2(g) \qquad [21]$$

$$Ca(s) + 2\,H_2O \longrightarrow Ca(OH)_2(aq) + H_2(g) \qquad [22]$$

$$Mg(s) + H_2O \xrightarrow{\Delta} MgO(s) \qquad [23]$$

$$2\ Al(s) + 6\ H_2O \longrightarrow 2\ Al\ (OH)_3(s) + 3\ H_2(g)$$ [24]

All alkali metals and the alkaline earth metals, along with Zn, Cd, Sn, and Al, react with dilute HCl(aq) and $H_2SO_4(aq)$ as in Equations [25] through [27]. The reactions of the alkali metals are quite violent. Pb reacts initially, but surface coatings of $PbCl_2(s)$ and $PbSO_4(s)$ prevent further reaction. Neither Hg nor Bi reacts with dilute HCl(aq) or $H_2SO_4(aq)$. The reactions of the alkali metals and the alkaline earth metals (except Be) with concentrated HCl(aq) and $H_2SO_4(aq)$, with $HNO_3(aq)$, and with NaOH(aq) are the same as their reactions with water.

$$Be(s) + H_2SO_4(aq) \longrightarrow BeSO_4(aq) + H_2(g) \qquad [25]$$

$$Cd(s) + 2\ HCl(aq) \longrightarrow CdCl_2(aq) + H_2(g) \qquad [26]$$

$$2\,Al(s) + 6\,HCl(aq) \longrightarrow 2\,AlCl_3(aq) + 3\,H_2(g) \qquad [27]$$

In their reactions with concentrated HCl(aq), the metals Be, Zn, Cd, Sn, Pb, and Al all form chloro complexes. The reactions of Be, Zn, Cd, and Sn are similar to that of Pb, Equation [28]. The reaction of Al is given in Equation [29]. Hg and Bi do not react with concentrated HCl(aq).

$$Pb(s) + 2\ H^+ + 4\ Cl^- \longrightarrow [PbCl_4]^{2-} + H_2(g) \qquad [28]$$

$$2\,Al(s) + 6\,H^+ + 12\ Cl^- \longrightarrow 2[AlCl_6]^{3-} + 3\ H_2(g) \qquad [29]$$

The reaction of Sn with concentrated $H_2SO_4(aq)$ is similar to its reaction with dilute $H_2SO_4(aq)$. Be, Zn, Cd, Hg, and Pb all react with concentrated $H_2SO_4(aq)$ to yield $SO_2(g)$ and the divalent cation in aqueous solution, as in Equation [30] for Hg. Bi does not react with concentrated $H_2SO_4(aq)$; Al does not either because of the formation of a protective coating of Al_2O_3.

$$Hg(l) + 2\ H_2SO_4(conc.) \longrightarrow HgSO_4(aq) + SO_2\ (g) + 2\,H_2O \qquad [30]$$

In reactions with dilute $HNO_3(aq)$, moderately active metals like Zn and Be yield $N_2O(g)$ and the divalent cation in aqueous solution, Equation [31]. Less active metals yield NO(g), Equation [32], with Hg giving Hg_2^{2+}, Equation [33]. With concentrated $HNO_3(aq)$, moderately active metals give N_2 (Equation [34]), less active metals give $NO_2(g)$ (Equation [35]), and Hg yields Hg^{2+}. Aluminum does not react with dilute or concentrated $HNO_3(aq)$ because of the formation of an oxide coating.

$$4\,Zn(s) + 10\ HNO_3(aq) \longrightarrow 4\,Zn(NO_3)_2(aq) + N_2O(g) + 5\ H_2O \qquad [31]$$

$$3\ Pb(s) + 8\ HNO_3(aq) \longrightarrow 3\ Pb(NO_3)_2\,(aq) + 2\ NO(g) + 4\ H_2O \qquad [32]$$

$$6\,Hg(l) + 8\,HNO_3(aq) \longrightarrow 3\ Hg_2(NO_3)_2(aq) + 2\,NO(g) + 4\ H_2O \qquad [33]$$

$$5\,Zn(s) + 12\ HNO_3(conc.) \longrightarrow 5\,Zn(NO_3)_2(aq) + N_2(g) + 6\ H_2O \qquad [34]$$

$$Cd(s) + 4\ HNO_3(conc.) \longrightarrow Cd(NO_3)_2(aq) + 2\,NO_2(g) + 2\ H_2O \qquad [35]$$

Be, Zn, Sn, Pb, and Al all react with NaOH(aq) by forming hydroxo complexes. The reactions of Be and Zn are similar to that of Sn, Equation [36]. Al forms $[Al(OH)_4]^-$ and Pb forms $[Pb(OH)_6]^{4-}$; other than that, their reactions are similar.

$$Sn(s) + 2\ OH^-(aq) + 2\ H_2O \longrightarrow [Sn(OH)_4]^{2-}(aq) + H_2(g) \qquad [36]$$

*** 6. Write equations to represent the amphoteric nature of the oxides and hydroxides of Al, Sn, and Zn.**

By *amphoteric*, we mean that a substance reacts as an acid when treated with a strong base, and as a base when treated with a strong acid. Consider the oxide and hydroxide of aluminum as examples. When treated with strong acid, these substances react as bases.

$$2\ Al(OH)_3(s) + 3\ H_2SO_4(aq) \longrightarrow Al_2(SO_4)_3(aq) + 6\ H_2O \qquad [37]$$

$$Al_2O_3(s) + H_2SO_4(aq) \longrightarrow Al_2(SO_4)_3(aq) + 3\ H_2O \qquad [38]$$

When treated with strong bases, $Al(OH)_3$ and Al_2O_3 react as acids, producing the aluminate anion AlO_3^{3-}.

$$Al(OH)_3(s) + 3\ NaOH(aq) \longrightarrow Na_3AlO_3(aq) + 3H_2O \qquad [39]$$

$$Al_2O_3(s) + 6\ NaOH(aq) \longrightarrow 2\ Na_3AlO_3(aq) + 3\ H_2O \qquad [40]$$

The anion produced depends on the oxide that reacts with the strong base. ZnO yields $Zn(OH)_4^{2-}$, SnO yields $Sn(OH)_3^-$, and SnO_2 yields $Sn(OH)_6^{2-}$. The oxides of Pb yield hydroxo complexes of the same formula as the oxides of Sn.

7. Cite examples of ways in which the first member of a group differs from other members of the same family, including a description of the *diagonal relationship*.

The *diagonal relationship* refers to the similarity between an element of the second period and the element with one more valence electron (residing in the third period). Thus, one notes similarities between Li and Mg, Be and Al, etc. It is reasonable that the behavior of the second-period element should be somewhat different from the other members of a family. For the second-period element, the electron shell just below the valence shell has the configuration $(n-1)s^2 = 1s^2$, while, for the other elements in the family, that configuration is $(n-1)s^2p^6$. Examples of diagonal behavior include the following:

(1) Li (Family 1A) reacts with N_2 directly to form a nitride, as does Mg (Family 2A). The other alkali metals (2A) do not react directly with N_2.

(2) Li_2CO_3 thermally decomposes to the metal oxide and $CO_2(g)$, as do the alkaline earth (2A) carbonates. The other alkali metal (1A) carbonates are thermally stable. These

relative thermal stabilities also are true of the hydroxides; NaOH and KOH are thermally stable, while LiOH, $(OH)_2$, and $Ca(OH)_2$ thermally decompose to produce the metal oxide and $H_2O(g)$.

(3) The fluorides, carbonates, and phosphates of Li (1A), Mg, and the other alkaline earth (2A) elements are relatively insoluble in water. The fluorides, carbonates, and phosphates of the other alkali metals are quite soluble.

(4) Both Be (2A) and Al (3A) are unreactive toward both cold water and steam. Mg (2A) reacts with steam and Ca (2A) reacts with cold water.

(5) BeO dissolves in alkaline solutions, as does Al_2O_3. MgO and CaO do not so dissolve.

(6) Molten BeF_2 and $BeCl_2$ (2A) are poor conductors of electricity, as is also true of $AlCl_3$ (3A), but not of $MgF_2, MgCl_2, CaF_2$, and $CaCl_2$ (2A).

(7) Boron (3A) has two allotropes: amorphous and crystalline; the latter not quite as hard as diamond (4A).

(8) Boron (3A) is a semiconductor, similar to Si (4A). Al (3A) is a metal.

In addition, the properties of a second-period element often differ from other elements in the same family. However, these are not similar to the properties of the elements in the next family. Some differences include:

(1) C (Family 4A) forms compounds with long chains of atoms, unlike any other element.

(2) N_2 (5A) exhibits a quite strong triple bond.

(3) HF is a weak acid, while HCl, HBr, and HI are strong (all Family 7A).

(4) HF, H_2O, and NH_3 display strong hydrogen bonding. The hydrides of other elements do not.

8. Explain ways in which Li differs from other members of Group 1A; Tl differs from other members of 3A; Pb differs from Sn; and Hg differs from Zn and Cd.

Lithium differs from the other alkali metals in that it forms a nitride; it forms a normal oxide (Li_2O) on reaction with $O_2(g)$, rather than a superoxide or a peroxide; both its carbonate and its hydroxide can be thermally decomposed to the oxide; and its carbonate, hydroxide, fluoride, and phosphate have low solubility in water. *Thallium* differs from the other metals of Group 3A in that its monovalent cation, Tl^{+}, is more stable than its trivalent cation, Tl^{3+}; $Tl^{3+}(aq)$ is a strong oxidizing agent; Tl^{+} behaves much like K^{+} in aqueous solution; and TlOH is a strong base. *Lead* differs from Sn in that Pb^{2+} is more stable than Pb^{4+}, while Sn^{2+} and Sn^{4+} are almost equally stable; Pb(IV) compounds are good oxidizing agents, while Sn(II) compounds are reducing agents; and Pb has only the metallic allotrope, while Sn exists in three allotropic forms, of which one (gray tin) is distinctly nonmetallic with a crystal structure similar to diamond. *Mercury* differs from Zn and Cd in that Hg has less tendency to combine with oxygen; it does not dissolve in nonoxidizing acids; it forms few water-soluble compounds; it forms many covalent compounds, including a wide variety of organic mercury compounds; mercury halides are only slightly ionized in aqueous solution; and it forms Hg_2^{2+}, the only common diatomic metal cation.

*** 9. State some of the principal uses of the representative metals. Describe some of the more important reactions and uses of compounds of the representative elements.**

Most of the reactions of the representative metals and their compounds have been presented in the preceding objectives of this chapter. The remaining important reactions are presented in Table 21-2. Those reactions that your instructor emphasizes should be thoroughly mastered.

TABLE 21-2 Some Reactions of the Representative Metals and Their Compounds

Group 1A	*Group 2A*
$X: 2M(s) + X_2(g) \longrightarrow 2\,MX(s)$	$M(s) + X_2(g) \longrightarrow MX_2(s)$
$O_2: 4Li(s) + O_2(g) \longrightarrow 2Li_2O(s)$	$2M(s) + O_2(g) \longrightarrow 2MO(s)$
$2Na(s) + O_2(g) \longrightarrow Na_2O_2(s)$	Ba forms BaO_2 above 500°C
$M''(s) + O_2(g) \longrightarrow M''O_2(g)$ $M'' = K, Rb, Cs$	
$N_2: 6Li(s) + N_2(g) \longrightarrow 2Li_3N(s)$	$3M'(s) + N_2(g) \longrightarrow M_3N_2(s)$
$H_2: 2M(s) + H_2(g) \longrightarrow 2MH(s)$	$M'(s) + H_2(g) \longrightarrow M'H_2(s)$ $M' = Ca,\ Sr,\ Ba$
$H^+: 2M(s)\ 2\,H^+ \longrightarrow 2M^+ + H_2(g)$	$M(s) + 2H^+ \longrightarrow M^{2+} + H_2(g)$
$H_2O: 2\,M(s) + 2\,H_2O \longrightarrow 2M^+ + 2OH^- + H_2$	$M'(s) + 2\,H_2O \longrightarrow M'^{2+} + 2OH^- + H_2$
	$Mg(s) + H_2O \xrightarrow{\Delta} MgO(s) + H_2(g)$
$M_2O(s) + H_2O \longrightarrow 2\,MOH(aq)$	$MO(s) + H_2O \longrightarrow M(OH)_2(s)$
$M_2O(s) + CO_2(g) \longrightarrow M_2CO_3(s)$	$MO(s) + CO_2(g) \longrightarrow MCO_3(s)$
$2LiOH(s) \xrightarrow{\Delta} Li_2O(s) + H_2O(g)$	$M(OH)_2)(s) \xrightarrow{\Delta} MO(s) + H_2O(g)$
$Li_2CO_3(s) \xrightarrow{\Delta} Li_2O(s) + CO_2(g)$	$MCO_3(s) \xrightarrow{\Delta} MO(s) + CO_2(g)$
$2MHCO_3(s) \longrightarrow M_2CO_3(s) + CO_2(g) + H_2O(g)$	$M(HCO_3)_2(s) \xrightarrow{\Delta} 2MO(s) + 2\,CO_2(g) + H_2$C
$2\,MNO_3(s) \xrightarrow{\Delta} 2\,MNO_2(s) + O_2(g)$ (not $LiNO_3$)	
$4LiNO_3(s) \xrightarrow{\Delta} 2\,Li_2O(s) + 4NO_2(g) + O_2(g)$	
$2\,MOH(aq) + SO_2(aq) \longrightarrow M_2SO_3(aq) + H_2O$	
$Na_2SO_3(aq) + S \xrightarrow{\Delta} Na_2S_2O_3\ (aq)$	
Group 3A	*Group 4A*
$4\,Al(s, powder) + 3\,O_2(g) \longrightarrow 2\,Al_2O_3(s)$	$NaNO_3(l) + Pb(s) \xrightarrow{\Delta} PbO(s) + NaNO_2(s)$
$4\,LiH + AlCl_3 \longrightarrow LiAlH_4(s) + 3\,LiCl$ (in ether soln)	

Many of the important uses of the representative metals and their compounds, arranged by groups in the periodic table, are given in Table 21-3. This material should be reviewed carefully. The portions that your instructor emphasizes should be committed to memory.

TABLE 21-3 Uses of Representative Metals and Some of Their Compounds

Substance	*Important or unique uses*
	Group 1A
H_2 gas	welding gas; production of $NH_3(g)$, synthetic methanol, and hydrogenated vegetable oils
Li metal	electrical batteries; future fusion fuel (Objective 26-12)
Li_2CO_3	medication for gout and mental illness
$LiAlH_4$	reducing agent for organic reactions
Na metal	reducing agent for chemical manufacture; in Na vapor lamps; coolant in some nuclear reactors
NaOH	(caustic soda or soda lye) paper manufacturing; aluminum ore purification; chemical manufacturing; petroleum refining; soap production
Na_2CO_3	(soda) in glass; water softener in detergents
Na_2SO_4	glass, soap, and paper pulp manufacturing; dyeing textiles
$Na_5P_2O_{10}$	water softener in detergents
NaCl	in chemical manufacturing; refrigeration; household use
$NaNO_3$	manufacturing of fertilizers, KNO_3, and HNO_3
$NaHCO_3$	(bicarbonate of soda) stomach antacid; baking soda
NaCN	Photographic process; in metal electroplating baths; rat poison
KOH	Manufacturing of soft soap
K_2CO_3	(potash: gave the element its name) in hard glass; to make textile dyes, bleaches, soft soap
KCl	fertilizer production (K^+ is a plant nutrient)
KNO_3	oxidizing agent in gun powder; dye production; fertilizer
$KMnO_4$	common laboratory and small-scale industrial oxidizing agent
$KClO_3$	strong oxidizing agent in fireworks, matches, and explosives
Rb metal	to degas vacuum tubes; in photoelectric cells
Cs metal	in photoelectric cells
Cs_2SO_4	with V_2O_5 to catalyze oxidation of SO_2 to SO_3
	Group 2A
Be metal	Windows in X-ray tubes; in copper alloys that must be flexible and conduct electricity; in alloys for non-sparking safety tools; in aircraft alloys for lightness and strength
Mg metal	Lightweight, strong alloys, often with Al; flashbulbs
MgO	Furnace liner because it has a high melting point; antacid
$MgSO_4$	in sizing paper, cotton, and leather; Epsom salts
Ca metal	to remove O_2 and N_2 from molten steel; hardener for Pb alloys in storage batteries; reducing agent in Cr metallurgy
CaO, lime	in plaster and mortar; in agriculture; as a cheap base in chemical industry; basic furnace lining

$CaCO_3$	(limestone, marble, chalk) glass manufacturing; building material; flux to remove SiO_2 in metallurgy: $CaCO_3 + SiO_2 \longrightarrow CaSiO_3 + CO_2(g)$; paper whitener
$CaCl_2$	by-product of Na_2CO_3 manufacturing; to melt ice on roads; to absorb water to settle dust on roads
$CaSO_4 \cdot 2\ H_2O$	(gypsum) reversibly dehydrates to plaster of Paris
$Ca(NO_3)_2 \cdot 2\ H_2O$ and $Ca_3(PO_4)_2$	fertilizers
CaC_2	portable source of acetylene: $CaC_2(s) + 2\,H_2O \longrightarrow Ca(OH)_2(s) + HCCH(g)$
$SrCl_2$, $Sr(NO_3)_2$	red color in flares and fireworks
Ba metal	in alloys for cathodes in vacuum tubes and spark plugs
$BaSO_4$	with ZnS in the white pigment lithopone; whitener and filler in paper, rubber, and flooring tile
$Ba(NO_3)_2$, $Ba(ClO_3)_2$	green color in fireworks
$BaSiF_6$	rat poison
$BaCO_3$	insecticide
$BaCrO_4$, $BaMnO_4$, BaU_2O_7	porcelain pigments: yellow, green, and yellow or orange, respectively
	Group 13
B	to harden steel; control rods for nuclear reactors
B_4C_3	industrial abrasive
Borax	$(Na_2B_4O_7.10H_2O)$ weak base; water softener
B_2O_3	Ingredient with SiO_2 of borosilicate glass (Pyrex)
H_3BO_3	antiseptic; for fireproofing fabric
Al	Structural metal, often alloyed with Cu, Mn, Si, Zn, and Fe; "silver" paint pigment; electric transmission lines; foil; in the thermite process $(Fe_2O_3 + 2\,Al \longrightarrow 2\ Fe + Al_2O_3)$ for welding and incendiary bombs
Al_2O_3	refractory (high-temperature resistant) furnace liner; industrial abrasive; synthetic jewels
Alum	$[KAl(SO_4)_2 \cdot 12H_2O]$ mordant to fix dye to cloth; to clarify water, size paper, and fireproof fabric
$Al_2(SO_4)_3$	in fire extinguishers; paper manufacturing ("sizing"); food additives; treatment of industrial wastes and municipal waste water
Na_3AlF_6	flux for Al electrolytic reduction
Ga metal	in high-temperature thermometers
GaAs, GaP, InSb, InAs	in semiconductors
In metal	in bearing alloys; to "silver" mirrors; in nuclear control rods
Tl metal	in semiconductors
Tl_2SO_4	rat poison

	Group 14
Si	semiconductor for microcircuitry
SiO_2	glass; Portland cement; sandpaper
Na_2SiO_3	(water glass) sizing of fabric and paper; industrial glue; laundry detergent water softener (partly replacing $Na_5P_3O_{10}$)
Silicates	water-softening equipment
SiC	Industrial abrasive
Sn metal	tin plating; in solders, pewter, and bronze
$SnCl_2$	reducing agent
SnF_2	anticavity additive in toothpaste
SnO_2	jewelry abrasive
SnS_2	bronze pigment; for imitation gilding
$SnCl_4$	mordant; durable ceramic glaze
$Sn(CrO_4)_2$	rose and violet ceramic paint
Pb metal	in storage batteries, solder, and type metal; plumbing for chemical laboratories; radiation shields
PbO	in cement; pottery glaze; additive to glass to increase brilliance
Pb_3O_4	corrosion-preventing paint pigment for structural steel
$Pb(C_2H_3O_2)_4$	oxidizing agent in organic synthesis
PbO_2	in explosives and matches
$PbCrO_4$	yellow paint pigment
$Pb(C_2H_5)_4$	formerly to increase the octane number of gasoline (little used presently)
$2PbCO_3 \cdot Pb(OH)_2$	white ceramic glaze and former paint pigment

10. Explain the origin of flame colors that are characteristic of some of the elements.

Flame colors of many elements are given in Table 8-2. These colors are emitted when an excited state gives off energy in returning to the ground state. The excited state is produced when the element is raised to a high temperature, as in a Bunsen burner flame. For the alkali metals, the excited state is produced by raising the outermost *s* electron to the *p* subshell. For example, an excited state of Na is $[Ne]3p^1$. For the alkaline earth metals, the excited states are energetic and, thus, unstable (molecules, such as MgOH, CaOH, etc). In all cases, the energy difference ΔE between the excited state energy (E^*) and the ground state energy (E_0) determines the frequency and, thus, the color of the emitted light. (Violet light has high frequency, while red light has low frequency.)

$$\Delta E = E^* - E_0 = h\nu \qquad [41]$$

12. Indicate the unique features of chemical bonding in the boron hydrides.

Attempts to draw a Lewis structure of diborane (B_2H_6) fail because there are 12 valence electrons and at least seven single bonds are needed (2 electrons per bond, totaling 14 electrons).

This apparent dilemma has been resolved by recognizing that a three-center, two-electron B—H—B bond is stable. Although the valence bond theory (Objectives 11-1 through 11-5)

explanation of such a bond is somewhat complex, the molecular orbital theory (Objectives 11-6 through 11-11) explanation is clearer. When three atoms combine, three molecular orbitals form: a bonding (*b*) orbital, a nonbonding (*n*) orbital, and an antibonding (*a*) orbital. These three orbitals are shown in Figure 21-5. The two electrons of the three atoms fill the bonding orbital and create the bond. A similar explanation applies to the B—B—B bond in higher boron hydrides, such as B_5H_9, B_9H_{15}, and $B_{10}H_{14}$.

FIGURE 21-5 Molecular Orbital Diagram for the B—H—B Two-Center Three-Electron Bond (*a* indicates the antibonding, *n* the nonbonding, and *b* the bonding orbital. ↑↓ is the pair of electrons.)

13. Write reactions for methods of producing compounds of nonmetals.

Many of these reactions have been given in previous objectives in this chapter and in the previous chapter. Some additional ones follow. Elemental silicon and phosphorus can be produced from their oxides by reduction with carbon.

$$SiO_2(s) + 2\,C(s) \longrightarrow Si(l) + 2\,CO(g)\ \text{[at 3000°C in an electric furnace]} \qquad [34]$$

$$2\,Ca_3(PO_4)_2(s) + 10\,C(s) + 6\,SiO_2(s) \xrightarrow{1500°C} 6\,CaSiO_3(l) + 10\,CO(g) + P_4(g) \qquad [35]$$

Phosphorus can be oxidized by oxygen, sulfur, or halogens. Whether phosphorus trichloride or phosphorus pentachloride is produced depends on the proportion of chlorine present.

$$P_4(s) + 5\,O_2(g) \longrightarrow P_4O_{10}(s) \quad \text{AND} \quad P_4(s) + 3\,S(s) \longrightarrow P_4S_3(s) \qquad [36]$$

$$P_4(s) + 10\,Cl_2(g) \longrightarrow 4\,PCl_5(g) \quad \text{AND} \quad P_4(s) + 6\,Cl_2(g) \longrightarrow 4\,PCl_3(g) \qquad [37]$$

Both silicon carbide and sodium cyanide are made by heating appropriate compounds with carbon.

$$NaNH_2(s) + C(s) \xrightarrow{\Delta} NaCN(s) + H_2(g) \qquad [38]$$

$$SiO_2(s) + 3\,C(s) \longrightarrow SiC(s) + 2\,CO(g) \qquad [39]$$

SELF-ASSESSMENT EXERCISES

78. SiO_2 will have the highest melting point because it has both a high ionic charge and a very small atomic radius.

79. Comparing reduction potentials, PbO_2 will have the greatest tendency to be reduced and, therefore, is the best oxidizing agent.

80. Na_2CO_3 cannot be used because hard water contains carbonates that form precipitates. The carbonate here will only increase the carbonate concentration.

81. (a) Al and Ba will both react with water to produce $H_2(g)$. (c) Na_2O_2 will produce oxygen, while sodium metal will produce H_2. (d) same as (c), and (e) $NaHCO_3$ will produce carbon dioxide.

82. (a) $Li_2O + CO_2(g)$

(b) $CaCl_2 + H_2O + CO_2(g)$

(c) $2NaAl(OH)_4 + 3H_2(g)$

(d) $Ba(OH)_2$ (s)

(e) Na_2CO_3 (aq)

84. (a) $K_2CO_3 + Ba(OH)_2 \rightarrow BaCO_3 + 2KOH$

(b) $Mg(HCO_3)_2 \rightarrow 2MgO\ (s) + 2\ CO_2\ (g) + H_2O\ (aq)$

(c) $SnO + C \rightarrow Sn + CO_2\ (g)$

(d) $CaF_2(s) + H_2SO_4\ (aq) \rightarrow CaSO_4 + H_2\ (g) + F_2\ (g)$

(e) $NaHCO_3 + HCl \rightarrow NaCl + H_2O + CO_2\ (g)$

(f) $PbO_2(s) + 4HBr(aq) \rightarrow PbBr_2 + 2H_2O + Br_2$

(g) $SiF_4 + 4Na\ (s) \rightarrow 4NaF + Si(s)$

85. $CaSO_4{\cdot}2H_2O + (NH_4)_2CO_3 \rightarrow (NH_4)_2SO_4 + CaCO_3 + H_2O$

86. (a) $SrO(s) + CO_2(g)$

(b) $Al(s) + O_2(g)$

(c) $Li_2O(s) + CO_2(g)$

89. One mole of CO_2 is produced for each mole of dolomite.

5.00×10^3 kg x 10^3g/kg x 1 mol/174.4 g = 2.87×10^4 mol

$$V = \frac{nRT}{P} = \frac{2.87x10^4 \text{ mol x } 0.08206 \text{ L atm mol}^{-1}\text{K}^{-1}\text{x } 288 \text{ K}}{102 \text{ kPa x } 1 \text{ atm}/101.32 \text{ kPa}}$$

$= 6.73 \times 10^5$ L x 10^3 cm^3L^{-1} x 1 $m^3/10^6$ cm^3 = 6.73×10^2 m^3

90. Calcium will displace two H^+ ions for each calcium. 77.5 ppm is 77.5g/10^6 g.

Find the moles of H^+: 77.5 g Ca^{+2} x $2H^+/Ca^{2+}$ x 1 mol/40.08 g = 3.86 mol H^+

Find the concentration: 3.86 mol/10^6 g x 1 g/mL x 10^3 mL/L = 3.86×10^{-3} mol/L

pH = 2.4.

DRILL PROBLEMS

1. (1) Rewrite each of the following lists of species in order of increasing (smallest to largest) value of the indicated property. Base your predictions on what you have learned thus far.

A. Ionization potential: Ca, Mg, Ba, C

B. Atomic radius: K, Br, Ca, Se

C. Ionic radius: Cs^+, Li^+, K^+, Be^{2+}

D. Hydration energy: Cs^+, Al^{3+}, Mg^{2+}, Ca^{2+}

E. Ionization potential: Cs, Al, F, Sr

F. Density of solid: S, Mg, Na, I

G. Electrical conductivity: Ca, Al, Cs, Rb

H. Melting point: Al, Mg, Cs, Ca

I. Melting point: Mg, Rb, Na, Cs

J. Ionic radius: Al^{3+}, Li^+, Mg^{2+}, Si^{4+}

K. Atomic radius: Br, I, Cs, Sr

L. Hydration energy: Na^+, Al^{3+}, Mg^{2+}, K^+

M. Electrical conductivity: Na, Cs, Rb, Li

N. Melting point: Ca, Cs, Sr, Mg

O. Density of solid: P, C, S, I

(2) Use the data below to predict the melting point (°C), boiling point (°C), density of the element (g/cm^3), and density of the oxide (g/cm^3) for cesium

	Element									
	K	*Ca*	*Sc*	*Rb*	*Sr*	*Y*	*Ba*	*La*	*Cs*	
Melting point	64	839	1539	39	769	1523	725	920	P. ____	°C
Boiling point	774	1484	2832	688	1384	3337	1640	3454	Q. ____	°C
Density	0.86	1.54	2.99	1.53	2.6	4.34	3.51	6.19	R. ____	g/cm^3
Oxide density	2.32	33.8	3.86	3.72	4.7	5.01	5.72	6.51	S. ____	g/cm^3

2. Write equations for the following processes:

A. production of Mg by electrolysis
B. production of Na by electrolysis
C. production of Al by electrolysis
D. concentration of aluminum ore
E. concentration of Mg from its natural source
F. calcining of limestone
G. production of a cheap strong base from limestone

3. Complete and balance the following equations. "No reaction" may be the correct answer.

A. $Be + H_2O \longrightarrow$
B. $K + HCl(aq) \longrightarrow$
C. $Zn + HCl(aq) \longrightarrow$
D. $Pb + HCl(dil) \longrightarrow$
E. $Al + NaOH(aq) \longrightarrow$
F. $Sn + HCl(conc.) \longrightarrow$
G. $Pb + H_2SO_4(conc.) \longrightarrow$
H. $Al + HCl(conc.) \longrightarrow$
I. $Cd + H_2SO_4(aq) \longrightarrow$
J. $Hg + HNO_3(dil.) \longrightarrow$
K. $Pb + HNO_3(conc.) \longrightarrow$
L. $Be + HNO_3(dil.) \longrightarrow$
M. $Be + NaOH(aq) \longrightarrow$
N. $Bi + H_2SO_4(conc.) \longrightarrow$

4. Complete and balance the following equations:

A. $ZnO + HCl(aq) \longrightarrow$
B. $SnO_2 + H_2SO_4 \longrightarrow$
C. $ZnO + NaOH(aq) \longrightarrow$
D. $SnO + HCl(aq) \longrightarrow$
E. $Zn(OH)_2 + H_2SO_4(aq) \longrightarrow$
F. $Sn(OH)_2 + NaOH(aq) \longrightarrow$
G. $Al_2O_3 + CaO(s) + H_2O \longrightarrow$
H. $Sn(OH)_4 + H_2SO_4(aq) \longrightarrow$
I. $Sn(OH)_4 + CaO(s) + H_2O \longrightarrow$
J. $SnO_2 + NaOH(aq) \longrightarrow$

5. Give at least two substances with each of these uses. Use Table 21-3 data.

A. in glass
B. in nuclear reactors
C. in soap
D. fertilizer
E. in fireworks
F. in vacuum tubes
G. to protect metals
H. in metallurgy
I. in photoelectricity
J. in explosives
K. in medications
L. fireproofing
M. in pigments
N. poisons, disinfectants
O. in paper manufacture
P. reducing agents
Q. oxidizing agent
R. in batteries
S. furnace liners
T. in textile manufacture
U. in ceramics

QUIZZES (15–20 minutes each) Choose the best answer for each question.

QUIZ A

1. What method is *not* used to treat hard water to make it soft? (a) add CaO; (b) add Na_2CO_3; (c) boil; (d) add $(NaPO_3)_n$; (e) pass through charcoal.

2. Sea water is the source of which metal? (a) Ca; (b) Na; (c) Au; (d) I_2; (e) Mg.

3. In the Solvay process, $NH_3(g)$ is recovered by adding which of these materials? (a) NaCl(aq); (b) $CaCO_3(s)$; (c) $CO_2(g)$; (d) $Ca(OH)_2(aq)$; (e) none of these.

4. In alkaline (basic) solution, the ionic form of tin(IV) is (a) $Sn^{4+}(aq)$; (b) $Sn(OH)_4(aq)$; (c) $Sn(OH)_6^{2-}(aq)$; (d) $Sn(OH)_5^{-}$; (e) none of these.

5. Which element's flame test color is not due to an excited atom? (a) sodium; (b) lithium; (c) magnesium; (d) potassium; (e) none of these.

6. One of the unique characteristics of Li among the alkali metals is its (a) basic hydroxide; (b) thermally stable carbonate; (c) nitride formation; (d) soluble chloride; (e) none of these.

7. Which of the following is used in nuclear reactors? (a) Sn; (b) CdO; (c) In; (d) $PbCrO_4$; (e) none of these.

8. Which two elements do not display a diagonal relationship? (a) Li and Mg; (b) Be and Al; (c) B and Si; (d) N and Si; (e) none of these.

9. Which has the highest electrical conductivity? (a) Al; (b) Na; (c) Cs; (d) Ba; (e) Ca.

10. Which has the largest atomic radius? (a) Rb; (b) Sr; (c) Mg; (d) Al; (e) Li.

Quiz B

1. Which of the following does not dissolve in concentrated $HNO_3(aq)$? (a) Zn; (b) Hg; (c) Cd; (d) Al; (e) Pb.

2. Which ion must be present in temporary soft water? (a) Ca^{2+}; (b) Mg^{2+}; (c) Fe^{2+}; (d) SO_4^{2-}; (e) CO_3^{2-}.

3. Pb differs from Sn in which of the following ways? (a) Pb(IV) compounds exist; (b) Pb(II) compounds are reducing agents; (c) Pb has two allotropes; (d) Sn(II) compounds are reducing agents; (e) none of these.

4. Which element does not dissolve in $H_2SO_4(aq)$? (a) Na; (b) Zn; (c) Mg; (d) Ag; (e) Fe.

5. Which of the following is insoluble in NaOH(aq)? (a) ZnO; (b) MgO; (c) PbO; (d) Al_2O_3; (e) none of these.

6. Which of the following is not used in explosives? (a) $Hg(CNO)_2$; (b) PbO_2; (c) $KClO_3$; (d) KNO_3; (e) none of these.

7. Which element readily forms a peroxide with $O_2(g)$? (a) Li; (b) Na; (c) Rb; (d) Mg; (e) Ba.

8. Which of the following metals does not react readily with water? (a) Na; (b) Cs; (c) Mg; (d) Li; (e) K.

9. Which has the smallest density? (a) Li; (b) Cs; (c) Ba; (d) Mg; (e) Na.

10. Which has the greatest ionization potential? (a) Al; (b) Cs; (c) Ca; (d) Na; (e) Ba.

Quiz C

1. Which of the following dissolves in dilute $H_2SO_4(aq)$? (a) Hg; (b) Pb; (c) Bi; (d) Zn; (e) none of these.

2. Permanent hard water (a) softens when boiled; (b) contains CO_3^{2-}; (c) forms from limestone; (d) forms boiler scale; (e) is softened by zeolites.

3. Thallium differs from other metals of Group 3A in what way? (a) Tl_2O_3 is amphoteric; (b) Tl has only one allotrope; (c) Tl_2O is a strong reducing agent; (d) TlOH is a strong base; (e) none of these.

4. Which element does not dissolve in $HNO_3(aq)$? (a) Ag; (b) Fe; (c) Al; (d) Cu; (e) Zn.

5. The diagonal relationship is thought to be due to which of the following? (a) 2 K-shell electrons; (b) equal numbers of isotopes; (c) approximately equal atomic weights; (d) similar flame colors; (e) none of these.

6. Which of these is used in paper production? (a) SnO_2; (b) $Al_2(SO_4)_3$; (c) CaO; (d) $CaCl_2$; (e) none of these.

7. Which element dissolves in both dilute and concentrated acids and bases but not in water? (a) Ca; (b) Cd; (c) Zn; (d) Pb; (e) Hg.

8. Which has the lowest melting point? (a) Al; (b) Cs; (c) Li; (d) Mg; (e) Ba.

9. Which has the largest radius? (a) Ba; (b) Mg; (c) Be; (d) Li; (e) Al.

10. When Zn reacts with $HNO_3(aq)$, one of the products is likely to be (a) $N_2O(g)$; (b) $NO_2^-(aq)$; (c) NO(g); (d) $NO_2(g)$; (e) $H_2(g)$.

QUIZ D

1. Which of the following does not dissolve in NaOH(aq)? (a) Mg; (b) Zn; (c) Be; (d) Sn; (e) Al.

2. After hard water is passed through a zeolite, the effluent has (a) high $[CO_3^{2-}]$; (b) high $[Na^+]$; (c) high $[H^+]$; (d) low $[CO_3^{2-}]$; (e) high $[Cl^-]$.

3. In which of the following ways does Hg differ from Zn and Cd? (a) Hg_2^{2+} has a covalent bond; (b) Hg is poisonous; (c) Hg has a low melting point; (d) Hg forms several useful alloys; (e) none of these.

4. Which element reacts readily with water? (a) Al; (b) Ca; (c) Fe; (d) Cu; (e) Zn.

5. Which of the following is insoluble in dilute HCl(aq)? (a) Na; (b) Mg; (c) Al; (d) Zn; (e) none of these.

6. Which of the following is not used in glass? (a) PbO; (b) Na_2CO_3; (c) $CaCO_3$; (d) K_2CO_3; (e) all are.

7. Which element is the most inert chemically? (a) Al; (b) Cd; (c) Sn; (d) Be; (e) Pb.

8. Which of the following melts at the highest temperature? (a) Na; (b) Ba; (c) Ca; (d) Al; (e) Cs.

9. Which property decreases as one goes from left to right in a row of the periodic table? (a) density of solid; (b) ionization potential; (c) hydration energy of cation; (d) electrical conductivity for metals; (e) atomic radius.

10. Which is used as a water softener? (a) lithium carbonate; (b) sodium nitrate; (c) sodium carbonate; (d) potassium chloride; (e) hydrochloric acid.

SAMPLE TEST (40 minutes)

1. Throughout the temperature range 0°C to 2000°C, $\Delta S° = 179.5\ \text{J mol}^{-1}\ \text{K}^{-1}$ and $\Delta H° = -221$ kJ/mol for $2\ C(s) + O_2(g) \longrightarrow 2\ CO(g)$; and $\Delta S° = -138.1\ \text{J mol}^{-1}\ \text{K}^{-1}$

   and $\Delta H° = -554$ kJ/mol for $2\ Fe(s) + O_2(g) \longrightarrow 2\ FeO(s)$. What is the minimum Kelvin temperature at which CO(g) spontaneously will reduce FeO(s) to Fe(s)?

2. Write equations for the synthesis of the following substances. Use the naturally occurring compounds (those in the ores) as sources of raw materials. In addition, you may use water, $O_2(g)$, and aqueous solutions (concentrated or dilute) of HCl, HNO_3, H_2SO_4, and NaOH. Use laboratory conditions (moderate ones) rather than industrial conditions when possible.

   A. $PbCl_2(s)$ B. $Hg_2SO_4(s)$ C. $Ca_3(AlO_3)_2(s)$

   D. $MgSO_4(s)$ E. $Na_2SnO_2(s)$ F. ZnO(s) without using $O_2(g)$

   G. $BaSO_4(s)$ from $BaCO_3$ ore H. $KNO_3(s)$ from KCl ore

SYNTHESES

Try these 14 syntheses. You should give the reaction or series of reactions that will produce the desired end product from the list of reagents given. The product must be pure unless it is an aqueous solution. If the product is a solid, indicate how the solid is recovered from solution (evaporation, precipitation, or some other means). You may make any intermediates or by-products that you wish, but indicate how the by-products are separated from the desired end product. Use large-scale industrial processes if necessary. You also may use heat, light, and electricity as you need. You do not need to use all of the reagents listed.

1. $Na_2O_2(s)$ from Na(s), $O_2(g)$, and $H_2O(l)$
2. NaCl(s) from Na(s), $H_2O(l)$, and HCl(aq)
3. $Li_2CO_3(s)$ from Li(s), $H_2O(l)$, and $CO_2(g)$
4. $CaSO_4(s)$ from Ca(s), $Cl_2(g)$, $H_2O(l)$, & H_2SO_4(ac
5. $Al_2(SO_4)_3$ from Al(s), $H_2SO_4(aq)$, and $O_2(g)$
6. $H_2(g)$ from Mg(s), $NH_3(g)$, and $H_2O(l)$
7. $Na_2CO_3(s)$ from $Na_2O(s)$ and $Li_2CO_3(s)$
8. $Ca(OH)_2(s)$: $CaCO_3(s)$, $H_2O(l)$, $H_2(g)$, & $O_2(g)$
9. $CaCO_3(s)$ from Ca(s), $O_2(g)$, and C(s)
10. $H_2(g)$ from Al(s), Ca(s), $NH_3(g)$, and $H_2O(l)$
11. $CaCO_3(s)$ from Ca(s), $O_2(g)$, $H_2O(l)$, and C(s)
12. $LiNO_3(s)$ from $HNO_3(aq)$ and any elements
13. $H_2S(g)$ from Ca(s), S(s), and any strong acid.
14. HCl(g) from Mg(s), $Cl_2(g)$, and any strong acid (specify which one) except HCl.

22 MAIN-GROUP ELEMENTS II: GROUPS 18, 17, 16, 15, AND HYDROGEN

CHAPTER OBJECTIVES

1. Know the characteristics of the Group 18 elements.

The noble gases, Group 18, are generally unreactive, but a few compounds of Xe, Kr, and Rn have been prepared. Predict the electron geometry and structure of XeF_2, XeF_4, and XeF_6.

*** 2. Discuss the oxidizing and/or reducing power of the halogen elements and their oxoacids and oxoanions, oxygen, ozone, hydrogen peroxide, the oxoacids and oxoanions of sulfur and nitrogen, and the noble gas compounds.**

An examination of reduction potential diagrams (Figures 22-1, 22-2, 22-3, and 22-4) reveals that oxoanions are more powerful oxidizing agents in acid than in base. The different half-equations reveal why this is so. Consider the reduction of ClO_4^- to ClO_3^- in acidic and basic solutions, Equations [30] and [31], respectively.

$$ClO_4^- + 2\,H^+ 2\,e^- \longrightarrow ClO_3^- + H_2O \qquad [30]$$

$$ClO_4^- + H_2O + 2\,e^- \longrightarrow ClO_3^- + 2\,OH^- \qquad [31]$$

If we consider these two equations in light of Le Châtelier's principle, we notice that, in acidic solution, an excess of reactant (H^+) is present, encouraging the forward reaction. In basic solution, on the other hand, an excess of product (OH^-) is present, favoring the reverse reaction and making the forward reaction less likely. A Nernst equation analysis gives a similar result. The Nernst equations for the acidic and basic reactions are given in Equations [32] and [33].

$$E_{\text{acidic}} = E° - \frac{0.0592}{2}\log\frac{[ClO_3^-]}{[ClO_4^-][H^+]^2} \quad \text{AND} \quad E_{\text{basic}} = E° - \frac{0.0592}{2}\log\frac{[ClO_3^-][OH^-]^2}{[ClO_4^-]} \qquad [32 \,\&\, 33]$$

TABLE 22-1 Standard Reduction Potentials for Nonmetallic Species and Oxoanions in Acidic Solution.

$I_2\vert I^-$ +0.535 V	$Br_2\vert Br^-$ +1.065 V	$Cl_2\vert Cl^-$ +1358 V	$F_2\vert F^-$ +2.886 V
$HClO\vert Cl_2$ +1.62 V	$HClO_2\vert Cl_2$ +1.63 V	$ClO_3^-\vert Cl_2$ +1.46 V	$ClO_4^-\vert Cl_2$ +1.38 V
$HBrO\vert Br_2$ +1.59 V		$BrO_3^-\vert Br_2$ +1.54 V	
$HIO\vert I_2$ +1.45 V		$IO_3^-\vert I_2$ +1.19 V	
$H_2O_2\vert H_2O$ +1.763 V	$O_2\vert H_2O$ +1.229 V	$O_2\vert H_2O_2$ +0.695 V	
$SO_4^{2-}\vert SO_2$ +0.17 V	$SO_2\vert S$ +0.449 V		
$HNO_2\vert NO$ +0.99 V	$NO_3^-\vert HNO_2$ +0.94 V	$NO_3^-\vert NO_2$ +0.81 V	

$XeF_2 \mid Xe, F^-$	$+2.2$ V

Notice that a high $[H^+]$ (or a low $[OH^-]$) increases the value of E, making the oxoanion a better oxidizing agent with a more positive half-cell potential.

To compare the oxidizing power of various species, we need to look at their reduction potentials. These reduction potentials must be those of reactions that actually occur (rather than those constructed from half-reaction potentials). In addition, it is most helpful if the end product is the same for each series of species that are compared. Such data is presented in Table 22-2. Reduction potentials in acidic solution for the halogens to the halides are presented in the first line. Those for the halogen oxoanions to the halogens are in the next three lines. The next line contains those of oxygen and hydrogen peroxide, while the next two lines are those of the common oxoanions of sulfur and nitrogen. The last entry is for XeF_2.

Careful examination of the data in this table reveals the following trends.

(1) The most electronegative halogens are the strongest oxidizing agents; they have the highest reduction potentials. Thus, oxidizing power decreases in the order $F_2 > Cl_2 > Br_2 > I_2$. XeF_2 is almost as strong an oxidizing agent as F_2.

(2) For oxoanions of a given element, the oxoanion with that element in the highest oxidation state is the weakest oxidizing agent; it has the smallest reduction potential. Thus, the oxidizing power of the chlorine oxoanions decrease in the order $HClO \approx HClO_2 > ClO_3^- > ClO_4^-$. In like fashion, O_2 is a better oxidizing agent than H_2O_2 when both are considered to be reduced to H_2O.

(3) For oxoanions of elements in the same period with similar formulas, the oxoanion of the more electronegative nonmetal is the better oxidizing agent. Thus, in terms of oxidizing power, $ClO_4^- > SO_4^{2-}$ and $ClO_3^- > SO_3^{2-}$.

***3. Use electrode (reduction) potential diagrams as a means of writing half-equations and predicting oxidation-reduction equations.**

In electrode potential diagrams, the oxidized and reduced forms of an element are connected with a line. The standard reduction potential is written above that line. The two forms of the element are called a *couple*. Thus, $SO_4^{2-} \mid SO_3^{2-}$ is a couple. The couple can be expanded into a complete half-equation in the same way we balanced half-equations (see Objective 5-8). In base: $SO_4^{2-} + H_2O + 2e^- \longrightarrow SO_3^{2-} + 2OH^- \quad E° = -0.93$ V.

When we wish to combine two couples (as $SO_4^{2-} \mid SO_3^{2-}$ and $SO_3^{2-} \mid S$) to obtain the potential for a third couple ($SO_4^{2-} \mid S$), we cannot simply add potentials, because the result is a half-equation. But we *can* add values of the free energy change, $\Delta G° = -nFE°$.

$$SO_4^{2-} + H_2O + 2e^- \longrightarrow SO_3^{2-} + 2OH^- \quad E_1^\circ\{SO_4^{2-} \mid SO_3^{2-}\} = -0.93\text{ V}; \Delta G_1^\circ = -2(-0.93)\text{ F}$$

$$SO_3^{2-} + 3H_2O + 4e^- \longrightarrow S + 6OH^- \quad E_2^\circ\{SO_3^{2-} \mid S\} = -0.67\text{ V}; \Delta G_2^\circ = -4(-0.67)\text{ F}$$

$$SO_4^{2-} + 4H_2O + 6e^- \longrightarrow S + 8OH^- \quad \Delta G_3^\circ = 2(0.93)\text{ F} + 4(0.67)\text{ F} = -6\,FE_3^\circ$$

Thus, $-6E_3^\circ = 4.54\,\text{V}$ *or* $E_3^\circ = -4.54\,\text{V}/6 = -0.76\,\text{V}$. Several electrode potential diagrams follow: Figure 22-1 for oxygen, Figure 22-2 for sulfur, Figure 22-3 for nitrogen, and Figure 22-4 for chlorine.

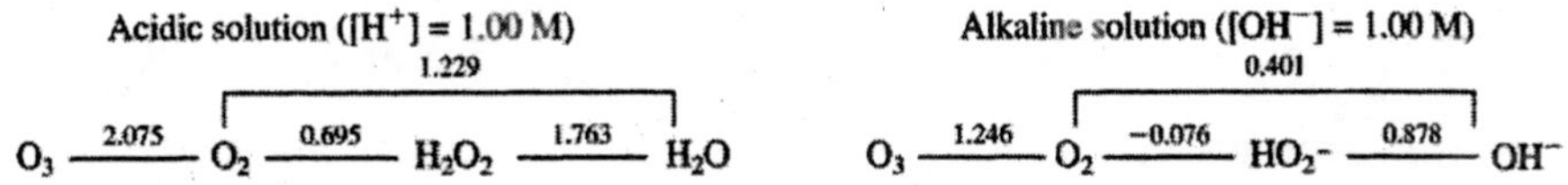

FIGURE 22-1 Standard Electrode Potential Diagrams for Oxygen

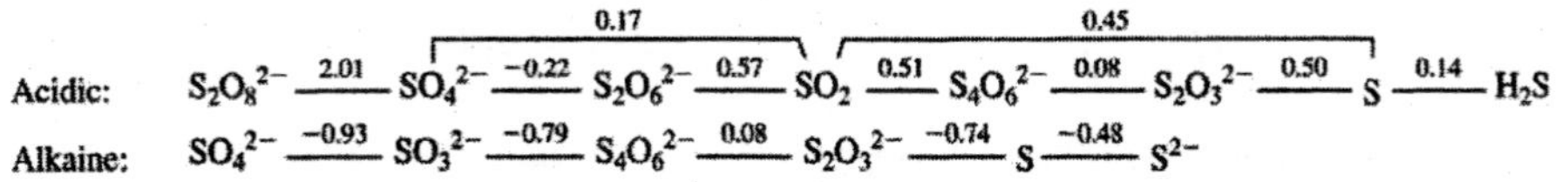

FIGURE 22-2 Standard Electrode Potential Diagrams for Sulfur

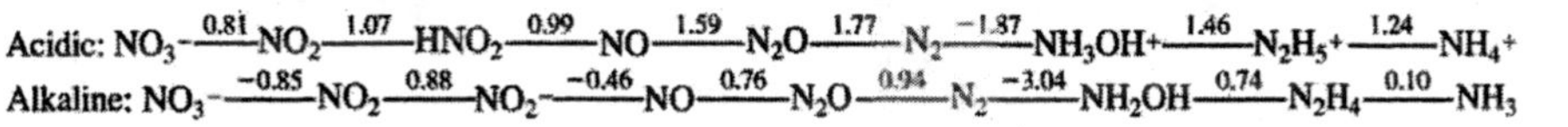

FIGURE 22-3 Standard Electrode Potential Diagrams for Nitrogen

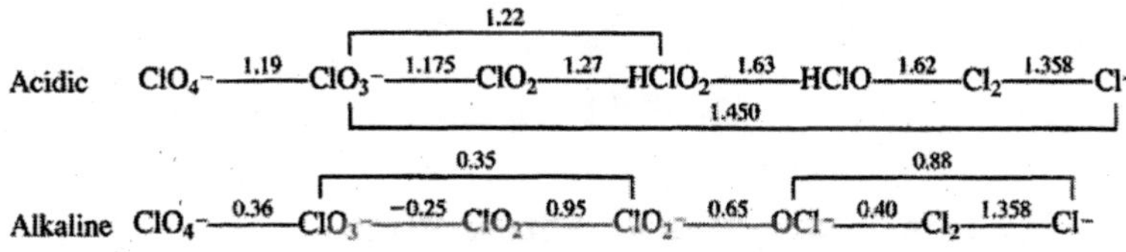

FIGURE 22-4 Standard Electrode Potential Diagrams for Chlorine

*** 4. Outline the principal methods of preparing the halogen elements, the hydrogen halides, and the oxoacids and oxoanions of the halogens.**

Methods of preparing F_2 and Cl_2 are given in Equations [1] and [21-1], respectively. *Bromine* is extracted from Br^- in seawater with Cl_2 as an oxidizing agent, Equation [2]. An abundant natural source of *iodine* is $NaIO_3$, from which it is extracted by a reducing agent, Equation [3], in which all species are aqueous.

$$2\,\text{HF(l, in molten KHF}_2) \xrightarrow{\text{electricity}} \text{H}_2(\text{g}) + \text{F}_2(\text{g}) \qquad [1]$$

$$\text{Cl}_2(\text{g}) + 2\,\text{Br}^-(\text{aq}) \longrightarrow \text{Br}_2(\text{l}) + 2\,\text{Cl}^-(\text{aq}) \qquad [2]$$

$$\text{IO}_3^- + 3\,\text{HSO}_3^- \longrightarrow \text{I}^- + 3\,\text{SO}_4^{2-} + 3\,\text{H}^+ \text{ AND}$$

$$5\,\text{I}^- + \text{IO}_3^- + 6\text{H}^+ \longrightarrow 3\,\text{I}_2 + 3\,\text{H}_2\text{O} \qquad [3]$$

The *hydrogen halides* can be prepared in several ways. Direct union of hydrogen with the halogen, Equation [4], occurs explosively with F_2, explosively with Cl_2 in the presence of light, at a moderate

rate with Br_2, and slowly with I_2. Combination of the halide ion with hydrogen ion is another method. For HF, almost any acid is suitable, Equation [5]. (This is why acids should not be stored near fluoride salts; HF(g) is exceedingly toxic.) For the other hydrogen halides, a rich source of protons (a concentrated strong acid) is needed. H_2SO_4(conc.) is suitable for HCl, Equation [6], but not for HBr and HI, as H_2SO_4 will oxidize Br^- or I^- to the element, Equation [7]. Thus, H_3PO_4(conc.) is used to produce HBr and HI from bromide and iodide salts, Equation [8]. Finally, many hydrogen halides can be produced by the hydrolysis of nonmetal halides, Equations [9] and [10].

$$H_2(g) + X_2(g) \longrightarrow 2\,HX(g) \qquad [4]$$

$$NaF(s) + HC_2H_3O_2(aq) \longrightarrow NaC_2H_3O_2(aq) + HF(g) \qquad [5]$$

$$NaCl(s) + H_2SO_4(conc.) \longrightarrow NaHSO_4(aq) + HCl(g) \qquad [6]$$

$$2\,HI + H_2SO_4 \longrightarrow 2\,H_2O + I_2(g) + SO_2(g) \qquad [7]$$

$$NaI(s) + H_3PO_4(conc.) \longrightarrow NaH_2PO_4(aq) + HI(g) \qquad [8]$$

$$PBr_5(g) + 4\,H_2O \longrightarrow H_3PO_4(aq) + 5\,HBr(g) \qquad [9]$$

$$SiCl_4(g) + 2\,H_2O \longrightarrow SiO_2(s) + 4\,HCl(g) \qquad [10]$$

The preparations of the *oxoacids and oxoanions* of chlorine are typical of the preparations of the other halogens. The hypohalous acids (HOX) are prepared by adding the element to water, Equation [11]. The equilibrium is shifted to the right and the hypohalite ion (OX^-) in aqueous solution is prepared if dilute cold base is added to the solution, Equation [12]. All of the hypohalites are thermally unstable. This instability is used to advantage in preparing the halates (XO_3^-); the element is added to a hot, concentrated solution of strong base, Equation [13]. All of the halic acids can be prepared in the same manner as chloric acid by reaction of $Ba(ClO_3)_2$ with H_2SO_4, Equation [14].

$$X_2 + H_2O \rightleftharpoons HOX(aq) + HX(aq) \qquad [11]$$

$$X_2 + OH^-(aq) \longrightarrow XO^-(aq) + X^-(aq) + H_2O \qquad [12]$$

$$3\,X_2 + 6\,OH^-(aq) \longrightarrow XO_3^-(aq) + 5\,X^-(aq) + 3\,H_2O \qquad [13]$$

$$Ba(ClO_3)_2(aq) + H_2SO_4 \longrightarrow 2\,HClO_3(aq) + BaSO_4(s) \qquad [14]$$

Chlorous acid can be prepared from a suspension of $Ba(ClO_2)_2$ by reaction with H_2SO_4, Equation [15]. Chlorites are formed by the disproportionation of $ClO_2(g)$ in water, Equation [16], and $ClO_2(g)$, in turn, is prepared by the reaction of $HClO_3$ with oxalic acid, Equation [17]. Finally, perchloric acid is produced by the electrolysis of perchlorate salts, half-equation [18], followed by distilling the resulting mixture in the presence of H_2SO_4 under reduced pressure (since $HClO_4$ explodes above 92°C), Equation [19].

$$Ba(ClO_2)_2(aq, susp.) + H_2SO_4(aq) \longrightarrow BaSO_4(s) + 2\,HClO_2(aq) \qquad [15]$$

$$2\,ClO_2 + 2\,OH^- \longrightarrow ClO_2^- + ClO_3^- + H_2O \qquad [16]$$

$$2\ HClO_3(aq) + H_2C_2O_4(s) \longrightarrow 2\ ClO_2(g) + 2\ CO_2(g) + 2\ H_2O \qquad [17]$$

$$ClO_3^-\ (aq) + H_2O \longrightarrow ClO_4^-(aq) + 2\ H^+(aq) + 2e^- \quad E° = -0.36\ V \qquad [18]$$

$$ClO_4^-\ (aq) + H_2SO_4(conc.) \longrightarrow HSO_4^-(aq) + HClO_4(g) \qquad [19]$$

***5. Describe bonding and structures of the oxoanions of the halogens, sulfur, and nitrogen. Predict the shapes of interhalogen compounds, polyhalide ions, and noble gas compounds.**

To predict the shapes of oxoanions, interhalogens, polyhalide ions, and noble gas compounds, one uses the VSEPR theory (recall Objective 10-12). This requires first drawing a plausible Lewis structure (recall Objectives 10-6 through 10-10). The total number of electron pairs (= ligands + lone pairs) determines the electron-pair geometry of the molecule and the hybridization of the central atom (see Table 11-1). The number of ligands and the number of lone pairs determines the molecular shape (see Table 10-4 and Figure 10-2). You should be aware of one additional hybridization: sp^3d^3 in IF_7 with a pentagonal bipyramid electron-pair geometry and molecular shape.

6. Discuss the significance of sulfur dioxide emissions.

Acid rain is produced by the following series of reactions:

$SO_2(g) + NO_2(g) \rightarrow SO_3(g) + NO(g)$ and

$SO_2(g) + \frac{1}{2}O_2(g) \rightarrow SO_3(g)$

The SO_3 formed react with H_2O:

$SO_3(g) + H_2O \rightarrow H_2SO_4(aq)$

The sulfuric acid forms as tiny droplets high in the atmosphere and falls to earth as acid rain. Acid rain may be harmful to vegetation and the environment.

7. Cite similarities and differences between the first and second members of a group in the periodic table.

Here we are comparing the second-period elements F, O, N, C, and B with the third-period elements Cl, S, P, Si, and Al. The first evident difference is that the second period elements are the more nonmetallic; they have the more positive reduction potentials and, thus, are the better oxidizing agents. Second, the third-period elements are able to form ions and compounds that have more ligands, largely because of the availability of *d* orbitals in those elements. For example, Cl_2, ClF, ClF_3, ClF_5, ClO^-, ClO_2^-, ClO_3^-, and ClO_4^- are all well characterized, while the only analogous fluorine species are F_2 and HFO (the latter having been prepared only during the 1970s). Third, the hydrides of the second-period nonmetals (with the exceptions of those of C and B) form strong hydrogen bonds while those of the third-period elements do not. Finally, the oxoanions of the third-period elements readily join together ($P_2O_7^{4-}$, $S_2O_7^{2-}$, and silicates); this is not so common with second-period elements.

These differences are not as important as the similarities of elements in the same family. Cl_2 behaves more like F_2 than it does like O_2, despite the weak diagonal relationship between O_2 and Cl_2. These similarities are most evident in the organization of how we study the elements. Both Chapters 21 and 22 in the *text* are organized by families in the periodic table.

8. List several methods of preparing oxygen.

Oxygen is prepared commercially by the fractional distillation of liquid air. This method requires a large capital investment and, in general, is not practical for laboratory work. Several methods that are suitable for laboratory preparations include: (1) heating oxides of metals of low reactivity, Equation [20]; (2) heating certain oxygen-containing salts, Equation [21]; (3) decomposition of water by a thermochemical cycle or by electrolysis, Equation [22]; (4) reaction of an ionic peroxide or a superoxide with water, Equation [23]; and (5) decomposition or oxidation of hydrogen peroxide, Equation [24].

$$2\,HgO(s) \xrightarrow{\Delta} 2\,Hg(l) + O_2(g) \quad AND \quad 2\,Ag_2O(s) \xrightarrow{\Delta} 4\,Ag(l) + O_2(g) \qquad [20]$$

$$2\,KClO_3(s) \xrightarrow{\Delta} 2\,KCl(s) + 3\,O_2(g) \qquad [21]$$

$$2\,H_2O \xrightarrow{\Delta} 2\,H_2(g) + O_2(g) \quad AND \quad 2\,H_2O \xrightarrow{electrolysis} 2\,H_2(g) + O_2(g) \qquad [22]$$

$$2\,O_2^{2-} + 2\,H_2O \longrightarrow 4\,OH^-(aq) + O_2(g) \quad AND \quad 4\,O_2^- + 2\,H_2O \longrightarrow 4\,OH^-(aq) + 3\,O_2(g) \qquad [23]$$

$$2\,H_2O_2(aq) \longrightarrow 2\,H_2(g) + O_2(g) \quad AND$$
$$5\,H_2O_2 + 2\,MnO_4^- + 6\,H^+ \longrightarrow 2\,Mn^{2+} + 8\,H_2O + 5\,O_2(g) \qquad [24]$$

9. Describe the acid-base and oxidation-reduction properties of NH_3, N_2H_4, and NH_2OH.

All three of these compounds are weak bases, Equations [25] through [28], and are built around an NH_2— core:

NH_2—H, NH_2—NH_2, and NH_2 — OH.

$$NH_3(aq) + H_2O \rightleftharpoons NH_4^+(aq) + OH^-(aq) \qquad K_b = 1.8 \times 10^{-5} \qquad [25]$$

$$N_2H_4(aq) + H_2O \rightleftharpoons N_2H_5^+(aq) + OH^-(aq) \qquad K_{b1} = 8.5 \times 10^{-7} \qquad [26]$$

$$N_2H_5^+(aq) + H_2O \rightleftharpoons N_2H_6^+(aq) + OH^-(aq \qquad K_{b2} = 8.9 \times 10^{-16} \qquad [27]$$

$$NH_2OH(aq) + H_2O \rightleftharpoons NH_3OH^+(aq) + OH^-(aq) \qquad K_b = 9.1 \times 10^{-9} \qquad [28]$$

NH_3 is always a reducing agent. NH_2OH acts as either an oxidizing agent ($E° = +1.35$ V for $NH_3OH^+ \mid NH_4^+$ in acidic solution and $E° = +0.42$ V in basic solution for $NH_2OH \mid NH_3$) or as a reducing agent ($E° = +1.87$ V for $NH_3OH^+ \mid N_2$ in acidic solution and $E° = +3.04$ V for $NH_2OH \mid N_2$ in basic solution). Likewise, N_2H_4 acts as either an oxidizing agent ($E° = +1.24$ V for $N_2H_5^+ \mid NH_4^+$ in acidic solution and $E° = +1.15$ V for $N_2H_4 \mid N_2$ in basic solution) or as a reducing agent ($E° = +0.20$ V for $N_2H_5^+ \mid N_2$ in acidic solution and $E° = +1.15$ V for $N_2H_4 \mid NH_3$ in basic solution). The ability of N_2H_4 (hydrazine) to act as a reducing agent accounts for its use as a rocket fuel in combination with hydrogen peroxide, Equation [29].

$$N_2H_4(l) + 2\,H_2O_2(l) \longrightarrow N_2(g) + 4\,H_2O(g) \qquad [29]$$

10. Outline methods of preparing the oxides of nitrogen.

Some methods of preparing the several oxides of nitrogen are summarized in Table 22-1. Notice that two of the nitrogen oxides are produced by thermal decomposition: $N_2O(g)$ from $NH_4NO_3(s)$ and $NO_2(g)$ from the nitrate of an inactive metal, $Pb(NO_3)_2$. Two of the nitrogen oxides are the products of nitric acid reactions: NO(g) from the reaction of moderately concentrated HNO_3 with Cu(s) and $N_2O_5(g)$ as a dehydration product of HNO_3 with $P_4O_{10}(s)$ as a dehydrating agent. Finally, three of the nitrogen oxides are formed by direct combination of two gases: $N_2O_3(g)$ from $NO_2(g)$ and NO(g), $NO_2(g)$ from NO(g) and $O_2(g)$, and $N_2O_4(g)$ from $NO_2(g)$ at low temperature and moderate pressures.

TABLE 22-2 Some Methods of Preparation of Nitrogen Oxides

N_2O	$NH_4NO_3(s) \xrightarrow{\Delta} N_2O(g) + 2\ H_2O(l)$		
NO	$3\ Cu(s) + 8\ HNO_3(aq) \longrightarrow 3\ Cu(NO_3)_2(aq) + 2\ NO(g) + 4\ H_2O$		
N_2O_3	$NO(g) + NO_2(g) \rightleftharpoons N_2O_3(g)$	$K_p = 0.48$ at 298 K	
NO_2	$2\ Pb(NO_3)_2 \xrightarrow{\Delta} 2\ PbO(s) + 4\ NO_2(g) + O_2(g)$		
	$2\ NO(g) + O_2(g) \rightleftharpoons 2\ NO_2(g)$	$K_p = 2.3 \times 10^{12}$ at 298 K	$\Delta H° = -114$ kJ/mol
N_2O_4	$2\ NO_2(g) \rightleftharpoons N_2O_4(g)$	$K_p = 0.48$ at 298 K	$\Delta H° = -57$ kJ/mol
N_2O_5	$4\ HNO_3(l) + P_4O_{10}(s) \longrightarrow 4\ HPO_3 + 2\ N_2O_5$		

11. Describe the different molecular or physical forms of helium, carbon, oxygen, sulfur, and phosphorus, as well as the physical behavior associated with them.

Above its critical point of 5.3 K, *He* is gaseous, with a boiling point of 4.2 K. Lowering the temperature of liquid, He at 1.00 atm pressure does not yield the solid, but rather a superfluid liquid phase with zero viscosity and a high thermal conductivity. Solid He is produced by applying 25 atm pressure to the initial normal liquid phase.

At normal pressure and temperature, carbon and oxygen each exist in two allotropic forms. In the case of *carbon*, these two forms—graphite (2.267 g/cm^3) and diamond (3.515 g/cm^3)—are solids with different crystal structures arising from different types of bonding. In graphite, sp^2 orbitals bond carbon atoms together into sheets, with the remaining valence electron on each carbon atom located in a $2p$ orbital perpendicular to the plane of atoms. These $2p$ orbitals combine to form molecular orbitals that extend over the entire sheet. This bonding scheme is reflected in the physical properties of graphite: good electrical conductivity in the direction of the sheets, but rather poor electrical conductivity perpendicular to them (the *p* electrons in the molecular orbitals are free to move); and good lubricating properties (the sheets can slide across each other). Diamond, in contrast, is composed of sp^3-hybridized carbon atoms. The crystal, thus, is a three-dimensional network of covalently bonded carbon atoms. The physical properties of diamond—high melting point, extreme hardness, inability to conduct an electrical current—reflect the fact that the crystal is held together entirely by covalent bonds, and there are no mobile electrons.

The two allotropes of *oxygen* are O_2, sometimes known as dioxygen and O_3 (ozone). O_3, produced from O_2 with either an electric discharge or ultraviolet light as an energy source, has a sharp, pungent odor and is an excellent oxidizing agent.

The two solid allotropes of *sulfur* are rhombic (2.08 g/cm^3) and monoclinic (1.96 g/cm^3). Rhombic crystals are somewhat squat, although we normally see this form as a powder—flowers of sulfur. Long, needle-shaped monoclinic crystals are unstable at room temperature, and slowly convert to rhombic crystals. The S_8 molecules of the solid allotropes also are present in the low-temperature melt, which is relatively fluid. Higher temperatures produce S_8 and longer chains, which entangle, producing a more viscous liquid. At still higher temperatures, the chains fragment, making the liquid less viscous.

Phosphorus is composed of tetrahedral P_4 molecules in white phosphorus, in the liquid, and in the vapor below 800°C. White phosphorus (1.828 g/cm^3) is a highly reactive, waxy solid, melting at 44.1°C. It burns spontaneously in air and is poisonous, very volatile, and soluble in CS_2(l). Red phosphorus (2.34 g/cm^3) is very much less reactive and less volatile than white phosphorus; it is insoluble in CS_2(l) and nonpoisonous. Black phosphorus is formed when white or red phosphorus is heated under pressure (200°C, 4000 atm) or in the presence of a catalyst. It is very unreactive, does not ignite in air below 400°C, is a semiconductor, and has a relatively high density (2.65 g/cm^3). The structure, consisting of pleated sheets of phosphorus atoms and black phosphorus, is somewhat flaky like graphite.

*** 12. Cite the uses of some of the compounds discussed in this chapter.**

Many of the important industrial uses of the representative elements and their compounds are given in Table 12-3. However, uses for some additional compounds have been introduced in this chapter. These compounds, arranged by groups in the periodic table, are given in Table 22-3. That material should be reviewed carefully and the portions that your instructor emphasizes should be committed to memory.

TABLE 22-3 Uses of Some of the Nonmetallic Elements and Some of Their Compounds.

Substance	*Important or Unique Uses*
	Group 7A
F_2	manufacture of fluorocarbons as lubricants and nonstick coatings (Teflon).
HF	etching glass ($SiO_2 + 4\,HF \longrightarrow SiF4 + 2H_2O$)
NaF	insecticide; rat poison; very low concentrations in drinking water prevent tooth decay.
Cl_2	water purification; plastic manufacturing; solvent production; pulp and paper bleach; poison gas (bertholite) in World War I
HCl	cleaning metals and masonry; removing boiler scale; refining some ores
ClF_3, BrF_3	fluorination reagents
UF_6	in the separation of uranium isotopes by gaseous diffusion
Ca(OCl)Cl	(bleaching powder) produces Cl_2(g) when moistened
NaClO	commercial bleach; swimming pool disinfectant
$NaClO_2$	textile bleach
$KClO_3$	oxidant in fireworks; weed killer
$HClO_4$	metal finishing
$KClO_4$	in detonators and explosives
NH_4ClO_4	in solid propellants

Br_2	to manufacture fire retardants
HBr	used to make inorganic and organic bromides
$C_2H_4Br_2$	gasoline additive; fumigant; pesticide
$AgBr$	light-sensitive agent in photographic film
I_2	antiseptic; in medications for the thyroid gland; to analyze laboratory reducing agents
ICl	organic iodination reagent
$NaIO_3$	natural source of iodine
AgI	photography
HCN	(pseudo hydrogen halide) poison gas; in the preparation of synthetic fibers (Orlon)

Group 6A

O_2	steel refining; high-temperature torches
O_3	industrial oxidant; water purification. [formed by electric discharge through O_2]
Na_2O_2	paper and fabric bleach
H_2O_2	germicide; bleach
S	Manufacturing of H_2SO_4; vulcanization of rubber; in fireworks, gunpowder, and matches
H_2SO_4	principal product of the chemical industry; make fertilizers and plastics; clean steel
SO_2	refrigerant; bleach; food preservative
Na_2SO_3	bleaches for natural fabrics and paper
$Na_2S_2O_3$	("hypo") photography; to remove excess Cl_2 used to bleach fabric and paper, analytical chemical
$Na_2S_2O_4$	bleach for dyes
$S_2O_8^{2-}$	production of H_2O_2 by hydrolysis
Se	in photoelectric devices and photocopying machines

Group 5A

N_2 gas	inert atmosphere for air-sensitive reactions and working active metals
NH_3	refrigerant, fertilizer
N_2O	(laughing gas) inhalation anesthetic
HNO_3	to make nitrates for fertilizers, plastics, dyes, and explosives
$NaNO_2$	to make dyes; as a meat preservative
KNO_3	gunpowder ingredient
$AgNO_3$	photography
NH_4NO_3	fertilizer; explosive (amatol is a mixture of NH_4NO_3 and TNT, trinitrotoluene)
NH_4Cl	as a flux to clean oxides from metal surfaces in galvanizing and soldering
N_2H_4	with H_2O_2 as a rocket propellant

$NaNH_2$	dehydrating agent; in synthesis of N-containing organic compounds
NO_2	in the preparation of explosives and organic nitrates
$Pb(N_3)_2$	in detonators
P	red allotrope in the striking strip of safety matches; smoke bombs; tracer bullets; burned in air to make P_4O_{10}
P_4S_3	in the tips of kitchen matches
P_4O_{10}	laboratory drying agent; with H_2O it forms H_3PO_4
$Ca(H_2PO_4)_2$	(triple superphosphate) fertilizer
H_3PO_4	fertilizer manufacture; in soft drinks; in dyeing
$Na_5P_3O_{10}$	cleaning agent, detergent builder, water softener; in cement manufacturing; in oil well drilling
$(NaPO_3)_n$	water softener
Bi metal	with Sn, Pb, and Cd in low-melting alloys; in casting alloys; to "silver" mirrors
$BiONO_3$	to treat gastrointestinal infections and ulcers, and skin infections such as eczema
BiOCl	white pigment; in face powders
Bi_2O_3	disinfectant; to fireproof fabrics and plastics
$(BiO)_2CO_3$	in white opalescent ceramic glazes; also pharmaceutically as for $BiONO_3$

TABLE 22-4 Equations for the Preparation and Uses of N_2, O_2, H_2, NH_3, CO, and CO_2

N_2 *uses*	$N_2(g) + 3H_2(g) \rightleftharpoons 2NH_3(g)$	400°C, 200 atm, catalyst. NH_3 *preparation*
	$N_2(g) + O_2(g) \longrightarrow 2NO(g)$	in high-temperature combustion; in lightning
NH_3 *uses*	$2NH_3(aq) + H_2SO_4(aq) \longrightarrow (NH_4)_2SO_4(aq)$	
	$NH_3(aq) + HCl(aq) \longrightarrow NH_4Cl(aq)$	
	$NH_3(aq) + HNO_3(aq) \longrightarrow NH_4NO_3(aq)$	
	$NH_4NO_3(s) \longrightarrow N_2O(g) + 2H_2O(g)$	200 – 260°C
	$NH_3(aq) + H_3PO_4(aq) \longrightarrow NH_4H_2PO_4$ and $(NH_4)_2HPO_4$	
	$2NH_3(g) + CO_2(g) \longrightarrow CO(NH_2)_2(s) + H_2O$	
HNO_3 *prep*	$4NH_3(g) + 5O_2(g) \longrightarrow 4NO(g) + 6H_2O(g)$	850°C, Pt catalyst
	$2NO(g) + O_2(g) \longrightarrow 2NO_2(g)$	
	$3NO_2(g) + H_2O(g) \longrightarrow 2HNO_3(aq) + NO(g)$	
O_2 *prep*	$2KClO_3(s) \longrightarrow 2KCl(s) + 3O_2(g)$	heated, MnO_2 catalyst.

	$4KO_2(g) + 2CO_2(g) \longrightarrow 2K_2CO_3(s) + 3O_2(g)$	$O_2(g)$ regeneration in submarines
	$2H_2O(l) \longrightarrow 2H_2(g) + O_2(g)$	electrolysis, dissolved $H_2SO_4(aq)$
O_2 *uses*	$3O_2(g) \longrightarrow 2O_3(g)$	electric discharge needed
	$2C_8H_{18}(l) + 25O_2(g) \longrightarrow 16CO_2(g) + 18H_2O(l)$	complete combustion example
CO_2 *prep*	$CaCO_3(s) \longrightarrow CaO(s) + CO_2(g)$	at about 900°C
	$C_6H_{12}O_6(aq) \longrightarrow 2C_2H_5OH(aq) + 2CO_2(g)$	fermentation
	$C(s) + O_2(g) \longrightarrow CO_2(g)$	
& H_2 *prep*	$CH_4(g) + 2H_2O(g) \xrightarrow{\Delta} CO_2(g) + 4H_2(g)$	steam reforming
	$CO(g) + H_2O(g) \xrightarrow{\Delta} CO_2(g) + H_2(g)$	a water gas reaction
CO & H_2	$CH_4(g) + 2H_2O(g) \xrightarrow{\Delta} CO(g) + 3H_2(g)$	steam reforming
prep	$C(s) + H_2O(g) \xrightarrow{\Delta} CO(g) + H_2(g)$	a water gas reaction
CO *uses*	$CO(g) + 2H_2(g) \longrightarrow CH_3OH(l)$	
	$Fe_2O_3(s) + 3CO(g) \xrightarrow{\Delta} 2Fe(l) + 3CO_2(g)$	also for other metal oxides
H_2 *prep*	$Zn(s) + 2H^+(aq) \longrightarrow Zn^{2+}(aq) + H_2(g)$	laboratory scale preparation
H_2 *uses*	$H_2(g) + Cl_2(g) \longrightarrow 2HCl(g)$	
	$2M(s) + H_2(g) \longrightarrow 2MH(s)$ OR $M'(s) + H_2(g) \rightarrow M'H_2(s)$ ($M = 1A$ metal, $M' = Ca, Sr, Ba$)	
	$2H^-(s) + 2H_2O(l) \longrightarrow 2OH^-(aq) + 2H_2(g)$	($M = 1A$ metal, M' = Ca, Sr, Ba)

***13. Describe the uses of the noble gases N_2, O_2, H_2, NH_3, CO, CO_2, and some of their compounds.**

The uses of these gases are taken from the *text* and listed in Table 8-1. The noble gases are primarily useful because of their inertness. The uses of nitrogen compounds are often based on either the many oxidation states displayed by nitrogen in its compounds or on the need of plants and animals for nitrogen, supplied in the form of fertilizers and feed supplements. Oxygen's utility depends on its oxidizing power, while those of H_2 and CO depend on their reducing ability.

***14. Write chemical equations for the preparation and uses of N_2, O_2, H_2, NH_3, CO, and CO_2.**

Both $N_2(g)$ and $O_2(g)$ are prepared by the fractional distillation of liquid air, although gaseous diffusion through molecular (very fine-holed) sieves is beginning to be used to prepare oxygen mixed with argon. The chemical equations for the preparation of the other gases and for the uses of all these gases are given in Table 8-2. Many of the techniques that are used industrially are impractical on a laboratory scale because they require high temperatures and/or pressures.

15. Write chemical equations for O_2, O_3, and HNO_3 acting as oxidizing agents, and H_2 and CO acting as reducing agents.

$O_2(g)$ and $O_3(g)$ both are powerful oxidizing agents, with $O_3(g)$ being the stronger of the two. The strength of $O_2(g)$ as an oxidant is not appreciated by us because we live in atmosphere that is approximately one-fifth $O_2(g)$, but it has been projected that if the concentration of $O_2(g)$ in the atmosphere were doubled to 40%, life as we know it could not exist. That is part of the reason why He is used in underwater breathing mixtures to dilute the $O_2(g)$. Oxygen will react with virtually every other element in the periodic table, with varying degrees of ease. We expect metals to produce oxides with the metals in their highest common oxidation state. Ozone will produce the same products, but more readily.

$$2\,Ca(s) + O_2(g) \longrightarrow 2\,CaO(s) \qquad [30]$$

$$4\,Fe(s) + 3\,O_2(g) \longrightarrow 2\,Fe_2O_3(s) \qquad [31]$$

Most metals react with nitric acid. Gold and platinum are among the few exceptions. For this reason, nitric acid is used in reprocessing nuclear fuels; virtually all of the material of the spent fuel reacts to form a solution that can be further treated. In all cases, the metal ion in its highest common oxidation state is produced in solution with nitrate ion. The nitrogen-containing reduction product of the reaction of $HNO_3(aq)$ with a metal depends on both the concentration of the nitric acid and the activity of the metal. With metals that are not very active, such as copper, silver, and nickel, dilute nitric acid produces NO(g) and concentrated nitric acid produces $NO_2(g)$. With more active metals, such as zinc, $N_2O(g)$ is formed.

$$3\,Cu(s) + 8\,HNO_3(aq, dil.) \longrightarrow 3\,Cu(NO_3)_2(aq) + 2\,NO(g) + 4\,H_2O(l) \qquad [32]$$

$$Ag(s) + 2\,HNO_3(aq, conc.) \longrightarrow AgNO_3(aq) + NO_2(g) + H_2O(l) \qquad [33]$$

$$4\,Zn(s) + 10\,HNO_3(aq) \longrightarrow 4\,Zn(NO_3)_2(aq) + N_2O(g) + 5\,H_2O(l) \qquad [34]$$

The principal applications of $H_2(g)$ and CO(g) as reducing agents are in metallurgical processes. Examples include the reduction of $WO_3(s)$ or NiO(s) with $H_2(g)$ and the reduction of $Fe_2O_3(s)$ or ZnO(s) with CO(g).

$$WO_3(s) + 3\,H_2(g) \xrightarrow{\Delta} W(s) + 3\,H_2O(g) \text{ OR } NiO(s) + H_2(g) \xrightarrow{\Delta} Ni(s) + H_2O(g) \qquad [35]$$

$$FeO_3(s) + 3\,CO(g) \xrightarrow{\Delta} 2\,Fe(l) + 3\,CO_2(g) \text{ OR } ZnO(s) + CO(g) \xrightarrow{\Delta} Zn(g) + CO_2(g) \qquad [36]$$

The CO(g) used is generated within the reaction mixture by the incomplete oxidation of C(s). This can be a problem, since some metals form metal carbides (compounds with carbon) that have properties that are undesirable in the finished metal. The main problem with using $H_2(g)$ as a reducing agent is that it is considerably more expensive than CO(g). It also is more difficult to use, since $H_2(g)$ must be generated outside the reaction mixture and transported to it, rather than being generated *in situ* as CO(g) is.

SELF-ASSESSMENT EXERCISES

100. Br_2 will be displaced by Cl_2. This is an oxidation-reduction reaction and the reduction potential will favor the reaction:

$$Cl_2(ag) + 2Br^- \rightarrow Br_2 + 2Cl^-$$

101. (a) The reaction will yield O_2: $2KClO_3 \rightarrow 2KCl + 3O_2$

(b) $2KClO_4 \rightarrow 2KCl + 4O_2$

(c) $2N_2O \rightarrow 2N_2 + O_2$

(d) $CaCO_3 \rightarrow CaO + CO_2$ (does not produce oxygen.)

(e) $Pb(NO_3)_2 \rightarrow Pb(NO_2)_2 + O_2$

102. (a) $H_2NNH_2 + H_2O \rightarrow N_2H_5^+ + OH^-$

(b) $NH_3 + H_2O \rightarrow NH_4^+ + OH^-$

(c) HN_3, not a base. It lacks the NH_2 core.

(d) $NH_2OH + H_2O \rightarrow NH_3OH + OH^-$

(e) $CH_3NH_2 + H_2O \rightarrow CH_3NH_3^+ + OH^-$

103. (a) H_2S. To be a reducing agent, the compound must be oxidized. Comparing reduction half-reactions, H_2S has the most favorable oxidation reaction.

104. Na_2CO_3 is not used in fertilizers.

105. (b) XeF_4 will have two unshared pairs of electrons and, therefore, will not be tetrahedral.

106. (a) $CaH_2(s) + 2H_3O^+ \rightarrow H_2(g) + Ca^{+2} + 2H_2O$ and (d) $Al(s) + 3H_3O^+ \rightarrow Al^{+3} + H_2(g) + 3H_2O$.

111. Need 4.5×10^7 tons H_2SO_4.

$$4.5 \times 10^7 \text{ tons} \times \frac{\mathbf{2x10^3}\text{ lb.}}{\text{ton}} \times \frac{1\text{kg}}{2.2\text{ lb.}} = 4.0 \times 10^{10} \text{ kg } H_2SO_4$$

$$4.0 \times 10^{10} \text{ kg } H_2SO_4 \times 1\text{kmol}/98.06 \text{ kg} = 4.0 \times 10^8 \text{ kmol } H_2SO_4$$

DRILL PROBLEMS

1. Write chemical reactions for the preparation of the following. Be sure to indicate how the final product is separated from the reaction mixture. [For example, if $KClO_3(s)$ is requested, and synthesis produces $KClO_3(aq)$, indicate that the solution should be gently evaporated to dryness.] Use sodium or potassium halides as the only source of the halogen elements. Use data in the text as necessary.

A. $Cl_2(g)$	B. $Br_2(l)$	C. $I_2(s)$	D. $KClO_3(s)$
E. HI(g)	F. $HClO_3(aq)$	G. $HClO_2(aq)$	H. $Ba(ClO_2)_2(s)$
I. $HClO_4(l)$	J. $ClF_3(g)$	K. $KI_3(aq)$	L. $F_2(g)$
M. HF(g)	N. HCl(g)	O. HBr(g)	P. NaOCl(aq,impure)

2. Give the hybridization on the central atom and the molecular shape of each of the following species:

A. ICl_5	B. SO_3^{2-}	C. IO_6^{5-}	D. NO_3^-
E. BrF_3	F. I_3^-	G. ClO_4^-	H. BrF_5
I. IF_7	J. ICl_3	K. NO_2^-	L. ClO_2^-
M. SO_4^{2-}	N. BrO_3^-	O. ICl_5	P. ClO_3^-

3. Complete and balance the following equations. A reaction does occur in each case.

A. $CaCO_3(s) \xrightarrow{\Delta}$
B. $ZnO(s) + CO(g) \xrightarrow{\Delta}$
C. $NH_3(aq) + H_3PO_4(aq) \longrightarrow$
D. $Zn(s) + H_2SO_4(aq) \longrightarrow$
E. $Na(s) + H_2(g) \longrightarrow$
F. $CdO(s) + H_2(g) \xrightarrow{\Delta}$
G. $N_2(g) + O_2(g) \xrightarrow{\Delta}$
H. $N_2(g) + H_2(g) \longrightarrow$
I. $NO(g) + O_2(g) \longrightarrow$
J. $CO_2(g) + NH_3(g) \longrightarrow$
K. $NH_3(aq) + HCl(aq) \longrightarrow$
L. $NH_3(g) + O_2(g) \longrightarrow$
M. $Fe_2O_3(s) + H_2(g) \longrightarrow$
N. $SnO_2(s) + CO(g) \longrightarrow$
O. $\longrightarrow NH_4H_2PO_4(aq)$
P. $Cu(s) + HNO_3(conc.) \longrightarrow$
Q. $Ag(s) + HNO_3(dil.) \longrightarrow$
R. $Al(s) + HNO_3(dil.) \longrightarrow$
S. $NaH(s) + H_2O(l) \longrightarrow$
T. $H_2(g) + O_2(g) \longrightarrow$
U. $NH_3(aq) + H_2SO_4(aq) \longrightarrow$
V. $KClO_3(s) \xrightarrow{\Delta}$
W. $NO_2(g) + H_2O(l) \longrightarrow$
X. $+ O_2(g) \longrightarrow CO(g)$
Y. $CH_4(g) + O_3(g) \longrightarrow$
Z. $MnO_2(s) + CO(g) \longrightarrow$

4. Use the principles of this objective to rearrange the following species in order of decreasing value (largest to smallest) of reduction potential:

A. Cl_2, I_2, Br_2, F_2
B. BrO_3^-, BrO_4^-, $HBrO$
C. ClO_4^-, HPO_4^{2-}, SO_4^{2-}
D. "H_2SO_3", SO_4^{2-}, SO_3^{2-}
E. NO_3^-, HNO_2, $NaNO_2$
F. HPO_4^{2-}, H_3PO_3, $H_2PO_3^-$
G. H_4SiO_4, HSO_4^-, ClO_4^-, HPO_4^{2-}
H. SeO_3^{2-}, H_4GeO_4, BrO_3^-, $H_2AsO_3^-$

5. Give the formulas of at least two substances that are used for each of the following purposes. Use information in Table 22-3.

A. photography	B. textile manufacturing	C. analytical reagent
D. detonators	E. explosives	F. water softener
G. medication	H. nitrate synthesis	I. halogenating agents
J. bleach	K. disinfectant	L. poison

6. Write the balanced half-equation and determine the standard reduction potential for each of the following couples. Use the data of Figures 22-1, 22-2, 23-3, and 22-4.

A. $O_2\|H_2O$ (acidic)	B. $O_2\|OH^-$ (basic)	C. $HO_2^-\|OH^-$ (basic)
D. $HClO\|Cl^-$ (acidic)	E. $ClO_4^-\|Cl_2$ (acidic)	F. $ClO_4^-\|Cl^-$ (acidic)
G. $HClO_2\|Cl^-$ (acidic)	H. $ClO_3^-\|Cl_2$ (basic)	I. $ClO_4^-\|Cl_2$ (basic)
J. $ClO_2^-\|Cl^-$ (basic)	K. $NO_3^-\|NO$ (acidic)	L. $NO_3^-\|NO$ (basic)

M. $NO_3^-|HNO_2$ (acidic) N. $NO_3^-|NO_2^-$ (basic) O. $NO_3^-|NH_4^+$ (acidic)
P. $HNO_2|N_2$ (acidic) Q. $SO_4^{2-}|S$ (acidic) R. $SO_4^{2-}|S^{2-}$ (basic)

QUIZZES (15-20 minutes each) Choose the best answer for each question.

QUIZ A

1. Trigonal pyramid is the molecular shape of (a) ClO_2^-; (b) NO_3^-; (c) IF_3; (d) BrO_3^-; (e) none of these.
2. Which is *not* found free in nature? (a) N_2; (b) O_2; (c) P_4; (d) S_8; (e) none of these.
3. In basic solution, $E° = 0.65$ V for $ClO_2^-|OCl^-$ and $E° = 0.40$ V for $OCl^-|Cl_2$. What is the value of $E°$ for $ClO_2^-|Cl_2$? (a) 1.05 V; (b) 0.52 V; (c) 0.48 V; (d) 0.57 V; (e) none of these.
4. Which of these is the strongest oxidizing agent? (a) Br_2; (b) HNO_2; (c) H_2SO_4; (d) $NaNO_3$; (e) HNO_3.
5. Which is used in textile manufacturing? (a) HBr; (b) $Na_2S_2O_3$; (c) HCN; (d) N_2H_4; (e) none of these.
6. Which reagent is not used to prepare HBr(g)? (a) $H_2(g)$; (b) H_2O; (c) H_2SO_4; (d) H_3PO_4; (e) none of these.
7. Which element must be produced by electrolysis? (a) Cl_2; (b) Na; (c) F_2; (d) K; (e) none of these.

QUIZ B

1. Which of the following has a linear molecular shape? (a) ClO_2^-; (b) XeF_2; (c) O_3; (d) H_2O; (e) none of these.
2. Which element has the highest melting point? (a) ozone; (b) sodium; (c) white phosphorus; (d) diamond; (e) sulfur.
3. In acidic solution, $E° = 1.229$ V for $O_2|H_2O$ and $E° = 1.77$ V for $H_2O_2|H_2O$. What is the value of $E°$ for $O_2|H_2O_2$? (a) 1.50 V; (b) 0.54 V; (c) 2.31 V; (d) 0.68 V; (e) none of these.
4. Which of the following is the weakest oxidizing agent? (a) $NaClO_3$; (b) $HClO_3$; (c) $HClO_2$; (d) XeF_2; (e) Cl_2.
5. Which of the following is not sensitive to light? (a) AgBr; (b) Se; (c) $H_2 + Cl_2$; (d) NO_2; (e) AgI.
6. Which reagent is not used in a laboratory preparation of $O_2(g)$? (a) BaO_2; (b) H_2O; (c) $KClO_3$; (d) Ag_2O; (e) $Na_2S_2O_8$.
7. Which central atom produces compounds with the largest number of ligands? (a) F; (b) I; (c) N; (d) Cl; (e) O

QUIZ C

1. Which has a T-shaped molecular shape? (a) ClO_3^-; (b) ClF_3; (c) NO_3^-;

   (d) SO_3^{2-}; (e) none of these.

2. Which produces more than one product when it reacts with $Cl_2(g)$? (a) Na;

   (b) Mg; (c) P; (d) H_2; (e) Be.

3. In acidic solution, $E° = 0.17$ V for $SO_4^{2-} | SO_2$ and $E° = 0.45$ V for $SO_2 | S$. What is the value of $E°$ for $SO_4^{2-} | S$? (a) 0.62 V; (b) 0.21 V; (c) 0.36 V;

   (d) 1.07 V; (e) none of these.

4. Which of these is the strongest oxidizing agent? (a) $HClO_2$; (b) $HClO_3$;

   (c) $HClO_4$; (d) $KClO_3$; (e) $KClO_2$.

5. Which of these is *not* used in connection with explosives? (a) HCN; (b) $Pb(N_3)_2$; (c) $KClO_4$; (d) NO_2; (e) none of these.

6. $N_2O_3(g)$ can be prepared from (a) HNO_3; (b) NH_4NO_3; (c) $NO_2 + NO$;

   (d) $NO + O_2$; (c) NO_2.

7. Which compound forms the weakest hydrogen bonds? (a) HF; (b) H_2O; (c) HCl; (d) NH_3; (e) H_2O_2.

QUIZ D

1. Which has a square planar molecular shape? (a) ClO_4^-; (b) SO_4^{2-}; (c) ICl_3;

   (d) ClF_5; (e) none of these.

2. Which is *not* composed of small (fewer than 10 atoms) molecules? (a) white phosphorus; (b) ozone; (c) monoclinic sulfur; (d) black phosphorus; (e) rhombic sulfur.

3. In basic solution, $E° = -3.04$ V for $N_2 | NH_2OH$ and $E° = 0.42$ V for $NH_2OH | NH_3$. What is the value of $E°$ for $N_2 | NH_3$? (a) -2.62 V;

   (b) -0.87 V; (c) -0.73 V; (d) -1.89 V; (e) none of these.

4. Which of these is the weakest oxidizing agent? (a) $KClO_4$; (b) $HClO_4$; (c) $KClO_3$;

   (d) $HClO_3$; (e) $HClO_2$.

5. Which of the following is *not* used for killing (insects, bacteria, and so forth)?

   (a) Br_2; (b) NaClO; (c) $C_2H_4Br_2$; (d) HCN; (e) I_2.

6. Which reagent is used to prepare H_2O_2? (a) N_2H_4; (b) $Na_2S_2O_8$;

(c) $Na_2S_2O_4$; (d) NH_2OH; (e) $H_4P_2O_7$.

7. Which element forms the largest variety of oxoanions? (a) S; (b) Cl; (c) P; (d) Br; (e) I.

SAMPLE TEST (60 minutes)

1. In acidic solution $E° = -1.07$ V for $Br^- \mid Br_2$, $E° = -1.59$ V for $Br_2 \mid HBrO$, and $E° = -1.49$ V for $HBrO \mid BrO_3^-$.

   A. Is HBrO spontaneous with respect to disproportionation to Br_2 and BrO_3^-?

   B. What is $E°$ for $Br_2 \mid BrO_3^-$?

2. Write an equation or a series of equations for each following process:

   A. $NaBrO_3$(aq) from Br_2(l), H_2O, and aOH(aq)
   B. HBr(g) by at least three methods
   C. O_2(g) by at least three methods
   D. $NaClO_3$(aq) from NaClO(aq)
   E. H_2S(g) and Cd^{2+}(aq) from CdS
   F. NO_2(g) by at least two methods
   G. N_2O_4(g) with HNO_3(aq) as the source of N
   H. N_2(g) from a nitrogen-containing compound
   I. $H_4P_2O_7$ from P_4O_{10} and H_2O
   J. H_2O_2 from a sulfur-containing compound
   K. $CaSO_3$(s) from Ca(s), O_2(g), H_2O(l), and S(s)
   M. H_3PO_4 from any elements; use no compounds
   L. HCl(g) from Na(s), Cl_2(g), and any strong acid (specify which one) except HCl
   N. NH_3(g) from Li(s), N_2(g), and H_2O(l)
   O. H_2(g) from Al(s), Ca(s), NH_3(g), and H_2O(l)
   P. $CaCO_3$(s) from Ca(s), O_2(g), H_2O(l), & C(s)
   Q. H_2O_2 from H_2SO_4(aq) and any elements
   R. $LiNO_3$(s) from HNO_3(aq) and any elements
   S. H_2S(g) from Ca(s), S(s), and any strong acid
   T. HCl(g) from Mg(s), Cl_2(g), and any strong acid (specify which one) except HCl
   U. H_3PO_3(aq) from any elements; use no compounds.

23 THE TRANSITION ELEMENTS

CHAPTER OBJECTIVES

*** 1. State ways in which the transition elements differ from the representative elements and know the trends in their properties.**

The transition elements correctly include those of Families 3B through 1B; those elements whose atoms or ions have partially filled *d* subshells. Often the zinc group (2B) is included in the transition elements, a practice that we follow here. Sometimes the copper group (1B), and occasionally the scandium group (3B), are excluded from the transition elements, because the chemistry of the elements in these two groups is similar to those of some representative elements. In particular, the chemistry of the elements of Group 3B is quite similar to that of the elements of Family 3A (Al, Ga, In, Tl). All of the transition elements (except those of Family 2B) have high melting and boiling points, as well as high heats of fusion and vaporization. Most are good conductors of heat and electricity, especially those of Group 1B (Cu, Ag, Au). Transition elements are denser than the representative elements of the same period. Many transition metal compounds are highly colored and paramagnetic, owing to partly filled underlying *d* orbitals.

Except for the elements of Group 3B (Sc, Y, La) and Zn, the transition elements exhibit more than one oxidation state. Higher oxidation states are common in combination with highly electronegative nonmetals: oxygen, fluorine, and chlorine. The highest oxidation state increases across each period, peaking in Family 8B and declining thereafter. Higher oxidation states are more important for heavier members of a family. Thus, Fe(II) and Fe(III) are most important for iron; Os(IV), Os(VI), and Os(VII) for osmium. For representative metals, on the other hand, lower oxidation states are more important for heavier members of a family. Thus, Tl(I) is more common than Tl(III), Pb(II) is more stable than Pb(IV), and Bi(III) is more stable than Bi(V). In all cases, higher oxidation states of a given element form more acidic oxides. Thus, CrO is basic, Cr_2O_3 is amphoteric, and CrO_3 is acidic.

One notes the following trends: A general but not constant decrease in atomic radius occurs across a period, with the smallest radii in Group 8B (as shown in Figure 9-3); Group 8B elements are the most dense. The elements are denser but softer toward the bottom of each group. Generally, across each period, the electronegativity increases, the reduction potential becomes more positive, and the first ionization potential increases.

2. Describe the lanthanide contraction and explain how it affects the properties of the transition elements.

The lanthanide contraction is the gradual decrease in atomic and ionic radii that occurs from La to Lu (Figure 9-4 and Objective 9-4). Because of the lanthanide contraction, the atoms and ions of each group of the second and third transition series are almost of the same size (Figure 9-3). Because they have the same valence electron configurations, their properties are virtually the same. The similarity is closest at the beginning of the transition series, as shown in Table 23-1.

*** 3. Describe the sources and uses of the elements of the first transition series (a 3×5-inch card or audio cassette project).**

Scandium is produced by electrolysis of a molten mixture of $ScCl_3$, LiCl, and KCl. It is too rare (about \$150 per gram) for important uses, although its density, strength, and chemical behavior are similar to those of Al.

Titanium ores are rutile (TiO_2), perovskite ($CaTiO_3$), ilmenite ($FeTiO_3$), and titanite ($TiSiO_4$ and Ca_2SiO_4). Its metallurgy is discussed in Objective 24-6. Titanium alloys are strong, lightweight, and heat and corrosion resistant. They are used for aircraft, jet engines, and pipes in chemical plants.

Vanadium is found as V_2O_3, $Pb(VO_4)Cl$, $Pb(VO_3)_2$, and $BiVO_4$. The ores are treated with concentrated HCl, and NH_4Cl is added to produce $(NH_4)_3VO_5$. Roasting $(NH_4)_3VO_5$ yields V_2O_5. Reduction of V_2O_5 with Si, Ca, or Mg produces the metal; if this reduction is carried out in the presence of Fe, the resulting mixture is known as *ferrovanadium*. Vanadium steel is used for automotive springs and axles and high-speed machine tools because of its toughness. Alloyed with Ti, V is both strong and heat resistant, and is used in rockets.

TABLE 23-1 Periodic Properties of Niobium, Tantalum, Silver, and Gold

	Nb	Ta	Ag	Au
ionic radius, pm	70	70	126	137(+1)
ionization potential, kJ/mol	664	761	731	890
oxidation states	$+5,+4$	$+5$	$+1$	$+3,+1$
heat of fusion, kJ/mol	27	28	11.3	12.7
electrical conductance, mho	0.080	0.081	0.616	0.42

Chromium's principal ore is chromite, $Fe(CrO_2)_2$. Reduction with C produces ferrochrome: $Fe(CrO_2)_2 + 4\,C \longrightarrow Fe + 2\,Cr + 4\,CO(g)$. Chromate compounds are produced by roasting with Na_2CO_3 in air: $4\,Fe(CrO_2)_2 + 8\,Na_2CO_3 + 7\,O_2 \longrightarrow 2\,Fe_2O_3 + 8\,Na_2CrO_4 + 8\,CO_2(g)$. Chrome steel is hard and tough. Stainless steel is 14% Cr. Nichrome (80% Ni and 20% Cr) is used for electrical resistance heaters. Chrome plating of steel adds beauty and provides some protection against corrosion.

Manganese's principal ore is pyrolusite, MnO_2. Reduction with Al (or C) produces the metal $3\,MnO_2 + 4\,Al \longrightarrow 4\,Al_2O_3 + 3\,Mn$. Steel alloy containing 12% Mn is used for armor plate and earth-moving equipment. Manganese also improves the ease of working molten steel.

Iron is found as Fe_2O_3, Fe_3O_4, *and* $FeCO_3$. Its metallurgy is discussed in Objective 23-6. Iron and steel are the most important structural metals.

Cobalt's common ores are $CoAs_2$, CoAsS, and Co_3S_4. A series of reactions produces Co_2O_3, which is reduced to the metal by C(s) or $H_2(g)$ at high temperature. Cobalt is alloyed with Fe and other metals to produce heat-resistant, metal-cutting tools and surgical instruments. Some alloys are magnetic.

Nickel occurs with Cu and Fe as NiS. To obtain pure nickel, the sulfides are separated by selective flotation. NiS then is roasted to NiO, which in turn is reduced with C to Ni:

$2\,NiS + 3\,O_2 \longrightarrow 2\,NiO + 2\,SO_2$; $2\,NiO + C \longrightarrow 2\,Ni + CO_2(g)$. The resulting 96% Ni is refined electrolytically. Often the mixed sulfide ore is not separated, but roasted to the oxides and reduced with C. The mixture of Ni, Fe, and Cu from the ore is called *Monel metal*; it is very resistant to chemical attack. Nickel formerly was separated from the mixture with CO(g) at 60°C in the Monel process: $Ni(s) + 4\,CO(g) \rightleftharpoons Ni(CO)_4(g)$. Above 200°C, $Ni(CO)_4(g)$ decomposes, releasing free Ni(s). Nickel plate also makes Fe corrosion resistant.

Copper is found as $Cu_2O, CuFeS_2$, and Cu_3FeS_3, often mixed together. Its metallurgy is discussed in Objective 23-6. Electrolytically refined copper is used for electrical wiring. Unrefined copper is used for plumbing and limited structural purposes (rain gutters, roof flashing).

Silver and *gold* often are found free in nature, although most of the rich deposits have been worked. The metallurgies of both metals are described in Objective 23-6. Both metals are used in jewelry and electrical circuits.

Zinc oxide is obtained by roasting the carbonate or sulfide. Reduction with coke yields the metal, which is refined either by fractional distillation of the molten impure metal or electrolytically. The latter process combines reduction and refining. ZnO(s) is dissolved in $H_2SO_4(aq)$, Equation [1]. Zn dust is added to the solution to displace Cd, and the solution is electrolyzed to produce pure Zn(s), Equation [2]. Zinc metal is used in Pb and Zn metallurgy to recover Ag and Cd, respectively; to galvanize Fe; in brass; in the manufacturing of dry cells; and in roofing materials. *Cadmium* is refined in a similar fashion to zinc. Cadmium metal is used in bearing alloys, low-melting solders, and copper alloys; in voltaic cells (Ni-Cd cell); as low-cost plating to replace Cr; and for control rods and shielding in nuclear reactors. Note that HgS is reduced when it is roasted, Equation [3]. *Mercury* is refined by first adding dilute HNO_3 to oxidize the impurities; the oxides float on the surface and are skimmed off. The mercury is further refined by distillation. The metal is used in thermometers, barometers; electrical relays, electrodes in electrochemical cells, fluorescent tubes, and dental amalgams.

$$ZnO(s) + H_2SO_4(aq) \longrightarrow Zn^{2+}(aq) + SO_4^{\;2-}(aq) + H_2O \qquad [1]$$

$$Zn^{2+}(aq) + SO_4^{\;2-}(aq) + H_2O \xrightarrow{\text{electrolysis}} Zn(s) + H_2SO_4(aq) + \tfrac{1}{2}O_2(g) \qquad [2]$$

$$HgS(s) + O_2(g) \xrightarrow{\Delta} Hg(l) + SO_2(g) \qquad [3]$$

*** 4. Describe uses of some of the important compounds of the transition elements.**

This is a 3×5-inch card project. Use the information in Table 23-2. You should be aware of how each element and its compounds are economically important. In addition, be aware that many of these metals are used extensively in alloys. The development and formulation of alloys with specific characteristics is a specialized branch of science.

TABLE 23-2 Important Compounds of Transition Elements

Sc	No important compounds
TiH_2	removes $O_2(g)$ and $N_2(g)$ from electronic tubes
$TiCl_3$	powerful reducing agent; laundry stain remover
TiO_2	high-quality white pigment for paper, plastics, and ceramics
$TiCl_4$	smoke screens as it produces dense clouds of TiO_2 on reaction with H_2O

TiC	edges for cutting tools
$BaTiO_3$	a dielectric for high-performance capacitors
VCl_2	mordant in textile dyeing
V_2O_3	industrial catalyst in oxidations
V_2O_5	catalyst for H_2SO_4 production; manufacturing of yellow glass; photography; dye mordant
$CrCl_2$	to absorb $O_2(g)$ in gas analysis; organic catalyst
Cr_2O_3	green pigment for paints, glass, and ceramics; abrasive; in semiconductors
Chrome alum	$[KCr(SO_4)_2 \cdot 12\,H_2O]$ dye mordant; waterproofing fabrics
CrO_3	used in chrome-plating baths; in glassware cleaning solutions
$PbCrP_4$, $ZnCrO_4$	yellow paint pigments
Na_2CrO_4	prevents corrosion in boilers and radiators
$FeCrO_4$	lining open-hearth furnaces for steel production
$Na_2Cr_2O_7$	tanning leather; photography; oxidizing agent
$K_2Cr_2O_7$	laboratory oxidizing agent
$MnCl_2$	disinfectant; drying agent in paints
MnO_2	decolorizing glass; drier for paints; oxidant in dry cells; source of Mn compounds
$KMnO_4$	versatile laboratory oxidizing agent
$FeSO_4$	reducing agent; pigment in ink, dye, and paint; weed killer; wood preservative; starting material for Fe compounds
$FeCl_3$	water clarification
$Fe_4[Fe(CN)_6]_3$	ink and paint pigment; blueprints (that is, plans for buildings)
Fe_2O_3	paint (Venetian red) and cosmetic (rouge) pigment
CoO	blue glass and ceramic pigment
$K_3Co(NO_2)_6$	yellow paint pigment for oil colors, watercolors, glass, and porcelain; rubber pigment
$NiSO_4$	in nickel plating; textile dye mordant
Ni_2O_3	Edison storage battery (nickel-cadmium battery)
Cu_2O	ruby glass pigment; fungicide for seeds
$CuSO_4$	to kill algae in drinking water; calico dyes; starting material for Cu compounds
$Cu_3(AsO_4)_2$	fungicide and insecticide
CuO	insecticide; absorbent for CO
$AgNO_3$	source of Ag compounds
AgX	(silver halides) photography
$HAuCl_4$	("gold chloride") in gold plating; ruby glass manufacture
ZnO	reinforcing agent and white pigment in rubber; in burn ointments; in cosmetics; dietary supplement; photoconductors in copying machines
ZnS	X-ray and television screen phosphor; in luminous paints;

	+ $BaSO_4$ = lithopone (white pigment)
$ZnSO_4$	rayon manufacturing; in animal feeds; wood preservative
$ZnCrO_4$	yellow corrosion-resistant paint pigment
$ZnCl_2$	fireproofing agent; its aqueous solutions dissolve cellulose to produce fiberboard when dried
$ZnHPO_4$	dental cement
CdO	in electroplating; in batteries; as catalyst; nematocide
CdS	yellow pigment in glass, textiles, paper, rubber, ceramics, and soap; in solar cells; photoconductor in xerography; phosphors
$CdSO_4$	electroplating of Cd, Cu, Ni; standard voltaic cells (Weston cell)
HgO	polishing compounds; in dry cells; antifouling paint for ships' bottoms; fungicide; red pigment
$HgCl_2$	manufacture of Hg compounds; disinfectant; fungicide; insecticide; wood preservative; formerly used to make felt for hats (from which "mad as a hatter" became a synonym for Hg poisoning)
Hg_2Cl_2	in electrodes; pharmaceuticals (cathartic and diuretic); fungicide
$Hg(CNO)_2$	fulminate of mercury: highly sensitive explosive for detonators and percussion caps
$C_6H_5HgC_2H_3O_2, C_2H_5HgCl$	fungicides for treating seeds stored for planting

*** 5. Describe the four fundamental processes of traditional extractive metallurgy—concentration, roasting, reduction, and refining—with specific reference to metals such as Sn, Pb, Zn, Cd, and Hg.**

The first step in processing metallic ores is *concentration* or removal of the unwanted material—the *gangue.* This gangue usually is silicate rock of low density (that of SiO_2 is $2.66\ g/cm^3$). One method of concentration is *washing*—allowing rapidly moving water to carry the lighter gangue away from the metal-containing mineral. (Some mineral densities are: $SnO_2, 6.95\ g/cm^3$; PbS, $7.5\ g/cm^3$; ZnS, $4.09\ g/cm^3$; $HgS, 8.10\ g/cm^3$; $Cu_2S, 5.6\ g/cm^3$; and CuS, $4.6\ g/cm^3$). A second method of concentration is *flotation.* The ore is mixed with water, a wetting agent, and a foaming agent. The mixture is agitated and air is blown through. The particles of the mineral stick to the foam on the surface, which is skimmed off. Concentration normally removes more than 99% of the gangue.

During *roasting*, the concentrated ore is heated strongly in the presence of air. This calcines carbonates (Equation [4]), transforms sulfides to oxides (Equation [5]), or oxidizes metallic impurities and converts nonmetallic ones to volatile oxides. $SO_2(g)$ is not vented to the atmosphere; it can be reduced to S(s) by reaction with red-hot C, Equation [6].

$$CuCO_3 \xrightarrow{\Delta} CuO(s) + CO_2(g) \qquad [4]$$

$$2\,MS(s) + 3\,O_2 \xrightarrow{\Delta} 2\,MO(s) + 2\,SO_2(g) \quad M = Pb, Zn, Cd \qquad [5]$$

$$SO_2(g) + 2\,C(s) \xrightarrow{\Delta} S(l) + 2\,CO(g) \qquad [6]$$

The oxide is transformed to the metal during *reduction.* Typical reducing agents are C(s), obtained from coking coal (heating it in the absence of air), and CO(g), obtained from coke (as in Equation [7]).

The reductions of SnO_2 and PbO are typical reactions; the reductions of ZnO and CdO are similar to that of PbO, except that the metal is produced as a vapor in each case. When heated, thermally unstable HgO decomposes to Hg(g) and $O_2(g)$.

$$2\,C(s) + O_2 \rightleftharpoons 2\,CO(g) \qquad [7]$$

$$SnO_2 + 2\,C(s) \xrightarrow{\Delta} Sn(l) + 2\,CO(g) \qquad [8]$$

$$PbO(s) + C(s) \xrightarrow{\Delta} Pb(l) + CO(g) \text{ AND } PbO(s) + CO(g) \xrightarrow{\Delta} Pb(l) + CO_2(g) \qquad [9]$$

Several other elements are obtained by reacting their oxides or halides with a reducing agent at high temperatures. This reducing element may be an active metal, as it is for beryllium, boron, and sodium.

$$BeF_2(l) + Mg(l) \longrightarrow MgF_2(l) + Be(l) \qquad [10]$$

$$B_2O_3(s) + 3\,Mg(l) \longrightarrow 2\,B(\text{amorphous, impure}) + 3\,MgO(s) \qquad [11]$$

$$Na(g) + KCl(l) \longrightarrow K(g) + NaCl(l) \qquad [12]$$

Refining removes the impurities from the metal. *Tin* is refined by remelting. Impurities are oxidized; the oxides float on the surface and are skimmed off. Metallic impurities remain unmelted and the molten pure tin is poured off. *Lead* also is refined by remelting; copper crystallizes out and remains behind when the molten lead is poured off. When air is blown through the melt, the nonmetallic impurities form lead oxoanion compounds that float on the surface and are skimmed off. Finally, 1–2% Zn is added to the melt. Ag impurity dissolves in the Zn(1) and floats on the surface; it is skimmed off and the Ag is recovered by distilling away the Zn.

*** 6. Outline the metallurgies of Ti, Fe and steel, Cu, Ag, and Au.**

Titanium is produced mainly from rutile ore, TiO_2. Reaction with HCl or Cl_2 produces the chloride, Equation [13]. $TiCl_4(g)$ is reduced with magnesium metal at high temperatures (800 °C) under an He atmosphere, Equation [14].

$$TiO_2(s) + 2\,Cl_2(s) + C(s) \longrightarrow TiCl_4(g) + 2\,CO(g) \qquad [13]$$

$$TiCl_4(g) + 2\,Mg(l) \longrightarrow Ti(s) + 2\,MgCl_2(l) \qquad [14]$$

Iron is mainly produced from the oxide ores Fe_2O_3 and Fe_3O_4. The oxides are reduced by CO(g) and $H_2(g)$, Equation [15]. $H_2(g)$ and CO(g), in turn, are made from coke (produced by heating coal to high temperatures in the absence of air), oxygen, and water (Equation [16]).

$$Fe_2O_3(s) + 3\,CO(g) \longrightarrow 2\,Fe(l) + 3\,CO_2(g) \quad \text{and} \quad Fe_2O_3(s) + 3\,H_2(g) \longrightarrow 2\,Fe(l) + 3\,H_2O(g) \qquad [15]$$

$$2\,C(s) + O_2(g) \longrightarrow 2\,CO(g) \text{ AND } C(s) + CO_2(g) \longrightarrow 2\,CO(g) \quad \text{and} \quad C(s) + H_2O(l) \longrightarrow CO(g) + H_2(g) \qquad [16]$$

Many impurities are removed in a liquid slag formed with $CaCO_3$ (equations in [17]).

$$CaCO_3(s) \longrightarrow CaO(s) + CO_2(g) \quad \text{then}$$
$$CaO(s) + SiO_2(s) \longrightarrow CaSiO_3(l) \quad \text{and} \qquad [17]$$
$$3\,CaO(s) \longrightarrow Ca_3(PO_4)_2(l)$$

The final product, pig iron, contains 4–5% C, along with impurities of about 1% each of Si, Mn, and P. It is transformed into steel by melting the mass and injecting $O_2(g)$ in the presence of CaO. The impurities are converted into their oxides, which either bubble out of the mixture (CO and CO_2) or combine to form slag [$FeSiO_3, MnSiO_3, Ca_3(PO_4)_2$]. Alloying metals (Cr, Ni, Mn, V, Mo, and W) are added to the molten iron to produce steel of the desired composition.

Copper ores are concentrated by flotation. Roasting in air at low temperatures produces Cu_2S and converts FeS to FeO, Equation [18], which combines with SiO_2 to form slag, Equation [19]. After the slag is separated, air is blown through the molten Cu_2S, yielding free copper (also called *blister copper* because of frozen bubbles of $SO_2(g)$), Equation [20]. This impure copper is refined electrolytically.

$$2\,CuFeS_2(s) + 4\,O_2(g) \longrightarrow Cu_2S(l) + 2\,FeO(s) + 3\,SO_2(g) \qquad [18]$$

$$FeO(s) + SiO_2(l) \longrightarrow FeSiO_3(l) \quad \text{AND THEN}$$
$$Cu_2S(l) + O_2(g) \longrightarrow 2\,Cu(l) + SO_2(g) \qquad [19]\ \&\ [20]$$

Both *silver* and *gold* are extracted from their low-grade ores with cyanide-ion-containing solutions (equations in [21]). The pure metals are recovered from these solutions with the addition of zinc (equations in [22]).

$$4\,Au(s) + 8\,CN^-(aq) + O_2(g) + 2\,H_2O \longrightarrow 4\,[Au(CN)_2]^-(aq) + 4\,OH^-(aq) \qquad [21a]$$

$$Ag_2S(s) + 4\,CN^-(aq) \longrightarrow 2\,[Ag(CN)_2]^-(aq) + S^{2-}(aq) \qquad [21b]$$

$$2\,[Au(CN)_2]^-(aq) + Zn(s) \longrightarrow 2\,Au(s) + [Zn(CN)_4]^{2-}(aq) \qquad [22a]$$

$$2\,[Ag(CN)_2]^-(aq) + Zn(s) \longrightarrow 2\,Ag(s) + [Zn(CN)_4]^{2-}(aq) \qquad [22b]$$

Ag and Au are also recovered from the anode "muds" produced during the electrolytic refining of Cu and Ni. Ag is recovered when impure Pb is refined.

7. State which are the most common oxidation states of the first transition series and Group 1B and 2B elements, which of these are stable, and which are unstable.

The common oxidation states of these metals, and an example of a compound of each, are given in Table 23-3.

TABLE 23-3 Common Oxidation States (and Examples) of the First Transition Series (**boldface** = most stable)

Scandium		**+3 Sc_2O_3**	
Titanium	+2 TiH_2	+3 $TiCl_3$	**+4 TiO_2**

Vanadium		+2 VCl_2	+3 V_2O_3	+4 VF_4	**+5 V_2O_5**	
Chromium		+2 $CrCl_2$	+3 Cr_2O_3		**+6 K_2CrO_4**	
Manganese		**+2 $MnCl_2$**	+3 MnF_3	+4 MnO_2	+6 K_2MnO_4	+7 $KMnO_4$
Iron		+2 $FeSO_4$	**+3 $FeCl_3$**			
Cobalt		**+2 CoO**	+3 $K_3[Co(NO_2)_6]$			
Nickel		**+2 $NiSO_4$**	+3 Ni_2O_3	[Silver	**+1 AgX**	+2 AgF_2]
Copper	+1 Cu_2O	**+2 $CuSO_4$**		[Gold	+1 $AuCl$	**+3 $HAuCl_4$**]
Zinc		**+2 $ZnSO_4$**		[Mercury	+1 Hg_2Cl_2	**+2 $HgSO_4$**]

*** 8. Use electrode potential data to predict the conditions under which certain compounds are likely to disproportionate.**

Electrode potential diagrams for the various oxidation states of vanadium and manganese are given together in Figure 23-1. Let us use this data to determine whether $MnO_4^{2-}(aq)$ is stable. We first write the balanced half-equations for the oxidation of $MnO_4^{2-}(aq)$ to $MnO_4^-(aq)$, Equation [23], and its reduction to $MnO_2(s)$, Equation [24]. Equation [23] is multiplied by 2 and added to Equation [24] to produce the net ionic equation, [25]. The positive value of $E°$ indicates that the overall reaction is spontaneous; $MnO_4^{2-}(aq)$ spontaneously disproportionates to $MnO_4^-(aq)$ and MnO_2.

$$MnO_4^{2-}(aq) \longrightarrow MnO_4^-(aq) + e^- \qquad -E°\{MnO_4^- \mid MnO_4^{2-}\} = -0.56\,V \qquad [23]$$

$$2\,e^- + MnO_4^{2-}(aq) + 4\,H^+(aq) \longrightarrow MnO_2(s) + 2\,H_2O \qquad E°\{MnO_4^{2-} \mid MnO_2\} = 2.26\,V \qquad [24]$$

$$3\,MnO_4^{2-}(aq) + 4\,H^+(aq) \longrightarrow MnO_2(s) + 2\,MnO_4^-(aq) + 2\,H_2O \qquad E° = +1.70\,V \qquad [25]$$

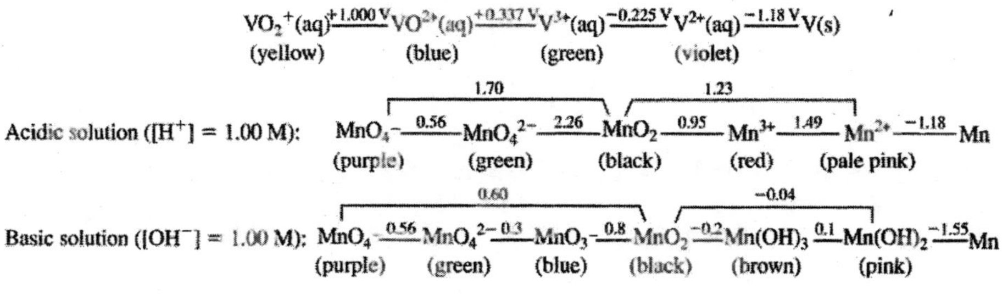

FIGURE 23-1 Electrode Potential Diagrams for Vanadium and Manganese

*** 9. Write equations for some important oxidation-reduction reactions, especially those involving permanganate and dichromate ions.**

Of the two oxoanions, MnO_4^- is the somewhat stronger oxidizing agent.

$$MnO_4^- + 8\,H^+ + 5\,e^- \longrightarrow Mn^{2+} + 4\,H_2O \qquad E° = +1.51\,V \qquad [26]$$

$$Cr_2O_7^{2-} + 14\,H^+ + 6\,e^- \longrightarrow 2\,Cr^3 + 7\,H_2O \quad E° = +1.33\text{ V} \qquad [27]$$

However, both ions are much better oxidizing agents in acid than in base.

$$MnO_4^- + 2\,H_2O + 3\,e^- \longrightarrow MnO_2 + 4\,OH^- \qquad E° = +0.60\text{ V} \qquad [28]$$

$$CrO_4^{2-} + 4\,H_2O + 3\,e^- \longrightarrow Cr(OH)_3 + 5\,OH^- \qquad E° = -0.11\text{ V} \qquad [29]$$

10. Discuss the chromate-dichromate equilibrium and the effect of pH on the concentrations of these ions in aqueous solution.

$Cr_2O_7^{2-}$ is a good oxidizing agent (see Equation [27]) and a poor precipitating agent (nearly all dichromates are soluble). CrO_4^{2-} is a poor oxidizing agent (see Equation [29]) but a good precipitating agent. The two ions are in equilibrium in aqueous solution.

$$2\,CrO_4^{2-} + 2\,H^+ \rightleftharpoons Cr_2O_7^{2-} \qquad K_c = 3.2\times10^{14} \qquad [30]$$

$$[Cr_2O_7^{2-}]/[CrO_4^{2-}]^2 = 3.2\times10^{14}[H^+]^2 \qquad [31]$$

If we know the total chromium(VI) concentration and the pH of the solution, we can find $[CrO_4^{2-}]$ and $[Cr_2O_7^{2-}]$.

EXAMPLE 23-1 Determine $[CrO_4^{2-}]$ when 0.200 mol $K_2Cr_2O_7$ is added to 1.00 L of a pH = 4.70 buffer. $[H^+] = 2.0\times10^{-5}$ M constantly, because of the buffer. The setup is based on the balanced chemical equation.

Rxn:	$2\,CrO_4^{2-}$ +	$2\,H^+$	$\rightleftharpoons$	$Cr_2O_7^{2-}$	$+H_2O$
Initial:		2.0×10^{-5} M		0.200 M	
Changes	$+2x$ M			$-x$ M	
:					
Equil:	$2x$ M	2.0×10^{-5} M		$(0.200-x)$ M	

$$[Cr_2O_7^{2-}]/[CrO_4^{2-}] = (0.200-x)/(2x)^2 = 3.2\times10^{14}(2.0\times10^{-5})^2 = 1.3\times10^5$$

Assuming $<< 0.200\text{ M}$, then $0.200/4x^2 \approx 1.3\times10^5$ $or\, x = 6.2\times10^{-4}$ M.

Thus, $[Cr_2O_7^{2-}] = 1.2\times10^3$ M and $[CrO_4^{2-}] = 0.199$ M.

With the a similar method to Example 23-1, $[Cr_2O_7^{2-}] = 5.1\times10^{-3}$ M and $[CrO_4^{2-}] = 0.400$ M in a solution buffered at pH = 9.000

11. Write chemical equations to illustrate the amphoteric nature of certain transition metal oxides and hydroxides, especially those of Cr(III).

An amphoteric substance behaves as a base in the presence of strong acid and as an acid in the presence of strong base. Amphoteric transition metal oxides include VO_2, V_2O_5, Cr_2O_3, FeO_3, FeO, Fe_2O_3, and CuO. The reactions with acid for the chromium(III) hydroxide and oxide are the equations in [32]; those with base are the equations in [33].

$$Cr(OH)_3(s) + 3\,H_3O^+(aq) \rightleftharpoons [Cr(H_2O)_6]^{3+}(aq) \qquad [32a]$$

$$Cr_2O_3(s) + 6\,H_3O^+(aq) + 3\,H_2O \rightleftharpoons 2\,[Cr(H_2O)_6]^{3+}(aq) \qquad [32b]$$

$$Cr(OH)_3(s) + OH^-(aq) \rightleftharpoons [Cr(OH)_4]^-(aq) \qquad [33a]$$

$$Cr_2O_3(s) + 2\,OH^-(aq) + 3\,H_2O \rightleftharpoons 2\,[Cr(OH)_4]^-(aq) \qquad [33b]$$

12. Describe ferromagnetism and the features of atomic structure that lead to it.

Ferromagnetism is the ability of a substance to become strongly and relatively permanently magnetic. For ferromagnetism to be important, the substance must already possess *magnetic domains*, or areas in which the magnetic moments are aligned. If the atoms are small, they will get close enough to pair up (↑↓) and no domains will form. (You can see this pairing with two small magnets. Place one free on a table and slide the other one up to it: side to side and north pole to north pole. When the moving magnet gets close, the free magnet suddenly will swing around.) If the atoms are too large, they will not feel each other's paramagnetism. Under the influence of a strong magnetic field, the domains line up in the same direction. The magnetism of the material is destroyed when this alignment is disrupted by heat or vibration.

13. Describe how Pb, Hg, and Cd function as poisons in the human body, some of their toxic effects, and some of the environmental sources of these poisons.

The serious nature of poisoning by lead, mercury, and cadmium is enhanced by their tenacity. Once these metals are absorbed by the body, they are excreted slowly, if at all. Both Pb and Hg are eliminated over a period of months. There is no evidence that Cd is eliminated in any way.

Lead poisoning seems to be due to interference with the synthesis of the heme group in hemoglobin. Initial symptoms of lead poisoning include listlessness, vomiting, and convulsions. Long-term symptoms include anemia, weakness, weight loss, and permanent brain damage. Sources of lead in the environment include old paint and automobile exhaust. $Pb(OH)_2 \cdot 2\,PbCO_3$ formerly was used as a white pigment; it is now banned. $Pb(C_2H_5)_4$ is present in leaded gasoline to improve its octane rating; blood levels of Pb have dropped since unleaded gasoline has been required for new automobiles.

Mercury functions as a poison by interfering with the action of sulfur-containing enzymes. The symptoms of mercury poisoning include personality changes and lack of coordination. Mercury attacks the central nervous system and causes brain damage, paralysis, and blindness. Because of the chemical inertness of Hg, it was originally thought that elemental Hg could safely be dumped in the environment. It has since been found that microorganisms produce $Hg(CH_3)_2$ from elemental mercury. Hg discarded in the past is being cleaned up because of the hazard it poses. Hg compounds also are being phased out as fungicides. One further source remains: discarded mercury batteries, such as those for hearing aids.

Cadmium functions biochemically as a poison by replacing Zn in enzymes. Symptoms of Cd poisoning include liver and kidney damage, lung disease, high blood pressure, and skeletal disorders. Environmental sources include metal plating, mining, cigarette smoke, and zinc metal and Zn compounds in which Cd is a contaminant. A major source of Cd is the dust produced from the wear of automobile tires in which ZnO (and, thus, low levels of CdO) is used as a reinforcing agent and a white pigment.

SELF-ASSESSMENT EXERCISES

79. Because an oxidizing agent must be reduced, we can examine the reduction half-reactions to determine which species is most easily reduced. Ag^+ will be most easily reduced.

80. (b) H_2S. All of the other complexes have similar solubility, but Fe_2S_3 does not readily form, so NiS will precipitate.

83. (a) First, write the half-reaction:

$$3S^{-2} \rightarrow 3S(s) + 6e^-$$

$$4e^- + H_2O + O_2 \rightarrow 3OH^-$$

Balance the second half-reaction:

$$12e^- + 6H_2O + 3O_2 \rightarrow 12OH^-$$

Balance the electrons by multiplying the first reaction by 2.

Overall equation:

$$2Fe_2S_3 + 6H_2O + 3O_2 \rightarrow 4Fe(OH)_3(s) + 6S(s)$$

(b) Half-reactions:

$$S_2O_8^{-2} + 2e^- \rightarrow 2SO_4^{-2}$$

$$Mn^{2+} \rightarrow MnO_4^- + 5e^-$$

Balance the electrons by multiplying the first reaction by 5 and the second by 2:

$$2Mn^{2+} + 5S_2O_8^{-2} + H_2O \rightarrow 2MnO_4^- + 10SO_4^{-2} + H^+$$

Use H_2O and H^+ to balance oxygen and hydrogen:

$$2Mn^{2+} + 5S_2O_8^{-2} + 8H_2O \rightarrow 2MnO_4^- + 10SO_4^{-2} + 16H^+$$

(c) The half-reactions are:

$$Ag(s) \rightarrow Ag^+ + 1e^-$$

$$\tfrac{1}{2} O_2 + 2e^- \rightarrow OH^-$$

Balancing electrons and combining the overall equation is:

$$2Ag(s) + 4CN^- + \tfrac{1}{2} O_2 + H_2O \rightarrow 2[Ag(CN)_2]^- + 2OH^-$$

DRILL PROBLEMS

1. Rewrite each of the following lists in order of increasing value (smallest to largest) of the indicated property:

A. atomic radius: Mn, Sc, Fe, Ti

B. density: Cr, Sc, Os, Fe

C. highest oxidation state: Cr, Sc, V, Mn

D. acidity: MnO, Mn_2O_7, MnO_3, MnO_2

E. first ionization energy: Zn, Sc, Mn, Fe

F. number of unpaired electrons: Fe, V, Sc, Mn

G. density: Ti, Pd, Nb, Pt
H. acidity: VO, V_2O_3, V_2O_5, VO_2
I. highest oxidation state: V, Ti, Mn, Y
J. number of unpaired elns: Cu^{2+}, Fe^{3+}, V^{3+}, Ni^{2+}
K. atomic radius: Ti, Cr, Fe, Ca

2. Give the chemical symbol of the transition element or elements with these properties, sources, or uses.

A. Mg is used to produce this metal
B. metal produced by electrolysis
C. metal refined by electrolysis
D. used in electrical wires
E. never used in structural alloys
F. many ores are sulfides
G. metal vaporizes during reduction
H. used to coat iron to protect it from corrosion
I. present in all steels
J. forms heat-resistant alloys
K. found in many magnets

3. Give the chemical formula of the transition metal compound or compounds with the following uses:

A. pigment for paper
B. fluorescent screens
C. glass pigment
D. paint pigment
E. insecticide
F. ingredient of burn ointments
G. in H_2SO_4 manufacture
H. oxidizing agent
I. used in dyeing
J. corrosion preventative
K. used in water treatment
L. used in electrical batteries

4. Write equations for the following processes:

A. roasting of $ZnCO_3$
B. roasting of ZnS
C. reduction of NiO with H_2
D. roasting of HgS
E. red'n of Fe_2O_3 with H_2
F. recovery of SO_2
G. red'n of SnO_2 with C
H. red'n of PbO with CO
I. roasting of PbS
J. red'n of CuO with H_2
K. roasting of CdS

5. Write the equation for each of the following processes:

A. reducing Cu_2S to metallic copper
B. generating to reducing agent for iron ore
C. initial treatment of titanium ore.
D. formation of silicate slag in iron metallurgy
E. production of titanium metal with C
F. roasting copper ore
G. treatment of Ag with $CN^-(aq)$
H. formation of silicate slag in Cu metallurgy

6. Predict whether or not the following disproportionations will occur. Use the data of Table 20-1 and Figure 23-1. Assume 1.00 M acidic aqueous solution unless otherwise stated.

A. Cu^{+} to Cu^{2+} and Cu° B. Fe^{2+} to Fe^{3+} and Fe°
C. Sn^{2+} to Sn^{4+} and Sn° D. H_2O_2 to $O_2(g)$ and H_2O
E. $MnO_2(s)$ to Mn^{2+} and MnO_4^- F. $MnO_2(s)$ to $Mn(OH)_2$ and MnO_4^-, 1.0 M OH^-
G. V^{3+} to V^{2+} and VO^{2+} H. MnO_4^{2-} to MnO_4^- and MnO_3^-, 1.0 M OH^-

7. Complete and balance the following equations in aqueous solution. Not all products are given

A. $K_2Cr_2O_7 + KBr + H_2SO_4 \longrightarrow Br_2(l)$ B. $KMnO_4 + Na_2C_2O_4 + H_2SO_4 \longrightarrow CO_2(aq)$
C. $H_2S + KMnO_4 \longrightarrow KOH + SO_2(g)$ D. $K_2Cr_2O_7 + SnSO_4 + H_2SO_4 \longrightarrow Sn(SO_4)_2$
E. $KMnO_2 + H_2O_2 \longrightarrow KOH + O_2(g)$ F. $Cd + KMnO_4 + H_2SO_4 \longrightarrow CdSO_4$
G. $K_2Cr_2O_7 + HNO_2 \longrightarrow KNO_3$ H. $NaMnO_4 + NaI \longrightarrow I_2(s)$
I. $Na_2CrO_4 + SnCl_2 + HCl \longrightarrow SnCl_4$ J. $HMnO_4 + AsH_3 + H_2SO_4 \longrightarrow H_3AsO_4$
K. $Cr(OH)_3 + Na_2O_2 \longrightarrow NaOH$ L. $KMnO_4 + FeSO_4 + H_2SO_4 \longrightarrow Fe_2(SO_4)_3$

QUIZZES (20 minutes each) Choose the best answer for each question.

QUIZ A

1. Which element displays ferromagnetism? (a) Co; (b) Cu; (c) Ti; (d) Cr; (e) none of these.
2. Which element has the greatest density? (a) Mn; (b) Ti; (c) Cr; (d) V; (e) Fe.
3. Which is the best oxidizing agent? (a) MnO_4^- in acid; (b) MnO_4^- in base; (c) $Cr_2O_7^{2-}$ in base; (d) $Cr_2O_7^{2-}$ in acid; (e) CrO_4^{2-} in acid.
4. MnO_4^- yields what manganese-containing species when used as an oxidizing agent in base? (a) MnO_4^{2-}; (b) Mn; (c) Mn^{2+}; (d) Mn_2O_3; (e) none of these.
5. Which is the most basic? (a) Cr_2O_3; (b) CrO_3; (c) CrO; (d) CO_2; (e) ClO_2.
6. Which ammonium sulfide group cation precipitates as a hydroxide in the group precipitate? (a) Fe^{2+}; (b) Fe^{3+}; (c) Cu^{2+}; (d) Ca^{2+}; (e) none of these.
7. The paramagnetism of transition element compounds is due to (a) paired electrons spinning in opposite direction; (b) unpaired electrons in *d* or *f* orbitals; (c) shared valence electrons; (d) unshared valence electron pairs; (e) unpaired electrons in *s* or *p* orbitals.
8. Which of the following is not a heavy metal poison? (a) Pb; (b) Cd; (c) Zn; (d) Hg; (e) all are poisons.

QUIZ B

1. Which metal gives high-temperature strength to steel? (a) Sc; (b) Zn; (c) Cu; (d) Co; (e) none of these.
2. Which element has the lowest first-ionization potential? (a) Fe; (b) Ni; (c) Co; (d) Mn; (e) Ti.
3. Which is a green pigment? (a) Cr_2O_3; (b) $PbCrO_4$; (c) ZnO; (d) CoO; (e) none of these.

4. $Cr_2O_7^{2-}$ yields what species when used as an oxidant in base? (a) CrO_3^{2-}; (b) CrO_3; (c) Cr^{3+}; (d) $Cr(OH)_3$; (e) none of these.

5. Which is most acidic? (a) Mn_2O_7; (b) MnO_2; (c) MnO; (d) MgO; (e) Mn_2O_3.

6. Which ion is NOT colored? (a) $[Co(SCN)_4]^{2-}$; (b) $[Fe(H_2O)_5SCN]^{2+}$; (c) $[Al(OH)_4]^{2-}$; (d) $Ni(DMG)_2$; (e) all are colored.

7. A property common to all transition metals is that each type of metal (a) is found in many different oxidation states; (b) produces colored compounds; (c) has a high electronegativity; (d) exhibits one oxidation state equal to the group number; (e) reacts with water to produce hydrogen gas.

8. For which metal's metallurgy is carbon an effective reducing agent? (a) Mg; (b) Sn; (c) Al; (d) Na; (e) Sr.

QUIZ C

1. Which metal is very resistant to chemical attack? (a) Fe; (b) Ni; (c) Cu; (d) Zn; (e) none of these.

2. Which has the largest atomic radius? (a) Ti; (b) Fe; (c) Co; (d) Ni; (e) Cr.

3. Which is *not* used in dyeing? (a) $CuSO_4$; (b) ZnO; (c) $FeSO_4$; (d) $KCr(SO_4)_2 \cdot 12H_2O$; (e) all are.

4. Which will enhance the oxidizing strength of MnO_4^- the most? (a) double $[H^+]$; (b) double $[MnO_4^-]$; (c) halve $[H^+]$; (d) halve $[Mn^{2+}]$; (e) halve $[MnO_4^-]$.

5. Which will not dissolve in acidic solution? (a) Al_2O_3; (b) P_2O_5; (c) MgO; (d) Sc_2O_3; (e) all will dissolve.

6. Which is a common test for Fe^{2+}? (a) brownish-red oxide; (b) red color with dimethyl glyoxime; (c) white precipitate with NaCl(aq); (d) red color with NH_3(aq); (e) none of these.

7. The last step in the production of copper metal used for electrical conduction is (a) smelting; (b) crushing; (c) roasting; (d) reduction; (e) electrolytic refining.

8. NaOH(aq) is used in the metallurgy of (a) Sn; (b) Zn; (c) Hg; (d) Al; (e) none of these.

QUIZ D

1. Which metal is produced in the smallest amounts industrially? (a) Fe; (b) Cr; (c) Mn; (d) Cu; (e) Sc.

2. Which has the highest oxidation state in some of its compounds? (a) Mn; (b) Sc; (c) Cu; (d) Zn; (e) V.

3. Which is not used to control insects, fungi, or algae? (a) $CuSO_4$; (b) $ZnCl_2$; (c) CuCl; (d) $Cu_3(AsO_4)_2$; (e) Cu_2O.

4. Under what conditions is Hg^{2+}(aq) most likely to be formed in a $0.100\ M\ K_2CrO_4, 0.100\ M\ Hg_2(NO_3)_2$ solution? (a) high temperature; (b) low pH; (c) addition of KOH; (d) none of these will work.

5. Which will dissolve in basic solution? (a) $Ni(OH)_2$; (b) MgO; (c) ZnO; (d) Fe_2O_3; (e) none will dissolve.

6. Which of the following ions precipitates as its hydroxide, rather than its sulfide, when its solution is treated with $NH_4Cl - NH_3$ and $(NH_4)_2S$ is added? (a) Fe^{2+}; (b) Co^{2+}; (c) Fe^{3+}; (d) Ni^{2+}; (e) Pb^{2+}.

7. The transition elements are *not* characterized by (a) tendency to form complexes; (b) ability to have several different oxidation states; (c) greater reactivity from left to right in a period; (d) tendency to form colored compounds; (e) none of these.

8. Which of the following is not produced or refined electrically? (a) Na; (b) Mg; (c) Al; (d) Pb; (e) all are.

SAMPLE TEST (20 minutes)

1. An alloy is suspected of containing Fe, Cu, and Ni. Describe how you would qualitatively analyze this alloy for these three elements, including how you would dissolve the alloy. Write the chemical equation for each reaction that occurs during the analysis.

2. (a) Write equations for the production of aluminum chloride and sodium aluminate from aluminum oxide.

   (b) What mass of oxide is needed to produce 100.0 g of each of the two products, with the reactions written in the first part of this question?

3. Use the couples $Cr^{2+} \mid Cr^{3+}, +1.51\ V$ and $Cr^{3+} \mid Cr_2O_7^{2-}, -1.33\ V$ to draw an electrode potential diagram for chromium and determine $E°$ of $Cr_2O_7^{2-} \mid Cr^{2+}$.

24 COMPLEX IONS AND COORDINATION COMPOUNDS

CHAPTER OBJECTIVES

1. Describe Werner's theory of coordination compounds.

Werner's theory explaned how two compounds that each contained three atoms of Cl^- reacted differently to $AgNO_3(aq)$. One compound released 3 moles of Cl^- per mole of compound and the other released 2 moles of Cl^- per mole of compound (see Section 24-1 in text). Werner proposed that certain metal ions, primarily those of transition metals, have two types of valence: the primary valence is based on the number of electrons the atom loses in forming an ion; the secondary valence is responsible for bonding of ligands to the central atom.

*** 2. Identify the central ions and the ligands, determine the coordination number and the oxidation state of the central ion, and establish the net charge on a complex ion.**

The formula of a coordination compound has the *complex ion* enclosed in brackets. Thus, the complex ion of $[Fe(NO_3)(NH_3)_5](NO_3)_2$ is $[Fe(NO_3)(NH_3)_5]^{2+}$, that of $K_2[Zn(CN)_4]$ is $[Zn(CN)_4]^{2-}$, and that of $[Cr(NH_3)_6]Cl_3$ is $[Cr(NH_3)_6]^{3+}$. The *net charge* on the complex ion is that needed to balance the total charge of the simple cations or anions in the compound. The *central ion* is the metal ion in the complex: Fe^{3+} in $[Fe(NO_3)(NH_3)_5]^{2+}$, Zn^{2+} in $[Zn(CN)_4]^{2-}$, and Cr^{3+} in $[Cr(NH_3)_6]^{3+}$. The *ligands* are the groups attached to the central ion: NH_3 and NO_3^- in $[Fe(NO_3)(NH_3)_5]$, CN^- in $[Zn(CN)_4]^{2-}$, and NH_3 in $[Cr(NH_3)_6]^{3+}$. The ligands are fairly common ions and simple molecules (see Table 24-2). The *oxidation state* of the central ion is determined by the charge on the complex and the total charges on the ligands.

$$\text{complex ion charge} = \text{total charge on ligands} + \text{central ion oxidation state} \qquad [1]$$

The *coordination number* is the number of places where ligands are bonded to the central ion. If all the ligands are *monodentate*—meaning that each forms only one bond to the central ion—the coordination number equals the number of ligands.

*** 3. Give the coordination numbers of some common metal ions and write the names and formulas of some common monodentate and polydentate ligands.**

The coordination numbers of common metal ions given in Table 24-1 should be committed to memory. The names, formulas, and charges of the various ligands in Table 24-2 should also be memorized.

*** 4. Write distinctive names based on formulas of complexes, and formulas based on names.**

The following rules are a restatement of those *in the text*:

(1) Name the cation first and the anion second.

(2) Within the complex ion, name the ligands first (in alphabetical order by ligand name), then the central ion. In writing formulas, this order is reversed, except that anion ligand formulas are written before neutral ligand formulas.

(3) Ligand names are those given in Table 24-2. Notice that the name of each anion ligand ends with "o".

(4) The number of ligands of a given type is indicated by the prefixes *mono-* (1, used for emphasis only), *di-* (2), *tri-* (3), *tetra-* (4), *penta-* (5), and hexa- (6) for monodentate ligands; *bis-* (2), *tris-* (3), and *tetrakis-* (4) for polydentate ligands. Enclose the names of polydentate ligands in parentheses.

(5) The central ion's oxidation state is given as a parenthesized Roman numeral after the complex's name.

(6) Complex *anion* names always end with *-ate*, followed by the appropriate Roman numeral. If the name of the atom ends in *-um*, *-ium*, or *-enum*, the ending is replaced with *-ate* in anion complexes.

Cr = chromate; Al = aluminate; Mo = molybdate

TABLE 24-1 Coordination Numbers of Common Metal Ions

Coordination number(s)	*Ion(s)*
2 only	Ag^+
2 or 4	Cu^+, Au^+
4 only	$Zn^{2+}, Au^{3+}, Cd^{2+}, Pt^{2+}$
4 or 6	$Co^{2+}, Ni^{2+}, Cu^{2+}, Al^{3+}$
6 only	$Ca^{2+}, Fe^{2+}, Sc^{3+}, Cr^{3+}, Fe^{3+}, Co^{3+}, Pt^{4+}$

TABLE 24-2 Formulas and Names of Common Ligands

Monodentate ligands—anions							
OH^-	hydroxo	F^-	fluoro	NH_2^-	amido	SO_3^{2-}	sulfito
H^-	hydro	Cl^-	chloro	NO_3^-	nitrato	SO_4^{2-}	sulfato
O^{2-}	oxo	Br^-	bromo	$—NO_2^-$	nitro	$S_2O_3^{2-}$	thiosulfato
O_2^{2-}	peroxo	I^-	iodo	$—ONO^-$	nitrito	CO_3^{2-}	carbonato
ClO_2^-	chlorito	ClO_3^-	chlorato	$—NC^-$	iso Cyano	$—CN^-$	cyano
$—NCS^-$	thiocyanato	$—SCN^-$	isothiocyanato			Ac^-	acetato ($C_2H_3O_2^-$)
Monodentate ligands—molecules							
H_2O	aqua	NH_3	ammine	CH_3NH_2	Methyl amine	CO	carbonyl

NO	nitrosyl	C_6H_5N	pyridine		
		Polydentate ligands[a]			
en	ethylenediamine (2)	dien	Diethylenetriamine (3)	trien	triethylenetetraamine (4)
$C_2O_4^{2-}$	oxalato (2) (or ox)	*o*-phen	*o*-phenanthroline (2)	EDTA	ethylenediaminetetraacetato (6)

The following metal names also are used in anion complexes:

Mn	manganate	Fe	ferrate	Co	cobaltate	Ni	niccolate	Zn	zincate
Cu	cuprate	As	arsenate	Ag	argentate	Pb	plumbate	Au	aurate
Sn	stannate	Sb	antimonate	W	tungstate				

Some examples follow:

$[Ag(NH_3)_2]Cl$	diamminesilver(I) chloride	$[CoCl_3(NH_3)_3]$	triamminetrichlorocobalt(III)
$[Ni(CO)_4]$	tetracarbonylnickel(II)	$K_4[Fe(CN)_6]$	potassium hexacyanoferrate(II)
$[Cu(en)_2]SO_4$	bis(ethylenediamine)copper(II) sulfate		
$[Pt(NH_3)_4][PtCl_6]$	tetraammineplatinum(II) hexachloroplatinate(IV)		

*** 5. Draw plausible structures for complex ions from their names and formulas.**

The coordination number (C.N.) of a metal ion is a guide to the geometry of the ligands around that ion. C.N. = 2 has a linear geometry, and that of C.N. = 6 is octahedral. C.N. = 4 can be either square planar, as it is for Ni^{2+}, Cu^{2+}, and Pt^{2+}; or tetrahedral, as it is for $Ni^{2+}, Co^{2+}, Zn^{2+}, Cd^{2+}$, and Al^{3+}. The square planar geometry is shown by the less active metals that have C.N. = 4, but Ni^{2+} shows both square planar and tetrahedral geometries. The structures of four complexes are drawn in Figure 24-1.

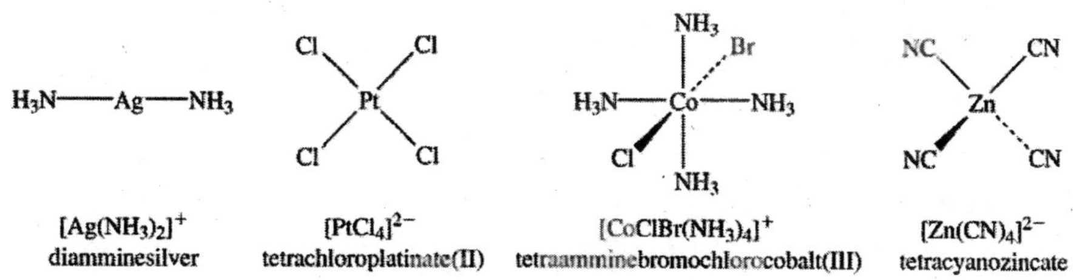

FIGURE 24-1 Structures of Four Complex Ions

*** 6. Describe the types of isomerism found among complex ions and identify the possible isomers in specific cases.**

---

[a] Numbers in parentheses are the number of monodentate ligands that each polydentate ligand replaces—the *denticity* of the ligand.

Isomers are compounds with the same formulas but different structures; thus, they have different chemical or physical properties. Coordination compounds show four types of isomerism: (1) Ionization isomers have different ions as ligands. For example, $[CoSO_4(NH_3)_5]Br$ is red and gives a positive test for Br^- ion when in solution, but $[CoBr(NH_3)_5]SO_4$ is green and gives a positive sulfate test. (2) Linkage isomerism occurs when the ligand attaches to the central ion in different ways. For example, pentaamminenitrocobalt(III) chloride, $[CoNO_2(NH_3)_5]Cl_2$, is yellow; while pentaamminenitritocobalt(III) chloride, $[Co(ONO)(NH_3)_5]Cl_2$, is red. Other possible (but not common) linkage isomer ligands are CN^-, SCN^-, and CO. (3) Coordination isomerism occurs when a given ligand is bonded to the cation in one isomer and bonded to the anion in the other. The pair of cobalt(III) complexes $[Co(NO_2)_3(NH_3)_3]$ and $[Co(NO_2)_2(NH_3)_4][Co(NO_2)_4(NH_3)_2]$ are one example. (4) Geometric isomers differ in the arrangement of the ligands around the central ion. This is most easily seen in square planar complexes, such as $[PtCl_2(NH_3)_2]$.

cis isomer dichlorodiammineplatinum(II) *trans* isomer

In the *trans* isomer, the ammines are across the Pt from each other, on opposite sides. In the *cis* isomer, they are next to each other. Geometric isomers also occur in octahedral complexes. The *fac* designation arises because the three common ligands are at the corners of a *face* of the octahedron. The *mer* designation signifies that the three common ligands are on the same *mer*idian.

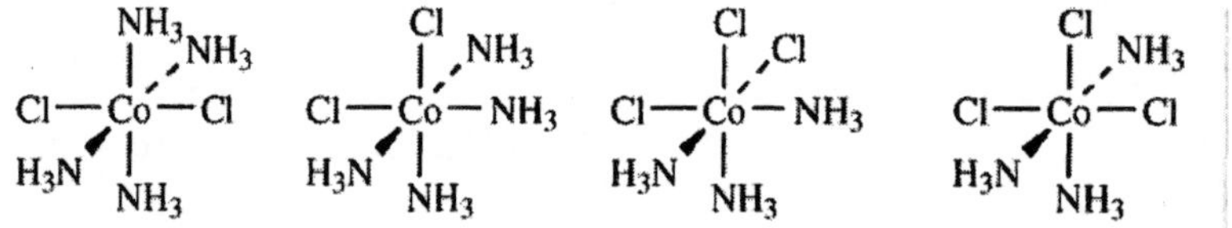

trans isomer *cis* isomer *cis* or *fac* isomer *trans* or *mer* isomer

7. **Explain the basis of the crystal field theory of bonding in complex ions. Use the spectrochemical series to make predictions about *d*-level splitting and the number of unpaired electrons in complex ions.**

In crystal field theory, a complex ion is held together by the attraction between the positive charge of the central ion and the negative electrons of the ligands. When they get close, the ligand electrons repel the valence shell electrons of the central ion. This makes the five *d* orbitals, which are the same energy in the isolated ion, of different energies in the complex. This difference in energies is the *crystal field splitting energy*. It is shown for octahedral and tetrahedral geometries in Figure 24-2. In that figure, the energy of the central ion surrounded by unorganized ligands (the ligands have not yet taken the octahedral or tetrahedral configuration) is called the excited ion's energy. Since the ligands place negatively charged electron pairs in the vicinity of the positively charged central ion, the ion is higher in energy (or less stable) than an isolated ion. The crystal field splitting energy is symbolized by $\Delta = 10$ Dq. Δ is much smaller for tetrahedral complexes than for octahedral ones. In addition, the size of Δ is determined by the ligands. Those that produce a large value of Δ are called *strong field ligands* and are strong Lewis bases. Those that produce a small value of Δ are *weak field ligands* and are weak Lewis bases. The relative strengths of ligands is summarized in the spectrochemical series.

$CN^- > —NO_2^- > en > py \approx —NH_3 > —NCS^- > H_2O > C_2O_4^{2-} >$ [2]
$OH^- > F^- > Cl^- > Br^- > I^-$

This series should be memorized. Nitrogen-bonding ligands are stronger than oxygen-bonding ones. Halides are the weakest ligands. We saw this displacement of weaker ligands by stronger ones before (Objective 18-9).

The strength of the crystal field determines the number of unpaired electrons in d^4, d^5, d^6, and d^7 octahedral complexes and in d^3, d^4, d^5, and d^6 tetrahedral complexes, as shown in Figure 24-3. The dividing line within the spectrochemical series between strong field ligands and weak field ligands is not the same for each cation.

*** 8. Explain the origin of color in aqueous solutions of complex ions.**

The size of Δ determines the color that a complex shows. When white light strikes a complex ion, electrons in the low energy set of d orbitals absorb light of energy Δ,and we see the remaining light—that which is not absorbed. Values of Δ and the resulting colors are given in Table 24-3.

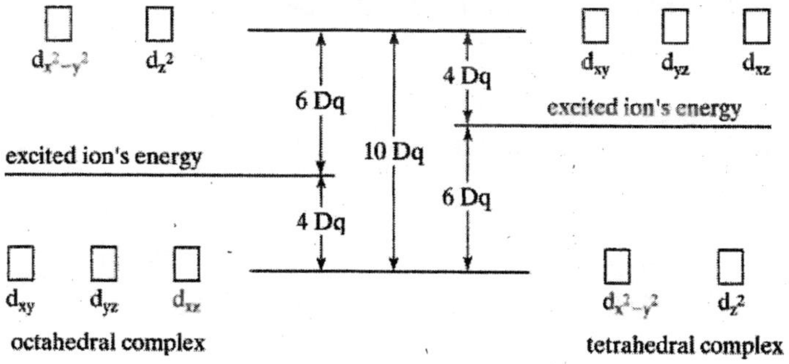

FIGURE 24-2 Ligand Field Splitting in Octahedral and Tetrahedral Complexes

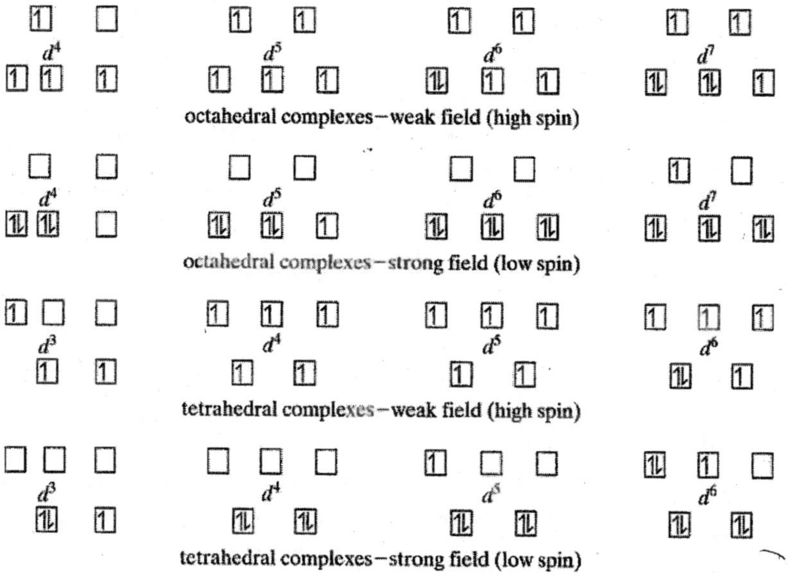

FIGURE 24-3 Strong and Weak Field Configurations in Octahedral and Tetrahedral Complexes

9. Write equations to show the formation of complex ions by the stepwise displacement of H_2O by other ligands, and relate the overall formation constant, K_f, to stepwise formation constants.

In earlier chapters, we have written equations for reactions in which complex ions were formed. Many examples are in Objective 23-13. One, for instance, is as follows:

$$Ni^{2+} + 6\ NH_3 \longrightarrow [Ni(NH_3)_6]^{2+} \qquad [3]$$

The Ni^{2+} ion in aqueous solution is more properly written $[Ni(H_2O)_6]^{2+}$. A reaction such a [3] is then revealed as one in which one Lewis base, NH_3, is displacing another one, H_2O, as in Equation [4]. The complex ion is properly viewed as an adduct between a Lewis acid (the central ion) and Lewis bases (the ligands).

$$[Ni(H_2O)_6]^{2+} + 6\ NH_3 \longrightarrow [Ni(NH_3)_6]^{2+} + 6\ H_2O \qquad [4]$$

The exchange of ligands does not take place in one step. In many cases, the equilibrium constant for the replacement of each aqua ligand by an ammine ligand has been measured. For example, the reaction of $Ag^+(aq)$ with NH_3 occurs in two steps:

$$[Ag(H_2O)_2]^+ + NH_3 \rightleftharpoons [AgNH_3H_2O]^+ + H_2O \quad K_1 = \frac{[[AgNH_3H_2O]^+]}{[[Ag(H_2O)_2]^+][NH_3]} = 2.0 \times 10^3 \qquad [5]$$

$$[AgNH_3H_2O]^+ + NH_3 \rightleftharpoons [Ag(NH_3)_2]^+ + H_2O \quad K_2 = \frac{[[Ag(NH_3)_2]^+]}{[[AgNH_3H_2O]^+][NH_3]} = 7.9 \times 10^3 \qquad [6]$$

The overall equation is a combination of Equations [5] and [6], and its equilibrium constant, K_f, is the product of the equilibrium constants of those two reactions, $K_f = K_1 \times K_2$

$$[Ag(H_2O)_2]^+ + 2\ NH_3 \rightleftharpoons [Ag(NH_3)_2]^+ + 2\ H_2O \quad K_f = \frac{[[Ag(NH_3)_2]^+]}{[[Ag(H_2O)_2]^+][NH_3]^2} = 1.6 \times 10^7 \qquad [7]$$

TABLE 24-3 Colors of Complex Ions with Different Values of 10 Dq

10 Dq = Δ (approx)		λ (approx),	Color	Complementary
J/molecule	kJ/mol	nm	absorbed	color seen
3.06×10^{-19}	185	650	red	green
3.31×10^{-19}	200	600	orange	blue
3.43×10^{-19}	206	580	yellow	violet
3.82×10^{-19}	230	520	green	red
4.23×10^{-19}	255	470	blue	orange
4.85×10^{-19}	292	410	violet	yellow

Some of the equations from previous chapters appear less artificial when aqua complex ions are written. For example, Equation [44] of objective 23-12 becomes

$$[Fe(H_2O)_6]^{3+} + SCN^- \rightleftharpoons [FeSCN(H_2O)_5]^{2+} + H_2O \qquad K_1 = \frac{[[FeSCN(H_2O)_5]^{2+}]}{[[Fe(H_2O)_6]^{3+}][SCN^-]} = 890 \qquad [8]$$

10. Describe how aqua complexes ionize as acid. Explain amphoterism from this viewpoint.

In addition to $Fe^{3+}(aq)$ mentioned in the text, $Co^{3+}(aq)$ also ionizes as an acid. In fact, the first ionization constant for $Co^{3+}(aq)$ is stronger than that for many of the weak acids listed in Table 16-1.

$$[Co(H_2O)_6]^{3+} + H_2O \rightleftharpoons [Co(H_2O)_5OH]^{2+} + H_3O^+ \qquad K = 0.018 \qquad [9]$$

Amphoterism also can be understood from the standpoint of the addition and removal of protons of aqua ligands. Consider the reactions of $Cr(OH)_3$ with strong acid and with strong base:

$$[Cr(H_2O)_3(OH)_3](s) + 3\,H_3O^+ \rightleftharpoons [Cr(H_2O)_6]^{3+} + 3\,H_2O \qquad [10]$$

$$[Cr(H_2O)_3(OH)_3](s) + 3\,OH^- \rightleftharpoons [Cr(OH)_6]^{3-} + 3\,H_2O \qquad [11]$$

11. Explain how complex ion formation can be used to stabilize oxidation states.

We normally think of copper(I) as unstable in aqueous solution. In fact, copper(I) spontaneously disproportionates to copper(II) and elemental copper.

$$2\,Cu^+ \rightleftharpoons Cu^{2+} + Cu \qquad E^\circ = +0.36\text{ V} \qquad [12]$$

From this value of E°, we can compute a value of $\Delta G^\circ (= -nfE^\circ)$, and from that a value of K_{eq} $(\Delta G^\circ = -RT \ln K_{eq})$.

$$K_{eq} = [Cu^{2+}]/[Cu^+]^2 = 1 \times 10^6 \qquad [13]$$

This ratio indicates that the value of $[Cu^+]$ must be quite small. For instance, if $[Cu^{2+}] = 1.0\,M$, then $[Cu^+] = 1 \times 10^{-3}$ M. We can achieve this small $[Cu^+]$ if we add cyanide ion to the solution and form $[Cu(CN)_4]^{3-}$. In this way, the formation of a complex ion will stabilize a normally unstable oxidation state.

$$Cu^+ + 4\,CN^- \rightleftharpoons [Cu(CN)_4]^{3-} \qquad K_f = 2.0 \times 10^{30} \qquad [14]$$

12. Cite ways in which complex ion equilibria are used in the qualitative analysis scheme.

Complex ion equilibria are often used to separate one ion or a group of ions from a larger group of ions or to confirm the presence of an ion by displaying a distinctive color. Some separations include: (1) Often Ag^+ is separated from the mixed precipitate of the chloride qualitative analysis group by adding $NH_3(aq)$. Neither $Hg_2{}^{2+}$ nor Pb^{2+} form a very stable ammine complex, but Ag^+ does (Equation [7]). (2) In the ammonium sulfide cation group, Al^{3+}, Cr^{3+}, and Zn^{2+} are kept in solution by adding OH^-; at the same time, $Mn(OH)_2, Co(OH)_2, Ni(OH)_2$, and $Fe(OH)_3$ precipitate. (Examples are Equations [30] through [33] of Objective 24-13.) (3) $NH_3(aq)$ keeps $[Ni(NH_3)_6]^{2+}$ and $[Co(NH_3)_6]^{3+}$ in solution, while $Fe(OH)_3(s)$ precipitates (Equations [38], [41], and [42] of Objective 23-13).

Some qualitative analysis confirmations that involve complexes are as follows: (1) The formation of $[Fe(H_2O)_5SCN]^{2+}$ confirms Fe^{3+}, Equation [8]. (2) The formation of $Ni(DMG)_2$ confirms Ni^{2+},

$$Ni(NH_3)_6]^{2+} + 2\ DMG^- \longrightarrow [Ni(DMG)_2] + 6\ NH_3 \qquad [15]$$

(3) $[Pb(C_2H_3O_2)_2]$ formation begins the confirmation of Pb,

$$PbSO_4(s) + 2\ C_2H_3O_2^-(aq) \longrightarrow [Pb(C_2H_3O_2)_2] + SO_4^{2-}(aq) \qquad [16]$$

13. Describe applications of complex ion formation in the photographic process, in electroplating, and in water treatment.

In developing *photographic* film, the unexposed AgBr is removed by forming a thiosulfato complex ion.

$$AgBr(s) + 2\ S_2O_3^{2-} \longrightarrow [Ag(S_2O_3)_2]^{3-} + Br^- \qquad [17]$$

In *electroplating*, the cation of the element to be plated often is in solution as a complex ion. This maintains a low concentration of free metal ion in solution, which produces a smoother, more adherent plating. Frequently, even low concentrations of some metal ions in *water* have undesired effects on chemical reactions and manufacturing processes. Often, these metal ions can be incorporated into a stable complex by chelation with a polydentate ligand.

SELF-ASSESSMENT EXERCISES

67. (e) 6. Each en (ethylenediamine) can donate two electron pairs, so there are four electron pairs from en and two from Cl.

68. (a) This molecule will be linear with no isomerism. (b) NO_2 can exhibit linkage isomerism. The NO_2 can be coordinated through the N or one of the O atoms. (c) Cis and trans isomers are possible. (d) and (e) will not form isomers.

69. (a) The mirror image is not superiposable.

70. Because NH_3 is a strong field ligand, there are two unpaired electrons.

71. (c) $[Fe(H_2O)_6]^{+3}$ because the small, highly charged Fe^{+3} attracts electrons away from an O-H bond in a ligand water molecule. H^+ is transferred to a solvent water molecule.

DRILL PROBLEMS

1. For each, give the central atom, its oxidation state, the coordination number, and the ligands' formulas.

A.	$[Ag(S_2O_3)_2]^{3-}$	B.	$[AlH_4]^-$
C.	$[Co(CN)_4(H_2O)_2]^-$	D.	$[PtCl_4(NH_3)_2]$
E.	$[Ni(NH_3)_4]^{2+}$	F.	$[Cr(SCN)_4(NH_3)_2]^-$
G.	$[Fe(en)_3]^{3+}$	H.	$[Pt(NO_2)_4]^{2-}$
I.	$[CoCl(NH_3)_5]^{2+}$	J.	$[AlOH(H_2O)_5]^{2+}$

K.	$[PtCl_3(NH_3)_3]^+$	L.	$[Fe(CO)_4]$
M.	$[Mn(CN)_5(NO)]^{3-}$	N.	$[AuF_4]^-$
O.	$[Co(NH_2)_2(NH_3)_4]^+$	P.	$[SbCl_5]^{2-}$
Q.	$[NiCl_2(en)_2]$	R.	$[Fe(CN)_6]^{4-}$
S.	$[NiBr_2(NH_3)_2]$	T.	$[Fe(OH)_2(H_2O)_4]^+$
U.	$[Zn(H_2O)_2(NH_3)_2]^{2+}$		

2. (1) List all the ions for each coordination number.

A. two (2) B. six (6) C. four (4) or six (6)

(2) Give the coordination number(s) of the following ions:

D. Cu^{2+}	E. Fe^{2+}	F. Fe^{3+}	G. Ca^{2+}	H. Co^{2+}
I. Au^+	J. Al^{3+}	K. Co^{3+}	L. Ag^+	

(3) Give the name or formula, as appropriate, and the denticity (1 for monodentate, 2 for bidentate, and so forth) of the following ligands (for example, Cl^-, chloro (1)):

M.	$—NO_3^-$	N.	EDTA	O.	triethylenetetraamine	P.	ammine
Q.	hydroxo	R.	carbonyl	S.	$—NO_2^-$	T.	thiocyanato
U.	NO	V.	N_2	W.	fluoro	X.	sulfato
Y.	cyano	Z.	$—ONO^-$				

3. (1) Name all of the ions listed in the drill problems for Objective 24-1.

(2) Give the name or formula, as appropriate, of each of the following compounds:

V.	$Na_3[Au(CN)_4]$	W.	sodium dithiosulfatoargentate
X.	$K_3[Co(NO_2)_6]$	Y.	diamminedibromoplatinum(II)
Z.	$[CoCl_2(NH_3)_4]Br$	Γ.	tetraaquadichlorochromium(III) bromide
Δ.	diaquatetrachloroplatinum(IV)	Θ.	$K_2[PtCl_6]$
Λ.	diamminesilver chloride	Ξ.	hexaamminenickel(II) sulfate
Π.	$Na_2[Cu(en)_2(SO_4)_2]$	Σ.	$[Al(H_2O)_6][Co(CN)_6]$
Υ.	$Ca[Cr(NH_3)_2(SCN)_4]_2$	Φ.	$[Co(NH_3)_6]ClSO_4$
Ψ.	sodium trioxalatochromate(III)	Ω.	bis(ethylenediamine)copper(II) chloride

4. Draw a reasonable structure for each of the complexes given in the drill problems for Objective 24-1.

5. Use these 10 coordination compounds in answering the questions below:

1.	$Li[Al(CN)_4]$	2.	$[Co(NH_3)_6][ClSO_4]$
3.	$[Co(CN)_2(H_2O)_4]Cl$	4.	$[Co(en)_2(NH_3)_2]SO_4$
5.	$[Al(H_2O)_6][Co(CN)_6]$	6.	$[Pt(NO_2)_2(H_2O)_2]$
7.	$[CrCl_2(H_2O)_4]Br$	8.	$[Pt(H_2O)_4][PtCl_6]$
9.	$[CoCl(NH_3)_5]SO_4$	10.	$[Mn(CN)_3(CO)_3]$

A. Identify all of the compounds that have ionization isomers, and give the names and formulas of two ionization isomers of each one of them.

B. Identify compounds that have linkage isomers and give the names of two linkage isomers of each compound.

C. Identify all of the compounds that have coordination isomers and give the formulas of two coordination isomers of each compound.

D. Identify all of the compounds that can display geometric isomerism. Draw the structures and designate each geometric isomer as *cis* or *trans*.

6. Predict whether each of the following complex ions will have a strong field or a weak field, and predict the number of unpaired electrons in each complex:

A.	$[Co(NH_3)_6]^{3+}$	B.	$[Cr(en)_3]^{3+}$	C.	$[PtCl_6]^{2-}$	D.	$[Co(ox)_3]^{4-}$
E.	$[Fe(CN)_6]^{4-}$	F.	$[CoCl_4]^{2-}$	G.	$[CoI_6]^{3-}$	H.	$[Ni(C_2O_4)_3]^{4-}$
I.	$[ScCl_3(H_2O)_3]$	J.	$[Co(NH_3)_4]^{2+}$				

QUIZZES (20 minutes each) Choose the best answer or fill in the blank.

QUIZ A

1. $[Fe(C_2O_4)_3]^{2-}$ has a coordination number of (a) 2; (b) 3; (c) 4; (d) 6; (e) none of these.

2. Au^{3+} shows a coordination number of (a) 2 only; (b) 2 or 4; (c) 4 only; (d) 4 or 6; (e) 6 only.

3. The nitrito ligand has the formula (a) $—NO_2^-$; (b) $—NO_3^-$; (c) $—ONO^-$; (d) NO; (e) none of these.

4. The name of $[AuCl_2(NH_3)_2]Br$ is ______________________________.

5. The formula of tris(ethylenediamine)cobalt(III) acetate is ________________.

6. $[Fe(CN)_6]^{4-}$ has how many unpaired electrons? (a) 2; (b) 0; (c) 4; (d) 3; (e) none of these.

7. Coordination isomerism could be shown by (a) $Li[AlH_4]$;

   (b) $[Ag(NH_3)_2][CuCl_2]$; (c) $[CoCl_2(NH_3)_4]Br$; (d) $[PtCl_2(H_2O)_4]$; (e) none of these.

8. The formation constant for the complex $[Zn(NH_3)_4]^{2+}$ is the equilibrium constant of the reaction represented by (a) $Zn^{2+}(aq) + 4\ NH_3(aq) \rightleftharpoons [Zn(NH_3)_4]^{2+}(aq)$; (b) $[Zn(NH_3)_3(H_2O)]^{2+}(aq) + NH_3(aq) \rightleftharpoons [Zn(NH_3)_4]^{2+}(aq) + H_2O$;

(c) $Zn(s) + 4\ NH_3(aq) \rightleftharpoons [Zn(NH_3)_4]^{2+}(aq) + 2\ e^-$;

(d) $[Zn(H_2O)_4]^{2+}(aq) + 4\ NH_3(aq) \rightleftharpoons [Zn(NH_3)_4]^{2+}(aq) + 4\ H_2O$

QUIZ B

1. $[CoCl_2(en)_2]^+$ has a coordination number of (a) 2; (b) 4; (c) 6; (d) 3; (e) none of these.

2. Cu^{2+} shows a coordination number of (a) 2 only; (b) 2 or 4; (c) 4 only; (d) 4 or 6; (e) 6 only.

3. The carbonyl ligand has the formula (a) CO; (b) CO_3^{2-}; (c) SCN^-; (d) CN^-; (e) none of these.

4. The name of $Ca[Cu(CN)_4]$ is ________________.

5. The formula of potassium pentachloroantimonate(III) is ________________.

6. $[CoCl_6]^{4-}$ has how many unpaired electrons? (a) 6; (b) 3; (c) 1; (d) 0; (e) none of these.

7. Ionization isomerism could be shown by (a) $[CoCl_2(en)_2]$; (b) $Na[Ag(CN)_2]$;

   (c) $[CrCl(NH_3)_5]I_2$; (d) $[PtCl_4][PtCl_2(NH_3)_4]$; (e) none of these.

8. The $[Fe(CN)_6]^{3-}$ complex ion (a) exhibits square planar geometry; (b) is diamagnetic; (c) should be very stable; (d) has two unpaired electrons; (e) none of these.

QUIZ C

1. $[Zn(en)Cl_2]$ has a coordination number of (a) 2; (b) 3; (c) 4; (d) 6; (e) none of these.

2. Au^+ shows a coordination number of (a) 2 only; (b) 2 or 4; (c) 4 only; (d) 4 or 6; (e) 6 only.

3. The oxalato ligand has the formula (a) $C_2O_4^{2-}$ (b) O^{2-}; (c) O_2^-; (d) CO_3^{2-};

   (e) none of these.

4. The name of $K_2[Pt(CN)_2Cl_2]$ is ________________.

5. The formula of calcium diamminetetrachlorochromate(III) is ________________.

6. $[Pt(NH_3)_6]^{4+}$ has how many unpaired electrons? (a) 0; (b) 3; (c) 2; (d) 4; (e) none of these.

7. Linkage isomerism can be shown by (a) $[PtCl_2(H_2O)_4]$; (b) $[CoCl_2(NH_3)_4]$;

   (c) $[Zn(H_2O)_4][CdCl_4]$; (d) $K[Ag(CN)_2]$; (e) none of these.

8. Which of the following is a chelating agent? (a) H_2O; (b) $H_2NCH_2CH_2NH_2$; (c) HCl; (d) NH_3; (e) $S_2O_3^{2-}$.

QUIZ D

1. $[Cr(en)(NH_3)(H_2O)_2]^{3+}$ has a coordination number of (a) 3; (b) 4; (c) 5; (d) 6; (e) none of these.

2. Al^{3+} shows a coordination number of (a) 2 only; (b) 2 or 4; (c) 4 only; (d) 4 or 6; (e) 6 only.

3. The hydroxo ligand has the formula (a) OH^-; (b) H_2O; (c) O^{2-}; (d) $C_2O_4^{\ 2-}$; (e) none of these.

4. The name of $Na_2[CoCl_4(NO)_2]$ is ______________________________.

5. The formula of potassium tetracyanodimethylamineferrate(II) is ____________.

6. A $[Co(OH)_2(H_2O)_4]^+$ ion has how many unpaired electrons? (a) 0; (b) 4; (c) 1; (d) 3; (e) none of these.

7. Geometric isomerism can be shown by (a) $[Ag(CN)(NH_3)]$; (b) $Na_2[Cd(NO_2)_4]$; (c) $[PtCl_4I_2]$; (d) $[Au(CN)_4][PtCl(NH_3)_3]$; (e) none of these.

8. One mole of a compound with empirical formula $CoCl_3 \cdot 4NH_3$ yields 1 mole of AgCl on treatment with excess $AgNO_3(aq)$. Ammonia is not removed by treatment with concentrated H_2SO_4. The formula of this compound is (a) $Co(NH_3)_4Cl_3$; (b) $[Co(NH_3)_4Cl_3$; (c) $[CoCl_3(NH_3)_3]NH_3$;

   (d) $[CoCl_2(NH_3)_4]Cl$; (e) none of these.

SAMPLE TEST

1. For each of the following complexes, draw the *d* orbital splitting diagram and predict how many unpaired electrons are in the complex.

   A. $[Co(CN)_4]^{2-}$ B. $W(CO)_6$ C. $[MnI_4]^{2-}$ D. $[Fe(CN)_6]^{3-}$

2. A compound contains 24.8% Na, 34.2% Cu, and 41.0% F. A 0.010 M solution of this compound has an osmotic pressure at 298 K of 559 mmHg. What is the formula and the name of this compound?

25 NUCLEAR CHEMISTRY

CHAPTER OBJECTIVES

1. Name the different types of radioactive decay processes and describe the characteristics of their radiation.

Alpha (α) *decay* involves nuclides with an atomic number larger than 83 ($Z > 83$) and a mass number larger than 200 ($A > 200$). It often leaves the nucleus in an excited state. The alpha particle is a ^{4}He nucleus, $^4He^{2+}$. Thus, the atomic number decreases by two units and the mass number by four units. The alpha particle has great ionizing power but poor penetrating power; it can be stopped by a sheet of paper.

Beta (β) *decay* occurs when electrons are emitted from the nucleus. The nuclide's mass number is unchanged, but its atomic number increases by one. Thus, β decay has the effect of changing a neutron to a proton. β particles are less ionizing but more penetrating than α particles. They are stopped by about 0.5 mm of aluminum, whereas an α particle is stopped by 0.015 mm of Al.

Gamma (γ) *decay* is the emission of high-energy photons. This is how the nucleus gets rid of excess energy. It occurs within 1 nanosecond of some other decay process. If the high-energy (excited) nucleus survives for more than a nanosecond, it is called an *isomer* and its gamma decay is an *isomeric transition* (IT). Gamma rays have relatively little ionizing power but great penetrating power. They are stopped by 5 to 11 mm of Al.

Positron (β^+) *emission* occurs only in artificial nuclides. A positron is a positively charged electron. Positron emission leaves the mass number unchanged and decreases the atomic number by 1. Thus, positron emission has the effect of changing a proton to a neutron. A positron has almost no penetrating power; it is destroyed as soon as it encounters an electron.

Electron capture (EC) occurs when a nucleus captures or absorbs one of its own electrons, usually one in the *K* or *L* shell. An *isomer* is formed that emits X-rays or γ rays to get rid of its excess energy.

***2. Complete nuclear equations for radioactive decay processes.**

Make sure that the sum of the mass numbers and the sum of the atomic numbers are the same on each side of the equation. Recall (Objective 2-8) that the mass number is the preceding superscript of the atomic symbol and the atomic number is the preceding subscript. A decay process has one nuclide on the left side (the "parent" nucleus) and another on the right side (the "daughter" nucleus), along with the emitted particle: alpha ($^4_2$He), beta ($^{\ 0}_{-1}$e), neutron ($^1_0$n), or positron ($^0_1$e).

$$\text{parent nucleus} \longrightarrow \text{daughter nucleus} + \text{emitted particle} \qquad [1]$$

EXAMPLE 25-1 Alpha decay of ^{210}Po yields what nuclide?

The incomplete decay equation is $^{210}_{84}\text{Po} \longrightarrow ? + ^4_2\text{He}$. The daughter has an atomic number of 82 ($84 = 82 + 2$) and a mass number of 206 ($210 = 206 + 4$). An atomic number of 82 is Pb. Thus, $^{206}_{82}$Pb is formed.

3. Describe the three natural radioactive decay series, with the uranium series as an example.

$^{238}_{92}U$, $^{232}_{90}Th$, and $^{235}_{92}U$ are the naturally occurring parents of three decay series: the uranium series, the thorium series, and the actinium series, respectively. These three series are shown in Figure 25-1. Note that the atomic numbers (on the horizontal axis) are offset so that the series are not plotted on top of each other. The thorium series is the $4n$ series (the mass numbers of its nuclides are multiples of 4). The uranium series is the $4n+2$ series, and the actinium series is the $4n+3$ series. Half-lives and isotope masses of the nuclides in these three series are given in Table 25-1.

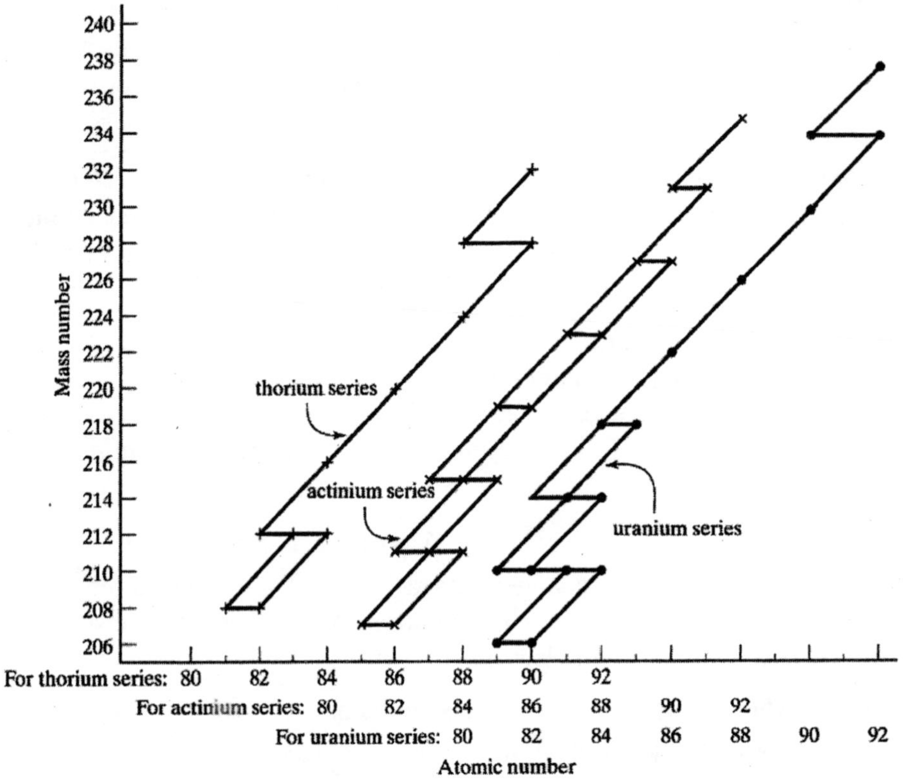

FIGURE 25-1 The Thorium, Actinium, and Uranium Decay Series.

Uranium series			Actinium series			Thorium series		
Nuclide	*Half-life*	*Mass, u*	*Nuclide*	*Half-life*	*Mass, u*	*Nuclide*	*Half-life*	*Mass, u*
^{238}U	7.13×10^{8} y	235.0439	^{235}U	4.51×10^{9} y	238.0508	^{232}Th	1.41×10^{10} y	232.0381
^{234}Th	24.1 d	234.0458	^{231}Th	25.5 h	231.0363	^{228}Ra	6.7 y	228.0311
^{234}Pa	6.75 h	234.0433	^{231}Pa	3.25×10^{4} y	231.0359	^{228}Ac	6.13 h	228.0310
^{234}U	2.47×10^{5} y	234.0409	^{227}Ac	21.6 y	227.0278	^{228}Th	1.91 y	228.0287
^{230}Th	8.0×10^{4}y	230.0331	^{227}Th	18.17 d	227.0278	^{224}Ra	3.64 d	224.0202
^{226}Ra	1602 y	226.0254	^{223}Fr	22 min	233.0198	^{220}Rn	53.3 s	220.0114
^{222}Rn	3.82 d	222.0175	^{223}Ra	11.4 d	223.1086	^{216}Po	145 ms	216.0019
^{218}Po	3.05 min	218.0089	^{219}At	54 s	219.0113	^{212}Pb	10.64 h	211.9919
^{218}At	2.0 s	218.0086	^{219}Rn	4.0 s	219.0095	^{212}Bi	60.6 min	211.9913
^{214}Pb	26.8 min	213.9998	^{215}Bi	8 min	215.0018	^{212}Po	0.30 µs	211.9889
^{214}Bi	19.7 min	213.9987	^{215}Po	1.8 ms	214.9995	^{208}Tl	3.10 min	207.9820
^{214}Po	150 µs	213.9952	^{215}Pb	100 µs	214.9987	^{208}Pb	stable	207.9766
^{210}Tl	1.32 min	209.9901	^{211}Pb	36.1 min	210.9887			
^{210}Pb	20.4 y	209.9842	^{211}Bi	2.15 min	210.9873			
^{210}Po	138 d	209.9829	^{207}Tl	4.78	206.9775			
^{206}Tl	4.19 min	205.9761	^{207}Pb	stable	206.9757			
^{206}Pb	stable	205.9745						

TABLE 25-1 Half-Lives and Masses of Nuclides in the Three Naturally Occurring Radioactive Series

***4. Write equations for the artificial production of nuclides.**

Many of these equations look like Equation [2].

$$\text{target nucleus} + \text{bombarding particle} \longrightarrow \text{emitted particle} + \text{product nucleus} \qquad [2]$$

Occasionally, more than one particle and/or product results. A shorthand notation for Equation [2] is Equation [3].

TABLE 25-2 Particles Involved in Nuclear Reactions

Name	*Complete symbol*	*Abbreviated symbol*	*Mass, u*
alpha particle	$^{4}_{2}He^{2+}$	α	4.001507
beta particle *or* electron	$^{0}_{-1}e$	β *or* $\beta-$	0.000548
neutron	$^{1}_{1}H^{+}$	n	1.008665
proton	$^{2}_{1}H^{+}$	p	1.007277
deuteron	$^{4}_{2}He^{0}$	d	2.013554
helium atom	$^{0}_{1}e$		4.002603
positron	$^{0}_{1}e$	β^{+}	
hydrogen atom	$^{1}_{1}H^{0}$		1.007825
deuterium atom	$^{2}_{1}H^{0}$		2.014102

$$\text{target nucleus (bombarding particle, emitted particle) product nucleus} \qquad [3]$$

Thus, the bombardment of ^{242}Cm by alpha particles is Equation [4] or Equation [5].

$$^{242}_{96}Cm + ^{4}_{2}He \longrightarrow ^{245}_{98}Cf + ^{1}_{0}n \qquad ^{242}_{96}Cm(\alpha,n)^{245}_{98}Cf \qquad [4 \;\&\; 5]$$

In the shorthand notation, we use the abbreviated symbols of the particles given in Table 25-2. Predicting products is not expected of general chemistry students. But by balancing mass numbers and atomic numbers, we can predict one of the four species in Equation [2], given the other three.

5. Name some of the transuranium elements and describe how they are made.

The transuranium elements are those with atomic numbers greater than 92 and are made by bombarding lighter nuclei with other particles. The equations for their production are given in Table 25-3. The names and discoverers of elements with atomic numbers higher than 103 have yet to be officially decided. There may be two sets of discoverers listed. In Table 25-3, those names preferred in the United States are in brackets. In the interim, element symbols and names are based on their atomic numbers, stringing together the syllables *nil* (0), *uni* (1), *bi* (2), *tri* (3), *quad* (4), *pent* (5), *hex* (6), *sept* (7), *oct* (8), and *enn* (9) and ending with *-ium*. Thus, element 108, which has been reported, is uniniloctium

Neptunium, plutonium, americium, curium, berkelium, and californium are available commercially in kilogram quantities. Where two reactions are given in Table 25-3 with only one set of discoverers, the first is the reaction of discovery. The second reaction is that of commercial preparation. You should be familiar with the commercial preparations for neptunium through californium.

6. Describe how a charged-particle accelerator operates.

A charged-particle accelerator produces beams of charged particles that have very high energies per particle. Often, these high-energy particles are atomic nuclei and, thus, are positively charged. Charged particles are accelerated by an electrical potential that exerts a force on them. In a circular accelerator, the polarity of the electric potential (whether it is positive or negative) in each half of the accelerator alternates as the particles move in a nearly circular path. The field alternates to always accelerate the particles. Such circular accelerators are also called *cyclotrons*.

In a synchrotron, the beam of particles travels through a donut-shaped tube. Magnetic fields are used to keep the beam in the center of the tube (focusing magnets) and to bend the beam into a circular path (deflection magnets). The particles are accelerated by many electrical fields spaced around the ring (or donut). The voltage of each field varies so that the particles are accelerated as they pass. As the particles move around the ring more rapidly, the field must vary more rapidly. (Think of a steel ball traveling around the outer track in a roulette wheel. If you want to speed up the ball by hitting it each time it passes, you have to hit it more often as it goes faster.) The strength of the deflection magnets also must increase to keep the particles in their circular path.

Linear accelerators also use electrical fields of increasing strength to accelerate particles. However, particles go around a synchrotron many times, being accelerated constantly. Thus, a synchrotron can accelerate a particle to a much higher speed than can a linear accelerator.

*** 7. Given two of the following quantities, calculate the third: rate of radioactive decay, half-life, and number of atoms in a sample of a radioactive nuclide.**

It is very helpful to realize that radioactive decay obeys first-order kinetics. Thus, radioactive decay problems can be solved with the methods of Chapter 14. The three quantities mentioned are related in Equations [6] and [7].

TABLE 25-3 Production of Transuranium Elements

Element	*Equation*	*Discoverers and date*
neptunium	$^{238}U(n,\beta)^{239}Np$	McMillan and Abelson
	$^{235}U(n,\gamma)^{236}U(n,\gamma)^{237}U(,\beta)^{237}Np$	Berkeley, CA (1940)
plutonium	$^{238}U(d,2n)^{238}Np(,\beta)^{238}Pu$	Seaborg, McMillan, Kennedy, Wahl
	$^{238}U(n,\gamma)^{238}Np(,\beta)^{238}Pu$	Berkeley, CA (1940)
americium	$^{239}Pu(n,\gamma)^{240}Pu(n,\gamma)^{241}Pu(,\beta)^{241}Am$	Seaborg, James, Morgan, Ghiorso
		Chicago, Il (1944)
curium	$^{239}Pu(\alpha,n)^{242}Cm$	Seaborg, James, Ghiorso
	$^{241}Am(n,\gamma)^{242m}Am(,\beta)^{242}Cm$	Berkeley, CA (1944)
berkelium	$^{241}Am(\alpha,2n)^{243}Bk$	Thomson, Ghiorso, Seaborg
		Berkeley, CA (Dec 1949)
californium	$^{242}Cm(\alpha,n)^{245}Cf$	Thomson, Street, Ghiorso, Seaborg
	$^{249}Bk(n,\gamma)^{250}Bk(,\beta)^{250}Cf$	Berkeley, CA (1950) & in nuclear reactors
einsteinium	$^{238}U(15\,n,7\,\beta)^{253}Es$	Ghiorso and colleagues
	products of first hydrogen bomb	Berkeley, CA (Dec 1952)
fermium	$^{238}U(^{16}O,4\,n)^{250}Fm$	Ghiorso and colleagues
	products of first hydrogen bomb	Berkeley, CA (Dec 1952)
mendelevium	$^{253}Es(\alpha,n)^{256}Md$	Ghiorso, Harvey, Choppin, Thompson, Seaborg Berkeley, CA (1955)
nobelium	$^{246}Cm(^{12}C,\ n)^{254}No$	Ghiorso, Sikkeland, Walton, Seaborg
		Berkeley, CA (Apr 1958)
lawrencium	$^{252}Cf(^{11}B,5\,n)^{258}Lw$	Ghiorso, Sikkeland, Larsh, Latimer
		Berkeley, CA (Mar 1961)
element 104	$^{242}Pu(^{22}Ne,4\,n)^{260}Ku$	Joint Nuclear Research Institute
(kurchatovium)		Dubna, Russia (1964)
[rutherfordium]	$^{249}Cf(^{12}C,4\,n)^{257}Rf$	Ghiorso and colleagues
		Berkeley, CA (1969)
element 105	$^{243}Am(^{22}Ne,4n)^{261}Unp$	Joint Nuclear Research Institute
	(no name proposed)	Dubna, Russia (early 1970)
[hanium]	$^{249}Cf(^{15}N,4\,n)^{263}Ha$	Ghiorso and colleagues
		Berkeley, CA (Mar 1970)
element 106	$^{249}Cf(^{18}O,4\,n)^{263}Sg$	Lawrence Berkeley Laboratories
[seaborgium]		Berkeley, CA (Sept 1974)
	$^{206-208}Pb(^{54}Cr,?)^{?}Ns$	Joint Nuclear Research Institute
	(no name proposed)	Dubna, Russia (June, 1974)
element 107	$^{204}Bi(^{54}Cr,?)^{?}Ns$	Joint Nuclear Research Institute

[nielsbohrium]	(no name proposed)	Dubna, Russia (1976)
element 108	^{265}Hs	Armbruster, *et al.* Heavy Ion Research Lab
[hassium]		Darmstadt, (West) Germany (1984)
element 109	$^{209}\text{Bi}(^{58}\text{Fe,n})^{266}\text{Mt}$	Armbruster, *et al.* Heavy Ion Research Lab
[meiterium]		Darmstadt, (West) Germany (1982)
element 110	$^{208}\text{Pb}(^{62}\text{Ni,n})^{269}\text{Uun}$	Hofmann, *et al.* Heavy Ion Research Lab
		Darmstadt, (West) Germany (1994)
	^{272}Uun	Organessian, *at al.* Joint Nuclear Res. Inst.
		Dubna, Russia (1987)
element 111	$^{209}\text{Pb}(^{64}\text{Zn,n})^{272}\text{Uuu}$	Hofmann, *et al.* Heavy Ion Research Lab
		Darmstadt, (West) Germany (1994)
element 112	$^{208}\text{Pb}(^{70}\text{Zn,n})^{269}\text{Uub}$	Hofmann, *et al.* Heavy Ion Research Lab
		Darmstadt, (West) Germany (1996)

$$t_{1/2} = 0.693/\lambda = \text{half - life} \quad [6]$$

$$\text{decay rate} = \lambda \text{N} \quad [7]$$

In these two equations, $t_{1/2}$ is the half-life of the nuclide, λ is the decay constant of the nuclide (a first-order rate constant), and N is the number of atoms in the sample. Decay rate is also known as *activity*.

EXAMPLE 25-2 A 1.00×10^{-6} g sample of a nuclide with an atomic weight of 231 g/mol decays at a rate of 1.00×10^{5} atoms/min. What are the half-life and the identity of the nuclide (refer to Table 25-1)?

$N = 1.00 \times 10^{-6}\ \text{g} \times \text{mol}/231\ \text{g} \times 6.022 \times 10^{23}\ \text{atoms/mol} = 2.61 \times 10^{15}\ \text{atoms}$

Now, we determine the decay constant and the half-life:

$1.00 \times 10^{5}\ \text{atoms/min} = \lambda\,(2.61 \times 10^{15}\ \text{atoms})\ \textit{or}\ \gamma = 3.83 \times 10^{-11}/\text{min}$

$t_{1/2} = 0.693/3.83 \times 10^{-11}/\text{min} = 1.81 \times 10^{10}\ \text{min} \times \text{h}/60\ \text{min} \times \text{d}/24\ \text{h} \times \text{y}/365\ \text{d} = 3.44 \times 10^{4}\ \text{y}$

This period of 3.44×10^{4} years is quite close to the half-life of ^{231}Pa.

*** 8. Determine the ages of rocks from the measured ratio of a stable nuclide to a radioactive one and the ages of carbon-containing materials from the decay rate of ^{14}C.**

The age of a rock can be found from the relative amounts of the parent and the final stable product of one of the three decay series.

$$\ln(\text{N}_t/\text{N}_0) = \ln N_t - \ln N_0 = -\lambda t \quad [8]$$

N_t is the number of atoms of parent nuclide present now, and N_0 is the number of atoms of parent present when the rock was formed.

EXAMPLE 25-3 A rock has 18.05 g of ^{235}U for every 1.00 g of ^{207}Pb. Assume that all the ^{207}Pb was produced by the decay of ^{235}U, and estimate the age of the rock.

The half-life of ^{235}U is 7.13×10^8 y; thus,

$\lambda = 0.693/t_{1/2} = 0.693/7.13 \times 10^8 \text{ y} = 9.72 \times 10^{-10}/\text{y}$

N_t is found from the mass of ^{235}U that is present now.

$18.05 \text{ g } ^{235}U \times \text{mol } ^{235}U/235 \text{ g} \times 6.022 \times 10^{23} \text{ atoms/mol} = 0.291 \times 10^{22} \text{ atoms } ^{235}U$

We also determine the number of ^{207}Pb atoms present:

$1.00 \text{ g } ^{207}Pb \times \text{mol } ^{207}Pb/207 \text{ g} \times 6.022 \times 10^{23} \text{ atoms/mol} = 4.63 \times 10^{22} \text{ atoms } ^{235}U$

Each atom of ^{235}U ultimately produces 1 atom of ^{207}Pb (Figure 25-1). The overall equation is

$^{235}U \longrightarrow ^{207}Pb + 7\,^4He + 4\,\beta^-$.

Thus, $N_0 = N_t +$ number of ^{207}Pb atoms present now $= 4.63 \times 10^{22} + 0.291 \times 10^{22}$ atoms

$= 4.92 \times 10^{22}$ atoms of ^{235}U

$\ln\,(N_t/N_0) = \ln\,(4.63 \times 10^{22}/4.92 \times 10^{22}) = -0.0608 = -\lambda t = -(9.72 \times 10^{-10}/\text{y})t$ or $t = 6.25 \times 10^7$ y.

Equation [8] also can be used when only the decay rate (λN) is known at two different times, as shown in Example 25-4.

EXAMPLE 25-4 A sample of radioactive nuclide has a decay rate of 1.000×10^6 β emissions/min. 100 min. later, the decay rate is 14.7% of the initial rate. What is the half-life of the nuclide? Use the data of Table 25-1 and Figure 25-1 to identify the nuclide.

rate at 100 min $= (1.000 \times 10^6/\text{min})(0.147) = 0.147 \times 10^6/\text{min}$

$\ln(N_t/N_0) = \ln\,(0.147 \times 10^6/\text{min} \div 1.000 \times 10^6/\text{min}) = -1.92 = -\lambda(100 \text{ min})$ or $\lambda = 0.0192/\text{min}$

$t_{1/2} = 0.693/\lambda = 0.693 \div 0.0192/\text{min} = 36.1 \text{ min}$

^{211}Pb is the only nuclide in Table 25-1 with a 36 1 min. half-life, and ^{211}Pb is a beta emitter (Figure 25-1).

*** 9. Calculate the energies associated with nuclear reactions.**

In spontaneous nuclear reactions, the products have less mass than the reactants. The mass difference appears as energy and is determined by Equation [9]. Energies are often expressed in millions of electron volts (MeV = megaelectron volts = 10^6 eV), where $1.000 \text{ MeV} = 1.602 \times 10^{-13}$ J.

$$E = mc^2 \qquad (c = 2.998 \times 10^8 \text{ m/s, the speed of light}) \qquad [9]$$

EXAMPLE 25-5 How much energy in MeV is produced by the alpha (4.0026 u) decay of each ^{235}U atom (235.0439 u)? The daughter is ^{231}Th (231.0363 u).

Mass difference = 235.0429 – (231.0363 + 4.0026) = 0.0050 u

$E = mc^2 = 0.0050\ \text{u} \times 1.660 \times 10^{-27}\ \text{kg/u} \times (2.998 \times 10^8\ \text{m}/s)^2 = 7.47 \times 10^{-13}\ \text{J} \times 1\ \text{MeV}/1.602 \times 10^{-13}\ \text{J}$

$= 4.7\ \text{MeV}$

In calculations of this sort, we must make sure that the masses of the electrons are taken into account. Although ^{235}U emits an alpha particle ($^4He^{2+}$), the calculation is performed as if it emits a helium-4 atom ($^4He^0$).

*** 10. Calculate the average nuclear binding energy per nucleon for a nuclide.**

A nuclide always weighs less than the sum of its particle masses. Consider a ^{12}C atom with a mass of 12.000 u. total particle mass = (6 × electron mass) + (6 × neutron mass) + (6 × proton mass).

$= (6 \times 0.00548\ \text{u}) + (6 \times 1.008665\ \text{u}) + (6 \times 1.007277\ \text{u})$

$= 12.09894\ \text{u}$

The mass defect is the difference between the total particle mass and the mass of the nuclide. The binding energy is the mass defect expressed in units of energy J or MeV.

mass defect = 12.09894 u – 12.00000 u = 0.09894 u

$$\text{binding energy} = 0.09894\ \text{u} \times \frac{1.000 \times 10^{-3}\ \text{kg}}{6.022 \times 10^{23}\ \text{u}} \times (2.998 \times 10^8\ \text{m/s})^2$$

$$= 1.478 \times 10^{-11}\ \text{J} \times 1\ \text{MeV}/1.602 \times 10^{-13}\ \text{J} = 92.26\ \text{MeV}$$

Often, binding energy is expressed as energy per nucleon:

binding energy/nucleon = 92.29 MeV/12 nucleons = 7.690 MeV/nucleon

The higher the binding energy per nucleon, the more stable the nucleus. A heavy nucleus of low stability can break apart or fission to produce more stable light nuclei. On the other hand, two very light nuclei can join together or fuse to produce a more stable heavy nucleus.

*** 11. Describe the factors that determine nuclear stability, establish whether a nuclide is likely to be stable or radioactive, and predict the type of decay process expected for a radioactive nuclide.**

Nuclei tend to be stable if they have a "magic number" of protons or neutrons (Table 25-4). Nuclei with an even number of protons and/or an even number of neutrons also tend to be stable. All nuclei with $Z > 83$ are unstable.

TABLE 25-4 Magic Numbers of Protons (Z) and Neutrons (A)

Z	2	8	20	28	50	82	114		
A	2	8	20	28	50	82	126	184	196

The unstable nuclei decay by α, β, or β^+ emission, or by electron capture. We can predict the type of decay: α decay if in the belt of stability with $Z > 83$, β decay if above and to the left of the belt of stability, and β^+ decay or electron capture if below and to the right of the belt of stability.

12. Describe the processes of nuclear fission and nuclear fusion, including the problems with using them as energy sources.

Nuclear fission is the splitting apart of large nuclei into more stable ones, a process often initiated by neutrons. The fission reaction of ^{235}U is not a simple process. Two possibilities are given in Equations [10] and [11]:

$$^{235}U + {}^{1}n \longrightarrow {}^{144}Xe + {}^{90}Sr + 2\,{}^{1}n \qquad [10]$$

$$^{235}U + {}^{1}n \longrightarrow {}^{140}Ba + {}^{93}Kr + 3\,{}^{1}n \qquad [11]$$

These reactions also occur in the atomic bomb. About 3.20×10^{-11} J or 200 MeV is produced by each ^{235}U fission. Once started by a stray neutron, the reaction continues because more neutrons are produced than are consumed. But the neutrons given off are moving too fast to efficiently split ^{235}U nuclei. They are slowed down by the graphite moderator in a nuclear reactor. The ^{1}H nuclei of the reactor's cooling water also serve as moderators. Boron control rods absorb neutrons to stop the reaction.

$$^{10}B + {}^{1}n \longrightarrow {}^{7}Li + {}^{4}He \qquad [12]$$

The ^{1}H nuclei of the cooling water also absorb neutrons. Because of this, ^{235}U fuel is enriched to 1–4% from a natural 0.7% abundance. If heavy water is used as cooling water, however, the fuel need not be enriched. ^{1}H nuclei absorb nuclei 600 times better than do the ^{2}H nuclei of heavy water. The CANDU reactor of Canada uses heavy water as a coolant. The savings are considerable; enriched fuel is about five times as expensive to produce as the fuel used by the CANDU reactor.

In breeder reactors, ^{239}Pu is produced from ^{238}U (natural abundance = 99.3%). See Table 25-3 for the reactions. Plutonium has the benefits of an abundant raw material (^{238}U) and no need for moderation because fast neutrons are more efficient than slow ones. Two disadvantages are plutonium's toxicity and its use in nuclear weapons.

In a fusion reaction, light nuclei bind together to form heavier, more stable ones. The overall reaction in the sun is:

$$4\,{}^{1}H \longrightarrow {}^{4}He + 2\,\beta^{-} + 26.7\text{ MeV} \qquad [13]$$

Two fusion reactions are being considered for commercial use:

$$^{6}Li + {}^{2}H \longrightarrow 2\,{}^{4}He + 22.4\text{ MeV} \qquad [14]$$

$$5\,{}^{2}H \longrightarrow {}^{4}He + {}^{3}He + {}^{1}H + 2\,{}^{1}n + 24.9\text{ MeV} \qquad [15]$$

Although the products of Reactions [14] and [15] are not radioactive, the high gamma and neutron radiation that accompany fusion reactions will cause the coolant and the reactor vessel to become radioactive.

All nuclear reactors have the advantage of producing large quantities of energy from small quantities of fuel. In addition, they do not contribute to pollution or to the CO_2 burden of the atmosphere, as do fossil fuel power plants. On the other hand, the reactors produce nuclear radiation while they are operating and leave the reactor and the spent fuel radioactive. The disposal of the spent fuel is a problem for which no final solution has been adopted.

13. Explain the effects of ionizing radiation on matter and describe several radiation-detection devices based on these effects.

When ionizing radiation passes through matter, it leaves a trail of ions in its path. It does this by knocking electrons off the atoms and molecules that it encounters. Of course, the bonds in molecules are severely disrupted by these ionizations. In the cloud chamber and the bubble chamber, the ions promote the formation of droplets of liquid and bubbles of vapor, respectively. In a Geiger-Muller counter, ions create a pathway for an electric spark. Because of their lack of charge, neutrons penetrate matter as far as gamma rays, but they produce few ions. They damage matter by combining with stable nuclei to produce new ones that are often radioactive, a process known as *transmutation.*

14. Discuss methods of expressing radiation dosages, some biological hazards of ionizing radiation, and sources of radiation to which the general population is exposed.

The units of radiation dosage are listed in Table 25-5. *Rem* is the abbreviation for "radiation equivalent—man" and depends on the amount of biological damage. Radiation can kill directly and quickly by ionizing molecules to stop an organism from functioning. In this way, bacteria can be killed and food is preserved by radiation processing. (Of course, the food does not become radioactive.) Radiation can disrupt individual cells, causing some of them to reproduce without limit. Cancer is one result of such uncontrolled growth. Radiation can change genetic material, causing mutations in offspring. Radiation also can destroy only part of an organism, as a serious burn from fire or chemicals will destroy tissue.

TABLE 25-5 Units of Radiation Dosage

Unit	Definition
Curie	An amount of radioactive material decaying at the same rate as 1 g of radium $(3.7\times10^{10}\ \text{dis/s})$.
Rad	A dosage of radiation able to deposit 1×10^{-2} J of energy per kilogram of matter.
Rem	A unit related to the rad, but taking into account the varying effects of different types of radiation of the same energy on biological matter. This relationship is through a "quality factor," which may be taken as equal to 1 for X-rays, γ rays, and β particles. For protons and slow neutrons, the factor has a value of about 5; for α particles, 10. Thus, an exposure to 1 rad of X-rays is about equal to 1 rem, but 1 rad of α particles is equal to 10 rem.

Sources of α radiation are relatively harmless when external to the body and extremely hazardous when taken internally, as in lungs and stomach. Other forms of radiation (X-rays, γ rays), because they are highly penetrating, are hazardous even when external to the body.

TABLE 25-6 Decay Modes, Half-lives, and Masses of Some Nuclides

			Nuclide mass, u	
Nuclide	Decay mode	Half-life	Parent	Daughter
^{3}H	β^-	12.26 y	3.016050	3.016030
^{6}He	β^-	0.81 s	6.018893	6.015125
^{7}Be	EC	53.37 d	7.016929	7.016004

^{8}Be	β^-	2×10^{-16} s	8.0053	4.002603
^{12}B	β^+	0.02 s	12.0143	8.022487
^{13}N		9.96 m	13.005738	13.003354
^{37}Ar	EC	35 d	36.966772	36.965898
^{61}Co	β^-	1.65 h	60.932440	60.931056
^{90}Sr	β^-	28 y	89.907747	89.907163
^{98}Tc	β^-	1.5×10^{6} y	97.907110	97.905289
^{109}In	β^+	4.3 h	108.907096	108.904928
^{161}Tm	EC	30 m	160.933730	160.929950
^{194}Au	β^+	39.5 h	193.965418	193.962725

The isotopes of the naturally occurring decay series are spread throughout the environment. Starting with the rock (such as granite) in which uranium and thorium commonly are found, the parents and daughters are spread by weathering processes. Certain plants concentrate these elements and, thus, are a source of these isotopes. Mining activities of all kinds speed up the dispersion of these elements by bringing rocks to the surface, where they are exposed to the weather and erode faster. In addition, a considerable fraction of the total radiation enters from outer space. All of these sources produce an annual background of 130 mrem per person. X-rays for dentistry and medicine also are a source of radiation, as are nuclear power plants.

15. Discuss some practical, beneficial uses of radioisotopes.

Radioisotopes are widely used in medicine to deliver radiation to the cancer cells to be destroyed without irradiating many normal cells. This is done by injecting the radioisotope or by using natural body processes. (For example, iodine will concentrate in the thyroid gland.) Radioisotopes are used as tracers to detect the fate of certain substances during plant and animal growth, as well as in industrial processes and chemical reactions. Certain radioisotopes concentrate in specific organs of the body. The organ then can be readily seen by a radiation-detecting camera, allowing us to see abnormalities. Radioisotopes also are used industrially in quality control, as in the production of aluminum foil or plastic film. The isotope is placed below the moving sheet and a detector above. The level of radiation reaching the detector depends on the thickness of the aluminum or plastic.

SELF-ASSESSMENT EXERCISES

80. The decay rate = λN, where λ is the decay constant and N is the number of atoms in the sample.

$$\lambda = \frac{0.693}{t_{1/2}} = \frac{0.693}{1\text{ hr.}} \times \frac{1\text{ hr.}}{3.6 \times 10^3\text{ s}} = 1.92 \times 10^{-4}\text{s}^{-1}$$

Decay Rate = 1000 atoms s^{-1} = $1.92 \times 10^{-4}\ s^{-1} \times N$

N = 5.15 x 10^6 atoms and $\ln\frac{N_t}{N_0} = -\lambda t$ = -1.92 x 10^{-4} s^{-1} x 1.08 x 10^4 s at three hours. N_t/N_0 = 0.125, answer (d) 125 atoms s^{-1}.

82. For 1% to be left, N_t/N_0 = 0.01. $\ln\frac{N_t}{N_0} = -\lambda t$, λ = 0.693/11.4 day = 0.0608 day^{-1}. Substituting into the equation and solving for t, ln(0.01)/-0.0608 day^{-1} = 75.7 days.

83. Decay rate = λN, λ = 0.693/87.9 d = 7.88 x 10^{-3} d^{-1}. Converting to min.,

7.88 x 10^{-3} d^{-1} x 1 day/24 hr. x 60 min./1 hr. = 5.47 x 10^{-6} $min.^{-1}$

Having found λ, we can now use the decay rate and λ to find the number of atoms:

Decay rate = 1.00 x 10^3 atoms $min.^{-1}$, N = Decay rate/λ

N = 1.00 x 10^3 atoms $min.^{-1}$/5.47 x 10^{-6} $min.^{-1}$ = 1.83 x 10^8 atoms.

To calculate the time:

(a) 253 dpm = λN = 5.46 x 10^{-6} $min.^{-1}$N. N = 4.63 x 10^7, $\ln\frac{4.63\times10^7}{1.83\times10^8} = -\lambda t$

t = 2.51 x 10^5 min.

(b) For 104 dpm, N = 104dpm/λ = 104 dpm/5.47 x 10^{-6} = 1.90 x 10^7, and solving for t as in part (a) t = 4.14 x 10^5 min.

(c) For 52 dpm, N = 9.50 x 10^6 and t = 5.41 x 10^5 min.

DRILL PROBLEMS

1. (1) Complete each of the following nuclear equations:

A. $sd^{210}Pb \longrightarrow \_\_\_\_ + ^{210}Bi$ B. $^{8}Be \longrightarrow ^{4}He + \_\_\_\_$ C. $^{205}Pb + \_\_\_\_ \longrightarrow ^{205}Tl$

D. $^{227}Pa \longrightarrow \_\_\_\_ + ^{223}Ac$ E. $\_\_\_\_ \longrightarrow ^{4}He + ^{228}Th$ F. $^{194}Hg + \_\_\_\_ \longrightarrow ^{194}Au$

G. $\_\_\_\_ \longrightarrow \beta- + ^{200}Hg$ H. $^{6}He \longrightarrow \beta- + \_\_\_\_$ I. $^{191}Os \longrightarrow \_\_\_\_ + ^{191}Ir$

J. $\_\_\_\_ \longrightarrow \beta- + ^{17}O$ K. $^{199}Pb \longrightarrow \_\_\_\_ + ^{199}Tl$ L. $\_\_\_\_ + \beta- \longrightarrow ^{37}Cl$

M. $^{210}Pb \longrightarrow ^{4}He + \_\_\_\_$ N. $^{191}Hg \longrightarrow \beta- + \_\_\_\_$ O. $\_\_\_\_ \longrightarrow ^{4}He + ^{218}Po$

(2) Write a nuclear equation for the reaction that occurs when each nuclide below decays as indicated:

P. ^{7}Be (EC) Q. ^{191}Hg (β^+) R. ^{177}Pt (α) S. ^{3}H (β^-) T. ^{177}W (EC) U. ^{109}In (β^+)

V. ^{186}Re (β^-) W. ^{235}U (α) X. ^{12}B (α) Y. ^{161}Tm (EC) Z. ^{98}Tc (β^-)

2. (1) Write full and abbreviated equations in each of the following cases:

A. ^{59}Co (p, n)____ B. ^{9}Be (^{6}Li, ____) ^{14}N C. ^{14}N (n, ____) ^{9}Be D. ____ (p, α) ^{12}C

E. ____(p, d) ^{4}He F. ____(d, α) ^{21}Ne G. ^{7}Li (α, ____) ^{10}B H. ^{9}Be (α, n)____

I. $^{35}Cl\,(n, p)$____ J. ____$(p, n)\ ^{44}Sc$ K. $^{27}Al\,(d, \alpha)$____ L. ____$(\alpha, p)\ ^{28}Al$

(2) Expand each of the abbreviated equations given in Table 25-3:

M. discovery of neptunium N. neptunium in reactors O. discovery of plutonium
P. americium Q. discovery of curium R. curium in reactors
S. berkelium T. californium U. mendelevium
V. nobelium W. lawrencium

3. (1) Find $t_{1/2}$ and λ of each nuclide and the activity of the given mass. Use Tables 25-1 and 25-6 as needed.

A. $^{223}Fr, 1.00\ \mu g$ B. $^{61}Co, 4.00\ pg$ C. $^{98}Tc, 7.00\ g$
D. $^{3}H, 1.31\ mg$ E. $^{194}Au, 1.00\ g$ F. $^{228}Ra, 1.32\ g$

(2) Find $t_{1/2}$ and λ of each nuclide, and then find the mass needed to produce the given activity:

G. $^{6}He, 3.3\times10^{15}/s$ H. $^{7}Be, 5.1\times10^{15}/d$ I. $^{194}Au, 19.4\times10^{16}/s$
J. $^{3}H, 187\times10^{20}/m$ K. $^{90}Sr, 8.46\times10^{13}/m$ L. $^{212}Bi, 34.5\times10^{15}/s$

(3) Find $t_{1/2}$ and λ of each nuclide based on mass number, mass, and activity:

M. $A = 214, 45.3\ \mu g, 3.3\times10^{15}/s$ N. $A = 109, 7.56\ pg, 1.87\times10^{9}/s$
O. $A = 212, 0.165\ g, 8.46\times10^{15}/s$ P. $A = 37, 4.20\ \mu g, 9.41\times10^{11}/m$
Q. $A = 215, 25.2\ g, 1.02\times10^{20}/s$ R. $A = 210, 37.2\mu g, 145\times10^{10}/m$

4. (1) Given the initial decay rate, the elapsed time, and the final decay rate (in that order), determine the half-life of each nuclide:

A. $1.43\times10^{8}/s, 20.4\ d, 6.73\times10^{7}/s$ B. $1.94\times10^{7}/m, 12.1\ h, 8.43\times10^{5}/m$
C. $7.93\times10^{12}/s, 1.41\ M, 5.73\times10^{11}/s$ D. $3.14\times10^{5}/s, 8.75\ m, 1.72\times10^{5}/s$
E. $1.77\times10^{12}/m, 7.63\ d, 1.82\times10^{11}/m$ F. $9.32\times10^{6}/s, 1.32\ d, 4.14\times10^{5}/s$
G. $4.17\times10^{10}/m, 4.00\ d, 1.00\times10^{10}/m$ H. $8.00\times10^{12}/s, 10.0\ d, 9.00\times10^{10}/s$

(2) In each part below, the mass number of the radioactive isotope is followed by the mass number of the stable isotope. Then, the weight ratio of the two isotopes (radioactive/stable) in a sample is followed by the half-life of the first isotope (the radioactive one). Assume that none of the stable isotope was present when the sample was formed, and determine the age of the sample.

I. $231, 227, 8.00/0.625, 3.4\times10^{4}\ y$ J. $290, 205, 6.50/1.25, 103\ y$
K. $233, 209, 0.262/1.00, 1.6\times10^{5}$ L. 210, 210, 1.13, 0.132, 19.4 y
M. $41, 41, 7.30/1.00, 8.0\times10^{4}\ y$ N. $41, 41, 1.26/5.00, 5700\ y$

(3) In each part below, the initial rate of decay of a sample is followed by the decay rate at some later time and the half-life (all assumed to be three significant figures) of the nuclide that is decaying. Determine the age of each sample:

O. $3.12\times10^{4}/s, 164/s, 7.20\ y$ P. $6.64\times10^{9}/m, 7.30\times10^{5}/m, 5700\ y$

Q. $1.72\times10^5/\text{s}, 4.44\times10^4/\text{d}, 80000\ \text{y}$ R. $8.52\times10^6/\text{h}, 1.66\times10^4/\text{h}, 5.27\ \text{y}$
S. $1.96\times10^4/\text{h}, 127/\text{d}, 16.0\ \text{y}$ T. $8.95\times10^6/\text{h}, 47.5/\text{s}, 2.60\ \text{y}$

5. Use the data of Tables 25-2 and 25-6 to determine the energy associated with the decay of each of the following nuclides. Determine each energy both in MeV/atom and in kJ/mol

A. ^{3}H B. ^{6}He C. ^{7}Be D. ^{8}Be E. ^{12}B
F. ^{13}N G. ^{37}Ar H. ^{61}Co I. ^{90}Sr J. ^{98}Tc
K. ^{109}In L. ^{161}Tm M. ^{194}Au

6. The nuclide mass in u of the most abundant stable isotope of an element is given in each part below. Compute the average binding energy per nucleon for each nuclide in MeV, then plot the energies you have computed against atomic number.

A. ^{1}H, 1.007825 B. ^{56}Fe, 55.934936 C. ^{4}He, 4.002603
D. ^{58}Ni, 57.935342 E. ^{9}Be, 9.012186 F. ^{64}Zn, 63.929146
G. ^{12}C, 12.000000 H. ^{80}Se, 79.916527 I. ^{16}O, 15.994915
J. ^{84}Kr, 83.911503 K. ^{20}Ne, 19.992440 L. ^{90}Zr, 89.904700
M. ^{24}Mg, 23.985042 N. ^{102}Ru, 101.9004348 O. ^{28}Si, 27.976928
P. ^{114}Cd, 113.903360 Q. ^{32}S, 31.972074 R. ^{130}Te, 129.906238
S. ^{40}Ar, 39.962384 T. ^{138}Ba, 137.905000 U. ^{40}Ca, 39.962589
V. ^{142}Nd, 41.907663 W. ^{48}Ti, 47.947960 X. ^{158}Gd, 157.924178
Y. ^{52}Cr, 51.940513 Z. ^{166}Er, 165.932060

7. Predict whether each of the following nuclides is likely to be stable or radioactive. If radioactive, predict the type of decay process that will occur.

A. ^{8}B B. ^{238}U C. ^{19}F D. ^{119}Cd E. ^{50}Sc
F. ^{212}Po G. ^{88}Sr H. ^{17}F I. ^{226}Ra J. ^{50}Mn
K. ^{106}In L. ^{197}Au M. ^{63}Co

QUIZZES (20 minutes each) Choose the best answer for each question.

QUIZ A

1. The alpha decay of ^{226}Ra produces (a) ^{3}He; (b) ^{4}Li; (c) ^{224}Rn; (d) ^{224}Po; (e) none of these.

2. Which of the following could be a member of the thorium series, which begins with ^{232}Th? (a) ^{223}Fr; (b) ^{216}Po; (c) ^{221}Rn; (d) ^{214}Pb; (e) none of these.

3. Neutron bombardment of ^{23}Na produces an isotope that is a beta emitter. After beta emission, the final product is (a) ^{24}Na; (b) ^{24}Mg; (c) ^{23}Ar; (d) ^{24}Ar; (e) none of these.

4. A nuclide has a decay rate of $2.00\times10^{10}/\text{s}$. 25.0 days layer, its decay rate is $6.25\times10^{8}/\text{s}$. What is the nuclide's half-life? (a) 25.0 d; (b) 12.5 d; (c) 50.0 d; (d) 5.00 d; (e) none of these.

5. A nuclide has a half-life of 35.0 h. What is the value of its decay constant? (a) 0.0285; (b) 0.0198; (c) 24.3; (d) 0.0412; (e) none of these.

6. The binding energy per nucleon is largest for (a) ^{3}He; (b) ^{59}Co; (c) ^{235}U; (d) ^{98}Tc; (e) ^{31}P.

7. The most massive particle is (a) alpha; (b) beta; (c) gamma; (d) positron; (e) neutron.

8. Based on magic numbers, which nuclide is the most stable? (a) ^{3}He; (b) ^{16}O; (c) ^{15}N; (d) ^{119}Sn; (e) ^{206}Pb.

QUIZ B

1. Electron capture by ^{41}Ca produces (a) ^{41}K; (b) ^{40}Ca; (c) ^{42}Ca; (d) ^{41}Sc; (e) none of these.

2. Which could be a member of the actinium series, which begins with ^{235}U? (a) ^{223}Ra; (b) ^{221}Ra; (c) ^{214}Ba; (d) ^{216}Po; (e) none of these.

3. Proton bombardment of ^{230}Th, followed by emission of two alpha particles, produces (a) ^{222}Rn; (b) ^{223}Fr; (c) ^{223}Ra; (d) ^{222}Fr; (e) none of these.

4. A nuclide has a half-life of 1.91 y. Its decay constant has a numerical value of (a) 1.32; (b) 2.76; (c) 0.363; (d) 0.524; (e) none of these.

5. The activity of a radioactive sample declines to 1.00% of its original value in 300.00 days. What is the half-life of the nuclide in this sample? (a) 3.35×10^{-5} d; (b) 0.0145 d; (c) 0.00667 d; (d) 3.00 d; (e) none of these.

6. The binding energy per nucleon is smallest for (a) ^{13}C; (b) ^{3}He; (c) ^{52}Cr; (d) ^{56}Fe; (e) none of these.

7. The most highly charged particle of these is (a) alpha; (b) beta; (c) gamma; (d) positron; (e) neutron.

8. Based on magic numbers, which nuclide is the least stable? (a) ^{96}Nb; (b) ^{119}Sn; (c) ^{40}K; (d) ^{15}O; (e) ^{40}Ca.

QUIZ C

1. The beta decay of ^{90}Sr produces (a) ^{90}Sr; (b) ^{91}Sr; (c) ^{89}Sr; (d) ^{89}Rb; (e) none of these.

2. Which could be a member of the uranium series, which begins with ^{238}U? (a) ^{215}Po; (b) ^{213}Po; (c) ^{210}Po; (d) ^{212}Po; (e) none of these.

3. Alpha-particle bombardment of ^{27}Al, followed by neutron emission, produces (a) ^{30}P; (b) ^{30}Si; (c) ^{31}Si; (d) ^{29}P; (e) none of these.

4. A nuclide's activity decreases 75% in 4.00 days. Its half-life is (a) 2.00 d; (b) 1.00 d; (c) 4.00 d; (d) 0.347 d; (e) none of these.

5. A nuclide is a beta emitter with a 28 y half-life. The ratio of its mass to the mass of its stable product in a sample is 1.00/8.25. If none of the stable product was present initially, what is the age of the sample? (a) 89.9 y; (b) 85.3 y; (c) 231 y; (d) 259 y; (e) none of these.

6. The binding energy per nucleon is largest for isotopes of (a) U; (b) Co; (c) Cs; (d) He; (e) Hg.

7. The most ionizing radiation is of what type? (a) alpha; (b) beta; (c) gamma; (d) electron; (e) neutron.

8. Based on magic numbers, which nuclide is the most stable? (a) ^{91}Nb; (b) ^{91}Zr; (c) ^{58}Co; (d) ^{13}C; (e) ^{20}Ne.

QUIZ D

1. The beta decay of ^{45}Ca produces (a) ^{45}K; (b) ^{45}Sc; (c) ^{44}K; (d) ^{44}Ca; (e) none of these.

2. Which could be a member of the uranium series, which begins with ^{238}U? (a) ^{236}U; (b) ^{217}At; (c) ^{216}Po; (d) ^{219}At; (e) none of these.

3. Deuteron (^{2}H) bombardment of ^{96}Mo, followed by neutron emission, produces (a) ^{98}Tc; (b) ^{97}Tc; (c) ^{97}Nb; (d) ^{98}Mo; (e) none of these.

4. A nuclide has a decay constant of $4.28 \times 10^{-4}/\text{h}$. If the activity of a sample is $3.14 \times 10^{5}/\text{s}$, how many atoms of the nuclide are present in the sample? (a) 2.64×10^{12}; (b) 7.34×10^{8}; (c) 2.04×10^{5}; (d) 4.40×10^{10}; (e) none of these.

5. The activity of a nuclide declines to 10.0 % of its original value in 145 d. What is the decay constant of this nuclide? (a) 0.0159/d; (b) 63.0 d; (c) 0.00690/d; (d) 0.00478/d; (e) none of these.

6. The binding energy per nucleon is smallest for isotopes of (a) Li; (b) Co; (c) Ge; (d) Sc; (e) S.

7. The least penetrating radiation is of what type? (a) alpha; (b) beta; (c) gamma; (d) positron; (e) neutron.

8. Based on magic numbers, which nuclide is the least stable? (a) ^{59}Ni; (b) ^{51}V; (c) ^{122}Sb; (d) ^{16}O; (e) ^{12}C.

SAMPLE TEST (30 minutes)

1. If there were no nuclear explosions, would the ratio of ^{12}C to ^{14}C in living material increase, decrease, or remain the same? Briefly explain why.

2. An atom of ^{253}Es decays to one of ^{237}Np. What kinds of particles are given off in this process, and how many of each are given off?

3. Write or complete, as appropriate, and balance each of the following nuclear equations:

   A. alpha emission by ^{243}Cm B. positron emission by ^{18}F C. electron capture by ^{88}Zr

   D. $^{10}B(\alpha, n)$____ E. $^{45}Sc(\alpha, p)$____ F. ____ $+ ^{2}H \longrightarrow 2\,^{1}n + ^{51}Cr$

4. ^{11}Be decays via the reaction $^{11}Be \longrightarrow \beta^- + ^{11}B + g$. What is the maximum energy of the gamma ray produced if atomic masses are $^{11}B = 11.00931\,u$, $^{11}Be = 11.0216\,u$, $\beta^- = 0.00055\,u$ $(1.000\,u = 932.8\,\text{MeV})$?

5. How many ^{238}U atoms would decay in a mole of ^{238}U during 995 million years? The half-life of ^{238}U is 4.51×10^{9} y.

26 ORGANIC CHEMISTRY

CHAPTER OBJECTIVES

1. Give examples of alkane, alkene, alkyne, and aromatic hydrocarbons.

Hydrocarbons are compounds that contain only carbon and hydrogen. Alkanes have only single bonds between carbon atoms; alkenes have at least one carbon-carbon double bond ($C{=}C$); and alkynes have at least one carbon-carbon triple bond ($C{\equiv}C$). Aromatic hydrocarbons have a benzene ring.

* 2. Draw structural and condensed formulas for hydrocarbons, given systematic (IUPAC) names; and name hydrocarbon molecules, given structural or condensed formulas.

Condensed formulas are simply a way of writing structural formulas all on one line. We have been writing condensed formulas for some time. For example, CH_4 is the condensed formula for methane, and its Lewis structure is its structural formula. The condensed formula of a straight chain hydrocarbon is fairly easy to write. Simply write down the carbons and, after each carbon, write the formula of the hydrogen atoms attached to it. Double and triple bonds are written between a carbon and the preceding carbon or hydrogen. Thus, $CH_3CH_2CH_3$ is propane, while $CH_2{=}CHCH_3$ is the structural formula of propene. If a multi-atom group replaces a hydrogen atom, its formula is enclosed in parentheses. Thus, if a methyl group $—CH_3$ replaces the hydrogen on the central carbon of propene, the structural formula of the resulting compound is $CH_2{=}C(CH_3)CH_3$, better written as $CH_2{=}C(CH_3)_2$.

For *noncyclic hydrocarbons* (those with no ring of carbon atoms), the IUPAC rules for naming follow:

(1) The longest hydrocarbon chain determines the base name (see Table 26-1).

(2) The longest chain must contain any double or triple bonds. After that requirement is satisfied, and there is more than one possible longest chain, choose the one that has the greatest number of side chains. (This ensures that the side chains—the *substituents*—are as simple as possible,)

(3) Add *-ane* to the base name if there are no multiple bonds, *-ene* for a $C{=}C$ bond, and -*yne* for a $C{\equiv}C$ bond.

(4) The position of the multiple bond is shown by a number that precedes the base name and is separated from it by a hyphen. The number must be as small as possible. There are always two ways to number the longest carbon chain, depending on which end you start with.

(5) *Alkyl groups* (side chains or branches) are named by changing their alkane names from -*ane* to *-yl* (see Figure 26-1). Their positions on the chain are indicated by the number (2, 3, 4, and so on) of the carbon to which they are attached. This number is separated from the alkyl group name by a hyphen. The alkyl group numbers must be as small as possible, after Rule (4) is satisfied. There must be a number prefix for *each* side chain even if the same number is repeated.

(6) The number of alkyl groups of one type is given as a prefix (*di-*, *tri-*, *tetra-*, *penta-*, *hexa-*, *hepta-*, and so on) to the alkyl group name. (This prefix follows the numbers that indicate positions on the main chain. Thus, one would say 2,2,3,4-tetramethylpentane).

(7) The side chains are named in alphabetical order.

(8) Separate numbers from each other by commas and from letters by hyphens, with no spaces in a name.

TABLE 26-1 The First Ten Alkanes

Chain length	*Formula*	*Name*	*Chain length*	*Formula*	*Name*
1	CH_4	methane	6	$CH_3(CH_2)_4CH_3$	hexane
2	CH_3CH_3	ethane	7	$CH_3(CH_2)_5CH_3$	heptane
3	$CH_3CH_2CH_3$	propane	8	$CH_3(CH_2)_6CH_3$	octane
4	$CH_3(CH_2)_2CH_3$	butane	9	$CH_3(CH_2)_7CH_3$	nonane
5	$CH_3(CH_2)_3CH_3$	pentane	10	$CH_3(CH_2)_8CH_3$	decane

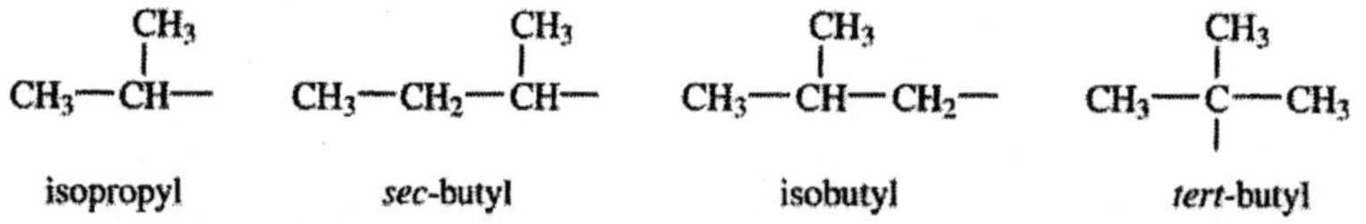

FIGURE 26-1 Names and Structures of Some Alkyl Groups

For *cyclic hydrocarbons,* the IUPAC rules for naming are as follows:

(9) The number of carbons in the ring determines the base name, which always begins with *cyclo-* for nonaromatic hydrocarbons.

(10) A double or triple bond in the ring is assumed to be between carbons 1 and 2. Numbering proceeds around a ring so that the side chains are attached at the lowest possible numbers.

(11) Cyclic hydrocarbons names then follow Rules 5, 6, 7, and 8 above.

Aromatic hydrocarbons are named as derivatives of benzene:

(12) Side-chain positions are designated by the smallest possible numbers, the largest side chain on carbon 1.

(13) For compounds with only two groups on the ring, the position of the second group can be named. Positions 2 and 6 are ortho (*o-*), 3 and 5 are meta (*m-*), and 4 is para (*p-*).

(14) If aromatic hydrocarbons are named as derivatives of toluene, the methyl group is on carbon 1.

(15) Aromatic hydrocarbon names then follow Rules 5, 6, 7, and 8 above.

*** 3. Determine all the possible skeletal isomers of simple hydrocarbons of given formulas.**

An abbreviated way of drawing hydrocarbon molecules is to draw only the bonds. All bonds are shown except those to hydrogen atoms. Abbreviated skeletons are drawn in Figure 26-2.

EXAMPLE 26-1 Use the abbreviated skeletal notation to draw and name all the isomers of formula C_5H_{10}. These isomers are the last nine in Figure 26-2, starting with cyclopentane.

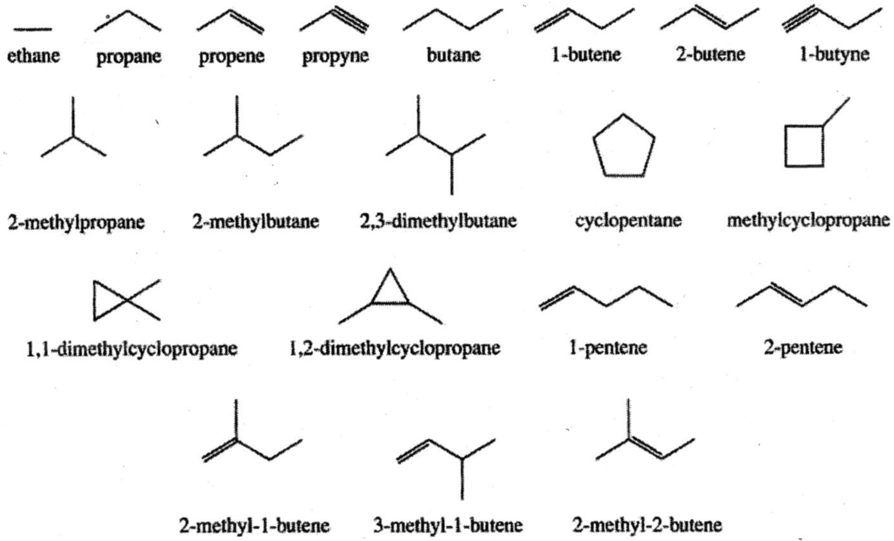

FIGURE 26-2 Abbreviated Skeletons of Some Hydrocarbons

*** 4. Discuss the physical properties of aliphatic and aromatic hydrocarbons in relation to their bonding, structures, and molecular weights.**

Within a series of hydrocarbons, such as the alkanes, boiling points increase with molecular weight. London forces between larger molecules are stronger, and it is more difficult to disrupt these stronger intermolecular attractions. The melting points and the molar heats of fusion and vaporization increase with mole weight for the same reason. A branched molecule boils at a lower temperature than does a straight-chain one with the same number of carbons. The branched molecule has less surface area and, hence, less area over which intermolecular forces can act. The melting points and the molar heats of fusion and vaporization also decrease as branching increases. Finally, we notice that the number of π electrons in a molecule increases the intermolecular forces. For molecules with the same number of carbons, those with more π electrons have larger molar heats of fusion and vaporization (although there is no clear trend in melting and boiling points). Data illustrating these predictions are given in Table 26-2. Predictions based on molecular weight are the best ones. Boiling points and molar heats of vaporization are predicted better than are melting points and molar heats of fusion.

TABLE 26-2 Physical Properties of Hydrocarbons

Effect of molecular weight		M,	m.p.,	b.p.,	ΔH_{fus}	ΔH_{vap}
Name	*Formula*	g/mol	°C	°C	kJ/mol	kJ/mol
methane	CH_4	16.04	−182.5	−164	0.937	8.908
ethane	C_2H_6	30.1	−183.3	− 88.6	2.862	15.648
propane	C_3H_8	44.1	−181.7	− 42.1	3.527	20.133
butane	C_4H_{10}	58.1	−138.3	− 0.5	4.661	24.271
pentane	C_5H_{12}	72.2	−129.7	36.1	8.427	27.593

hexane	C_6H_{14}	86.2	– 95.3	69.0	13.079	31.911
heptane	C_7H_{16}	100.2	– 90.6	98.4	14.163	35.187
octane	C_8H_{18}	114.2	– 56.8	125.7	20.652	38.581

Effect of branching—heptanes		m.p.,	b.p.,	ΔH_{fus}	ΔH_{vap}
Name	*Structure*	°C	°C	kJ/mol	kJ/mol
n-heptane		– 90.6	98.4	14.163	35.187
2-methylhexane	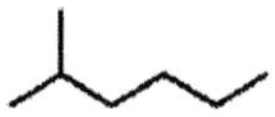	–118.2	90.0	8.870	35.727
2,2-dimethylpentane	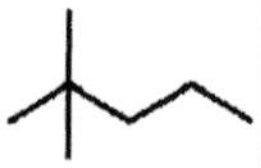	– 123.8	79.2	5.862	33.920
2,4-dimethylpentane	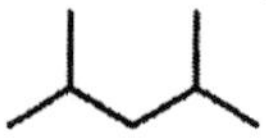	– 119.9	89.8	6.686	34.171
3,3-dimethylpentane	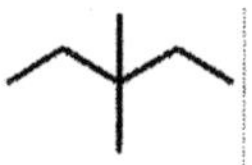	– 134.9	80.5	7.067	34.079
3-ethylpentane	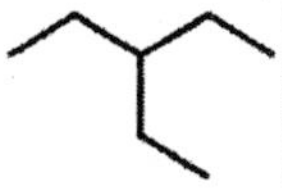	–118.6	93.5	9.552	36.162
2,2,3-trimethylbutane	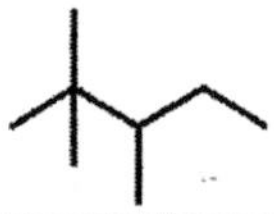	– 25.0	80.9	2.201	32.497

Effect of π bonding[b]		m,	m.p.,	b.p.,	ΔH_{fus}	ΔH_{vap}
Name	*Structure*	g/mol	°C	°C	kJ/mol	kJ/mol
n-pentane		72.2	–129.7	36.1	8.427	27.593
1-pentene		70.4	– 166.2	30.0	5.837	28.999
2-pentene		70.4	– 140.2	36.4	8.389	n.a.[c]
1,3-pentadiene		68.1	–148.8	42.0	6.142	30.548
cyclohexane	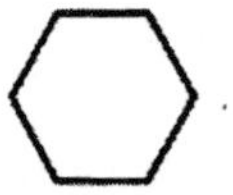	84.16	6.6	80.7	2.632	32.765
cyclohexene	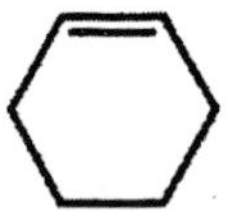	82.15	–103.5	82.9	3.293	n.a.[c]
benzene		78.12	5.53	80.1	9.954	42.903

---

[b] five-carbon chains and six-carbon rings

[c] n.a. means data are not available.

[c] n.a. means data are not available.

straight-chain alkanes [b]five-carbon chains and six-carbon rings [c]n.a. menas data are not available

5. **Write equations for several reactions used in the preparation of alkanes, alkenes, and alkynes.**

These equations are summarized below. Those indicated by your instructor should be memorized. R represents any alkyl group.

(a) Hydrogenation of multiple bonds.

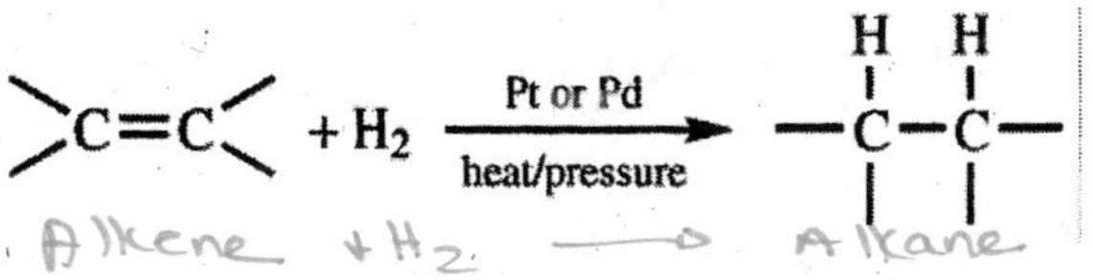

$$>C=C< \; + H_2 \xrightarrow[\text{heat/pressure}]{\text{Pt or Pd}} \; -\overset{H}{\underset{|}{C}}-\overset{H}{\underset{|}{C}}- \tag{1}$$

Alkene + H$_2$ ⟶ Alkane

$$-C\equiv C- \; + H_2 \xrightarrow[\text{heat/pressure}]{\text{Pt or Pd}} \; -\overset{H}{\underset{H}{C}}-\overset{H}{\underset{H}{C}}- \tag{2}$$

Alkyne + H$_2$ $\xrightarrow[\Delta,]{Pt\,Pd}$ Alkane.

b) Wurtz reaction: chain length is doubled.

$$2\,R{-\!}Br + 2\,Na \longrightarrow 2\,NaBr + R{-\!}R \tag{3}$$

c) Alkali carbonates (such as Na_2CO_3, K_2CO_3) fused with alkali hydroxides (such as NaOH, KOH).

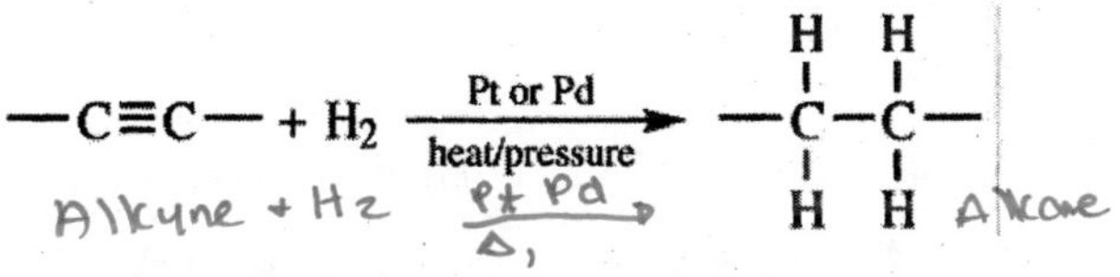

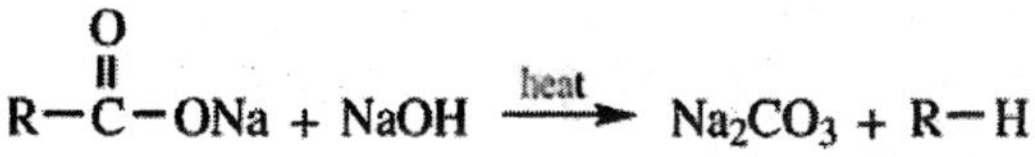

$$R-\overset{O}{\overset{\|}{C}}-ONa + NaOH \xrightarrow{\text{heat}} Na_2CO_3 + R{-}H \tag{4}$$

d) Dehydration of (removing the elements of water from) an alcohol.

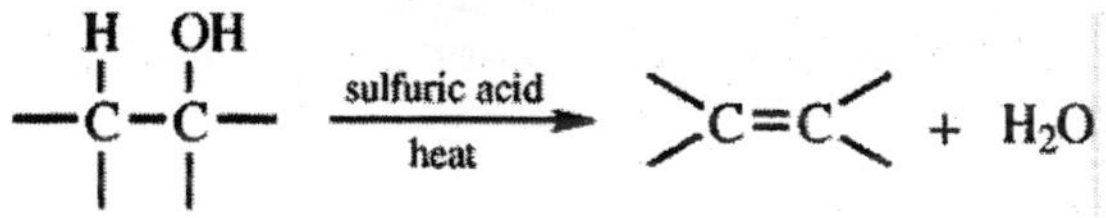

$$-\overset{H}{\underset{|}{C}}-\overset{OH}{\underset{|}{C}}- \xrightarrow[\text{heat}]{\text{sulfuric acid}} \; >C=C< \; + H_2O \tag{5}$$

e) Dehydrohalogenation of (removing the elements of HX from) an alkyl halide.

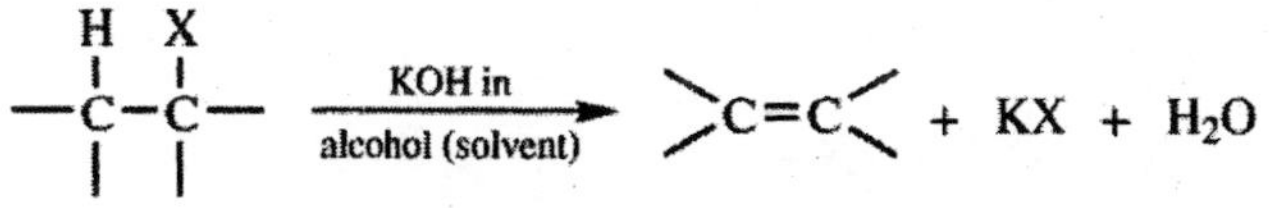

$$-\overset{H}{\underset{|}{C}}-\overset{X}{\underset{|}{C}}- \xrightarrow[\text{alcohol (solvent)}]{\text{KOH in}} \; >C=C< \; + KX + H_2O \tag{6}$$

f) Extension of alkyne chain length with sodium amide.

$$H{-\!}C\equiv C{-\!}H + NaNH_2 \longrightarrow NH_3 + H{-\!}C\equiv C\text{- }Na^+$$
$$H{-\!}C\equiv C\text{- }Na^+ + R{-\!}CH_2Br \longrightarrow H{-\!}C\equiv C{-\!}CH_2R + NaBr \tag{7}$$

g) Double dehydrohalogenation of an alkyl halide.

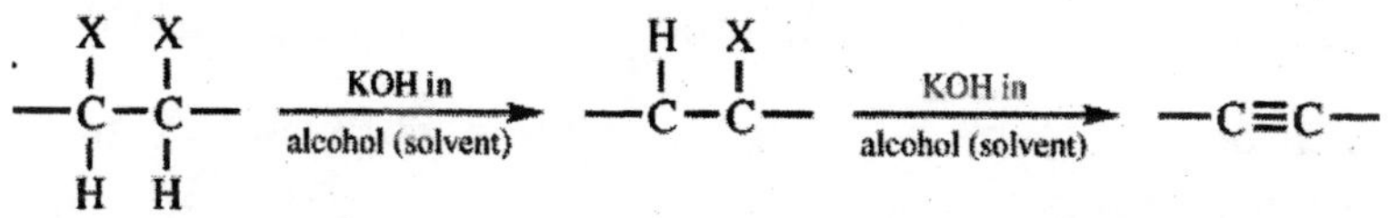

$$-\overset{X}{\underset{H}{C}}-\overset{X}{\underset{H}{C}}- \xrightarrow[\text{alcohol (solvent)}]{\text{KOH in}} -\overset{H}{\underset{|}{C}}-\overset{X}{\underset{|}{C}}- \xrightarrow[\text{alcohol (solvent)}]{\text{KOH in}} -C\equiv C- \tag{8}$$

6. **Explain why alkanes and aromatic hydrocarbons react by substitution, and why alkenes and alkynes react by addition.**

For alkanes, substitution is the only possible type of reaction, because carbon is capable of forming only four bonds. For alkenes and alkynes, addition reactions produce more energy than do substitution reactions. Consider the hydrogenation of a double bond in Equation [1]. The net result is that a C═C bond (611 kJ/mol) and an H—H bond (435 kJ/mol) are broken, and a C—C bond (347 kJ/mol) and two C—H bonds (414 kJ/mol) are formed. Thus, the enthalpy change for hydrogenation of a double bond is -127 kJ/mol [= 611 + 435 − 347 − 2(414)].

Aromatic hydrocarbons are considerably more stable than we might expect based on the number of single and double bonds present (see Objectives 11-10 and 11-11). This extra stability is the resonance stabilization energy. If an aromatic hydrocarbon participates in an addition reaction, the resonance stabilization energy is lost. For example, when benzene is hydrogenated to cyclohexane, the enthalpy change is -206.1 kJ/mol.

$$C_6H_6(g) + 3\ H_2(g) \xrightarrow[\text{heat/pressure}]{\text{Pt or Pd}} C_6H_{12}(g) \qquad [9]$$

This -206.1kJ/mol is far short of the -381 kJ/mol (3×-127 kJ/mol) that would result from hydrogenating 3 moles of C═C bonds. The difference, 175 kJ/mol, is the resonance stabilization energy, the energy by which benzene is more stable than a non-resonance structure of C_6H_6 that contains three noninteracting double bonds.

7. Write equations for the reactions of alkanes with halogens and with oxygen, emphasizing the radical chain nature of the alkane-halogen reaction.

When an alkane reacts with a halogen (F_2, Cl_2, Br_2, or I_2, symbolized as X_2), halogen atoms substitute for some of the hydrogens, producing the halogenated hydrocarbon and HX. The reaction proceeds by a free-radical chain mechanism consisting of initiation, propagation, and termination steps. A free radical is one of the species produced when a bond breaks and the two electrons in the bond are split between the two fragments, one electron going with each. This is homolytic cleavage of the bond. (Heterolytic cleavage produces a cation and an anion.) Free-radical halogenation of an alkane involves the following steps.

$$\text{Initiation:}\quad X{-}X \xrightarrow{\text{heat or light}} 2\ X. \qquad [10]$$

$$\text{Propagation:}\quad R{-}H + X\cdot \longrightarrow R\cdot + H{-}X \quad \text{then}$$

$$R\cdot + X{-}X \longrightarrow R{-}X + X \quad . \qquad [11]$$

$$\text{Termination:}\quad X\cdot + \cdot \longrightarrow X{-}X,\ R\cdot + \cdot X \longrightarrow R{-}X \ \text{and}$$

$$R\cdot + \cdot R \longrightarrow R{-}R \qquad [12]$$

The relative strengths of the R—X and H—X bonds compared with R—H and X—X determine the energy and speed of the reaction. In general, reactions of F_2 are faster and more energetic than those of Cl_2, Br_2, or I_2. Any number of hydrogens may be substituted, producing a mixture of products.

Of course, all hydrocarbons react with oxygen to produce carbon dioxide and water. Many of these combustion reactions are free-radical in nature.

*** 8. Predict the products of an addition reaction at a multiple bond.**

The more positive fragment (usually a hydrogen atom) of an unsymmetrical addition reagent attaches to the carbon atom of a multiple bond that has the largest number of hydrogen atoms. For alkynes with the triple bond at the end of the molecule, the last carbon has the most hydrogen atoms (one). For alkenes, there are three possibilities. They are shown in Figure 26-3. The carbon to which the positive fragment would add is shown by an arrow. R is some group other than hydrogen.

$$R-C\equiv C-H \qquad R-\overset{\overset{H}{|}}{C}=\overset{\overset{H}{|}}{C}-H \qquad R-\overset{\overset{R}{|}}{C}=\overset{\overset{H}{|}}{C}-H \qquad R-\overset{\overset{R}{|}}{C}=\overset{\overset{R}{|}}{C}-H$$

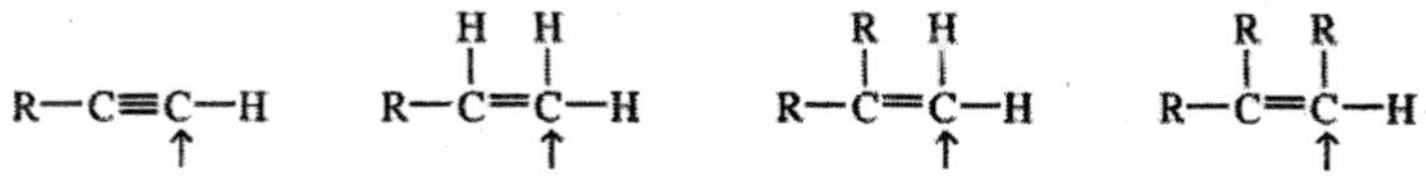

FIGURE 26-3 Addition Sites for Positive Fragments

9. Name the common functional groups, and give examples of compounds containing them.

Although the descriptive chemistry of inorganic compounds is based on the periodic table, that of organic compounds is based on functional groups. Organic compounds with the same functional group react in similar ways. Some of the common classes of organic compounds are given in Table 26-4; characteristics and endings for some names appear in Table 26-3. R represents any alkyl group.

TABLE 26-3 Characteristics of Common Classes of Organic Compounds and Endings for Some Names

Class	Structural Group Present	Ending
alkane	only C—H and C—C single bonds	-ane
alkene	at least one C=C bond	-ene
alkyne	at least one C≡C bond	-yne
arene	an aromatic ring, ⬡	
alkyl halide	an alkyl group bonded to a F, Cl, Br, or I atom	-yl halide (as -yl iodide)
alcohol	an —OH group bonded to a C atom	-ol
ether	two alkyl group bonded to an O atom	-yl ether
aldehyde	an alkyl group bonded to a $-\overset{\overset{O}{\|\|}}{C}-H$ group	-anal (as ethanal)
ketone	two alkyl groups bonded to a $-\overset{\overset{O}{\|\|}}{C}-$ group	-one
acid	an alkyl group bonded to a $-\overset{\overset{O}{\|\|}}{C}-OH$ group	-anoic acid
ester	two alkyl groups bonded to each end of a $-\overset{\overset{O}{\|\|}}{C}-O-$ group	-anoate
amine	an alkyl group bonded to an —NH$_2$ group	-yl amine

TABLE 26-4 Some Common Classes of Organic Compounds

Class of compound	*General formula*		*Examples with names*	
alkane	R—H	CH_3CH_3 ethane	CH_3 (branch) $CH_3CC\ CH_3$ 2-methylpropane	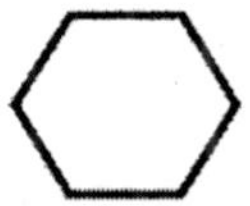 cyclohexane
alkene	$R_2C{=}CR_2$	$H_2C{=}CHCH_3$ 1-propene	$(CH_3)_2C{=}C(CH_3)_2$ 2,3-dimethyl-2-butene	 cyclopentene
alkyne	R—C≡C—R	$CH_3C{\equiv}CCH_3$ 2-butyne	CH_3 (branch) $CH_3CH\ C{\equiv}CH_2CH_3$ 2-methyl-3-hexyne	
arene (aromatic hydrocarbon)	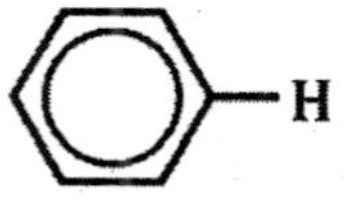 or Ar—H	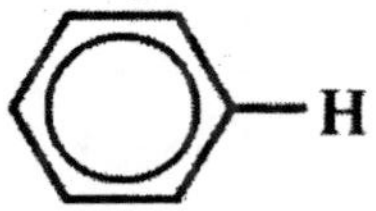 benzene	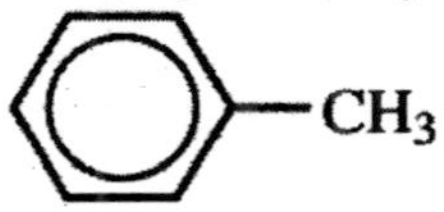 methylbenzene or toluene	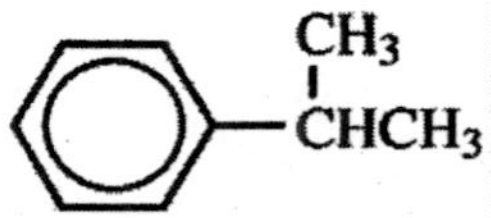 isopropylbenzene
alkyl halide	R—X	CH_3CH_2Br bromoethane	$CH_3CF_2CH_2CH_2$ 2,2-difluorobutane	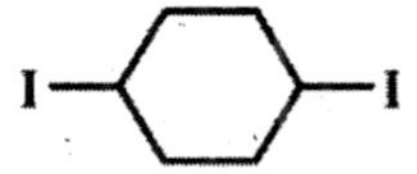 1,4-diiodocyclohexane
alcohol	R—OH	OH (branch) $CH_3CH_2CHCH_2CH_3$ 3-pentanol	CH_3 (branch) $CH_3C—CH_2CH_3$ OH (branch) 2-methyl-2-butanol	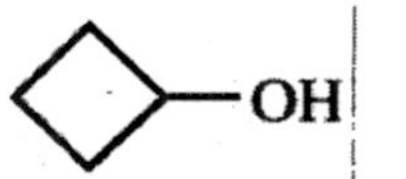 cyclobutanol

ether	$R-O-R'$	CH_3-O-CH_3 dimethyl ether	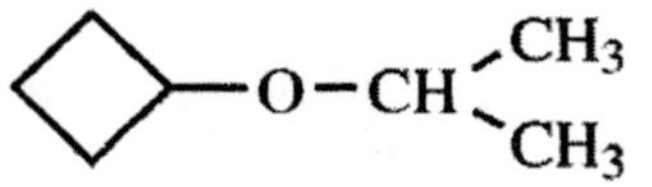 cyclobutyl isopropyl ether	
aldehyde	$R-C(=O)-H$	$CH_3C(=O)-H$ ethanal *or* acetaldehyde	$CHCH_2CH_2C(=O)-H$ butanal *or* butryraldehyde	
ketone	$R-C(=O)-R'$	$CH_3C(=O)CH_3$ propanone	$CH_3CH(CH_3)CH_2C(=O)CH_3$ 4-methyl-2-pentanone	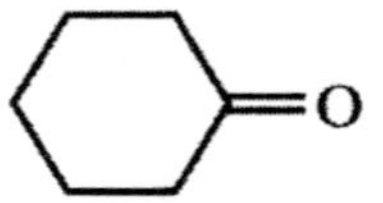 cyclohexanone
acid	$R-C(=O)-OH$	$CH_3C(=O)-OH$ ethanoic acid *or* acetic acid	$CH_3CH(CH_3)CH_2C(=O)-OH$ 3-methylbutanoic acid	
ester	$R-C(=O)-OR'$	$CH_3C(=O)-OCH_3$ methyl ethanoate *or* methyl acetate	$CH_3CH(CH_3)CH_2C(=O)-OCH_2CH_3$ ethyl 3-methylbutanoate	
amine	$R-NH_2$	$CH_3CH_2NH_2$ ethylamine	$CH_3CH_2C(CH_3)-NH_2$ 2-aminobutane	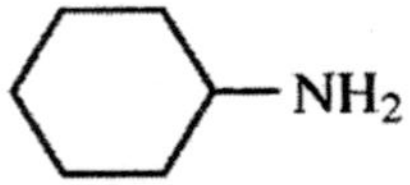 cyclohexylamine

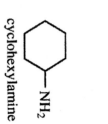

10. Write structural formulas, identify possible isomers, and name organic compounds containing functional groups.

First, one needs to thoroughly learn the structural formulas in Table 26-3. The summary in Table 26-4 may help in writing structural formulas. Different structural isomers are found by drawing all the possible carbon skeletons and then placing the functional group in all possible positions. Nomenclature depends on the class of the compound. Alcohols, alkyl halides, ketones, and amines have names obtained from the corresponding alkane, with the position of the group on the main chain indicated with the smallest possible number. This number precedes the base name and is separated from it by a dash. The -*ane* ending of the base name is replaced by the ending given in the far-right column in Table 26-4. For both aldehydes and carboxylic acids, the functional group is at one end of the longest chain, and the carbon of the functional group is numbered 1.

Esters are derived from carboxylic acids by replacing the H atom in the —COOH group with an alkyl group. The alkyl group name precedes that of the acid, and the -*oic acid* ending is replaced by -*oate*. In ethers the two alkyl group names are followed by *ether*. In order to determine the numbering of substituents on alkyl groups in ethers, the oxygen-bonded carbon atoms are numbered 1.

1. Describe methods of preparing alcohols, ethers, aldehydes, ketones, acids, and amines. Write equations for some typical reactions of these classes of organic compounds

Many methods of preparation are summarized below in Equations [13] through [25]. Following them are some of the reactions that compounds in each class of compounds undergoes.

a. Alcohols

Hydration (adding water to) of alkenes:

$$\ce{>C=C< + H2O ->[sulfuric acid] -\underset{|}{\overset{H}{C}}-\underset{|}{\overset{OH}{C}}-} \qquad [13]$$

Hydrolysis (attack by water) of alkyl halides:

$$\ce{-\underset{|}{C}-\underset{|}{\overset{X}{C}}- + OH^- -> -\underset{|}{C}-\underset{|}{\overset{OH}{C}}- + X^-} \qquad [14]$$

Production of methanol:

$$\ce{CO(g) + 2 H2(g) ->[350^\circ C,\ 200\ atm][ZnO, Cr2O3] CH3OH(g)} \qquad [15]$$

Ethers Elimination of water between two alcohol molecules, a condensation reaction:

$$\ce{R-OH + HO-R' ->[conc\ sulfuric\ acid] R-O-R' + H2O} \qquad [16]$$

Aldehydes and ketones. Produced by oxidation of alcohols.

Primary alcohols form aldehydes:

$$RCH_2OH \xrightarrow{Cr_2O_7^{2-},\ H^+} R-\overset{\overset{\displaystyle O}{\|}}{C}-H \xrightarrow{Cr_2O_7^{2-},\ H^+} R-\overset{\overset{\displaystyle O}{\|}}{C}-OH \qquad [17]$$

$$RCH_2OH \xrightarrow{Cu.\ 200\text{–}300°C} R-\overset{\overset{\displaystyle O}{\|}}{C}-H \qquad [18]$$

2. Secondary alcohols form ketones:

$$R-\overset{\overset{\displaystyle OH}{|}}{CH}-R' \xrightarrow[or\ Cu,\ 200\text{–}300°C]{KMnO_4\ or\ K_2Cr_2O_7} R-\overset{\overset{\displaystyle O}{\|}}{C}-R' \qquad [19]$$

d *Carboxylic acids*

1. Oxidation of a primary alcohol:

$$RCH_2OH \xrightarrow{KMnO_4,\ OH^-\ or\ Cr_2O_7^{2-},\ H^+} R-\overset{\overset{\displaystyle O}{\|}}{C}-OH \qquad [20]$$

2. Oxidation of an aldehyde:

$$R-\overset{\overset{\displaystyle O}{\|}}{C}-H \xrightarrow{KMnO_4,\ OH^-\ or\ Cr_2O_7^{2-},\ H^+} R-\overset{\overset{\displaystyle O}{\|}}{C}-OH \qquad [21]$$

e. *Amines* Reduction of nitro compounds:

$$R—NO_2 \xrightarrow{Fe,\ HCl} R—NH_3^+\ Cl^- \xrightarrow{NaOH} R—NH_2 \qquad [22]$$

Reactions of these classes of compounds are summarized in the following general equations ([5], [23] - [33]).

f. *Alcohols*

1. Dehydration to form alkenes (see Equation [5]).

2. Reactions with acid halides to form esters.

$$R'-OH + R-\overset{\overset{\displaystyle O}{\|}}{C}-X \longrightarrow R-\overset{\overset{\displaystyle O}{\|}}{C}-OR' + HX \qquad [23]$$

3. Reactions with an carboxylic acids to form esters.

$$R'-OH + R-\overset{\overset{\displaystyle O}{\|}}{C}-OH \xrightleftharpoons{H^+} R-\overset{\overset{\displaystyle O}{\|}}{C}-OR' + H_2O \qquad [24]$$

4. Reactions with nitric acid.

$$R—OH + HONO_2 \longrightarrow R—ONO_2 + H_2O \qquad [25]$$

5. Reactions with alkali hydroxides.

$$R—OH + NaOH \longrightarrow R—C^- + Na^+ + H_2O \qquad [26]$$

6. Reaction with sodium.

$$R{-}OH + Na^0 \longrightarrow R{-}O^- Na^+ + \tfrac{1}{2} H_2(g) \qquad [27]$$

g. *Ethers*

1. Stable when attacked by most oxidizing and reducing agents and dilute acids and alkalis.

2. Cleavage to alkyl halides by strong acids.

$$R{-}O{-}R' + 2\,HCl \longrightarrow R{-}Cl + R'{-}Cl + H_2O \qquad [28]$$

h *Aldehydes*

1. Oxidation to acids (see Equation [17]).

2. Reduction to alcohols with hydrogen.

$$R{-}\overset{\overset{\large O}{\|}}{C}{-}H \xrightarrow{H_2,\,Ni} R{-}CH_2{-}OH \qquad [29]$$

i. *Ketones* Reduction to alcohols with hydrogen.

$$R{-}\overset{\overset{\large O}{\|}}{C}{-}R' \xrightarrow{H_2,\,Ni} R{-}\overset{\overset{\large OH}{|}}{CH}{-}R' \qquad [30]$$

j. *Acids*

1. Reaction with ammonia, followed by heating.

$$R{-}\overset{\overset{\large O}{\|}}{C}{-}OH + NH_3 \longrightarrow R{-}\overset{\overset{\large O}{\|}}{C}{-}O^-NH_4^+ \xrightarrow{\Delta,\,P_2O_5} R{-}C{\equiv}N + H_2O \qquad [31]$$

2. Reaction with an alcohol to give an ester (see Equation [24]).

3. Elimination of water between two acid molecules.

$$2\,R{-}\overset{\overset{\large O}{\|}}{C}{-}OH \xrightarrow{\Delta,\,P_2O_5} R{-}\overset{\overset{\large O}{\|}}{C}{-}O{-}\overset{\overset{\large O}{\|}}{C}{-}R \qquad [32]$$

k. *Amines*

Reaction with an alkyl halide to form secondary amines.

$$RNH_2 + R'X \longrightarrow R{-}NH{-}R' + HX \qquad [33]$$

*** 12. Propose schemes for synthesizing some simple organic compounds.**

You should be able to combine the reactions given in Equations [1] through [8], and [13] through [33] to synthesize simple organic compounds. Often it is easier to work backward: determine what reactants are needed to produce the desired compound. Then figure out how to make those reactants, always keeping in mind what the starting materials are.

EXAMPLE 26-2 Produce propyl butanoate from simple alcohols.

$$CH_3CH_2CH_2CH_2OH \xrightarrow{KMnO_4} CH_3CH_2CH_2COOH$$

$$CH_3CH_2CH_2COOH + HOCH_2CH_2CH_3 \xrightarrow{H+} CH_3CH_2CH_2COOCH_2CH_2CH_3 + H_2O$$

13. Predict the products of an aromatic substitution reaction, including ortho, para and meta direction.

Aromatic compounds react by substitution rather than by addition, as discussed in Objective 26-6. If other groups are attached to the aromatic ring, they will influence the position where the substitution takes place. One group of substituents direct substitution to occur mainly in the *ortho* and *para* (that is, the 1, 3, and 5) positions on the ring. These are called ortho, para directors. Listed from strongest to weakest, they are as follows:

$$—NH_2 \text{ (an amine)} > —OR > —OH \text{ (an alcohol)}$$

$$> — OCOR \text{ (an ester)} > —R \quad > —X \text{ (a halide)}$$

Another group of substituents direct substitution to occur mainly in the *meta* (the 2 and 4) positions on the aromatic ring. These are called meta directors. They are listed from strongest to weakest:

$$—NO_2 > —CN > —SO_3H > —CHO \text{ (an aldehyde)}$$

$$> —COR \text{ (an ether)} > —COOH \text{ (an acid) (an acid)} > —COOR \text{ (ester)}$$

(Notice how the different orientation of the ester changes it from being an ortho, para to a meta director.) We should emphasize that the direction of aromatic substitution is not absolute. Rather it is an emphasis on, for instance, the ortho and para isomers at the expense of the meta isomers in the mix of final products.

14. Illustrate the principal methods of polymer formation—free-radical addition and condensation—through structural formulas of monomers and polymers.

Free-radical polymerization consists of initiation, propagation, and termination steps, as does free-radical halogenation (Objective 26-7). There are, however, two major differences. First, there is very little initiator present. In halogenation, there are about equal amounts of halogen and alkane, while here the mixture is principally alkene. Second, propagation consists of one reaction rather than two. The propagating free radical is an ever-lengthening molecule. Organic peroxides easily form free radicals and often are used as chain initiators, Equation [34]. Propagation depends on the alkene monomer; three examples are given in Equations [35] through [37]. Termination occurs when any two free radicals combine.

$$\text{Initiation: } RO—OR \longrightarrow 2\ RO\cdot \qquad [34]$$

Polyethylene propagation (ethylene, $CH_2{=}CH_2$, monomer).

$$RO\cdot + CH_2{=}CH_2 \longrightarrow ROCH_2CH_2\cdot$$

$$ROCH_2CH_2 + CH_2{=}CH_2 \longrightarrow RO(CH_2)_3CH_2\cdot$$

$$RO(CH_2)_3CH_2\cdot + CH_2{=}CH_2 \longrightarrow RO(CH_2)_5CH_2\cdot \qquad [35]$$

Polyvinyl chloride propagation (vinyl chloride, $CH_2{=}CHCl$, monomer).

$$RO\cdot + CH_2{=}CHCl \longrightarrow ROCH_2CHCl\cdot$$

$$ROCH_2CHCl\cdot + CH_2{=}CHCl \longrightarrow RO(CH_2CHCl)_2\cdot$$

$$RO(CH_2CHCl)_2\cdot + CH_2{=}CHCl \longrightarrow RO(CH_2CHCl)_3\cdot \qquad [36]$$

Polystyrene propagation (styrene, $CH_2{=}CH\phi$, monomer; $\phi = C_6H_5$, a phenyl group).

$$RO\cdot + CH_2{=}CH\phi \longrightarrow ROCH_2CH_2\phi$$

$$ROCH_2CH\phi\cdot + CH_2{=}CH\phi \longrightarrow RO(CH_2CH\phi)_2\cdot$$

$$RO(CH_2CH\phi)_2\cdot + CH_2{=}CH\phi \longrightarrow RO(CH_2CH\phi)_3\cdot \qquad [37]$$

Condensation polymerization occurs when a small molecule such as water is eliminated between the two growing molecules. This reaction is promoted by the presence of a substance that consumes the small molecule as it is formed. For the removal of water, often an acid anhydride is used to increase the yield, for example, acetic anhydride $[(CH_3CO)_2O + H_2O \longrightarrow 2\ CH_3COOH]$. This also can speed up the reaction, since many of these polymerization reactions are acid-catalyzed. The reaction can be represented by the generalized Equation [38]. Note that there must be two functional groups per molecule of monomer, one on each end.

$$HO{-}R{-}OH + HO{-}R{-}OH \longrightarrow HO{-}R{-}O{-}R{-}OH + H_2O \qquad [38]$$

15. Distinguish between natural and synthetic polymers and among elastomers, fibers, and plastics.

Many natural polymers are discussed in Chapter 27. These include polysaccharides (Objectives 27-4 and 27-6), which have monosaccharides as monomers; proteins (Objective 27-9), which have amino acids as monomers; and nucleic acids (Objective 27-13), which have nucleotides as monomers. In addition, rubber also is a natural polymer. Rubber is composed of isoprene monomers, as shown in Figure 26-4. Synthetic polymers include polyethylene, polystyrene, polyvinyl chloride, dacron, nylon, neoprene, teflon, and lucite.

Elastomers are polymers that can deform or stretch under stress. Fibers are long single- or double-stranded chains. Plastics are crosslinked chains that form a somewhat rigid two- or three-dimensional structure. Thus, they can be formed into films such as mylar (backing for magnetic tapes for audio cassettes and VCRs) or solids such as polyvinyl chloride (plumbing pipe and phonograph records).

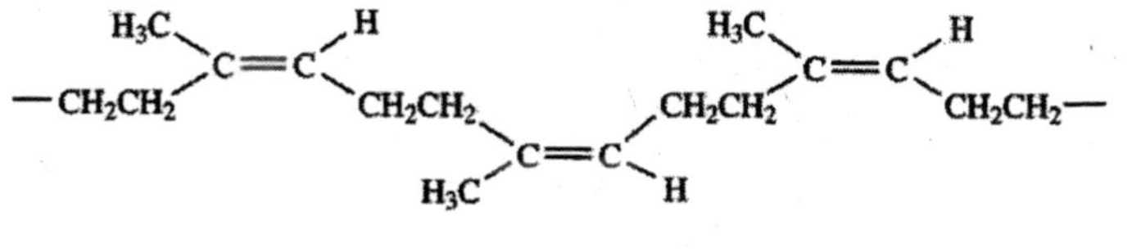

FIGURE 26-4 Rubber—Polyisoprene

SELF-ASSESSMENT EXERCISES

128. C_4H_8 will have more isomers because the double bond will allow different structural isomers.

129. Cyclobutane has the C_4H_8, which is the same as $CH_3CH=CHCH_3$.

130. (a) The unsaturated hydrocarbon, C_6H_6 will have a higher boiling point.

(b) The smaller alcohol will be more soluble. There are more H-bonds for the size of the molecule.

(c) C_6H_5COOH is benzoic acid, and it will be more acidic.

DRILL PROBLEMS

1. (1) Draw structural formulas for each of the following compounds. (You may draw either the *cis* or the *trans* isomer for those compounds that exhibit geometric isomerism.)

A. 3-methyl-3-hexene	B. 2-pentene	C. 2-methylbutane
D. 2,2,4-trimethylpentane	E. 2-butyne	F. 2-methylethylbenzene
G. toluene	H. *m*-xylene	I. 3-ethylcyclopentene
J. 2,3,4-trimethylpentane	K. 2,3-dimethylbutane	L. 2,4-dimethyl-4-ethylheptane
M. 2,3-dimethyl-2-butene	N. 3-methyl-4-ethyl-3-hexene	O. 2-pentyne
P. 1,3-dimethylcyclohexane	Q. 3,5-dimethylcyclopentene	R. 3-methylcyclopentene
S. cyclohexylcyclohexane	T. 1,1-dimethyl-4-ethylcycloheptane	
U. methylcyclopentene	V. 1,2,3,4-tetramethylcyclobutene	W. 1,3,5-trimethylbenzene

(2) Give the systematic (IUPAC) name of each of the following compounds:

A.	$CH_3CH_2CH-CH_3$ \| CH_3	B.	H, C=C ,H with CH₂CH₃	C.	H_3C, $C=C$, CH_2CH_3 / H_3C, CH_2CH_3
D.	CH_3 CH_3 \| \| $CH_3-CH--CHCH_2CH_3$	E.	$CH_2-CH-CH_2CH_3$ \ / CH_2	F.	(cyclohexane) CH_3 / CH_3
G.	(benzene) CH_3 / CH_3	H.	(cyclohexane)$-CH_2CH_2CH_3$	I.	(benzene)$-\overset{CH_3}{\underset{CH_3}{C}}-CH_3$
J.	CH_3 \| $CH_3CH_2CH_2CCH_2CH_2CH_3$	K.	$CH_3CH_2C{\equiv}CH$	L.	$CH_3CH=CHCH_3$
M.	$CH_3CH_2CH_2CH=CH_2$	N.	$CH{\equiv}CCH_3$	O.	C_3H_8

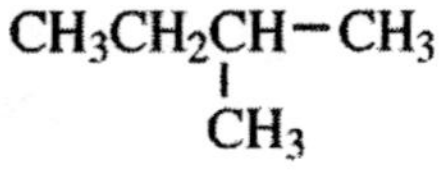

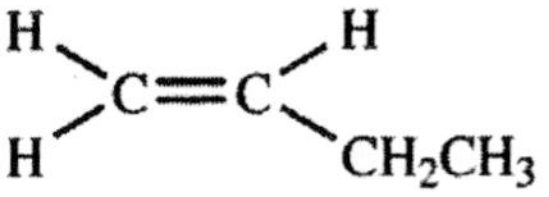

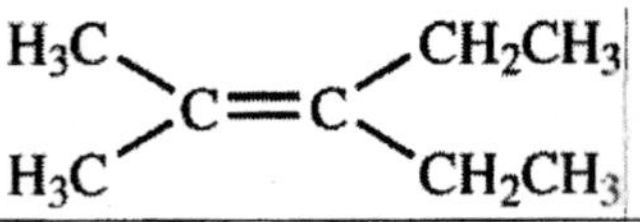

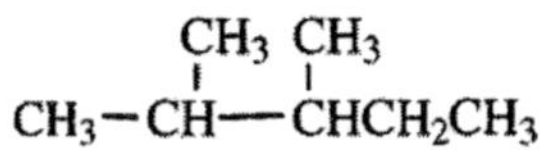

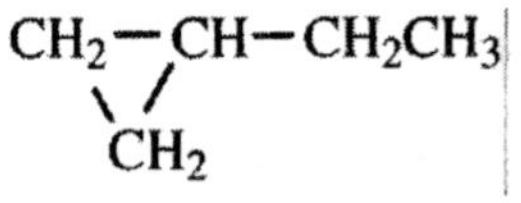

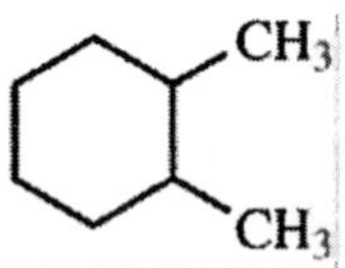

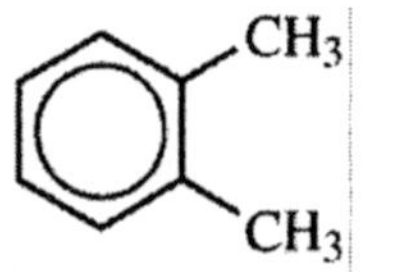

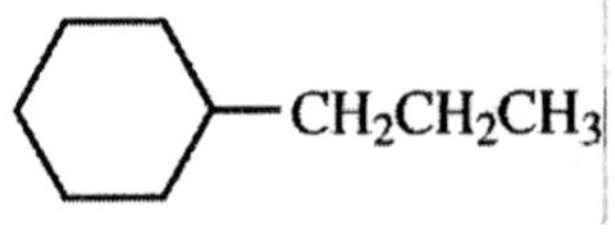

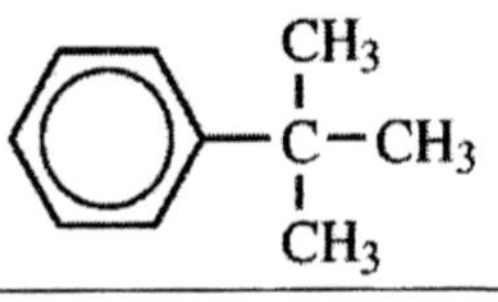

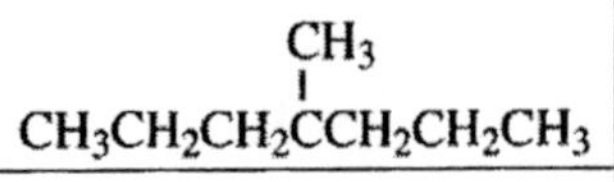

P.	CH$_3$ \| CH$_3$CHCH$_3$	Q.	CH$_3$ CH$_3$ \| \| CH$_3$—C——CHCH$_3$ \| CH$_3$	R.	CH$_3$ CH$_2$CH$_3$ \| \| CH$_3$CH—CCH$_2$CH$_2$CH$_3$ \| CH$_2$CH$_3$
S.	CH$_3$ CH$_2$CH$_3$ \| \| CH$_3$CH—CHCH$_2$CH$_3$	T.	CH$_3$ \| CH$_3$CHCH$_2$CH$_2$CH$_3$	U.	CH$_3$ CH$_3$ \| \| CH$_3$CH$_2$CH—CCH$_3$ \| CH$_3$
V.	CH$_3$ \| CH$_3$CH$_2$CH$_2$CH$_2$CCH$_2$CH$_2$CH$_3$ \| CH$_3$			W.	—CH$_3$

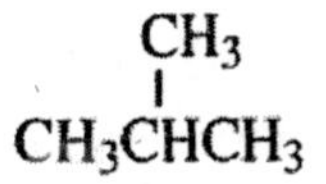

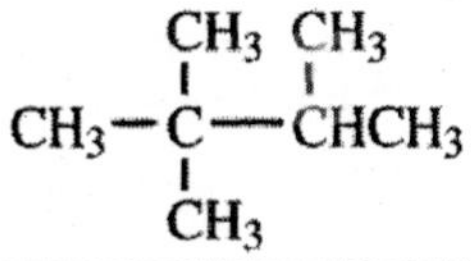

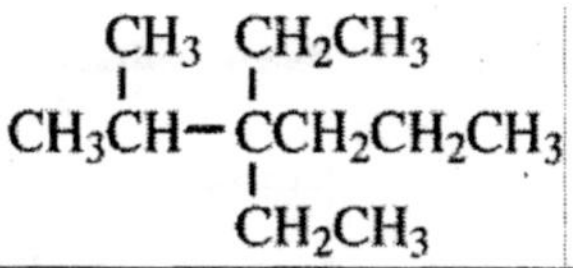

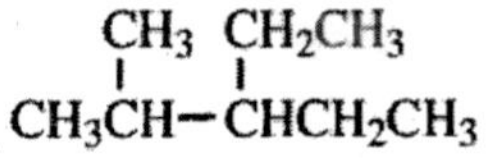

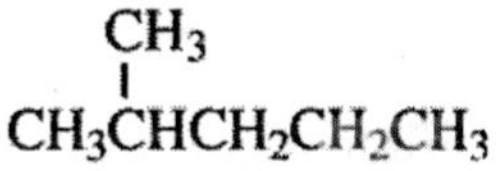

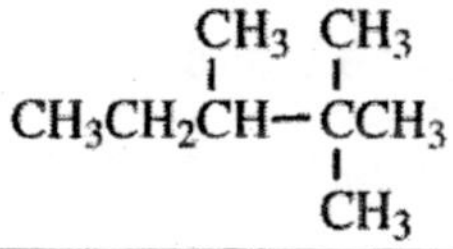

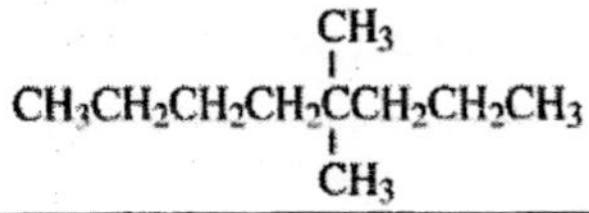

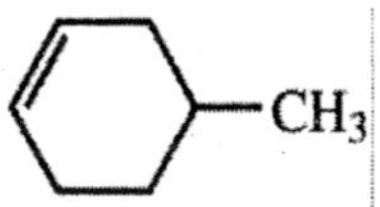

2. Draw all the possible skeletal isomers of each of the following hydrocarbons, and then name each isomer.

A. aromatic compounds of formula C$_9$H$_{12}$ B. C$_4$H$_8$

C. cyclic compounds of formula C$_6$H$_{10}$ D. C$_6$H$_{14}$

E. C$_4$H$_{10}$ F. C$_5$H$_{10}$

G. hydrocarbons of molecular weight 86

3. Predict which compound in each group below has the highest value of the listed property and which one has the lowest value:

Melting point

A. cyclohexane; hexane; benzene B. pentane; 2,2-dimethylpropane; 2-methylbutane

C. ethyne; ethane; ethene D. benzene; 1,4-dimethylbenzene; toluene

Boiling point

E. butane; propane; hexane F. 3,3-dimethylpentane; 2-methylhexane; n-heptane

G. cyclohexane; benzene; cyclohexene H. butane; cyclobutane; methylcyclopropane

Molar heat of fusion

I. methane; ethane; ethene J. ethyne; propyne; butyne

K. cyclohexane; hexane; benzene L. n-hexane; 2,2-dimethylbutane; 3-methylpentane

Molar heat of vaporization

M. xylene; benzene; toluene N. 3,3-dimethylpentane; 2-methylhexane; n-heptane

O. cyclohexane; hexane; benzene P. butane; pentane; propane

4. Write the complete equation for each of the following reactions:

A. 1-bromohexane with KOH in alcohol B. 1-propanol ⟶ Propylene

C. ethanol with H$_2$SO$_4$ and heat D. chloroethane ⟶ butane

E. 1,1-dichloropropane ⟶ propyne

F. 2-chloropropane ⟶ propylene

G. 2-chlorobutane ⟶ NaCl

H. 1-butene ⟶ butane

I. 2-chloro-2,3-dimethylpentane with KOH in alcohol

5. Predict the product of each of the following reactions:

A. $1-\text{butene}+H_2O\xrightarrow{H_2SO_4,HgSO_4}$

B. $3-\text{methyl}-2-\text{pentene}+HCl\longrightarrow$

C. $1-\text{butene}+HI\longrightarrow$

D. $2-\text{methyl}-2-\text{butene}+HBr\longrightarrow$

E. $2-\text{methyl}-2-\text{pentene}+HCl\longrightarrow$

F. $2-\text{methylcyclohexene}+HCl\longrightarrow$

6. (1) Give the name of each compound:

A. $CH_3CH_2CHBrCH_2$

B. $(CH_3)_2CH_2CHOH(CH_3)_3$

C. $CH_3CH_2CH_2COCH_2CH_3$

D. $CH_3CH_2CH_2CH_2COOH$

E. $(CH_3)_3CCHO$

F. $CH_3OCH_2CH_2CH_3$

G. $CH_3CH_2CH_2CH_2CH_2NH_2$

H. HCOOH

I. $CH_3CH_2COOCH_2CH_3$

J. $(CH_3)_3COH$

K. $(CH_3)_3COCH_2CH_3$

L. $CH_3CCl_2CHClCH_2CH_3$

M. $CH_3CH_2COCH_2CH_3$

N. $CH_3CH_2COOCH_2CH_2CH_3$

O. $CH_3CH_2CH_2CH_2CHO$

(2) Give the condensed structural formula of each compound:

P. butanone

Q. 2-methylbutanal

R. 2-methylpentanoic acid

S. 3-bromopentane

T. methylbutanoate

U. propyl isopropyl ether

V. 3-methylpentanal

W. 2,2-dichloropropane

X. 2-amino-3-methylpentane

Y. 3-hexanol

Z. 2-propanol

Γ 3-aminopentane

Δ. 2-methylpropanoic acid

Θ 1-iodo-2-methylbutane

Λ 2-pentanone

7. Detail all the steps in synthesizing the following compounds from the given starting material:

A. propyne from ethyne

B. ethene from ethanal

C. Propene from propane

D. 2-butene from ethane

E. propanone from propane

F. Diethyl ether from ethane

G. propanal from propane H. cyclohexane from hexane I. ethyl butanoate from ethane

QUIZZES (20 minutes each) Choose the best answer for each question.

QUIZ A

1. The IUPAC name of $(CH_3)_2CHCH_2CH_3$ is (a) isopentane; (b) *sec*-pentane; (c) 3-methylbutane; (d) 2-methylbutane; (e) none of these.

2. The formula of 2-methyl-1-butanol is (a) $CH_3CH_2COH(CH_3)_2$; (b) $CH_3CH_2CH(CH_3)CH_2OH$; (c) $CH_3CH(CH_3)CH_2OH$; (d) $CH_3COH(CH_3)_2$; (e) none of these.

3. C_4H_8 has how many noncyclic structural (skeletal) isomers, including geometric isomers? (a) 2; (b) 3; (c) 4; (d) 5; (e) none of these.

4. Which compound is the lowest boiling? (a) methane; (b) ethene; (c) ethane; (d) cyclopropane; (e) ethyne.

5. Which monomer would be suitable for condensation polymerization? (a) CH_3CH_2OH; (b) $CH_3CH{=}CH_2$; (c) $CH_2{=}CHCl$; (d) $HOCH_2CH_2OH$; (e) none of these.

6. What reactant could *not* be used to produce an alkene in fewer than three steps? (a) bromoethane; (b) bromobutane; (c) methane; (d) propyne; (e) none of these.

7. Which class of compounds is the least reactive? (a) ethers; (b) alcohols; (c) ketones; (d) alkyl halides; (e) aldehydes.

8. To produce an organic ester, one uses a carboxylic acid and an (a) alkyl halide; (b) alkane; (c) alcohol; (d) aldehyde; (e) none of these.

QUIZ B

1. The IUPAC name of $H_2C{=}C(CH_3)CH(Cl)CH_3$ is (a) 4-chloro-2-methylbutene; (b) 2-chloro-2-methyl-4-butene; (c) 4-chloro-2-methylpentene; (d) 2-chloro-4-methyl-4-butene; (e) none of these.

2. The formula of 2-methylbutanal is (a) $CH_3CH_2CH(CH_3)CHOH$; (b) $CH_3CH(CH_3)CHOH$; (c) $CH_3CH_2CH(CH_3)CHO$; (d) $CH_3CH(CH_3)CHO$; (e) none of these.

3. C_5H_{12} has how many skeletal isomers? (a) 2; (b) 3; (c) 4; (d) 5; (e) none of these.

4. Which compound has the lowest molar heat of vaporization? (a) hexane; (b) cyclohexane; (c) benzene; (d) heptane; (d) 2-heptene.

5. Which step does not occur during polymerization? (a) condensation; (b) termination; (c) propagation; (d) dehydrohalogenation; (e) initiation.

6. What could be used to produce an alkane in one step? (a) an alkene; (b) an alcohol; (c) an acid; (d) an alkali metal carbonate; (e) none of these.

7. A free-radical chain reaction is used to prepare (a) methanol; (b) methanal; (c) methyl chloride; (d) dimethyl ether; (e) none of these.

8. To produce an alcohol from a ketone, one uses (a) Cu, heat; (b) HCl; (c) H_2, Ni; (d) NH_3; (e) none of these.

QUIZ C

1. The formula of 2-chloro-1-butene is (a) $CH_2{=}CClCH_2CH_3$; (b) $CH_3CCl{=}CHCH_3$; (c) $CH_2{=}CClCH_3$; (d) $CH_3CCl{=}CH_2$; (e) none of these.

2. The name of $CH_3CH_2COCH_2CH_3$ is (a) diethyl ether; (b) ethyl propyl ether; (c) butanone; (d) 3-pentanone; (e) none of these.

3. C_4H_9Cl has how many skeletal isomers? (a) 1; (b) 2; (c) 3; (d) 4; (e) none of these.

4. Which compound is the highest melting? (a) propanone; (b) propene; (c) cyclopropane; (d) propyne; (e) methylpropane.

5. Which is the formula of a straight-chain alkene? (a) C_2H_6; (b) C_2H_2; (c) C_6H_{10}; (d) C_6H_{14}; (e) none of these.

6. To add to a double bond, which reactant(s) is *not* used? (a) HCl; (b) H_2, Pt, (c) Na; (d) H_2O, H_2SO_4; (e) none of these.

7. A free-radical chain reaction is *not* used to produce (a) bromoethane; (b) polystyrene; (c) polyethylene; (d) ethanol; (e) none of these.

8. To produce an ether from methanol, one uses (a) heat; (b) concentrated H_2SO_4; (c) KOH in alcohol; (d) Cu, heat; (e) none of these.

QUIZ D

1. The name of $CH_3CCl{=}CBr_2$ is (a) 1,1-dibromo-2-chloropropane; (b) 3,3-dibromo-2-chlorethane; (c) 1,1-dibromo-2-chloropropene; (d) 3,3-dibromo-2-chloropropene; (e) none of these.

2. The formula of 2-methylbutanal is (a) $CH_3CH_2CH(CH_3)CHO$; (b) $(CH_3)_2CHCH_2CHO$; (c) $(CH_3)_2CHCHO$; (d) $CH_3CH(CH_3)CHO$; (e) none of these.

3. C_3H_8O has how many skeletal isomers? (a) 1; (b) 2; (c) 3; (d) 4; (e) none of these.

4. Which compound has the highest molar enthalpy of fusion? (a) butane; (b) 2-butane; (c) cyclobutane; (d) methylcyclopropane; (e) methylcyclobutane.

5. Which of these could be used as a reactant with HBr to produce 1-bromobutane? (a) $CH_2{=}CHCH_2CH_3$; (b) $CH{\equiv}CCH_2CH_3$; (c) $CH_3CH_2CH_2CH_3$; (d) $CH_3CH{=}CHCH_3$; (e) none of these.

6. Which could produce an alkene in one step? (a) an amine; (b) an alkyl halide; (c) an alkyl dihalide; (d) an alkane; (e) all of these could be used.

7. Which class of compounds is the most oxidized? (a) alkanes; (b) alcohols; (c) acids; (d) aldehydes; (e) ketones.

8. To produce an aldehyde from 2-butanol, what reactant is used? (a) MnO_4^-; (b) Cu, heat; (c) $Cr_2O_7^{2-}$; (d) H_2SO_4; (e) impossible with any of these reagents.

SAMPLE TEST (20 minutes)

1. Draw all of the ethers of the formula $C_4H_{10}O$. Name each compound.

2. Explain why propanol is soluble in water but hexanol is not.

3. Consider the dehydrohalogenation of 2-bromobutane in the presence of KOH. Draw the structures of all the products and name them.

4. Give the names and structures of compounds A, B, C, and D formed in the following sequence of reactions:

$$2-\text{methylpropanal}\xrightarrow{H_2,\,Pt}A\xrightarrow{H_2SO_4,\,200°C}B\xrightarrow{HBr}C\xrightarrow{KOH,\,alcohol}D$$

Carbohydrates are composed of carbon, hydrogen, and oxygen, with the general formula of $C_x(H_2O)_y$.

27 CHEMISTRY OF THE LIVING STATE

CHAPTER OBJECTIVES

1. List the four principal types of substances found in cells—lipids, carbohydrates, proteins, and nucleic acids—and describe the chemical composition of each.

Lipids readily dissolve in nonpolar solvents. They are nonpolar or of very low polarity. Lipids are subdivided into triglycerides (fats and oils), phosphatides or phospholipids, and waxes. See Objectives 27-2, 27-2a, and 27-3.

Polysaccharides are carbohydrate polymers composed of more than 10 monosaccharides units. A monosaccharide is an aldehyde or a ketone with hydroxyl groups (—OH) on all other carbons. Aldehyde monosaccharides are aldoses; ketone ones are ketoses. See Objectives 27-4 and 27-6.

Proteins are polymers of amino acids. An α-amino acid is a carboxylic acid with an amino group ($—NH_2$) attached to the α-carbon, the carbon bonded to the carboxylic group (—COOH). These acids have the general formula shown in Figure 27-1. R groups of common amino acids are given in Table 27-1. See Objectives 27-8 and 27-9.

TABLE 27-1 Some Common Amino Acids

Symbol	*Name*	*Formula of R* (* designates formula of entire amino acid)	p*I*
		NEUTRAL AMINO ACIDS	
Gly	glycine	—H	5.97
Ala	alanine	$—CH_3$	6.00
Val	valine	$—CH(CH_3)_2$	5.96
Leu	leucine	$—CH_2CH(CH_3)_2$	6.02
Ile	isoleucine	$—CH(CH_3)CH_2CH_3$	5.98
Ser	serine	$—CH_2OH$	5.68
Thr	threonine	$—CH(OH)CH_3$	5.6
Phe	phenylalanine	$—CH_2C_6H_5$	5.48
Met	methionine	$—CH_2CH_2—S—CH_3$	5.74
Cys	cysteine	$—CH_2—SH$	5.05
$(Cys)_2$	cystine	$—CH_2—S—CH_2—$(another chain)	4.8
Tyr	tyrosine	$—C_6H_4—OH$ (—OH in the *para* position)	5.66
Trp	tryptophan	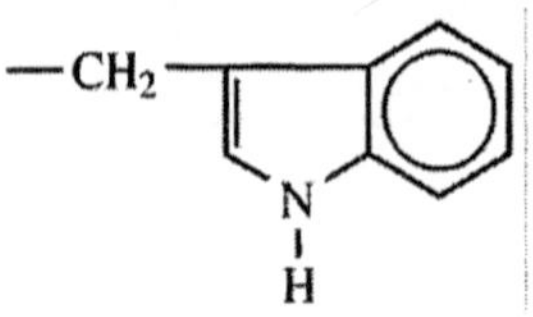	5.89

Pro	proline* (left structure)	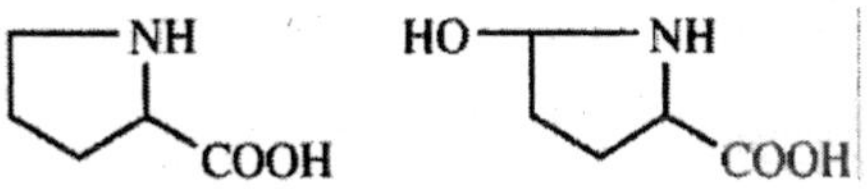	6.30
Hyp	hydroxyproline*		5.82
		ACIDIC AMINO ACIDS	
Asp	aspartic acid	$—CH_2COOH$	2.77
Glu	glutamic acid	$—CH_2CH_2COOH$	3.22
		BASIC AMINO ACIDS	
Lys	lysine	$—CH_2CH_2NH_2$	9.74
Arg	arginine	$—CH_2CH_2CH_2—NH—C(=NH)—NH_2$	10.76
			7.64
His	histidine	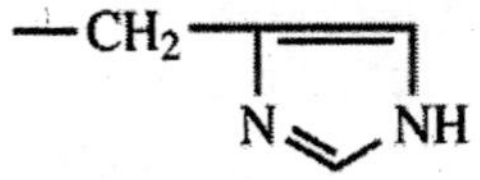	

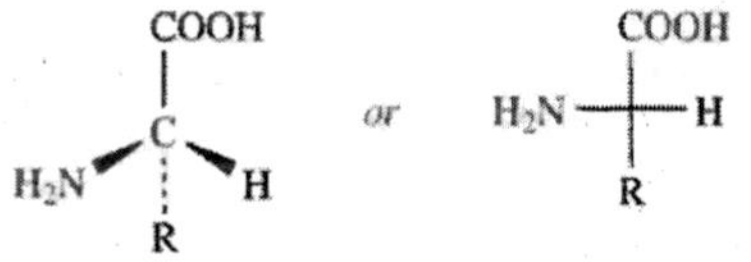

FIGURE 27-1 General Formula for Amino Acids

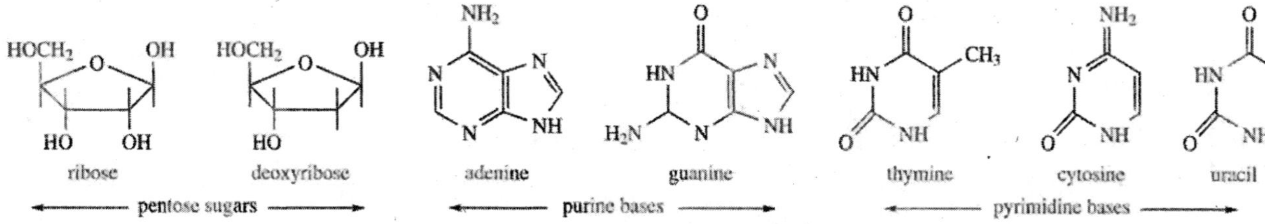

FIGURE 27-2 Components of Nucleic Acids

Nucleic acids are composed of phosphoric acid, pentose sugars (ribose and 2-deoxyribose), the purine bases (adenine and guanine), and the pyrimidine bases (thymine, cytosine, and uracil); the purine and pyrimidine bases are shown in Figure 27-2. See Objective 26-13.

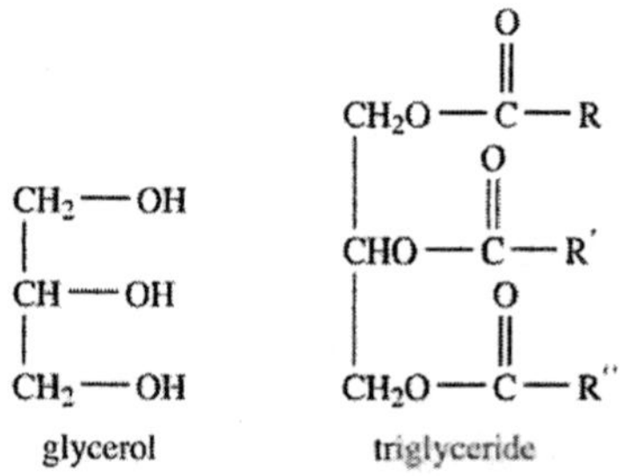

FIGURE 27-3 General Formula of Triglycerides

*** 2. Write structural formulas and the names of triglycerides, and indicate whether their constituent fatty acids are saturated or unsaturated.**

Triglycerides are esters of glycerol and carboxylic acids with long hydrocarbon chains as R groups (Figure 28-3). When the carboxylic acids are saturated, meaning they contain no double $(C = C)$ bonds, the triglyceride is a saturated fat. Saturated fats are waxy solids, such as lard. When the R groups are unsaturated, containing one or more double bonds, the triglyceride is an unsaturated fat. Unsaturated fats are liquids, commonly called oils. Because of the reactivity of the double bonds, oils react with oxygen in the air to become rancid.

Fats are named beginning with *glyceryl,* followed by the names of the fatty acid anions. (The anion of stearic acid, $C_{17}H_{35}COOH$, is the stearate anion, $C_{17}H_{35}COO^-$.) A prefix indicates when there is more than one anion of a given type, such as *distearate*. In a mixed triglyceride—where the three fatty acids are not the same—the anions are named in alphabetical order and the *-ate* endings of the first two are changed to *-o*. All three names are written together with no spaces. The names of some common fatty acids are given in Table 27-2.

*** 2a. Relate the structures, the saponification values, and the iodine numbers of triglycerides.**

The saponification value is the mass in milligrams of KOH that reacts with 1 gram of triglyceride. We know that 3 moles of KOH reacts with 1 mole of triglyceride, as shown in Equation [1]. Thus, we can use the saponification value to determine the molar mass of a triglyceride, as in Example 27-1.

TABLE 27-2 Some Common Fatty Acids—Saturated and Unsaturated

Common Name	IUPAC Name	Formula	COMMON NAME	IUPAC Name	Formula
	SATURATED ACIDS			UNSATURATED ACIDS	
Lauric acid	dodecanoic acid	$C_{11}H_{23}COOH$	oleic acid	9-octadecenoic acid	$C_{17}H_{33}COOH$
myristic acid	tetradeca noic acid	$C_{13}H_{27}COOH$	linoleic acid	9,12-octadecadiienoic acid	$C_{17}H_{31}COOH$
palmitic acid	hexadecanoic acid	$C_{15}H_{31}COOH$	linolenic acid	9,12,15-octadecatrienoic acid	$C_{17}H_{29}COOH$
stearic acid	octadecanoic acid	$C_{17}H_{35}COOH$	eleostearic acid	9,11,13-octadecatrienoic acid	$C_{17}H_{29}COOH$

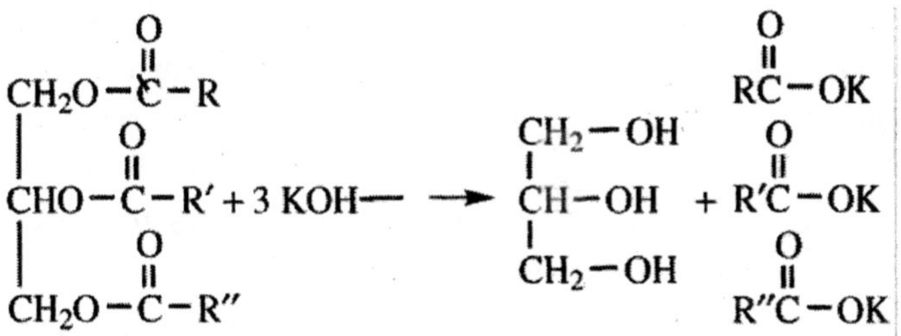

[1]

$$CH_2O-\overset{\overset{\displaystyle O}{\|}}{C}-R$$
$$CHO-\overset{\overset{\displaystyle O}{\|}}{C}-R' + 3\ KOH \longrightarrow$$
$$CH_2O-\overset{\overset{\displaystyle O}{\|}}{C}-R''$$

$$\begin{array}{l} CH_2-OH \\ CH-OH \\ CH_2-OH \end{array}$$

$$+ \begin{array}{l} R\overset{\overset{\displaystyle O}{\|}}{C}-OK \\ R'\overset{\overset{\displaystyle O}{\|}}{C}-OK \\ R''\overset{\overset{\displaystyle O}{\|}}{C}-OK \end{array}$$

EXAMPLE 27-1 What is the molar mass of triglyceride with a saponification value of 185.7?

$$\text{amount} = 185.7 \text{ mg KOH} \times \frac{1.000 \text{ g}}{1000 \text{ mg}} \times \frac{\text{mol KOH}}{56.11 \text{ g}} \times \frac{\text{mol triglyceride}}{3 \text{ mol KOH}} = 1.103 \times 10^{-3} \text{ mol}$$

$$M = 1.00 \text{ g}/ 1.103 \times 10^{-3} \text{ mol} = 906.3 \text{ g/mol}$$

The iodine number is the mass in grams of I_2 that reacts with 100 g of triglyceride by addition of I_2 across the double bonds. Thus, the iodine number can be used to establish the number of double bonds in a triglyceride molecule.

EXAMPLE 27-2 How many double bonds are there in the triglyceride of Example 28-1 if its iodine number is 168 g of I_2 (per 100 g triglyceride, remember)?

$$\begin{array}{l} \text{no. double} \\ \text{bonds} \end{array} = \frac{168 \text{ g } I_2}{100 \text{ g triglyceride}} \times \frac{906.3 \text{ g triglyceride}}{\text{mol triglyceride}} \times \frac{\text{mol } I_2}{253.8 \text{ g } I_2} = 6.00 \ \frac{\text{mol } I_2}{\text{mol triglyceride}}$$

There are six double bonds in this triglyceride molecule.

$$CH_2O-\overset{\overset{\displaystyle O}{\|}}{C}-R$$
$$CHO-\overset{\overset{\displaystyle O}{\|}}{C}-R'$$
$$CH_2O-\overset{}{\underset{\underset{\displaystyle O^-}{|}}{P}}O-A$$
phospholipid

$$R\overset{\overset{\displaystyle O}{\|}}{C}-OR'$$
wax

$$CH_3(CH_2)_{24}\text{-}CO(CH_2)_{30}CH_3$$
carnauba wax

$$CH_3(CH_2)_{12}\text{-}\overset{\overset{\displaystyle O}{\|}}{C}O(CH_2)_{25}CH_3$$
beeswax

$$CH_3(CH_2)_{14}\text{-}\overset{\overset{\displaystyle O}{\|}}{C}O(CH_2)_{15}CH_3$$
spermaceti wax

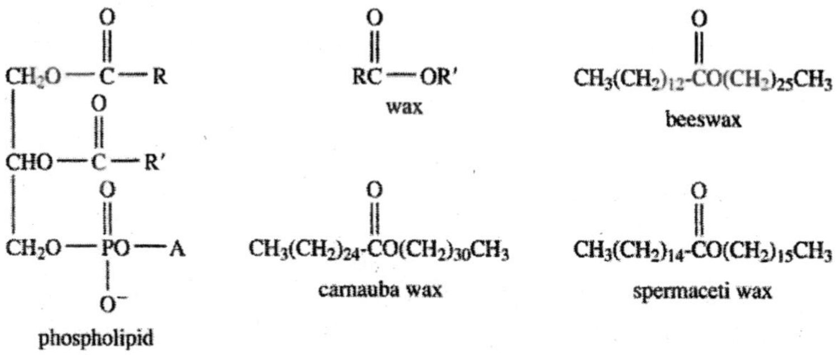

FIGURE 27-4 Structures of Phospholipids and Waxes

3. Explain how phosphatides and waxes, which are also lipids, differ from triglycerides.

Phosphatides (or phospholipids) are esters of glycerol with two fatty acids and a derivative of phosphoric acids (Figure 28-4). There are two major classes of phosphatides: lecithins, in which —A is $-CH_2CH_2N(CH_3)_3^+$; and cephalins, in which —A is $-CH_2CH_2NH_3^+$. A wax is an ester of a long-chain *mono*hydric alcohol (an alcohol with *one* OH group) with a fatty acid.

4. Classify carbohydrates as monosaccharides, oligosaccharides, and polysaccharides, and give examples of each.

The general formula of a monosaccharide is $(CH_2O)_n$ where $n = 3, 4, 5, 6$, etc. The total number of carbon atoms in the molecule determines its name: triose ($n = 3$), tetrose ($n = 4$), pentose ($n = 5$), hexose ($n = 6$), and so on. Whether the monosaccharide is an aldose or a ketose is indicated by a prefix: aldo- or keto-. In addition, the position of the C=O in a ketose is indicated by a preceding number (as in Figure 27-5). With the exception of ketotriose, there are several optical isomers of each monosaccharide (see Objective 27-5); the number of optical isomers is given in Figure 27-5.

Oligosaccharides are composed of from two to 10 monosaccharide units. The number of saccharide units is given by a numerical prefix: *di*saccharide, *tri*saccharide, *tetra*saccharide, and so forth. Polysaccharides contain more than 10 monosaccharide units. Common disaccharides are sucrose (cane sugar), lactose (milk sugar), maltose, and cellobiose. Common polysaccharides are starch (used to store energy in plants), cellulose (used as a building material by plants), glycogen (used to store energy in animals), and amylopectin. All of these are polymers of D-glucose.

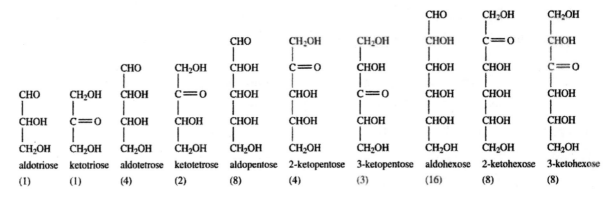

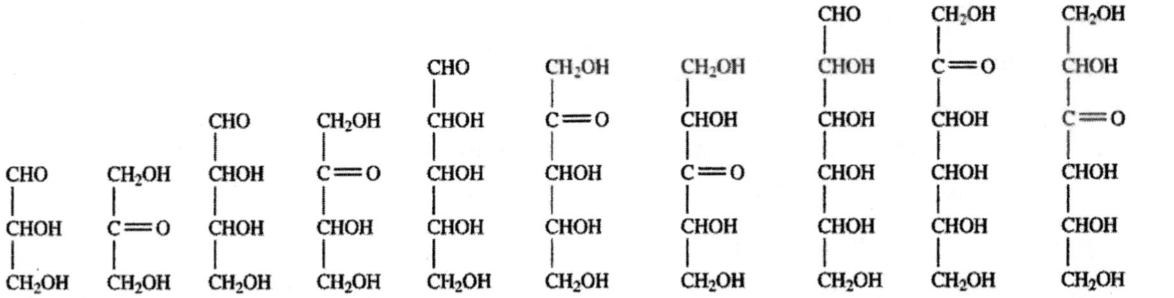

aldotriose (1) ketotriose (1) aldotetrose (4) ketotetrose (2) aldopentose (8) 2-ketopentose (4) 3-ketopentose (3) aldohexose (16) 2-ketohexose (8) 3-ketohexose (8)

FIGURE 27-5 Names of Monosaccharides (parenthesized numbers are number of optical isomers.)

*** 5. Describe the phenomenon of optical activity and the structural features of a molecule that produces it.**

An optically active molecule is one that rotates the plane of polarized light passed through it. Light is accompanied by both an electric field and a magnetic field, which vibrate at right angles to the direction of travel. If we could see this vibration, light traveling at us would look like Figure 27-6a. When light is plane polarized, the electric or magnetic field vibrates in one plane (Figure 27-6b). Light can be polarized in two ways: by passing it through polarizing material or by reflecting it off a flat surface. You probably have seen light polarized by a flat surface. It looks like shimmering water (a mirage) at a distance and often is seen on the highway on bright days.

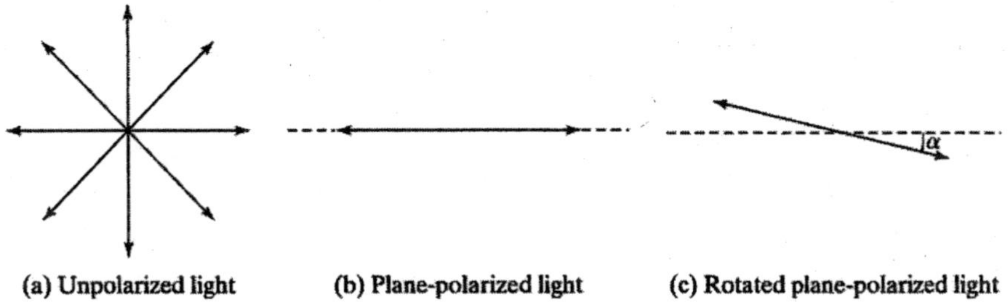

FIGURE 27-6 The Polarization of Light

Optically active molecules rotate the plane of polarized light, as shown in Figure 27-6c, where the plane is rotated by an angle α. A molecule is optically active if it cannot be superimposed on its mirror image. If there are four different groups bonded to a carbon atom in a molecule, that carbon atom is termed a chiral carbon or a chiral center. (A chiral carbon formerly was referred to as an asymmetric carbon.) A chiral carbon often is present in an optically active molecule. A molecule with only one chiral carbon must be optically active. The easiest way to see optical activity is to build a model of the molecule and of its mirror image and see if they can be superimposed. After some practice with models, you can use sketches. There are two types of sketches that you can draw: wedge and dash sketches, and Fischer projections. Both types are shown in Figure 27-7 for CHFClBr. In Fischer projections, the groups above and below the chiral carbon are assumed to point away from you and the groups on either side point toward you. Two compounds that differ only in their optical activity are called enantiomers. You should build models of all these compounds to convince yourself that they are truly non-superimposable. Building models will also help you visualize sketches of structures and will help you sketch structures more easily.

The presence of a chiral carbon does not go hand-in-hand with optical activity. A molecule may *lack* a chiral carbon and yet be optically active, that is, possess a nonsuperimposable mirror image. An example is 1-bromo-1-chloro-2-fluoro-1,2-propadiene, shown in Figure 27-8.

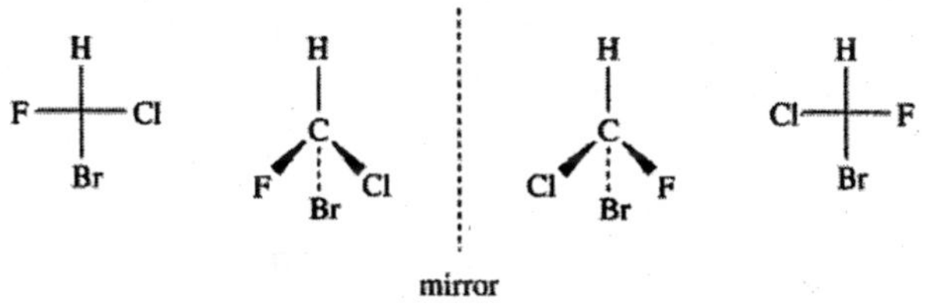

FIGURE 27-7 Wedge and Dash Sketches (in center) and the Corresponding Fischer Projections (on the outside) of the Two Enantiomers of CHFClBr.

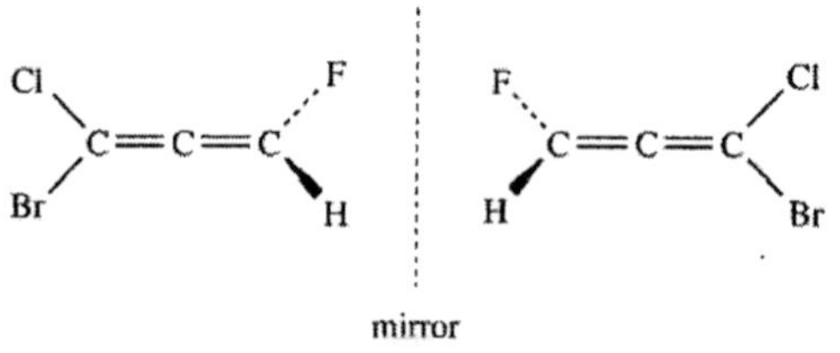

FIGURE 27-8 Wedge and Dash Drawings of the Two Enantiomers of 1-bromo-1-chloro-2-fluoro-1,2-propadiene.

On the other hand, a molecule may possess two or more chiral carbons and yet its mirror image *is* superimposable, as exemplified by 2,3-dichlorobutane in Figure 27-9.

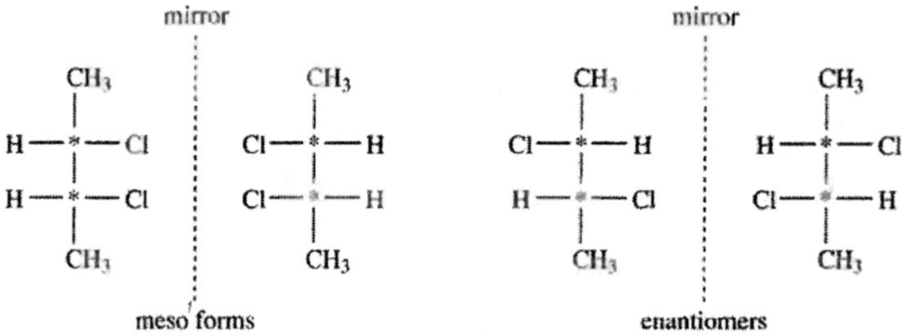

FIGURE 27-9 Fischer Projections of the Meso Forms and Enantiomers of 2,3-dichlorobutane (* designates a chiral carbon).

6. Describe the conversion of a straight-chain monosaccharide into its cyclic form and the joining of monosaccharide units into polysaccharides.

The aldohexoses do not exist in the straight-chain form shown in Figure 27-5 but rather as six-membered rings. The principal evidence for the ring is that two forms of D-glucose exist: α-D-glucose and β-D-glucose. Whereas the straight-chain form of glucose has four chiral carbons (numbered 2–5 in Figure 28-10), the ring form has five (1–5 in Figure 27-10). α-D-glucose and β-D-glucose also are known as pyranoses, because of their similarity to the pyran ring structure. The glucose ring opens and closes readily in aqueous solution and thus the α and β forms of D-glucose interconvert. At 25°C, the equilibrium mixture contains about two molecules of β-D-glucose for every α-D-glucose molecule. The chair conformations of α-D-glucose and β-D-glucose in Figure 27-10 show that the —OH group on carbon 1 is axial in α-D-glucose and equatorial in β-D-glucose. (That is what distinguishes α from β.) The equatorial position is favored for a large group (since there is more room available there than at the axial position), and thus β-D-glucose is more stable.

Monosaccharides join together to form disaccharides, other oligosaccharides, and polysaccharides, with the loss of one water molecule for each bond that is formed. The linkage reaction is given in Equation [2]. Note that it is the —OH groups on the two molecules that react in forming the bond. This is a condensation polymerization reaction, like those discussed in Objective 26-14.

[2]

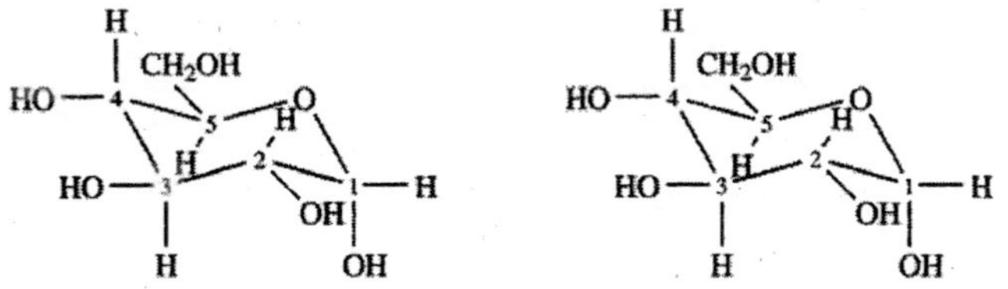

α-D-glucose β-D-glucose

FIGURE 27-10 α- and β-D-glucose

Hexoses can join between carbon 1 on one molecule, a $(1 \longrightarrow 4)$ linkage, or they can join by a $(1 \longrightarrow 6)$ linkage. These linkages may be either α or β (axial or equatorial) from carbon 1. The monomers (or subunits) and the linkages for common disaccharides are given in Table 27-1.

TABLE 27-1 Joining of Monosaccharides

Carbohydrate name	*Name of subunit(s)*	*Linkage(s)*
maltose	glucose	$(1 \longrightarrow 4)$
lactose	galactose, glucose	$(1 \longrightarrow 4)$
cellobiose	glucose	$(1 \longrightarrow 4)$
α-amylose (starch)	glucose	$(1 \longrightarrow 4)$
amylopectin (a component of starch)	glucose	$(1 \longrightarrow 4)$ backbone
		$\alpha(1 \longrightarrow 6)$ side chain branches
glycogen (in liver and muscle)	glucose	same as amylopectin, but more highly branched
cellulose	glucose	$\beta(1 \longrightarrow 4)$

7. Show how the form adopted by an amino acid in solution varies with the pH.

Two of the groups on an amino acid change form depending on acidity: the amino group ($—NH_2$) and the carboxylic acid group ($—COOH$). We have shown both of these groups as uncharged (Figure 28-1). Since the forces between uncharged molecules are relatively low, we would expect solid amino acids to be low melting. Yet most crystalline amino acids melt above 200°C, a temperature more characteristic of ionic compounds. ($LiNO_3$ melts at 264°C.) This is one of the many data that indicate that solid amino acids exist as dipolar ions called zwitterions, which are depicted in Figure 27-11. (The name comes from the German word for two, *zwei.*) The zwitterion also exists in neutral solution or, more correctly, at the isoelectric point, a pH intermediate between acidic and basic. The pH of the isoelectric point is given as p*I* in the right-hand column of Table 27-1. The isoelectric point depends on the acidic character of the specific amino acid. In basic solution, a proton is donated by (and thus removed from) the nitrogen of the ammonium group ($—NH_3^+$), yielding an anion. In acidic solution, a proton is accepted by (and thus added to) an oxygen of the carboxylate group ($—COO^-$), yielding a cation.

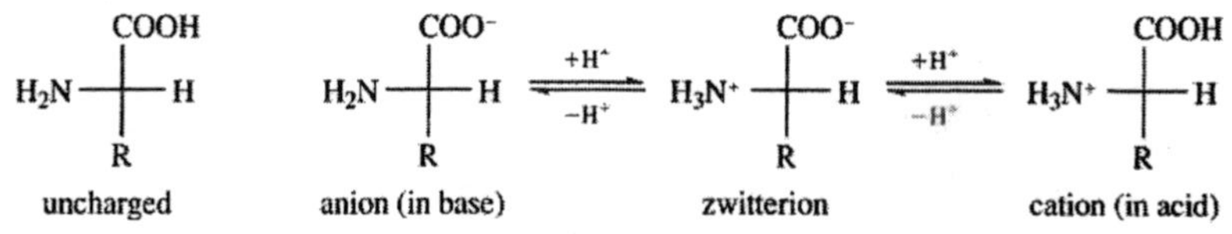

FIGURE 27-11 Different Forms of Amino Acids

8. Describe the formation of a peptide bond between two amino acids, and assign structures and systematic names to polypeptides.

We have already seen several cases of a bond being formed between two molecules by the elimination of water between them. Among these are the formation of pyrophosphates and pyrosulfates, the formation of ethers and esters in Equations [16] and [23] of Objective 26-11, and condensation polymerization in Equation [38] of Objective 26-14. An analogous reaction occurs between two amino acids, equation [3].

A polypeptide is named by naming the amino acids in order starting at the N-terminal end. Except for the last amino acid (the one at the C-terminal end) the *-an* or *-ine* ending of each amino acid is changed to *-yl*. For large polypeptides, more than ten amino acids, only the peptide sequence is given. This also starts the the N-terminal end and uses the three-letter symbols given in Table 27-1. The symbols, names, and structures in that table should be thoroughly learned.

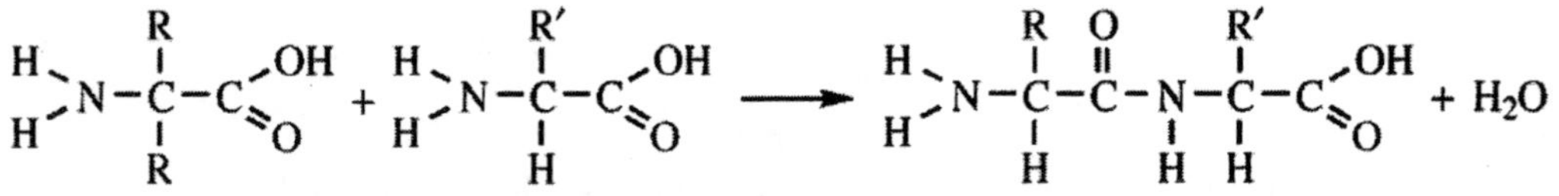

[3]

***9. Determine the sequence of amino acids in a polypeptide chain from information acquired in the hydrolysis of the polypeptide.**

With regard to polypeptide analysis, degradation is the process of breaking the molecule apart, identifying the fragments, and then deducing the structure. In determining the structure—the sequence of amino acids—a large "marker" group is attached to the N-terminal end (the end with the amino group, $-NH_2$). The polypeptide is gently hydrolyzed by heating gently in a mildly acidic or alkaline solution. (A more precise method of hydrolysis is treatment with selective enzymes. Some of these enzymes are so specific that they only hydrolyze the peptide joining two distinct amino acids as, for example, an alanine-glycine bond.) Note that hydrolysis is the reverse of peptide bond formation, Equation [3]. After the initial hydrolysis, the N-terminal marker is placed on each fragment. This process—of gentle hydrolysis followed by marking the N-terminal group—is repeated until the chain has been degraded to di-, tri-, and tetrapeptides. The identities of these small molecules are determined by conventional means: melting point, solubility, spectral analysis, and so forth. Putting together the sequence of the polypeptide is a matter of piecing together these fragments.

10. Explain the meaning of primary, secondary, tertiary, and quaternary protein structure.

The *primary* structure of a protein is the sequence of amino acids in the molecule, starting at the N-terminal end.

The *secondary* structure refers to how the backbone of the polypeptide chain is arranged in space. Most often the polypeptide chain has the form of an α-helix. Hydrogen bonds are formed between the carboxyl oxygen atoms of one amino acid and the amide nitrogen of the amino acid four positions further along the chain.

The *tertiary* structure describes how different parts of the α-helix are held together. The R groups of the amino acids point outward from the helix and therefore can attract each other. Thus the helix will bend into a three-dimensional coil, rather like a twisted worm. The forces that hold the tertiary structure together are those between the atoms of the R groups. They are of four types, and are illustrated in Figure 27-12.

FIGURE 27-12 Interactions that Influence Tertiary Structure

(a) *Hydrogen bonds* link $-NH_2$ or $-OH$ groups of one amino acid's R group and $C=O$ groups of another's. The amino acids with $-NH_2$ groups are asparagine, glutamine, tryptophan, lysine, arginine, and histidine. Those with $-OH$ groups are serine, 4-hydroxyproline, threonine, and tyrosine. Those with $C=O$ groups are asparagine, glutamine, aspartic acid, and glutamic acid.

(b) *Salt linkages* occur between acidic and basic R groups. Acidic amino acids are aspartic acid and glutamic acid. Basic ones are lysine, arginine, and histidine.

(c) *Disulfide bonds* join the two "halves" of cystine.

(d) *Hydrophobic interactions* occur because nonpolar R groups, such as those in glycine, alanine, valine, leucine, phenylalanine, and proline, tend to point inward in the molecule, away from the polar solvent.

In addition to these attractions between R groups, the protein may be wrapped around and bound to another type of molecule. Thus in hemoglobin the proteins (called globins) are bound to a heme molecule.

The quaternary structure of a protein is how two or more individual polypeptide chains are packed together to form the final protein. Many proteins possess only one polypeptide chain and thus have no quaternary structure. But in hemoglobin, for example, four globin chains and four heme molecules combine to produce the final protein.

11. In general terms, explain how lipids, polysaccharides, and proteins are metabolized in the human body and how the energy released by metabolism is stored.

The body breaks down lipids, polysaccharides, and proteins by hydrolysis of the bond between the parts. Thus, lipids break down into glycerol and fatty acids, polysaccharides into monosaccharides, and proteins into amino acids.

Lipids, polysaccharides, and proteins are not rapidly hydrolyzed. The human body uses protein catalysts, called enzymes, to speed up these reactions. Both fatty acids and the monosaccharides are further broken down within the body, resulting ultimately in carbon dioxide and water. Some are stored as fat deposits in the body, and some as carbohydrate deposits, principally glycogen in liver and muscle tissue. Amino acids can be further metabolized to provide energy, but they also are used as raw material for protein synthesis.

12. Explain enzyme action in terms of the "lock-and-key" model.

Enzymes are protein catalysts. Each enzyme speeds up a certain type of chemical reaction. For example, hydrolases speed up hydrolysis reactions. Other enzymes control the rate of peptide bond formation, disulfide bond formation or breakage, and so forth. Enzymes can be quite specific. There are enzymes for triglyceride hydrolysis that will not affect the rate of polysaccharide hydrolysis. Enzymes are also stereospecific. An enzyme may catalyze a reaction involving α-D-glucose but not affect a reaction involving α-L-glucose.

The molecule on which the enzyme acts is known as the substrate. The involved structure of an enzyme and its high specificity are strong evidence that the substrate fits almost perfectly into a pocket in the enzyme, called the active site. The active site often is only a very small portion of the large enzyme molecule. The substrate fits into the active site much as a key fits into a lock. While the substrate is held in place, the enzyme causes the desired reaction.

13. Name the principal constituents of nucleic acids and indicate how these constituents are linked together into chains and, in the case of DNA, a double helix.

Nucleic acids consist of phosphate groups, the purine and pyrimidine bases, and the pentose sugars (shown in Figure 27-2). The purine and pyrimidine bases bond to the $1'$ carbon of the pentose sugar, thus forming *nucleosides*. Guanine bonds to the pentose through the same nitrogen as does adenine. Cytosine and uracil bond through the same nitrogen as does thymine. A *nucleotide* consists of a phosphate group bound to a nucleoside. The phosphate group can bond to the $5'$ or the $3'$ carbon of either pentose, and also to the $2'$ carbon of ribose. However, all naturally occurring nucleotides have the phosphate group bound to the $5'$ carbon.

Nucleotides form chains by a bond between the phosphate group on one nucleotide and the $3'$ carbon on the next nucleotide. This is known as a $3'$-$5'$ linkage.

The single-stranded DNA forms a double-stranded helix by formation of hydrogen bonds. A purine (the larger molecules: adenine and guanine) always is hydrogen-bonded to a pyrimidine (the smaller molecules: thymine and cytosine). This base pairing holds the two strands of the chain together.

DRILL PROBLEMS

1. (1) Name the following triglycerides.

A.

$$\begin{array}{l} CH_2OOC(CH_2)_7CH=CH(CH_2)_7CH_3 \\ | \\ CHOOC(CH_2)_7CH=CH(CH_2)_7CH_3 \\ | \\ CH_2OOC(CH_2)_7CH=CH(CH_2)_7CH_3 \end{array}$$

B.

$$\begin{array}{l} CH_2OOC(CH_2)_8CH_3 \\ | \\ CHOOC(CH_2)_8CH_3 \\ | \\ CH_2OOC(CH_2)_8CH_3 \end{array}$$

C.

$$\begin{array}{l} CH_2OOC(CH_2)_{10}CH_3 \\ | \\ CHOOC(CH_2)_7{=}\,CHCH_2CH_3 \\ | \\ CH_2OOC(CH_2)_8CH_3 \end{array}$$

D.

$$\begin{array}{l} CH_2OOC(CH_2)_7(CH{=}CH)_3(CH_2)_3CH_3 \\ | \\ CHOOC(CH_2)_7(CH{=}CH)_3(CH_2)_3CH_3 \\ | \\ CH_2OOC(CH_2)_7(CH{=}CH)_3(CH_2)_3CH_3 \end{array}$$

2 Draw the condensed formula of the following triglycerides.

E. glyceryl dipalmitoleate F. glyceryl lauromyristopalmitate G. glyceryl dilaurolinoleate

H. glyceryl trilinolenate I. glyceryl triricinoleate J. glyceryl dioleolaurate

3. (1) Determine the average mole weights and the average number of double bonds per molecule for each of the following oil by using the given average value of the saponification value and the iodine number (in that order in parentheses):

A. butter (220, 27)

B. coconut oil (256, 8.8) C. soybean oil (192, 132) D. safflower oil (191, 148)

(2) Determine the saponification value and the iodine number of each of the following triglycerides:

E. glyceryl trioleate F. glyceryl eleosterate

G. glyceryl tristearate H. glyceryl dilaurolinoleate

4. Draw the enantiomers in Fischer projection for each of the following molecules that is optically active:

A. 2-chloropentane B. 3-chloro-1-pentene C. $CH_3CH(NH_2)COOH$
D. $CH_2CH_2CHOHCH_3$ E. 2-chlorobutane F. 1-chloro-3-methylpentane
G. 1,2-dichlorobutane H. 1,4-dichlorobutane I. 2,2-dichloropentane
J. 3-chloropentanol K. 1,3-dibromopentane L. 1-bromo-3-chloro-1-butene
M. 3-chloro-4-methyl-1-pentane

5. (1) Based on the hydrolysis fragments that follow, determine the sequence of each polypeptide, starting with the N-terminal end. Each fragment starts with the N-terminal end.

A. Lys-Asp-Gly, Glu-Ser-Gly, Ala-Ala-Glu, Asp-Gly-Ala, Ala-Glu-Ser

B. Lys-Phe-Ile, Arg-Glu-Lys, Ala-Ala-His, Glu-Lys-Phe, Ala-His-Arg-Glu

C. Cys-Lys-Ala, Tyr-Cys-Lys, Arg-Arg-Gly, Ala-Arg-Arg

D. Phe-Ala, Ser-Ala-Gly, Glu-Ser-Ala, Ala-Glu

E. Val-Ala-Lys-Glu, Met-Gly-Gly-Phe, Val-Met-Tyr-Cys, Glu-Glu-Phe-Val, Tyr-Cys-Glu, Glu-Phe-Val, Glu-Trp-Met, Gly-Gly-Phe

(2) Name the results of A through D above.

QUIZZES (15–20 minutes each) Choose the best answer for each question.

QUIZ A

1. Which function is not regulated by a hormone? (a) pregnancy; (b) vision; (c) metabolism; (d) growth; (e) reproduction.

2. The molar mass of a triglyceride can be determined from its (a) iodine number; (b) oxygen content; (c) unsaturation; (d) saponification value; (e) none of these.

3. All lipids contain (a) nitrogen atoms; (b) long-chain alcohols; (c) glycerol; (d) a phosphate group; (e) none of these.

4. Which of the following is not a monosaccharide? (a) glucose; (b) ribose; (c) glyceraldehyde; (d) sucrose; (e) fructose.

5. Which molecule is not optically active? (a) 1,2-dichlorobutane; (b) 1,4-dichlorobutane; (c) 1,3-dichlorobutane; (d) 1,2-dichloropropane; (e) none of these.

6. Which symbol does not specify how a compound rotates polarized light? (a) d; (b) +; (c) *dl*; (d) D; (e) none of these.

7. Amino acids are joined into proteins by (a) a 3'–5' linkage; (b) a $\beta(1 \longrightarrow 4)$ linkage; (c) DNA; (d) an ester; (e) none of these.

8. The R group $—CH_2CH_2NH_2$ is that of what amino acid? (a) alanine; (b) serine; (c) threonine; (d) arginine; (e) none of these.

9. Glycine is (a) a sugar; (b) an amino acid; (c) a nucleoside; (d) a lipid; (e) none of these.

QUIZ B

1. Which of these is not present in DNA? (a) purine; (b) pentose; (c) phosphate; (d) heme; (e) pyrimidine.

2. The number of double bonds in a triglyceride molecule can be determined from its (a) mole weight; (b) saponification value; (c) iodine number; (d) oxygen content; (e) none of these.

3. Lipids always contain (a) glycerol; (b) phosphate groups; (c) double bonds; (d) fatty acids; (e) none of these.

4. Which of these is a monosaccharide? (a) sucrose; (b) glycogen; (c) starch; (d) cellulose; (e) none of these.

5. Which molecule is optically active? (a) 1,2-dichlorobutane; (b) 1,4-dichlorobutane; (c) *cis*-1,2-dichlorocyclobutane; (d) *cis*-2-butene; (e) none of these.

6. D refers to (a) rotation of plane polarized light; (b) a meso form; (c) configuration of sugars; (d) racemic mixtures; (e) none of these.

7. What is present in all amino acid molecules? (a) a chiral carbon; (b) two nitrogen atoms; (c) a peptide bond; (d) a $—COOH$ or $—COO^-$ group; (e) none of these.

8. Glycine has what R group? (a) $—CH_3$; (b) $—H$; (c) $—CH_2CH_2COOH$; (d) $—CH_2SH$; (e) none of these.

9. 2'-Deoxyribose is a (a) nucleoside; (b) sugar; (c) wax; (d) nucleotide; (e) purine.

QUIZ C

1. All of the following are major components of every cell except (a) triglycerides; (b) carbohydrates; (c) proteins; (d) nucleic acids; (e) hormones.

2. Which of the following fatty acids is unsaturated? (a) palmitic acid; (b) oleic acid; (c) lauric acid; (d) stearic acid; (e) none of these.

3. Biological waxes are (a) the same as triglycerides; (b) the same as lecithins; (c) soluble in water; (d) esters of long-chain alcohols; (e) none of these.

4. Which of the following does *not* contain six carbon atoms? (a) deoxyribose; (b) glucose; (c) aldohexose; (d) methylcyclopentane; (e) benzene.

5. Which molecule is optically active? (a) 2-chlorobutane; (b) 1-chlorobutane; (c) *trans*-propene; (d) *cis*-propene; (e) none of these.

6. Which of the following is optically active? (a) racemate; (b) *dl*; (c) an enantiomer; (d) DL; (e) none of these.

7. In strongly alkaline solution, an amino acid has what charges? (a) + and –; (b) – only; (c) + only; (d) no charges; (e) none of these.

8. The —R group of serine is (a) $—CH_2OH$; (b) $—CH_2COOH$; (c) $—CH_2CH_2CH_3$; (d) $—CH_2CH_2CH_2CH_2NH_2$; (e) none of these.

9. Cytosine is (a) an amino acid; (b) a nucleoside; (c) a purine; (d) a sugar; (e) none of these.

Quiz D

1. All of the following contain both carbon and nitrogen atoms except (a) nucleosides; (b) nucleotides; (c) pyrimidines; (d) purines; (e) none of these.

2. A fatty acid usually has (a) many branched chains; (b) many double bonds; (c) no double bonds; (d) an odd number of carbon atoms; (e) none of these.

3. Which of the following is *false* about lipids? (a) some are triglycerides; (b) they are soluble in water; (c) they are soluble in hexane; (d) some are lecithins; (e) all are true.

4. Which of these is *not* a monosaccharide? (a) ribose; (b) fructose; (c) glucose; (d) glycogen; (e) pentose.

5. Enantiomers *always* (a) have an asymmetric carbon; (b) have different physical properties; (c) change the color of light; (d) rotate polarized light; (e) none of these.

6. Which of the following is *not* broken by hydrolysis with dilute acid? (a) DNA; (b) peptide bonds; (c) polysaccharides; (d) ring forms of hexoses; (e) triglycerides.

7. At the isoelectric point, an amino acid has what charges? (a) + and –; (b) none; (c) + only; (d) – only; (e) none of these.

8. 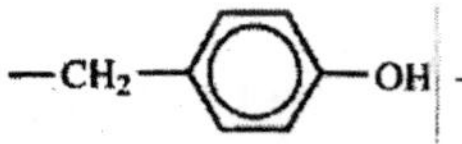

— is the R group of (a) tryptophan; (b) tyrosine; (c) serine; (d) phenylalanine; (e) none of these.

9. Which of these is a pyrimidine? (a) adenine; (b) glycine; (c) phenylalanine; (d) thymine; (e) none of these.

Sample Test (20 minutes)

1. Draw the cyclic forms of threose, $OHC(CHOH)_2CH_2OH$, remembering that five- or six-membered rings are more stable than larger or smaller ones.

2. Describe three structural features of proteins that influence their tertiary structures. Give an example of each.

3. On gentle hydrolysis, a protein gives the following polypeptides, with the N-terminal end written first for each fragment. Give the amino acid sequence of this protein:

Gly-His-Leu, Phe-Val-Asp, Cys-Gly-Ser-His, His-Leu-Cys-Gly, Asp-Gly-His, Gly-Ser-His, Phe-Val.

Answers and Solutions

CHAPTER 1

Drill Problems

1. (1) A. 4 B. 6 C 4 D. 4 E. 2 F. 4–6 G. 3 H. 4 I. 8 J. 5 K. 6 L. 7 M. 6 N. 4 O. 5 P. 5 Q. 2–4 R. 1–5 S. 4 T. 6 U. 6 V. 3 W. 1–4 X. 3

(2) Y. 6.168 Z. 213.3 Γ. 1.200×10^3 Δ. 3136 Θ. 6.196

Λ. 14.16 Ξ 0.002246 Π 152.0 Σ. 0.001987 Υ. 3.024×10^5

Φ. 14.16 Ψ. 3.141×10^4

2. A. 0.592 B. 1.701×10^{-23} C. 56 D. 2×10^4 or 1.7×10^4

or 1.75×10^4 E 8.9×10^2 F. 3.0×10^1 G. 13.6 H. 0.21 or

0.208 or 0.2078 I. 15.3 J. 3.60×10^7 K. 123 L. 87.6

M. -91 N. 108.8 O. 4.362 P. 696 Q. 133.7 R. 75.9

S. 1.4×10^3 T. 412 U. -1.86 V. 3.8 W. 2.03

3. A. 3.74×10^{-3} B. 1.2×10^3 C. 4.06389×10^3

D. 1.751×10^5 E. 6.4604×10^{10} F. 6.627×10^{-27}

G. 9.475×10^{-3} H. 3.74×10^4 I. 1.42×10^{-4}

J. 1.7645×10^4 K. 2.12×10^8 L. 2.66×10^{-3}

M. 8.43×10^7 N. 9.400×10^7 O. 4.963×10^{-8}

P. 8.43214×10^{-1} Q. 2.12×10^9 R. 8.39×10^{-1}

S. 8.9413×10^2 T. 8.314×10^{-7} U. 4.90006×10^4

V. 9.204×10^8 W. 8.7012×10^{27} X. 1.413×10^{-7}

Y. 1.7645×10^{-11} Z. 2.66 Γ. 8.134×10^{-5}

4. (1) A. 88 in. B. 204 fl oz C. 7.244 mi D. 2.346 lb

E. 165.4 mm F. 3.74×10^4 μm G. 0.146 L. H. 0.03754 m^3

I. 2.72×10^7 g J. 0.454 lb K. 8.46 fl oz L. 2.23 m M 0.355

L. N. 84.8 kg O. 22 cm P. 3.78 L Q. 236 cm^3 .R. 14.2 g

S. 57.6 kg T. 5.92 gal U. 0.998 oz V. 3.38 fl oz W. 16.4 yd

X. 157 cm

(2) Y. $76.5\,m^3$ Z. $1.31\,yd^3$ Γ. 1.7×10^3 in.^{3} Δ. $6.45\,cm^3$

Θ. $929\,cm^3$ Λ. 202 gal Ξ. $7.35\,ft^3$ Π. 4.10 L

5. A. 5.00 g, 0.276 oz B. 96.5 g, 3.40 oz C. 56.7 g, 2.00 oz D. 197 g E. 250 g F. $4.41\,cm^3$ G. $5.61\,cm^3$ H. $57.8\,cm^3$ I. $23.5\,cm^3$ J. $837\,cm^3$ K. 2.09 lb L. 746 g M. 17.6 in.^{3} N. 7.18 in.^{3}

6. A. 2.86 g B. 0.13 ton Al, 87 L Fe C. 3.7×10^5 L N_2, 3.2×10^3 L air D. 71.4 lb solder E. 20 lb bleach, 2.3 gal bleach F. 1.22 qt whiskey G. 1.99 L vinegar H. 2.29×10^3 g sulfuric acid I. \$0.50, 2.0×10^2 pennies J.\$33.60, 3.72 g silver nitrate

7. A. 0°C B. 37.8°C C. 209°F D. –3.9°C E. 212°F F. 39.2°F G. 10°C H. 260°C I. –76°C J. –459.67°F

8. A. $x = 0.602$ B. $r = (3V/4\pi)^{1/3}$ C. $x = 14/15$ D. $h = 1.15H/13.6$ E. $T = pvt/PV$ F. $R = PV/nT$ G. $n/V = P/RT$ H. $M = mRT/PV$ I. $m/V = PM/RT$ J. $u = (3RT/\mathrm{m})^{1/2}$ K. $n = \{1/[(1/4) - (v/Rc)]\}^{1/2}$ L. $v = h/m\lambda$ M. $x = 4/5$

Quizzes

A: 1. (d) 2. (a) 3. (e) 4. (e) 5. (d) 6. (d) 7. (b) 8. (b) 9. (a) 10. (e) 90°F 11. (c) 12. (d)

B: 1. (a) 2. (a) 3. (a) 4. (e) 5. (a) 6. (d) 7. (c) 8. (c) 9. (e) 10. (a) 11. (a) 12. (d)

C: 1. (c) 2. (b) 3. (c) 4. (d) 5. (e) 6. (c) 7. (e) 8. (c) 9. (e) 10. (e) 11. (e) 12. (b)

D: 1. (b) 2. (d) 3. (b) 4. (b) 5. (d) 6. (a) 7. (c) 8. (b) 9. (a) 10. (c) 11. (c) 12. (b)

Sample Test

1. $10.72\,\text{g}\times\dfrac{0.417\,\text{g Au}}{1\,\text{g ring}}\times\dfrac{1\,\text{lb}}{454\,\text{g}}\times\dfrac{16\,\text{oz}}{1\,\text{lb}}\times\dfrac{\$652.50}{1\,\text{oz}} = \$103$

2. a. physical—no change of identity, only of physical state b. physical—separation of a mixture c. chemical—two new substances are produced d. chemical—a compound is produced from elements e. physical—separation of a mixture Air is a mixture that later is separated physically; pressure and temperature are not forms of matter; natural gas is possibly a mixture, more likely a compound, since it is later used in a reaction; steam and carbon dioxide are compounds; hydrogen is one of the elements; and ammonia gas is a compound.

3\. a. $12.5\frac{\text{lb}}{\text{in.}^2}\times\frac{16\text{ oz}}{\text{lb}}\times\left(\frac{\text{in.}}{2.54\text{ cm}}\right)^2 = 31.0\text{ oz/cm}^2$

$$31.0\frac{\text{oz}}{\text{cm}^2}\times\frac{\text{lb}}{16\text{ oz}}\times\frac{\text{kg}}{2.205\text{ lb}}\times\left(\frac{100\text{ cm}}{\text{m}}\right)^2$$

$$= 8.79\times10^3\text{ kg/m}^2$$

b. $(77.4°\text{F} - 32.0)\tfrac{5}{9} = 25.2°\text{C}$ c. $5.2\text{ m}^3\times\frac{1000\text{ L}}{\text{m}^3}$

$$= 5.2\times10^3\text{ L}\times\frac{\text{qt}}{0.946\text{ L}} = 5.5\times10^3\text{ qt}$$

CHAPTER 2

Drill Problems

1. (1) A. 8.21 g products, 1.80 g water B. 8.15 g products, 6.35 g Cu, 79.9% Cu C. 10.8 g water, 60.0% water D. 2.23 g Fe, 69.9% Fe, 0.592 g carbon dioxide, 0.592 ton carbon dioxide E. 14.36 g products, 7.96 g copper(II) oxide F. 10.8 g water, 19.2 g oxygen, 4.27 lb oxygen

(2) G. 28.7 g S, 28.5 g oxygen H. 27.3 g C, 72.7 g oxygen I. 8.33 g hydrogen, 66.7 g oxygen J. 21.1 g oxygen, 53.0 g Ca K. 23.1 g table salt, 9.1 g Na L. 182 g ammonia, 150 g nitrogen

2. A. $^{81}\text{Br}^-$, -1, 46 n B. $^{59}\text{Co}^{3+}$, $Z = 27$, 24 e^-

C. +2, $A = 43$, $Z = 20$, 18 e^-, 23 n D. $^{15}\text{N}^{3-}$, $A = 15$, 10 e^-

E. ^{20}Ne, 10 e^-, 10 n F. $^{127}\text{I}^-$, −1, 74 n G. $^{23}\text{Na}^+$, $Z = 11$, 10 e^-

H. +4, $A = 192$, $Z = 76$, 72 e^-, 116 n

I. impossible, since $A < A - Z$ J. $^{52}\text{Cr}^{2+}$, 22 e^-, 28 n

K. $^{60}\text{Co}^{2+}$, +2, 33 n L. $^{17}\text{F}^-$, $Z = 9$, 10 e^-

M. −2, $A = 80$, $Z = 34$, 36 e^-, 46 n N. $^{14}\text{C}^{4-}$, $Z = 6$, 10 e^-

O. $^{118}\text{Sn}^{4+}$, 46 e^-, 68 n

3. A.246.5314 B. 1.11435 C. 5.06364 D. 3.62797

E. 162.219 F. 254.293 G. 249.140 H. 159.868 I. 10.6212 J. 41.6073 K. 54.2572 L. 0.383873

4. A. Li, 6.941, 92.58% B. 10.81, 19.9%, 80.1% C. 13.0, 1.11% D. Ne, 20.19, 9.8% E. Cu, 69.5%, 30.5% F. 35.453, 36.947, 24.47% G. K, 41 (40.964), 6.9% H. Ga, 69.7, 39.6% I. 79.904, 50.4%, 49.6% J. 85.47, 86.92, 27.85%

5. A. 1.01, 0.416 g, 2.48×10^{23} atoms B. S, 2.72 mol,

1.64×10^{24} atoms C. 16.0, 25.4 g, 9.58×10^{23} atoms

D. 12.0, 6.02 g, 0.501 mol E. Cl, 1.52 g, 2.57×10^{22} atoms

F. N, 0.153 mol, 9.21×10^{22} atoms G. 24.3, 4.08 mol,

2.46×10^{24} atoms H. P, 155 g, 5.00 mol I. 79.9,

4.89×10^{-3} g, 3.69×10^{19} atoms J. 39.1, 5.64×10^{-3},

1.44×10^{-4} mol K. 9.01, 4.52×10^{11} g, 3.02×10^{34} atoms

L. F, 1.59×10^{-5} mol, 9.57×10^{-18} atoms

Quizzes

A: 1. (a) 2. (c) 3. (a) 4. (b) 5. (c) 6. (c) 7. (c) 8. (c) 9. (b)

B: 1. (c) 2. (c) 3. (c) 4. (a) 5. (e) 6. (b) 7. (a) 8. (a) 9. (c)

C: 1. (a) 2. (e) 3. (b) 4. (d) 5. (d) 6. (d) 7. (b) 8. (d) 9. (d)

D: 1. (a) 2. (a) 3. (e) 4. (b) 5. (b) 6. (d) 7. (e) 8. (e) 9. (e)

Sample Test

1. $\frac{16.00\,\text{g O}}{\text{mol O}} \times \frac{\text{mol O}}{\text{mol X}} \times \frac{46.7\,\text{g X}}{53.5\,\text{g O}} = 14.0 \frac{\text{g X}}{\text{mol}}$

2. a. isotope b. radioactivity c. atomic number d. Dalton e. cathode tube emission f. Millikan g. Rutherford h. fundamental particles i. Thomson

3. $[10.013 + (11.009 \times 4)]/5 = 10.810$ amu (boron)

CHAPTER 3

Drill Problems

1. (1) A. 472.2 g/mol B. 58.5 g/mol C. 260.7 g/mol

D. 106.3 g/mol E. 84.0 g/mol F. 213.0 g/mol G. 208.3 g/mol

H. 34.0 g/mol I. 27.0 g/mol J. 300.9 g/mol K. 152.0 g/mol

L. 27.67 g/mol

(2) M. 152.0 g/mol, 1.14 mol N. 93.9 g/mol, 232 g

O. 58.5 g/mol, 3.98 mol P. 159.4 g/mol, 11.1 g

Q. 162.3 g/mol, 0.346 mol R. 34.0 g/mol, 66.7 mol

S. 40.0 g/mol, 58.8 g T. 76.3 g/mol, 0.638 g U. 324.6 g/mol,

0.0127 mol V. 99.0 g/mol, 789 g W. 85.0 g/mol,

7.64 mol X. 144.9 g/mol, 316 g

2&3. A. 1.05×10^{23} C atoms B. 4.01 g NH_3

C. 19.8 kg $CsIO_3$ D. 1.57×10^{24} O atoms

E. 2.70×10^{24} HCl molecules F. 5.83×10^{24} atoms

G. 1.21 Mg H_2SO_4 H. 6.88 kg $(NH_4)_2Cr_2O_7$ I. 2.82×10^{24}

atoms J. 0.0266 g $BaCl_2$ K. 2.06×10^{25} O atoms

L. 7.36 mol C M. 0.500 mol H N. 111 mol C O. 2.70 mol Al

4. Mass ratios are given in the same order as the compounds are listed, followed by the whole number ratios.

A. $gC/gO = 0.751:1.123:0.376$ or 2.00:2.99:1.00 or 2:3:1

B. $g\,K/g\,O = 4.88:2.45:1.63$ or 1.99:1.50:1.00 or about 6:3:2

C. $g\,C/g\,H = 12.0:3.00:4.00:5.99$ or about 12:3:4:6

D. $g\,Cl/g\,O = 4.43:2.22:0.634$ or 6.99:3.50:1.00 or $\approx 14:7:2$

E. $g\,Cl/g\,S = 1.11:2.22:4.43$ or 1.00:2.00:3.99 or about 1:2:4

F. $g\,N/g\,H = 4.68:7.00:42.5:13.9$ or 1.00:1.50:9.08:2.97 or

about 2:3:18:6 G. $g\,P/g\,H = 10.2:10.2:10.4$ or about 1:1:1;

$g\,O/g\,P = 2.07:1.54:1.03$ or 2.01:1.50:1.00 or about 4:3:2

H. $g\,S/g\,H = 16:16:16$ or $1:1:1$; $g\,O/g\,S = 1.50;1.99:2.49$ or about 3:4:5

5. & 6. A. H_2SO_3, 82.06 g/mol, 58.54% O B. 40.0 g/mol,

57.5% Na, 40.0% O, 2.5% H C. $MgCO_3$, 84.3 g/mol,

28.9% Mg D. 213.0 g/mol, 12.7% Al, 19.7% N, 67.6% O

E. $LiClO_4$, 106.4 g/mol, 33.3% Cl

F. 153.4 g/mol, 42.6% Zn,

15.6% C, 41.7% O G. K_2MnO_4, 197.1 g/mol, 32.4% O

H. 76.3 g/mol, 31.8% Mg, 31.4% C, 36.7% N I $CuCO_3$,

123.5 g/mol, 9.7% C J. 98.0 g/mol, 3.1% H, 31.6% P,

65.3% O K. $Hg(BrO_3)_2$, 456.4 g/mol, 44.0% Hg

7. A. 0.406 g C, 0.102 g H, 0.812 g O; $C_2H_6O_3$ B. 61.72 g C, 12.87 g H, 0 g O; C_2H_5 C. 3.05 g C, 1.03 g H, 4.06 g O; CH_4O D. 0.0140 g C, 0.00118 g H, 0 g O, CH E. 0.658 g C, 0.110 g H, 0.175 g O; $C_5H_{10}O$ F. 6.11 g C, 1.53g H, 4.06g O; C_2H_6O G. 7.77 g C, 1.30 g H, 10.33 g O; CH_2O H. 3.23 g C, 0.6501 g H, 3.44 g O; $C_5H_{12}O_4$ I. 6.65 mg C, 1.12 mg H, 17.7 mg O; CH_2O_2 J. 0.106 g C, 0.0178 g H, 0.432 g O; CH_2O_3 K. 60.6 g C, 5.04 g H, 80.4 g O; CHO L. 86.62 g C, 5.82 g H. 0 g O; C_5H_4

8a. A. 67.49% Ba B. 10.92% Mg C. 73.20% Pb D. 0.668% Ca E. 29.42% Al F. 24.7% Cr G. 4.27% Li H. 52.20% F I. 15.44% S J. 95.45% C

8b. A. 58.71 g/mol B. 55.87 g/mol C. 15.6 g/mol D. 23.00 g/mol E. 140.1 g/mol F. 26.97 g/mol G. 39.11 g/mol H. 144.2 g/mol

9. (1) A. $Hg = +1$, B. $Na = +1, Cl = -1$ C. $Sn = +4, Cl = -1$ D. $Ca = +2, H = +1, S = -2$

E. $Na = +1, H = +1, O = -2, C = +4$ F. $Al = +3, O = -2, N = +5$ G. $Ba = +2, Cl = -1$

H. $H = +1, O = -1$ I. $H = +1, N = -3, C = +2$ J. $Cl = -1, S = -2, C = +5$
K. $H = +1, N = -3, O = -2, Cr = +6$ L. $H = +1, B = -3$ (2) M. $H = +1, N = -3, O = -2, Cr = +6$
N. $Li = +1, O = -2, S = +4$ O. $Na = +1, Cl = -1$,

P. $Ag = +1, O = -2, Cl = +1$

Q. $Cl = -1, Fe = +3$ R. $H = +1, O = -1$ S. $Na = +1, H = +1, O = -2$ T. $Mg = +2, N = -3, C = +2$
U. $Hg = +2, O = -2, N = +5$ V. $Cl = -1, Cu = +1$ W. $Na = +1, O = -2, N = +5$
X. $H = +1, O = -2, Br = +7$

10. A. sodium chloride B. aluminum sulfate C. strontium chloride D. chromium(II) hydrogen phosphate E. sodium thiosulfate F. copper(I) chloride G. iron(III) phosphate H. lithium sulfite I. ammonium chromate J. potassium hydrogen sulfate K. calcium acetate L. copper(II) hydrogen sulfide M. magnesium carbonate N. chromium(III) cyanide O. barium sulfate P. $KMnO_4$ Q. $NaC_2H_3O_2$ R. $FeCl_3$ S. $Sr(NO_3)_2$ T. $AgClO_3$ U. Cu_2O V. $Hg_2(NO_3)_2$ W. $Fe(NO_3)_2$ X. $Sr(HCO_3)_2$ Y. $Zn_3(PO_4)_2$ Z. K_2CrO_4 Γ. $Mg(CN)_2$ Δ. Al_2O_3 Θ. NaOH Λ. $Ca(OCl)_2$

11. A. water B. hydrogen sulfide C. carbon dioxide D. dinitrogen trioxide E. diphosphorus pentoxide F. iodine pentachloride G. nitrogen trichloride H. sulfur tetrachloride I. chlorine dioxide J. sulfur trioxide K. NO L. P_2O_5 M. NH_3 N. HF O. SiF_4 P. XeF_6 Q. PBr_3 R. SiS_2 S. CH_4 T. BCl_3 U. N_2O_3 V. Sb_2S_5

12. A. sodium perchlorate: $Na+1, Cl+7, O-2$ B. H_2SeO_4 : $H+1, Se+6, O-1$ C. ammonium sulfide: $H+1, N-3, S-2$ D. copper(I)iodate: $Cu+1, I+5, O-2$ E. carbonic acid: $H+1, C+4, O-2$ F. H_3PO_4 : $H+1, P+5, O-2$ G. Na_3BO_3 or $NaBO_2$: $Na+1, B+3, O-2$ H. sodium nitrate: $Na+1, N+5, O-2$ I. Lithium hypochlorite: $Li+1, Cl+1, O-2$ J. H_2SO_2 : $H+1, S+2, O-2$ K. aluminum phosphate: $Al+3, P+5, O-2$ L. H_3BP_3 : $H+1, B+3, O-2$ M. perchloric acid: $H+1, Cl+7, O-2$ N. HNO_2 : $H+1, N+3, O-2$ O. $Mg(ClO_2)_2$: $Mg+2, Cl+3, O-2$ P. scandium aluminate: $Sc+3, Al+3, O-2$ Q. magnesium germanate: $Mg+2, Ge+4, O-2$ R. ammonium nitrate: $N-3, H+1, N+5, O-2$ S. hypofluorous acid: $H+1, F+1, O-2$ or $H+1, F-1, O\,0$ T. HNO: $H+1, N+1, O-2$ U. tellurous acid: $H+1, O-2, Te+4$

13. A. 9.38 B. 0.113 C. 0.116 D. 0.122 E. 0.832 F. 360. G. 0.884 H. 121 I. 1.58 J. 331 K. 0.498 L. 0.173

14. a. carboxylic acid, ethanoic acid; b. chlorohydrocarbon, 2-chloropropane; c. bromohydrocarbon, 2-bromopentane; d. alcohol, propanol; e. carboxylic acid, butanoic acid

15. a. CH_3CH_2COOH, b. CH_3OH, c. $CH_3CH_2CH_2CH_2CH_2CH_3$ d. $FICH_2CH_3$ e. $CH_3CH_2CH = CHCH_2CH_2CH_3$ f. C_6H_{12}

Quizzes

A: 1. (d) 2. (c) 3. (e) 4. (c) 5. (e) 6. (b) 7. (c) 8. (d) 9. (d) 10. (b) 11. (c)

B: 1. (a) 2. (a) 3. (c) 4. (e) 5. (d) 6. (b) 7. (b) 8. (d) 9. (b) 10. N_2O_3 11. (b)

C: 1. (d) 2. (d) 3. (d) 4. (b) 5. (c) 6. (c) 7. (d) 8. (c) 9. (a) 10. dichlorine heptoxide 11. (d)

D: 1. (a) 2. (a) 3. (b) 4. (e) 5. (e) 6. (e) 7. (c) 8. (a) 9. (d) 10. P_2O_3 11. (b)

Sample Test

1. $16.88\text{ g } CO_2 \times \dfrac{12.01\text{ g C}}{44.01\text{ g } CO_2} = 4.606\text{ g C} \times \dfrac{\text{mol C}}{12.01\text{ g}} = 0.3835\text{ mol C}$

$$9.21\text{ g } H_2O \times \frac{2.016\text{ g H}}{18.01\text{ g } H_2O} = 1.03\text{ g H} \times \frac{\text{mol H}}{1.008\text{ g H}} = 1.023\text{ mol H}$$

$$\text{mass O} = 9.72\text{ g} - 4.606\text{ g C} - 1.03\text{ g H} = 4.08\text{ g O} \times \frac{\text{mol O}}{16.0\text{ g O}} = 0.255\text{ mol O}$$

$0.384\text{ mol C}/0.255 = 1.51\text{ mol C}$

$1.023\text{ mol H}/0.0255 = 4.01\text{ mol H}$

$0.255\text{ mol O}/0.255 = 1.00\text{ mol O}$ $C_3H_8O_2$

2. $$\frac{2.9\text{ g Fe} \times \dfrac{\text{mol Fe}}{55.8\text{ g}} \times \dfrac{6.02\times10^{23}\text{ atoms}}{\text{mole Fe}}}{2.6\times10^{13}\text{ red blood cells}} = 1.2\times10^{9}\ \frac{\text{Fe atoms}}{\text{red blood cell}}$$

3. a. $CuSO_4$: Cu + 2, S + 6, O − 2 b. sodium thiosulfate: Na + 1, S + 2, O − 2
c. $Mg(ClO_2)_2$: Mg + 2, Cl + 3, O − 2 d. sodium perborate: Na + 1, B + 5, O − 2
e. $(NH_4)_2O$: N − 3, H + 1, O − 2 f. nitrogen triiodide: N + 3, I − 1 g. calcium xenate: Ca + 2, Xe + 6, O − 2 h. BaO_2: Ba + 2, O − 1 i. Sb_2S_3: Sb + 3, S − 2

CHAPTER 4

Drill Problems

1. A. $2\,Al + 3\,MgO \longrightarrow 3\,Mg + Al_2O_3$

B. $AlCl_3 + 3\,NaOH \longrightarrow Al(OH)_3 + 3\,NaCl$

C. $3\,AgNO_3 + Na_3PO_4 \longrightarrow Ag_3PO_4 + 3\,NaNO_3$

D. $Cl_2 + 2\,KI \longrightarrow I_2 + 2\,KCl$

E. $Fe_2(SO_4)_3 + 3\,Ca(OH)_2 \longrightarrow 2\,Fe(OH)_3 + 3\,CaSO_4$

F. $Ba(NO_3)_2 + (NH_4)_2CO_3 \longrightarrow BaCO_3E + 2\,NH_4NO_3$

G. $2\,KOH + H_2SO_4 \longrightarrow K_2SO_4 + 2\,H_2O$

H. $Fe(OH)_2 + 2\,HCl \longrightarrow FeCl_2 + 2\,H_2O$

I. $CuSO_4 + H_2S \longrightarrow CuS + H_2SO_4$

J. $2\,AgNO_3 + K_2CrO_4 \longrightarrow Ag_2CrO_4 + 2\,KNO_3$

1a. (1) A. potassium dichromate B. silver bromate C. magnesium bisulfite D. potassium hypoiodite E. sodium perchlorate F. potassium thiocyanate G. lead(IV) carbonate H. copper(II) oxalate I. potassium oxide J. thallium(I) sulfide K. tin(IV) manganate *or* tin(II) permanganate L. manganese(II) iodate M. nickel(II) cyanide N. manganese(II) cyanate O. cadmium sulfate P. manganese(II) sulfite Q. ammonium dichromate R. cobalt(II) manganate S. silver thiocyanate T. lithium peroxide U. mercury(I) acetate V. thallium(III) sulfide W. silver dichromate X. gold(III) sulfate Y. magnesium iodate Z. cadmium bisulfite

(2) A. SrO_2 B. $MnSO_3$ C. PbC_2O_4 D. $RbBrO_3$ E. $Mg(CNO)_2$ F. NaHS G. $CaI(IO)_2$ H. FeO_2 I. Ag_2S J. $Ba(IO_3)_2$ K. $CdSO_4$ L. K_2MnO_4 M. $AsCl_3$ N. $Cr(BrO_2)_2$ O. $Sn(CO_3)_2$ P. As_2S_3 Q. $Mn(HCO_3)_2$ R. $Pb(CO_3)_2$ S. $Au_2Cr_2O_7$ T. $K_2C_2O_4$ U. MnS_2O_3 V. H_2O_2 W. $Au(IO_3)_3$ X. NaCN Y. CuCNO Z. SnF_2

2. A. $P_4 + 5\,O_2 \longrightarrow 2\,P_2O_5$

B. $Na + O_2 \longrightarrow Na_2O_2$

C. $2\,Al + 6\,HCl \longrightarrow 2\,AlCl_3 + 3\,H_2$

D. $Ca + 2\,H_2O \longrightarrow Ca(OH)_2 + 3\,H_2$

E. $2\,FeCl_3 + 3\,Ca(OH)_2 \longrightarrow 2\,Fe(OH)_3 + 3\,CaCl_2$

F. $2\,Al + N_2 \longrightarrow 2\,AlN$

G. $6\,HCl + Fe_2O_3 \longrightarrow 2\,FeCl_3 + 3\,H_2O$

H. $3\,Cl_2 + 3\,H_2O \longrightarrow 5\,HCl + HClO_3$

I. $Al(OH)_3 + 3\,HCl \longrightarrow AlCl_3 + 3\,H_2O$

J. $CaSO_3 + H_2SO_4 \longrightarrow CaSO_4 + H_2O + SO_2$

K. $2\,NaCl + H_2SO_4 \longrightarrow Na_2SO_4 + 2\,HCl$

L. $2\,Pb(NO_3)_2 \longrightarrow 2\,PbO + 4\,NO + 3\,O_2$

M. $3\,HNO_2 \longrightarrow HNO_3 + 2\,NO + H_2O$

N. $3\,Ca(OH)_2 + 2\,H_3PO_4 \longrightarrow Ca_3(PO_4)_2 + 6\,H_2O$

O. $SiF_4 + 2\,H_2O \longrightarrow 4\,HF + SiO_2$

2a. A. $3\,Cu + 2\,Fe^{3+} \longrightarrow 3\,Cu^{2+} + 2\,Fe$

B. $Zn + 2\,H^{+} \longrightarrow Zn^{2+} + H_2$

C. $2\,H^+ + S^{2-} \longrightarrow H_2S(g)$

D. $Pb^{2+} + 2\,Hg^{2+} \longrightarrow Pb^{4+} + Hg_2{}^{2+}$

E. $2\,Fe^{3+} + Sn^{2+} \longrightarrow 2\,Fe^{2+} + Sn^{4+}$

F. $3\,H^+ + N^{3-} \longrightarrow NH_3\,(g)$

G. $6\,CN^- + Fe^{2+} \longrightarrow Fe(CN)_6{}^{4-}$

H. $2\,Bi^{3+} + 3\,S^{2-} \longrightarrow Bi_2S_3\,(s)$

3. A. $2\,CH_3OH + 3\,O_2 \longrightarrow 2\,CO_2 + 4\,H_2O$

B. $2\,C_6H_6 + 15\,O_2 \longrightarrow 12\,CO_2 + 6\,H_2O$

C. $C_5H_{12} + 8\,O_2 \longrightarrow 5\,CO_2 + 6\,H_2O$

D. $C_{12}H_{22}O_{11} + 12\,O_2 \longrightarrow 12\,CO_2 + 11\,H_2O$

E. $2\,CO + O_2 \longrightarrow 2\,CO_2$

F. $HC_2H_3O_2 + 2\,O_2 \longrightarrow 2\,CO_2 + 2\,H_2O$

G. $C_{10}H_8 + 12\,O_2 \longrightarrow 10\,CO_2 + 4\,H_2O$

H. $2\,C_9H_5O_4 + 17\,O_2 \longrightarrow 18\,CO_2 + 6\,H_2O$

I. $2\,C_7H_6O_2 + 15\,O_2 \longrightarrow 14\,CO_2 + 6\,H_2O$

J. $C_9H_{28}O_2 + 25\,O_2 \longrightarrow 19\,CO_2 + 14\,H_2O$

3a. C = combination, De = decomposition, Di = displacement, M = metathesis

A. C: $Al + 3\,Br_2 + \longrightarrow 2\,AlBr_3$

B. Di: $2\,Al + Fe_2O_3 \longrightarrow 2\,Fe + Al_2O_3$

C. M: $AlCl_3 + 3\,KOH \longrightarrow Al(OH)_3 + 3\,KCl$

D. M: $BaCl_2 + Na_2CO_3 \longrightarrow BaCO_3 + 2\,NaCl$

E. M: $BaCl_2 + 2\,AgNO_3 \longrightarrow 2\,AgCl + Ba(NO_3)_2$

F. C: $BaO + SO_3 \longrightarrow BaSO_4$

G. M: $Ba(OH)_2 + H_2CO_3 \longrightarrow BaCO_3 + 2\,H_2O$

H. C: $Br_2 + H_2 \longrightarrow 2\,HBr$

I. M: $Cd(NO_3)_2 + (NH_4)_2S \longrightarrow CdS + 2\,NH_4NO_3$

J. Di : $Ca + 2\,H_2O \longrightarrow Ca(OH)_2 + H_2$

K. Di : $6\,HCl + 2\,Al \longrightarrow 2\,AlCl_3 + 3\,H_2$

L. C : $H_2O + Li_2O \longrightarrow 2\,LiOH$

M. M : $3\,Ba(OH)_2 + 2\,H_3PO_4 \longrightarrow Ba_3(PO_4)_2 + 6\,H_2O$

N. C : $BaO + H_2O \longrightarrow Ba(OH)_2$

O. M : $CaO + 2\,HCl \longrightarrow CaCl_2 + H_2O$

P. C : $CaO + CO_2 \longrightarrow CaCO_3$

Q. M : $Ca(OH)_2 + 2\,HNO_3 \longrightarrow Ca(NO_3)_2 + 2\,H_2O$

R. M : $CaO + 2\,HNO_3 \longrightarrow Ca(NO_3)_2 + H_2O$

S. C : $NH_3(g) + HCl(g) \longrightarrow NH_4Cl(s)$

T. C : $2\,SO_2 + O_2 \longrightarrow 2\,SO_3$

U. Di : $2\,AgNO_3 + Cu \longrightarrow 2\,Ag + Cu(NO_3)_2$

V. C : $Cl_2 + 2\,FeCl_2 \longrightarrow 2\,FeCl_3$

W. Di : $Cl_2 + 2\,KI \longrightarrow 2\,KCl + I_2$

X. C : $N_2 + 3\,H_2 \longrightarrow 2\,NH_3$

Y. M : $2\,Al(OH)_3 + 3\,H_2SO_4 \longrightarrow Al_2(SO_4)_3 + 6\,H_2O$

4. & 4a. A. 35.4 g HCl, 52.4 g $FeCl_3$, 17.4 cm³ H_2O

B. 40.5 g $Fe(OH)_3$, 20.5 g H_2O

C. 17.8 g $Fe(OH)_3$, 27.0 g $FeCl_3$

D. 50.8 g O_2, 43.7 g CO_2, 21.5 cm³ H_2O

E. 0.808 g C_5H_{12}, 1.212 g H_2O

F. 0.410 g C_5H_{12}, 0.640 L CO_2, 0.616 cm³ H_2O

G. 0.314 g H_2, 1.77 g NH_3

H. 11.0 g N_2, 2.37 g H_2

I. 89.8 g N_2, 175 L NH_3

J. 10.6 g HNO_3, 6.05 g H_2O, 3.76 L NO

K. $13.3\ g\ HNO_3, 6.34\ g\ NO, 7.61\ g\ H_2O$

L. $43.4\ L\ HCl,\ 23.3\ g\ H_2O, 40.7\ g\ HNO_3$

M. $8.30\ g\ HNO_3, 0.988\ g\ NO$

N. $9.20\ g\ Cu, 27.1\ g\ Cu(NO_3)_2$

O. $460\ g\ HNO_3, 513\ g\ Cu(NO_3)_2, 40.8\ L\ NO$

P. $0.802\ cm^3\ Cu, 1.69\ L\ NO, 2.71\ g\ H_2O$

5. A. 51.26 g. 0.8535 mol

B. 46.30 g, 3.175 M

C. 73.2 g, 200 mL (0.200 L)

D. 50.9 g, 0.244 mol

E. 0.244 mol, 167 mL

F. 0.492 mol, 2.83 M

G. 21.30 g, 0.2128 mol

H. 0.1155 mol, 0.3301 M

I. 781 g, 2.06 L $(2.06 \times 10^3\ mL)$

J. 33.7 g, 166 mL

K. 511.0 g, 5.210 mol

L. 0.223 mol, 0.745 M

6. A. $[KNO_3] = 0.0526\ M, 570\ mL$

B. 240 mL HCl soln, 360 mL H_2O

C. 64.8 ml water, 152 mL total

D. $[NaCl] = 0.271\ M$, 620 mL total

E. 31.3 mL $MgSO_4$ soln, 269 mL water

F. 840 mL total, 700 mL water

G. $[K_2SO_4] = 0.0583\ M$, 720 mL total

H. 133 mL AlI_3 soln, 267 mL water

I. $[NaCl] = 0.228\ M$, 300 mL total

J. 135 mL total 85.0 mL KCl soln

K. $[NaNO_3] = 0.456\ M$, 300 mL total

L. 305 mL soln A, 195 mL soln B

M. 300 Ml HCl soln, $[HCl]=1.90$ M

N. $[NaNO_3]=3.00$ M, 500 mL total

7. A. 4.19 g Ag_3PO_4, $[NaNO_3]=0.300$ M

B. 31.5 mL Na_3PO_4 soln, $[NaNO_3]=0.473$ M

C. 131 mL $AgNO_3$ soln, 36.6 g Ag_3PO_4

D. 0.178 M Na_3PO_4 soln, 32.0 mL $NaNO_3$ soln,

E. 9.48 g HCN, 176 mL $MgSO_4$ soln

F. 13.4 g $Mg(CN)_2$, 29.3 mL H_2SO_4 soln

G. 45.9 g $Mg(CN)_2$, 3.50 M, 32.5 g HCN

H. 26.7 g $Mg(CN)_2$, 18.9 g HCN

I. 11.3 mL H_2SO_4 soln, 1.22 g H_2O

J. 56.4 g Al_2O_3, 29.3 g H_2O

K. $[Al_2(SO_4)_2]=0.333$ M, 3.53 g H_2O

L. 14.0 g Al_2O_3, 412 mL H_2SO_4 soln

M. $[KOH]=7.06$ M, 42.5 g CO_2, $[K_2CO_3]=3.53$ M

N. 4.27 g CO_2, 194 mL K_2CO_3 soln, 1.75 g H_2O

O. $[KOH]=0.429$ M, 0.548 g H_2O

P. 3.81 g CO_2, 1.56 g H_2O

Q. $[H_2SO_4]=1.82$ M, 121 mL K_2SO_3 soln, 3.28 g H_2O

R. 217 mL KOH soln, 5.15 g H_2O

S. 88.3 mL H_2SO_3 soln, 124 mL K_2SO_3 soln

8. A. 64.8 g Ag_3PO_4 produced, 23.1 g $AgNO_3$ left

B. 22.1 g H_2O produced, 23.0 g H_2SO_4 left

C. 6.33 g CO produced, 9.69 g C left

D. 0.220 g H_2 produced, 6.2 g HCl left

E. 34.1 g MgS produced, 6.6 g Mg left

F. 68.8 g CuCl produced, 6.7 g HCl left

G. 32.8 g H_2O produced, 29.5 g SiO_2 left

H. 46.6 g H_2O produced, 175 g $Al(OH)_3$ left

I. 57.8 g H_2SO_4 produced, 121 g HNO_3 left

J. 212 g $FeCl_3$ produced, 100 g Fe_2O_3 left

K. 70.3 g H_2O produced, 64.1 g H_2S left

L. 92.0 g NaCl produced, 5.8 g HCl left, 112 g Na_2TeO_4 left

M. 217 g $ZnCl_2$ produced, 137 g H_3AsO_4 left, 12 g HCl left

N.42.0 g K_2SO_4 produced, 7 g H_2SO_4 left, 48 g Hg left

O. 19.9 g Cl_2 produced, 48.2 g C_2H_3OCl left, 45 g H_2SO_4 left

9. A. 91.6% B. 63.6 g PCl_3 C. 29.0 g P_4 D. 79.7% E. 59.4 g CaO F. 98.1 g $CaCO_3$ G. 84.9% H. 249 g O_2 I. 23.1 g H_2O J. 64.0% K. 330 g Fe L. 2.01 g H_2 M. 89.8% N. 5.53 g H_2O O. 3.29 g H_2S

10. A. 5.03 g NaCl, 172 mL NaCl soln

B. 8.45 g CO_2, 2.70×10^3 mL HCl soln

C. 61.6 g CO_2, 110.6 g solid

D. 1.23 L KOH soln, 13.8 g H_2O

E. 0.563 g H_2, 2.79 L HCl soln

F. 33.4 g H_2O, 535 mL H_2SO_4 soln

G. 20.4 g $NaNO_3$, 37.7 g solid

11. A. 19.9 g H_2SO_4 B. 17.9 g O_2 C. 153 g NaCl D. 152 mL H_2SO_4 soln
E.15.3 g Cu F. 255 g HNO_3 G. 1.03 g H_2 H. 21.6 g $CaCl_2$ I. 25.1 g NH_3
J. 240 g $CaCO_3$ K. 301 g AgCl L. 283 mL HNO_3 soln M. 1.97 g P_4, 6.78 g Cl_2

12. $CuS + 2\,O_2 + H_2O \longrightarrow CuO + H_2SO_4$

$6\,NaCl + 3\,H_2SO_4 + 2\,Fe + 3\,CuO \longrightarrow 3\,Na_2SO_4 + 2\,FeCl_3 + 3\,Cu + 3\,H_2O$

$6\,N_2 + 18\,H_2 + 21\,O_2 \longrightarrow 14\,H_2O + 8\,HNO_3 + 4\,NO$

$CaCO_3 + 2\,NH_4Cl \longrightarrow CO_2 + CaCl_2 + 2\,NH_3 + H_2O$

$P_4 + 6\,Cl_2 + 12\,H_2O + 12\,AgNO_3 \longrightarrow 4\,H_3PO_3 + 12\,AgCl + 12\,HNO_3$

Quizzes

A: 1. sodium nitrite 2. H_2SO_4 3.(d) 4. (e) (sum = 9) 5. (a) 6. (b) 7. $Na_2S + 2\,HCl \longrightarrow 2\,NaCl + H_2S(g)$ 8. (b) 9. (d) 10. (a)

B: 1. magnesium chlorate 2. NaFO 3. (b) 4. (b) 5. (b) 6. (d) 7. $Al_2O_3 + 3\,H_2SO_4 \longrightarrow Al_2(SO_4)_3 + 3\,H_2O$ 8. (c) 9. (e) 10. (c)

C: 1. rubidium perbromate 2. HNO_3 3. (c) 4.(c) . 5. (b) 6 (b) 7. $Ba(OH)_2 + H_2SO_4 \longrightarrow BaSO_4 + 2\,H_2O$ 8. (d) 9 (a) 10. (b)

D: 1. sulfuric acid 2. $NaHCO_3$ 3. (c) 4. (e) (sum = 9) 5. (c) 6. (b) 7. $CaCO_3 + 2\,HCl \longrightarrow CaCl_2 + CO_2 + H_2O$ 8. (d) 9. (b) 10. (a)

Sample Test

1. a.lead(II) thiocyanate b. $Pb(S_2O_3)_2$ c. As_2S_3

d. manganese(II)hypobromite e. nitrous acid f. HIO_4

g.SnS h. NH_4HCO_3

2. $200.0\text{ mL} \times 0.240\text{ M} \times 2\text{ mmol K}^+ / \text{mmol K}_2\text{SO}_4 = 96.0\text{ mmol K}^+$

Let V be the volume of KNO_3 solution added, in mL.

$$0.400\text{ M} = \frac{96.0\text{ mmol} + (01.60\text{ M} \times V)}{200.0 + V}$$

or $V = 66.7$ mL

3. a. $BaCl_2 + 2\,AgNO_3 \longrightarrow Ba(NO_3)_2 + 2\,AgCl$

b. $Ca(OH)_2 \xrightarrow{\text{heat}} CaO + H_2O$ c. $NH_3 + HCl \longrightarrow NH_4Cl$

d. $Zn + CuSO_4 \longrightarrow ZnSO_4 + Cu$

e. $2\,C_2H_6 + 7\,O_2 \longrightarrow 4\,CO_2 + 6\,H_2O$

4. $55.2\text{ g CuO} \times \frac{\text{mol CuO}}{79.5\text{ g CuO}} \times \frac{1\text{ mol H}_2\text{O}}{2\text{ mol HCl}} = 0.213\text{ mol H}_2\text{O}$

$0.213\text{ mol H}_2\text{O} \times 18.0\text{ g H}_2\text{O/mol} = 3.84\text{ g H}_2\text{O}$

5. $26.4\text{ g} \times \frac{0.400\text{ g CaO}}{\text{g mixture}} \times \frac{\text{mol CaO}}{56.1\text{ g CaO}} \times \frac{\text{mol CaCl}_2}{\text{mol CaO}} \times \frac{111.1\text{ g CaCl}_2}{\text{g mixture}} = 20.9\text{ g CaCl}_2$

$26.4\text{ g} \times \frac{0.600\text{ g NaOH}}{\text{g mixture}} \times \frac{\text{mol NaOH}}{40.0\text{ g NaOH}} \times \frac{\text{mol NaCl}}{\text{mol NaOH}} \times \frac{58.5\text{ g NaCl}}{\text{mol NaCl}} = 23.2\text{ g NaCl}$

Total mass of solid $= 20.9\text{ g CaCl}_2 + 23.2\text{ g NaCl} = 44.1\text{ g}$

CHAPTER 5

Drill Problems

1. Electrolytes are designated as follows: N = nonelectrolyte, W = weak electrolyte, and S = strong electrolyte. Also a = acid, b = base, and s = salt. Of course, for instance, a weak electrolyte that is a base is also a weak base.

A.Sa B.N C.Wb D.Ss E.Sa F.Wa G.Ss H.N I.Ss J.Ss K.Ss L.Sb M.Ss N.Wa O.N P.Ss Q.Ss R.Ss S.Ss T.Wb U.N V.Ss W.Ss X.Wa Y.Ss Z.Ss Γ. N Δ. Ss Θ. Ss Λ. Wa Ξ. N Π. N Σ. Ss Υ. Ss Φ. Ss Ψ. N

1a. A.sulfuric acid B. HNO_3 C. H_2SO_4 D.acetic acid E. $HC_2H_3O_2$ F. hydrochloric acid G.phosphoric acid H.perchloric acid I.iodic acid J.HF K. HNO_2 L.hypofluorous acid M.oxalic acid N. HIO_4 O.manganic acid P.HBrO Q.HCN R.hypoiodous acid S. H_2CO_3 T. $HMnO_4$ U. $HClO_4$ V.chlorous acid

2. (1) A. 6.48 M B. 1.22 M C. 0.466 M D. 0.630 M E. 4.80 M F. 2.23 M G. 0.269 M H. 0.214 M I. 0.169 M J. 0.203 M K. 0.324 M L. 0.665 M

(2) M.31.3 mL $MgSO_4$ soln, 269 mL water N. 840 mL total, 700 mL water O. $[K^+] = 0.117$ M, 720 mL total

P. 44.4 mL $Al(NO_3)_3$ soln, 356 mL water Q. $[Na^+] = 0.228$ M, 300 mL total R. $[Na^+] = 0.456$ M, 300 mL total

S. 305 mL $MgSO_4$ soln, 195 mL Na_2SO_4 soln T. 300 mL KCl soln, [KCl] = 1.90 M

U. $[Cl^-] = 3.00$ M, 500 mL total

3. i = insoluble, s = soluble, m = moderately soluble, p = sparingly soluble A. i B. i C. s D. i E. s F. s G. s H. m I. s J. s K. i L. s M. i N. i O. s P. s Q. s R. s S. s T. i U. s V. s W. s X. i Y. s Z. m Γ i Δ.s Θ . i Λ s Ξ . s Π i. Σ . m Υ . i Φ . i Ψ . s

4. E. L. and P are the same: $OH^- + H^+ \longrightarrow H_2O$

A. $2\,Cl^- + Pb^{2+} \longrightarrow PbCl_2(s)$ B. $3\ Ca^{2+} + 2\ PO_4^{3-} \longrightarrow Ca_3(PO_4)_2(s)$

C. $Ba^{2+} + CO_3^{2-} \longrightarrow BaCO_3(s)$ D. $Cl^- + H^+ \longrightarrow HCl(g)$ F. $S^{2-} + 2\ H^+ \longrightarrow H_2S(g)$

G. $Ba^{2+} + SO_4^{2-} \longrightarrow BaSO_4(s)$ H. $2\ Ag^+ + SO_4^{2-} \longrightarrow Ag_2SO_4(s)$

I $CO_3^{2-} + 2\,H^+ \longrightarrow H_2O + CO_2(g)$ J $NH_4^+ + OH^- \longrightarrow H_2O + NH_3(g)$ K $Cd^{2+} + S^{2-} \longrightarrow CdS(s)$

M. $Fe^{3+} + 3\ OH^- \longrightarrow Fe(OH)_3(s)$ N. $3\ Ag^+ + PO_4^{3-} \longrightarrow Ag_3PO_4(s)$ O $3\ Sr^{2+} + 2\ PO_4^{3-} \longrightarrow Sr_3(PO_4)_2(s)$ and $H^+ + OH^- \longrightarrow H_2O$

5. If the reaction goes to completion, the reason will be included in brackets: [p] = precipitate forms, [n] = nonelectrolyte forms, [g] = gas escapes from solution, [eq] reaction does not go to completion. A. [p] $3\,Hg_2(NO_3)_2(aq) + 2\,H_3PO_4(aq) \longrightarrow Hg_6(PO_4)_2(s) + 6\,HNO_3(aq$

B [p] $Pb(NO_3)_2(aq) + 2\,HCl(aq) \longrightarrow PbCl_2(s) + 2\ HNO_3(aq)$

C [p] $Sr(NO_3)_2(aq) + H_2SO_4(aq) \longrightarrow SrSO_4(s) + 2\,HNO_3(aq)$

D [n] $2\ RbOH(aq) + H_2SO_3(aq) \longrightarrow Rb_2SO_3(aq) + 2\,H_2O(l)$

E [p] $2\ AsCl_3(aq) + 3\ H_2S(aq) \longrightarrow As_2S_3(s) + 6\ HCl(aq)$

F [eq] $MnCl_2(aq) + H_2S(aq) \longrightarrow MnS(s) + 2\ HCl(aq)$

G [n, p] $Ba(OH)_2(aq) + H_2SO_4(aq) \longrightarrow BaSO_4(s) + 2\ H_2O(l)$

H [n] $Zn(OH)_2(aq) + 2\ HCl(aq) \longrightarrow ZnCl_2(aq) + 2\ H_2O(l)$

I [n] $Al_2O_3(aq) + 6\ HNO_3(aq) \longrightarrow 2\ Al(NO_3)_3(aq) + 3\ H_2O(l)$

J [n] $Cu_2O(s) + 2\ HCl(aq) \longrightarrow 2\ CuCl(s) + H_2O(l)$

K [n] $Cr_2O_3(s) + 3\ H_2SO_4(aq) \longrightarrow Cr_2(SO_4)_3(aq) + 3\ H_2O(l)$

L [n] $2\ KOH(aq) + CO_2(g) \longrightarrow K_2CO_3(aq) + H_2O(l)$

M [eq] $CuSO_4(aq) + H_2S(aq) \longrightarrow CuS(s) + H_2SO_4(aq)$

N [eq] $ZnCl_2(aq) + H_2S(aq) \longrightarrow ZnS(s) + 2\ HCl(aq)$

O [n] $3\ ZnO(s) + 2\ H_3PO_4(aq) \longrightarrow Zn_3(PO_4)_2(aq) + 3\ H_2O(l)$

P [n] $Al_2O_3(s) + 3\ H_2SO_4(aq) \longrightarrow Al_2(SO_4)_3(aq) + 3\ H_2O(l)$

Q [g] $2\ HCl(aq) + Na_2S(s) \longrightarrow 2\ NaCl(aq) + H_2S(g)$

R [g] $K_2CO_3(aq) + H_2SO_4(aq) \longrightarrow K_2SO_4(aq) + H_2O(l) + CO_2(g)$

S [g] $NH_4Cl(aq) + NaOH(aq) \longrightarrow NaCl(aq) + H_2O(l) + NH_3(g)$

T [g] $CuSO_3(s) + 2\ HBr(aq) \longrightarrow CuBr_2(aq) + H_2O(l) + SO_2(g)$

6. Oxidizing agent given first, followed by reducing agent.

A. HNO_3 or NO_3^-, Ag B. PbO_2, HCl or Cl^- C. HBr or H^+, Al

D. $BaSO_4$ or SO_4^{2-}, C E. Br_2, SO_2 F. Br_2, Br_2 G. HNO_3 or NO_3^-, C

H. $Ca_3(PO_3)_2$ or PO_3^{3-}, C I. ClO_2, H_2O_2 J. HNO_3 or NO_3^- Cu

K. $HClO_3$, $HClO_3$ L. I_2, H_3AsO_3 M. Na_2O_2 or O_2^{2-}, $Cr(OH)_3$

N. H_2SO_4 or SO_4^{2-}, FeI_2 O. HNO_3 or NO_3^-

7. (1) A. $Ag \rightarrow Ag^+ + e^-$; B. $NO_3^- + 4\,H^+ + 3\,e^- \rightarrow NO + 2\ H_2O$

C. $2\,Cl^- \rightarrow Cl_2 + 2\,e^-$; D. $PbO_2 + 4\,H^+ + 2\ e^- \rightarrow Pb^{2+} + 2\,H_2O$

E. $Br_2 + 6\,H_2O \rightarrow 2\,BrO_3^- + 12\,H^+ + 10\,e^-$

F. $Br_2 + 12\,OH^- \rightarrow 2\,BrO_3^- + 6\,H_2O + 10\,e^-$

G. $NO_3^- + 2\,H^+ + e^- \rightarrow NO_2 + H_2O$

H. $H_2O_2 + 2\ OH^- \rightarrow O_2 + 2\,H_2O + 2\,e^-$

I. $ClO_3^- + 2\,H^+ + e^- \rightarrow ClO_2 + H_2O$

J. $ClO_3^- + H_2O \rightarrow ClO_4^- + 2\,H^+ + 2\,e^-$

K. $Cr(OH)_3 + 5\,OH^- \rightarrow CrO_4^{2-} + 4\,H_2O + 3\,e^-$

L. $O_2^{2-} + 2\,H_2O + 2\,e^- \rightarrow 4\ OH^-$

M. $SO_4^{2-} + 4\,H^+ + 2\,e^- \rightarrow SO_2 + 2\,H_2$

N. $NO_2^- + 2\,OH^- \rightarrow NO_3^- + H_2O + 2\,e^-$

O. $MnO_4^{2-} + 2\,H_2O + 2\,e^- \rightarrow MnO_2 + 4\,OH^-$

(2) A. $H_2O_2 \rightarrow 2\,H^+ + O_2 + 2\,e^-; Cu^{2+} + 2\,e^- \rightarrow Cu;$

$CuSO_4 + H_2O_2 \longrightarrow Cu + O_2 + H_2SO_4$ B. $Mg \rightarrow Mg^{2+} + 2\,e^-;$

$Ag^+ + e^- \rightarrow Ag; Mg + 2\,AgNO_3 \longrightarrow 2\,Ag + Mg(NO_3)_2$

C. $Al \rightarrow Al^{3+} + 3\,e^-; 2\,H^+ + 2\,e^- \rightarrow H_2;$

$2\,Al + 3\,H_2SO_4(aq) \longrightarrow Al_2(SO_4)_3(aq) + 3\,H_2(g)$

D. $Pb \rightarrow Pb^{2+} + 2\,e^-; Fe^{3+} + 3\,e^- \rightarrow Fe;$

$Fe_2(SO_4)_3 + 3\,Pb \longrightarrow 3\,PbSO_4 + 2\,Fe$ E. $2\,I^- \rightarrow I_2 + 2\,e^-;$

$F_2 + 2\,e^- \rightarrow 2\,F^-; 2\,NaI + F_2 \longrightarrow I_2 + 2\,NaF$

F. $H_2 + 2\,OH^- \rightarrow 2\,H_2O + 2\,e^-;$

$ClO^- + H_2O + 2\,e^- \rightarrow Cl^- + 2\,OH^-; H_2 + KClO \longrightarrow KCl + H_2O$

G. $2\,Cl^- \rightarrow Cl_2 + 2\,e^-; MnO_2 + 4\,H^+ + 2\,e^- \rightarrow Mn^{2+} + 2$

$H_2O; MnO_2 + 4\,HCl \longrightarrow MnCl_2 + 2\,H_2O + Cl_2$

H. $I_2 + 6\,H_2O \rightarrow 2\,IO_3^- + 12\,H^+ + 10\,e^-; Br_2 + 2\,e^- \rightarrow 2\,Br^-;$

$I_2 + 6\,H_2O + 5\,Br_2 \longrightarrow 10\,HBr + 2\,HIO_3$ I. $Ag \rightarrow Ag^+ + e^-;$

$NO_3^- + 4\,H^+ + 3\,e^- \rightarrow NO + 2\,H_2O;$

$3\,Ag + 4\,HNO_3 \longrightarrow 3\,AgNO_3 + NO + 2\,H_2O$

J. $S + 3\,H_2O \rightarrow SO_3^{2-} + 6\,H^+ + 4\,e^-; Pb^{2+} + 2\,e^- \rightarrow Pb;$

$S + 3\,H_2O + 2\,Pb(NO_3)_2 \longrightarrow 2\,Pb + H_2SO_3 + 4\,HNO_3;$

$K. O_2 + H_2O \rightarrow O_3 + 2\,H^+ + 2\,e^-; O_2 + 2\,H^+ + 2\,e^- \rightarrow H_2O_2;$

$2\,O_2 + H_2O \longrightarrow O_3 + H_2O_2;$ $L. H_2S \rightarrow S + 2\,H^+ + 2\,e^-;$

$Br_2 + 2\,e^- \rightarrow 2\,Br^-; H_2S + Br_2 \longrightarrow S + 2\,HBr;$

$M. Fe^{2+} \rightarrow Fe^{3+} + e^-; Sn^{2+} + 2\,e^- \rightarrow Sn;$

$SnSO_4 + 2\,FeSO_4 \longrightarrow Sn + Fe_2(SO_4)_3;$

$N. S + 3\,H_2O \rightarrow H_2SO_3 + 4\,H^+ + 4\,e^-;$

$SO_4{}^{2-} + 4\,H^+ + 2\,e^- \rightarrow H_2SO_3 + H_2O;$

$2\,H_2SO_4 + S + H_2O \longrightarrow 3\,H_2SO_3;$

$O. I_2 + 6\,H_2O \rightarrow 2\,IO_3{}^- + 12\,H^+ + 10\,e^-; I_2 + 2\,e^- \rightarrow 2\,I^-;$

$3\,I_2 + 3\,H_2O \longrightarrow 5\,HI + HIO_3$

8. $A. 3\,Ag + 4\,HNO_3 \longrightarrow 3\,AgNO_3 + 2\,H_2O + NO$

$B. 4\,HCl + PbO_2 \longrightarrow 2\,H_2O + Cl_2 + PbCl_2$

$C. 2\,Al + 6\,HBr \longrightarrow 2\,AlBr_3 + 3\,H_2$ $D. BaSO_4 + 4\,C \longrightarrow BaS + 4\,CO$

$E. Br_2 + 2\,H_2O + SO_2 \longrightarrow 2\,HBr + H_2SO_4$

$F. 3\,Br_2 + 6\,KOH \longrightarrow 5\,KBr + KBrO_3 + 3\,H_2O$

$G. C + 4\,HNO_3 \longrightarrow CO_2 + 2\,H_2O + 4\,NO_2$

$H. 3\,Ca(PO_3)_2 + 10\,C \longrightarrow Ca_3(PO_4)_2 + 10\,CO + P_4$

$I. 2\,ClO_2 + H_2O_2 + 2\,KOH \longrightarrow 2\,H_2O + 2\,KClO_2 + O_2$

$J. Cu + 4\,HNO_3 \longrightarrow Cu(NO_3)_2 + 2\,NO_2 + 2\,H_2O$

$K. 3\,HClO_3 \longrightarrow HClO_4 + 2\,ClO_2 + H_2O$

$L. H_3AsO_3 + H_2O + I_2 \longrightarrow 2\,HI + H_3AsO_4$

$M. 2\,Cr(OH)_3 + 3\,Na_2O_2 \longrightarrow 2\,Na_2CrO_4 + 2\,NaOH + 2\,H_2O$

$N. 2\,FeI_2 + 6\,H_2SO_4 \longrightarrow Fe_2(SO_4)_3 + 2\,I_2 + 3\,SO_2\ 6\,H_2O$

$O. FeS + 6\,HNO_3 \longrightarrow Fe(NO_3)_3 + S + 3\,NO_2 + 3\,H_2O$

9. (4-2) A. c, o B. c, o C. di, o D. di, o E. m, p F. c, o G. m, a H. ?, o I. m, a J. m, p K. m, g L. de, o M. de, o N. m, (p, a) O. m, (p [$SiO_2(s)$], g)

(4-4) A. c, o B. di, o C. m, p D. m, p E. m, p F. c, a G. m, (a,p) H. c, o I. m, (p,g) J. di, o K. di, o L. c, a M. m, (p, a) N. c, a O. m, a P. c, a Q. m, a R. m, a S. c, a T. c, o U. di, o V. c, o W. di, o X. c, o Y. m, a Z. c, o

10. A. 0.0750 M B. 0.165 M C. 1.25 M D. 1.18 M E. 0.0510 M F. 3.10 M G. 0.0388 M H. 19.4% I. 0.0697 M J. 0.384 M K. 0.0919 M L. 41.3 % M. 4.44% N. 62.5% O. 0.142 M

11. A. NH_4NO_3 B. $FeSO_4$ C. KCl D. $AgClO_3$ E. $MgSO_4$

F. FeS G. $FeSO_4$ H. $CoCl_2$ I. MgS J. $NaC_2H_3O_2$ K. Na_2CO_3 L. $MgCO_3$

Quizzes

A: 1. (d) 2. (d) 3. (a) 4. (e) 5. (b) 6. (a) 7. (e) (5) 8. (c)

9. $2\,AgClO_3(aq) + H_2SO_4(aq) \longrightarrow Ag_2SO_4(s) + 2\,HClO_3(aq)$

10. $2\,LiOH(aq) + H_2C_2O_4(aq) \longrightarrow Li_2C_2O_4(aq) + 2\,H_2O(l)$

B: 1. (c) 2. (a) 3. (d) 4. (b) 5. (b) 6. (b) 7. (c) 8. (c)

9. $MnSO_3(aq) + 2\ HCl(aq) \longrightarrow MnCl_2(aq) + H_2O + SO_2(g)$

10. $ZnSO_4(aq) + Bas(aq) \longrightarrow ZnS(s) + BaSO_4(s)$

C: 1. (c) 2. (b) 3. (e) 4. (b) 5. (d) 6. (d) 7. (d) 8. (a)

9. $FeS(s) + H_2SO_4(aq) \longrightarrow FeSO_4(aq) + H_2S(g)$

10. $2\ Al(OH)_3(s) + 3\ H_2SO_4(aq) \longrightarrow Al_2(SO_4)_3(aq) + 6\ H_2O(l)$

D: 1. (b) 2. (d) 3. (b) 4. (c) 5. (d) 6. (a) 7. (d) 8. (c)

9. $BaCO_3(s) + 2\ HNO_3(aq) \rightarrow Ba(NO_3)_2(aq) + H_2O(l) + CO_2(g)$

10. $Ba(OH)_2(aq) + H_2SO_4(aq) \longrightarrow BaSO_4(s) + 2\ H_2O(l)$

Sample Test

1. The abbreviations are the same as those for objective 5-9.

a. $Sr(NO_3)_2(aq) + H_2SO_4(aq) \rightarrow SrSO_4(s) + 2\ HNO_3(aq)$ [m, p]

b. $Na_2CO_3(s) + 2\ HCl(aq) \longrightarrow 2\ NaCl(aq) + H_2O(l) + CO_2(g)$

[m, g] c. $Zn(s) + CuSO_4(aq) \longrightarrow ZnSO_4(aq) + Cu(s)$ [di, o] d. $CaO(s) + 2\ HNO_3(aq) \longrightarrow Ca(NO_3)_2(aq) + H_2O$ [m, a] e. $2\ NaI(aq) + Cl_2(aq) \longrightarrow 2\ NaCl(aq) + I_2(s)$ [di, o] f. $BaBr_2(aq) + 2\ AgNO_3(aq) \longrightarrow Ba(NO_3)_2(aq) + 2\ AgBr(s)$ [m, p] g. $CaO(s) + CO_2(g) \longrightarrow CaCO_3(s)$ [c, ?] h $2\,Al(s) + 3\ Cl_2(g) \longrightarrow 2\ AlCl_3(s)$ [c, o]

2. a. $3\ Ag(s) + 4\ HNO_3(aq) \rightarrow 3\ AgNO_3(aq) + NO(g) + 2\ H_2O$ b. $Cl_2(aq) + 2\ KOH(aq) \longrightarrow KCl(aq) + KClO(aq) + H_2O(l)$ c.

$Fe_2(SO_4)_3(aq) + SnSO_4(aq) \longrightarrow Sn(SO_4)_2(aq) + 2\ FeSO_4(aq)$ d. $5\ K_2C_2O_4(aq) + 2\ KMnO_4(aq) + 8\ H_2SO$
$\longrightarrow 2\ MnSO_4(aq) + 6\ K_2SO_4(aq) + 10\ CO_2(g) + 8\ H_2O(l)$

3. The pungent odor with NaOH(aq) indicates the presence of NH_4^+ ion. The compound thus is expected to be soluble. The absence of a precipitate with $Cu^{2+}(aq)$ eliminates the generally insoluble ions: carbonate, chromate, phosphate, oxalate, sulfite, hydroxide, oxide, and sulfide anions. The formation of a precipitate with $Ba^{2+}(aq)$ eliminates chloride, bromide, and iodide anions. This leaves sulfate and thiosulfate anions, of which the latter forms $SO_2(g)$ on the addition of acid. The unknown is ammonium sulfate, $(NH_4)_2SO_4$.

CHAPTER 6

Drill Problems

1. A. 441 lb/in.^{2} = 3.04×10^3 kPa B. 2.75 lb/in.^{2} = 1.89×10^4 Pa C. 16.5 atm = 1.67×10^3 kPa D. 3.80×10^5 mmHg = 5.07×10^7 Pa E. 0.658 atm = 6.67×10^4 Pa F. 4.93×10^{-3} atm = 3.75 mmHg G. 34.0 atm = 3.45×10^6 Pa H. 0.511 lb/in.^{2} = 26.4 mmHg

2. A. 6.13 L B. 39.5 atm C. 1.16×10^{-4} L D. 23.0 atm E. 12.6 L F. 0.334 atm G. 1.51×10^3 ft^3 H. 257 kPa I. 187 L J. 756 mmHg

3. A. 8.73 L B. 1.59×10^3 K C. 170 L D. −24.5°C E. 24.5 L F. 1.22×10^3 K G. 13.6 L H. −214°C

4. A. 49.7 L B. 187 K C. 0.277 mol D. 22.4 atm E. 31.9 qt F. −72.0°C G. 19.9 mol H. 6.57 L I. 112 mmHg J. 8.50 L K. 23.0 g H_2O L. 0.337 lb/in.^{2} M. 284°F N. 0.418 L O. 7.22 g CH_4 P. 44.5 mmHg

5. (1) A. 215 K B. 21.9 L C. 2.24 atm D. 22.78 K E. 1.32 L F. 0.342 atm G. 559 K H. 143 L I. 0.618 atm J. 715 K K. 10.92 L L. 12.1 atm

(2) M. 10.7 mol N. 25.1 atm O. 0.634 mol P. 685°C Q. 2.90 mol R. 67.8 L

6. A. 2.03 mol, 15.4 g/mol, 31.3 g B. 3.02 L, 32.3 g/mol, 24.8 g/L C. 19.4 g/mol, 30.7 K, 0.474 g/L D. 0.511 atm, 14.5 g, 0.418 g/L E. 1.11×10^3 K , 42.0 g, 4.04 g/L F. 422 mol, 33.4 kg, 249 g/L G. 13.2 L, 181 g, 13.7 g/L H. 15.5 L, 40.0 g/mol, 4.92 g/L I. 24.2 L, 5.15 mol, 17.2 g/mol J. 95.9 L, 1.08 atm, 205 g K. 333 L, 0.0853 atm, 82.5 g/mol L. 95.2 L, 2.72 atm, 9.00 mol M. 0.0881 L, 0.0116 mol, 409 g/mol N. 0.546 mol, 24.5 g/mol, 13.4 g O. 99.7 g/mol, 203 K, 542 g P. 0.770 mol, 204 K, 6.49 g

7. 0.459 mol, 42.1 g/mol, C_3H_6 B. 0.788 mol, 128 g/mol $C_4H_4F_4$ C. 0.330 mol, 186 g/mol, C_6F_6 D. 0.0302 mol, 32.0 g/mol, N_2H_4 E. 0.109 mol, 84.4 g/mol, C_6H_{12} F. 3.50 mol, 48.0 g/mol, O_3 G. 0.0244 mol, 52.1 g/mol, C_4H_4

H. 21.5 mol, 46.0 g/mol, NO_2 I. 0.213 mol, 40.0 g/mol, C_2H_2N J. 2.72 mol, 58.0 g/mol, $C_2H_2O_2$ K. 0.723 mol, 60.0 g/mol, $C_2H_4O_2$ L. 0.570 mol, 30.0 g/mol, CH_2O_2

8. (1) A. 59.3 L O_2, 37.1 L CO_2, B. 141 L O_2, 105 L H_2O C. C_5H_{12} limiting: 33.8 L CO_2, 40.5 L H_2O. D. 81.6 L H_2, 54.4 L NH_3 E. H_2 limiting: 16.8 L NH_3 F. 10.7 L N_2, 32.1 L H_2 G. 6.09 L H_2O, 3.05 L NO H. 3.26 L. HCl, 1.09 L NO, 1.63 L Cl_2 I. 0.973 L H_2O, 0.487 L NO J. 4.61 L NO,

6.92 L H_2O K. O_2 limiting: 33.7 L NO, 50.0 L H_2O L. 3.43 L O_2, 4.11 L H_2O M. 38.4 L NH_3, 28.8 L O_2 N. O_2 limiting: 12.4 L N_2, 37.2 L H_2O O. 9.73 L NH_3, 7.30 L O_2, 4.87 L N_2

(2) P. 12.9 L CO_2 Q. O_2 limiting: 22.6 L H_2O R. 101 L O_2, 67.1 L CO_2 S. 7.15 L O_2, 4.29 L CO_2 T. 2.82 g O_2, 1.59 g H_2O, 8.16 L CO_2 U. C_6H_6 limiting: 3.07 L H_2O V. CH_4 limiting: 12.3 L CO_2 W. $C_2H_6O_2$ limiting: 2.07 L CO_2 X. 8.96 L CO_2 Y. 14.8 L CO_2, 12.2 g H_2O

9. A. 48.2 L B. 2.53 atm C. 308 K D. 584 atm E. 139 K F. 7.41 atm G. 15.5 L H. 6.13 atm I. 160 K

10. A. He 1356 m/s, Ne 608.8 m/s, H_2 1927 m/s, N_2 516.8 m/s, O_2 483.6 m/s, CO_2 412.3 m/s, Ar 432.8 m/s, H_2O 644.5 m/s, NH_3 662.8 ms, CH_4 682.9 m/s B. He 14.4 K, Ne 72.8 K, H_2 7.3 K, N_2 101.1 K, O_2 115.5 K, CO_2 158.8 K, Ar 144.1 K, CH_4 57.9 K, NH_3 61.5 K, H_2O 65.0 K C. 2.65 times faster D. 1.069 times faster E. 25.3 hr F. 536 hr

Quizzes

A. 1. (b) 2. (a) 3. (d) 4. (b) 5. (a) 6. (c) 7. (c) 8. (e) 9. (c) 10. (c)

B. 1. (c) 2. (d) 3. (b) 4. (d) 5. (a) 6. (a) 7. (c) 8. (a) 9. (b) 10. (c)

C. 1. (d) 2. (a) 3. (e) 4. (a) 5. (b) 6. (b) 7. (e) 8. (c) 9. (b) 10. (d)

D. 1. (b) 2. (a) 3. (d) 4. (d) 5. (d) 6. (b) 7. (a) 8. (a) 9. (e) 10. (a)

Sample Test

1. a. 1.80 atm = 182 kPa b. 0.0566 m^3 = 56.6 L c. 33.3°C = 306.5 K

2. $n = PV/RT = \dfrac{(1.80\text{ atm})(56.6\text{ L})}{[(0.08206\text{ L atm mol}^{-1}\text{ K}^{-1})(306.5\text{ K})]} =$

4.05 mol $\quad m = \dfrac{(1.305\text{ lb} \times 453.5\text{ g/lb})}{4.05\text{ mol}} = 146\text{ g/mol}$

3. $$\frac{0.08206\text{ L atm}}{\text{mol K}} \times \frac{1000\text{ mL}}{\text{L}} \times \frac{760\text{ mmHg}}{\text{atm}} \times \frac{\text{mol}}{1000\text{ mmol}}$$

$$= 62.37\text{ mL mmHg mmol}^{-1}\text{ K}^{-1}$$

4. $n_A = PV/RT = \dfrac{(1.067\text{ atm})(14.20\text{ L})}{[(0.08206\text{ L atm mol}^{-1}\text{ K}^{-1})(303.1\text{ K})]}$

$= 0.6092$ mol

$n_B = PV/RT = \dfrac{(26.42\text{ atm})(1.251\text{ L})}{[(0.08206\text{ L atm mol}^{-1}\text{ K}^{-1})(327.5\text{ K})]}$

$= 1.230$ mol $\quad n_{total} = 1.839$ mol

a. $P = \dfrac{(1.839\text{ mol})(0.08206\text{ L atm mol}^{-1}\text{ K}^{-1})(291.0\text{ K})}{(8.78\text{ atm})}$

$= 2.91$ atm

b. $P_A = (n_A / n_{total})P = (0.6092\ \text{mol}/1.839\ \text{mol})(8.78\ \text{atm})$

$= 2{:}91$ atm $V_B = (n_B / b_{total})V = (1.230\ \text{mol}/1.839\ \text{mol})(5.00\ \text{L})$

5. a. Ne b. He c. He d. equal pressures

CHAPTER 7

Drill Problems

1. (1) A. 30.5°C B. 441 J C 0.62 J g^{-1} °C^{-1} D. 117 g

E. 29.9°C F. −1.18 kJ G. 0.0862 J g^{-1} °C^{-1} H. 157 g

I. 21.2°C J. 1.5 mol K. 24.2 J mol^{-1} deg^{-1} L. −0.27 kJ

M. 20.0°C

(2) N. 1.8 J mol^{-1} deg^{-1} O. 20.4°C P. 153.5°C Q. 522 g

R. 6.53 g S. 0.208 J g^{-1} °C^{-1} T. 72.2°C U. 46.0°C V. 22.2°C

2. (1) Values in calories: A. 4.216 C. 586.8 E. 74.66

G. 150.0 I. 1619 K. 956.5 Values in joules: B. 2394 D. 2268 F. 306.3 H. 5.288 J. 10.88 L. 144.4

(2) M. −68 J N. −1207 J O. −55 J P. +1663 J Q. +1.96 kJ

R. −3.88 kJ S. −0.68 kJ T. +5.28 kJ U. −0.314 kJ

V. +2131 J W. −1382 J X. 272 J

3. A. 0.47 kJ/°C B. 40.7 kJ/g C. 21.3°C D. 0.79 g

E. 0.716 kJ/°C F. 48 kJ/g G. 1.09 g H. 63.2°F

4. A. +39.7 kJ/mol B. 21.8°C. C. 246 g H_2O

D. + 24.6 kJ/mol E. 40.3°C F. 18.4°C G. + 25.1 kJ/mol

5. All values are in kJ/mol A. +107.6 B. − 406.8 C. −112.3

D. −411.0 E. −167.4 F. −217.5 G. +406 H. −529 I. − 25.5

J. + 226.8 K. −155.2

6. A. $2\ \text{C(graphite)} + 3\ H_2(g) + \frac{1}{2}O_2(g) \longrightarrow C_2H_5OH(l)$

B. $N_2(g) + 4\ H_2(g) + \text{Sb(c, III)} + \frac{5}{2}\ Cl_2(g) \longrightarrow (NH_4)_2SbCl_5(s)$

C. $2\ \text{Hg(l)} + Cl_2(g) \longrightarrow Hg_2Cl_2(s)$

D. $\frac{3}{2}\ H_2(g) + \text{P(s, white)} + 2\ O_2(g) \longrightarrow H_3PO_4(s)$

E. $\text{Na(s)} + \frac{1}{2}\ I_2(g) + \frac{3}{2}\ O_2(g) \longrightarrow NaIO_3(s)$

F. $Sn(s, white) + F_2(g) \longrightarrow SnF_2(s)$ G. $Xe(g) + \frac{3}{2}O_2(g) \longrightarrow XeO_3(G)$

H. $\frac{1}{2}H_2(g) + \frac{1}{2}N_2(g) + \frac{3}{2}O_2(g) \longrightarrow HNO_3(l)$

I. $K(s) + \frac{1}{2}Br_2(l) + \frac{3}{2}O_2(g) \longrightarrow KBrO_3(s)$

J. $\frac{1}{2}I_2(s) + \frac{3}{2}Cl_2(g) \longrightarrow ICl_3(g)$

7. All values are in kJ/mol A. +178 B. −176.0 C. −124.0 D. −261.9 E. +224.5 F. −86.1 G. −1530.6 H. −1169.2 I. −114.14 J. −71.1 K. −305.9 L. +88.1 M −117.2 N. +869.3 O. −65.2

Quizzes

A. 1. (a) 2. (c) 3. (d) 4. (d) 5. (e) 6. (b) 7. (e)(−103 kJ/mol) 8. (b)

B. 1. (c) 2. (b) 3. (c) 4. (a) 5. (c) 6. (b) 7. (e)(108.8 kJ/mol) 8. (d)

C. 1. (b) 2. (b) 3. (a) 4. (c) 5. (b) 6. (c) 7. (a) 8. (e)(−1125.6 kJ/mol)

D. 1. (d) 2. (d) 3. (e)(+24.8 J) 4. (b) 5. (e) 6. (d) 7. (d) 8. (b)

Sample Test

1. a. $Sn(s, white) + 2\ Cl_2(g) \rightarrow SnCl_4(s)$

b. $7\ C(graphite) + 3\ H_2(g) + O_2(g) \longrightarrow C_6H_5COOH(s)$

c. $C(graphite) + \frac{1}{2}O_2(g) + Cl_2(g) \longrightarrow COCl_2(g)$

2. a. −185 kJ/mol b. −807 kJ/mol

3. $$5.00\ g \times \frac{mol\ LiCl}{42.4\ g} \times \frac{37.2\ kJ}{mol} = 4.39\ kJ \times \frac{1000\ J}{kJ}$$

$$\times \frac{g\,°C}{4.00\ J} = 1097\ g°C$$

$\Delta t = 1097\,g°C/115.0\,g = 9.54°C$. Thus $t_f = 29.5°C$.

CHAPTER 8

Drill Problems

1. & 2. A. 1.1×10^3 nm, ir B. 1.9×10^{18} Hz, γ ray C. 12 m, TV D. 200 kHz, radio E. 1.71×10^{-5} nm, γ ray F. 4.13×10^{10} Hz, microwave G. 5.83 km, radio H. 242 MHz, radar I. 4.04×10^{-8} cm, x-ray J. 5.95×10^{11} kHz, visible K. 95.2 nm, UV L. 2.410×10^7 Hz, TV M. 3.55×10^{-15} m x-ray N. 2.21×10^5 MHz, microwave O. 5.98×10^3 Å, visible P. 3.64×10^9 MHz, UV Q. 2.05×10^{-5} km, microwave R. 2.27×10^{21} Hz, x-ray S. 0.355 cm, microwave T. 6.82×10^{19} MHz, x-ray U. 0.571 m, radio

3. A. 2, Balmer, visible B. 3, Lyman, UV C. 379.9 nm, Balmer, visible D. 1875 nm, Paschen, ir E. 2, Balmer, visible F. 5, Balmer, visible G. 1945 nm, Brackett, ir H. 12372 nm, Humphreys, ir I. 5, Pfund, ir J. 2, Lyman, UV K. 389.0 nm, Balmer, visible L. 92.3 nm, Lyman, UV

4. A. 1.95×10^{18} Hz, 1.29×10^{-8} erg, 7.78×10^{5} kJ/mol

B. 6.94×10^{-12} erg, 1.05×10^{15} Hz, 286 nm

C. 2.86 m, 6.96×10^{-19} erg, 4.19×10^{-5} kJ/mol

D. 0.0644 kJ/mol, 1.61×10^{11} Hz, 1.85×10^{6} nm

E. 5.69×10^{14} Hz, 3.77×10^{-12} erg, 227 kJ/mol

F. 4.73×10^{-13} erg, 7.14×10^{13} Hz, 4.20×10^{-4} cm

G. 4.90×10^{4} nm, 4.06×10^{-14} erg, 2.44 kJ/mol

H. 4.90×10^{-6} kJ/mol, 1.23×10^{7} Hz, 24.5 m

I. 2.24×10^{5} Hz, 1.48×10^{-21} erg, 8.94×10^{-8} kJ/mol

J. 8.73×10^{-15} erg, 1.32×10^{15} Hz, 228 nm

K. 0.586 km, 3.39×10^{-21} erg, 2.04×10^{-7} kJ/mol

L. 1.89×10^{4} kJ/mol, 4.74×10^{16} Hz, 6.33 nm

M. 4.88×10^{9} Hz, 3.23×10^{-17} erg, 1.95×10^{-3} kJ/mol

N. 1.20×10^{-14} erg, 1.82×10^{12} Hz, 0.165 mm

O. 8.77 nm, 2.27×10^{-10} erg, 1.37×10^{4} kJ/mol

P. 7.47×10^{7} kJ/mol, 1.87×10^{20} Hz, 1.60 pm

5. (1) A. $m_l = 0$ B. $3p_x, 3p_y, 3p_z$,; $m_l = -1, 0, +1$

C. $4d_{xy}, 4d_{yx}, 4d_{xz}, 4d_{x^2-y^2}, 4d_{z^2}$; $m_l = -2, -1, 0, +1, +2$

D. $m_l = 0$ E. $m_l = -3, -2, -1, 0, +1, +2, +3$

F. $4p_x, 4p_y, 4p_z$; $m_l = -1, 0, +1$ G. $m_l = 0$

H. $2p_x, 2p_y, 2p_z$; $m_l = -1, 0, +1$

I. $5d_{xy}, 5d_{yx}, 5d_{xz}, 5d_{x^2-y^2}, 5d_{z^2}$; $m_l = -2, -1, 0, +1, +2$

J. $3d_{xy}, 3d_{yx}, 3d_{xz}, 3d_{x^2-y^2}, 3d_{z^2}$; $m_l = -2, -1, 0, +1, +2$

(2) K. 1*s*; 2 electrons L. 3*s*, 3*p*, 3*d*; 18 electrons M. 2*s*, 2*p*; 8 electrons N. 4*s*, 4*p*, 4*d*, 4*f*; 32 electrons O. 5*s*, 5*p*, 5*d*, 5*f*, 5*g*; 50 electrons

6. See Figure A-1.

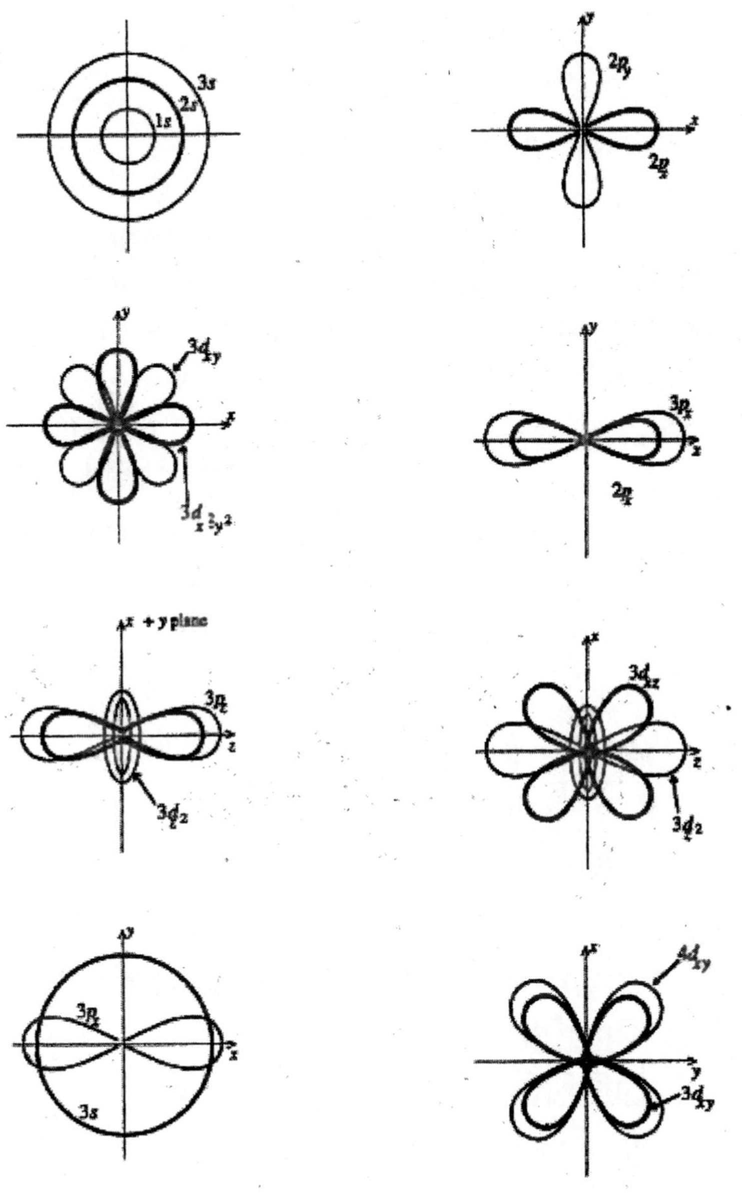

FIGURE A-1 Answers to 8-6 Drill Problem

7. A. (3); $1s^2 2s^2 2p_x^1 2p_y^1 2p_z^1$ B. (5); $1s^2 2s^2 2p^6 3s^2 3p^1$ C. (5); $1s^2 2s^2 2p^1$ D. $1s^2 2s^2 2p^6 3s^2 3p^3$ E. (4); [Ar]$4s^1 3d^{10}$ F (5); $1s^2 2s^2$ G. (2); [Ne] [↑↓] H. (1); $1s^2 2s^2 2p_x^1 2p_y^1$ I. (1); $1s^2 2s^2 2p^6 3s^2 3p^6 4s^2 3d^3$ J. (3); [Ne]$3s^2 3p_x^2 3p_y^1 3p_z^1$ K. (4); [Ar]$4s^2 3d^5$ L. (5); $1s^2 2s^2 2p^3$ M. (4); [Kr]$5s^1 4d^{10}$ N. (3); [Kr]$5s^2 4d_{xy}^2 4d_{xz}^2 4d_{yz}^1 4d_{z^2}^1 4d_{x^2-y^2}^1$ O. (5); $1s^2 2s^2 2p^6 3s^1$ P. (1); [Ne]$3s^2 3p^6 4s^2 3d^1$ Q.(5); $1s^2 2s^2 2p^6 3s^2 3p^5$ R. (3); $1s^2 2s^2 2p_x^1 2p_y^1$ S. (5); [Ne] [↑↓] [↑↓] [↑↓] [↑↓] T. (1); $1s^2 2s^2 2p_x^1$ U. When two subshells are very close in energy, an electron can move from one to the other if that move results in an increase in the number of empty, half-filled, or full electron shells. Cr is [Ar]$4s^1 3d^5$ Ag is [Kr]$5s^1 4d^{10}$.. V. Each orbital of a subshell must contain one electron before electrons are paired in the orbitals of that subshell. N is $1s^2 2s^2 2p_x^1 2p_y^1 2p_z^1$ Si is [Ne]$3s^2 3p_x^1 3p_y^1$. Electrons fill orbitals in the order given in expression [15], Figure 8-4, and Figure 8-5. K is $1s^2 2s^2 2p^6 3s^2 3p^6 4s^1$; Fr is $1s^2 2s^2 2p^6 3s^2 3p^6 4s^2 3d^{10} 4p^6 5s^2 4d^{10} 5p^6 6s^2 4f^{14} 5d^{10} 6p^6 7s^1$. X. No two electrons may have the same set of four quantum numbers. C is $1s^2 2s^2$ [↑][↑][ ], not $1s^2 2s^2$ [↑↓][ ][ ] and not $1s^3 2s^3$.

8. (1) A. Sc [Ar]$4s^2 3d^1$ B. C $1s^2 2s^2 2p_x^1 2p_y^1$ C. F $1s^2 2s^2 2p_x^2 2p_y^2 2p_z^1$ D. Ti [Ar]$4s^2 3d^2$ E. Al [Ne]$3s^2 3p^1$ F. Cl $1s^2 2s^2 2p^6 3s^2 3p^5$ G. Cr [Ar]$4s^1 3d^5$ H. V [Ar]$3d^3 4s^2$ I. Sc $1s^2 2s^2 2p^6 3s^2 3p^6 4s^2 3d^1$ J. B $1s^2 2s^2 2p^1$ K. N $1s^2 2s^2 2p_x^1 2p_y^1 2p_z^1$

(2) L. S M. Sc N. Zr O. Tl P. Ce Q. Mo, Tc R. V S. Se T. Sr

(3) U. [He]$2s^2 2p_x^2 2p_y^2 2p_z^1$; [Ne] $_{3s}$ [↑↓] $_{3p}$ [↑] [↑] [↑];

V. [Ne]$3s^2 3p_x^2 3p_y^2 3p_z^1$; [Ne) $_{3s}$ [↑↓] $_{3p}$ [↑↓] [↑] [↑];

W. [Ne]$3s^2 3p_x^2 3p_y^2 3p_z^1$; [Ne] $_{3s}$ [↑↓] $_{3p}$ [↑↓] [↑↓] [↑];

X. [Ne]$3s^2 3p_x^1 3p_y^1$; [Ne] $_{3s}$ [↑↓] $_{3p}$ [↑] [↑]; Y. [Ar]$4s^1 3d^{10}$;

[Ar] $_{4s}$ [↑] $_{3d}$ [↑↓] [↑↓] [↑↓] [↑↓] [↑↓] Z. [Ne]$3s^1$; [Ne] $_{3s}$ [↑] [Ne]$3s^1$; Γ. [Xe]$6s^1$; [Xe] $_{6s}$ [↑] Δ. [Xe]$6s^2 5d^1 4f^1$;

[Xe]$6s^2 5d^1 4f^1$; [Xe] $_{6s}$ [↑↓] $_{5d}$ [↑][ ][ ][ ][ ] $_{4f}$ [↑][ ][ ][ ][ ][ ][ ]

Θ. [Ar]$4s^2 3d^1$; [Ar] $_{4s}$ [↑↓] $_{3d}$ [↑][ ][ ][ ][ ]

Λ. [He]$2s^2 2p_x^2 2p_y^1 2p_z^1$; [He] $_{2s}$ [↑↓] $_{2p}$ [↑↓] [↑] [↑]; Π. [Ar]$4s^2 3d^6$ [Ar] $_{4s}$ [↑↓] $_{3d}$ [↑↓] [↑] [↑] [↑] [↑] [↑]

S. [Kr]$6s^2$; [Kr] $_{6s}$ [↑↑]

9. (1) A. $ns^2(n-1)d^x$ B. ns^2 C. ns^2np^5 D. ns^2np^2 E. ns^2np^1 F. ns^2np^4 G. ns^1 H. ns^2np^6 I. ns^2np^4 J. $ns^2(n-1)d^1$ K. $ns^2(n-1)d^2$ L. ns^2np^3

(2) M. s^2p^3 N. s^2p^2 O. s^2p^6 P. s^2p^3 Q. s^1 R. s^2p^4 S. s^2p^1 T. s^2p^5 U. s^1 V. s^2p^6 W. s^2 X. s^2p^1 Y. s^2p^4 Z. s^2p^2 Γ. s^2p^5 Δ. s^2

(3) Θ. [Ar]$4s^1 3d^{10}$ Δ. [Kr]$5s^1 4d^5$ Ξ. [Xe]$6s^2 4f^{14} 5d^6$ Π. [Xe]$6s^2 5d^0 4f^1$ Σ. [Xe]$6s^2 5d^1$ Υ. [Xe]$6s^2 4f^{14} 5d^6$ Φ. [Ar]$4s^2 3d^6$ Ω. [Kr]$5s^2 4d^3$ Ø. [Kr]$5s^2 4d^1$

Further Questions A. $[Ar]4s^1$ B. $[He]2s^2 2p^2$ C. $[Ar]4s^2 3d^5$ D. $[Ar]4s^2 3d^1$ E. $[Ar]4s^1 3d^5$ F. $[Xe]6s^1 4f^{14} 4d^{10}$ G. $[Ne]3s^2 3p^2$ H. $[Ar]4s^2 3d^{10}$ I. $[Ne]3s^2$ J. $[He]2s^2 2p^5$ K. $[Ne]3s^2 3p^3$ L. $[Ar]4s^2 3d^6$ M. $[Rn]7s^2 6d^1 5f^3$ N. $[He]2s^2 2p^6$ O. $[Xe]6s^2 5d^1 4f^2$, but actually $[Xe]6s^2 5d^0 4f^3$ P. $[Kr]5s^1 4d^{10}$ Q. $[Ne]3s^2 3p^1$ R. $[Kr]5s^2 4d^{10} 5p^2$ S. $[Ar]4s^2 3d^{10} 4p^2$ T. $[Kr]5s^1 4d^5$ U. $[Ar]4s^2 3d^{10} 4p^1$ V. $[Ar]4s^2$ W. $[Kr]5s^2 4d^{10} 5p^5$ X. $[Xe]6s^1 4f^{14} 5d^5$, but actually $[Xe]6s^2 4f^{14} 5d^4$ Y. $[He]2s^2 2p^1$ Z. $[Ar]4s^1 3d^{10}$

Quizzes

A: 1. (a) 2. (c) 3. (b) 4. (a) 5. (c) 6. (d) 7. (c) 8. (a) 9. (a) 10. (b) 11. (c) 12. (e)(Sc)

B: 1. (c) 2. (d) 3. (e) 4. (d) 5. (c) 6. (d) 7. (b) 8. (c) 9. (c) 10. (d) 11. (b) 12. (d)

C: 1. (d) 2. (c) 3. (d) 4. (d) 5. (b) 6. (e) 7. (c) 8. (d) 9. (b) 10. (c) 11. (b) 12. (a)

D: 1. (b) 2. (b) 3. (a) 4. (b) 5. (d) 6. (b) 7. (c) 8. (a) 9. (d) 10. (d) 11. (c) 12. (b)

Sample Test

1. $$\varepsilon = \frac{hc}{\lambda} = \frac{(6.63\times10^{-34}\ \text{J}\cdot\text{s})(3.00\times10^{8}\ \text{m/s})}{3.28\ \text{m}}$$
$$= 6.06\times10^{-26}\ \text{J/photon}$$

$$E = \frac{6.06\times10^{-26}\ \text{J}}{\text{photon}} \times \frac{6.022\times10^{23}\ \text{photon}}{\text{mole}} \times \frac{\text{kJ}}{1000\ \text{J}}$$
$$= 3.65\times10^{-5}\ \text{kJ/mol}$$

2. See Figures8–2 and8–3.

3. a. $1s^2 2s^2 2p^6 3p_x^1 3p_y^1 3p_z^1$ b. $1s^2 2s^2 2p^6 3s^2 3p^6 4s^2 3d^5$ c. $1s^2 2s^2 2p^6 3s^2 3p_x^1 3p_y^1$

4. a. [Ar] 4s [↑↓] 3d [↑↓][↑][↑][↑][↑] b. [Ne] 3s [↑↓] 3p [↑][ ][ ] c. [He] 2s [↑↓] 2p [↑↓][↑][↑]

CHAPTER 9

Drill Problems

1. A. Be < Mg < Ca < Sr B. O < S < Te < Po C. He < Ar < Kr < Rn D. Si < Ge < Sn, Pb E. B < Al < Ga < In F. O < C < B < Li G. Bi < Tl < Ba < Cs H. Te < Sn < In < Rb I. Br < As < Ge < K J. Cl < P < Al, Mg K. Fe < V < Ti < K L. Os < Hf < La < Ba M. $W^{3+} < Tm^{3+} < Eu^{3+} < Ce^{3+}$ N. Os < W < U < Ac O. $Ca^{2+} < Ar < P^{3-} < Si^{4-}$ P. $Na^+ < Ne < O^{2-} < C^{4-}$ Q. $Sc^{3+} < K^+ < Cl^- < S^{2-}$ R. $Y^{3+} < Sr^{2+} < Br^- < Ga^{5-}$ S. Ne < O < Si < Ge T. $Be^{2+} < Mg^{2+} < F^- < C^{4-}$ U. N < Al < Na < K V. $Na^+ < Ne < N < Li$ W. Ge < Ga < Tl < Ba

2. A. Cs < Rb < K < Na B. Sb < As < P < N C. Tl < In < Al < B D. Sr < Ca < Mg < Be E. N < Al < Cl < Ar F. Ca < Ge < As < Kr G. Rb < In < Te < I H. Li < B < N < Ne I. Ba < Ga < Se < F J. Sr < Sb < P < Ne K. Cs < Sn < As < Cl L. Rb < In < Al < F

3. A. F < Cl < Br < I B. N < P < Sb < Bi C. Li < Na < K < Rb D. Br < Se < Ge < K E. I < Sb < In < Rb F. F < P < In < Cs G. Cl < Se < Sb < Tl H. B < Al < In < Tl I. Cl < As < In < Sr

4. A. Na < K < Rb < Cs B. F < P < In < Cs C. Br < As < Ca < K D. F < O < N < C E. Cl < Al < Mg < Na F. Se < Ge < Fe < Ce G. Be < Mg < Ca < Sr H. C < B < Be < Li I. Al < Ga < Sr < Rb

5. A. $K^+ < Ce^{3+} < C < P$ B. $(He, Cs^+) < Sc < S$ C. $B < (Si, S) < P$ D. $Ca < (K, Sc) < Ti$ E. $Cu < V < Mn < Cr$ F. $((Zn, Ca) < Ti < As$ G. $Na^+ < (Na, S^-) < S$ H. $Be < (Li, B) < C$ I. $Cu^+ < Cu^{2+} < Fe^{2+} < Fe^{3+}$

Quizzes

A: 1. (c) 2. (d) 3. (d) 4. (c) 5. (b) 6. (b) 7. (e) 8. (c)

B: 1. (c) 2. (c) 3. (e) 4. (a) 5. (b) 6. (c) 7. (a) 8. (a)

C: 1. (e) 2. (e) 3. (c) 4. (b) 5. (c) 6. (b) 7. (a) 8. (a)

D: 1. (a) 2. (b) 3. (d) 4. (b) 5. (b) 6. (c) 7. (c) 8. (a)

Sample Test

1. Se_8 or Se(metal), both allotropes exist. 2. 34 3. $4s^2 4p^4$ 4. 78.96 (vertical trends predict 78.4) 5. 2 6. 116 pm (horizontal trends predict 116 pm or 117 pm) 7. 16.5 mL/mol (period 5 trend predicts 16.1 mL/mol, halogen group trend predicts 18.9 mL/mol) 8. −195 kJ/mol (period 4 trend predicts −201 chalcogen group trend predicts −192 kJ/mol) 9. 9.8 eV/atom, same as predicted by group trend 10. $0.35\ J\ g^{-1}\ °C^{-1}$ (group trends predict $0.30\ J\ g^{-1}\ °C^{-1}$)

CHAPTER 10

Drill Problems

1.

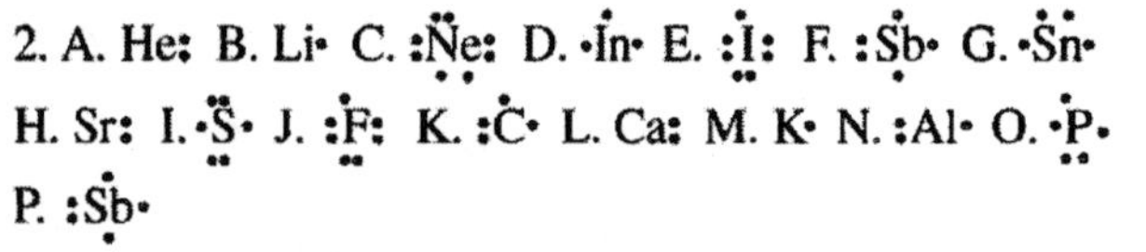
2. A. He: B. Li• C. :Ne: D. •In• E. :I: F. :Sb• G. •Sn•
H. Sr: I. •S• J. :F: K. :C• L. Ca: M. K• N. :Al• O. •P•
P. :Sb•

2. A. $Li^+ [H:]^-$ B. $Na^+[|\overline{\underline{S}}|]^{2-}$ C. $Ca^{2+} [|\overline{\underline{Br}}|]^-$ D. $Sr^{2+} [|\overline{\underline{O}}|]^{2-}$ E. $Mg^{2+} [|\overline{\underline{N}}|]^{3-}$ F. $Sc^{3+} [|\overline{\underline{S}}|]^{2-}$ G. $Na^+ [|\overline{\underline{F}}|]^-$ H. $Ti^{4+} [|\overline{\underline{O}}|]^{2-}$ I. $Mg^{2+} [|\overline{\underline{S}}|]^{2-}$ J. $La^{3+} [|\overline{\underline{Cl}}|]^-$ K. $K^+ [|\overline{\underline{Se}}|]^{2-}$ L. $Mg^{2+} [|\overline{\underline{C}}|]^{4-}$ M. $Ce^{4+} [|\overline{\underline{Cl}}|]^-$ N. $Al^{3+} [|\overline{\underline{N}}|]^{3-}$ O. $Sr^{2+} [|\overline{\underline{O}}|]^{2-}$ P. $Rb^+ [|\overline{\underline{P}}|]^{3-}$

3. A. C < N < O < F B. K < Ca < As < Br C. I<Br<Cl<F D. Cs < Rb < K < Li E. Rb < Ru < In < Se F. Na < Li < B < F G. Cs < As < Cl < F H. Ge < Si < N < O I. Cs < Br < Tl < Bi

4. A.	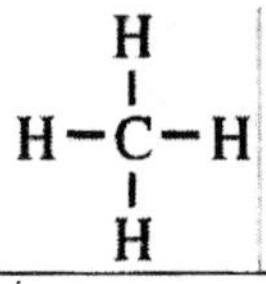	B.	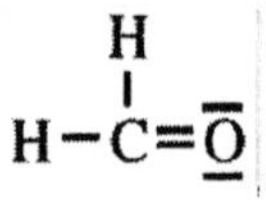	C.	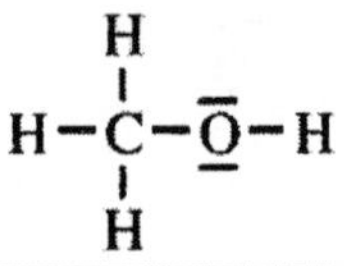

D.	H H \| \| H–N–N–H	E.	\|F̄–Ō–F̄\|	F.	F̄ \| \|F̄–N–F̄\|
G.	⎛ Ō̄ ⎞⁴⁻ ⎜ \| ⎜\|Ō–Si–Ō\| ⎜ \| ⎝ Ō̄ ⎠	H.	Ō=N–C̄l\|	I.	C̄l\| \| \|C̄l–Ge–C̄l\| \| C̄l\|
J.	H H \| \| H–C=C–H	K.	C̄l\| \| \|C̄l–Si–C̄l\| \| C̄l\|	L.	C̄l\| \| \|C̄l–C=Ō
M.	H–Ō–C̄l–Ō\|	N.	H–Ō–C≡N\|	O.	H–Ō–F̄\|
P.	H–Ō–N̄=Ō\|	Q.	H–Ō–Ō–H	R.	H–C≡C–H
S.	\|F\| \| \|F̄–As–F̄\|	T.	⎛ Ō̄ ⎞⁻ ⎜ \| ⎜\|Ō–Br–Ō\| ⎜ \| ⎝ Ō\| ⎠	U.	⎛ Ō̄ ⎞³⁻ ⎜ \| ⎜\|Ō–P–Ō\| ⎜ \| ⎝ Ō\| ⎠
V.	\|B̄r–S̄e–Br\|	W.	⎛ Ō\| ⎞³⁻ ⎜ \| ⎝\|Ō–P–Ō\| ⎠	X.	⎛ F̄\| ⎞⁻ ⎜ \| ⎜\|F̄–B–F̄\| ⎝ F̄\| ⎠
Y.	[\|Ō–F̄\|]⁻	Z.	⎛ Ō\| ⎞⁻ ⎜ \| ⎝\|Ō–C̄l–Ō\| ⎠		
Γ.	[H–Ō–Ō\|]⁻	Δ.	[\|C≡N\|]⁻	Θ.	H–Ī\|
Λ.	H \| \|C̄l–C–C̄l\| \| H	Π.	H \| H–Ō–C=Ō	Σ.	H H \| \| H–C–C–H \| \| H H
Φ.	[\|Ō–Ō\|]²⁻				

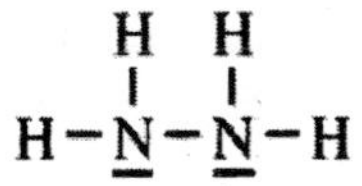

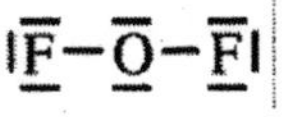

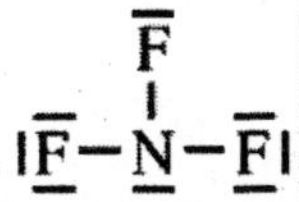

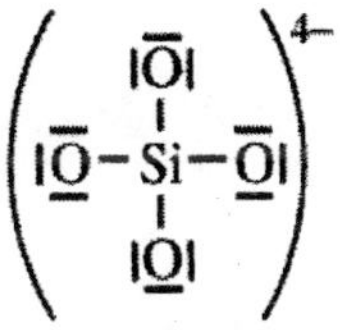

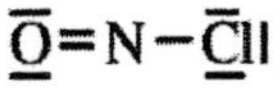

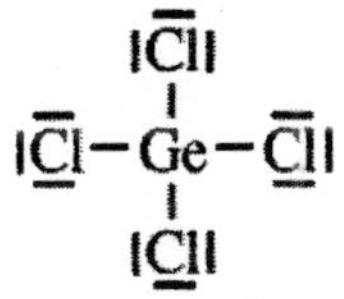

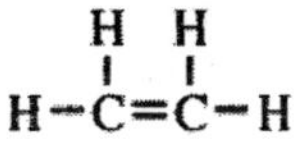

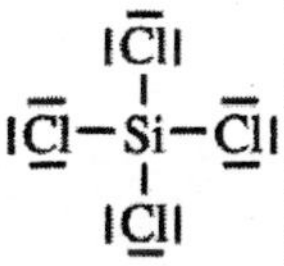

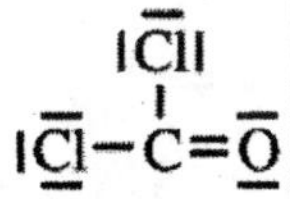

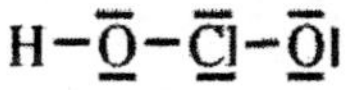

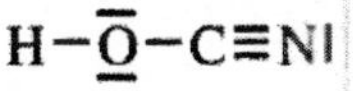

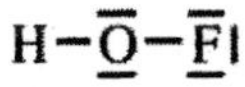

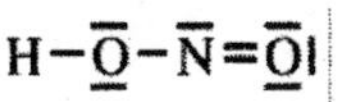

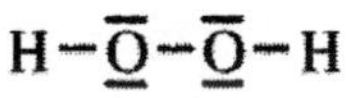

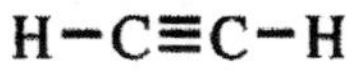

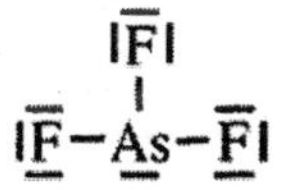

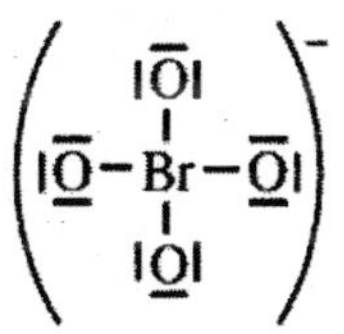

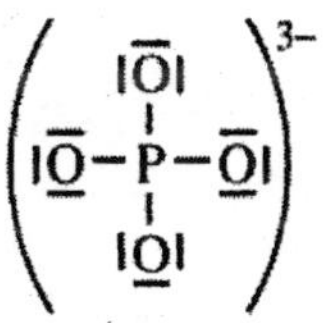

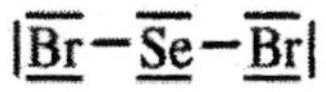

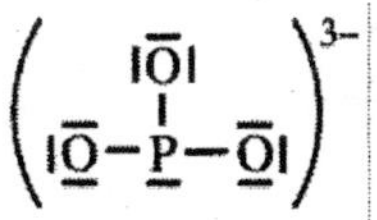

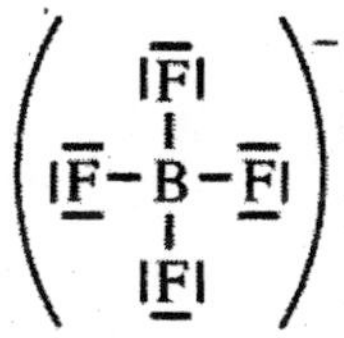

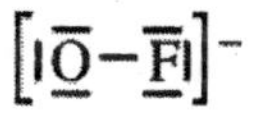

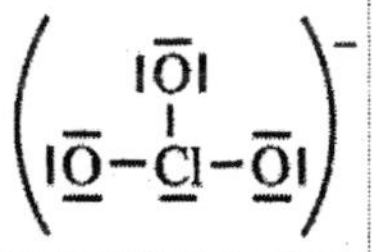

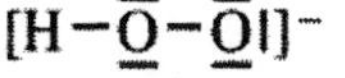

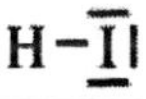

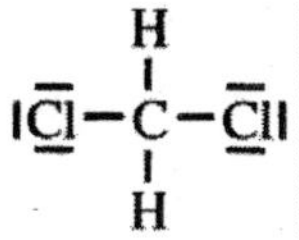

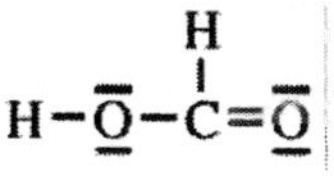

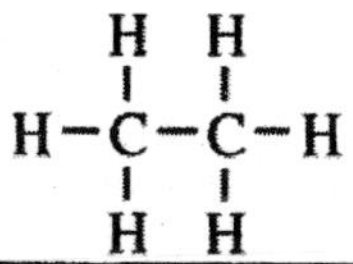

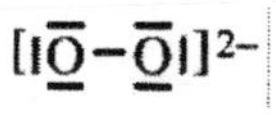

5. Formal charges are given in parentheses after each structure, when they are different from zero. The "best" structure is drawn first in each case.

A. H — Ō — C ≡ N\|; H — Ō — N ≡ Cl (N, +1; C, –1)

B. $H—C≡N|; H—N≡Cl\,(N,+1; C,-1)$

C.

$$H-\underline{\overline{O}}-\underset{\underset{|\underline{O}|}{|}}{\overset{\overset{|\overline{O}|}{|}}{C}l}-\underline{\overline{O}}|\ (Cl,+2; O,-1)\qquad H-\underset{\underset{|\underline{O}|}{|}}{\overset{\overset{|\overline{O}|}{|}}{C}l}-\underline{\overline{O}}\ |\ \begin{array}{l}(Cl,+3)\\(O,-1)\end{array}$$

D. $|C≡O|\,(C,-1; O,+1); |\overline{C}=\overline{O}\,(C,-2; O,+2)$

E. $\overline{O}=\underline{N}—\overline{Cl}\,|; \underline{\overline{N}}=O—\overline{Cl}\,|\,(N,-1; O,+1)$

F. $[|\underline{N}=C=\underline{S}\,|]^-\,(N,-1); [|\underline{C}=N=\underline{S}\,|]^-\,(C,-2; N,+1)$

G. $H—\underline{N}=C=\underline{S}\,|; H—\underline{C}=N=\underline{S}\,|\,(C,-1; N,+1)$

H.

$$|\overline{\underline{C}l}-\underset{\underset{}{B}}{\overset{\overset{|\overline{\underline{C}}l|}{|}}{}}-\overline{\underline{C}l}|; \quad \overline{\underline{C}l}=\underset{\underset{}{\underline{B}}}{\overset{\overset{|\overline{\underline{C}}l|}{|}}{}}-\overline{\underline{C}l}|\ (=Cl,+1; B,-1)$$

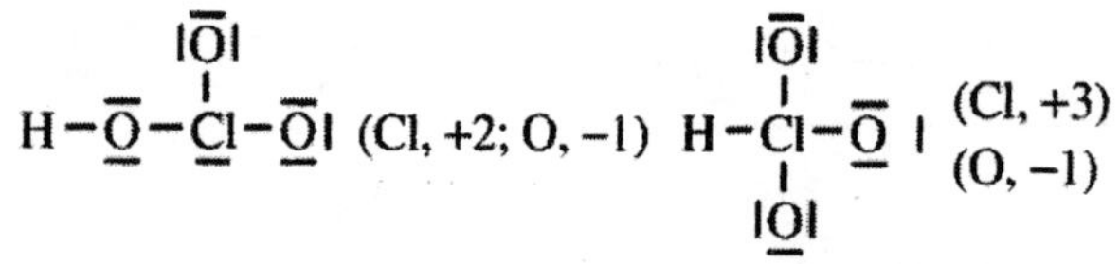

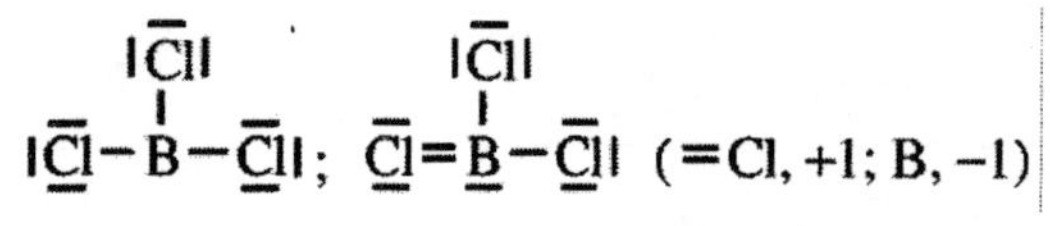

I. $H—\underline{N}=\underline{O}\,|; H—\underline{O}=\underline{N}\,|\,(O,+1; N,-1)$

J. $|\overline{Cl}—Be—\overline{Cl}\,|; \overline{Cl}=Be—\overline{Cl}\,|\,(=Cl,+1; B,-1)$

K.

$$|\overline{\underline{F}}-\underline{\overline{O}}-\underset{\underset{}{\underline{Se}}}{\overset{\overset{}{}}{}}-\overline{F}|; \quad |\underline{\overline{O}}-\underset{\underset{}{\underline{Se}}}{\overset{\overset{|\overline{F}|}{|}}{}}-\overline{F}|\ (O,-1; Se,+1);$$

$$|\underline{\overline{Se}}-\underset{\underset{}{\underline{O}}}{\overset{\overset{|\overline{F}|}{|}}{}}-\overline{F}|\,(Se,-1; O,+1)$$

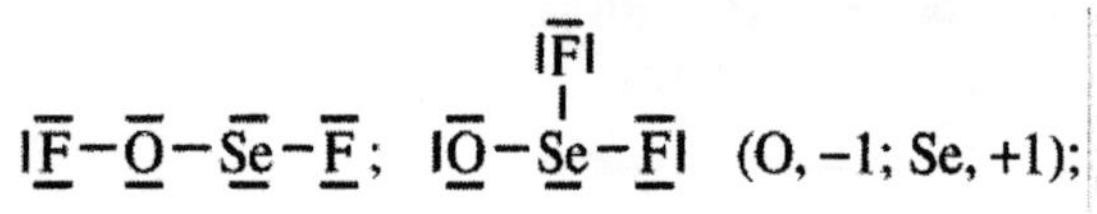

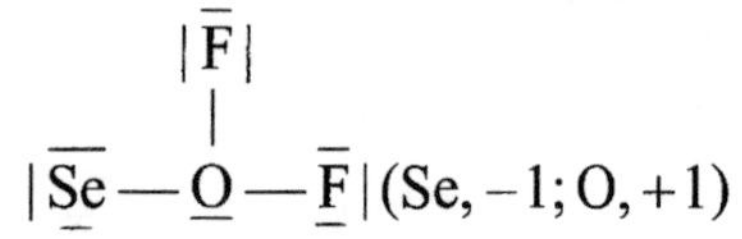

L. $[|\underline{N}=C=\underline{N}\,|]^{2-}\,(=N,-1); [|\underline{C}=N=\underline{N}\,|]^{2-}\,(C,-2\ N,+1; N,-1)$

M.

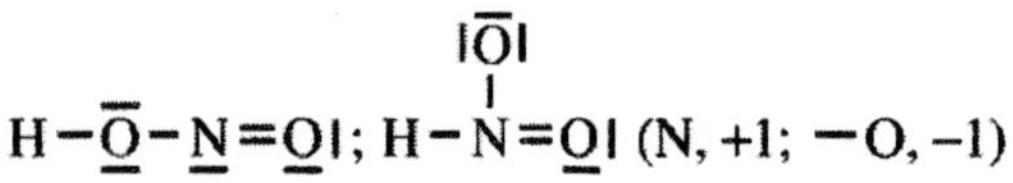

$$H-\underline{\overline{O}}-\underline{N}=\underline{O}\,|; H-\underset{\underset{}{N}}{\overset{\overset{|\overline{O}|}{|}}{}}=\underline{O}\,|\,(N,+1; -O,-1)$$

N.

$$|\underline{\overline{O}}-\underset{\underset{|\underline{C}l|}{|}}{\overset{\overset{|\overline{\underline{C}}l|}{|}}{P}}-\overline{\underline{C}l}|\ (P,+1; O,-1)\qquad |\underset{\underset{|\underline{C}l|}{|}}{\overset{\overset{|\overline{\underline{C}}l|}{|}}{P}}-\underline{\overline{O}}-\overline{\underline{C}l}|$$

O. $|\overline{Cl}—C≡N\,|; |\overline{Cl}—N≡Cl\,(N,+1; C,-1)$

P. $H—\underline{N}=N=\underline{\overline{N}}\,(=N=,+1; =N,-1); H—N≡N—\underline{\overline{N}}\,|\,(both\ —N≡,+1; —\underline{\overline{N}}\,|,-2)$

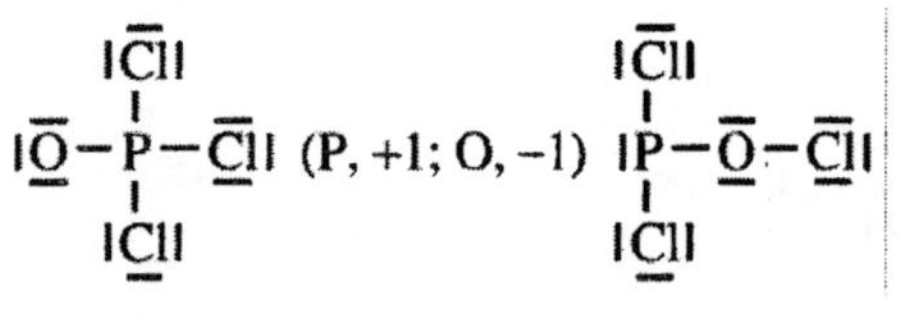

Q.

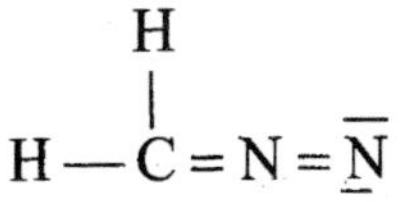

$$H-\overset{\overset{\displaystyle H}{|}}{C}=N=\overline{N} \quad (=N=,+1, \ =\overline{N}. \ -1)$$

$$H-\overset{\overset{\displaystyle H}{|}}{N}=C=\overline{N} \quad (-N=,+1; \ =\overline{N}, -1$$

R. $|\overline{Cl}-\overline{O}-N=\overline{O};\ |\overline{Cl}-\overline{O}-O=\overline{N}(-N=,+1;N,-1)$

6. A.

$$\left(H-\overset{\overset{\displaystyle I\overline{O}I}{|}}{\underline{C}}=\overline{O}I\right)^{-} \longleftrightarrow \left(H-\overset{\overset{\displaystyle IOI}{\|}}{\underline{C}}=\overline{O}I\right)^{-}$$

B. $[\overline{O}=\underline{N}-\overline{O}\,|]^{-} \longleftrightarrow [|\overline{O}-\underline{N}=\overline{O}]^{-}$

C. $[\overline{N}=N=\overline{N}]^{-} \longleftrightarrow [|N\equiv N-\overline{N}\,|]^{-} \longleftrightarrow [|\overline{N}-N\equiv N\,|]^{-}$

D. $O=\underline{S}-\overline{O}\,|\longleftrightarrow|\overline{O}-\underline{S}=\overline{O}$

E.

$$\left(\overline{O}=\overset{\overset{\displaystyle I\overline{O}I}{|}}{\underline{C}}-\overline{O}I\right)^{2-} \longleftrightarrow \left(|\overline{O}-\overset{\overset{\displaystyle I\overline{O}I}{|}}{\underline{C}}=\overline{O}I\right)^{2-} \longleftrightarrow \left(|\overline{O}-\overset{\overset{\displaystyle IOI}{\|}}{\underline{C}}-\overline{O}I\right)^{2-}$$

F.

$$\left(\overline{O}=\overset{\overset{\displaystyle I\overline{O}I}{|}}{\underline{B}}-\overline{O}I\right)^{3-} \longleftrightarrow \left(|\overline{O}-\overset{\overset{\displaystyle I\overline{O}I}{|}}{\underline{B}}=\overline{O}I\right)^{3-} \longleftrightarrow \left(|\overline{O}-\overset{\overset{\displaystyle IOI}{|}}{\underline{B}}=\overline{O}I\right)^{3-}$$

G.

$$\left(\overline{O}=\overset{\overset{\displaystyle I\overline{O}I}{|}}{\underline{N}}-\overline{O}I\right)^{-} \longleftrightarrow \left(|\overline{O}-\overset{\overset{\displaystyle I\overline{O}I}{|}}{\underline{N}}=\overline{O}I\right)^{-} \longleftrightarrow \left(|\overline{O}-\overset{\overset{\displaystyle IOI}{\|}}{\underline{N}}-\overline{O}I\right)^{-}$$

H.

$$\left(\overline{O}=\overset{\overset{\displaystyle I\overline{O}I}{|}}{\underline{P}}-\overline{O}I\right)^{-} \longleftrightarrow \left(|\overline{O}-\overset{\overset{\displaystyle I\overline{O}I}{|}}{\underline{P}}=\overline{O}I\right)^{-} \longleftrightarrow \left(|\overline{O}-\overset{\overset{\displaystyle IOI}{\|}}{\underline{P}}-\overline{O}I\right)^{-}$$

I. $\overline{O}=\underline{C}-\underline{O}=C=\overline{O}\longleftrightarrow\overline{O}=C=\underline{O}-\underline{C}=\overline{O}$

J.

$$\overline{O}=\overset{\overset{\displaystyle I\overline{O}I}{|}}{\underline{S}}-\overline{O}I \longleftrightarrow I\overline{O}-\overset{\overset{\displaystyle I\overline{O}I}{\|}}{\underline{S}}=\overline{O}I \longleftrightarrow I\overline{O}-\overset{\overset{\displaystyle IOI}{\|}}{\underline{S}}-\overline{O}I$$

K.

$$\left(\overline{O}=\overset{\overset{\displaystyle I\overline{O}I}{|}}{\underline{C}}-\overset{\overset{\displaystyle I\overline{O}I}{|}}{\underline{C}}=\overline{O}\right)^{2-} \longleftrightarrow \left(I\overline{O}-\overset{\overset{\displaystyle I\overline{O}I}{\|}}{C}-\overset{\overset{\displaystyle I\overline{O}I}{\|}}{C}-\overline{O}I\right)^{2-}$$

L. $\overline{O}=\underline{O}-\overline{O}\,|\longleftrightarrow|\overline{O}-\underline{O}=\overline{O}$

7. All formal charges equal zero unless otherwise indicated.

A. $\underline{N}=\overline{O}$ B. $|\overline{O}-N=\overline{O}\longleftrightarrow\overline{O}=N-\overline{O}\,|(-O,-1;N,+1)$

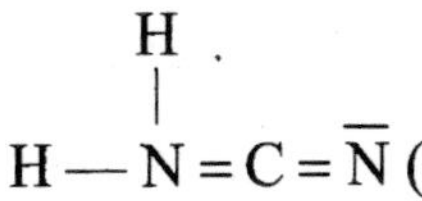

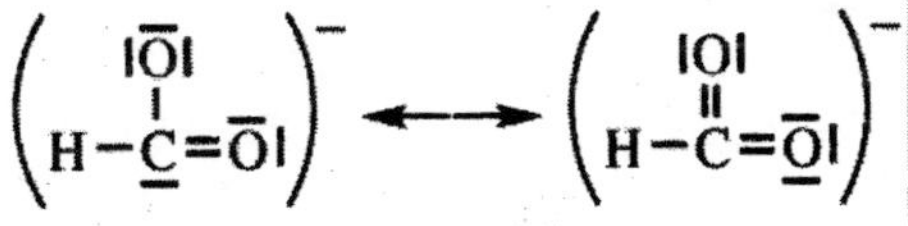

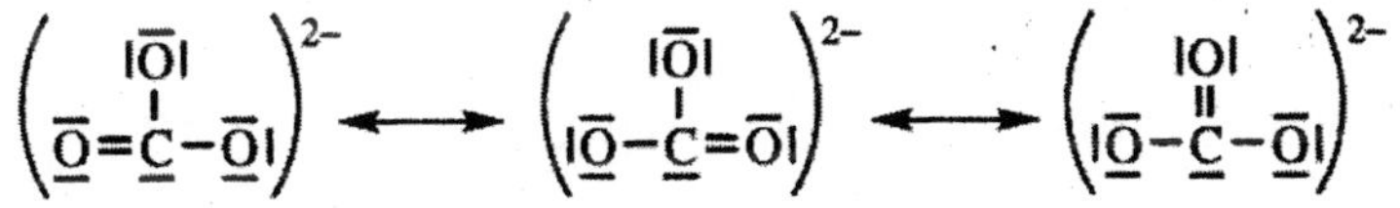

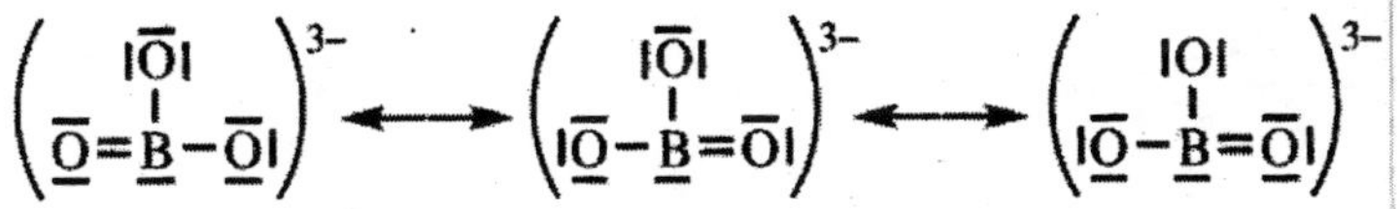

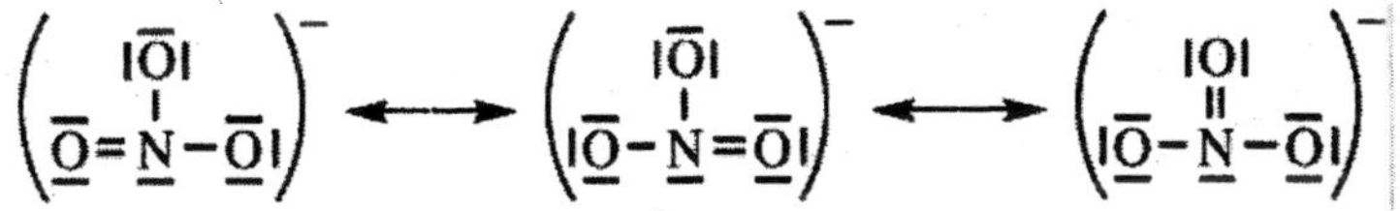

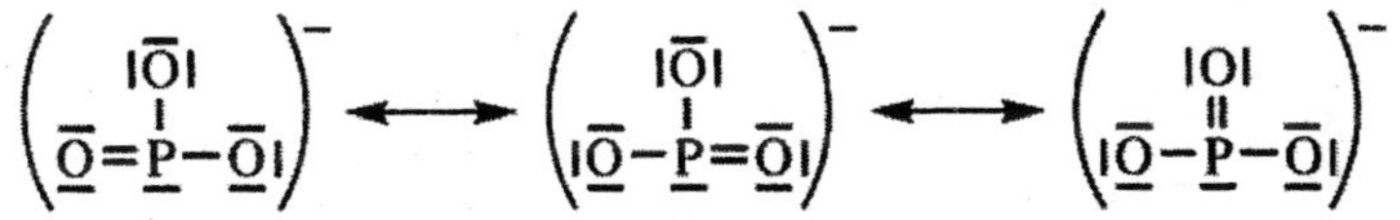

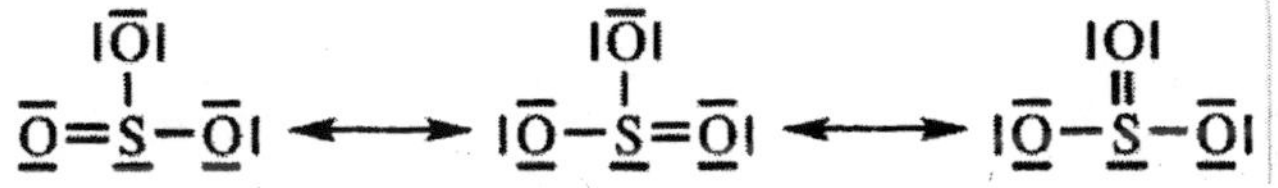

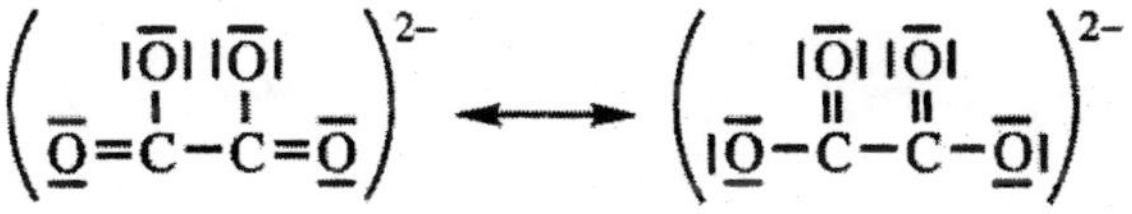

C. $|\overline{\underline{Br}} - \overline{Sn} - \overline{\underline{Br}}|$

D. $|\overline{\underline{O}} - \underline{Cl} = \overline{O} \longleftrightarrow \overline{O} = \underline{Cl} - \overline{\underline{O}}|$ $(-O, -1; Cl, +1)$

E. $|\overline{\underline{I}} - \overline{Pb} - \overline{\underline{I}}|$ F. $|\overline{\underline{Cl}} - Be - \overline{\underline{Cl}}|$

G.
$$\overline{\underline{O}} = P - \overline{\underline{O}} \cdot \longleftrightarrow \cdot \overline{\underline{O}} - P = \overline{O}| \quad (|\overline{\underline{O}}-, -1; P, +1)$$
with $|\overline{\underline{O}}|$ atoms above P, and four other resonance forms.

H. H — Be — H I. $[|N \equiv C - \overline{\underline{N}} \cdot]^- \longleftrightarrow [\cdot \overline{\underline{N}} - C \equiv N|]^-$ $[N, -1]$

J. $B \equiv N|$ K. $|\overline{\underline{I}} - B - \overline{\underline{I}}|$ with $|\overline{\underline{I}}|$ above B L. $H - \underset{\cdot\cdot}{\overset{H}{C}} - H$

M. $[\cdot \overline{\underline{O}} - \overline{S} = \overline{O}]^+ \longleftrightarrow [\overline{O} = \overline{S} - \overline{\underline{O}} \cdot]^+$ $(S, +1)$

N. $|\overline{\underline{Cl}} - B - \overline{\underline{Cl}}|$ with $|\overline{\underline{Cl}}|$ above B O. $[\cdot \overline{\underline{O}} - \overline{\underline{O}}|]^- \longleftrightarrow [|\overline{\underline{O}} - \overline{\underline{O}} \cdot]^-$ $(|\overline{\underline{O}}-, -1)$

8. All formal charges equal zero unless otherwise indicated.

A. $|\overline{\underline{F}} - \overset{\frown}{Br} - \overline{\underline{F}}|$ with $|\overline{\underline{F}}|$ below Br

B. $\left[\begin{array}{c} |\overline{\underline{Cl}}\quad \overline{\underline{Cl}}| \\ |\overline{\underline{Cl}} - Sb - \overline{\underline{Cl}}| \\ |\overline{\underline{Cl}}| \end{array}\right]^{2-}$ $(Sb, -2)$

C. $\left(\begin{array}{c} |\overline{O}| \\ |\overline{\underline{O}} - \underline{Br} - \overline{\underline{O}}| \end{array}\right)^-$ (all O, –1; Br, +2)

D. $[|\overline{\underline{I}} - \overline{\underline{I}} - \overline{\underline{I}}|]^-$ (central I, –1) E. $\begin{array}{c} |\overline{\underline{Cl}}\quad \overline{\underline{Cl}}| \\ |\overline{\underline{F}} - P - \overline{\underline{F}}| \\ |\overline{\underline{F}}| \end{array}$

F. $|\overline{\underline{Cl}} - \underline{Pb} - \overline{\underline{Cl}}|$ G. $H - \overline{\underline{O}} - \overset{|O|}{\underset{||}{S}} - \overline{\underline{O}} - H$

H. $\begin{array}{c} |\overline{\underline{Cl}}\quad \overline{\underline{Cl}}| \\ |\overline{\underline{Cl}} - Sb - \overline{\underline{Cl}}| \\ |\overline{\underline{Cl}}| \end{array}$ I. $\left(\begin{array}{c} |\overline{\underline{Cl}}| \\ |\overline{\underline{Cl}} - I - \overline{\underline{Cl}}| \\ |\overline{\underline{Cl}}| \end{array}\right)^-$ $(I, -1)$

J. $\left[\begin{array}{c} |\overline{\underline{Cl}}\quad \overline{\underline{Cl}}| \\ |\overline{\underline{Cl}} - Sn - \overline{\underline{Cl}}| \\ |\overline{\underline{Cl}}\quad \overline{\underline{Cl}}| \end{array}\right]^{2-}$ $(Sn, -2)$ K. $\begin{array}{c} |\overline{\underline{F}}\quad \overline{\underline{F}}| \\ |\overline{\underline{F}} - I - \overline{\underline{F}}| \\ |\overline{\underline{F}}| \end{array}$

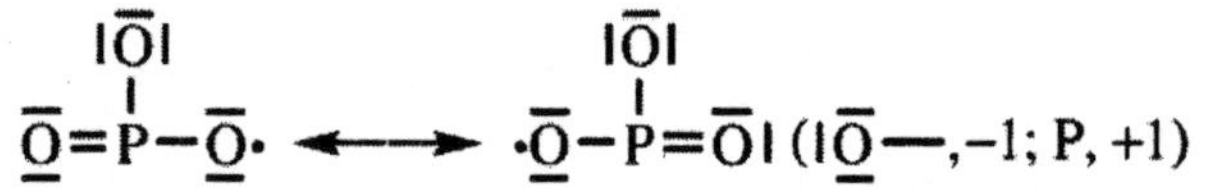

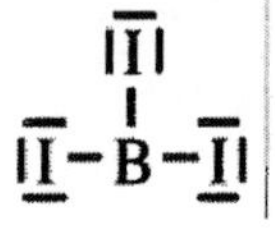

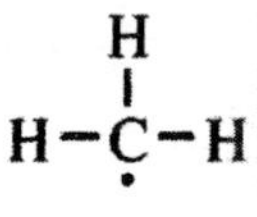

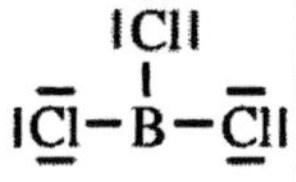

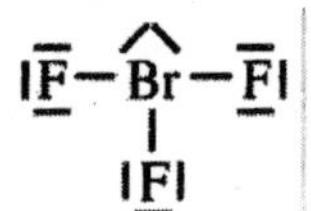

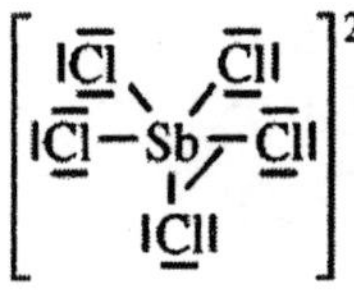

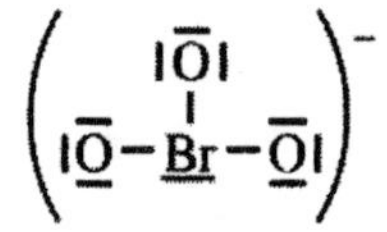

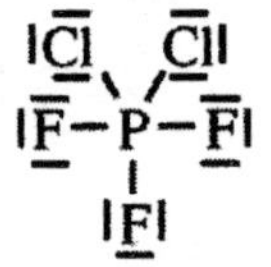

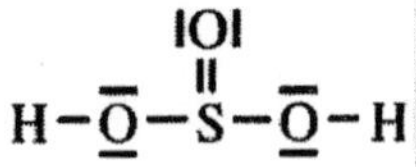

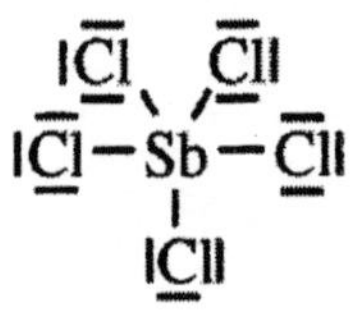

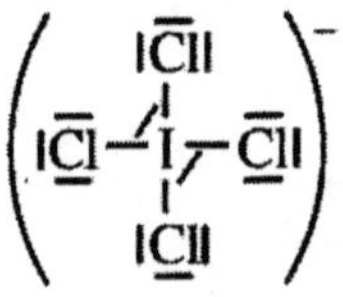

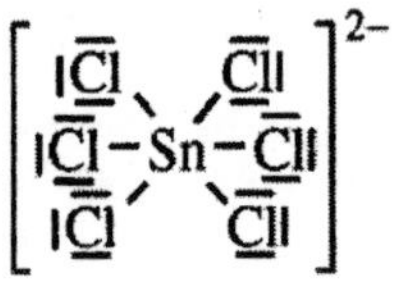

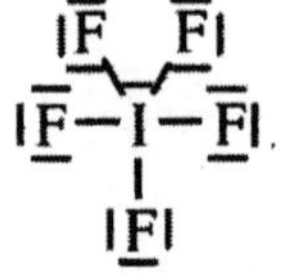

L. 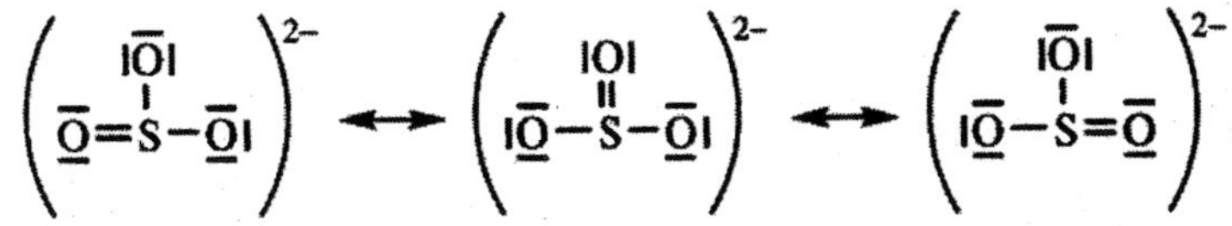 (—O, −1)

M. 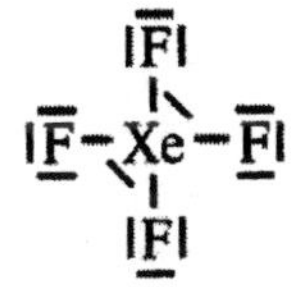

N. 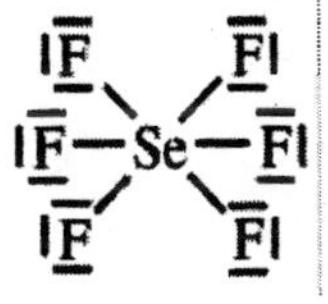

O. 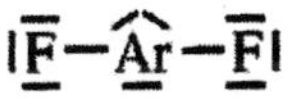

P. 

Q. 

R. 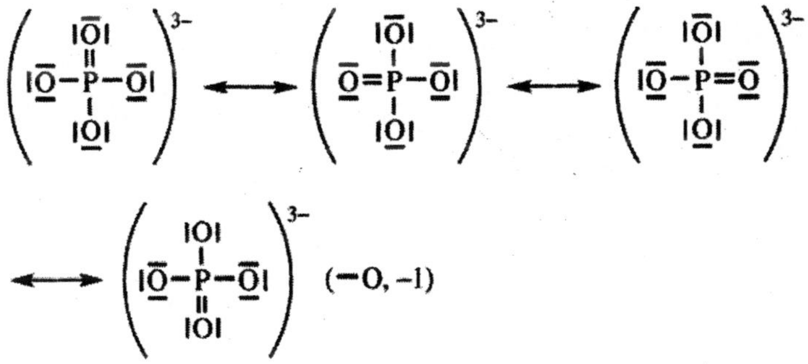 (—O, −1)

S. [|S—I—S|]$^-$ (S, −1; I, +1) T. |F—Xe—F| U. 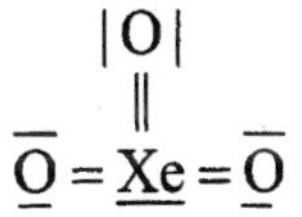

9. All values are in kJ/mol A. −114 B. −676 C. −338 D. −821 E. −374 F. −80 G. −80 H. −1926 I. +134 J. −2612 K. −132 L. −993 M. −35 N. −8 O. +497 P. −1632

10. Electron pair geometry is given first, and molecular shape is given second, abbreviated as follows: lin, linear; tripl, trigonal planar; tet, tetrahedron; tripyr, trigonal pyramid; tbp, trigonal bipyramid; irtet, irregular tetrahedron; T, T-shaped; oct, octahedron; sqpyr, square pyramid; sqpl, square planar. A. tripl, tripl B. tet, tet C. tet, bent D. tbp, T

A tripl, tripl

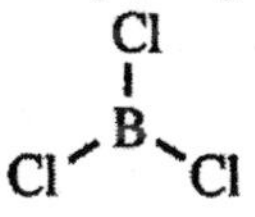

B. tet, tet

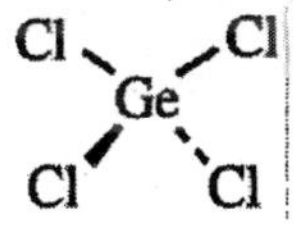

C. tet, bent

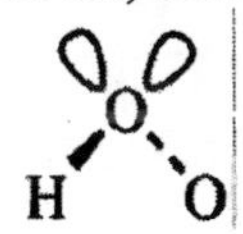

D. tbp, T

E. tbp, tbp

F. tripl, bent

G. tripl, tripl

H. lin, lin

Cl—Be—Cl

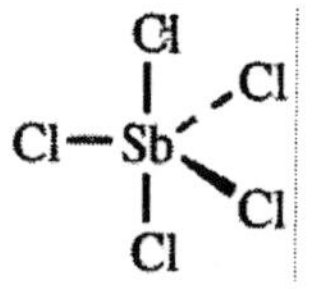
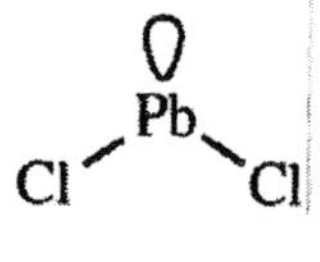
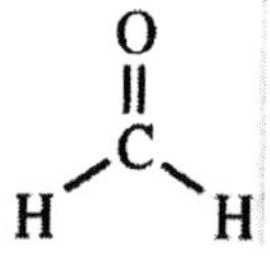

I. oct, sqpl

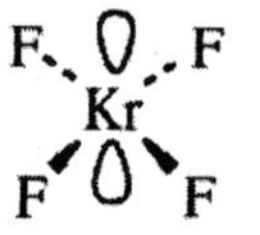

J. tbp, lin

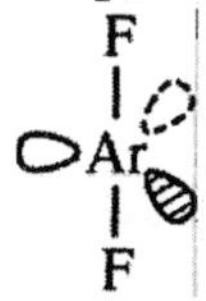

K. tbp, irtet

L. tripl, bent

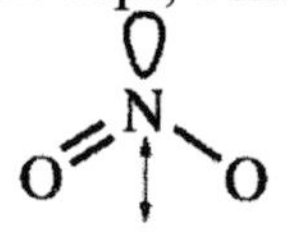

M. tet tripyr

N. tbp, T

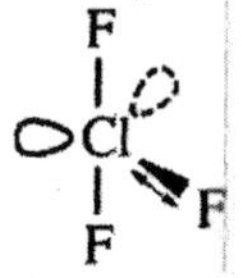

O. oct, sqpyr

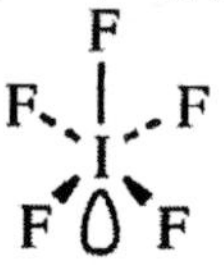

P. tbp, tbp

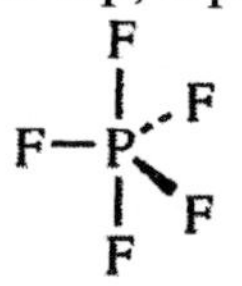

Q. tet, tet

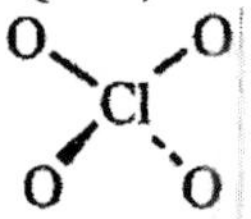

R. tet, tripyr

S. oct, sqpl

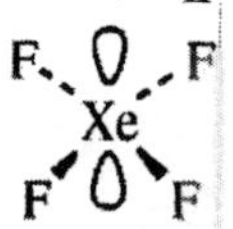

T. tripl, tripl

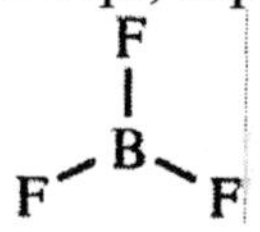

U. tet, bent

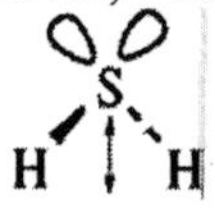

V. tet, tet

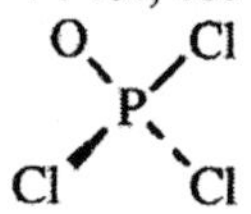

W. tbp, irtet

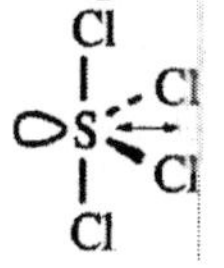

X. tripl, tripl

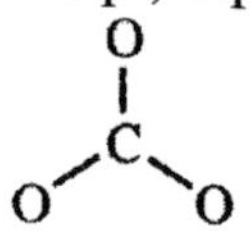

Y. tripl, tripl

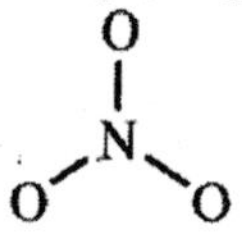

Z. lin, lin

N—N—O

Γ. tet, tet

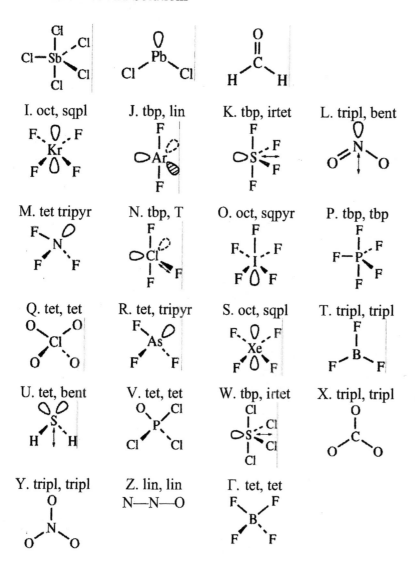

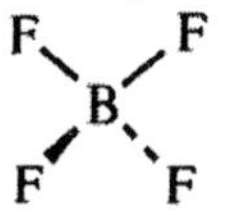

11. Most of the molecular polarities are indicated on the sketches that are part of the answers to the drill problems of objective 11–12. The following molecular polarities are described because they do not appear clearly on a sketch. C. from the midpoint of the H's to the O F from the Pb to the midpoint of the Cl's G. from C to O M. from N to the center of the 3-F triangle O. from I to the top F R. from As to the center of the 3-F triangle V. along the P—O bond Z. from N to O.

Quizzes

A: 1. (d) 2. (a) 3. (d) 4. (e) 5. (c) 6. (b) 7. (c) 8. (d) 9. (c) 10. (e) (+177)

B: 1. (b) 2. (d) 3. (c) 4. (e) 5. (d) 6. (c) 7. (a) 8. (a) 9. (a) 10. (c)

C: 1. (b) 2. (b) 3. (b) 4. (d) 5. (a) 6. (d) 7. (e) (linear) 8. (b) 9. (d) 10. (d)

D: 1. (c) 2. (c) 3. (d) 4. (b) 5. (d) 6. (d) 7. (b) 8. (a) 9. (d) 10. (d)

Sample Test

1. Formal charge is given only when not zero.

a.

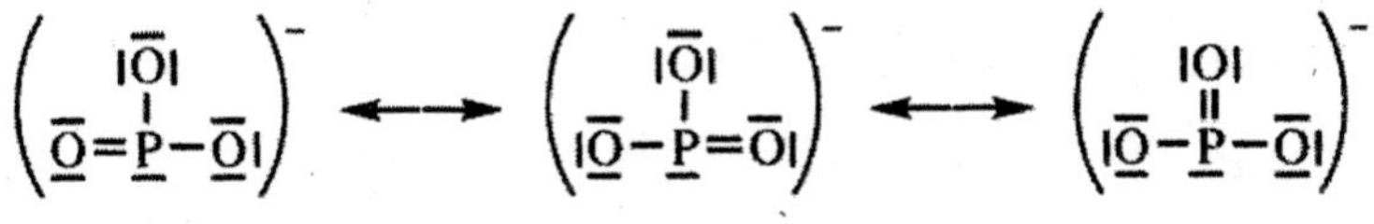

$$\left(\begin{array}{c} |\overline{\text{O}}| \\ | \\ \overline{\text{O}}=\text{P}-\overline{\text{O}}| \end{array}\right)^{-} \longleftrightarrow \left(\begin{array}{c} |\overline{\text{O}}| \\ | \\ |\overline{\text{O}}-\text{P}=\overline{\text{O}}| \end{array}\right)^{-} \longleftrightarrow \left(\begin{array}{c} |\text{O}| \\ \| \\ |\overline{\text{O}}-\text{P}-\overline{\text{O}}| \end{array}\right)^{-}$$

b. $[\overline{\text{O}}=\text{N}=\overline{\text{O}}]^{+}$ (N, +1) c.

$$\begin{array}{c} \text{H} \\ | \\ \text{H}-\text{C}-\overline{\underline{\text{Cl}}} \\ | \\ \text{H} \end{array}$$

d.

$$\begin{array}{c} |\overline{\text{F}}-\overline{\text{I}}-\overline{\text{F}}| \\ | \\ |\underline{\text{F}}| \end{array}$$

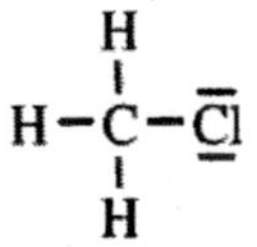

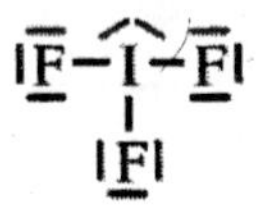

2. Electron pair geometry is given first, followed by molecular shape, abbreviated as in the answers for objective 11–10.

a.

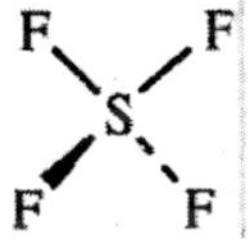

tet, tet

b. tet, tripyr, M c. tbp, irtet, W

d. oct, sqpyr, O (For questions b, c, and d, the sketches are given in the answers to objectives 11–12 and 11–13, to which the capital letters [M, W, O] refer.)

3. $\text{CH}_4(\text{g}) + 2\,\text{O}_2(\text{g}) \longrightarrow \text{CO}_2(\text{g}) + 2\,\text{H}_2\text{O}(\text{g})$ (C O.S. $= -4$ in CH_4) Bonds broken: 4
C—H + 2 O=O = 4(414) + 2(498) = 2642 kJ Bonds formed: 2
C=O + 4 H—O = 2(707) + 4(464) = 3270 kJ Energy/g = (2642 − 3270) kJ/16 g = −38.6 kJ/g.

$2\,\text{CH}_3\text{OH}(\text{g}) + 3\,\text{O}_2(\text{g}) \longrightarrow 2\,\text{CO}_2(\text{g}) + 4\,\text{H}_2\text{O}(\text{g})$ (C O.S $= -2$ in CH_3OH) Bonds broken:
6 C—H + 2 C—O + 2 O—H + 3 O=O = 6(414) + 2(360) + 2(464) + 3(498) = 5626 kJ/mol. Bonds
formed: 4 C=O + 8 H—O = 4(707) + 8(464) = 6540 kJ/mol.

Energy/g = (5626 − 6540) kJ/64.0 g = −14.3 kJ/g. $\text{H}_2\text{CO}(\text{g}) + \text{O}_2(\text{g}) \longrightarrow \text{CO}_2(\text{g})$
$+ \text{H}_2\text{O}(\text{g})$ (C O.S. = 0 in H_2CO) Bonds broken 2 C—H + 1 C=O + 1 O=O
= 2(414) + 1(707) + 1(498) = 2033 kJ. Bonds formed: 2 C=O + 2 H—O = 2 (707) + 2(464) = 2342 kJ.
Energy/g = (2033 − 2342) kJ/30.0 g = −10.3 kJ/g.

$2\,\text{HCOOH}(\text{g}) + \text{O}_2(\text{g}) \longrightarrow 2\,\text{CO}_2(\text{g}) + 2\,\text{H}_2\text{O}(\text{g})$ (C O.S. $= +2$ in HCOOH) Bonds broken:
2 H—O + 2 H—C + 2 C=O + 2 C—O + O=O =
2(464) + 2(414) + 2(707) + 2(360) + 1(498) + 4388 kJ Bonds formed:
4 C=O + 4 H—O = 4(707) + 4(464) = 4684 kJ. Energy/g = (4388 − 4684) kJ/92.0 g = −3.22 kJ/g.
The more negative the oxidation state of carbon, the more energy given off per gram of combusted material.

CHAPTER 11

Drill Problems

Abbreviations are those used in answers for objective 11-10.

1. (1)A. sp, lin, lin B. sp^3, tet, tet C. sp^2, tripl, bent D. sp^3 tet, tripyr E. sp^3, tet, tripyr F. sp^3, tet, bent G. sp^3, tet, bent H. sp^3, tet, bent I. sp^2, tripl, tripl J. sp^2, tripl, bent K. sp^3, tet, tet L. sp^3, tet, tet M. sp^3, tet, tripyr N. sp^3, tet, tet O. sp^2, tripl, tripl P. sp^3, tet, bent

(2)Q. sp^3d, tbp, irtet R. sp^3d^2, oct, oct S. sp^3d, tbp, T T. sp^3d, tbp, tbp U. sp^3d, tbp, T V. sp^3d, tbp, lin W. sp^3d^2, oct, sqpyr X. sp^3d, tbp, lin Y. sp^3d, tbp, tbp Z. sp^3d, tbp, lin Γ. sp^3d, tbp, tbp Δ. sp^3d^2, oct, spqyr Θ. sp^3d^2, oct, sqpl Λ. sp^3d, tbp, irtet Ξ. sp^3d^2, oct, sqpyr Π. sp^3d, tbp, irtet Σ. sp^3d^2 oct, sqpl Υ. sp^3d tbp, T Φ. sp^3d^2 oct, oct Ψ. sp^3d^2, oct, sqpl Ω. sp^3d^2, oct, oct

2. All bonds are σ bonds unless specified as π bonds. The number of bonds of a given type is given before the abbreviated description of the bond. "l.p." means lone pair; "l.e." means lone or unpaired electron. Only the lone pairs and lone electrons of the central atom are indicated. In the cases where there is some doubt, "hyb" precedes the description that assumes the central atom to be hybridized, and "unhyb" precedes the description in which the central atom is not assumed to be hybridized.

A. 2σ Be(sp)–H$(1s)$ B. 3σ C(sp^3)–H$(1s)$, l.e. in C(sp^3) C. 4σ Kr(d^2sp^3)–F$(2p)$, 2 l.p. in Kr(d^2sp^3) D. 4σ B(sp^3)–F$(2p)$ E. hyb: 2σ Sn(sp^2)–Cl$(3p)$, 1 l.p. in Sn (sp^2) unhyb: 2σ Sn$(5p)$–Cl$(3p)$, 1 1 .p. in (Sn$(5s)$ F. 4σ N(sp^3)–H$(1s)$ G. 4σ I(sp^3d^2)–Cl$(3p)$, 2 l.p. in I(sp^3d^2) H. 3σ C(sp^2)–H$(1s)$ I. hyb: 2σ S(sp^3)–I$(5p)$, 2 l.p. in S(sp^3); unhyp: 2σ S$(3p)$–I$(5p)$, l.p. in S$(3p)$, l.p. in S$(3s)$ J. 4σ Se(sp^3d)–F$(2p)$, l.p. in Se (sp^2d) K. 2σ Xe(sp^3d)–F$(2p)$, 3 l.p. in Xe(sp^3d) L. hyb: 2σ C(sp^2)–Cl$(3p)$, l.p.in C(sp^2); unhyb: 2σ C$(2p)$–Cl$(3p)$, l.p. in C$(2s)$ M. 2σ N(sp^2)–H$(1s)$, 2 l.p. in N(sp^3) N. 3σ Br(sp^3d)–F$(2p)$, 2 l.p. in Br(sp^3d) O. 5σ I(sp^3d^2)–F$(2p)$, 1 l.p. in I(sp^3d^2) P. 2σ Cl(sp^3)–O$(2p)$, 2 l.p. in Cl(sp^3) Q. 3σ B(sp^2)–Cl$(3p)$ R. Sb(sp^3d)–Cl$(3p)$ S. 6σ Se(sp^3d^2)–F$(2p)$ T. 3σ O(sp^3)–H$(1s)$ 1 l.p. in O(sp^3)

3. Refer to the paragraph at the beginning of the answers for objective 11-3. These bonding descriptions are based on the Lewis structure that is drawn first in each part.

A. $[\overline{\text{N}}\ _a\!=\!\text{C}=\overline{\text{N}}^b]^{2-}$ 2σ : C(sp)–N$(2p)$, π C$(2p_y)$–N$^a(2p_y)$, π C$(2p_z)$–N$b(2p_z)$

B.
$$\text{H}^b\!-\!\overset{\overset{\displaystyle \text{H}^a}{|}}{\text{C}}\!=\!\underline{\text{S}}$$
3σ : C(sp^2)–H$^a(1s)$, C(sp^2)–H$^b(1s)$, C(sp^2)–S$(3p_y)$; π C$(2p_z)$–S$(3p_z)$

C.
$$|\overline{\text{C}}|^b\!-\!\overset{\overset{\displaystyle |\overline{\text{Cl}}|^a}{|}}{\text{C}^a}\!=\!\overset{\overset{\displaystyle \text{H}^a}{|}}{\text{C}}\!-\!\text{H}^b$$
5σ : C$^a(sp^2)$–Cl$^a(3p)$, C$^a(sp^2)$–Cl$^a(3p)$, C$^a(sp^2)$–Cl$^b(3p)$, C$^b(sp^2)$–H$^a(1s)$, C$^b(sp^2)$–H$^b(1s)$ C$^a(sp^2)$–C$^b(sp^2)$;

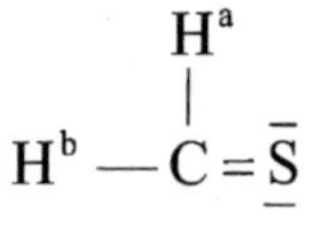

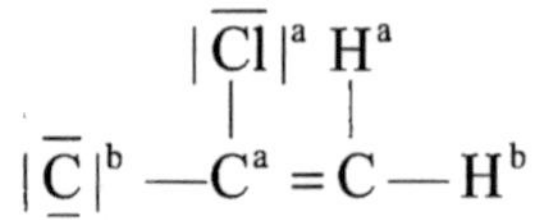

D. $|\overline{\text{Cl}}|^{a}$ over $|\overline{\text{Cl}}|^{b}-\text{C}=\overline{\text{S}}$ 3σ: $\text{Cl}^{a}(3p)-\text{C}(sp^2)$, $\text{Cl}^{b}(3p)-\text{C}(sp^2)$, $\text{C}(sp^2)-\text{S}(3p)$, $\pi\,\text{C}(2p_z)-\text{S}(3p_z)$

E. $[\overline{\text{S}}=\text{C}=\overline{\text{N}}]^{-}$ $\sigma\,\text{C}(sp)-\text{S}(3p_y)$, $\sigma\,\text{C}(sp)-\text{N}(2p_z)$, $\pi\,\text{C}(2p_y)-\text{N}(2p_y)$, $\pi\,\text{C}(2p_z)-\text{S}(3p_z)$

F. $|\text{N}^{a}\equiv\text{N}^{b}-\overline{\text{O}}|$ $\sigma\,\text{N}^{a}(2p_x)-\text{N}^{b}(sp)$, $\sigma\,\text{N}^{b}(sp)-\text{O}(2p_z)$, $\pi\,\text{N}^{a}(2p_y)-\text{N}^{b}(2p_y)$, $\pi\,\text{N}^{a}(2p_z)-\text{N}^{b}(2p_z)$

G. $\overline{\text{O}}^{a}=\text{C}=\overline{\text{O}}^{b}$ $\sigma\,\text{O}^{a}(2p_y)-\text{C}(sp)$, $\sigma\,\text{O}^{b}(2p_z)-\text{C}(sp)$, $\pi\,\text{O}^{a}(2p_z)-\text{C}(2p_z)$, $\pi\,\text{O}^{b}(2p_y)-\text{C}(2p_y)$

H. $\text{H}-\overline{\text{O}}-\text{C}\equiv\text{N}|$ 3σ: $\text{H}(1s)-\text{O}(sp^3)$, $\text{O}(sp^3)-\text{C}(sp)$, $\text{C}(sp)-\text{N}(2px)$; 2 l.p. in $\text{O}(sp^3)$; $\pi\,\text{C}(2p_y)-\text{N}(2p_y)$, $\pi\,\text{C}(2p_z)-\text{N}(2p_z)$

I. $\text{H}-\text{N}=\overline{\text{O}}$ $\sigma\,\text{H}(1s)-\text{N}(sp^2)$, $\sigma\,\text{N}(sp^2)-\text{O}(2p_y)$, l.p. in $\text{N}(sp^2)$, $\pi\,\text{N}(2p_z)-\text{O}(2p_z)$

J. $\text{H}-\text{C}\equiv\text{N}|$ $\sigma\,\text{H}(1s)-\text{C}(sp)$, $\sigma\,\text{C}(sp)-\text{N}(2p_x)$, $\pi\,\text{C}(2p_y)-\text{N}(2p_y)$, $\pi\,\text{C}(2p_z)-\text{N}(2p_z)$

K. $\text{H}^{a}-\text{C}^{a}\equiv\text{C}^{b}-\text{H}^{b}$ 3σ: $\text{H}^{a}(1s)-\text{C}^{a}(sp)$, , $\text{C}^{a}(sp)-\text{C}^{b}(sp)$, $\text{C}^{b}(sp)-\text{H}^{b}(1s)$; $\pi\,\text{C}^{a}(2p_y)-\text{C}^{b}(2p_y)$, $\text{C}^{a}(2p_z)-\text{C}^{b}(2p_z)$

L. $[\overline{\text{O}}^{a}=\text{N}=\overline{\text{O}}^{b}]^{+}$ $\sigma\,\text{O}^{a}(2p_y)-\text{N}(sp)$, $\sigma\,\text{N}(sp)-\text{O}^{b}(2p_z)$, $\pi\,\text{O}^{a}(2p_z)-\text{N}(2p_z)$, $\pi\,\text{N}(2p_y)-\text{O}^{b}(2p_y)$

M. $\overline{\text{O}}=\text{C}=\overline{\text{S}}$ $\sigma\,\text{O}(2p_y)-\text{C}(sp)$, $\sigma\,\text{C}(sp)-\text{S}(2p_z)$, $\pi\,\text{O}(2p_z)-\text{C}(2p_z)$, $\pi\,\text{C}(2p_y)-\text{S}(2p_y)$

N. $[|\overline{\text{F}}-\text{P}=\overline{\text{O}}$ $\sigma\,\text{F}(2p_z)-\text{P}(sp^2)$, $\sigma\,\text{P}(sp^2)-\text{O}(2p_y)$, l.p. in $\text{P}(sp^2)$, $\pi\,\text{P}(3p_z)-\text{O}(2p_z)$

O. $|\text{O}|^{a}$ (double-bonded) over $\text{H}^{a}-\text{C}-\overline{\text{O}}^{b}-\text{H}^{b}$ 4σ $\text{H}^{a}(1s)-\text{C}(sp^2)$, $\text{C}(sp^2)-\text{O}^{a}(2p_y)$, $\text{C}(sp^2)-\text{O}^{b}(sp^3)$,

$\text{O}^{b}(sp^3)-\text{H}^{b}(1s)$; 2 l.p. in $\text{O}^{b}(sp^3)$, $\pi\,\text{C}(2p_z)-\text{O}^{a}(2p_z)$

4. Simplified molecular orbital diagram given first (in the order $\sigma_{2s}^{b}\ \sigma_{2s}^{*}, \pi_{2p}^{b}, \sigma_{2p}^{b}, \pi_{2p}^{*}, \sigma_{2p}^{*}$), followed by answers (1)-(6).

A. KK [↑↓] [↑↓] [↑↓][↑↓] [↑↓] [][] []; 8, 2, 3, 6, 4, 0
B. KK [↑↓] [↑↓] [↑↓][↑↓] [↑↓] [↑↓][↑↓] []; 8, 6, 1, 6, 8, 0
C. KK [↑↓] [↑↓] [↑↓][↑↓] [↑↓] [↑][] []; 8, 3, ½, 6, 5, 1
D. KK [↑↓] [↑↓] [↑↓][↑↓] [] [][] []; 6, 2, 2, 4, 4, 0
E. KK [↑↓] [↑↓] [↑↓][↑↓] [] [][] []; 6, 2, 2, 4, 4, 0
F. KK [↑↓] [↑↓] [↑↓][↑↓] [] [][] []; 6, 2, 2, 4, 4, 0
G. KK [↑↓] [↑] [][] [] [][] []; 2, 1, ½, 3, 0, 1
H. KK [↑↓] [↑↓] [↑↓][↑↓] [↑] [][] []; 7, 2, ½, 5, 4, 1
I. KK [↑↓] [↑↓] [↑↓][↑↓] [↑↓] [↑↓][↑↓] [↑]; 8, 7, ½, 7, 8, 1
J. KK [↑↓] [↑↓] [↑↓][↑↓] [↑↓] [↑↓][↑↓] [↑]; 8, 7, ½, 7, 8, 1
K. KK [↑↓] [↑↓] [↑↓][↑↓] [↑↓] [↑↓][↑] []; 8, 5, ½, 6, 7, 1
L. KK [↑↓] [↑↓] [↑↓][↑↓] [↑↓] [↑][↑] []; 8, 4, 2, 6, 6, 2
M. KK [↑↓] [↑↓] [][] [] [][] []; 2, 2, 0, 4, 0, 0

5. Refer to the paragraph at the beginning of objective 12-3. These descriptions are based on the resonance structure drawn first in each part.

A. $\bar{O} = N - \bar{O}\cdot^b \longleftrightarrow \cdot\bar{O} - N = \bar{O}$ $\sigma\, O^a(2p_y) - N(sp^2)$, $\sigma\, N(sp^2) - O^b(2p_y)$, l.p. in $N(sp^2)$,

$\pi\, O^a(2p_z) - N(2p_z) - O^b(2p_z)$ (3-eln, 3-center π bond)

B.

$$|\bar{O}|^b \qquad |\bar{O}| \qquad |O|$$
$$\bar{O}^a = S - \bar{O}|^c \longleftrightarrow |\bar{O} - S = \bar{O}| \longleftrightarrow |\bar{O} - S = \bar{O}|$$

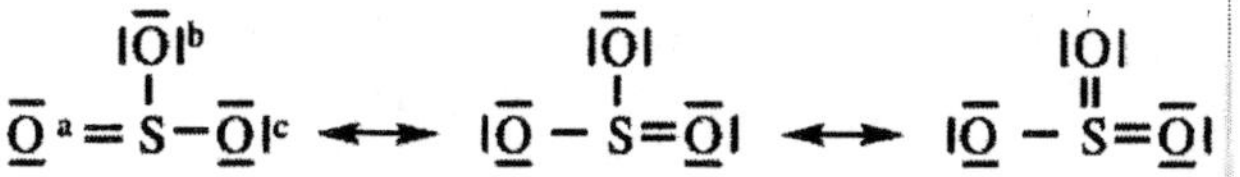

$3\sigma : S(sp^2) - O^a(2p_y)$, $S(sp^2) - O^b(2p_y)$, $S(sp^2) - O^c(2p_y)$,

$\pi\, S(3p_z) - O^a(2p_z) - O^b(2p_z) - O^c(2p_z)$ (6-electron, 4-center π bond)

C.

$$|\bar{Cl}|^b \qquad |\bar{Cl}| \qquad |Cl|$$
$$\bar{Cl}^a = B - \bar{Cl}|^c \longleftrightarrow |\bar{Cl} - B = \bar{Cl}| \longleftrightarrow |\bar{Cl} - B - \bar{Cl}|$$

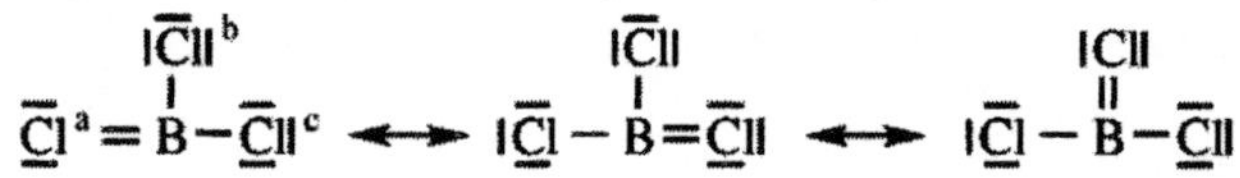

$3\sigma : B(sp^2) - Cl^a(3p_y)$, $B(sp^2) - Cl^b(3p_y)$, $B(sp^2) - Cl^c(3p_y)$,

$\pi\, B(2p_z) - Cl^a(3p_z) - Cl^b(3p_z) - Cl^c(3p_z)$ (6-electron, 4-center π bond)

D. $\bar{O} = \underline{O}^b - \bar{O}|^c \longleftrightarrow |\bar{O} - \underline{O} = \bar{O}$ $\sigma\, O^a(2p_y) - O^b(sp^2)$, $\sigma\, O^b(sp^2) - O^c(2p_y)$, l.p. in

$O^b(sp^2)$, $\pi\, O^a(2p_z) - O^b(2p_z) - O^c(2p_z)$ (4-center, 3-electron π bond)

E.

$$\left(\begin{array}{c} |O|^a \\ \| \\ H^a - C - \bar{O}|^b \end{array}\right)^{-} \longleftrightarrow \left(\begin{array}{c} |\bar{O}| \\ | \\ H - C = \bar{O}| \end{array}\right)^{-}$$

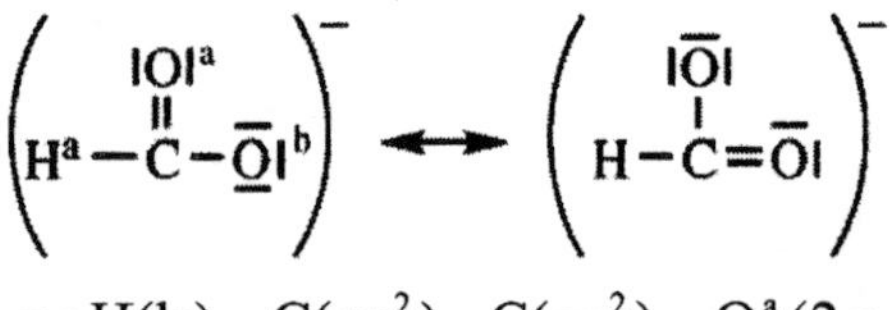

$3\sigma : H(1s) - C(sp^2)$, $C(sp^2) - O^a(2p_y)$, $C(sp^2) - O^b(2p_y)$; $\pi\, C(2p_z) - O^a(2p_z) - O^b(2p_z)$ (4-center, 3-electron π bond)

F. $[\bar{O} = \underline{C} - \bar{F}|]^{-} \longleftrightarrow [|\bar{O} - \underline{C} = \bar{F}]^{-}$ $\sigma\, O(2p_y) - C(sp^2)$, $\sigma\, C(sp^2) - F(2p_y)$, l.p. in $C(sp^2)$,

$\pi\, O(2p_z) - C(2p_z) - F(2p_z)$ (4-center, 3-electron π bond)

G.

$$\left(\begin{array}{c} |\bar{O}| \\ | \\ \bar{O} = C - \bar{O}| \end{array}\right)^{2-} \longleftrightarrow \left(\begin{array}{c} |\bar{O}| \\ | \\ |\bar{O} - C = \bar{O}| \end{array}\right)^{2-} \longleftrightarrow \left(\begin{array}{c} |O| \\ \| \\ |\bar{O} - C - \bar{O}| \end{array}\right)^{2-}$$

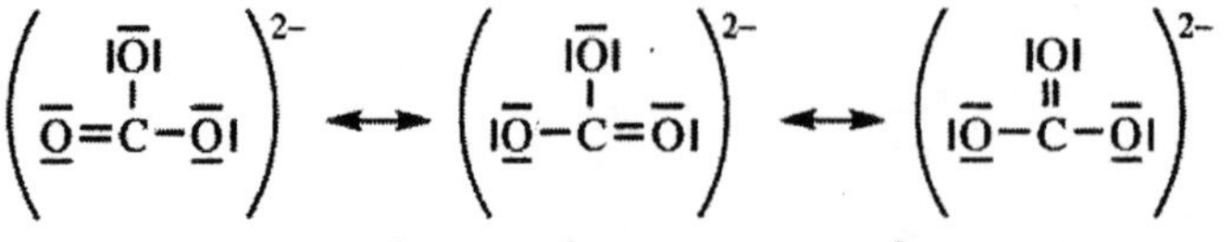

$\sigma\, O^a(2p_y) - C(sp^2)$, $\sigma\, O^b(2p_y) - C(sp^2)$, $\sigma\, O^c(2p_y) - C(sp^2)$

H. Same as BCl_3 except $2p$ orbitals used on ligands.

Quizzes

A: 1. (c) 2. (c) 3. (b) 4. (a) 5. (b) 6. (a) 7. (e) 8. (e)
B: 1. (a) 2. (b) 3. (c) 4. (b) 5. (b) 6. (b) 7. (a) 8. (c)
C: 1. (a) 2. (b) 3. (d) 4. (b) 5. (b) 6. (b) 7. (d) 8. (a)
D: 1. (a) 2. (a) 3. (c) 4. (d) 5. (c) 6. (c) 7. (d) 8. (c)

Sample Test

1. a. 2, 3, 2 b. 0, 0, 2 c. linear, triangular planar, tetrahedral d. linear, triangular planar, bent e. none, none sp^3 f. sp, sp^2, sp^3 g. $N(2p_x)-C^a(sp), C^a(sp)-C^b(sp^2)$, $C^b(sp^2)-S(3p_x), C^b(sp^2)-O(sp^3)$, $O(sp^3)-H(1s)$ h. $N(2p_y)-C^a(2p_y)$, $N(2p_z)-C^a(2p_z)$, $C^b(2p_z)-S(3p_y)$ i. 180° j. 120° k.120° l. 109.5°

2. a. $5/2$ b. NO^+ because it has a bond order of 3 compared to $5/2$ for O_2^+ c. 1, 2, 0

CHAPTER 12

Drill Problems

2. A. $200.\,g \times mol/32.0\ g \times 0.0816\ kJ\ mol^{-1}K^{-1} \times (64.96°C - 50.0°C) = 7.63\ kJ$; $(184.7\,kJ - 7.63\,kJ) \times mol/39.2\,kJ \times 32.0\ g/mol = 145\ g$ vaporizes; $m_l = 55\ g$; $t_f = 64.96°C$

B. $q = 204\,g \times mol/46.0\,g \times [0.113\,kJ\ mol^{-1}\ K^{-1}(78.5°C - 20.2°C) + 40.5\,kJ/mol + 0.0657\,kJ\ mol^{-1}K^{-1}(100.0°C - 78.5°C)] = 215\,kJ.$

C. $186\,g \times mol/64.1\,g \times 0.0852\,kJ\ mol^{-1}K^{-1}(-10.0°C + 51.0°C) + (186\,g - 46\,g)\,mol/64.1\,g \times 26.8\,kJ/mol = 68.7\,kJ;$

D. $20.2\,kJ = m \times mol/27.0\,g[0.0707\,kJ\ mol^{-1}\ K^{-1}(26.0°C + 14.0°C) + 30.7\,kJ\ mol + 0.0360\,kJ\ mol^{-1}\ K^{-1}(36.5°C - 26.0°C)]; m = 16.1\,g$

E. same method as A; $t_f = 118.5°C$, $m_l = 177\,g$ F. same method as B; $q = 126\,J\ m$ G. same method as A; $t_f = 34.6°C, m_l = 54.5\,g$ H.

$q = 14.0\ g \times mol/154.0\,g \times 0.133\,kJ\ mol^{-1}\ K^{-1}\ (76.8°C - 25.0°C) = 0.626\,kJ;$
$(16.4 - 0.626\,kJ) \times mol/\{[0.133\,kJ(76.8°C - 25.0°C)] + 34.4\,kJ\} \times 154.0\,g/mol = 58.9\,g;$
total mass $= 72.9\,g$

I. $314\,g \times mol/119.5\,g \times 0.115\,kJ\ mol^{-1}\ K^{-1}\ (t_f - 0.0°C) = 14.3\,kJ; t_f = 47.3°C$

J. Same method as B; 92.1 kJ K. Same method as C; 206 kJ;

L. $t_f = 58.78°C$; $q = 124\,g \times mol/159.8\,g \times 0.0757\,kJ\ mol^{-1}\ K^{-1}\ (58.78°C - 20.40°C) + 100.\ g \times mol/159.8\,g \times 33.9\,kJ/mol = 23.5\,kJ$ M. Same method as I; $t_f = -34.35°C$

4. (1) A. 11.5 mmHg B. 35.0 mmHg C. 115 mmHg D. 45.0 mmHg E. 101 mmHg F. 8.01 mmHg G. 1.10 mmHg
(2) H. 450 mmHg, 6.8 g liquid I. 205 mmHg, 5.73 g liquid

J. 370 mmHg, no liquid K. 110 mmHg, 2.4 g liquid
L. 570 mmHg, no liquid M. 400 mmHg, 4.6 g liquid
N. 650 mmHg, no liquid
5. (1) A. 31.4 kJ/mol, 41.7°C B. kJ/mol, 5.0° C
C. 50.2 kJ/mol, 149°C D. 46.0 kJ/mol, 139.7° C
E. 32.1 kJ/mol, 70.3°C F. 19.6 kJ/mol, -98.9° C
G. 78.5 kJ/mol, 678° C
(2) H. 113 mmHg, -9.3° C I. 1386 mmHg, 31°C J. 143 mmHg. -15° C K. 315 mmHg, 36° C L. 84.6 mmHg, -18.7° C M. 25.9 mmHg, 37° C N. 237 mmHg, -54°C O. 350 mmHg, 48° C
7. The enthalpies (in kJ/mol) are, in order $\Delta H_{fus}, \Delta H_{vap}, \Delta H_{sub}, \Delta H_{sol}, \Delta H_{cond}, \Delta H_{dep}$.
A. 6.01, 44.0, 50.0, -6.01, -44.0, -50.0
B. 10.7, 232.8, 243.5, -10.7, -232.8, -243.5
C. 27.2, 176.9, 204.1, -27.2, 176.9, -204.1
D. 47.2, 181.8, 229.0, -47.2, -181.8, -229.0
E. 201., 158.6, 178.7, -20.1, -158.6, -178.7
F. 19.0, 41.4, 60.4, -19.0, -41.4, -60.4
G. 56.6, 668.5, 725.1, -56.6, -668.5, -725.1
5. A. {{Figure}}
B. Yes, at about 14.00 K C. No; the critical pressure is 12.8 atm. D. No; the solid-liquid line slopes upward to the right.
E. {{Figure}}
F. Yes, about 49.6°C G. Yes, about 218° C H. Yes; the solid-liquid line slopes upward to the left. I. Begins a vapor; turns solid at about 5 mmHg, turns liquid between 300 and 400 mmHg and remains liquid. J. Begins as solid; turns liquid at about 273.15 K; vaporizes at 131° C (from the Clausius-Clapeyron equation). K. Begins as solid; sublimes about -70° C.
L. Begins as a solid; melts at about -56.7° C; vaporizes at perhaps 0° C. M. Begins as a solid; melts at about -55° C; never vaporizes as 80.0 atm is above the critical pressure. N. Begins as a solid; melts at 14.01 K; boils at 20.38 K. O. Starts as a vapor; becomes liquid at 0.12 atm; turns solid at about 5 atm and remains so. P. Begins as a vapor; becomes a solid at 0.40 atm; turns liquid at about 1.40 atm and remains so. Q. Begins as a solid; melts at about 49.6° C; vaporizes at about 225° C.
6. A. HI, Xe B. PF_3, BCl_3 C. HBr, HCl D. CH_4, GeH_4 E. PF_3, AlF_3 F. CH_3OH, C_3H_8OH G. H_2O, HF H. I_2, Cl_2 I. CH_4, CCl_4 J. CCl_3 F, Br_2
7. A. SrS, MgS B. NaH, NaBr C. SrI_2, $CaCl_2$ D. MgS, Na_2S E. $CaCl_2$, CaO F. Na_2CO_3, $Al_2(CO_3)_3$ G. LiF, LiI H. K_2S, MgS I. $Mg(NO_3)_2$, $Pb(NO_3)_2$ J. $InCL_3$, RbCl (Lattice energies are negative, exothermic.)

Quizzes

A: 1. (e) 2. (b) 3. (e) 4. (b) 5. (b) 6. (d) 7. (a) 8. (e)

B: 1. (b) 2. (b) 3. (d) 4. (d) 5. (a) 6. (e) 7. (d) 8. (b)
C: 1. (c) 2. (b) 3. (a) 4. (d) 5. (e) 6. (a) 7. (a) 8. (b)
D: 1. (e) 2. (b) 3. (b) 4. (e) 5. (c) 6. (a) 7. (e) 8. (e)

Sample Test

1. $-8.3145\frac{\text{J}}{\text{mol}\cdot\text{K}}\ln\frac{50.0}{760.0}=32.0\times10^3\left(\frac{1}{T}-\frac{1}{329.4}\right);$

$T=-6.0°\text{C}$

2. See Figure 13-1. 3. Review objective 13-9.

4. MgO. The lattice energy depends on $Z_A Z_C/(r_A+r_C)$. Large lattice energy means highly charged ions with the smallest radii.

CHAPTER 13

Drill Problems

1. A. 30.00 cm^3, 8.91 cm^3 21.80 g, 31.16%, 29.70%
B. 50.00 g, 0.9912 g/cm^3, 5.00%, 4.96%, 6.26%

C. 4.28 cm^3, 39.87 g, 8.50%, 8.47%, 10.71%
D. 91.22 cm^3,48.00 g, 120.00 g, 60.00%, 67.81%
E. 6.995 cm^3, 0.870 cm^3, 6.18 g, 9.81%, 12.43%
F. 60.00 g, 1.0094 g/cm^3, 9.00%, 8.08%, 8.19%
G. 4.35 cm^3, 120.46 g, 4.00%, 4.02%
H. 39.34 cm^3, 102.00 g, 150.00 g, 32.00%, 28.51%
I. 36.00 cm^3, 7.50 cm^3, 28.96 g, 25.41%, 20.83%
J. 100.00 g, 1.044 g/cm^3, 19.00%, 19.89%, 15.77%
K. 31.99 cm^3, 84.07 g, 48.00%, 53.80%
L. 74.30 cm^3, 11.20 g, 70.00 g, 84.00%, 88.80%
M. 71.40 cm^3, 53.75 cm^3, 20.02 g, 59.58%, 75.29%
2. A. 0.8535 mol, 4.163 *m*, 19.99%
B. 53.5 g/mol, 0.667 mol, 1.406 *m*
C. 0.224 mol, 1.156 g/mL, 0.9148 *m*
D. 3.175 M, 3.383 *m*, 10.98%
E. 42.40 g/mol, 500.0 mL, 2.620 *m*
F. 200. mL, 1.12 *m*, 27.0%
G. 42.5 g/mol, 0.963 M, 0.981 *m*
H. 0.2128 mol, 1.065 g/mL, 1.110 *m*
I. 0.9768 mol, 1.302 M, 1.368 *m*
J. 74.54 g/mol, 0.3084 mol, 0.9326 *m*
K. 0.2007 mol, 0.7434 M, 0.7440 *m*
L. 165.7 mL, 1.29 m, 20.0%
M. 3.782 mol, 1.261 g/mL, 6.060 *m*
N. 5.210 mol, 32.31 *m*, 76.06%
O. 420. mL, 14.3 M, 16.7 *m*
P. 58.44 g/mol, 6.012 *m*, 1.598 mol
Q. 400.0 mL, 6.803 *m*, 40.00%
R. 0.205 mol, 0.8548 M, 0.8873 *m*
3. A.0.06977, 93.023% B. 0.0247, 97.53% C. 0.0166, 98.34%
D. 0.05744, 94.256% E. 0.04507, 95.493% F. 0.0208, 97.92%

G. 0.0172, 98.28% H. 0.01959, 98.041% I. 0.02404, 97.596%
J. 0.0165, 98.35% K. 0.0135, 98.65% L. 0.0227, 97.73%
M. 0.09846, 90.154% N. 0.3677, 63.23% O. 0.231, 76.9%
P. 0.09769, 90.231 Q. 0.1092, 89.08% R. 0.01573, 98.427%
4. A. 0.0355 *m* B. 7.67 atm C. 0.000234 *m* D. 0.493 atm
E. 0.00144 *m* F. 4.57 atm G. 0.00278 *m* H. 4.58 atm
I. 1.12×10^{-4} *m* J. 11.7 atm
5. A. $\chi_B = 0.25, P = 87.5$ B. $\chi_A = 0.542, \chi_B = 0.458$
C. $P_A = 340, \chi_B = 0.75$ D. $P_B = 254, \chi_B = 0.70$
E. $P_A = 125, \chi_A = 0.86$ F. $\chi_A = 0.65, P = 218$
G. $\chi_A = 0.125, \chi_B = 0.875$ H. $\chi_A = 0.200, \chi_B = 0.800$
6. A. 106 g/mol, 15.83°C B. 4.88°C/*m*, 0.217 *m*, 80.64°C
C. 7.00°C/*m*, 23.1 g, 212.09°C D. 3.56°C/*m*, 83.8 g/mol, 49.91°C
E. 0.325 *m*, 100.17°C F. 14.2 g, 0.369 *m*, – 0.69°C
7. A. 0.0398 M, 0.749 g B. 0.00740 M, 512 g/mol
C. 488 mmHg, 0.0269 M D. 0.0808 M, 7.75 g E. 0.0258 M, 1214 g/mol
F. 579 mmHg, 0.0322 M G. 0.0968 M, 8.02 g
H. 0.0275 M, 99.8 g/mol I. 455 mmHg, 0.0231 M
J. 0.0256 M, 1.73 g K. 0.0414 M, 164 g/mol
L. 351 mmHg, 0.0187 M. 0.0524 M, 1.04 g N. 0.0350 M, 154 g/mol
O. 47.2 mmHg, 0.00250 M
8. A. $[Li^+] = 0.0709\,M, [Ca^{2+}] = 0.436\,M, [NO_3^-] = 0.943\,M$
B. $[Cl^-] = 0.0788\,N, [SO_4^{2-}] = 0.0966\,M, [H^+] = 0.2720\,M$
C. $[Na^+] = 0.124\,M, [Al^{3+}] = 0.305\,M, [Br^-] = 1.040\,M$
D. $[Cl^-] = 0.0611\,M, [PO_4^{3-}] = 0.0916\,M, [Na^+] = 0.3360\,M$
E. $[H^+] = 0.829\,M, [Na^+] = 0.457\,M, [NO_3^-] = 1.286\,M$
F. $[Ca^{2+}] = 0.0894\,M, [Mg^{2+}] = 0.120\,M, [Cl^-] = 0.420\,M$
G. $[Na^+] = 0.112\,M, [Al^{3+}] = 0.144\,M, [SO_4^{2-}] = 0.272\,M$
H. $[SO_4^{2-}] = 0.0946\,M,\ [PO_4^{3-}] = 0.184\,M,\ [Na^+] = 0.741\,M$
J. $[Cl^-] = 0.536\,M, [SO_4^{2-}] = 0.0848\,M, [Li^+] = 0.706\,M$

Quizzes

A: 1. (e) 2. (e) (concentrated) 3. (a) 4. (a) 5. (d) 6. (d) 7. (e) 8. (a)
B: 1. (b) 2. (a) 3. (c) 4. (e) 5. (e) (8/100) 6. (c) 7. (e) 8. (b)
C: 1. (e) 2. (d) 3. (d) 4. (c) 5. (e) 6. (e) 7. (a) 8. (c)
D: 1. (b) 2. (c) 3. (d) 4. (e) 5. (c) 6. (d) 7. (d) 8. (a)

Sample Test

1. a. $(750\,g + 85\,g)/810\,mL = 1.03\,g/mL$
b. $[86\,g/(750\,g + 85\,g)] \times 100\% = 10.2\%$

c. $\dfrac{85.0\text{ g} \times \text{mol}/180.0\text{ g}}{985.0\text{ g} \times \text{mol}/180.0\text{ g}} + (750\text{ g} \times \text{mol}/46.0\text{ g})) = 0.0281$

$= \chi_{\text{sucrose}}$

d. $\dfrac{875.0\text{ g} \times \text{mol}/180.0\text{ g}}{750\text{ g} \times \text{kg}/1000\text{ g}} = 0.630\ m$

e. $\dfrac{85.0\text{ g} \times \text{mol}/180.0\text{ g}}{810\text{ mL} \times \text{L}/1000\text{ mL}} = 0.583\text{ M}$

2. $914.0°\text{F} - 32)(5/9) = -10.°\text{C}; \Delta t_f = 10.0°\text{C} = K_F m;$

$m = 10.0°\text{C}/(1.86\,°\text{C}/m) = 5.38\ m$

mols of solute $= 20.0\text{ L H}_2\text{O} \times (1.00\text{ kg/L})\,(5.38\ m) = 108\text{ mol}$

$108\text{ mol} \times \dfrac{62.0\text{ g}}{\text{mol}} \times \dfrac{\text{cm}^3}{1.12\text{ g}} \times \dfrac{\text{L}}{1000\text{ cm}^3}$

$= 5.95\text{ L ethylene glycol}$

CHAPTER 14

Drill Problems

1. A. rate $= \Delta[O_2]/\Delta t = \Delta[NO_2]/4\Delta t = -\Delta[N_2O_5]/2\Delta t$

B. rate $= \Delta[CH_4]/\Delta t = \Delta[CO]/\Delta t = -\Delta[CH_3CHO]/\Delta t$

C. rate $= \Delta[CO_2]/\Delta t = \Delta[NO]/\Delta t = -\Delta[NO_2]/\Delta t$

D. rate $= \Delta[N_2]/\Delta t = \Delta[H_2O]/2\Delta t = -\Delta[NO]/2\Delta t = -\Delta[H_2]/2\Delta t$

E. rate $= \Delta[Cl_2]/\Delta t = \Delta[NO]/2\Delta t = -\Delta[NOCl]/\Delta t$

F. rate $= \Delta[FClO_2]/2\Delta t = -\Delta[F_2]/\Delta t = -\Delta[ClO_2]/2\Delta t$

G. rate $= \Delta[Cl^-]/\Delta t = \Delta[CH_3F]/\Delta t = -\Delta[F^-]/\Delta t = -\Delta[CH_3Cl]/\Delta t$

H. rate $= \Delta[Br_2]/3\Delta t = \Delta[H_2O]/3\Delta t = -\Delta[BrO_3^-]/\Delta t = -\Delta[Br^-]/5\Delta t = -\Delta[H^+]/6\Delta t$

I. rate $= \Delta[N_2]/\Delta t = \Delta[H_2O]/2\Delta t = -\Delta[NH_4^+]/\Delta t = -\Delta[NO_2^-]/\Delta t$

J. rate $= \Delta[HPO_3^{2-}]/\Delta t = \Delta[H_2]/\Delta t = -\Delta[OH^-]/\Delta t = -\Delta[H_2PO_4^-]/\Delta t$

K. rate $= \Delta[N_2O]/\Delta t = \Delta[H_2O]/\Delta t = -\Delta[NO]/2\Delta t = -\Delta[H_2]/\Delta t$

2. (1) A. $2.9 \times 10^{-3}\text{ mol L}^{-1}\text{ h}^{-1}$

B. $9.3 \times 10^{-4}\text{ mol L}^{-1}\text{ h}^{-1}$ C. $3.4 \times 10^{-4}\text{ mol L}^{-1}\text{ h}^{-1}$

D. $2.1 \times 10^{-3}\text{ mol L}^{-1}\text{ h}^{-1}$ E. $7.0 \times 10^{-4}\text{ mol L}^{-1}\text{ h}^{-1}$

F. $1.4 \times 10^{-3}\text{ mol L}^{-1}\text{ h}^{-1}$ G. $2.2 \times 10^{-4}\text{ mol L}^{-1}\text{ h}^{-1}$

H. $0.14\text{ mol L}^{-1}\text{ h}^{-1}$ I. $0.091\text{ mol L}^{-1}\text{ r}^{-1}$ J. $0.051\text{ mol L}^{-1}\text{ r}^{-1}$

K. $0.010\text{ mol L}^{-1}\text{ h}^{-1}$ L. $0.031\text{ mol L}^{-1}\text{ h}^{-1}$

M. $0.018\text{ mol L}^{-1}\text{ h}^{-1}$ N. $0.028\text{ mol L}^{-1}\text{ h}^{-1}$

(2) O. 0.44 min, $2.3 \times 10^{-3}\text{ mol L}^{-1}\text{ min}^{-1}$

P. 320 min, $1.1 \times 10^{-3}\text{ mol L}^{-1}\text{ min}^{-1}$

Q. 484 min, $7.6 \times 10^{-4}\text{ mol L}^{-1}\text{ min}^{-1}$

R. 688 min, 5.1×10^{-4} mol L^{-1} min^{-1}
S. 1.6 h, 0.28 mol L^{-1} h^{-1}
T. 21.2 h, 0.044 mol L^{-1} h^{-1}
U. 60.6 h, 0.010 mol L^{-1} h^{-1}
V. 40.7 h, 0.018 mol L^{-1} h^{-1}

3. A. $k[N_2O_5]$, first B. $k[CH_3CHO]^2$, second
C. $k[CO][NO_2]$, second D. $k[NO]^2[H_2]$, third
E. $k[NOCl]^2$, second
F. $k[F_2][ClO_2]$, second G. $k[CH_3Cl][F^-]$, second
H. $k[BrO_3^-][Br^-][H^+]^2$, fourth I. $k[NH_4^+][NO_2^-]$, second
J. $k[H_2PO_2^-]^2[OH^-]^2$, fourth K. $k[NO]^2[H_2]$, third

4. A. 0.294/h, 0.0588 mol L^{-1} h^{-1}, 0.294 mol L^{-1} h^{-1}
B. 9.00×10^{-5} L mol^{-1} s^{-1}, 2.03×10^{-6} mol L^{-1} s^{-1}, 1.02×10^{-5} mol L^{-1} s^{-1}
C. 1.2 L mol^{-1} s^{-1}, 0.30 mol L^{-1} s^{-1}, 0.86 mol L^{-1} s^{-1}
D. 1.35 L^2 mol^{-2} s^{-1}, 7.15×10^{-3} mol L^{-1} s^{-1}, 0.0335 mol L^{-1} s^{-1}
E. 4.0×10^{-8} L mol^{-1} s^{-1}, 1.0×10^{-8} mol L^{-1} s^{-1}, 2.1×10^{-9} mol L^{-1} s^{-1}
F. 2.4 L^2 mol^{-2} s^{-1}, 0.084 mol L^{-1} s^{-1}, 0.077 mol L^{-1} s^{-1}
G. 4.8×10^{11} L mol^{-1} s^{-1}, 0.39 mol L^{-1} s^{-1}, 0.21 mol L^{-1} s^{-1}
H. 2.11 L^3 mol^{-3} s^{-1}, 0.00601 mol L^{-1} s^{-1}, 2.32×10^{-4} mol L^{-1} s^{-1},
I. 3.9×10^{-7} L mol^{-1} s^{-1}, 7.7×10^{-9} mol L^{-1} s^{-1}
J. 2.1×10^{-4} L^3mol^{-2} s^{-1}, 1.9×10^{-5} mol L^{-1}, s^{-1}, 1.2×10^{-6} mol^{-2} s^{-1}
K. 1.9×10^{-5} mol L^{-1} s^{-1}, 1.2×10^{-6} mol L^{-1} s^{-1}

5. A. second 2.0×10^{-3} mol L^{-1} s^{-1} B. first, 0.0207/min
C. first, 0.0902/h D. second, 0.0581 L mol^{-1} min^{-1}
E. second, 3.33×10^{-3} L mol^{-1} min^{-1}
F. first, $5.86\times10^{-3}/min$ G. First, 0.0523/min
H. first, $2.01\times10^{-3}/min$ I. first, 0.0103/min
J. first, $1.974\times10^{-2}/min$

6. A. 2.5×10^2 s B. 33.5 min C. 7.68 h D. 43.0 min
E. 150 s F. 118 min G. 13.3 min H. 345 min
I. 67.3 min J. 35.11 min

7. A. 24.91 kJ/mol, 0.479/s B. 134.01 kJ/mol, $6.0\times10^{-14}/s$
C. 97.10 kJ/mol, $4.9\times10^{-39}/s$ D. 269.36 kJ/mol, 8.09/s
E. 80.96 kJ/mol, 0.797 L $mol^{-1}s^{-1}$

8. (1) A. rate $= k[CO][NO_2]$ B. rate $= k[NO_2]^2[CO]^2$
C. rate $= k[NO]^2$ D. rate $= k[NO_2]^2$ Since C involves an unlikely fast termolecular step, D is more plausible. They both give the same total reaction.
(2) E. rate $= k[Cl_2]^{3/2}[CO]$ F. rate $= k[Cl_2]^{3/2}[CO]$

G. rate $= k[Cl_2][CO]$ H. rate $= k[Cl_2]^{3/2}[CO]$ E, F, and H all sum to the total reaction and give the correct rate law, but H does so in an unlikely way, and E contains Cl_3, a very unlikely intermediate. Thus F seems the most plausible.

(3) I. rate $= k[H_2][Br_2]$ J. rate $= k[Br_2][H_2]^{1/2}$

K. rate $= k[H_2][Br_2]^{1/2}$ if the first step in J and K is disregarded as occurring to only a small extent, all three sum to the overall reaction. Only K gives the correct rate law; it is the most plausible.

(4) L. rate $= k[NO]^2[O_2]$ M. rate $= k[NO]^2[O_2]$ N. rate $= k[NO]^2[O_2]$ All three sum to the overall reaction and give the correct rate law. However, both L and N have unlikely termolecular steps. Thus M is the most plausible.

(5) O. rate $= k[C_4H_9Br][OH^-]/[H_2O]$ P. rate $= k[C_4H_9Br]$ Q. rate $= k[C_4H_9Br][OH^-]$ All sum to the overall reaction, but only Q gives the correct rate law.

(6) R sums to the overall reaction if the first step is disregarded as proceeding only to a minor extent. No termolecular collisions. rate $= k[CH_3Cl][Cl]^{1/2}$

(7) S sums to the overall reaction and contains no termolecular collisions. rate $= k[O_3][NO]$

Quizzes

A: 1. (d) 2. (d) 3. (b) 4. (d) 5. (e) (all change the *rate*) 6. (e) (zero order) 7. (b) 8. (d)
B: 1. (d) 2. (e) 3. (e) (0.012) 4. (b) 5. (a) 6. (e, mol L^{-1} s^{-1}) 7. (b) 8. (e)
C: 1. (b) 2. (b) 3. (a) 4. (d) 5. (b) 6. (d) 7. (c) 8. (d)
D: 1. (c) 2. (c) 3. (e) 4. (b) 5. (d) 6. (a) 7. (d) 8. (e) (changes the *rate*)

Sample Test

1. $-8.314\ \text{J mol}^{-1}\ \text{K}^{-1} \ln[(7.00\times10^{-9})/(1.00\times10^{-4})] =$
$E_a(1/400 - 1/500); E_a = 159.1\ \text{kJ/mol}$

2. Inspection of the data reveals a nearly constant half life of 280 seconds. This suggests a first-order reaction. Check this by calculating several values of k.
$kt = \ln([B]_0/[B]_t)$ $k = \ln(3.116/2.291) \div 120\ \text{s} = 2.56\times10^{-3}\ \text{s}^{-1}$ and
$k = \ln(2.542/1.223) \div 320 = 2.29\times10^{-3}\ \text{s}^{-1}$. We use the latter value to determine the rate.
Rate $= k[B] = 2.29\times10^{-3}\ \text{s}^{-1} \times 0.00504\ \text{M} = 1.15\times10^{-5}\ \text{M s}^{-1}$.

CHAPTER 15

Drill Problems

1. A. $[CH_4] = 0.0300\ \text{M}, [CCl_4] = 0.138\ \text{M}, [CH_2Cl_2] = 0.02392\ \text{M}$
B. $[CH_4] = [CCl_4] = 0.0390\ \text{M}, [CH_2Cl_2] = 0.0121\ \text{M}$
C. $[CH_4] = 0.1830\ \text{M}, [CCl_4] = 0.1004\ \text{M}, [CH_2Cl_2] = 0.0392\ \text{M}$
D. $[CH_4] = [CCl_4] = 0.06063\ \text{M}, [CH_2Cl_2] = 0.01874\ \text{M}$
E. $[PCl_5] = 0.0038\ \text{M}, [PCl_3] = [Cl_2] = 0.0462\ \text{M}$
F. $[PCl_5] = 0.0229\ \text{M}, [PCl_3] = 0.1381\ \text{M}, [Cl_2] = 0.1211\ \text{M}$
G. $[PCl_5] = 0.00100\ \text{M}, [PCl_3] = 0.0340\ \text{M}, [Cl_2] = 0.0170\ \text{M}$
H. $[HCHO] = 0.600\ \text{M}, [H_2] = [CO] = 0.400\ \text{M}$

I. $[HCHO] = 0.750\ M, [H_2] = 0.800\ M, [CO] = 0.250\ M$

J. $[HCHO] = 0.167\ M, [H_2] = 0.133\ M, [CO] = 0.333\ M$

K. $[NO] = 1.00\ M, [Br_2] = 2.50\ M, [NOBr] = 4.00\ M$

L. $[NO] = 0.188\ M, [Br_2] = 2.50\ M, [NOBr] = 0.750\ M$

M. $[NO] = 0.0781\ M, [Br_2] = 1.41\ M, [NOBr] = 0.234\ M$

N. $[H_2] = 0.0573\ M, [N_2] = 0.0185\ M, [NH_3] = 0.698\ M$

O. $[H_2] = 0.0697\ M, [N_2] = 0.0640\ M, [NH_3] = 1.742\ M$

2. (1) A. $[NO]^2[Br_2]/[NOBr]^2$ B. $[H_2O]^2[Cl_2]^2/[HCl]^4[O_2]$

C. $[NO]^2/[N_2][O_2]$ D. $[CH_2Cl_2]^2/[CH_4][CCl_4]$

E. $[PCl_3][Cl_2]/[PCl_5]$ F. $[H_2][CO]/[HCHO]$

G. $[NH_3]^2/[N_2][H_2]^3$ H. $[H_2O]^2[SO_2]^2/[H_2S]^2[O_2]^3$

I. $[NF_3]^2[HF]^6/[NH_3]^2[F_2]^6$

(2) J. 0.0263 M K. 1.42 M L. 0.348 M M. 6.59 M
N. 0.769 M O. 0.397 M P. 0.00155 M Q. 0.0386 M
R. 0.0560 M S. 0.199 M T. 0.775 U. 0.0396 M

3. A. $[HCHO]/[H_2][CO] = 3.75$ B. $[PCl_5]/[PCl_3][Cl_2] = 1.784$

C. $[N_2][H_2]^3/[NH_3]^2 = 7.14\times10^{-6}$ D. $[Br_2][Cl_2]/[BrCl]^2 = 0.159$

E. $[CH_4][CCl_4]/[CH_2Cl_2]^2 = 10.48$

F. $\sqrt{[CH_4][CCl_4]}/[CH_2Cl_2] = 3.24$ G. $[NOBr]/[NO]\sqrt{[Br_2]} = 2.529$

H. $[NH_3]/[N_2]^{1/2}[H_2]^{1/2}[H_2]^{3/2} = 374$ I. $[BrCl]/\sqrt{[Br_2][Cl_2]} = 2.51$

J. $\sqrt{[N_2][O_2]}/[NO] = 3.96\times10^7$

K. $([HCl]^2/[H_2O][Cl_2])\sqrt{[O_2]} = 0.0259$

L. $[H_2O][SO_2]/[H_2S][O_2]^{3/2} = 4.17\times10^{22}$

4. A. $P(PCl_5)/P(PCl_3)P(Cl_2), K_p = 0.0362$

B. $P(BrCl)^2/P(Br_2)P(Cl_2), K_c = 6.30$

C. $P(H_2O)^2P(SO_2)^2/P(H_2S)^2P(O_2)^3, K_p = 2.12\times10^{43}$

D. $P(NH_3)^2/P(N_2)^3P(H_2)^3, K_p = 92.9$

E. $P(NH_3)^2/P(N_2O_4), K_c = 4.62\times10^{-3}$

F. $P(NO)^2P(Br_2)/P(NOBr)^2, K_p = 4.489$

G. $P(CH_3OH)/P(H_2)^2P(CO), K_c = 1.23\times10^7$

H. $P(CH_2Cl_2)^2/P(CH_4)P(CCl_4), K_p = 0.0956$

I. $P(NO)^2P(Cl_2)^2/P(NOCl)^2, K_c = 2.69\times10^{-3}$

J. $P(H_2O)^2P(Cl_2)^2/P(HCl)^4P(O_2), K_p = 56.7$

K. $P(H_2)^2P(S_2)/P(H_2S)^2 = 4.64\times10^{-5}$

L. $P(NO)^2/P(N_2)P(O_2) = 6.37\times10^{-16}$

5. A. $1/[CO_2][NH_3]^2$ B. $[SO_2]/[CO]^2$ C. $[NH_3][HCl]$
D. $[H_2]/[NH_3]$ E. $[SO_2][Cl_2]$ F. $1/[H_2]^2[O_2]$ G. $[N_2O][H_2O]^2$
H. $[NaHS]/[H_2S][NaOH]$ I. $[H_3PO_3][HCl]^3/[PCl_3]$
J. $[AgNO_3]^3[NO]/HNO_3]^4$ K. $[H_2SO_4][NO_2]^6/[HNO_3]^6$
L. $[CO_2]$ M. $[NH_3]^2$

6. (1) A. 0.0954 B. 0.562 C. 0.267 D. 6.4 E. 1.40×10^5
(2) F. 1.16×10^{-3} G. 2.05×10^4 H. 1.95×10^{-5}
I. 1.21×10^{-4}

7. A. $Q_c = 298 < K_c$, equilibrium B. $Q_c = 1.62 < K_c$, right
C. $Q_c = 0.423 > K_c$, left D. $Q_c = 1.189 > K_c$, left E. $Q_c = 0.0956 = K_c$ equilibrium
F. $Q_c = 0.0624 < K_c$, right G. $Q_c = 0.0844 < K_c$, right
H. $Q_c = 0.346 > K_c$, left I. $Q_c = 7.87\times$
$10^4 < K_c$, right J. $Q_c = 7.87 < K_c$, right K. $Q_c = 17.3 > K_c$, left L. L. $Q_c = 1.67 < K$, right M. $Q_p = 0.0889 < K_p$, right N. $Q_p = 0.398 > K_p$, left O. $Q_p = 1.98 < K_p$, right P. $Q_p = 8.01 > K_p$, left Q. $Q_p = 0.168 < K_p$, right R. $Q_p = 0.957 > K_p$, left S. $Q_p = 0.306 > K_p$, left T. $3.18\times10^{-3} < K_p$, right U. $Q_p = 0.744 < K_p$, right V. $Q_p = 2.89 > K_p$, left

8. A. $K_p = 0.112$ B. $K_p = 0.334$ C. $K_p = 91.0$ D. $K_p = 0.556$ E. $K_p = 2.69$ F. $K_p = 8.4\times10^{-3}$ G. $K_p = 0.675$

9. A. R B. L C. R D. R E. U F. U G. L H. R I. R J. L K. R L. L M. R N. R O. U P. L

10. C. $[CH_3ONO] = 0.1203$ M, $[HCl] = 0.2623$ M, $[CH_3OH] = 0.0987$ M, $[NOCl] = 0.0778$ M
D. $[CH_3ONO] = 0.418$ M, $[HCl] = 0.174$ M, $[CH_3OH] = 0.140$ M, $[NOCl] = 0.126$ M
E. is at eqilibrium
F. $[CH_4] = 0.2057$ M, $[CCl_4] = 0.0143$ M, $[CH_2Cl_2] = 0.0168$ M
K. $[PCl_3] = [Cl_2] = 0.201$ M, $[PCl_5] = 0.0702$ M
L. $[PCl_5] = 0.1079$ M, $[PCl_3] = 0.2001$ M, $[Cl_2] = 0.3021$ M
M. $P(N_2O_4) = 0.245$ atm, $P(NO_2) = 0.167$ atm
N. $P(N_2O_4) = 0.185$ atm, $P(NO_2) = 0.144$ atm
O. $P(Br_2) = P(Cl_2) = 0.163$ atm, $P(BrCl) = 0.410$ atm
P. $P(Br_2) = 0.157$ atm, $P(Cl_2) = 0.171$ atm, $P(BrCl) = 0.410$ atm
Q. $P(PCl_5) = 0.089$ atm, $P(PCl_3) = 0.319$ atm, $P(Cl_2) = 0.190$ atm
R. $P(PCl_5) = 0.058$ atm, $P(PCl_3) = 0.290$ atm, $P(Cl_2) = 0.135$ atm
S. $P(H_2) = P(CO_2) = 0.293$ atm, $P(H_2O) = P(CO) = 0.027$ atm
T. $P(H_2) = 0.795$ atm, $P(CO_2) = 0.495$ atm, $P(H_2O) = 0.081$ atm, $P(CO) = 0.040$ atm
U. $P(SO_2) = 2.44$ atm, $P(Cl_2) = 1.10$ atm
V. $P(SO_2) = P(Cl_2) = 1.64$ atm

W. $P(\text{HCHO}) = 0.121\text{ atm},\ P(\text{H}_2) = P(\text{CO}) = 0.179\text{ atm}$

X. $P(\text{HCHO}) = 0.180\text{ atm},\ P(\text{H}_2) = P(\text{CO}) = 0.220\text{ atm}$

11. Since the stoichiometric coefficients are the same on the left- and right-hand side of each equation, there is no change in the position of equilibrium by changing the volume in parts J, K, L, M, R, or S. T. shift left, $[\text{Cl}_2] = 0.220$ M U. shift right, $[\text{PCl}_3] = 0.505$ M.

Quizzes

A: 1. $[\text{NaClO}_2]^2[\text{SO}_2]/[\text{Na}_2\text{SO}_3][\text{HClO}_3]^2$ 2. (e) (1.40) 3. (b) 4. (d) 5. (a) 6. (b) 7. (d)

B: 1. $P(\text{SO}_2)P(\text{Cl}_2)^2/P(\text{O}_2)$ 2. (c) 3. (c) 4. (b) 5. (e) 6. (e) 7. (a)

C: 1. $[\text{Cu(NO}_3)_2]^3[\text{NO}]^2/[\text{HNO}_3]^8$ 2. (d) 3. (b) 4. (c) 5. (e) 6. (e) 7. (d)

D: 1. $P(\text{CO}_2)/P(\text{SO}_2)$ 2. (c) 3. (b) 4. (a) 5. (d) 6. (c) 7. (e)

Sample Test

1. $\text{PCl}_5(g) \rightleftharpoons \text{PCl}_3(g) + \text{Cl}_2(g)$

	2.00 atm	1.00 atm
x	$-x$	$-x$
x	$2.00 - x$	$1.00 - x$

$1.19 = (2.00 - x)(1.00 - x)/x$

$x = (4.19 \pm \sqrt{17.56 - 8.00})/2 = 0.549\text{ atm} = P(\text{PCl}_5)$

2.

Reaction:	$\text{HCHO}(g) \rightleftharpoons$	$\text{H}_2(g) +$	$\text{CO}(g)$
Initial:	0.900	0.000	0.600
Changes:	-0.400	$+0.400$	$+0.400$
1st Eq:	0.500	0.400	1.000
$\text{V} \div 3$	1.50	1.20	3.00
Changes	$+x$	$-x$	$-x$
2nd Eq:	$1.50 + x$	$1.20 - x$	$3.00 - x$

The information in the first three lines are used to establish the value of the equilibrium constant:

$$K_c = \frac{(0.400\text{ mol}/3.00\text{ L})(1.000\text{ mol}/3.00\text{ L})}{1.500\text{ mol}/3.00\text{ L}} = 0.0889$$

Then, the set-up in the last three lines is used to determine $[\text{H}_2]$ once equilibrium is re-established:

$$K_c = \frac{(1.20 - x)(3.00 - x)}{1.50 + x} = 0.0889$$

$3.60 - 4.20x + x^2 = 0.0889x + 0.133$ *or* $x^2 - 4.29 + 3.47 = 0$

$$x = \frac{4.29 \pm \sqrt{18.4 - 13.9}}{2} = 1.08$$

Thus, $[\text{H}_2] = 1.20 - 1.08 = 0.12$ M

3. a. left b. no effect c. right d. right e. right f. left g. right h. right i. no effect j. no effect

CHAPTER 16

Drill Problems

1. **(1) In these equations, the species appear in the order:** acid + base $\rightleftharpoons$ acid + base.

(2) The products are followed by one conjugate pair in parentheses, conjugate base written first.

K. $Cl^- + NH^+ \ (NH_3, NH_4^+)$

L. $H_2O + H_3PO_4 \ (H_2O, H_3O^+)$

M. $C_2H_3O_2^- + H_2S \ (HS^-, H_2S)$

N. $N_3^- + CH_3NH_3^+ \ (N_3^-, NH_3)$

O. $CH_3NH_2 + H_2O \ (OH^-, H_2O)$

P. $NO_2^- + HN_3 \ (NO_2^-, HNO_2)$

2. The base is drawn first in each case.

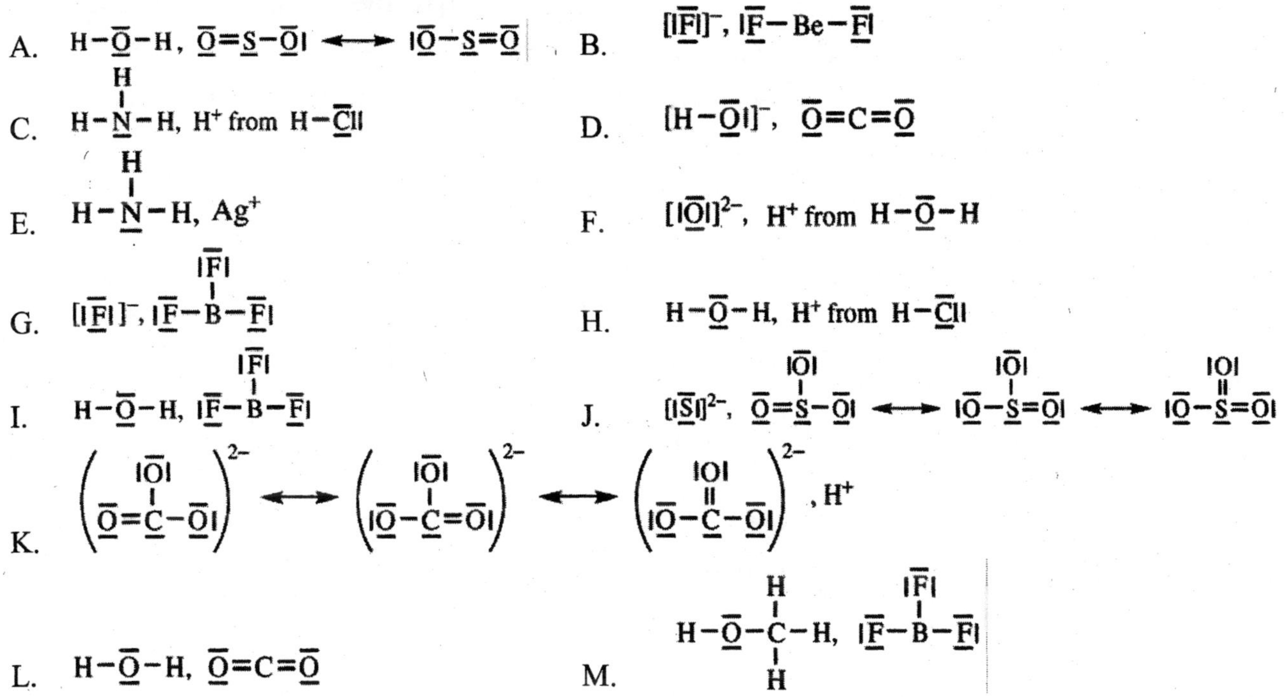

3. A. 1.6 g/L, 2.5×10^{-13} M, 0.040 M, 12.60, 1.40

B. 6.25×10^{-4} M, 0.0613 g/L, 8.00×10^{-12} M, 2.903, 11.097

C. 1.67×10^{-2} M, 1.23 g/L, 3.00×10^{-13} M, 12.522, 1.478

D. 0.0375 M, 3.84 g/L, 2.67×10^{-13} M, 12.574, 1.426

E. 0.0382 M, 0.0382 M, 2.62×10^{-13} M, 1.418, 12.582

F. 2.995×10^{-3} M, 1.67×10^{-12} M, 5.99×10^{-3} M, 11.777, 2.223

G. 0.788 g/L, 0.0125 M, 8.00×10^{-13} M, 1.903, 12.097

H. 0.045 M, 2.5 g/L, 2.2×10^{-13} M, 12.65, 1.35

I. 6.7×10^{-6} M, 8.2×10^{-4} g/L, 7.4×10^{-10} M, 1.3×10^{-5} M, 4.87

J. 2.2×10^{-5} M, 3.4×10^{-3} g/L, 4.5×10^{-10} M, 2.2×10^{-5} M, 9.35

K. 7.1×10^{-4} M, 0.057 g/L, 7.1×10^{-4} M, 1.4×10^{-11} M, 10.85

L. 0.056 M, 7.2 g/L, 0.056 M, 1.8×10^{-13} M, 1.25

4. A. $HOBr + H_2O \rightleftharpoons H_3O^+ + OBr^-$; $K_a = [H_3O][OBr^-]/HOBr]$

B. $(CH_3)_2NH + H_2O \rightleftharpoons (CH_3)_2NH_2^+ + OH^-$; $K_b = [(CH_3)_2NH_2^+][OH^-]/[(CH_3)_2NH]$

C. $HC_6H_7O_2 + H_2O \rightleftharpoons C_6H_7O_2^- + H_3O^+$; $K_a = [C_6H_7O_2^-][H_3O^+]/HC_6H_7O_2]$

D. $N_2H_4 + H_2O \rightleftharpoons N_2H_5^+ + OH^-$; $K_b = [N_2H_5^+][OH^-]/[N_2H_5^+]$

E. $HC_3H_5O_3 + H_2O \rightleftharpoons H_3O^+ + C_3H_5O^-$; $K_a = [H_3O^+][C_3H_5O_3^-]/HC_3H_5O_3]$

F. $(CH_3)_3N + H_2O \rightleftharpoons (CH_3)_3NH^+ + OH^-$; $K_b = [(CH_3)_3NH^+][OH^-]/[(CH_3)_3N]$

G. $HC_3H_5O_2 + H_2O \rightleftharpoons H_3O^+ + C_3H_5O_2^-$; $K_a = [H_3O^+][C_3H_5O_2^-]/[HC_3H_5O_2]$

H. $HC_2HCl_2O_2 + H_2O \rightleftharpoons H_3O^+ + C_2HCl_2O_2^-$; $K_a = [H_3O^+][C_2HCl_2O_2^-]/[HC_2HCl_2O_2]$

I. $H_3BO_3 + H_2O \rightleftharpoons H_3O^+ + H_2BO_3^-$; $K_a = [H_3O^+][H_2BO_3^-]/[H_3BO_3]$

J. $HC_2Cl_3O_2 + H_2O \rightleftharpoons H_3O^+ + C_2Cl_3O_2^-$; $K_a = [H_3O^+][C_2Cl_3O_2^-]/[HC_2Cl_3O_2]$

5. (1) A. 2.10×10^{-9} B. 7.10×10^{-4} C. 8.02×10^{-5} D. 9.92×10^{-7} E. 8.38×10^{-4} F. 7.41×10^{-5} G. 1.40×10^{-5} H. 3.32×10^{-2} I. 5.81×10^{-10} J. 0.210

(2) K. 2.68 L. 2.06 M. 1.92 N. 1.84 O. 1.92 P. 0.98 Q. 11.66 R.12.56 S. 12.67 T. 2.51 U. 9.83 V. 8.90

6. A. $pH = 4.36, [H_2CO_3] = 4.6 \times 10^{-3}$ M, $[HCO_3^-] = 4.4 \times 10^{-5}$ M, $[CO_3^{2-}] = 4.7 \times 10^{-11}$ M

B. $pH = 2.97, [H_2Se] = 8.9 \times 10^{-3}$ M, $[HSe^-] = 1.1 \times 10^{-3}$ M, $[Se^{2-}] = 1.0 \times 10^{-11}$

C. $pH = 0.64, [H_2C_2O_4] = 0.97$ M, $[HC_2O_4^-] = 0.23$ M, $[C_2O_4^{2-}] = 5.2 \times 10^{-5}$ M

D. $pH = 1.20, [H_3PO_4] = 0.686$ M, $[H_2PO_4^-] = 0.064$ M, $[HPO_4^{2-}] = 6.2 \times 10^{-8}, [PO_4^{3-}] = 4.7 \times 10^{-19}$ M

E. $pH = 0.72, [H_3PO_3] = 0.711$ M; $[H_2PO_3^-] = 0.209$ M; $[HPO_3^{2-}] = 2.5 \times 10^{-7}$ M

F. $pH = 0.97, [H_2SO_3] = 0.89$ M, $[HSO_3^-] = 0.11$ M; $[SO_3^{2-}] = 6.3 \times 10^{-8}$ M

7. Only the products of the dissolving reactions are written. All ions are (aq).

A. $Na^+ + Cl^-$; neutral; no hydrolysis

B. $NH_4^+ + Cl^-$; acidic; $NH_4^+ H_2O \rightleftharpoons NH_3(aq) + H_3O^+$

C. $N_2H_5^+ + Br^-$; acidic; $N_2H_5^+ + H_2O \rightleftharpoons N_2H_4(aq) + H_3O^+$

D. $K^+ + NO_3^-$; neutral; no hydrolysis

E. $C_2H_5NH_3^+ + H_2O$; acidic; $C_2H_5NH_3^+ + H_2O \rightleftharpoons C_2H_5NH_2(aq) + H_3O^+$

F. $2\ Na^+ + CO_3^{2-}$; alkaline; $CO_3^{2-} + H_2O \rightleftharpoons HCO_3^- + OH^-$

G. $C_6H_5NH_3^+ + I^-$; acidic; $C_6H_5NH_3^+ + H_2O \rightleftharpoons C_6H_5NH_2(aq) + OH^-$

H. $Ca^{2+} + 2\ ClO^-$; alkaline; $ClO^- + H_2O \rightleftharpoons HClO(aq) + OH^-$

I. $Rb^+ + ClO_2^-$; alkaline; $ClO_2^- + H_2O \rightleftharpoons HClO_2(aq) + OH^-$

J. $2\ HONO_3^+ + SO_4^{2-}$; acidic; $HONO_3^+ + H_2O \rightleftharpoons HONH_2(aq) + H_3O^+$

K. $Sr^{2+} + 2\ F^-$; alkaline; $F^- + H_2O \rightleftharpoons HF(aq) + OH^-$

L. $Cs^+ + C_2H_3O_2^-$; alkaline; $C_2H_3O_2^- + H_2O \rightleftharpoons HC_2H_3O_2(aq) + OH^-$

M. $NH_4^+ + CN^-$; basic; $NH_4^+ + H_2O \rightleftharpoons NH_3(aq) + H_3O^+$; $CN^- + H_2O \rightleftharpoons HCN(aq) + OH^-$

N. $HONH_3^+ + F^-$; acidic;

$$HONH_3^+ + H_2O \rightleftharpoons HONH_2(aq) + H_3O^+; \quad F^- + H_2O \rightleftharpoons HF(aq) + OH^-$$

O. $CH_3NH_3^+ + ClO_2^-$; acidic;

$$CH_3NH_3^+ + H_2O \rightleftharpoons CH_3NH_2(aq) + H_3O^+; \quad ClO_2^- + H_2O \rightleftharpoons HClO_2(aq) + OH^-$$

8. (1) A. 7.00 B. 4.62 C. 4.00 D. 7.00 E. 5.32 F. 12.16 G. 2.32 H. 10.92 I. 7.96 J. 2.76 K. 8.74 L. 9.37 (2) M. 9.03 N. 8.75 O. 8.87 P. 8.40 Q. 8.06 R. 8.00 S. 4.76 T. 5.48 U. 5.50 V. 4.00 W. 3.37 X. 5.06

9. Acids: $HCl > H_3O^+ > HSO_4^- > H_3PO_4 > HNO_2 > HN_3 > HC_2H_3O_2 > HCO_3^-$ $> H_2S > HCN > NH_4^+ > CH_3NH_3^+ > H_2O - HS^-$

Bases: $S^{2-} > OH^- > CH_3NH_2 > NH_3 > CN^- > HS^- > HCO_3^- > C_2H_3O_2^- > N_3^- >$ $NO_2^- > H_2PO_4^- > SO_4^{2-} > H_2O > Cl^-$

The products of each reaction follow, acids first. A. $HN_3 + OH^-$, left B. $H_2O + NO_2^-$, right C. $HCN + H_2O$, right D. $HCN + NH_2^-$, left E. $H_3PO_4 + CN^-$, left F. $NH_3 + H_2PO_4^-$, left G. $H_2O + H_2PO_4^-$, right H. $NH_3 + H_2O$, right I. $HC_2H_3O_2 + CO_3^{2-}$, left J. $H_2S + SO_4^{2-}$, right *or* $H_2SO_4 + S^{2-}$, left K. $H_2O + HS^-$, right L. $HC_2H_3O_3 + OH^-$, left M. $HSO_4^- + CO_3^{2-}$, left N. $H_2O + SO_4^{2-}$, right

10. (1) A. basic B. acidic C. amphoteric D. acidic E. acidic F. acidic G. amphoteric H. acidic I. amphoteric J. basic K. amphoteric L. basic M. basic N. acidic O. basic
(2) P. $CaO + SiO_2 \longrightarrow CaSiO_3$ Q. $Al_2O_3 + 3\,SO_2 \longrightarrow Al_2(SO_3)_2$ R. no reaction S. no reaction T. $BaO + SO_2 \longrightarrow BaSO_3$ U. $3\,PbO + P_2O_5 \longrightarrow Pb_3(PO_4)_2$ V. $3\,Na_2O + Al_2O_3 + \longrightarrow 2\,Na_3AlO_3$

11. A. H_2CO_3 more electronegative central atom B. H_2SO_4 higher oxidation state of sulfur C. H_3PO_4 more electronegative central atom D. $HClO_4$ higher oxidation state of chlorine E. HClO more electronegative central atom F. HNO_3 higher oxidation state of nitrogen G. H_2SO_4 more electronegative central atom H. CF_3OH greater inductive effect by F than by H I. FC_6H_4OH greater inductive effect by F than by Cl J. $HBrO_3$ more double-bonded oxygens on central atom K. $HClO_3$ more double-bonded oxygens on central atom L. H_2CO_3 more electronegative central atom

Quizzes

A: 1. (c) 2. (d) 3. (b) 4. (e) 5. (d) 6. (c) 7. (d) 8. (c) 9. (a)
B: 1. (b) 2. (a) 3. (a) 4. (c) 5. (d) 6. (d) 7. (c) 8. (b) 9. (b)
C: 1. (d) 2. (e) 3. (e) 4. (a) 5. (c) 6. (a) 7. (c) 8. (e) 9. (e)
D: 1. (b) 2. (a) 3. (c) 4. (b) 5. (a) 6. (d) 7. (d) 8. (d) 9. (e)

Sample Test

1. $[HA] = \frac{3.050\text{ g}}{0.250\text{ L}} \times \frac{\text{mol}}{122.0\text{ g}} = 0.100\text{ M}$

or $[H_3O^+] = 2.5 \times 10^{-3}$ M and pH = 2.60

2. a. Brønsted-Lowry; acids: HF, $N_2H_5^+$; bases: N_2H_4, F^- b. Brønsted-Lowry; acids: H_2O, OH^-; bases O^{2-}, OH^- c. Lewis; acid: SOI_2; base: $BaSO_3$ d. Lewis; acid: $HgCl_3^-$; base: Cl^-

3. a. $H_2SO_3 > HF > N_2H_5^+ > CH_3NH_3^+ > H_2O$

b. $OH^- > CH_3NH_2 > N_2H_4 > F^- > HSO_3^-$ (c) (i) to the right; (ii) to the left

4. $[ClO_2] = 2 \times 3.00$ mol/2.50 L $= 2.40$ M;

$K_b = 1.00 \times 10^{-14} / 1.1 \times 10^{-2} = 9.1 \times 10^{-13}$;

$[OH^-] = 1.5 \times 10^{-6}$ M *or* pH = 8.17

CHAPTER 17

Drill Problems

1. A. 2.68, 4.44 B. 2.68, 0.85 C. 2.06, 3.54 D. 1.92, 0.62 E. 1.78, 2.67 F. 1.98, 3.48 G. 0.98, 1.20 H. 11.66, 13.51 I. 12.01, 10.92 J. 11.89, 13.33 K. 11.77, 9.62 L. 11.47, 8.76 M. 11.84, 12.31 N. 12.22, 10.16

2. A. pH = 4.03

B. $[HC_7H_5O_2] = 0.660$ M $[C_7H_5O_2^-] = 0.640$ M; pH = 4.19

C. $[HCHO_2] = 0.105$ M, $[CHO_2^-] = 0.710$ M, pH = 4.57

D. pH = 9.72

E. $[C_2H_5NH_2] = 1.29$ M, $[C_2H_5NH_3^+] = 1.02$, pH = 10.74

F. $[CH_3NH_2] = 0.114$ M, $[CH_3NH_3^+] = 0.441$ M, pH = 10.03

G. pH = 3.08

H. $[HC_2H_2ClO_2] = 0.79$ M, $[C_2H_2ClO_2^-] = 2.14$ M, pH = 3.30

I. $[HNO_2] = 1.302$ M, $[NO_2^-] = 0.550$ M, pH = 2.76

J. pH = 10.97

K. $[(CH_3)_2N] = 0.922$ M, $[(CH_3)_3NH^+] = 0.314$ M, pH = 10.27

L. $[NH_3] = 0.0731$ M, $[NH_4^+] = 0.1549$ M, pH = 8.92

M. $[CHO_2^-] = 0.208$ M, pH = 2.77

N. $[HC_2H_3O_2] = 0.0337$ M, $[C_2H_3O_2^-] = 0.0224$ M, pH = 4.58

O. $[HC_7H_5O_2] = 0.512$ M, $[C_7H_5O_2^-] = 0.231$ M, pH = 3.86

P. $[(CH_3)_3N] = 0.218$ M, $[(CH_3)_3NH^+] = 0.613$ M, pH = 9.35

Q. pH = 10.77

R. $[C_2H_5NH_2] = 0.148$ M, $[C_2H_5NH_3^+] = 1.001$ M, pH = 9.80

3. The pH after acid addition is given first, followed by the pH after addition of base. A. 3.88, 4.15 B. 4.17, 4.21 C. 4.54, 4.61 D. 9.52, 9.96 E. 10.53, 10.95 F. 9.99, 10.20 G. 2.91, 3.23 H. 3.22, 3.39 I. 2.63, 2.92 J. 10.92, 11.02, K. 9.95, 10.59 L. 8.781, 9.035 M. 2.67, 2.86 N. 4.52, 4.64 O. 3.60, 4.01 P. 9.23, 9.46 Q. 10.23, 11.66 R. 9.27, 10.07

4. Answers are given in the order: buffer range, acid capacity (mol H^+), base capacity (mol OH^-). A. 3.76–5.76, 0.0852, 0.927 B. 3.20–5.20, 0.417, 0.433 C. 2.75–4.75, 0.636, 0.0309 D. 8.24–10.24, 0.0798, 0.0193 E. 9.63–11.63, 2.20, 1.65 F. 9.70–11.70, 0.376, 2.31 G. 2.17–4.17, 0.999, 1.30 H. 1.87–3.87, 21.4, 5.97 I. 2.29–4.29, 3.32, 9.87 J. 9.70–11.70, 0.519, 0.189 K. 8.87–10.87, 0.721, 0.179 L. 8.24–10.24, 3.98, 106 M. 2.75–4.75, 0.00683, 0.982 N. 3.76–5.76, 1.78, 2.95 O. 3.20–5.20. 0.248, 0.676 P. 8.87–10.87, 0.534, 2.02 Q. 9.62–11.62, 11.3, 7.29 R. 9.63–11.63, 0.0435, 0.897

5. Answers are given in the order: pH at 0%, 10%, 50%, 90%, 100%, and 110% titration.
A. 0.61, 0.71, 1.12, 1.93, 7.00, 12.02
B. 13.40, 13.26, 12.74, 11.89, 7.00, 2.18
C. 14.40, 14.26, 13.74, 12.89, 7.00, 1.18
D. 0.92, 1.02, 1.43, 2.24, 7.00, 11.71
E. 0.60, 0.74, 1.25, 2.11, 7.00, 11.82
F. 13.80, 13.63, 13.08, 12.21, 7.00, 1.85
G. 12.30, 12.24, 11.85, 11.07, 7.00, 2.97
H. 1.89, 2.18, 3.14, 4.09, 8.09, 12.02
I. 12.00, 11.57, 10.62, 9.67, 6.47, 2.18
J. 12.52, 11.58, 10.63, 9.68, 6.13, 1.18
K. 2.06, 2.22, 3.17, 4.12, 7.46, 11.71
L. 2.17, 2.80, 3.75, 4.70, 8.30, 11.82
M. 12.20, 11.57, 10.62, 9.67, 8.28, 1.85
N. 8.74, 6.13, 5.18, 4.23, 3.57, 2.97

6. Suitable indicators: A. though G. phenol red for all seven titrations. H. phenophthalien I. methyl red J. methyl red K. phenophthalein L. phenophthalein M. methyl red N. indistinct equivalence point; no indicator possible.

7. A. 8.32 B. 9.00 C. 4.06 D. 2.77 E. 4.60 F. 2.76 G. 9.24 H. 4.18 I. 9.2 J. 10.4 K. 10.7 L. 4.52

Quizzes

A: 1. (b) 2. (e) (2.0) 3. (d) 4. (e) 5. (b) 6. (e) (pH = pK_a) 7. (d)

B: 1. (a) 2. (c) 3. (e) 4. (a) 5. (c) 6. (b) 7. (e) (no OH^- present)

C: 1. (c) 2. (c) 3. (d) 4. (a) 5. (e) [(c) = endpoint] 6. (b) 7. (b)

D: 1. (b) 2. (b) 3. (e) 4. (d) 5. (d) 6. (a) 7. (e) (pOH = pK_b)

Sample Test

1. $pK_a = 4.602$ is $K_a = 2.50 \times 10^{-5}$; pH = 2.716 is $[H^+] =$ 1.92×10^{-3} M. $(1.92 \times 10^{-3})^2/[HA] = 2.50 \times 10^{-5}$ *or* [HA] = 0.148 M. Thus $[HA]_i = [HA] + [H^+] = 0.150$ M.
0.500 L × 0.150 M = 0.0750 mol HA = 7.500 g *and* m = 7.500 g/0.0750 mol = 100. g/mol
2. a. weak base
b. 50 mL HCl × L/1000 mL × 0.100 mol/L = 5.0×10^{-3} mol HCl m = 0.400 g/5.0×10^{-3} mol = 80. g/mol
c. pH = 8.0 at midpoint *or* pOH = 6.0 = pK_b. Thus $K_b = 1 \times 10^{-6}$
d. pH = 5 to pH = 3, methyl orange

CHAPTER 18

Drill Problems

1. A. $AgCN(s) \rightleftharpoons Ag+(aq) + CN^-(aq); [Ag^+][CN^-]$

B. $SrF_2(s) \rightleftharpoons Sr^{2+}(aq) + 2\,F^-(aq); K_{sp} = [Sr^{2+}][F^-]^2$

C. $TlBr(s) \rightleftharpoons Tl^+(aq) + Br^-(aq); K_{sp} = [Tl^+][Br^-]$

D. $Zn_3(AsO_4)_2(s) \rightleftharpoons 3\,Zn^{2+}(aq) + 2\,AsO_4^{3-}(aq); K_{sp} = [Zn^{2+}]^3[AsO_4^{3-}]^2$

E. $Hg_2Br_2(s) \rightleftharpoons Hg_2^{2+}(aq) + 2\,Br^-(aq); K_{sp} = [Hg_2^{2+}][Br^-]^2$

F. $Ag_3AsO_4(s) \rightleftharpoons 3\,Ag^+(aq) + AsO_4^{3-}(aq); K_{sp} = [Ag^+]^3[AsO_4^{3-}]$

G. $Li_2CO_3(s) \rightleftharpoons 2\,Li^+(aq) + CO_3^{2-}(aq); K_{sp} = [Li^+]^2[CO_3^{2-}]$

H. $PbC_2O_4(s) \rightleftharpoons Pb^{2+}(aq) + C_2O_4^{2-}(aq); K_{sp} = [Pb^{2+}][C_2O_4^{2-}]$

I. $La(IO_3)_3(s) \rightleftharpoons La^{3+}(aq) + 3\,IO_3^-(aq); K_{sp} = [La^{3+}][IO_3^-]^3$

J. $Ce_2(C_2O_4)_3(s) \rightleftharpoons 2\,Ce^{3+}(aq) + 3\,C_2O_4^{2-}(aq); K_{sp} = [Ce^{3+}]^2[C_2O_4^{2-}]^3$

K. $CdSO_3(s) \rightleftharpoons Cd^{2+}(aq) + SO_3^{2-}(aq); K_{sp} = [Cd^{2+}][SO_3^{2-}]$

L. $AlPO_4(s) \rightleftharpoons Al^3(aq) + PO_4^{3-}(aq); K_{sp} = [Al^{3+}][PO_4^{3-}]$

M. $Al_2S_3(s) \rightleftharpoons 2\,Al^{3+}(aq) + 3\,S^{2-}(aq); K_{sp} = [Al^{3+}]^2[S^{2-}]^3$

N. $Ca(IO_3)_2(s) \rightleftharpoons Ca^{2+}(aq) + 2\,IO_3^-(aq); K_{sp} = [Ca^{2+}][IO_3^-]^2$

O. $CoS(s) \rightleftharpoons Co^{2+}(aq) + S^{2-}(aq); K_{sp} = [Co^{2+}][S^{2-}]$

P. $FeAsO_4(s) \rightleftharpoons Fe^{3+}(aq) + AsO_4^{3-}(aq); K_{sp} = [Fe^{3+}][AsO_4^{3-}]$

2. A. 1.16×10^{-12}, 1.08×10^{-6} M, 1.79×10^{-5} g

B. 2.2×10^{-5}, 4.7×10^{-3} M, 0.096 g.

C. 0.031, 0.18 M, 0.18 M

D. 4.5×10^{-5} M, 4.5×10^{-5} M, 2.2×10^{-3} g

E. 6.2×10^{-19}, 7.9×10^{-10} M, 9.6×10^{-9} g

F. 3.4×10^{-6}, 9.5×10^{-3} M, 0.37 g

G. 1.7×10^{-16}, 3.5×10^{-6} M, 7.0×10^{-6} M

H. 5.6×10^{-3} M, 1.1×10^{-2} M, 0.22 g

I. 2.5×10^{-48}, 8.5×10^{-17} M, 1.4×10^{-15} g

J. 1.0×10^{-12}, 1.3×10^{-4} M, 3.3×10^{-3} g

K. 3.4×10^{-5}, 4.1×10^{-2} M, 2.0×10^{-2} M

L. 1.3×10^{-5} M, 3.9×10^{-5} M, 7.8×10^{-4} g

M. 1.3×10^{-16}, 4.8×10^{5} M, 2.0×10^{-3} g

N. 6.5×10^{-12}, 6.9×10^{-4} M, 0.046 g

O. 1.9×10^{-19}, 2.1×10^{-4} M, 1.4×10^{-4} M
P. 0.036 M, 0.054 M, 0.27 g
3. A. 2.6×10^{-6} B. 1.2×10^{-11} C. 1.3×10^{-6} M
D. 2.1×10^{-4} M E. 6.5×10^{-7} M F. 1.2×10^{-9} M
G. 1.4×10^{-12} M H. 6.1×10^{-10} M I. 2.2×10^{-3} M
J. 5.1×10^{-9} M
4. (1) A. 2.5×10^{-3} M B. 4.2×10^{-7} M C. 7.9×10^{-8} M
D. 1.4×10^{-3} M E. 1.8×10^{-6} M F. 0.033 M G. 0.56 M
H. 1.1×10^{-3} M I. 4.2×10^{-7} M J. 0.085 M
(2) Values of Q are given, followed by *yes* if precipitation should occur. K. 5.12×10^{-13}, yes L. 3.3×10^{-8}, yes M. 1.10×10^{-12}, yes N. 1.1×10^{-6}, yes O. 2.5×10^{-11}, no P. 1.5×10^{-8}, yes Q. 1.3×10^{-8}, yes R. 2.9×10^{-10}, no S. 1.7×10^{-2}, yes T. 1.7×10^{-25}, no
5. The concentrations of the two ions are followed by the formula(s) of the ion(s) for which precipitation is complete (or by neither). (1) K. $[Co^{2+}] = 3.2 \times 10^{-5}$ M, $[CO_3^{2-}] = 4.4 \times 10^{-9}$ M, CO_3^{2-} L. $[Pb^{2+}] = 4.6 \times 10^{-3}$ M, $[SO_4^{2-}] = 3.5 \times 10^{-6}$ M, neither M. $[Ag^+] = 1.2 \times 10^{-5}$ M, $[I^-] = 7.1 \times 10^{-12}$ M, I^-
N. $[Ba^{2+}] = 6.4 \times 10^{-2}$ M, $[SO_4^{2-}] = 1.7 \times 10^{-9}$ M, SO_4^{2-} O. no ppt.
P. $[Pb^{2+}] = 1.7 \times 10^{-10}$ M, $[CrO_4^{2-}] = 1.7 \times 10^{-3}$ M, Pb^{2+} Q. $[Ag^+] = 0.017$ M, $[CrO_4^{2-}] = 3.8 \times 10^{-9}$ M, CrO_4^{2-} R. no ppt, S. $[Li^+] = 1.6$ M, $[PO_4^{3-}] = 7.8 \times 10^{-10}$ M, PO_4^{3-}
T. $[Mg^{2+}] = 1.3 \times 10^{-6}$ M, $[PO_4^{3-}] = 6.0 \times 10^{-6}$ M, neither.
(2) A. $[Zn^{2+}] = 1.1 \times 10^{-3}$ M, $[CO_3^{2-}] = 9.1 \times 10^{-18}$ M, neither
B. $[Hg_2^{2+}] = 3.4 \times 10^{-4}$ M, $[Cl^-] = 6.2 \times 10^{-8}$ M, neither
C. $[Al^{3+}] = 6.6 \times 10^{-23}$ M, $[OH^-] = 2.7 \times 10^{-4}$ M, Al^{3+}
D. $[Pb^{2+}] = 1.0$ M, $[Cl^-] = 4.0 \times 10^{-3}$ M,, neither
E. $[Ag^+] = 1.5 \times 10^{-10}$ M, $[Br^-] = 3.3 \times 10^{-3}$ M, Ag^+
F. $[Ca^{2+}] = 4.6 \times 10^{-8}$ M, $[PO_4^{3-}] = 4.6 \times 10^{4}$ M, neither
6. The formula of the precipitating ion is given first, followed by its concentrations needed: to just cause precipitation, and for complete precipitation of each ion; "yes" means that the two ions can be separated by fractional precipitation.

A. Ag^+ : Cl^- 1.3×10^{-4} M; Br^- 5.6×10^{-10} M, 5.6×10^{-7} M; yes

B. SO_4^{2-}; Ba^{2+} 6.9×10^{-6} M, 6.9×10^{-3} M; Pb^{2+} 1.2×10^{-3} M, 1.2 M; no (barely)

C. Cl^- : Pb^{2+} 0.20 M, 6.4 M; Hg_2^{2+} 5.8×10^{-8} M, 1.8×10^{-6} M; yes

D. SO_4^{2-} : Sr^{2+} 20 M, 2.0×10^{4} M; Pb^{2+} 4.7×10^{-4} M, 0.47 M; yes

E. Mg^{2+} : OH^- 2.3×10^{-5} M, 23 M; F^- 2.2×10^{-6} M, 2.2 M; no

F. CrO_4^{2-} : Ag^+ 5.2×10^{-8} M 0.052 M; Pb^{2+} 1.2×10^{-11} M, 1.2×10^{-8} M; no

G. OH^-; Zn^{2+} 0.012 M, 0.39 M; Sn^{2+} 6.9×10^{-9} M, 2.2×10^{-7} M; yes

H. PO_4^{3-}; Li 1.6×10^{-8} M, 11 M, Mg^{2+} 1.6×10^{-8} M, 5.0×10^{-4} M; no

7. Concentrations are given at pH = 2, pH = 7, and pH = 10, in that order.

A. 1300 M, 1.3×10^{-12} M, 1.3×10^{21} M

B. 6.3×10^{5} M, 6.3×10^{-10} M, 6.3×10^{-19} M

C. 1.8×10^{13} M, 1.8×10^{3} M, 1.8×10^{-3} M

D. 1.6×10^{-3} M, 1.6×10^{-13} M, 1.6×10^{-19} M

E. 5×10^{21} M, 5×10^{11} M, 5×10^{5} M

F. $0.04\times$ M, 4×10^{17} M, 4×10^{-26} M

G. 5.5×10^{18} M, 5.5×10^{8} M, 550 M

H. 2×10^{4} M, 0.2 M, 0.0002 M

8. A. $Ag^+ + 2\,S_2O_3^{2-} \rightleftharpoons [Ag(S_2O_3)_2]^{3-}$ $S_2O_3^{2-} + H_2O \rightleftharpoons HS_2O_3^- + OH^-$; $Ag^+ + Cl^- \rightleftharpoons AgCl(s)$ decreases precipitate B. $Ag^+ + I^- \rightleftharpoons AgI(s)$; $Hg^{2+} + 4\,I^- \rightleftharpoons [HgI_4]^{2-}$; $AgI(s) + I^- \rightleftharpoons [AgI_2]^-$ decreases precipitate C. $Zn^{2+} + S^{2-} \rightleftharpoons ZnS(s)$; $Zn^{2+} + 4\,CN^- \rightleftharpoons [Zn(CN)_4]^{2-}$; $CN^- + H_2O \rightleftharpoons HCN(aq) + OH^-$ decreases precipitate D. $Fe^{2+} + 2\,OH^- \rightleftharpoons Fe(OH)_2(s)$; $Fe^{2+} + 6\,CN^- \rightleftharpoons [Fe(CN)_6]^{4-}$; $CN^- + H_2O \rightleftharpoons HCN(aq) + OH^-$ decreases precipitate E. $Al^{3+} + 3\,F^- \rightleftharpoons AlF_3(s)$; $Ca^{2+} + 2\,F^- \rightleftharpoons CaF_2(s)$; $AlF_3(s) + 3\,F^- \rightleftharpoons [AlF_6]^{3-}$ $F^- + H_2O \rightleftharpoons HF(aq) + OH^-$ decreases precipitate F. $Cu^+ + Cl^- \rightleftharpoons CuCl(s)$; $Ag^+ + Cl^- \rightleftharpoons AgCl(s)$ $Cu^+ + 2\,CN^- \rightleftharpoons [Cu(CN)_2]^-$ $Ag^+ + 2\,CN^- \rightleftharpoons [Ag(CN)_2]^-$; $CN^- + H_2O \rightleftharpoons HCN(aq) + OH^-$ decreases precipitate G. $Al^{3+} + 3\,OH^- \rightleftharpoons Al(OH)_3(s)$; $Al^{3+} + 6F^-$; $\rightleftharpoons [AlF_6]^{3+}$ $F^- + H_2O \rightleftharpoons HF(aq) + OH^-$ decreases precipitate H. $2Co^{3+} + 3S^{2-}$; $\rightleftharpoons Co_2S_3(s)$ $Co^{3+} + 3\,OH^- \rightleftharpoons Co(OH)_3(s)$; $Co^{3+} + 6\,NH_3 \rightleftharpoons [Co(NH_3)_6]^{3+}$; $S^{2-} + H_2O \rightleftharpoons HS^- + OH^-$; $NH_3 + H_2O \rightleftharpoons NH_4^+ + OH^-$ decreases precipitate I. $Cu^{2+} + 2\,OH^- \rightleftharpoons Cu(OH)_2(s)$; $Cu^{2+} + 4\,NH_3 \rightleftharpoons [Cu(NH_3)_4]^{2+}$; $NH_3 + H_2O \rightleftharpoons NH_4^+ + OH^-$ decreases precipitate

9. Free ion concentration given first, followed by that of complex ion.

A. 6.4×10^{-17} M, 0.002 M

B. 3.4×10^{-6} M, 8.2×10^{-5} M

C. 4.5×10^{-15} M, 0.0067 M

D. 9.7×10^{-22} M, 5.0×10^{-4} M

E. 4.8×10^{-34} M, 5.0×10^{-4} M

F. 1.4 $\times 10^{-40}$, 5.0×10^{-4} M

G. 1.4×10^{-45}, 5.0×10^{-4} M

H. 2.5×10^{-40}, 1.4×10^{-5} M

I. 3.8×10^{-13} M, 1.4×10^{-5} M

J. 3.5×10^{-36} M, 2.4×10^{-6} M

K. 2.0×10^{-21} M, 2.4×10^{-6} M

10. A. $[Ag^+] = 2.5 \times 10^{-11}$ M, $Q = 2.5 \times 10^{-13}$, precipitate forms B. $[Ag^+] = 2.5 \times 10^{-11}$ M, $Q = 2.5 \times 10^{-13}$, no precipitate C. $[Zn^{2+}] = 3.3 \times 10^{-18}$ M, $Q = 3.3 \times 10^{-22}$, precipitate forms D. $[Ag^+] = 1.9 \times 10^{-14}$ M,, $Q = 1.9 \times 10^{-16}$, no precipitate E. $[Cu^{2+}] = 1.9 \times 10^{-20}$ M, $Q = 2.2 \times 10^{-20}$, precipitate forms

11. a. Na^+, Sr^{2+}, Zn^{2+}; b. K^+, Fe^{3+}; c. Ba^{2+}, Ni^{2+}

Quizzes

A: 1. (a) 2. (d) 3. (e) 4. (d) 5. (d) 6. (c) 7. (e) 8. (a)
B: 1. (a) 2. (e) 3. (a) 4. (d) 5. (d) 6. (b) 7. (c) 8. (c)
C: 1. (d) 2. (b) 3. (b) 4. (d) 5. (a) 6. (c) 7. (c) 8. (b)
D: 1. (c) 2. (c) 3. (a) 4. (e) 5. (c) 6. (c) 7. (b) 8. (e)

Sample Test

1. The concentration of PO_4^{3-} removed is

$\Delta[PO_4^{3-}] = 8.06 \times 10^{-4}$ M $- 1.00 \times 10^{-12}$ M $= 8.06 \times 10^{-4}$ M The volume of the pond is $100.$ m $\times$ $200.$ m $\times$ 8.00 m $\times$ 1000 L/m^3 $= 1.60 \times 10^8$ L The $[Ca^{2+}]$ that must be present in solution when precipitation is complete is given by $[Ca^{2+}]^3[PO_4^{3-}]^2 = [Ca^2]^3(1.00 \times 10^{-12})^2 = 1.30 \times 10^{-32}$ *or* $[Ca^{2+}] = 2.35 \times 10^{-3}$ M The $Ca(NO_3)_2$ is needed both (1) to form precipitate and (2) to maintain the needed $[Ca^{2+}]$.

(1) 1.60×10^8 L $\times$ (3 mol Ca^{2+} / 2 mol PO_4^{3-}) $\times 8.06 \times 10^{-4}$ M
$= 1.93 \times 10^5$ mol $Ca(NO_3)_2$

(2) 1.60×10^8 L$\times$ 2.35×10^{-3} M $= 3.76 \times 10^5$ mol $Ca(NO_3)_2$.

Total $= 5.69 \times 10^5$ mol $Ca(NO_3)_2$ $\times$ 164.1 g/mol $\times$ 1 Mg/10^6 g =93.4 Mg

2. $[Ag^+]^2(2.0 \times 10^{-3}$ M$) = 2.4 \times 10^{-12}$ *or* $[Ag^+] = 3.5 \times 10^{-5}$ M for CrO_4^{2-} pptn
$[Ag^+](1.0 \times 10^{-5}$ M$) = 1.6 \times 10^{-10}$ *or* $[Ag^+] = 1.6 \times 10^{-5}$ M for Cl^- pptn
AgCl precipitates first. When Ag_2CrO_4 begins to precipitate,
$(3.5 \times 10^{-5}$ M$)[Cl^-] = 1.6 \times 10^{-10}$ *or* $[Cl^-] = 4.6 \times 10^{-6}$ M, 46% of its initial value.

3. Initially: $[Ag^+] = 0.0700$ M, $[SO_4^{2-}] = 0.0600$ M, $[S_2O_3^{2-}] = 0.200$ M

$$\frac{[[Ag(S_2O_3)_2]^{3-}]}{[Ag^+][S_2O_2^{2-}]^2} = \frac{0.0700 - x}{x(0.0600 + 2x)^2} = 1.7 \times 10^{13} \; or \; x = 1.1 \times 10^{-12}\,\text{M} = [Ag^+]$$

$Q = [Ag^+]^2[SO_4^{2-}] = 7.8 \times 10^{-26} < 1.4 \times 10^{-5} = K_{sp}$
It all dissolves

CHAPTER 19

Drill Problems

1. In order, the values are *q, w*, and ΔE. A. −196 J, −123 J, −319 J B. −189 J, +247 J, +58 J C. +647 J, −92.4 J, +555 J D. +1362 J, −312 J, +1041 J E. −142 J, +254 J, +112 J F. +802 J, +897 J, +1699 J G. −1432 J, +937 J, −495 J H. −972 J, −432 J, −1404 J

2. ΔE is given first, followed by ΔH (all in kJ/mol). A. −2667, −2672 B. −3061, −3063 C. −96.91, −91.83 D. −547.0, −563.5 E. −72.47, −72.47 F. −8981.2, −9015.9 G. −332.2, −338.1 H. +177.20, +184.73

3. + for $\Delta S > 0$, − for $\Delta S < 0$, u for unclear. A. − B. + C. + D. + E. + F. u G. − H. − I. u. J. − K. − L. - M. u N. + O. u P. − Q. − R. + S. − T. + U. − V. u W. u X. +

4. In order, the numbers are $\Delta G°$ (in kJ/mol), $\Delta H°$ (in kJ/mol), and $\Delta S°$ (in J mol^{-1} K^{-1}) A. –90.2, –176.8, –284.5, S, (2) B. +176.5, +225.4, +160.8, N (3) C. +18.1, +124.0, +354.8, N (2) D. +4.76, +57.20, +175.8, N (3) E. +130.6, +178., +160.5, N (3) F. –216, –225, –29.3, S (2) G. –37.2. –87.9, –170.2, S (2) H. –141.8, –197.8, –187.8, S (2) I. –5.10, –1.63, +12.0, S (1) J. –1007.5, –1123.9, –390.4, S (2) K. –32.96, –92.92, –198.7, S(2) L. –24.8, –90.2, –219.1, S (2) M. –20.43, –7.2, –44.3, S (4) N. +40.98, +77.08, +121.2, N (3) O. +173.1, +180.5, +24.9, N (3) P. –7.1, –133.6, –423.7, S (2) Q. –25.8, –75.8, –167.4, S (2) R. +19.8, +67.2, +159.3, N (3) S. +8.4, –20.3, –96.0, N (2) T. –326.4, –285.4, +137.4, S (1) U. –76.0, –114.4, –128.8, S (2) V. +28.6, +41.2, +42.1, S (3) W. +15.92, +9.48, –21.8, N (4) X. –34.7, –1.9, +109.5, S (3)

5. Value for $\Delta G°$ (in kJ/mol) given first, followed by K_{eq}. 200 K values given first, followed by 400 K values, separated by a semicolon.

A. $-119.90, 2.1 \times 10^{31}; -63.00, 1.7 \times 10^{8}$

B. $+193.24, 3.4 \times 10^{-51}; +161.08, 5.8 \times 10^{-26}$

C. $+53.04, 1.8 \times 10^{-14}; -17.92, 219$

D. $+22.04, 1.2 \times 10^{-6}; -13.12, 51.7$

E. $+145.9, 7.9 \times 10^{-39}; +113.8, 1.0 \times 10^{-15}$

F. $-219.1, 1.7 \times 10^{57}; -213.3, 7.1 \times 10^{27}$

G. $-53.9, 1.2\times10^{14}; -19.82, 387$

H. $-160.2, 7.1\times10^{41}; -122.7, 1.1\times10^{16}$

I. $-4.03, 11.3; -6.43, 6.9$

J. $-1045.8, 10^{273}; -976.7, 2.4\times10^{126}$

K. $-53.18, 7.7\times10^{13}; -13.44, 56.9$

L. $-46.38, 1.3\times10^{12}; -2.56, 2.16$

M. $-16.06, 1.6\times10^{4}; -24.92, 1.8\times10^{3}$

N. $+52.84, 1.6\times10^{-14}; +28.60, 1.8\times10^{-4}$

O. $+175.52, 1.5\times10^{-46}; +170.54, 5.4\times10^{-23}$

P. $-48.86, 5.8\times10^{12}; +35.88, 2.1\times10^{-5}$

Q. $-42.3, 1.1\times10^{11}; -8.84, 14$

R. $+35.34, 5.9\times10^{-10}; -3.48,\ 0.35$

S. $-1.10, 1.94; +18.1, 4.3\times10^{-3}$

T. $-132.9, 5.2\times10^{81}; -340.4, 2.8\times10^{44}$

U. $-88.64, 1.4\times10^{23}; -340.4, 2.8\times10^{44}$

V. $+32.78, 2.8\times10^{-9}; 24.36, 6.6\times10^{-4}$

W. $+13.84, 2.4\times10^{-4}; +18.2, 4.2\times10^{-3}$

X. $-23.80, 1.60\times10^{6}; -45.70, 9.3\times10^{5}$

6. Transitions are all from the first to the second substance. Values given are ΔH°_{tr} (in kJ/mol), ΔS°_{tr} (in J mol^{-1} K^{-1}), and T_{tr} in order unless otherwise indicated.

A. $-67.2, -152, 442$ K

B. $-22, -96, +231$ J mol^{-1} K^{-1} = S° of AsF_3 (l)

C. $+660.3, +183.3,\ 3602$ K

D. $+23.4, +80.9, 289$ K

E. $-24.8, -93.3, 24.8$ kJ/mol = ΔH°_f of Br_2 (g)

F. $-12, -5, 2400$ K

G. $-62.25, -143.9, 432.6$ K

H. 1.55, 0.5, 3100 K

I. 0.50, 1.25, 71.95 J mol^{-1} K^{-1} = S° for HgO (yellow).

7. (1) Expression for K_{eq} is followed by value of K_{eq} at 298 K.

A. $1/P(NH_3)P(HCl) = 1.3\times10^{16}$

B. $P(CO_2) = 8\times10^{-32}$

C. $P(NH_3)P(CO_2) = 8.2\times10^{-3}$

D. $P(NO_2)^2/P(N_2O_2) = 0.113$

E. $P(CO_2) = 1.4\times10^{-45}$

F. $P(CO_2)/P(SO_3) = 6.0\times10^{37}$

G. $P(PCl_5)/P(PCl_3)(Cl_2) = 5.2\times10^{6}$

H. $P(SO_2)^2/P(SO_2)^2 P(O_2) = 3.5\times10^{24}$

I. $P(BrCl)^2/P(Br_2)P(Cl_2) = 7.23$

J. $P(SO_2)^2/P(H_2S)^2 P(O_2)^3 = 7\times10^{176}$

K. $P(NH_3)^2/P(N_2)P(H_2)^3 = 6.8\times10^{5}$

L. $P(CH_3OH)/P(CO)P(H_2)^2 = 2.1\times10^{4}$

M. $P(CH_2Cl_2)^2/P(CH_4)P(CCl_4) = 2.63$

N. $P(NO)^2/P(Cl_2)P(NOCl)^2 = 5.7\times10^{-14}$

O. $P(\mathrm{NO})^2/P(\mathrm{N_2})P(\mathrm{O_2}) = 4.1\times10^{-31}$
P. $1/P(\mathrm{CO_2})P(\mathrm{NH_3})^2 = 1.1\times10^{-9}$
Q. $P(\mathrm{SO_2})/P(\mathrm{CO})^2 = 3.3\times10^{-5}$
R. $P(\mathrm{SO_2})P(\mathrm{Cl_2}) = 3.5\times10^{-3}$
S. $[\mathrm{CO_2}]/P(\mathrm{CO_2}) = 0.037$
T. $P(\mathrm{O_2})^3/P(\mathrm{O_3})^2 = 1.9\times10^{57}$
U. $P(\mathrm{H_2O})^2\,P(\mathrm{Cl_2})^2/P(\mathrm{HCl})^4\,P(\mathrm{O_2}) = 2.2\times10^{13}$
V. $P(\mathrm{H_2O})\,P(\mathrm{CO})/P(\mathrm{H_2})\,P(\mathrm{CO_2}) = 6.0\times10^{-6}$
W. $P(\mathrm{H_2})\,P(\mathrm{I_2})/P(\mathrm{HI})^2 = 1.1\times10^{-3}$
X. $P(\mathrm{H_2})\,P(\mathrm{CO})/P(\mathrm{HCHCO}) = 6.1\times10^{4}$
(2) Values of $\Delta G°$ (in kJ/mol): A. 0.00 B. 0.00 C. 17.1
D. 11.5 E. 28.7 F. 11.5 G. −17.1 H. −28.7
8. A. 216.9 kJ/mol B. 0.147, 355 K C. 3.57×10^4, 385 K
D. −17.32 kJ/mol E. 22.2, 297 K F. 7.81×10^{-72}, 509 K

Quizzes

A: 1. (a) 2. (b) 3. (e) (−4.0) 4. (d) 5. (e) 6. (b) 7. (d) 8. (a)
B: 1. (e) (−80.66) 2. (c) 3. (c) 4. (e) 5. (d) 6. (c) 7. (a) 8. (c)
C: 1. (d) 2. (b) 3. (e) (−1796.9) 4. (a) 5. (a) 6. (a) 7. (b) 8. (b)
D: 1. (c) 2. (a) 3. (d) 4. (a) 5. (d) 6. (c) 7. (b) 8. (a)

Sample Test

1. a. $P(\mathrm{NH_3})^2[\mathrm{Mg(OH)_2}]^3$ b. $P(\mathrm{SO_2Cl_2})/P(\mathrm{SO_2})\,P(\mathrm{Cl_2})$
c. $1/P(\mathrm{CO})\,P(\mathrm{H_2})^2$ d. $1/[\mathrm{H_2SO_4}]$
2. a. $\Delta G° = -RT \ln K_{eq}$
$= -(8.314\ \mathrm{J\ mol^{-1}\ K^{-1}})(298\ \mathrm{K})\ln(1.87\times10^{-7})$
$= 38.38\times10^3$ J/mol. The reaction is non-spontaneous since $\Delta G° > 0$ and $K_{eq} < 1$.

b. $\Delta G° = \Delta H° - T\Delta S° = 38.38\ \mathrm{kJ/mol} = \Delta H° - 298\ \mathrm{K}(0.18192\ \mathrm{kJ\ mol^{-1}\ K^{-1}})$ or $\Delta H° = 92.60$ kJ/mol
c. As temperature decreases, the $-T\Delta S$ factor becomes less negative. Thus $\Delta G°$ becomes more positive, making K_{eq} smaller. Let's check.

$\Delta G° = 92.60\ \mathrm{kJ/mol} - 200\ \mathrm{K}\ (0.18192\ \mathrm{kJ\ mol^{-1}\ K^{-1}})$
$= 56.211\ \mathrm{kJ/mol} = -RT \ln K_{eq}$ or $K_{eq} = 2.08\times10^{-15}$

This is in accord with LeChatelier's principle. Lowering the temperature of an endothermic reaction shifts the equilibrium left.
d. $\Delta G° = \Delta H° - T\Delta S° = 92.60\ \mathrm{kJ/mol} = 320\ \mathrm{K}(0.18192\ \mathrm{kJ\ mol^{-1}\ K^{-1}}) = 34.39$ kJ/mol. We assumed that $\Delta H°$ and $\Delta S°$ remain constant with temperature.

3. CaO(s) is a solid and thus more ordered than any gas. The remaining gases have more and more atoms. There are increasingly many mays in which their molecules can bend and stretch, and thus their molecules are more disordered.

4. $\ln(1.456/14.2) = -(\Delta H°/8.314\ \text{J mol}^{-1}\ \text{K}^{-1})(1/273 - 1/298)$; $\Delta H° = +61.6\ \text{kJ/mol}$

CHAPTER 20

Drill Problems

1. Sketches of the cells are in Figure A-2. The anode is the left half cell of each sketch. Cell diagrams follow.

A. $Pt(s) \mid H_2O_2(aq) \mid H^+(aq) \parallel Cu^{2+}(aq) \mid Cu(s)$

B. $Mg(s) \mid Mg^{2+}(aq) \parallel Ag^+(aq) \mid Ag(s)$

C. $Al(s) \mid Al^{3+}(aq) \parallel H^+(aq) \mid H_2(g) \mid Pt(s)$

D. $Pb(s) \mid Pb^{2+}(aq) \parallel Fe^{3+}(aq) \mid Fe(s)$

E. $Pt(s) \mid I_2(s) \mid I^-(aq) \parallel F^-(aq) \mid F_2(g) \mid Pt(s)$

F. $Pt(s) \mid H_2(g) \mid OH^-(aq) \parallel Cl^-(aq) \mid ClO^-(aq) \mid Pt(s)$

G. $Pt(s) \mid Cl_2(g) \mid Cl^-(aq) \parallel Mn^{2+}(aq) \mid MnO_2(s) \mid Pt$

H. $Pt(s) \mid I_2(s) \mid IO_3^-(aq) \parallel Br^-(aq) \mid Br_2(l) \mid Pt(s)$

I. $Ag(s) \mid Ag^+(aq) \parallel NO_3^-(aq) \mid NO(g) \mid Pt(s)$

J. $Pt(s) \mid S(s) \mid SO_3^{2-}(aq) \parallel Pb^{2+}(aq) \mid Pb(s)$

2. Spontaneous reactions have positive cell voltages.
A. −0.358 V B. +1.556 V C. +1.676 V D. +0.089 V
E. +2.351 V F. +1.718 V G. −0.128 V H. −0.135 V
I. +0.156 V J. −0.574 V K. −1.375 V L. +0.925 V
M. −0.908 V N. −0.279 V O. −0.665 V

3. A. −0.320 V B. +1.494 V C. +1.729 V D. +0.234 V
E. +2.304 V F. +1.494 V G. −0.025 V H. −0.064 V
I. +0.182 V J. −0.601 V

4. In order $\Delta G°, K_{eq}, \Delta G$ for each part. Free energies are in kJ/mol. A. $+69.08, 7.9 \times 10^{-13}, +61.75$
B. $-300.26, 4.1 \times 10^{52}, -288.3$ C. $-970.3, 10^{170}, -1000.9$
D. $-51.52,\ 1.1 \times 10^{9}, -135.5$ E. $-453.7,\ 3.1 \times 10^{79}, -444.6$
F. $-331.5, 1.2 \times 10^{58}, -335.2$ G. $24.70, 4.7 \times 10^{-5}, 4.82$
H. $130.25,\ 1.5 \times 10^{-11},\ 61\ 75$ I. $-45.15, 8.2 \times 10^{7}, -52.68$
J. $-221.5,\ 6.3 \times 10^{38}, -231.9$

5. A. $Cl_2, H_2, -1.330$ V B. Cl_2, Cu, −1.003 V
C. $Cl_2, H_2, -0.971$ V D. $O_2, H_2, -1.229$ V
E. Cl_2, Cu, −1.018 V F. I_2, Fe, −1.069 V

6. A. 0.560 *F* B. 226 g Sn C. 0.318 g Cu D. 49.3 g Cu
E. 1.28×10^{3} g Hg F. 1.01×10^{5} s = 28.0 hr G. 701 g Au

H. 28.0 L O_2 I. 82.1 g Al J. 5.85 g Cr K. 61.1 g Au
L. 40.2 g Ag

Quizzes

A: 1. (e) (5) 2. (c) 3. (d) 4. (d) 5. (b) 6. (b) 7. (b) 8. (c) 9. (d) 10. (a)
B: 1. (c) 2. (c) 3. (e) (9) 4. (b) 5. (a) 6. (d) 7. (a) 8. (e) (Nernst) 9. (c) 10. (c)
C: 1. (d) 2. (a) 3. (b) 4. (d) 5. (e) 6. (b) 7. (d) 8. (a) 9. (e) (1.04 M) 10. (d)
D: 1. (d) 2. (c) 3. (a) 4. (a) 5. (c) 6. (e) 7. (b) 8. (e) (11.7) 9. (e) 10. (a)

Sample Test

1. a. $Al(s) \mid Al^{3+}(aq) \parallel Pb^{2+}(aq) \mid Pb(s)$

b. $3(Pb^{2+} + 2e^- \longrightarrow Pb^\circ) \qquad E^\circ = -0.126\ V$

$2(Al \longrightarrow Al^{3+} + 3e^-) \qquad E^\circ = +1.66\ V$

$3Pb^{2+} + 2Al \longrightarrow 2Al^{3+} + 3Pb \quad E^\circ = +1.53\ V$

c. $E_{cell} = E^\circ_{cell} - \dfrac{0.0592}{n}\log Q = 1.53 - \dfrac{0.0592}{6}\log\dfrac{(0.0200)^2}{(4.00)^3}$

$= 1.53 + 0.05 = 1.58\ V$

2. $6.00 \times 10^{-4}\ F$ used

a. $Cu^{2+} + 2e^- \longrightarrow Cu$ $6.00 \times 10^{-4}\ F \times (\text{mol Cu}/2\ F) \times$
$(1000\ \text{mmol Cu/mol Cu}) = 0.300\ \text{mmol Cu}$

$50.0\ mL \times 0.100\ M = 5.00\ \text{mmol}\ Cu^{2+}$

$(5.00 - 0.30)\ \text{mmol} = 4.70\ \text{mmol}\ Cu^{2+}$ left in solution

$[Cu^{2+}] = 4.70\ \text{mmol}/50.0\ mL = 0.094\ M$

b. $2H_2O \longrightarrow O_2 + 4H^+ + 4e^-$

$6.00 \times 10^{-4}\ F \times (\text{mol}\ H^+/F) \times (1000\ \text{mmol}\ H^+/\text{mol}\ H^+)$

$= 0.600\ \text{mmol}\ H^+$

$[H^+] = 0.600\ \text{mmol}/50.0\ mL = 0.0120\ M$ or $pH = 1.921$

c. $2KMnO_4 + 3H_2S \longrightarrow 2MnO_2 + 3S + 2KOH + 2H_2O$

$3NiS + KClO_3 + 6HCl \longrightarrow 3NiCl_2 + KCl + 3S + 3H_2O$

CHAPTER 21

Drill Problems

1. A. Ba < Ca < Mg < C B. Br < Se < Ca < C C. $Be^{2+} < Li^+ < K^+ < Cs^+$
D. $Cs^+ < Ca^{2+} < Mg^{2+} < Al^{3+}$ E. Cs < Sr < Al < F F. Na < Mg < S < I
G. Cs < Rb < Ca < Al H. Cs < Ca < Mg < Al predicted, Cs < Mg < Al < Ca actual
I. Cs < Rb < Na < Mg
J. $Si^{4+} < Al^{3+} < Mg^{2+} < Li^+$ K. Br < I < Sr < Cs
L. $Al^{3+} < Mg^{2+} < Na^+ < K^+$

M. $Cs < Rb < Na < Li$ predicted, $Cs < Rb < Li < Na$ actual
N. $Cs < Sr < Ca < Mg$ predicted, $Cs < Mg < Sr < Ca$ actual
O. $C < P < S < I$ predicted, $P < S < C < I$ actual

(Notice how often predictions are incorrect because a second period element is "out of place.")
(2) P. 14.7°C (28.7°C) Q. 602°C (690°C) R. $2.1\,g/cm^3$ ($1.90\,g/cm^3$)
S. $4.7\,g/cm^3$ ($4.25\,g/cm^3$)
2. A. $MgCl_2(l) \xrightarrow{\text{electrolysis}} Mg(l) + Cl_2(g)$
B. $2\,NaCl(l) \xrightarrow{\text{electrolysis}} 2\,Na(l) + Cl_2(g)$
C. $2\,Al_2O_3(l) + 3\,C(s) \xrightarrow{\text{electrolysis}} 4\,Al(l) + 3\,CO_2(g)$
D.
$Al_2O_3(s) + 2\,OH^-(aq) + 3\,H_2O \longrightarrow 2\,Al(OH)_4^-(aq)$; $Al(OH)_4^-(aq) + H_3O^+ \longrightarrow Al(OH)_3(s) + 2\,H_2O$
E. $Mg^{2+}(aq) + 2\,OH^-(aq) \longrightarrow Mg(OH)_2(s)$
F. $CaCO_3(s) \xrightarrow{\Delta} CaO(s) + Co_2(g)$
G. from question F plus $CaO(s) + H_2O \longrightarrow Ca(OH)_2(aq)$
3. A. no reaction B. $2\,K(s) + 2\,HCl(aq) \longrightarrow 2\,KCl(aq) + H_2(g)$
C. $Zn(s) + 2\,HCl(aq) \longrightarrow ZnCl_2(aq) + H_2(g)$
D. no reaction
E. $2\,Al(s) + 2\,NaOH(aq) + 6\,H_2O \longrightarrow 2\,NaAl(OH)_4(aq) + 3\,H_2(g)$
F. $Sn(s) + 2\,H^+(aq) + 4\,Cl^-(aq) \longrightarrow H_2(g) + SnCl_4^{2-}(aq)$
G. $Pb(s) + 2\,H_2SO_4(conc.) \longrightarrow PbSO_4(s) + SO_2(g) + 2\,H_2O$
H. $2\,Al(s) + 6\,HCl(aq) \longrightarrow 2\,AlCl_3(aq) + 3\,H_2(g)$
I. $Cd(s) + H_2SO_4(aq) \longrightarrow CdSO_4(aq) + H_2(g)$
J. $6\,Hg(l) + 8\,HNO_3(dil.) \longrightarrow 3\,Hg_2(NO_3)_2(aq) + 2\,NO(g) + 4\,H_2O$
K. $Pb(s) + 4\,HNO_3(conc.) \longrightarrow Pb(NO_3)_2(aq) + 2\,NO_2(aq) + 2\,H_2O$
L. $5\,Be(s) + 12\,HNO_3(conc.) \longrightarrow 5\,Be(NO_3)_2(aq) + N_2(g) + 6\,H_2O$
M. $Be(s) + 2\,NaOH(aq) + 2\,H_2O \longrightarrow Na_2Be(OH)_4(aq) + H_2(g)$
N. no reaction
4. A. $ZnO(s) + 2\,HCl(aq) \longrightarrow ZnCl_2(aq) + H_2O$
B. $SnO_2(s) + 2\,H_2SO_4(aq) \longrightarrow Sn(SO_4)_2(aq) + 2\,H_2O$
C. $ZnO(s) + H_2O + 2\,NaOH \longrightarrow Na_2Zn(OH)_4(aq)$
D. $SnO(s) + 2\,HCl \longrightarrow SnCl_2(aq) + H_2O$
E. $Zn(OH)_2(s) + H_2SO_4(aq) \longrightarrow ZnSO_4(aq) + 2\,H_2O$
F. $Sn(OH)_2(s) + 2\,NaOH(aq) \longrightarrow Na_2Sn(OH)_4(aq)$
G. $Al_2O_3(s) + CaO(s) + 4\,H_2O \longrightarrow Ca[Al(OH)_4]_2$
H. $Sn(OH)_4(s) + 2\,H_2SO_4 \longrightarrow Sn(SO_4)_2(aq) + 4\,H_2O$

I. $Sn(OH)_4(s) + CaO(s) + H_2O \longrightarrow CaSn(OH)_6$

J. $SnO_2(s) + 2\ NaOH(aq) + 2\ H_2O \longrightarrow Na_2Sn(OH)_6$

5. A. $K_2CO_3, Na_2CO_3, PbO, CaCO_3$ B. Li, In, Pb

C. KOH, K_2CO_3, NaOH

D. $KNO_3, Ca(NO_3)_2 \cdot 2\ H_2O, Ca_3(PO_4)_2$

E. $KClO_3, SrCl_2, Sr(NO_3)_2, Ba(NO_3)_2, Ba(ClO_3)_2$

F. Rb, Ba

G. Sn, Pb_3O_4

H. $Na_3AlF_6, CaCO_3$, Zn

I. Rb, Cs J. $PbO_2, PbO_2, KClO_3$ K. MgO, $NaHCO_3$,

$Li_2CO_3, BiONO_3, (BiO)_2CO_3, SnF_2$ L. Bi_2O_3, Alum

M. $BaCrO_4, BaMnO_4, BaU_2O_7, PbCrO_4, BiOCl, SnS_2, Sn(CrO_4)_2, BaSO_4$

N. $Tl_2SO_4, Bi_2O_3, BaSiF_6, BaCO_3$

O. $Al_2(SO_4)_3, MgSO_4, CaCO_3, BaSO_4$

P. $SnCl_2$, Ca

Q. $KNO_3, KClO_3, KMnO_4, Pb(C_2H_3O_2)_4$

R. Li, Pb S. MgO, Al_2O_3, CaO,

T. $K_2CO_3, MgSO_4$, Alum, $SnCl_4$

U. $(BiO)_2CO_3, 2PbCO_3 \cdot Pb(OH)_2$, PbO

Quizzes

A: 1. (e) 2. (e) 3. (d) 4. (c) 5. (c) 6. (c) 7. (c) 8. (d) 9. (a) 10. (a)
B: 1. (d) 2. (e) 3. (c) 4. (d) 5. (b) 6. (a) 7. (b) 8. (c) 9. (a) 10. (a)
C: 1. (d) 2. (e) 3. (d) 4. (c) 5. (a) 6. (b) 7. (c) 8. (b) 9. (a) 10. (a)
D: 1. (a) 2. (a) 3. (a) 4. (b) 5. (e) 6. (e) 7. (e) 8. (d) 9. (e) 10. (c)

Sample Test

1. $2\ C(s) + O_2(g) \longrightarrow 2\ CO(g)$

$2\ FeO(s) \longrightarrow 2\ Fe(s) + O_2(g)$

$2\ C(s) + 2\ FeO(s) \longrightarrow 2\ Fe(s) + 2\ CO(g)$

$\Delta H^\circ_{rxn} = -221 - (-544) = +323\ kJ/mol$

$\Delta S^\circ_{rxn} = +179.5 - (-138.1) = +317.6\ J\ mol^{-1}\ K^{-1}$

The reaction becomes spontaneous where $\Delta G^\circ = 0 = \Delta H^\circ - T\Delta S^\circ$.
We solve for $T = \Delta H^\circ / \Delta S^\circ = (323\ kJ/mol)/(0.3176\ kJ\ mol^{-1}\ K^{-1}) = 1017\ K = 744°C$.

2. Only one method (of perhaps several) is given in each part.

a. $3\ PbS(s) + 8\ HNO_3(aq, dil) \longrightarrow 3\ Pb(NO_3)_2(aq) + 2\ NO(g) + 3\ S(s)$ [filter solution];

$Pb(NO_3)_2(aq) + 2\ HCl(aq) \longrightarrow PbCl_2(s) + 2\ HNO_3(aq)$ [filter solution]

b. $HgS(s) + O_2(g) \xrightarrow{\Delta} Hg(l) + SO_2(g)$;

$6\,Hg(l) + 8\,HNO_3(aq, dil) \longrightarrow 3\,Hg_2(NO_3)_2(aq) + 2\,NO(g) +$
$4\,H_2O; 2\,NaOH(aq) + H_2SO_4(aq) \longrightarrow Na_2SO_4(aq) + 2\,H_2O;$
$Hg_2(NO_3)_2(aq) + Na_2SO_4(aq) \longrightarrow Hg_2SO_4(s) + 2\,NaNO_3(aq)$
c. $Al_2O_3(s) + 3\,H_2O \longrightarrow 2\,Al(OH)_3(aq); CaCO_3(s) \xrightarrow{\Delta}$
$CaO(s) + CO_2(g); CaO(s) + H_2O \longrightarrow Ca(OH)_2(aq);$
$2\,Al(OH)_3(s) + 3\,Ca(OH)_2(aq) \longrightarrow Ca_3(AlO_3)_2(s) + 6\,H_2O$
[filter solutions; all solutions are saturated solutions of sparingly soluble compounds];
d. $Mg^2 + (aq, seawater) + 2\,OH^-$ (aq, from NaOH) $\longrightarrow$
$Mg(OH)_2(s)$ [filter off solid]; $Mg(OH)_2(s) + H_2SO_4(aq) \longrightarrow$
$MgSO_4(aq) + 2\,H_2O$ [evaporate solution to dryness]
e. $SnO_2(s) + 2\,C(s) \longrightarrow Sn(l) + 2\,CO(g)$ [solidify metal];
$Sn(s) + 2\,HCl(aq) \longrightarrow SnCl_2(aq) + H_2(g);$
$SnCl_2(aq) + 2\,NaOH(aq) \longrightarrow 2\,NaCl(aq) + Sn(OH)_2(s)$ [filter off solid];
$Sn(OH)_2(s) + NaOH(aq) \longrightarrow NaSn(OH)_3(aq)$ [evaporate solution to dryness];
$NaSn(OH)_3(s) + NaOH(s) \xrightarrow{\Delta} Na_2SnO_2(s) + 2\,H_2O(g)$
f. $ZnS(s) + 2\,HCl(aq) \longrightarrow ZnCl_2(aq) + H_2S(g)$;
$ZnCl_2(aq) + 2\,NaOH(aq) \longrightarrow Zn(OH)_2(s) + 2\,NaCl(aq)$; $Zn(OH)_2(s) \xrightarrow{\Delta} ZnO(s) + H_2O(g)$
g. $BaCO_3(s) + H_2SO_4(aq) \longrightarrow BaSO_4(s) + H_2O + CO_2(g)$ [filter off solid]
h. $2\,KCl(l) \xrightarrow{\text{electrolysis}} 2\,K(l) + Cl_2(g)$ [solidify metal]; $2\,K(s) + 2\,H_2O \longrightarrow 2\,KOH(aq) + H_2(g)$;
$KOH(aq) + HNO_3(aq) \longrightarrow KNO_3(aq) + H_2O$ [evaporate solution to dryness]

Syntheses

Only one possible sequence of reactions is given, except for the second question in which several are provided to give some indication of the possible diversity of answers. ESTD means "evaporate solution to dryness."

1. $2\,Na(s) + O_2(g) \longrightarrow Na_2O_2(s)$

2. $2\,Na(s) + 2\,H_2O(l) \longrightarrow 2\,NaOH(aq) + H_2(g)$; $NaOH(aq) + HCl(aq) \longrightarrow NaCl(aq) + H_2O(l)$ ESTD *or* $2\,Na(s) + 2\,HCl(aq) \longrightarrow 2\,NaCl(aq) + H_2(g)$, but this would be a very violent reaction. *or* $2\,HCl(aq) \xrightarrow{\text{electricity}} H_2(g) + Cl_2(g)$; gases separate naturally. Cl_2 at anode, H_2 at cathode; then 2
$2\,Na(s) + Cl_2(g) \longrightarrow 2\,NaCl(s)$ (very violent)

3. $2\,Li(s) + 2\,H_2O(l) \longrightarrow 2\,LiOH(s) + H_2(g)$ ESTD carefully;
$2\,LiOH(s) \longrightarrow Li_2O(s) + H_2O(g); Li_2O(s) + CO_2(g) \longrightarrow Li_2CO_3(s)$

4. $Ca(s) + 2\,H_2O(l) \longrightarrow Ca(OH)_2(aq) + H_2(g);$
$Ca(OH)_2(aq) + H_2SO_4(aq) \longrightarrow CaSO_4(s) + 2\,H_2O(l)$
$Ca(OH)_2(aq) + H_2SO_4(aq) \longrightarrow CaSO_4(s) + 2\,H_2O(l)$ filter off solid

5. $2\,Al(s) + 3\,H_2SO_4(aq) \longrightarrow Al_2(SO_4)_3(aq) + 3\,H_2(g)$ ESTD

6. $Mg(s) + H_2O(l) \xrightarrow{\Delta} MgO(s) + H_2(g)$

7. $Li_2CO_3(s) \xrightarrow{\Delta} Li_2O(s) + CO_2(g)$; $Na_2O(s) + CO_2(g) \longrightarrow Na_2CO_3(s)$

8. $CaCO_3(s) \xrightarrow{\Delta} CaO(s) + CO_2(g)$; $CaO(s) + H_2O(l) \longrightarrow Ca(OH)_2(aq)$ ESTD carefully

9. $2\,Ca(s) + O_2(g) \longrightarrow 2\,CaO(s)$ at low temps;

$C(s) + O_2(g) \longrightarrow CO_2(g)$; $CaO(s) + CO_2(g) \longrightarrow CaCO_3(s)$

10. $Ca(s) + 2\,H_2O(l) \longrightarrow Ca(OH)_2(aq) + H_2(g)$

11. $Ca(s) + 2\,H_2O(l) \longrightarrow Ca(OH)_2(aq) + H_2(g)$ ESTD;

$Ca(OH)_2(aq) \xrightarrow{\Delta} CaO(s) + H_2O(g)$; $C(s) + O_2(g) \longrightarrow CO_2(g)$; $CaO(s) + CO_2(g) \longrightarrow CaCO_3(s)$ **12.**

$2\,H_2(g) + O_2(g) \longrightarrow 2\,H_2O(l)$; $2\,Li(s) + 2\,H_2O(l) \longrightarrow 2\,LiOH(aq) + H_2(g)$;

$LiOH(aq) + HNO_3(aq) \longrightarrow LiNO_3(aq) + H_2O(l)$ ESTD

13. $Ca(s) + S(s) \longrightarrow CaS(s)$; $CaS(s) + 2\,HCl(aq) \longrightarrow CaCl_2(aq) + H_2S(g)$

14. $Mg(s) + Cl_2(g) \longrightarrow MgCl_2(s)$;

$3\,MgCl_2(s) + 2\,H_3PO_4(aq) \longrightarrow Mg_3(PO_4)_2(aq) + 6\,HCl(g)$

CHAPTER 22

Drill Problems

1. A. $2\,NaCl(aq) + 2\,H_2O \xrightarrow{\text{electroylsis}} 2\,NaOH(aq) + H_2(g) + Cl_2(g)$ [in a Downs cell]

B. $2\,KBr(aq) + Cl_2(g, \text{from A}) \longrightarrow 2\,KCl(aq) + Br_2(l)$ [separate denser $Br_2(l)$]

C. $2\,KI(s) + Cl_2(g, \text{from A}) \xrightarrow{\Delta} 2\,KCl(s) + I_2(g)$ [deposit $I_2(s)$ on a cool surface]

D. $6\,Cl_2(g, \text{from A}) + 6\,Ba(OH)_2(aq) \xrightarrow{\Delta} 5\,BaCl_2(aq) + Ba(ClO_3)_2(aq) + 6\,H_2O$

[separate salts by fractional crystallization];

$Ba(ClO_3)_2(aq) + H_2SO_4(aq) \longrightarrow BaSO_4(s) + 2\,HClO_3(aq)$ [filter off solid];

$HClO_3(aq) + KOH(aq) \longrightarrow KClO_3(aq) + H_2O$ [carefully evaporate solution to dryness]

E. $NaI(s) + H_3PO_4(aq) \longrightarrow NaH_2PO_4(aq) + HI(g)$

F. first two reactions of question D

G. $2\,HClO_3(aq, \text{from F}) + H_2C_2O_4(s) \longrightarrow 2\,ClO_2(g) + 2\,CO_2(g) + 2\,H_2O$;

$4\,ClO_2(g) + 2\,Ba(OH)_2(aq) \longrightarrow Ba(ClO_2)_2(s) + Ba(ClO_3)_2(aq) + 2\,H_2O$ [filter off solid];

$Ba(ClO_2)_2(s) + H_2SO_4(aq) \longrightarrow BaSO_4(s) + 2\,HClO_2(aq)$ [filter off solid]

H. First two eqations of question G

I. $ClO_3^-(aq, \text{ from F}) + H_2O \xrightarrow{\text{electrolysis}} ClO_4^-(aq) + H_2(g)$ [in acidic solution]

J. $2\,NaF(l) \xrightarrow{\text{electrolysis}} 2\,Na(l) + F_2(g)$ [in a Downs-type cell];

$Cl_2(g, \text{from A}] + 3\,F_2(g) \longrightarrow 2\,ClF_3(g)$ [controlled ratio of reactants];

$KI(aq) + I_2(s, \text{ from C}) \longrightarrow KI_3(aq)$

L. First reaction of question J

M. $NaF(s) + H_2SO_4(aq) \longrightarrow NaHSO_4(aq) + HF(g)$

N. $NaCl(s) + H_2SO_4(aq) \longrightarrow NaHSO_4(aq) + HCl(aq)$

O. $KBr(s) + H_3PO_4(aq) \longrightarrow NaH_2PO_4(aq) + HBr(g)$

P. $Cl_2(g) + 2\ NaOH(aq) \longrightarrow NaCl(aq) + NaOCl(aq) + H_2O$

2. Shape abbreviations are as in the answers for objective 11-12.

A. d^2sp^3, sqpyr B. sp^3, tet C. d^2sp^3, oct D. sp^2, tripl E. dsp^3, T F. dsp^3, lin
G. sp^3, tet H. d^2sp^3, sqpyr I. d^3sp^3, pentagonal bipyramid J. dsp^3, T K. sp^2, bent
L. sp^3, bent M. sp^3, tet N. sp^3, tripyr O. d^2sp^3, sqpyr P. sp^3, tripyr

3. A. $CaCO_3(s) \xrightarrow{\Delta} CaO(s) + CO_2(g)$
B. $ZnO(s) + CO(g) \xrightarrow{\Delta} Zn(g) + CO_2(g)$ C. $2\,NH_3(aq) + H_3PO_4(aq) \longrightarrow (NH_4)_2HPO_4$ D. $Zn(s) + H_2SO_4(aq) \longrightarrow ZnSO_4(aq) + H_2(g)$ E. $2\ Na(s) + H_2(g) \longrightarrow 2\ NaH(s)$ F. $CdO(s) + H_2(g) \xrightarrow{\Delta} Cd(s) + H_2O(g)$ G $N_2(g) + O_2(g) \xrightarrow{\Delta} 2\ NO(g)$ H. $N_2(g) + 3\ H_2(g) \rightleftharpoons 2\ NH_3(g)$ I. $2\ NO(g) + O_2(g) \longrightarrow 2\ NO_2(g)$ J. $CO_2(g) + NH_3(g) \longrightarrow CO(NH_2)_2(s) + H_2O(l)$ K. $NH_3(aq) + HCl(aq) \longrightarrow NH_4Cl(aq)$ L. $4\ NH_3(g) + 5\ O_2(g) \longrightarrow 4\ NO(g) + 6\ H_2O(g)$ M. $Fe_2O_3(s) + H_2(g) \longrightarrow 2\ Fe(l) + 3\ H_2O(g)$ N. $SnO_2(s) + 2\ CO(g) \longrightarrow Sn(l) + 2\ CO_2(g)$ O. $NH_3(aq) + H_3PO_4(aq) \longrightarrow NH_4H_2PO_4(aq)$ P. $Cu(s) + 4\ HNO_3(conc.) \longrightarrow Cu(NO_3)_2(aq) + 2\ NO_2(g) + 2\ H_2O(l)$ Q. $Ag(s) + 2\ HNO_3(dil.) \longrightarrow AgNO_3(aq) + NO_2(g) + H_2O(l)$ R. $8\ Al(s) + 30\ HNO_3\ (dil.) \longrightarrow 8\ Al(NO_3)_3(aq) + 3\ N_2O(g) + 15\ H_2O(l)$
S. $2\ NaH(s) + H_2O(l) \longrightarrow 2\ NaOH(aq) + 2\ H_2(g)$ T. $2H_2(g) + O_2(g) \longrightarrow 2\ H_2O(l)$
U. $2\ NH_3(aq) + H_2SO_4(aq) \longrightarrow (NH_4)_2SO_4(aq)$ V. $2\ KClO_3(s) \xrightarrow{\Delta} 2\ KCl(s) + 3\ O_2(g)$ W. $3\,NO_2(g) + H_2O(l) \longrightarrow 2\ HNO_3(aq) + NO(g)$
X. $2\,C(s) + O_2(g) \longrightarrow 2\,CO(g)$
Y. $3\,CH_4(g) + 4\,O_3(g) \longrightarrow 3\,CO_2(g) + 6\,H_2O(g)$
Z. $MnO_2(s) + 2\,CO(g) \longrightarrow Mn(s) + 2\,CO_2(g)$

4. A. $F_2 > Cl_2 > Br_2 > I_2$
B. $HBrO > BrO_3 > BrO_4^-$
C. $ClO_4^- > SO_4^{2-} > HPO_4^{2-}$
D. "H_2SO_3"$> Na_2SO_3 > SO_4^{2-}$
E. $HNO_2 > NaNO_2 > NO_3^-$
F. $H_3PO_3 > H_2PO_3^- > HPO_4^{2-}$
G. $ClO_4^- > HSO_4^- > HPO_4^{2-} > H_4SiO_4$
H. $BrO_3^- > SeO_3^{2-} > H_2AsO_3^- > H_2GeO_4$

5. A. AgBr, AgI B. $NaClO_2$, HCN C. I_2, $Na_2S_2O_3$ D. $KClO_4$, $Pb(N_3)_2$
E. $KClO_4$, NO_2 F. $(NaPO_3)_n$, Na_2CO_3 G. I_2, Li_2CO_3 H. NO_2, HNO_3

I. ClF_3, BrF_3, HBr, ICl J. NaClO, $NaClO_2$, $Na_2S_2O_4$ K. NaClO, I_2
L. HCN, NaF

6. A. $O_2 + 4\ H^+ + 4\ e^- \longrightarrow 2\ H_2O$,+1.229 V

B. $O_2 + 2\ H_2O + 4\ e^- \longrightarrow 4\ OH^-$,+0.401 V

C. $HO_2^- + H_2O + 2\ e^- \longrightarrow 3\ OH^-$,+0.878 V

D. $HClO + H^+ + 2\ e^- \longrightarrow Cl^- + H_2O$,+1.49 V

E. $2\ ClO_4^- + 16\ H^+ + 14\ e^- \longrightarrow Cl_2 + 8\ H_2O$,+1.39 V

F. $ClO_4^- + 8\ H^+ + 8\ e^- \longrightarrow Cl^- + 4\ H_2O$,+1.39 V

G. $HClO_2 + 3\ H^+ + 4\ e^- \longrightarrow Cl^- + 2\ H_2O$, +1.56 V

H. $2\ ClO_3^- + 6\ H_2O + 10\ e^- \longrightarrow Cl_2 + 12\ OH^-$, +0.48 V

I. $2\ ClO_4^- + 8\ H_2O + 14\ e^- \longrightarrow Cl_2 + 16\ OH^-$, +0.45 V

J. $ClO_2^- + 2\ H_2O + 4\ e^- \longrightarrow Cl^- + 4\ OH^-$, 0.77 V

K. $NO_3^- + 4\ H^+ + 3\ e^- \longrightarrow NO + 2\ H_2O$, +0.96 V

L. $NO_3^- + 2\ H_2O + 3\ e^- \longrightarrow NO + 4\ OH^-$, −0.14 V

M. $NO_3^- + 3\ H^+ + 2\ e^- \longrightarrow HNO_2 + H_2O$, +0.94 V

N. $NO_3^- + H_2O + 2\ e^- \longrightarrow NO_2^- + 2\ OH^-$, +0.02 V

O. $NO_3^- + 10\ H^+ + 8\ e^- \longrightarrow NH_4^+ + 3\ H_2O$, +0.88 V

P. 2. $HNO_2 + 6\ H^+ + 6\ e^- \longrightarrow N_2 + 4\ H_2O$, +1.45 V

Q. $SO_4^{2-} + 8\ H^+ + 6\ e^- \longrightarrow S + 4\ H_2O$, +0.36 V

R. $SO_4^{2-} + 4\ H_2O + 8\ e^- \longrightarrow S^{2-} + 8\ OH^-$, −0.68 V

Quizzes

A: 1. (d) 2. (c) 3. (d) 4. (b) 5. (b) 6. (c) 7. (c)
B: 1. (b) 2. (d) 3. (d) 4. (a) 5. (d) 6. (e) 7. (c)
C: 1. (b) 2. (c) 3. (c) 4. (a) 5. (a) 6. (c) 7. (c)
D: 1. (e) 2. (d) 3. (c) 4. (a) 5. (a) 6. (b) 7. (b)

Sample Test

1. a. $2\ HBrO + 2\ H^+ + 2\ e^- \longrightarrow Br_2 + 2\ H_2O$ $E° = +1.59$ V

$HBrO + 2\ H_2O \longrightarrow BrO_3^- + 5\ H^+ + 4\ e^-$ $E° = -1.49$ V

$5\ HBrO \longrightarrow 2\ Br_2 + BrO_3^- + 2\ H_2O + 2\ H^+$ $E° = +0.10$ V

Yes, it is spontaneous

b. $Br_2 + 2\ H_2O \longrightarrow 2\ HBrO + 2\ H^+ + 2\ e^-$ $E° = -1.59$ V

$2\ HBrO + 4\ H_2O \longrightarrow 2\ BrO_3^- + 10\ H^+ + 8\ e^-$ $E° = -1.49$ V

$Br_2 + 6\ H_2O \longrightarrow 2\ BrO_3^- + 12\ H^+ + 10\ e^-$

$E° = [2(-1.59) + 8(-1.49)]/10 = -1.51$ V

2. a. $Br_2(l) + 6\ NaOH(aq) \xrightarrow{\Delta} 2\ NaBrO_3(aq) + 5\ NaBr(aq) + 3\ H_2O$

b. $H_2(g) + Br_2(g) \longrightarrow 2\ HBr(g)$,

$PBr_3(g) + 3\ H_2O \longrightarrow H_3PO_3(aq) + 3\ HBr(g)$;

$NaBr(s) + H_3PO_4(aq) \longrightarrow NaH_2PO_4(aq) + HBr(g)$

c. $2\ H_2O \xrightarrow{\text{electrolysis}} 2\ H_2(g) + O_2(g)$, $2\ H_2O_2(\text{l or aq}) \longrightarrow 2\ H_2O + O_2(g)$;

$2\ HgO(s) \xrightarrow{\Delta} 2\ Hg(l) + O_2(g)$

d. $3\ NaClO(s) \xrightarrow{\Delta} 2\ NaCl(s) + NaClO_3(s)$

e. $CdS(s) + 2\ HCl(aq) \longrightarrow CdCl_2(aq) + H_2S(g)$

f. $2\ NO(g) + O_2(g) \longrightarrow 2\ NO_2(g)$; $N_2O_4(g)\ \$\$\$\ 2\ NO_2(g)$;

$2\ Pb(NO_3)_2(s) \xrightarrow{\Delta} 2\ PbO(s) + 4\ NO_2(g) + O_2(g)$

g. $2\ Cu(s) + 8\ HNO_3(aq) + 2\ NO(g) + 4\ H_2O$, $2\ NO(g) + O_2(g)$

$\longrightarrow 2\ NO_2(g) \rightleftharpoons N_2O_4(g)$ [favored by high P and low T]

h. $N_2H_4(l) + 2\ H_2O_2(l) \longrightarrow N_2(g) + 4\ H_2O(g)$

i. $P_4O_{10}(s) + 6\ H_2O \longrightarrow 4\ H_3PO_4(l) \xrightarrow{\Delta} 2\ H_4P_2O_7(l) + 2\ H_2O(g)$

j. $Na_2S_2O_8(s) + 2\ H_2O \longrightarrow 2\ NaHSO_4(aq) + H_2O_2(aq)$

k. $2\ Ca(s) + O_2(g) \longrightarrow 2\ CaO(s)$ at low temperatures;

$S(s) + O_2(g) \longrightarrow SO_2(g)$; $CaO(s) + SO_2(g) \longrightarrow CaSO_3(s)$

l. $2\ Na(s) + Cl_2(g) \longrightarrow 2\ NaCl(s)$;

$3\ NaCl(s) + H_3PO_4(aq) \longrightarrow Na_3PO_4(aq) + 3\ HCl(g)$. $H_2SO_4(aq)$ also could be used as the strong acid.

m. $2\ H_2(g) + O_2(g) \longrightarrow 2\ H_2O(l)$;

$P_4(s) + 5\ O_2(g) \longrightarrow P_4O_{10}(s)$;

$P_4O_{10}(s) + 6\ H_2O(l) \longrightarrow 4\ H_3PO_4(aq)$

n. $6\ Li(s) + N_2(g) \longrightarrow 2\ Li_3N(s)$;

$Li_3N(s) + 3\ H_2O(l) \longrightarrow 3\ LiOH(aq) + NH_3(g)$

o. $Ca(s) + 2\ H_2O \longrightarrow Ca(OH)_2(s) + H_2(g)$

p. $2\ Ca(s) + O_2(g) \longrightarrow 2\ CaO(s)$ at low temperatures; $C(s) + O_2(g) \longrightarrow CO_2(g)$;

$CaO(s) + CO_2(g) \longrightarrow CaCO_3(s)$

q. $Ba(s) + O_2(g) \xrightarrow{600°C} BaO_2(s)$; $BaO_2(s) + H_2SO_4(aq) \longrightarrow BaSO_4(s) + H_2O_2(aq)$

r. $2\ H_2(g) + O_2(g) \longrightarrow 2\ H_2O(l)$; $2\ Li(s) + 2\ H_2O(l) \longrightarrow 2\ LiOH(aq) + H_2(g)$;

$LiOH(aq) + HNO_3(aq) \longrightarrow LiNO_3(aq) + H_2O(l)$ ESTD

s. $Ca(s) + S(s) \longrightarrow CaS(s)$; $CaS(s) + 2\ HCl(aq) \longrightarrow CaCl_2(aq) + H_2S(g)$

t. $Mg(s) + Cl_2(g) \longrightarrow MgCl_2(s)$;

$3\ MgCl_2(s) + 2\ H_3PO_4(aq) \longrightarrow Mg_3(PO_4)_2(aq) + 6\ HCl(g)$

u.
$P_4(s) + 6\ Br_2(l) \longrightarrow 4\ PBr_3(l)$; $2\ H_2(g) + O_2(g) \longrightarrow 2\ H_2O(l)$;
$PBr_3(l) + 3\ H_2O(l) \longrightarrow H_3PO_3(aq) + 3\ HBr(g)$. Use water sparingly.

CHAPTER 23

Drill Problems

1. A. $Fe < Mn < Ti < Sc$ B. $Sc < Cr < Fe < Os$ C. $Sc < Ti < Cr < Fe$
D. $MnO < MnO_2 < MnO_3 < Mn_2O_7$ E. $Sc < Mn < Fe < Zn$
F. $Sc < V < Fe < Mn$ G. $Ti < Nb < Pb < Pt$ H. $VO < V_2O_3 < VO_2 < V_2O_5$
I. $Y < Ti < V < Mn$ J. $Cu^{2+} < V^{3+} < Ni^{2+} < Fe^{3+}$ K. $Fe < Cr < Ti < Ca$
2. A. Ti, V B. Sc C. Cu D. Cu E. Sc F. Cu, Co, Ni G. Zn H. Zn, Ni, Cr
I. Fe J. Ti, V K. Fe, Co, Ni
3. A. TiO_2 B. ZnS C. CoO, Cu_2O, CuO, Cr_2O_3
D. Cr_2O_3, $PbCrO_4$, $FeSO_4$, $Fe_4[Fe(CN)_6]_3$, Fe_2O_3, $ZnCrO_4$
E. CuCl, $Cu_3(AsO_4)_2$ F. ZnO G. V_2O_5 H. $KMnO_4$, $Na_2Cr_2O_7$
I. $KCr(SO_4)_2 \cdot 12H_2O$, $FeSO_4$, $CuSO_4$ J. Na_2CrO_4, $ZnCrO_4$
K. $FeCl_3$, $CuSO_4$ L. MnO_2, Ni_2O_3

4. A. $ZnCO_3(s) \xrightarrow{\Delta} ZnO(s) + CO_2$
B. $2\,ZnS(s) + 3\,O_2(g) \xrightarrow{\Delta} 2\,ZnO(s) + 2\,SO_2(g)$
C. $NiO(s) + H_2(g) \xrightarrow{\Delta} Ni(s) + H_2O(g)$
D. $HgS(s) + O_2(g) \xrightarrow{\Delta} Hg(l) + SO_2(g)$
E. $Fe_2O_3(s) + 3\,H_2(g) \xrightarrow{\Delta} 2\,Fe(s) + 3\,H_2O(g)$
F. $SO_2(g) + 2\,C(s) \xrightarrow{\Delta} S(s) + 2\,CO(g)$
G. $SnO_2(s) + 2\,C(s) \xrightarrow{\Delta} Sn(l) + 2\,CO(g)$
H. $PbO(s) + CO(g) \xrightarrow{\Delta} Pb(l) + CO_2(g)$
I. $PbS(s) + 3\,O_2(g) \xrightarrow{\Delta} 2\,PbO(s) + 2\,SO_2(g)$
J. $CuO(s) + H_2(g) \xrightarrow{\Delta} Cu(s) + H_2O(g)$
K. $2\,CdS(s) + 3\,O_2 \xrightarrow{\Delta} 2\,CdO(s) + SO_2(g)$
5. A. $Cu_2S(l) + O_2(g) \xrightarrow{\Delta} 2\,Cu(l) + SO_2(g)$
B. $2\,C(s) + O_2(g) \longrightarrow 2\,CO(g)$
C. $TiO_2(s) + 2\,Cl_2(g) + C(s) \longrightarrow TiCl_4(g) + 2\,CO(g)$
D. $CaO(s) + SiO_2(l) \longrightarrow CaSiO_3(l)$
E. $TiO_2(s) + C(s) \longrightarrow Ti(l) + CO_2(g)$
F. $2\,CuFeS_2 + 4\,O_2(g) \longrightarrow Cu_2S(l) + 2\,FeO(s) + 3\,SO_2(g)$
G. $4\,Ag(s) + 8\,CN^-(aq) + O_2(g) + 2\,H_2O \longrightarrow 4[Ag(CN)_2^-](aq) + 4\,OH^-(aq)$
H $FeO(s) + SiO_2(l) \longrightarrow FeSiO_3(l)$

6. Standard cell potentials follow; spontaneous reactions have positive $E°$ values. A. $+0.36$ V B. -0.291 V C. -1.211 V D. $+1.068$ V
E. -0.47 V F. -0.10 V G. -0.592 V H. -0.3 V

7. A. $K_2Cr_2O_7 + 6\,KBr + 7\,H_2SO_4 \longrightarrow 3\,Br_2 + 4\,K_2SO_4 + Cr_2(SO_4)_3 + 7\,H_2O$

B. $2\,KMnO_4 + 5\,Na_2C_2O_4 + 8\,H_2SO_4 \longrightarrow 2\,MnSO_4 + 10\,CO_2 + 5\,Na_2SO_4 + K_2SO_4 + 8\,H_2O$

C. $H_2S + 2\,KMnO_4 \longrightarrow 2\,KOH + SO_2 + 2MnO_2$

D. $K_2Cr_2O_7 + 3\,SnSO_4 + 7\,H_2SO_4 \longrightarrow 3\,Sn(SO_4)_2 + K_2SO_4 + Cr_2(SO_4)_3 + 7\,H_2O$

E. $2\,KMnO_4 + 3\,H_2O_2 \longrightarrow 3\,O_2 + 2\,MnO_2 + 2\,H_2O + 2\,KOH$

F. $5\,Cd + 2\,KMnO_4 + 8H_2SO_4 \longrightarrow 5\,CdSO_4 + 2\,MnSO_4 + K_2SO_4 + 8\,H_2O$

G. $K_2Cr_2O_7 + 3\,HNO_2 + 5\,HCl \longrightarrow 2\,KNO_3 + 4\,H_2O + CrCl_3 + CrCl_2NO_3$

H. $2\,NaMnO_4 + 6\,NaI + 4\,H_2O \longrightarrow 3\,I_2(s) + 2\,MnO_2 + 8\,NaOH$

I. $2\,Na_2CrO_4 + 3\,SnCl_2 + 16\,HCl \longrightarrow 3\,SnCl_4 + 2\,CrCl_3 + 4\,NaCl + 8\,H_2O$

J. $8\,HMnO_4 + 5\,AsH_3 + 8\,H_2SO_4 \longrightarrow 5\,H_3AsO_4 + 8\,MnSO_4 + 12\,H_2O$

K. $2\,Cr(OH)_3 + 3\,Na_2O_2 \longrightarrow 2\,Na_2CrO_4 + 2\,NaOH + 2\,H_2O$

L. $2\,KMnO_4 + 10\,FeSO_4 + 8\,H_2SO_4 \longrightarrow 5\,Fe_2(SO_4)_3 + 2MnSO_4 + K_2SO_4 + 8\,H_2O$

Quizzes

A: 1. (a) 2. (e) 3. (a) 4. (e, MnO_2) 5. (c) 6. (b) 7. (b) 8. (c)
B: 1. (d) 2. (e) 3. (a) 4. (d) 5. (a) 6. (c) 7. (d) 8. (b)
C: 1. (b) 2. (a) 3. (b) 4. (a) 5. (b) 6. (e) 7. (e) 8. (d)
D: 1. (e) 2. (a) 3. (b) 4. (b) 5. (c) 6. (c) 7. (c) 8. (d)

Sample Test

1. a. Dissolve in nitric acid. $Fe(s) + 2\,H^+ \longrightarrow Fe^{2+} + H_2(g)$ $Ni(s) + 2\,H^+ \longrightarrow Ni^{2+} + H_2(g)$;
   $3\,Cu(s) + 2\,NO_3^- + 8\,H^+ \longrightarrow 3\,Cu^{2+} + 2\,NO(g) + 4\,H_2O$

b. Treat with $NH_3{}^-NH_4Cl$, $(NH_4)_2S$. $Cu^{2+} + 4\,NH_3(aq) \longrightarrow [Cu(NH_3)_4]^{2+}$;

$Fe^{2+} + (NH_4)_2S(aq) \longrightarrow FeS(s) + 2\,NH_4^+$; $Ni^{2+} + (NH_4)_2S(aq) \longrightarrow NiS(s) + 2\,NH_4^+$

c. Treat remaining solution with Zn(s). $Zn(s) + [Cu(NH_3)_4]^{2+} \longrightarrow Cu(s) + [Zn(NH_3)_4]^{2+}$

d. Treat precipitate from step b with aqua regia.

$FeS(s) + NO_3^- + 4\,H^+ \longrightarrow Fe^{3+} + NO(g) + S(s) + 2\,H_2O$;

$3\,NiS(s) + 2\,NO_3^- + 8\,H^+ \longrightarrow 3\,Ni^{2+} + 2\,NO(g) + 3\,S(s) + 4\,H_2O$

e. Treat solution with

$NH_3(aq)$ $Fe^{3+} + 3\,NH_3(aq) + 3\,H_2O \longrightarrow Fe(OH)_3(s) + 3\,NH_4^+$

$Ni^{2+}(aq) + 6\,NH_3(aq) \longrightarrow [Ni(NH_3)_6]^{2+}$

f. Test solution from e with dimethylglyoxime.

$[Ni(NH_3)_6]^{2+} + 2\,DMG^- \longrightarrow Ni(DMG)_2\,(scarlet, s) + 6\,NH_3(aq)$

g. Dissolve solid from e in HCl(aq), then test with KSCN(aq). $Fe(OH)_3(s) + 3\,H^+ \longrightarrow Fe^{3+} + 3\,H_2O$;
$Fe^{3+} + SCN^- + 5\,H_2O \longrightarrow [Fe(H_2O)_5SCN]^{2+}$ (wine red)

2. a.

$Al_2O_3 + 6\,HCl(aq) \longrightarrow 3\,H_2O + 2\,AlCl_3(aq)$; $Al_2O_3(s) + 6\,NaOH(aq) \longrightarrow 2\,Na_3AlO_3(aq) + 3\,H_2O$

b. $100.0\text{ g } AlCl_3 \times \frac{\text{mol } AlCl_3}{133.5\text{ g}} \times \frac{\text{mol } Al_2O_3}{2\text{ mol } AlCl_3} \times \frac{102.0\text{ g}}{\text{mol } Al_2O_3} = 38.20\text{ g } Al_2O_3$

$100.0\text{ g } Na_3AlO_3 \times \frac{\text{mol } Na_3AlO_3}{144.0\text{ g}} \times \frac{\text{mol } Al_2O_3}{2\text{ mol } Na_3AlO_3}$

$\times \frac{102.0\text{ g}}{\text{mol } Na_3AlO_3} = 35.42\text{ g } Al_2O_3$

3.

$Cr_2O_7^{2-} \xrightarrow{+1.33\text{ V}} Cr^{3+} \xrightarrow{-1.51\text{ V}} Cr^{2+}$ (overall: $+0.62$ V)

$E° = [3(1.33) - 1.51] + 4 = 0.62\text{ V}$

CHAPTER 24

Drill Problems

1. A. Ag, +1, 2, $S_2O_3^{2-}$ B. Al, +3, 4, H^- C. Co, +3, 6, H_2O & CN^-
D. Pt, +4, 6, NH_3 & Cl^- E. Ni, +2, 4, NH_3
F. Cr, +3, 6, NH_3, & SCN^- G. Fe, +3, 6, en H. Pt, +2, 4, NO_2^-
I. Co, +3, 6, NH_3 & Cl^- J. Al, +3, 6, H_2O & OH^-
K. Pt, +4, 6, NH_3 & Cl^- L. Fe, 0, 4, CO M. Mn, +2, 6, NO & CN^-
N. Au, +3, 4, F^- O. Co, +3, 6, NH_3 & NH_2– P. Sb, +3, 5, Cl^-
Q. Ni, +2, 6, Cl– & en R. Fe, +2, 6, CN– S. Ni,. +2, 4, NH_3 & Br^-
T. Fe, +3, 6, H_2O & OH^- U. Zn, +2, 4, H_2O & NH_3

2. (1) A. Ag^+ B. Ca^{2+}, Fe^{2+}, Sc^{3+}, Cr^{3+}, Fe^{3+}, Co^{3+}, Pt^{4+}
C. Co^{2+}, Ni^{2+}, Cu^{2+}, Al^{3+}
(2) D. 4,6 E. 6 F. 6 G. 6 H. 4,6 I. 2,4 J. 4,6 K. 6 L. 2
(3) M. nitrato (1) N. ethylenediamminetetraacetato (6) O. trien (4) P. NH_3 (1) Q. OH^- (1) R. CO (1) S. nitro (1) T. —NCS– (1) U. nitrosyl (1) V. nitrogeno (1) W. F– (1) X. SO_4^{2-} (aq) Y. CN^- (1) Z. nitrito (1)

3. (1) A. dithiosulfatoargentate(I) ion
B. tetrahydroaluminate(III) ion
C. diaquatetracyanocobaltate(III) ion
D. diamminetetrachloroplatinum(IV)
E. tetraamminenickel(II) ion

F. diamminetetrathiocyanatochromate(III) ion
G. tris(ethylenediamine)iron(III) ion H. tetranitroplatinate(II) ion I. pentaamminechlorocobalt(III) ion
J. pentaaquahydroxoaluminum(III) ion
K. triamminetrichloroplatinum(IV) ion L. tetracarbonyliron(0)
M. pentacyanonitrosylmanganate(II) ion
N. tetrafluoroaurate(III) ion
O. diamidotetraamminecobalt(III) ion
P. pentachloroantimonate(III) ion
Q. dichlorobis(ethylenediamine)nickel(II)
R. hexacyanoferrate(II) ion S. diamminedibromonickel(II)
T. tetraaquadihydroxoiron(III) ion
U. diamminediaquazinc(II) ion
(2) V. sodium tetracyano aurate(I) W. $Na_3[Ag(S_2O_3)_2]$ X. potassium, hexanitrocobaltate(III) ion Y. $[Pt(NH_3)_2Br_2]$ Z. tetraamminedichlorocobalt(III) bromide Γ. $[Cr(H_2O)_4Cl_2Br_2]Br$. Δ .$[Pt(H_2O)_2]_4$ Θ.potassium hexachloroplatinate(IV) Λ. $[Ag(NH_3)_2Cl$ Ξ. $[Ni(NH_3)_6SO_4$ Π. sodium bis(ethylenediamine)disulfatocuprate(II) Σ. hexaaquaaluminum(III) hexacyanocobaltate(III)ϒ. calcium diamminetetraisothiocyanatochromate(III) Φ. hexaamminecobalt(III) chloride sulfate Ψ. $Na_3[Cr(C_2O_4)_3]$
Ω. $[Cu(en)_2Cl_2$

4. A. $[S_2O_3\text{—}Ag\text{——}S_2O_3]^{3-}$

B.
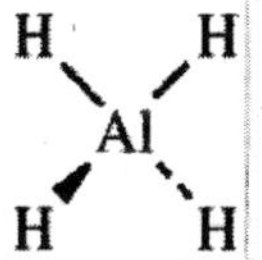

C.
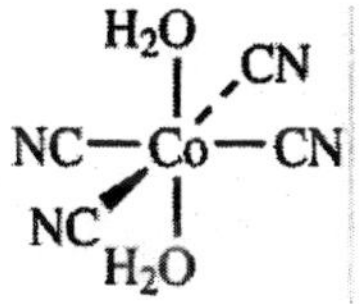

D.
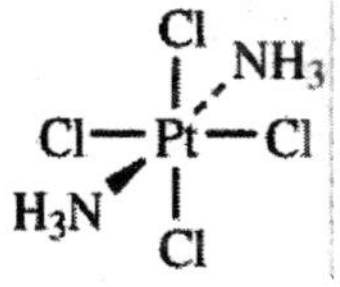

E.
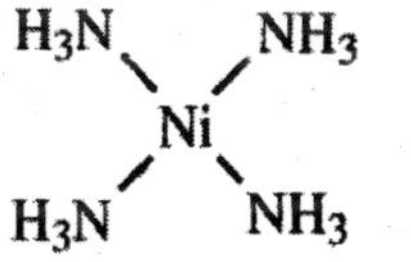

F.
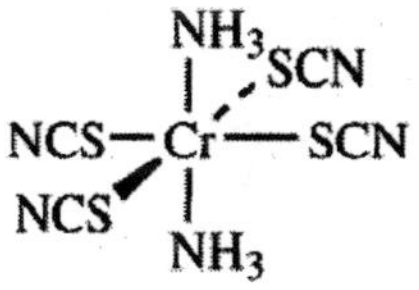

G.
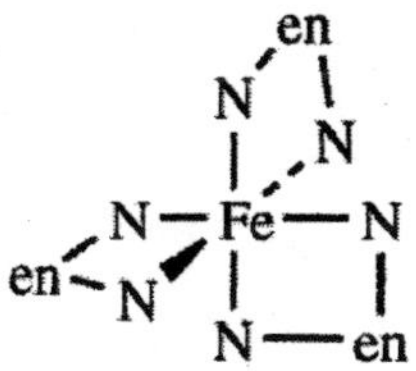

H.
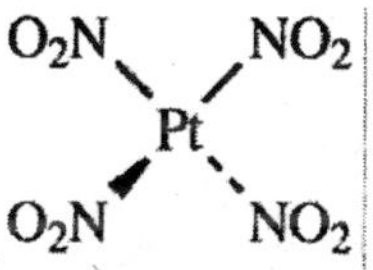

I.
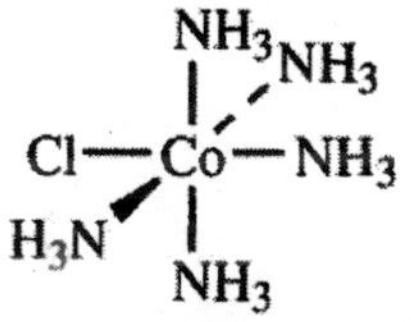

J.
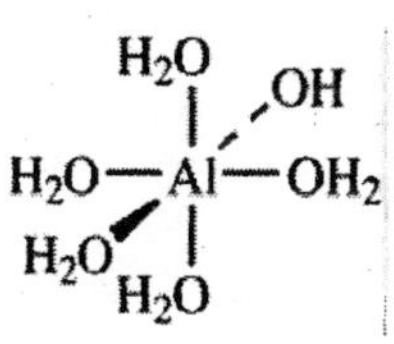

K.
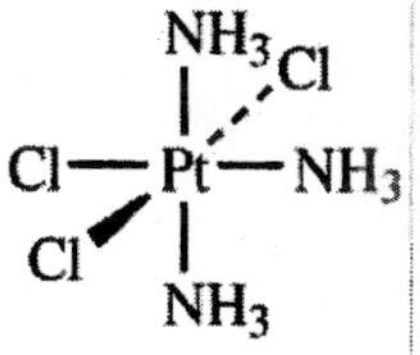

L. 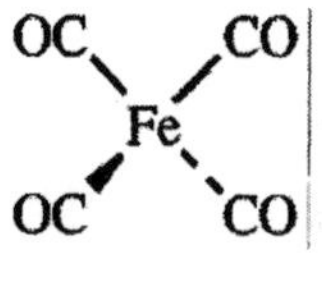

M. 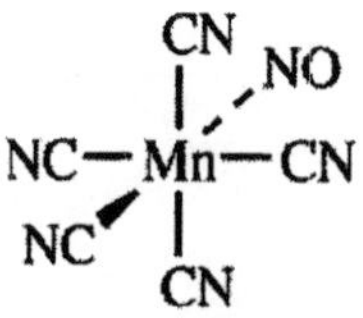

N. 

O. 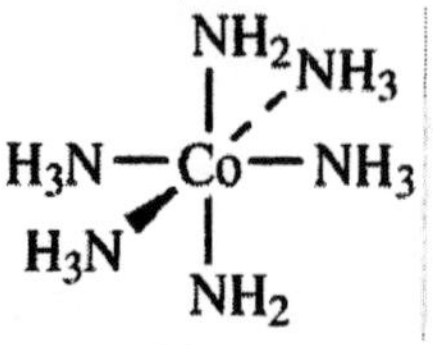

P. 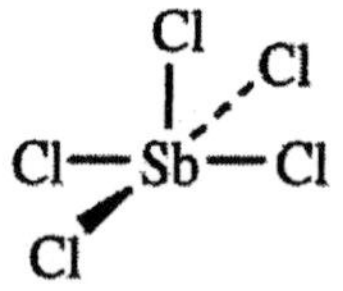

Q. 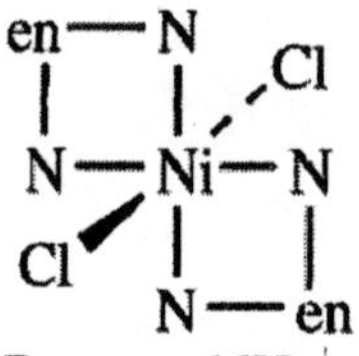

R. 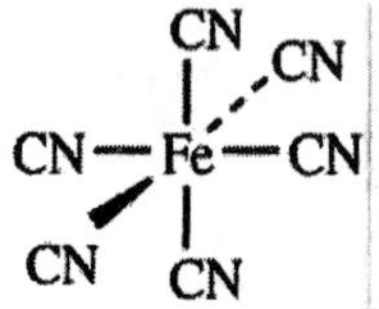

S. 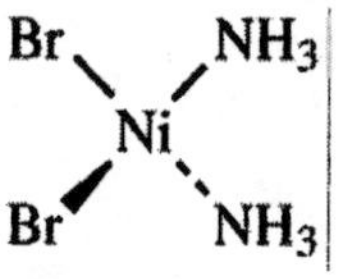

T. 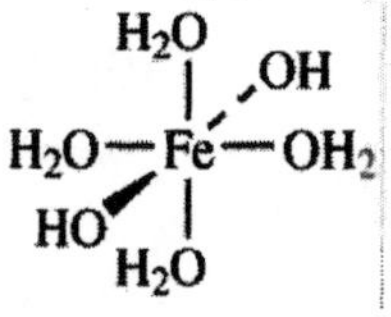

U. 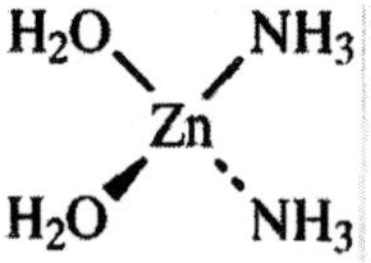

5.A.3: $[Co(H_2O)_4(CN)_2Cl$, tetraaquadicyanocobalt(III) chloride; $[Co(H_2O)_4(CN)Cl]CN$, tetraaquachlorocyanocobalt(III) cyanide 7: $[Cr(H_2O)_4Cl_2]Br$, tetraaquadichlorochromium(III) bromide; $[Cr(H_2O)_4ClBr]Cl$, tetraaquabromochlorochromium(III) chloride 9: $[Co(NH_3)_5Cl]SO_4$, pentaamminechlorocobalt(III) sulfate; $[Co(NH_3)_5SO_4]Cl$, pentaamminesulfatocobalt(III) chloride B. 1: lithium tetracyanoaluminate, lithium tetraisocyanoaluminate 3: tetraaquadicyanocobalt(III) chloride, tetraaquadiisocyanocobalt(III) chloride 5: hexaaquaaluminum hexacyanocobaltate(III), hexaaquaaluminum hexaisocyanocobaltate(III) 6: diaquadinitroplatinum(II), diaquadinitritoplatinum(II) 10: tricyanotrinitrosylmanganese(III), triisocyanotrinitrosylmanganese(III)

C. 5: $[Al(H_2O)_6][Co(CN)_6]$, $[Co(H_2O)_6[Al(CN)_6]$

8: $[Pt(H_2O)_4[PtCl_6]$, $[Pt(H_2O)_4Cl_2][PtCl_4]$

D. 3: *trans cis*

D. *trans* *cis*

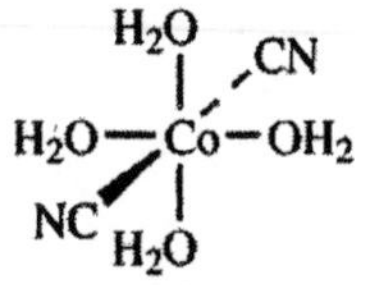 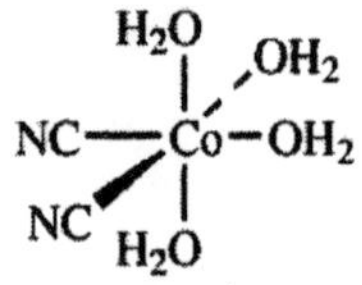

trans *cis*

4: 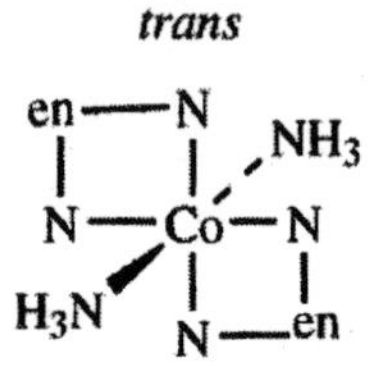 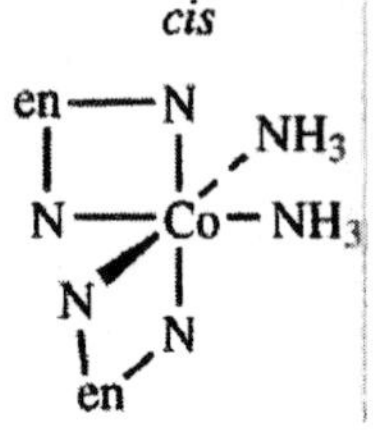

6:

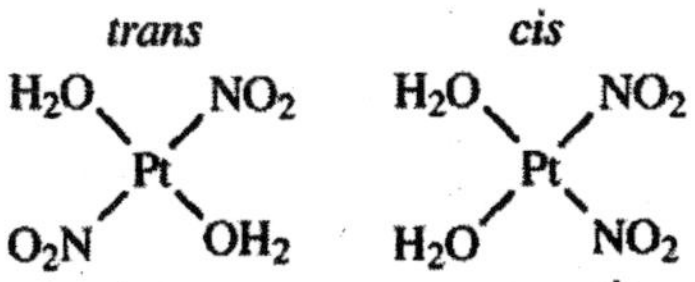

7:

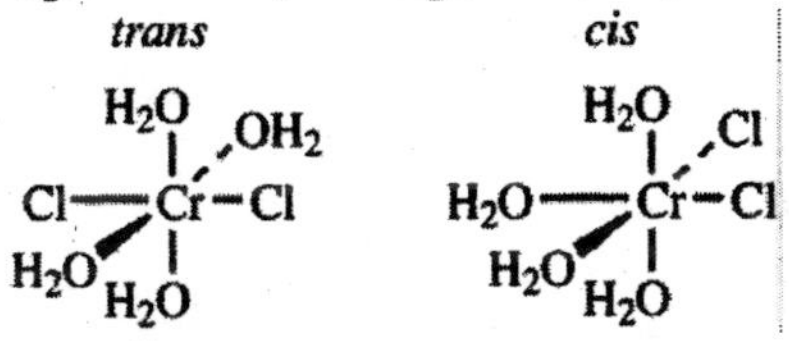

10:

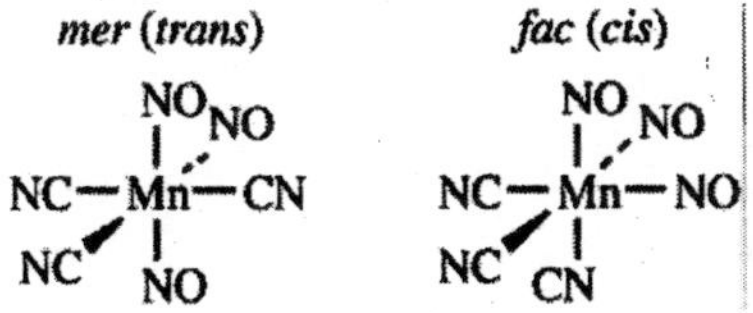

6. A. strong, 0 B. strong, 3 C. weak, 4 D. weak, 3 E. strong, 0 F. weak, 3 G. weak 4 H. weak, 2 I. weak, 0 J. strong, 1

Quizzes

A: 1. (d) 2. (c) 3. (c) 4. diamminedichlorogold(III) bromide 5. $K_2[SbCl_5]$ 6. (b) 7. (b) 8. (a)

B: 1. (c) 2. (d) 3. (a) 4. calcium tetracyanocuprate(II) 5. $Ca[Cr(NH_3)_2Cl_4]_2$ 6. (b) 7. (c) 8. (c)

C: 1. (c) 2. (b) 3. (a) 4. potassium dichlorodicyanoplatinate(II) 5. $Ca[Cr(NH_3)_2Cl_4]_2$ 6. (a) 7. (d) 8. (b)

D: 1. (c) 2. (d) 3. (a) 4. sodium tetrachlorodinitrosylcobaltate(II) 5. $K_2[Fe(CH_3NH_2)_2(CN)_4]$ 6. (b) 7. (c) 8. (d)

Sample Test

1. Number of unpaired electrons is given in parentheses.

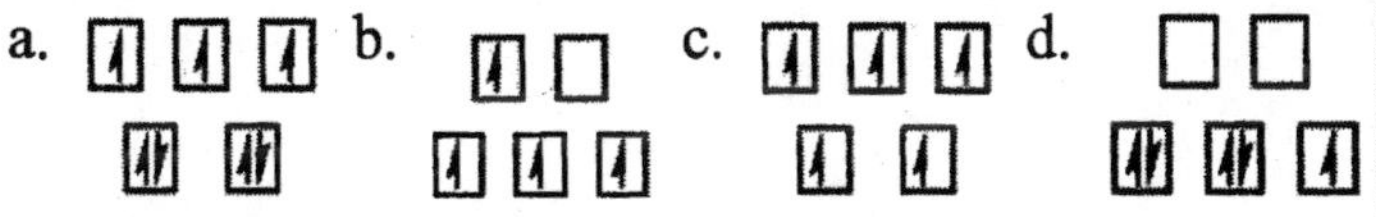

strong (3) weak (4) weak (5) strong (1)

2. 24.8 g Na × mol/23.0 g = 1.08 g Na

41.0 g F ×mol/19.0 g = 2.16 mol F

i(0.010 M)(0.0821 L atm mol^{-1} K^{-1}) (298 K)

= 559 mmHg ×atm/760 mmHg

i = 3.00. . Thus, the compound is $Na[CuF_4]$, sodium tetrafluorocuprate(II)

CHAPTER 25

Drill Problems

1. (1) A. β- B. ^{4}He C. β- D. ^{4}He E. ^{232}U F. β-
G. ^{200}Au H. ^{6}Li I. β- J. ^{17}F K. β+ L. ^{37}Ar M. ^{206}Hg N. ^{191}Au
O. ^{222}Rn

(2) P. $^7Be + \beta^- \longrightarrow {}^7Li$ Q. $^{191}Hg \longrightarrow \beta^+ + {}^{191}Au$

R. $^{177}Pt \longrightarrow {}^4He + {}^{173}Os$ S. $^3H \longrightarrow \beta^- + {}^3He$
T. $^{177}W + \beta^- \longrightarrow {}^{177}Ta$ U. $^{109}In \longrightarrow \beta^+ + {}^{109}Cd$
V. $^{186}Re \longrightarrow \beta^- + {}^{186}Os$ W. $^{235}U \longrightarrow {}^4He + {}^{231}Th$
X. $^{12}B \longrightarrow {}^4He + {}^8Li$ Y. $^{161}Tm + \beta^+ \longrightarrow {}^{161}Er$
Z. $^{98}Tc \longrightarrow \beta^- + {}^{98}Ru$

2. (1) A. $^{59}Co + {}^1H \longrightarrow {}^1n + {}^{59}Ni$; $^{59}Co(p,n)^{59}Ni$
B. $^9Be + {}^6Li \longrightarrow {}^1n + {}^{14}N$; $^9Be(^6Li,n)^{14}N$
C. $^{14}N + {}^1n \longrightarrow {}^9Be + {}^6Li$; $^{14}N(n,^6Li)^9Be$
D. $^{15}N + {}^1H \longrightarrow {}^4He + {}^{12}C$; $^{15}N(p,\alpha)^{12}C$
E. $^5He + {}^1H \longrightarrow {}^2H + {}^4He$; $^5He(p,d)^4He$
F. $^{23}Na + {}^2H \longrightarrow {}^4He + {}^{21}Ne$; $^{23}Na(d,\alpha)^{21}Ne$
G. $^7Li + {}^4He \longrightarrow {}^1n + {}^{10}B$; $^7Li(\alpha,n)^{10}B$
H. $^9Be + {}^4He \longrightarrow {}^1n + {}^{12}C$; $^9Be(\alpha,n)^{12}C$
I. $^{35}Cl + {}^1n \longrightarrow {}^1H + {}^{35}S$; $^{35}Cl(n,p)^{35}S$
J. $^{44}Ca + {}^1H \longrightarrow {}^1n + {}^{44}Sc$; $^{44}Ca(p,n)^{44}Sc$
K. $^{27}Al + {}^2H \longrightarrow {}^4He + {}^{25}Mg$; $^{27}Al(d,\alpha)^{25}Mg$
L. $^{25}Mg + {}^4He \longrightarrow {}^1H + {}^{28}Al$; $^{25}Mg(\alpha,p)^{28}Al$
(2) M. $^{238}U + {}^1n \longrightarrow \beta^- + {}^{239}Np$
N. $^{235}U + {}^1n \longrightarrow {}^{236}U + \gamma$; $^{236}U + {}^1n \longrightarrow {}^{237}U + \gamma$; $^{237}U \longrightarrow {}^{237}Np + \beta^-$
O. $^{238}U + {}^2H \longrightarrow {}^{238}Np + 2\,{}^1n$; $^{238}Np \longrightarrow {}^{238}Pu + \beta^-$
P. $^{239}Pu + {}^1n \longrightarrow {}^{240}Pu + \gamma$, $^{240}Pu + {}^1n \longrightarrow {}^{241}Pu + \gamma$; $^{241}Pu \longrightarrow {}^{241}Am + \beta^-$
Q. $^{239}Pu + {}^4He \longrightarrow {}^{242}Cm + {}^1n$
R. $^{241}Am + {}^1n \longrightarrow {}^{242}Am + \gamma$; $^{242}Am \longrightarrow {}^{242}Cm + \beta^-$
S. $^{241}Am + {}^4He \longrightarrow {}^{243}Bk + 2\,1n$
T. $^{242}Cm + {}^4He \longrightarrow {}^{245}Cf + {}^1n$
U. $^{253}Es + {}^4He \longrightarrow {}^{256}Md + {}^1n$
V. $^{246}Cm + {}^{12}C \longrightarrow {}^{254}No + 4\,{}^1n$
W. $^{252}Cf + {}^{11}B \longrightarrow {}^{258}Lw + 5\,{}^1n$

3. (1) Given in the order: $t_{1/2}$, λ, activity. A. 22 min, 0.0315/min, 8.5×10^{13} /min B. 1.65 h, 0.42/h, 1.7×10^{10} /h C. 1.5×10^6 y, 4.6×10^{-7}/y. 2.0×10^{16} /y D. 12.26 y, 0.05653/y, 1.49×10^{19} /y E. 39.5 h, 0.0175/h, 5.45×10^{19} /h F. 6.7 y, 0.10/y, 3.6×10^{20} /y

(2) Given in the order: $t_{1/2}$, λ, mass. G. 0.81 s, 0.86/s, 38 ng H. 53.37 d, 0.01298/d, 4.6 μg I. 39.5 h, 0.0175/h, 12.9 g J. 12.26 y, 0.05653/y, 866 kg K. 28 y, 0.025/y, 268 mg L. 60.6 min, 0.0114/min, 63.9 mg

(3) Given in the order: $t_{1/2}, \lambda$. M. 27 s, 0.026/s N. 15.5 s, 0.45/s O. 10.7 h, 1.81×10^{-5} /s P. 14.0 h, 1.38×10^{-6} /s Q. 8.00 min, 1.44×10^{-3} /s R. 14.2 h, 1.36×10^{-5} /s

4. (1) A. 18.8 d B. 2.67 h C. 0.372 min D. 10.1 min E. 2.32 d F. 0.293 d G. 1.94 d H. 1.54 d

(2) I. 3.8×10^3 y J. 26.6 y K. 3.81×10^5 y L. 3.09 y M. 1.5×10^4 y N. 1.32×10^4 y

(3) O. 54.5 y P. 7.50×10^4 y Q. 1.56×10^5 y R. 12.4 y S. 19.0 y T. 14.9 y

5. A. 2.0×10^{-5} amu, 1.8×10^9 kJ/mol, 0.019 MeV

B. 0.003768 amu, 3.162×10^{11} kJ/mol, 3.514 MeV

C. 0.000377 amu, 3.16×10^{10} kJ/mol, 0.352 MeV

D. 0.000094 amu, 7.9×10^9 kJ/mol, 0.088 MeV

E. 0.0143 amu, 1.20×10^{12} kJ/mol, 13.3 MeV

F. 0.001288 amu, 1.081×10^{11} kJ/mol, 1.201 MeV

G. 0.000326 amu, 2.74×10^{10} kJ/mol, 0.304 MeV

H. 0.001384 amu, 1.161×10^{11} kJ/mol, 1.291 MeV

I. 0.000548 amu, 4.90×10^{10} kJ/mol, 0.545 MeV

J. 0.001821 amu, 1.528×10^{11} kJ/mol, 1.698 MeV

K. 0.001072 amu, 8.996×10^{10} kJ/mol, 0.9997 MeV

L. 0.003232 amu, 2.712×10^{11} kJ/mol, 3.014 MeV

M. 0.001597 amu, 1.340×10^{11} kJ/mol, 1.489 MeV

6. Binding energy per nucleon is given first, then atomic number. A. 0.000 MeV, 1 B. 8.801 MeV, 26 C. 7.082 MeV, 2 D. 8.742 MeV, 28 E. 6.470 MeV, 4 F. 8.746 MeV, 30 G. 7.689 MeV, 6 H. 8.721 MeV, 34 I. 7.985 MeV, 8 J. 8.728 MeV, 36 K. 8.042 MeV, 10 L. 8.720 MeV, 40 M. 8.270 MeV, 12 N. 8.617 MeV, 44 O. 8.458 MeV, 14 P. 8.542 MeV, 48 Q. 8.503 MeV, 16 R. 8.440 MeV, 56 S. 8.605 MeV, 18 T. 8.405 MeV, 56 U. 8.561 MeV, 20 V. 8.356 MeV, 60 W. 8.733 MeV, 22 X. 8.211 MeV, 64 Y. 8.786 MeV, 24 Z. 8.142 MeV, 68 Results (except for parts A and E) are plotted below.

7. A. $^{8}B \longrightarrow {}^{8}Be + \beta^+$

B. $^{238}U \longrightarrow {}^{234}Th + {}^{4}He$

C. ^{19}F stable

D. $^{119}Cd \longrightarrow {}^{119}In + \beta^-$

E. $^{50}Sc \longrightarrow {}^{50}Ti + \beta^-$

F. $^{212}Po \longrightarrow {}^{208}Pb + {}^{4}He$

G. ^{88}Sr stable

H. $^{17}F \longrightarrow {}^{17}O + \beta^+$

I. $^{226}Ra \longrightarrow {}^{222}Rn + {}^{4}He$

J. $^{50}Mn \longrightarrow {}^{50}Cr + \beta^+$

K. $^{106}In \longrightarrow {}^{106}Cd + \beta^+$

L. ^{197}Au stable

M. $^{63}Co \longrightarrow ^{63}Ni + \beta^-$

Quizzes

A: 1. (e) $\left(^{222}Rn\right)$ 2. (b) 3. (b) 4. (d) 5. (b) 6. (b) 7. (a) 8. (b)
B: 1. (a) 2. (a) 3. (b) 4. (c) 5. (e) (45 d) 6. (b) 7. (a) 8. (a)
C: 1. (e) $\left(^{90}Y\right)$ 2. (c) 3. (a) 4. (a) 5. (a) 6. (b) 7. (a) 8. (e)
D: 1. (b) 2. (e) 3. (b) 4. (a) 5. (a) 6. (a) 7. (d) 8. (b)

Sample Test

1. Remain the same. ^{14}C is not produced by nuclear explosions but by cosmic rays in the upper atmosphere:

$$^{14}N + {}^1n \longrightarrow {}^{14}C + {}^1H$$

2. Only $\alpha(^4He)$ and $\beta(e^-)$ particle are given off; $^{253}Es \longrightarrow {}^{237}Np + 4\ {}^4He + 2\,e^-$. Thus four α particles and two β particles are given off.

3. a. $^{243}Cm \longrightarrow {}^{239}Pu + {}^4He$ b. $^{18}F \longrightarrow \beta^+ + {}^{18}O$

c. $^{88}Zr + \beta^- \longrightarrow {}^{88}Y$ d. $^{10}B + {}^4He \longrightarrow {}^1n + {}^{13}N$

e. $^{45}Sc + {}^4He \longrightarrow {}^1H + {}^{48}Ti$ f. $^{51}V + {}^2H \longrightarrow 2\,{}^1n + {}^{51}Cr$

4. $(11.0216\ \text{amu} - 11.00931\ \text{amu} - 0.00055\ \text{amu})$
$\times 932.8\ \text{MeV/amu} = 10.95\ \text{MeV}$

5. $\ln(N_t/N_0) = -0.693t/t_{1/2} - 0.693(995 \times 10^6\ \text{y}) \div (4.51 \times 10^9\ \text{y}) = -0.1529$

$N_t = N_0 e^{-0.1529} = 0.858 \times 6.022 \times 10^{23}$ atoms number decayed

$= N_0 - N_t = (6.022 - 5.17) \times 10^{23} = 8.5 \times 10^{22}$ atoms

CHAPTER 26

Drill Problems

1. (1)

$$\underset{\qquad\qquad CH_3CH_2\overset{\displaystyle CH_3}{\underset{|}{C}}=CHCH_2CH_3}{}$$

B. $CH_3CH=CHCH_2CH_3$ C. $CH_3\overset{\displaystyle CH_3}{\underset{|}{C}}HCH_2CH_3$

D. $CH_3\overset{\displaystyle CH_3}{\underset{\underset{|}{\underset{CH_3}{|}}}{C}}CH_2\overset{\displaystyle CH_3}{\underset{|}{C}}HCH_3$ E. $CH_3C\equiv CCH_3$

F. G.

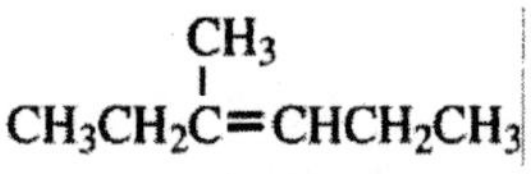

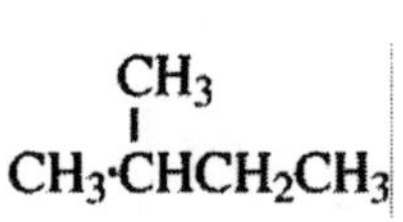

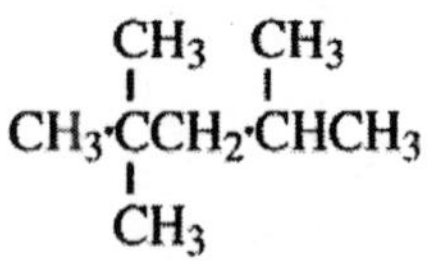

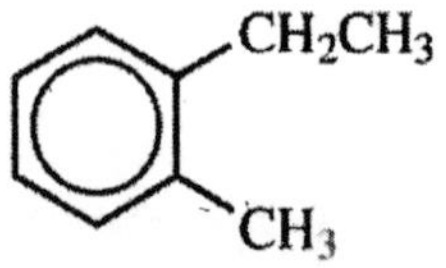

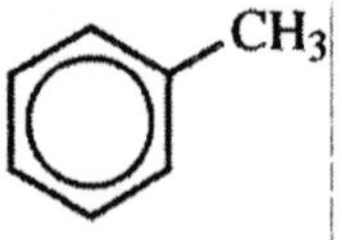

H. I.

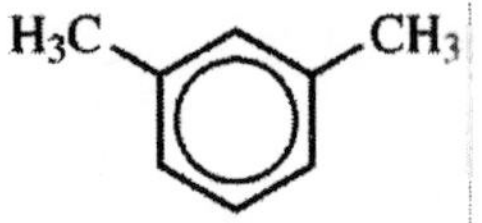

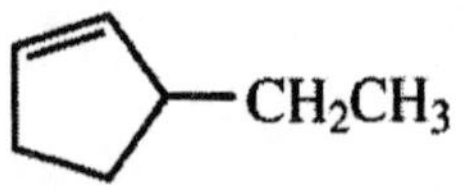

J. 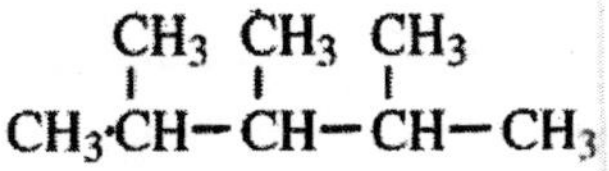

K. 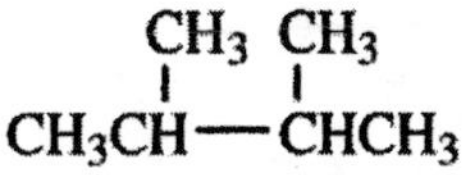

L. 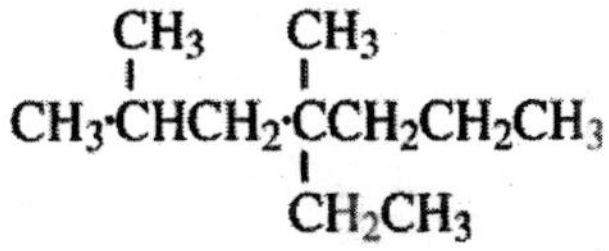

M. 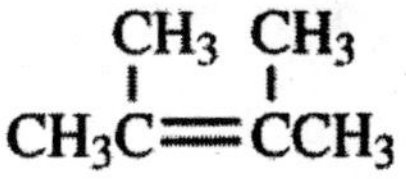

N. 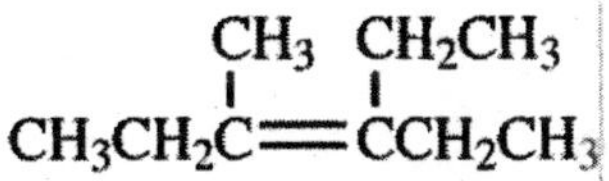

O. $CH_3C{=}CCH_2CH_3$

P. 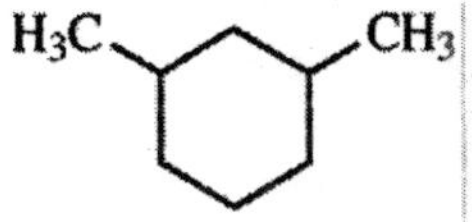

Q. 

R. 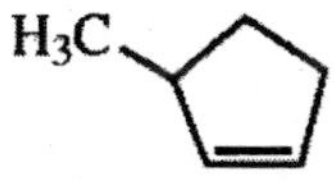

S. 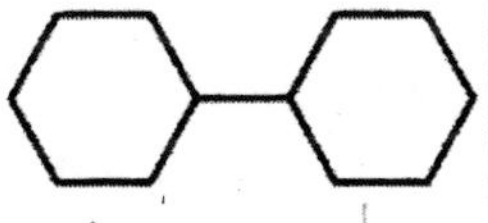

T. 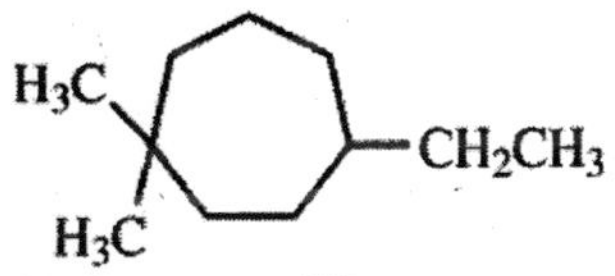

U. 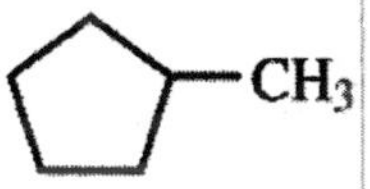

V. 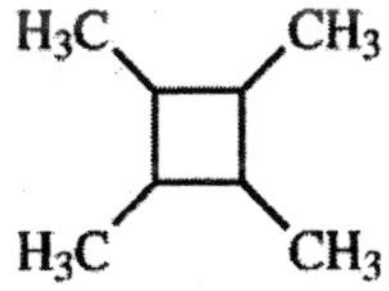

W. 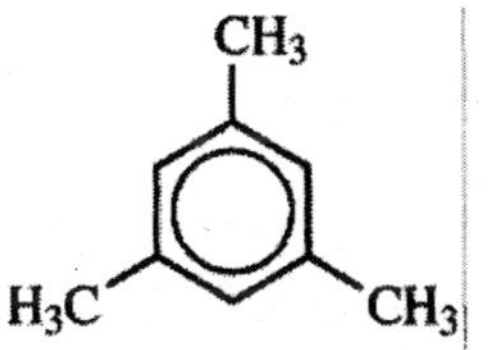

(2) A. 2-methylbutane B. 1-butene C. 3-ethyl-2-methyl-2-pentene D. 2,3-dimethylpentane E. ethylcyclopropane F. 1,2-dimethylcyclohexane G. *o*-xylene H. *n*-propylcyclohexane I. *t*-butylbenzene J. 4-methylheptane K. 1-butyne L. 2-butene M. 1-pentene N. 1-propyne *or* propyne O. propane P. 2-methylpropane Q. 2,2,3-trimethylbutane R. 3,3-diethyl-2-methylhexane S. 3-ethyl-2-methylpentane T. 2-methylpentane U. 2,2,3-trimethylpentane V. 4,4-dimethyloctane W. 4-methylcyclohexene

2. A.

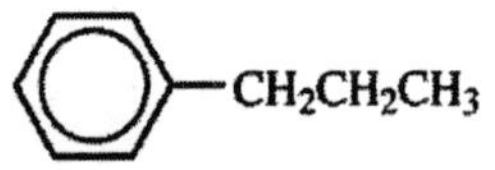

n-propylbenzene

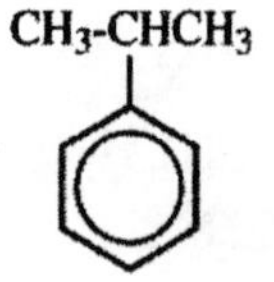

isopropylbenzene

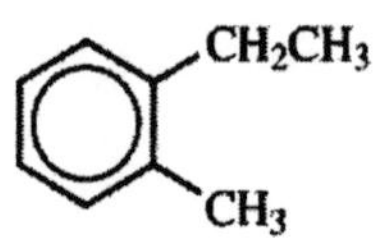

o-ethyltoluene

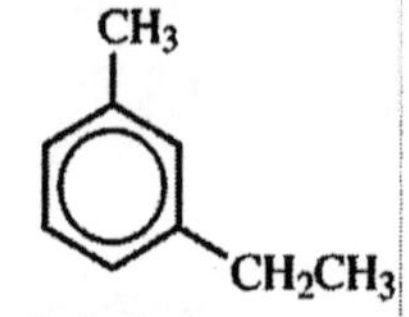

m-ethyltoluene

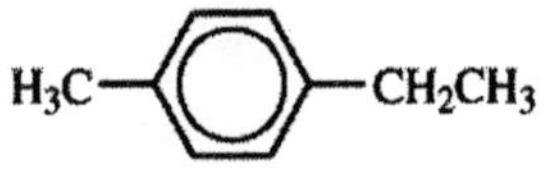

p-ethyltoluene

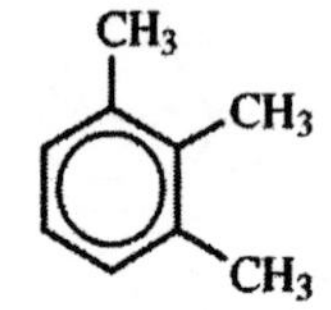

1,2,3-trimethylbenzene

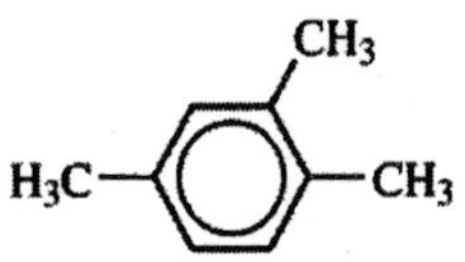

1,2,4-trimethylbenzene

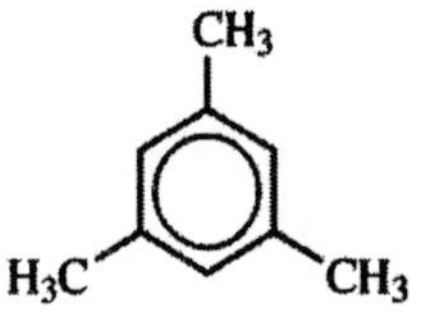

1,3,5-trimethylbenzene

B.

$H_2C=CHCH_2CH_3$
1-butene

$CH_3CH=CHCH_3$
2-butene

cyclobutane methylcyclopropane

$$CH_2=\overset{\overset{\displaystyle CH_3}{|}}{C}CH_3$$

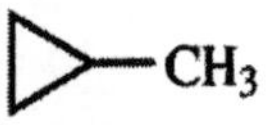

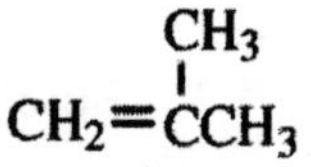

2-methyl-1-propene

c.

cyclohexane

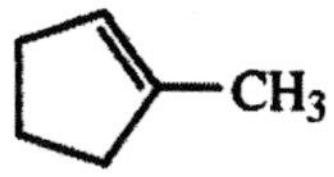

1-methylcyclopentene

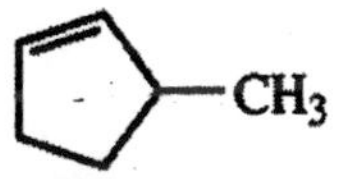

3-methylcyclopentene

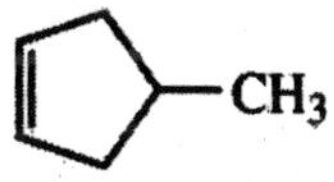

4-methylcyclopentene

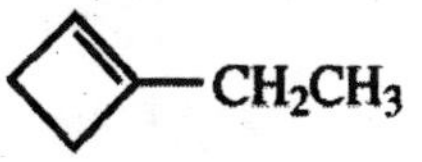

2-ethylcyclobutene

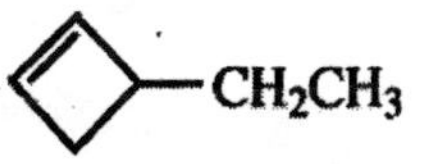

3-ethylcyclobutene

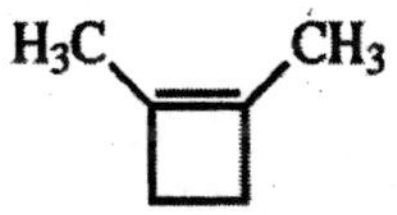

1,2-dimethylcyclobutene

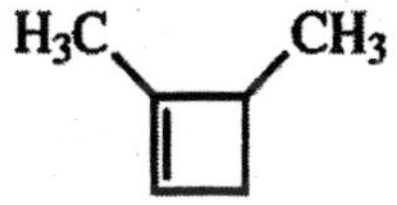

2,3-dimethylcyclobutene

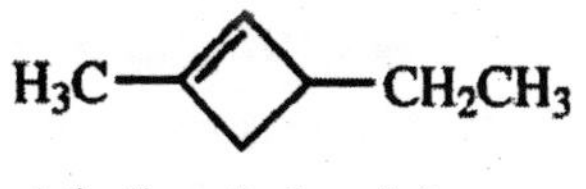

1,3-dimethylcyclobutene

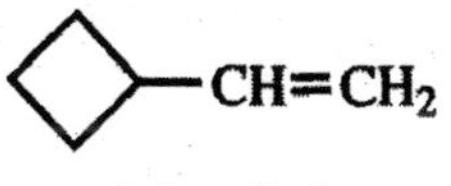

cyclobutylethene

and other unusual ones such as:

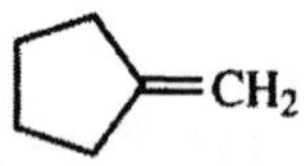

methylenecyclopentane

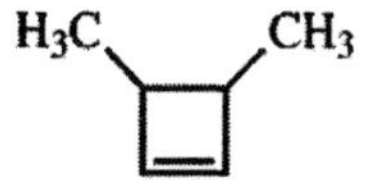

3,4-dimethylcyclobutene

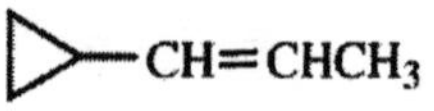

1-cyclopropylpropene

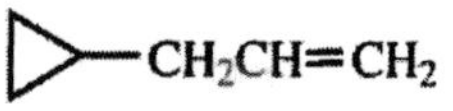

3-cyclopropylpropene

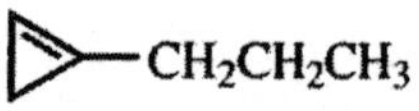

1-propylcyclopropene

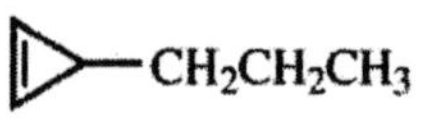

3-propylcyclopropene

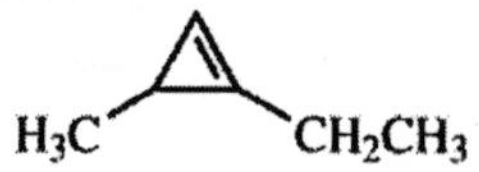

3-ethyl-1-methylcyclopropene

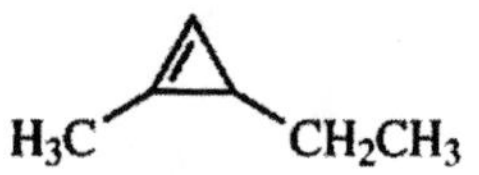

1-ethyl-3-methylcyclopropene

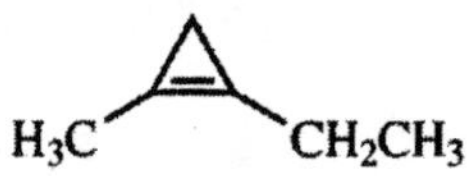

1-ethyl-2-methylcyclopropene

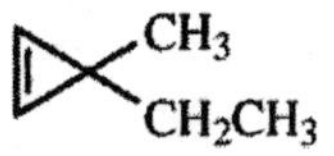

3-methyl-3-ethylcyclopropene

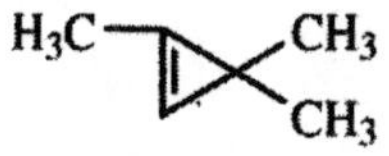

1,3,3-trimethylcyclopropene

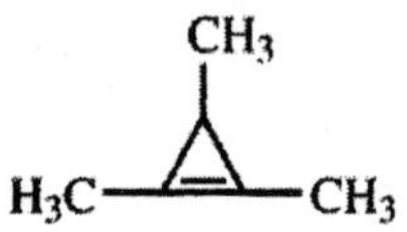

1,2,3-trimethylcyclopropene

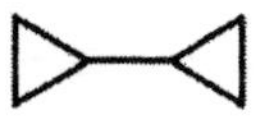

cyclopropylcyclopropane

D.

$CH_3CH_2CH_2CH_2CH_2CH_3$

hexane

$$\begin{array}{c} CH_3 \\ | \\ CH_3\text{-}CHCH_2CH_2CH_3 \end{array}$$

2-methylpentane

$$\begin{array}{c} CH_3 \\ | \\ CH_3CH_2CHCH_2CH_3 \end{array}$$

3-methylpentane

$$\begin{array}{c} CH_3CH_3 \\ |\quad | \\ CH_3\text{-}CH-CHCH_3 \end{array}$$

2,3-dimethylbutane

$$\begin{array}{c} CH_3 \\ | \\ CH_3\text{-}CCH_2CH_3 \\ | \\ CH_3 \end{array}$$

3,3-dimethylbutane

E.

$CH_3CH_2CH_2CH_3$

butane

$$\begin{array}{c} CH_3 \\ | \\ CH_3\text{-}CHCH_3 \end{array}$$

2-methylpropane

F. $CH_2{=}CHCH_2CH_2CH_3$
1-pentene

$CH_3CH{=}CHCH_2CH_3$
2-pentene

$CH_2{=}CH{-}CH(CH_3)CH_3$
3-methyl-1-butene

$CH_2{=}C(CH_3)CH_2CH_3$
2-methyl-1-butene

$CH_3C(CH_3){=}CHCH_3$
2-methyl-2-butene

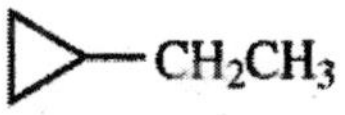
ethylcyclopropane

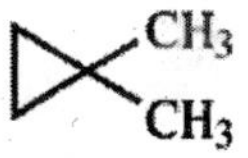
1,1-dimethylcyclopropane

1,2-dimethylcyclopropane

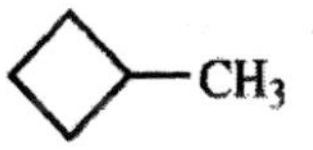
methylcyclobutane

G. Such a hydrocarbon could be C_7H_2 (unlikely) or C_6H_{14}. Thus, the answers are the same as for D.

3. H for highest value; L for lowest value. A. benzene H, hexane L B. pentane H, 2,2-dimethylpropane L C. ethyne H, ethane L D. 1,4-dimethylbenzene H, benzene L E. hexane H, butane L F. *n*-heptane H, 3,3-dimethylpentane L G. benzene H, cyclohexane L H. butane H, cyclobutane L I. ethene H, methane L J. butyne H, ethyne L K. benzene H, hexane L L. *n*-hexane H, 2,2-dimethylbutane L M. xylene H, benzene L N. *n*-heptane H, 3,3-dimethylpentane O. hexane H, benzene L P. pentane H, propane L

4. A. $BrCH_2CH_2CH_2CH_2CH_2CH_3 \xrightarrow{\text{KOH, alcohol}} CH_2{=}CHCH_2CH_2CH_2CH_3 + KBr + H_2O$

B. $HOCH_2CH_3 \xrightarrow{\text{sulfuric acid, } \Delta} CH_2{=}CHCH_3 + H_2O$

C. $HOCH_2CH_3 \xrightarrow{\text{sulfuric acid, } \Delta} H_2C{=}CH_2 + H_2O$

D. $2\ ClCH_2CH_3 + 2\ Na \longrightarrow CH_3CH_2CH_2CH_3 + 2\ NaCl$

E. $Cl_2CHCH_2CH_3 \xrightarrow{\text{KOH, alcohol}} HC{\equiv}CCH_3 + 2\ KCl + 2\ H_2O$

F. $CH_3CHClCH_3 \xrightarrow{\text{KOH, alcohol}} CH_2{=}CHCH_3 + KCl + H_2O$

G. $2\ CH_3CHClCH_2CH_3 + 2\ Na \longrightarrow CH_3\cdot CHCH_2CH_3 \,|\, CH_3\cdot CHCH_2CH_3$ (3,4 - dimethylhexane)

H. $CH_2 = CHCH_2CH_3 \xrightarrow{H_2,\ pt} CH_3CH_2CH_2CH_3$

I.

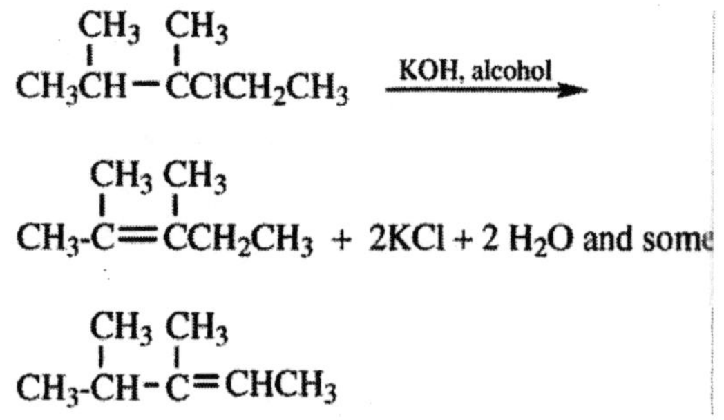

$$CH_3\underset{|}{CH}-\underset{|}{CCl}CH_2CH_3 \xrightarrow{\text{KOH, alcohol}}$$
(with CH_3, CH_3 substituents)

$$CH_3-\underset{|}{C}=\underset{|}{C}CH_2CH_3 + 2KCl + 2H_2O \text{ and some}$$
(with CH_3, CH_3 substituents)

$$CH_3-\underset{|}{CH}-\underset{|}{C}=CHCH_3$$
(with CH_3, CH_3 substituents)

5. A. $CH_3-\underset{\underset{OH}{|}}{CH}CH_2CH_3$

B. $CH_3CH_2-\underset{\underset{CH_3}{|}}{CCl}CH_2CH_3$

C. $CH_3-\underset{|}{CH}CH_2CH_3$

D. $CH_3-\underset{\underset{CH_3}{|}}{CBr}CH_2CH_3$

E. $CH_3-\underset{\underset{CH_3}{|}}{CCl}CH_2CH_2CH_3$

F. (cyclohexane ring with CH_3 and Cl)

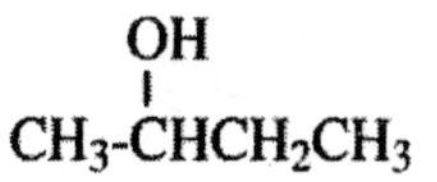

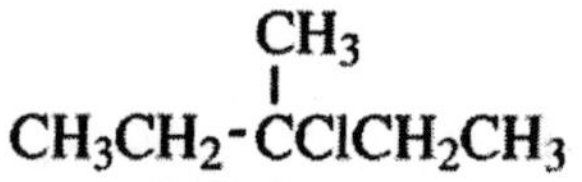

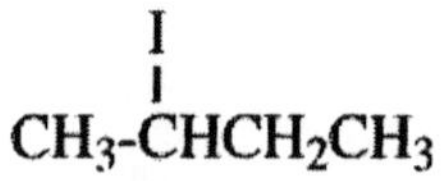

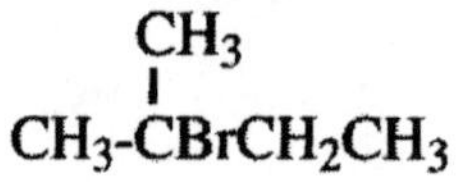

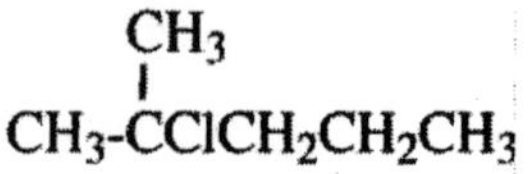

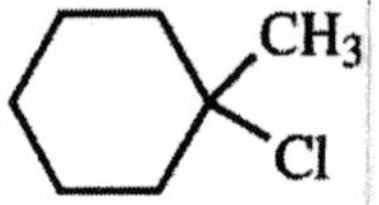

6. (1) A. 2-bromobutane B. 2,2,4-trimethyl-3-pentanol C. 3-hexanone D. pentanoic acid E. 2,2-dimethylpropanal
F. methyl *n*-propyl ether G. *n*-pentylamine H. methanoic acid *or* formic acid I. ethyl propanoate J. 2-methyl-2-propanol
K. ethyl isopropyl ether. L. 2,2,3-trichloropentane
M. 3-pentanone N. *n*-propyl propanoate O. pentanal

(2) P. $CH_3COCH_2CH_3$, Q. $CH_3CH_2CH(CH_3)CHO$

R. $CH_3CH_2CH_2CH(CH_3)COOH$ S. $CH_3CH_2CHBrCH_2CH_3$

T. $CH_3CH_2CH_2COOCH_3$ U. $CH_3CH_2CH_2OCH(CH_3)_2$

V. $CH_3CH_2CH(CH_3)CH_2CHO$ W. $CH_3CCl_2CH_3$

X. $CH_3CH(NH_2)CH(CH_3)CH_2CH_3$

Y. $CH_3CH_2CHOHCH_2CH_2CH_3$ Z. $CH_3CHOHCH_3$

Γ. $CH_3CH_2CH(NH_2)CH_2CH_3$ Δ. $CH_3CH(CH_3)COOH$

Θ. $CH_2ICH(CH_3)CH_2CH_3$ Λ. $CH_3COCH_2CH_2CH_3$

7. A. $CH \equiv CH + NaNH_2 \longrightarrow NH_3 + HC \equiv C^- \, Na^+$;

$HC \equiv C^- \, Na^+ + CH_3Br \longrightarrow HC \equiv CCH_3 + NaBr$

B.

$$CH_3\overset{\overset{O}{\|}}{C}-H \xrightarrow{H_2/Ni,\, H_2SO_4} CH_3CH_2OH;$$

$$CH_3CH_2OH \xrightarrow{H_2SO_4,\, heat} CH_2 = CH_2$$

C. $CH_3CH_2CH_3 \xrightarrow{Cl_2} CH_3CH_2CH_2Cl + CH_3CHClCH_3$;

either chloropropane $\xrightarrow{\text{KOH, alcohol}}$

$$CH_3CH = CH_2 + KCl + H_2O$$

D. $CH_3CH_3 + Cl_2 \longrightarrow CH_3CH_2Cl + HCl;$

$2\,CH_3CH_2Cl + 2\,Na \longrightarrow CH_3CH_2CH_2CH_3 + 2\ NaCl;$

$CH_3CH_2CH_2CH_3 + Cl_2 \longrightarrow$

$\quad CH_3CH_2CH_2CH_2Cl + CH_3CH_2CHClCH_3 + HCl;$

$CH_3CH_2CHClCH_3 \xrightarrow{\text{KOH, alcohol}}$

$\quad CH_3CH_2CH = CH_2 + CH_3CH = CHCH_3 + KCl + H_2O$

E. $CH_3CH_2CH_3 + Cl_2 \longrightarrow$

$\quad CH_3CHClCH_3 + CH_3CH_2CH_2Cl + HCl;$

$$CH_3CHClCH_3 + OH^- \longrightarrow CH_3-\overset{\overset{\displaystyle OH}{|}}{C}HCH_3 + Cl^-;$$

$$CH_3-\overset{\overset{\displaystyle OH}{|}}{C}HCH_3 \xrightarrow{K_2Cr_2O_7} CH_3\overset{\overset{\displaystyle O}{\|}}{C}CH_3$$

F. $CH_3CH_3 + Cl_2 \longrightarrow CH_3CH_2Cl + HCl;$

$CH_3CH_2Cl + OH^- \longrightarrow CH_3CH_2OH + Cl^-;$

$CH_3CH_2OH \xrightarrow{H_2SO_4,\text{ conc}} CH_3CH_2OCH_2CH_3$

G. $CH_3CH_2CH_3 + Cl_2 \longrightarrow CH_3CHClCH_3 + CH_3CH_2CH_2Cl$

$CH_3CH_2CH_2Cl + OH^- \longrightarrow CH_3CH_2CH_2OH + Cl^-$

$CH_3CH_2CH_2OH \xrightarrow{\text{Cu, 200-300°C}} CH_3CH_2CHO + H_2$

H. $CH_3CH_2CH_2CH_2CH_2CH_3 + Cl_2 \longrightarrow$ many isomers

Including $ClCH_2CH_2CH_2CH_2CH_2CH_2Cl;$

$ClCH_2CH_2CH_2CH_2CH_2CH_2Cl + Na \longrightarrow$ 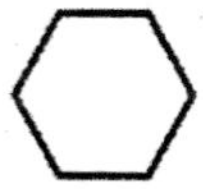 $+ NaCl$

I. $CH_3CH_3 + Cl_2 \longrightarrow CH_3CH_2Cl;$

$CH_3CH_2Cl + OH^- \longrightarrow CH_3CH_2OH + Cl^-;$

$CH_3CH_2Cl + Na \longrightarrow CH_3CH_2CH_2CH_3 + NaCl;$

$CH_3CH_2CH_2CH_3 + Cl_2 \longrightarrow$

$\quad CH_3CH_2CHClCH_3 + CH_3CH_2CH_2CH_2Cl + HCl;$

$CH_3CH_2CH_2CH_2Cl + OH^- \longrightarrow CH_3CH_2CH_2CH_2OH + Cl;$

$CH_3CH_2CH_2CH_2OH \xrightarrow{KMnO_4} CH_3CH_2CH_2COOH;$

$CH_3CH_2CH_2COOH + CH_3CH_2OH \longrightarrow$

$$CH_3CH_2CH_2\overset{\overset{\displaystyle O}{\|}}{C}-OCH_2CH_3$$

Quizzes

A: 1. (d) 2. (b) 3. (b) 4. (a) 5. (d) 6. (c) 7. (a) 8. (c)
B: 1. (e) 2. (c) 3. (b) 4. (a) 5. (d) 6. (a) 7. (c) 8. (c)
C: 1. (a) 2. (d) 3. (d) 4. (a) 5. (e) 6. (c) 7. (d) 8. (b)
D: 1. (c) 2. (a) 3. (c) 4. (e) 5. (a) 6. (b) 7. (c) 8. (b)

Sample Test

1. $CH_3CH_2OCH_2CH_3$ diethyl ether $CH_3OCH_2CH_2CH_3$ methyl propyl ether

$$\begin{array}{c} \quad CH_3 \\ \quad | \\ CH_3OCHCH_3 \end{array}$$

methyl isopropyl ether

2. Propanol contains an —OH group in a three-carbon molecule. This —OH group will hydrogen bond to water molecules, and the remaining $CH_3CH_2CH_2$ — propyl group is small enough that this hydrogen bonding is not significantly disrupted. Hence, a solution forms. In the case of hexanol, however, the relatively long six-carbon hexyl chain will disrupt the hydrogen bonds significantly and no solution forms.

3. $CH_2 = CHCH_2CH_3$

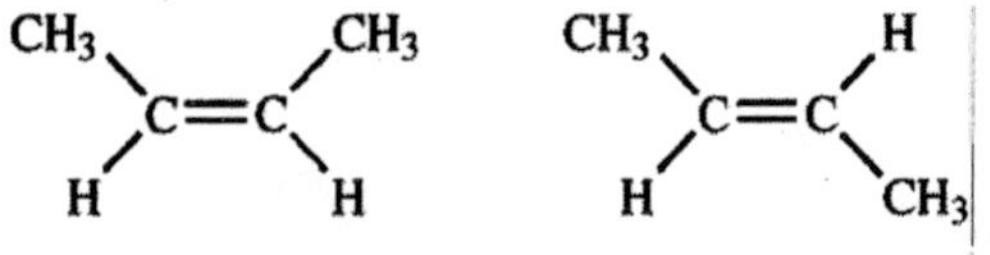

cis-2-butene   *trans*-2-butene

4.

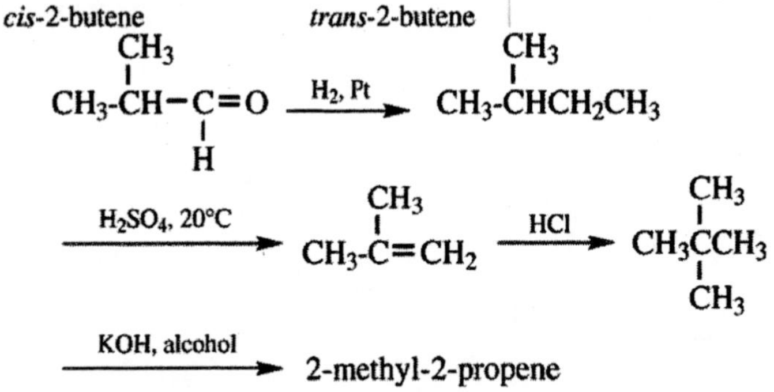

A is 2-methylpropanol, B is 2-methyl-2-propene, and C is 2-bromo-2-methylpropane.

CHAPTER 27

Drill Problems

1.(1) A. glyceryl trioleate B. glyceryl tricaprate C. glyceryl caprolaurolinolenate D. glyceryl trieleostearate

2. E. CH_2—$OOC(CH_2)_{14}CH_3$

|
$CH—OOC(CH_2)_7CH = CH(CH_2)_7CH_3$
|
$CH_2—OOC(CH_2)_{14}CH_3$

F. $CH_2—OOC(CH_2)_{10}CH_3$
|
$CH—OOC(CH_2)_{12}CH_3$
|
$CH_2—OOC(CH_2)_{14}CH_3$

G. $CH_2—OOC(CH_2)_{10}CH_3$
|
$CH—OOC(CH_2)_{10}CH_3$
|
$CH_2—OOC(CH_2)_7(CH=CHCH_2)_2(CH_2)_3CH_3$

H. $CH_2—OOC(CH_2)_7(CH=CHCH_2)_3CH_3$
|
$CH—OOC(CH_2)_7(CH=CHCH_2)_3CH_3$
|
$CH_2—OOC(CH_2)_7(CH=CHCH_2)_3CH_3$

I. $CH_2OOC(CH_2)_7CH=CHCH_2CHOH(CH_2)_5CH_3$
|
$CHOOC(CH_2)_7CH=CHCH_2CHOH(CH_2)_5CH_3$
|
$CH_2OOC(CH_2)_7CH=CHCH_2CHOH(CH_2)_5CH_3$

J. $CH_2OOC(CH_2)_7CH=CH(CH_2)_7CH_3$
|
$CHOOC(CH_2)_7CH=CH(CH_2)_7CH_3$
|
$CH_2OOC(CH_2)_{10}CH_3$

3. (1) A. 765.0 g/mol; 0.81 double bond/triglyceride B. 657.4 g/mol; 0.23 double bond/triglyceride C. 876.6 g/mol; 4.56 double bonds/triglyceride D. 881.2 g/mol; 5.14 double bonds/triglyceride
(2) E. 885.40 g/mol; SV =190.08; 3 C=C/molecule; IN = 86.0
F. 873.31 g/mol; SV = 192.72; 9 C_C/molecule; IN = 261.56
G. 891.46 g/mol; SV = 188.79; no C=C bonds; IN = 0.00
H. 719.11 g/mol; SV = 234.04; 2C=C/molecule; IN = 70.59

4. A.
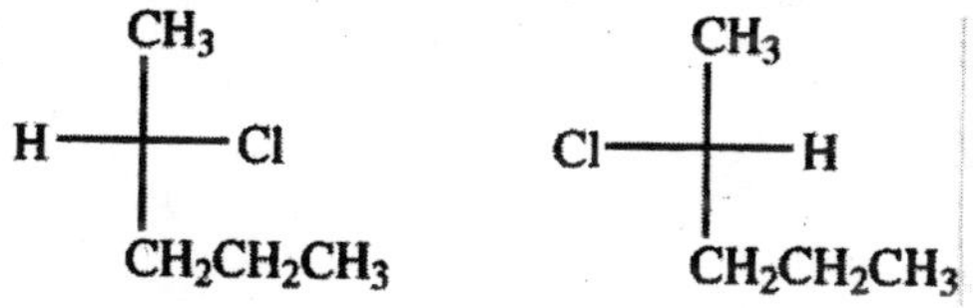

B.
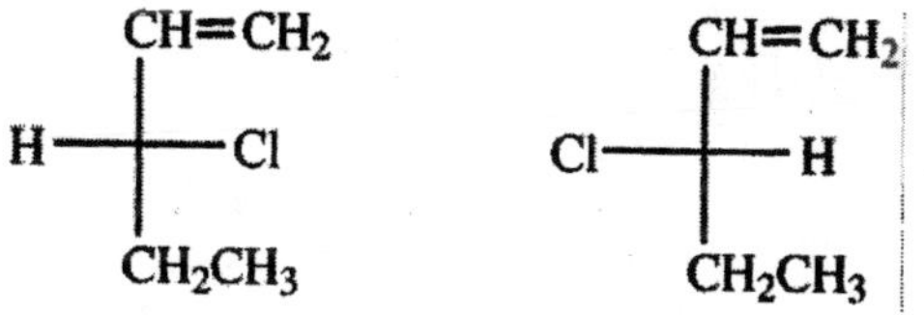

C.

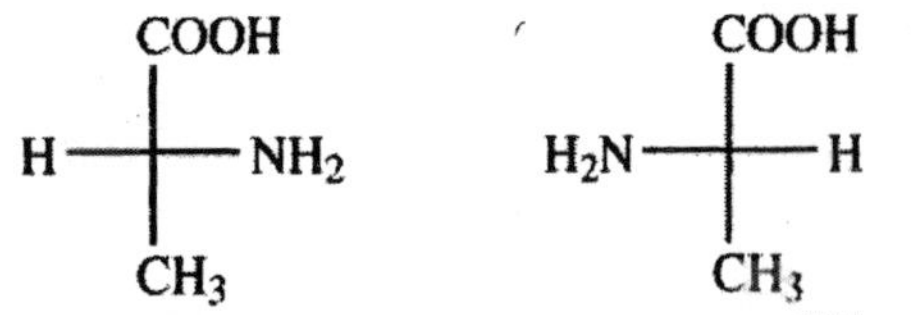

```
      COOH                      COOH
       |                          |
  H————NH₂              H₂N————H
       |                          |
      CH₃                       CH₃
```

D.

```
      CH₃                       CH₃
       |                          |
  H————OH              HO————H
       |                          |
    CH₂CH₃                    CH₂CH₃
```

E.

```
      CH₃                       CH₃
       |                          |
  H————Cl               Cl————H
       |                          |
    CH₂CH₃                    CH₂CH₃
```

F.

```
   CH₂CH₂Cl                 CH₂CH₂Cl
       |                          |
  H————CH₃             H₃C————H
       |                          |
    CH₂CH₃                    CH₂CH₃
```

G.

```
     CH₂Cl                     CH₂Cl
       |                          |
  H————Cl               Cl————H
       |                          |
    CH₂CH₃                    CH₂CH₃
```

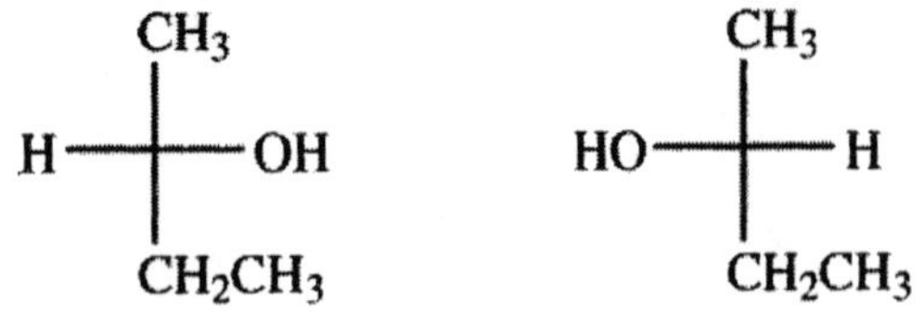

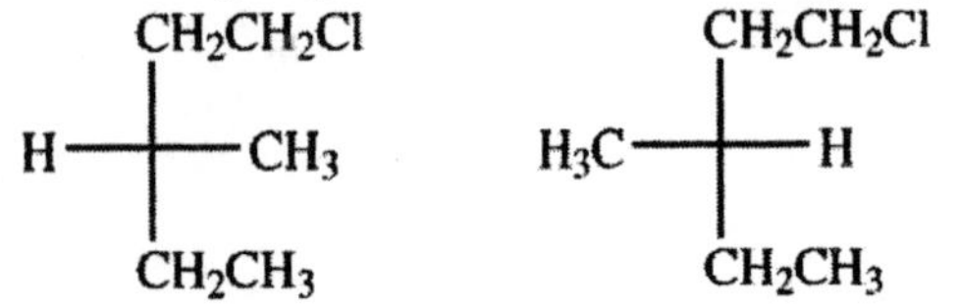

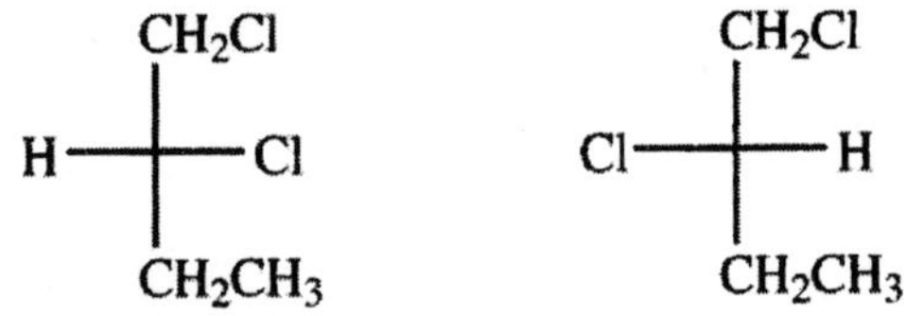

H. I., J. not optically active

K.

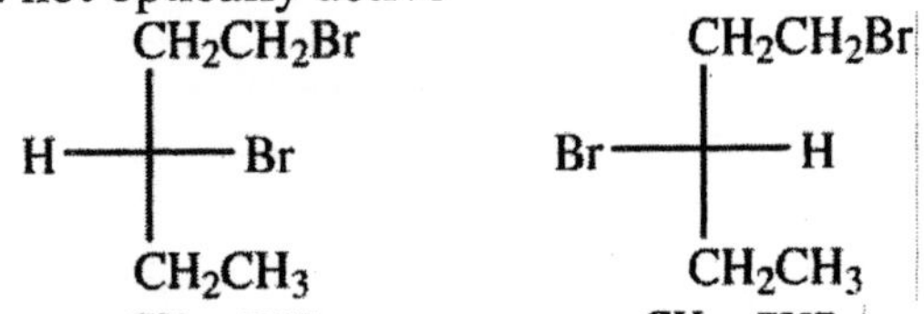

```
   CH₂CH₂Br                 CH₂CH₂Br
       |                          |
  H————Br               Br————H
       |                          |
    CH₂CH₃                    CH₂CH₃
```

L.

```
   CH=CHBr                   CH=CHBr
       |                          |
  H————Cl               Cl————H
       |                          |
      CH₃                       CH₃
```

M.

```
    CH=CH₂                    CH=CH₂
       |                          |
  H————Cl               Cl————H
       |                          |
  CH₃-CHCH₃                 CH₃-CHCH₃
```

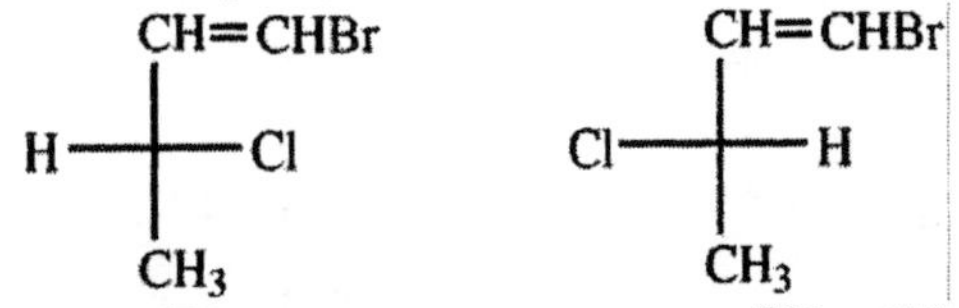

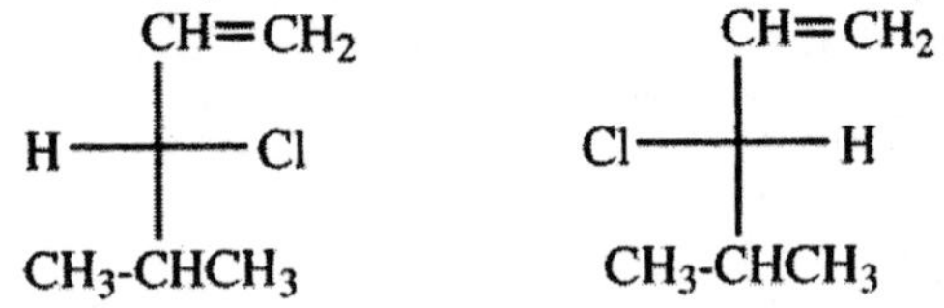

5. (1) A. Lys-Asp-Gly-Ala-Ala-Glu-Ser-Gly
B. Ala-Ala-His-Arg-Glu-Lys-Phe-Ile
C. Tyr-Cys-Lys-Ala-Arg-Arg-Gly
D. Phe-Ala-Glu-Ser-Ala-Gly
E. Val-Ala-Lys-Glu-Glu-Phe-Val-Met-Tyr-Cys-Glu-Trp-Met-Gly-Phe
(2) A. Lysylaspartylglycylalanylalanylglutylserylglycine
B. Alanylalanylhistidylarginylglutyllysylphenylalanylisoleucine
C. Tyrosylcysteyllysylalanylarginylarginylglycine
D. Phenylalanylglutylserylalanylglycine

Quizzes

A: 1. (b) 2. (d) 3. (c) 4. (d) 5. (b) 6. (d) 7. (e) 8. (e) 9. (b)
B: 1. (d) 2. (c) 3. (a) 4. (e) 5. (a) 6. (c) 7. (d) 8. (b) 9. (b)
C: 1. (e) 2. (b) 3. (d) 4. (a) 5. (a) 6. (c) 7. (b) 8. (a) 9. (e) (a pyrimidine)
D: 1. (e) 2. (e) 3. (b) 4. (d) 5. (d) 6. (d) 7. (a) 8. (b) 9. (d)

Sample Test

1. 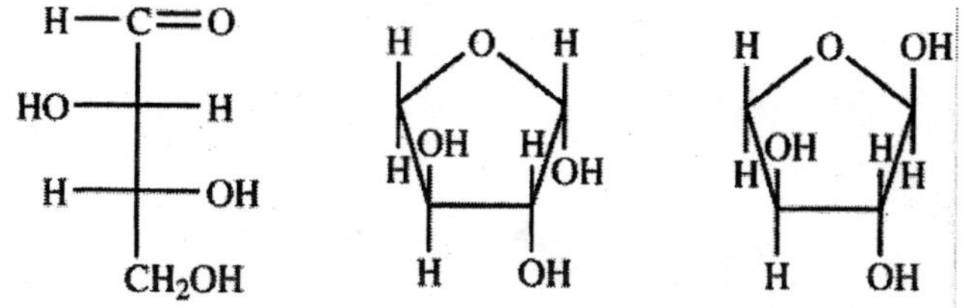

2. Refer to objective 27-10 and Figure 27-12
3. Phe-Val-Asp-Gly-His-Leu-Cys-Gly-Ser-His